新能源技术应用系列

Bioenergy

生物能源

EDITED BY
Judy D. Wall
University of Missouri-Columbia
Caroline S. Harwood
University of Washington, Seattle
Arnold Demain
Drew University, Madison, NJ

科学出版社
北京

图字:01-2011-2228 号

This is an annotated version of
Bioenergy
by Judy D. Wall, Caroline S. Harwood and Arnold Demain

ISBN: 978-1-55581-554-7

图书在版编目(CIP)数据

生物能源=Bioenergy:英文/(美)华尔(Wall,J. D.)等编. —北京:科学出版社,2011
ISBN 978-7-03-030841-2

Ⅰ.①生… Ⅱ.①华… Ⅲ.①生物能源-研究-英文 Ⅳ.①TK6

中国版本图书馆 CIP 数据核字(2011)第 070458 号

责任编辑:贾明月/责任印制:钱玉芬
封面设计:耕者设计工作室

科学出版社 出版
北京东黄城根北街 16 号
邮政编码:100717
http://www.sciencep.com

北京佳信达欣艺术印刷有限公司印刷
科学出版社发行 各地新华书店经销
*
2011 年 6 月第 一 版 开本:787×1092 1/16
2011 年 6 月第一次印刷 印张:30 插页:4
印数:1—2 000 字数:710 000

定价:128.00 元
(如有印装质量问题,我社负责调换)

序

《生物能源》一书简明扼要地介绍了通过微生物转化的各类可再生能源。每一章都由从事相关研究的资深专家撰写，这些专家同时也是未来研究方向的奠基人。本书并没有打算将生物能源领域全部囊括，新的科技前沿应由创造力无限的读者自己去发掘。

在编写本书的过程中，这些微生物学者意识到，我们有义务为能减少化石燃料需求的碳中立可再生能源发展开拓新道路。我们当中的许多人还能清楚记得1973～1974年美国的石油危机，当时油价狂飙，加油站前等着加油的汽车排成巨龙长队。而现在，油价又再次飙升，但这次并非政治原因，而是由于对石油的需求增加，同时供应量减少。如果任由其发展下去，其经济影响会导致社会结构的崩坏甚至大混乱，而且这次危机不会仅限于美国，会波及全球。我们现在迫切需要快速开发替代燃料以度过过渡期，从而避免因燃料严重短缺造成的不稳定性。

毫无疑问，单一的替代能源不可能满足所有需求，所以任何一种可用的能源都不应被忽视。在漫长的进化史中，微生物通过几乎每个化学反应来生成养分；而现在，我们须重点关注这些数量非常丰富的资源，以有效地满足对能源的庞大需求。由光合作用生成的生物质是其中一种理想的资源，而微生物转化生物质的过程已被证实颇有成效。

因此，这本书首先以10章来介绍比玉米乙醇更加经济可行且可持续的纤维素乙醇；紧接着，我们将探讨在开发中或仍在构思中的各个阶段的能源现状，包括甲烷、甲醇、氢能、电力、丁醇等。这些章节都反复强调，所有的这些能源都亟需进一步开发。我们希望这份关于所有进行中的研究的汇总，能给拥有先见之明的科学家一点启发和灵感，进而帮助突破一些颇具影响力的技术问题。大概浏览一下目录，对此感兴趣的业外人士就会发现，众多值得注意的微生物活动都具有为生物燃料生产作贡献的潜力。

我们要感谢美国微生物科学院在2006年3月举办的一次名为"微生物能源转化"的关键问题座谈会，本书的三个编辑都出席了此会。我们特别感谢科学院院长Carol Colgan，其卓越的组织能力保证了整个讨论会的圆满成功。本次座谈会同意编写一本能概括微生物在生物燃料制造中的参与和作用，并指出未来发展方向的著作，《生物能源》便是这次初步商谈的结晶。

具有讽刺意义的是，1973～1974年的石油危机也促成了一次专业座谈会的举办，当时顶尖的微生物学家和工程师都有出席。主办方是联合国训练研究所(UNITAR)和联邦德国研究技术部，主旨是回顾生物质生产和将生物质转化为能源中包含的微生物反应过程。这个座谈会的讨论结果也汇总成一本值得推荐阅读的书，名为《微生物能源转换》(*Microbial Energy Conversion*)，编者是H. G. Schlegel和J. Barnea，在1977年经Gamon Press出版。两本书的主题相近，均强调了过去对支持生物能源发展的疏忽，我们由于能源的暂时满足而缺少发展替代能源的紧迫感。

对替代能源的忽视只是过去专属的奢侈，这种情况绝不会重现了。

（李十中 译）

目　录

第一卷　生物乙醇

第一章　从玉米、甘蔗生产燃料乙醇

出于经济、能源安全和环境等方面的考虑，燃料乙醇的生产和使用在全世界范围内扩大增产。在美国，乙醇被用作充氧物以减少汽车尾气排放、对进口石油的依赖以及国内石油生产干扰的风险。使用乙醇的另外一个优势是拉动农村经济。巴西和美国是世界上两个最大的乙醇生产国和消费国。以10%～22%的比率掺混的乙醇可被传统的汽车发动机兼容，更高比率的燃料可在“灵活燃料车”中使用。

本章从乙醇发酵的生物学理论入手，介绍了从玉米淀粉转化成可发酵糖（葡萄糖和蔗糖），再从可发酵糖转化成乙醇的过程，特别介绍了能够参与其中的酵母菌。

其后描述了玉米乙醇工艺，这项传统工艺可分成干法研磨和湿法研磨两种。前者分为淀粉转化、糖化作用和发酵、蒸馏和脱水、釜馏物加工和动物饲料生产几个方面；后者着重于浸泡、油/纤维/谷蛋白分离和淀粉转化过程。最后介绍了玉米乙醇工艺的未来发展动向：如用转基因方法可针对性改良玉米品种以方便提取糖分；在低温下固态淀粉水解，无需蒸汽液化以达到发酵最佳温度；以及结合了湿法研磨和分馏技术的干法研磨，以提高副产品产量等。

甘蔗乙醇被巴西大规模使用，本章介绍了甘蔗在发酵前的加工处理、原料制备、发酵过程、蒸馏和脱水、釜馏物加工、蔗渣处理等环节。

燃料乙醇的生产和使用可持续增长。改良作物品种、酶制剂和原料等可提高乙醇工艺的生产力，工艺的改进也可为产品增值。未来，能够加工整株玉米（淀粉、纤维和秸秆）或甘蔗的生物炼制厂可同时生产液体燃料、食用油或糖、动物饲料、粉末、聚合物或化学中间产物等。即使木质纤维素原料也进入生产环节，淀粉和糖分会一直在乙醇生产中发挥重要作用。

第二章　通过 *Saccharomyces* 酵母菌转化的木质纤维素生物质

包括木质纤维素材料和农林残留物在内的可再生生物质是生物乙醇生产的廉价原料。其中，农业残留物（玉米、小麦和水稻等的秸秆）、农业加工副产品（玉米纤维、稻壳和蔗渣等）、能源作物（如柳枝稷）都可用来当作糖分的廉价来源，用于生产生物燃料。目前，木质纤维素生物质到可发酵糖的转化存在着一定的技术和经济难题，其成功相当程度上取决于高效的预处理程序，随后的木质纤维素底物也需要高效的酶制剂转化成可发酵糖；另外还取决于具胁迫耐受性的微生物催化剂。木质纤维素生物质在预处理基础上能自动或在酶水解作用下生成一系列的五碳糖和六碳糖。传统的 *Saccharomyces cerevisiae* 酿酒酵母能将葡萄糖高速高效地发酵成乙醇，但在五碳糖（木糖和阿拉伯糖）中它产出乙醇的功能就很有限。虽然有些酵母菌种（*Pachysolen tannophilus*、*Pichia stipilis* 和 *Candida shehatae*）拥有将木糖转化成乙醇的能力，但大多数菌种耐受性较差，乙醇发酵率低。经基因改造的 *S. cerevisiae* 被证实能够将木糖转化成乙醇（Kötter et al.，1990）。由于它的乙醇生产能力和对抑制物的高耐受性，它被认作是从木质纤维素衍生的水解物中所含六碳糖的最佳发酵酵母菌种。

具成本效益的可持续木质纤维素生物质转化成乙醇的工艺在未来主要存在两项挑战：异源糖的利用，和用于生物乙醇生产的催化发酵菌的胁迫耐受性基因工程。在这一章里，我们回顾了木质纤维素生物质的组成和结构的现有知识，它的预处理，酶参与的糖化作用（把纤维素和半纤维素转成单糖），把玉米纤维转化成乙醇，*S. cerevisiae* 利用木糖、葡萄糖和异源糖的发酵，原位脱毒的原理，通过定向进化性适应的菌种改良，产乙醇酵母的功能基因组学等。

为了实现具成本效益的稳定的生物质-乙醇工业，需要一支跨领域的团队在研究开发上不懈努力；同时也需要及时到位的研究资源和资金。基因工程是充实产乙醇菌种 *S. cerevisiae* 的综合遗传学背景知识的关键所在。从基因上改造的定向进化性适应，聚焦在特定分子功能改良和在实验室设定下的代谢动态，是一个改良和开发理想菌种的强大工具。这项技术与常规的遗传学研究知识结合，能为我们开拓产乙醇酵母 *S. cerevisiae* 的新视野。一个全面的基因组工程可使我们在往后 10 年或更远的未来更有把握迎接高效木质纤维素生物质转化乙醇技术的挑战。

第三章　用于半纤维素水解物发酵的毕赤酵母基因组工程

毕赤酵母（*Pichia stipitis*）是一种半子囊类真菌，与几种 内共生的酵母相近，这种甲虫生长在腐烂的白色硬木中，并靠分解木头为生。这些真菌貌似非常适应这种环境，因为 *P. stipitis* 的基因组序列中显示了许多能在木栖昆虫的体内存活和生长的特征。因此它能够在木聚糖上生长并让其发酵，且能够利用木材里存在的所有主要糖种（木糖和纤维素多糖）。它能够同化并发酵纤维二糖。另外，据研究报告，在有氧培养下，它能够将低分子量的木质素相关的芳香醛，如 3,4-二甲氧基苯甲醛和 3-甲氧基-4-羟基苯甲醛转化成相对应的醇类和酸类。

本章着重分析毕赤酵母的遗传系统、遗传转化、代谢调节、基因组结构，对麦芽糖、半乳糖和甘露糖的利用，糖同化途径以及葡萄糖和木糖运输等。这种酵母的乙醇产率比较高且能够代谢所有常见的糖种。科学家们开发了一种特殊的遗传系统，包含营养缺陷型标记 *URA3* 和 *LEU3*，还有抗腐草霉素 D1 标记的修改型 *Sh ble* 和 *Cre* 重组酶。这个系统被用于扰乱真菌呼吸能力，从而提高发酵率。基于表达序列标签和基因芯片的表达分析可指导进一步的基因改造。*P. stipitis* 的许多基因位于功能性相关的基因簇上。我们需要进一步的代谢工程和菌种选育以增加总体发酵率和 *P. stipitis* 对乙醇的耐受性，以实现半纤维素水解物的商业性生物转化。

第四章　乙醇，生物质和梭菌

近 30 年来，以乙醇为代表的生物燃料与石油的竞争和替代问题已成为全世界经济和政治争论的重点。然而，如果要继续满足运输业对燃料的需求增长，目前的生物燃料产业必须从谷物粮食原料转移到替代的可再生原料，如非粮食木质纤维素生物质。非粮生物燃料拥有众多优点，如原料成本低廉、可持续性较高、加强能源安全、减少温室气体等。但目前在纤维素水解反应解开可发酵糖分的过程中存在一定困难。现研究者正积极开发高效的预处理过程，以破坏木质素外层的晶体结构，增加气孔数量，使半纤维素和纤维素溶解，且使酶制剂易于接近目标聚合物。

目前有一种基于热纤梭菌(*Clostridium thermocellum*)的无氧共培养技术，可以生产木质纤维素乙醇，它可以将糖化和发酵结合到一起。热纤梭菌含有的独特的纤维体(cellulosome)是一个蛋白质复合体，能够增强酶与纤维素的亲和力。这个合并的生物工艺的特色包括：①嗜热厌氧菌 *C. thermocellum* 能同时产纤维素酶和半纤维素酶；②将纤维素转化成纤维寡糖和纤维二糖；③将包括阿拉伯木聚糖的半纤维素转化成单体木聚糖、木二糖、木寡糖和五碳糖；④*C. thermocellum* 从纤维素分解产物转化到乙醇；⑤结合嗜热厌氧解糖菌(*Thermoanaerobacterium saccharolyticum*)将半纤维素分解产物转化成乙醇。

这个无氧嗜热共培养系统的优点在于：①单步式实现糖化和乙醇生产，且减免了真菌酶制剂的成本；②不需要昂贵的氧气运输；③减少污染问题；④最低的冷却需求；⑤微生物的嗜热性方便了乙醇的转移与回收；⑥细胞产量少，相对更多的底物就能转化成乙醇；⑦利用纤维素的生产率和代谢率都相当高；⑧能够利用纤维素、半纤维素、葡萄糖、淀粉、木聚糖、木糖、甘露糖、半乳糖和阿拉伯糖。

第五章　利用嗜热性细菌纤维素生物质联合生物工艺产乙醇

阻挡生物燃料产业发展的直接原因其实是加工费用，而非原料费用(纤维素生物质原料价格仅为 3 美元/GJ，相当于 17 美元/桶的石油价格)。在降低从生物质转化成糖分的工艺成本方面，无添加糖分解酶的一步是微生物转化，又称作联合生物工艺(CBP)，因其拥有革命性突破的潜力，已经得到了越来越多的认可。当高级同步糖化与共发酵过程被成熟的 CBP 技术代替，每生产 1 加仑的乙醇所需酶成本仅需 0.10 美元，CBP 在生物加工环节中开支可为原来的 1/4，总体运行开支为原来的一半。

但这一工艺亟需一种配套的微生物，能够从经过预处理的生物质中迅速利用纤维素等成分，同时产出大量高浓度的乙醇。目前在自然界中尚未发现拥有双重特点的微生物，于是科学家们在能够产乙醇或利用纤维素的微生物身上进行基因工程，企图使其同时具备两种能力。

由于与温度成正相关的微生物水解结晶纤维素作用，嗜热菌成为科学家在 CBP 的研究兴趣之一。代谢需走发酵途径而非呼吸途径才能大量产出乙醇，热纤梭菌(*Clostridium thermocellum*)正是表征最详细的分解纤维素的发酵嗜热性微生物，且其纤维素利用率是目前观测的最高值之一。除此之外，一些嗜温性细菌(最佳生长环境接近 60℃)和一些极耐高温的分解纤维素细菌也可用作潜在的 CBP 宿主。

这一章在 CBP 的配置下关注嗜热性细菌产乙醇，重点是作为开发 CBP 配套微生物中的最大技术挑战——纤维素转化，特别是微生物分解纤维素过程近年来的进展。本章强调了非葡萄糖组分的糖分转化，因这类转化是 CBP 的必要缓解；对近年用嗜热性微生物在这方面的进展也作了概括。在 CBP 背景下，本章回顾了嗜热性细菌的多样性和生态学、对非葡萄糖成分的利用及对乙醇的耐受性；介绍了微生物纤维素转化的基础知识，包括底物同化作用和生物能量学、酶-微生物协同作用、底物捕获、对合成纤维素酶的调节和纤维素利用率等。接着回顾了与 CBP 配套微生物有关的非特异性基因位点突变、遗传系统开发和代谢工程的新近进展。最后，我们对 CBP 的可行性、可取性和障碍、不确定性等发表了最新的观点。

第六章　纤维体增强型生物质转化:通往生物乙醇的道路

作为热纤梭菌(*Clostridium thermocellum*)的纤维素酶系统优点之一的纤维体(*cellulosome*)在对付顽抗纤维素方面的更大潜力被重点提出。本章首先介绍了 *C. thermocellum* 的原型,接着介绍纤维体的趋异、组成部分及这些组成部分在各种微生物里的调节和表达,包括 *C. thermocellum*、*C. cellulovorans*、*C. cellulolyticum* 等。本章还探讨了共培养发酵的潜力,*C. thermocellum* 和其他分解糖分产乙醇的细菌一起培养,可共享纤维二糖、纤维寡糊精等,达到更高的生长率和生产力。未来,纤维体在生物质转化中的应用有好几种可能性,如设计人造纤维体(从基因重组的纤维体组分上构成不同纤维素酶和其他相关酶的复合体),以及在不能自然降解纤维素但可产生合意生物质产品的微生物细胞表面植入人造纤维体。

第七章　中温菌的纤维体

中温菌的细胞体外壁也镶嵌有纤维素酶复合体——纤维体。这一章只讨论表征完毕的拥有两个组成部分的中温菌纤维体。这些组分包括:①一种或几种含有酶结合位点(cohesin)的支架蛋白(scaffoldin);②含有对接域(dockerin domain)的纤维体酶。支架蛋白的酶结合位点和纤维体酶的对接域的互动作用,让细胞体外的多亚单位酶复合体(纤维体)组装起来。讨论主要针对的是纤维体的酶和催化特性。

对中温菌的纤维体的基础研究又如何联系到我们社会的能源需求呢?为此,纤维体的研究应至少提供以下信息:①纤维体酶和非纤维体酶对生物质降解的协同作用;②拥有特定功能的人为设计纤维体的构建;③利用纤维体酶对植物细胞壁进行自溶作用;④用作生产理想产品的纤维体酶异源表达。

第八章　酶结合位点-对接域的相互作用和纤维体蛋白折叠

这一章详细叙述了两种蛋白的特定位点间的相互作用如何生成纤维体,介绍了对接域(dockerin)和酶结合位点 (cohesin)的初级结构、结合特异性、亲和力、对接域的 ST 基序在不同菌种的特异性、酶结合位点的分子结构以及酶结合位点-对接域复合体的分子结构等。

通过众多研究小组约 20 年来对纤维体的研究,纤维体组装的机理谜团终于被揭开。在 *C. thermocellum* 中,I 型和 II 型的酶结合位点-对接域复合体给予了不同菌种的结合特异性,实现了纤维体的有序布局。纤维体的催化性亚基通过在各亚基上的 I 型对接域和 CipA 的 I 型酶结合位点的相互作用,组装成纤维体。另一方面,纤维体通过 CipA 的 II 型对接域和表层蛋白如 SdbA 等的 II 型酶结合位点的相互作用锚定在细胞表面。值得注意的是,酶结合位点和对接域序列的轻微变化就可造成种内和种间的特异性,为纤维体组装奠定了基础。

第九章　功能域相互作用和钙结合对碳水化合物活性酶的稳定性的作用

这一章概述了碳水化合物活性酶的模块结构，并分析了功能域间的相互作用对这些既有趣又重要的蛋白质的结构、稳定性和功能性的影响。虽然这些蛋白质中的大多数都是较为复杂的分子机器，但当中的催化域被不同的非催化性的辅助模块所包围，它们位于蛋白质内的一个特定位置，常与旁边特定的模块相关联。模块长度不等，拥有各种不同的功能，参与复杂的通信和互动作用。一系列的特定域间的相互作用大大稳定了整个蛋白质分子。大型的纤维体催化酶（如 *C. thermocellumde* 的 CbhA）的模块间的结构和通信由钙结合来调节。域间相互作用和串扰的复合体性质已被结构和构象分析并确定了。甚至域间的接头对于碳水化合物活性酶的功能性也是非常重要的，这些接头可当做“分子弹簧”，让催化部位能接触继而水解糖苷键。这些研究结果的结论可概括为：模块式的碳水化合物水解酶中的域间相互作用增强了催化部位的活性。

第十章　产乙醇细菌在燃料乙醇生产上的开发

最近在能源界发生的事件、气候变化和环保需求等，刺激了利用植物衍生材料作为液体燃料生产燃料的各项技术的发展。其中影响最大的事件当属石油的成本上升，其原因不仅是供应的不稳定性和不确定性，也因为全世界对内燃机燃料需求量的急剧增加。

目前对于开发可再生的国产燃料供应的关注影响着政府和金融机构的行动，其背后所需的技术也已经开发了多年。本章的重点放在了两种已证实可在将植物原料转化成液体燃料的化学和物理过程中发挥作用的高效微生物：大肠杆菌（*Escherichia coli*）和产酸克雷伯菌（*Klebsiella oxytoca*）。

能高效地将生物质来源里的所有主要糖分高效转化的生物催化剂的开发，是可再生资源生产乙醇产业化的关键一步。尽管如此，在评估这种未来技术的经济价值时，还是必须全局考虑。有些科学家首先选定一种工艺，然后再设计合适的生物催化剂以优化预设的工艺。本章所介绍的是另一种方式，即先选择一种高效的微生物，然后建起一套能利用该微生物的特性优势的工艺。作者相信这条途径是实现首创木质纤维素乙醇工厂的最快方法。

第十一章　金孢子菌属纤维素酶和木聚糖酶产纤维素生物燃料

纤维素酶和半纤维素酶是非常活跃的解聚酶系统，通过合作和动态协同作用，将生物质多糖转化成可发酵的单糖，如葡萄糖、木糖和其他相关的六碳糖、五碳糖等寡糖。纤维素酶和半纤维素酶因此在许多行业有广泛的应用，如食品、饮料、纺织、饲料和造纸等。

金孢子菌属 *Chrysosporium lucknowense* 是纤维素酶和半纤维素酶的真菌性过剩生产者。在每升的发酵液中，*C. lucknowense* 能产出超过 100 g 的蛋白质，分泌的蛋白质的主要成分正是纤维素酶和半纤维素酶。*C. lucknowense* 的酶配制剂在纤维素糖化过程中效率较高，总的来说，*C. lucknowense* 的纤维素酶和半纤维素酶在降解木质纤维素材料方面有很大的潜力。

本章分别介绍了 *C. lucknowens* 的这两种酶，特别是纤维二糖水解酶、内切葡聚糖酶、纤维素酶之间的协同作用和 *C. lucknowens* 纤维素酶在水解纤维素的潜力。

细丝状真菌是自然界中最常见的木质纤维素生物质分解者。这些真菌细胞工厂是各种糖基水解酶的复杂组合，如纤维素酶、半纤维素酶和其他辅助酶。科学家证明了真菌多酶复合体在纤维素生物质糖化过程中的水解效率，这个协同作用机理很大程度上取决于每个单一酶组分的性质和它们在混合酶配制剂中的比例。预处理、酶糖化和发酵过程等的低成本综合运用，在纤维素乙醇和其他替代性发动机生物燃料及化学品的商业化进程上开拓了一条有希望的道路。所有这些努力的结果会最终促进石油经济成功过渡到生物质燃料经济。

第十二章　从合成气产乙醇

在木质纤维素生物质发酵的研究中，另有一种称作间接发酵的方法，即生物质首先在气化炉中经热裂解转化成主要由一氧化碳、二氧化碳和氢气（如在气化炉中使用空气而非氧气，还含有氮气）组成的合成气。合成气可送入发酵过程，或转化成一系列其他所需产品，包括乙醇。另外一个有吸引力的潜在用途是把一氧化碳充入生物燃料电池中用来发电。

用厌氧性细菌把一氧化碳发酵成乙醇是仅仅不久前才发现的生化过程。这些细菌全都是产乙酸菌，它们的产乙醇功能是后来才发现的。一氧化碳本身的存在对于产乙酸菌生产乙醇代替醋酸是关键。目前的猜测是，一氧化碳充当了一种有效的还原剂。这一章里介绍了三种被确定能从一氧化碳生产乙醇的微生物：*Butyribacterium methylotrophicum*、*Clostridium ljungdahlii* 和 *Clostridium carboxidivorans*。

间接发酵对比起直接发酵有几个优点。生物质到合成气的热裂解反应被很广泛认为是一种成熟技术，如煤炭的气化工艺。热裂解是一种快速的放热过程，一个相对较小的单位可以处理大量的生物质。例如，一个直径才 25 cm 的气化炉中试装置每天能处理重达 1000 公斤的生物质。相比之下，直接发酵需要大发酵罐和相对较长的处理时间才能糖化生物质。即使完成了糖化，生物质的木质素依然保留；但在气化炉内，这些木质素和纤维素、半纤维素一起统统转化成合成气。这就诠释了气化炉为何能把任何可燃物甚至通常认为不可燃的原料产出合成气，如废水中的淤泥等。但由于生产率较低，而且产物在液体中的浓度较低，所以其发展受到了限制。

第二卷　甲 烷 生 产

第十三章　乙酸产甲烷

地球的生物圈每年大概产出10亿公吨的甲烷。甲烷(沼气)是厌氧环境下的复杂有机物分解后的最终产物,这个过程是全球碳循环的重要一环。

生物性制沼法被用于处理有机废物和生产甲烷生物燃料。市政污泥处理厂开发了这一工艺,用于民用和工业废物的大型处理,减少了75%的可挥发固体数量,与此同时用沼气发电支持工厂电力运作。垃圾填埋场生产的甲烷也为当地能源需求作出重大贡献。在发展中国家,该工艺用于家居废物的小型处理和家庭的供热做饭等。因此,可持续生产植物生物质的生物性制沼法具有替代化石燃料的潜质。甲烷的能量含量是等重氢气的三倍,且甲烷易于存放运输;而且,有可能通过催化重整技术,将甲烷转化成甲醇原料,进一步减少对石油的依赖性。这一章发掘了乙酸转化成甲烷过程的微生物学和生物化学,并为促进改良该工艺成为化石燃料的有力竞争替代品的目标而作一个基本的背景介绍。

通过发酵性和产乙酸厌氧菌的作用从复杂的生物质产乙酸,以及随后从乙酸转化成甲烷的过程,是生物性制沼法最具重要性的部分。乙醇分解型甲烷生成(aceticlastic methanogenesis)是控制该过程的速度和可靠性的主要因素;因此,对这些产烷生物的全面了解对于开发一套高效工艺,用于从可再生和废弃生物质生产生物燃料的生物性制沼法是首要任务。虽然甲烷八叠球菌属(*Methanosarcina*)的从乙酸到甲烷的反应酶化学已被充分了解,但只有少数几个研究报告针对能量守恒和基因表达调控的机理。全球蛋白质组学和基因芯片分析已经识别了甲烷八叠球菌属的一组蛋白质和基因,其中许多拥有未知但也许重要甚至是必需的功能。所幸的是,科学家控制了甲烷八叠球菌属的健全的遗传系统,对纠正这些缺陷拥有相当的把握,同时也为工序改进提供了基因操作的可能性。另一方面,人们对甲烷鬃菌(*Methanosaeta*)的微生物学、生物化学、分子生物学和遗传学了解甚少,因此为控制将生物质转化成甲烷生物燃料的工艺并将产量最大化,有必要将研究兴趣放在深入认识乙醇分解型甲烷的生成上。

第十四章　产甲烷给料网络的能量学

从燃料总产出的角度上看,有机质的产甲烷型降解,似乎是将大多数有机废料高效转化成能源燃料的绝佳反应过程。在此方面,产甲烷型降解比其他发酵产氢气、乙醇、丁醇或丙酮等的过程更加高效。活体生物质和有机废料的所有主要构成部分都可被转化成甲烷加CO_2,唯独木质素和木质纤维素除外。高转化效率是通过许多不同类型的发酵型细菌和产乙醇古细菌的协作实现的;它们完成大量特定反应,重新安排代谢功能模块,将多样性极高的主要生物质组成部分转化成越来越简单的中间产物。由于产甲烷型发酵的能量总产出较小,每个在这个交互共生网络内的生物体只有很少的可用能量用以催化特定的反应步骤。过程中需要好几个步骤才能产出刚好足够的能量供给生物体的催化反应,并满足合成ATP和生长所需的最低能量需求,即每摩尔反应的能量约为$-15\sim-20$ kJ。

只有以如此高的能量效率才能使整个过程变得有可能，但同时它限制了动力的用途。

第十五章　优化混合培养生物工艺将废料转化成生物能源

本章从介绍两种自然界中能够将顽固的植物大部分聚合物分解的模式群落开始，向读者展示通过分泌胞质或胞外的酶把聚合物分解成水溶物的微生物混养。人们利用这个优势设计出了一系列能产出各种化工产品，如甲烷、氢气、电力、溶剂和液体燃料等的生物工艺，其大部分利用无光厌氧发酵过程。将有机废料转化成生物燃料的效率和经济可行性由废料的特点决定，尤其是可转化成能源产品的化学成分和相关浓度等。所利用的原料包括低强度废料、高强度废料和固体废料等。强度高低取决于生化需氧量（BOD）或化学需氧量（COD）。

因为混养发酵涉及庞大的微生物群落，所以只能生产某几种化合物。有一些产物无法得到，因为它们以比形成速度更高的速度被混合培养转化成其他化合物。当清楚确定合适微生物群落的挑选标准并通过生物反应器设计改良后，可优化某种特定的理想产品产量。从理论上来说，应该应用对群落成员有利的直接选择压力来催化理想反应，但这种对理想产品（如能作为生物燃料的醇类）的标准还未被验明，包括甲烷、燃料醇类和电力等生物衍化能源载体的生产工艺在本章里会一一介绍。

被深入研究的厌氧消化罐内的食物链随不同营养类群的有用的相互作用而定。虽然微生物燃料电池中的食物链未被详细了解，但微生物燃料电池很明显包含能够比单一菌种系统产生更多电力的混合培养。这一章也概述了混养系统的优点。

过去，生物工程师并未意识到工艺改进效果能影响微生物群落的结构。在缺乏对这些系统的微生物学的基本认识的情况下，这些工程师试图通过优化反应器配置或反应器操作来提高转化率。在这些选择中，后者包括了电子受体的适当选择，加上了指向某特定营养类群（如产甲烷生物）的抑制剂。另外，近年分子生物学发生了不少进步，例如发现了能操作微生物群落组成的新方法，监控关于群落组成的工艺改进效果，选择特定的发酵产品，以及优化转化率和产量。这一章最后以介绍两种能够操作微生物群落构成的方法结束，包括水平基因转移的生物强化（bioaugmentation）和噬菌体生物防治。

第十六章　来自生物质、生物废料和生物燃料的生物甲烷

从农业、城市、工业废料生产沼气是制造生物燃料的一个好法子，可用于过程加热、发电和汽车燃料。沼气可被升级并注入天然气管道中，促使现有的配送基础设施的改变。液体、淤泥、固体和气体垃圾都可送进厌氧消化罐加工成沼气。人们开发了好几种消化罐设计，以优化不同原料的加工工艺。消化罐的尺寸可根据应用来订造，集中生产设施和混合废物共消化能够帮助实现规模经济效益，且改进性能。由于生产生物乙醇、生物柴油、生物氢气和合成气的副产品也同样适用于厌氧发酵，甲烷生产也可整合到生物炼制厂中，能提高能量净产出和为农作物生产再回收有价值的养分。最后，用陆生或海生的能源作物加工成生物甲烷可比其他生物燃料产出更高的能量。考虑到原料丰富和产物回收轻便，从能源作物和有机废料生产的甲烷提供了一种主要的可持续能源解决方案，并具有可再生、二氧化碳中立和就地取材等优点，可把能源产生有效分散和分配到当地市场，能保

护环境，创造就业并促进当地经济发展。

第十七章 来自气体水化物的甲烷

在本书的这一卷里，甲烷水合物被认为是一个棘手的难题。水合物在生产方面更类似于化石燃料；它们存在于自然界，多出现在有限的地下矿床，和天然气或石油类似，必须先经过勘探，然后钻洞，再从钻井中开采出来。但是，它们可以被归类成生物能源资源，因其形成主要归因于近期(以地质期计算)和当前活跃的生物过程。

就像从生物质产甲烷一样(但规模更大)，困在气体水合物中的甲烷主要是细菌对有机物的自然分解物。从陆地上冲进水中，并沉积到古代和现代的海盆，有机物碎屑沉积物形成各种活跃的地表下细菌群落的基本食料。在无氧条件下，通过各种中间产物代谢途径，不同的细菌种类互相合作，逐渐将复杂的有机大分子还原成甲烷气。在深海和极地永久冻土的低温高压环境中，这些甲烷与沉积物间隙水混合，形成了笼形包含水合物(clathrate hydrate)，即由水分子“笼子”困住外来气体所组成的冰晶状化合物。这种水合物在沉积物空隙中增长，相当数量的甲烷可积聚成可开采的潜力矿藏。

如果能捕获并储存那些释放出来的CO_2(通过燃烧或重整产氢气)，如在地层中(例如陈化石油储备、地下含盐蓄水层，或者作为海床的CO_2水合物等)，那么甲烷水合物便能作为一个清洁、潜在储量庞大的待开发的替代品。低碳沼气可用来产生能量，其比传统储油量能维持更久时间。

这一章的目的是概述自然界中存在的甲烷水合物：它们的物理性质、形成、在自然环境中的分布、目前的储量估计以及开采潜力等。

第三卷　甲　　醇

第十八章　来自生物质的甲醇

本章的作者提议采用甲醇作为乙醇的部分替代产品，利用如下反应链：二氧化碳/水/太阳能→生物质→甲烷→甲醇。首先介绍了在控制气候变化的背景下甲醇的生产和使用，包括甲醇的性质、制造，和将含碳原料(如煤炭、石油、天然气、二氧化碳等)转化成甲醇的实践方法。随后介绍了甲醇生产的未来潜在渠道，即从生物质和煤炭、气化工艺、合成甲醇、生物甲烷以及近期形成的二氧化碳等生产甲烷。接着是甲醇生产的生物反应实施，涉及从可再生原料产沼气，从生物质产氢气等。最后，作者介绍了甲烷转化成甲醇的生物途径的前景和考虑因素，包括甲烷转甲醇的氧化反应，从有机物产甲醇的还原-发酵反应，还有还原-碳酸盐呼吸作用产甲醇。

第十九章　甲醇生产的前景展望

能源与环境是改善生活品质的最关键的先决条件。全球变暖和能源危机令那些能提供高效碳中立燃料来源的创新技术增值，这些技术于是成为了维持全球单碳种类的氧化和还原反应途径平衡的绝对必要因素。持续增长的人口和工业化造成能源需求扩大，自然资源迅速耗尽。同样地，对化石能源的利用增加，释放过量的二氧化碳到环境中，也造成了全球变暖现象。在可预见的未来，预计化石能源的利用仍会持续一段时间，这是因为目前代用燃料的经济可行性不高，且只占总能源经济的一小部分。减少全球 CO_2 排放的长期战略在于可再生能源资源，尤其是它们的碳中立性。同样地，面临化石资源耗尽的趋势，我们急需替代能源解决方案。从长期看来，随着世界通过利用替代燃料资源以减少 CO_2 水平，替代燃料逐渐取代化石燃料目前的地位，人们迫切需要创新型可持续碳管理策略。然而这很有可能要首先树立一个未来能源技术过渡示范。考虑到燃料缺乏和全球变暖的严重威胁，我们需要颠覆性的技术革新，而非生物能源产生的增量式改进。

长期的环境和能源可持续性的问题涉及复杂的相互依赖的变量平衡，今后的关键技术将利用协同增效的方式维持和谐大局。改良的可再生燃料资源对持续经济增长和清洁、安全、可靠且具竞争力而无须牺牲环境的能源供应方式进展尤为重要。在这个前提下，本章描述的仿生法(biomimetic method)有望提供必需的解决方案，满足能源需求，无须损害环境的完整性。温室气体 CO_2 转化成高价值燃料(如甲醇等)能成为有前景的一个途径，不但能回收再利用温室气体，还能生产代用燃料资源。CO_2 到甲醇的酶转化类似于能将 CO_2 转化成糖分的光合作用。包含在溶胶基体中的三种不同的脱氢酶对二氧化碳相继的还原反应能提高甲醇的产量。甲醇的高效生产可建立一个便捷的渠道，不仅能从现成的资源就地生产甲醇，还可联系到相关的有潜力的能源技术利用，以及环境中的二氧化碳固定等。酶工艺对比起其他工艺而言非常重要的优势，在于它的特定性、选择性，还有能量强度较小的外界温度-压力操作条件。微生物或酶制剂产燃料的大型应用目前仍处于初级阶段，重大进展和突破仍需普遍的认可以及生产能源的可替代策略方案。新型绿色燃料技术要想在长期获得成功，那它们必须具有和目前的工艺方法匹敌的竞争力。新式仿生法要求具有长期经济可持续性，以取代当前的制造工序。仿生能源转化中利用酶的关键推动力在于资本和其他资源的投资，以大型生产达到必需的技术框架以及具成本效益的可持续发展为目的。在今后长期内，随着适当的资源分配，酶作用的生物甲醇生产可望实现自我持续的技术实践发展。

第四卷　氢能生产

第二十章　氢能生产中的氢化酶工程

未来可取代排放温室气体的化石燃料中，一个引起人们关注的选项是分子氢，其前提是以清洁和可再生的方式生产。目前科学家对其生产、使用和储存进行了大量研究。新的生物技术的工艺发展是为满足未来的能源需求而设计，运用能在生物进化初期便能够利用 H_2 的微生物。许多生物体(包括某些细菌、古生菌和低等真核生物)都拥有活跃有效的氢代谢，利用 H_2 分子的键裂，在其变成两个电子和两个质子的同时得到能量，或释放还原能力。在生物系统中，该反应由含金属的氢化酶催化进行。

目前找到三类按所含金属划分的氢化酶：①含两个铁原子的活性位点、对氧敏感的[FeFe]氢化酶；②含有异源双原子核镍-铁活性位点、更健全的[NiFe]氢化酶；③亚甲基四氢甲醇蝶呤(methylenetetrahydromethanopterine)氢化酶。虽然[FeFe]氢化酶貌似最适合用来生产氢气，但因为它们对氧化性损害极为敏感，关于这类酶的工程研究还十分有限，其生化特质不能够确定。这就是为什么这一章的重点放在大多数研究中稳定性较强的[NiFe]氢化酶上。

目前来说，氢化酶的构效关系还处于基础研究阶段，目的在于了解酶的催化机理。但这些研究只局限于少数几个酶模型，因为使用常规的异源宿主[如大肠杆菌(*E. coli*)]制造基因重组氢化酶存在巨大的困难。虽然[NiFe]氢化酶操纵子是一个在进化史上高度保守的序列，也表现出高度的相似性，但每个蛋白质成熟系统都仅限于相对应的酶。这种特定性障碍造成了氢化酶无法在异源宿主内成熟。然而，这个基本研究的深层发掘是许多生物技术应用的主要挑战，拥有最优氢化酶的微生物因此可在氢能燃料生产中发挥重要作用。

为使酶的效率达到满意的水平以及技术的完善，氢化酶必须从机能上优化，其需通过改进从电子供体的电子传递(即改变酶的动力学参数)或改进对 O_2 的耐受性来实现。这些待攻克的难题可通过基因工程的两种策略达到：在细胞水平上，通过代谢工程，可为底物还原创造适宜条件，或避免接触 O_2；在酶水平上，通过蛋白质工程，目的是改进氢化酶，使其可以比其他酶在争夺底物上更有优势，或使用对热力学有利的底物，或变得 O_2 耐受性更强，或生产 H_2 的效率比摄取 H_2 的更高等等。在这一章中，我们概括了氢化酶的构效关系研究，改进的酶的新特性可作为生物技术工艺中氢化酶合理最优化的灵感源泉。

蛋白质副产物可参与新型氢能经济发展。如能有利于分子研究的[NiFe]氢化酶，为促进仿生催化剂发展而研究催化原理、机体工程等，其他问题也是当前研究的重点对象。

第二十一章　紫色无硫光合细菌用固氮酶催化产氢气

氢气在作为石油的替代品方面，得到了相当的重视。它是一种完全燃烧燃料载体，可在氢燃料电池中转化成电能，用作汽车动力或其他电力驱动的进程。推广氢气的部分原因在于它可从一系列的原料制造出来，包括天然气、煤炭、化石燃料、生物质和太阳能。这一一章着重于氢能生产的生物过程、利用环保的生物质和太阳能。

一种被称作紫色无硫细菌(PNSB)的不生氧光合作用细菌在正常的生长条件下产生大量的氢气,利用固氮酶而非氢化酶。目前已发现超过20个属的PNSB,它们全部都能够在固氮实验条件下生长并释放氢气,作为这种生长模式的副产物。PNSB使用无氧光合作用产生ATP,供应固氮酶催化的氢气生产。PNSB通常获取低电势电子,通过氧化有机物产氢气,但它们也能够氧化选定的几种无机化合物达到此目的。大多数菌种能氧化发酵工艺废物流中的化合物。因此,PNSB可作为双重用途生物体,在废水处理与伴随的氢气生产中有应用潜力。

这一章回顾了PNSB固氮酶催化产氢气的基础生物学.作者还介绍了几种改进和稳定这类细菌产氢气的基本策略,还有构建高效低成本产氢气生物反应器的策略。科学家相当了解产氢气所需的固氮酶调控、代谢过程、光合磷化作用,但他们普遍尚未在深入了解产氢气的系统生物学的前提下研究这类课题。从实验台到大型反应器,菌种改良和生物反应器的设计需要有计划有步骤。在实施时须以整个细菌细胞的生理特征为出发点,而不能只关注产氢气的几个小部分。

第二十二章　光合水裂解产氢气

生物产氢是本章的主要话题,重点是与绿藻相关的太阳能直接转化过程(阳光加上水通过光合作用直接联系到氢化酶的功能)。本章作者认为藻类所含的[FeFe]氢化酶优于固氮酶和存在于蓝藻及光合细菌中的[NiFe]氢化酶,水可作藻类的反应底物而无需其他有机物参与;藻类的氢气生产理论效率比蓝藻的固氮酶系统高出好几倍。而且同样在很多细菌体内也出现的[FeFe]氢化酶,能够比[NiFe]氢化酶更容易地表达在基因重组或非原生的生物体中,[NiFe]氢化酶通常需要好几种某生物体特定的蛋白质成熟基因的联合表达。最后,藻类中的[FeFe]氢化酶的最大活性与含氧光合作用的最高速度相匹配。

这一章着重介绍产氢气的生物体的光生物学,和能够将光合作用的水氧化直接联系到[FeFe]氢化酶催化产氢气功能的过程。首先介绍一系列背景知识,包括产氢气的生物种类、产氢酶、含氧光合作用、产氢途径、在铁氧化还原蛋白层面上对还原剂的竞争、产氢气的效率和能量学、影响因素等。其后重点描述[FeFe]氢化酶,包括它的遗传学、基因表达,以及蛋白质成熟、结构和建模等。另外,多个产氢气的物种的基因组的测定,也为藻类产氢研发相关的转录物组、蛋白质组、代谢物组以及计算系统生物学等奠定了实验检测的基础。最后作者点明了在实际运用中需要注意的一些问题,譬如[FeFe]氢化酶对氧气非常敏感,所以需要从反应混合物中移走O_2以维持光生物原理的H_2生产。一个比较有效的办法是剥夺绿藻的含硫养分,其对含氧光合作用有可逆性抑制作用,因此氧气消耗率比生产率高,造成绿藻封闭培养内的缺氧环境,因而启动氢化酶途径。还有就是减少每个细胞对阳光的吸收度,以增加绿藻大量培养中的透射度和照射利用率,因为绿藻的光合作用在阳光全面照射的五分之一强度便已经饱和。

第五卷　燃料电池：电力

第二十三章　产电生物生产电能

“微生物燃料电池”(MFC)这个词组用于形容差异较大的一些在微生物的帮助下产电的系统。近年来人们对 MFC 又慢慢产生兴趣，部分原因是 MFC 的工程改进证明能增加电力输出。另外，如这一章所述，最近，一种“产电微生物”(electricigen)的发现清楚表明，高效、可持续及可广泛应用的 MFC 雏形渐渐形成，它能利用废弃有机物和再生生物质产生电能。这一章的目的是回顾 MFC 的微生物学，尤其是由产电生物推动的燃料电池。

产电微生物与其他被利用于 MFC 的微生物的主要区别在其产电原理。产电微生物通过将有机化合物完全氧化至二氧化碳，把电子直接传输到 MFC 的阳极上，并将能量用到自身生长上。产电微生物可作为纯培养物，为使 MFC 有效将有机燃料转化成电能，至少绝对需要一种产电微生物的存在。然而，一些非产电微生物的菌种也可有助于产生电流，在某些情况下，甚至是必需品；譬如在将某些有机燃料转化成更简单的燃料以供产电微生物氧化利用方面。随后本章介绍了按产电微生物来划分的 MFC 的种类。接着作者详述了电子传输到阳极的原理，以帮助材料设计，使之更好地与细胞和阳极等更合适的电接触点互动，取代目前大多数 MFC 所应用的石墨导板。另外，这类知识也可帮助阳极表面的设计，以确保燃料输送和最佳电子传输等。

在改善 MFC 设计方面，有很多选择可能，如提高阴极的反应率，或克服阳极和阴极隔间之间的质子交换率潜在的有限性等等。但在了解了电子传输到阳极的原理之后，阳极材料的设计或模式结构以及与产电微生物和阳极间的电连接相关的基因工程等方法就变得尤为引人注目。

第二十四章　生态系统工程在微生物燃料电池中的应用

对能量的高追求决定了微生物菌落的性质。菌落发展成各种各样的生态系统，微生物在底物转化过程中最大化地获得能量，决定了它们生存、生长以及在一个群落里获得主导地位的几率。在这种情况下，不但底物和养分成为了必需品，合适的电子受体也是关键。MFC 的阳极和阴极的隔间变成了人工改造的生态系统，其中微生物群落发挥机能，与固体传导面紧密地相互作用。由于阳极和阴极都有干预微生物的能量代谢，它们因此施加了自然环境生态系统中并不存在的选择压力。对于在 MFC 中生长的菌落，底物和养分的可用度变成了取决于空间的因素。而且，电极的固体性质迫使细菌发展出一套专门的对策以回传输电子到电极。这些因素与那些依附在电极表面的生物膜的细菌存在着因果关系。在这一章里我们讨论了细菌产生和消耗电能的动力，大致介绍 MFC 里发现的微生物菌落、使其成为菌落结构的相互作用、菌落里进行的反应过程以及基因工程如何影响 MFC 里的微生物资源。

燃料电池里的微生物菌落并不比其他技术系统(如活化污泥处理池或厌氧消化器)中更复杂。就像对待这些系统一般，我们必须了解相关的工程参数，实现高效转化率和有效发电量。

目前,控制着 MFC 输出的因素是开始的足量的微生物多样性,即设定电极电势,在细菌新陈代谢过程中获得和输送到系统的电力间得到一个适当的平衡,避免反应淬火,困住电子用于其他反应(如甲烷产生等),最后是通过增加小段距离或者促进电子传导基体来提供稳定的电子流量。

第六卷 有机溶剂

第二十五章 丁醇发酵的分子状态

丁醇是生物基产品中很重要的一员，它在生物燃料用途上有很大的发展潜力。对比起乙醇，丁醇有更优越的化学性能，例如挥发性和亲水性较低、与其他烃类化合物的混合度高，以及每单位质量的能量值较高等。丁醇并不是生物基化学品和燃料大家庭的新生儿，实际上，在经济价值较高的化工法盛行的20世纪50年代到来之前的40多年中，厌氧丙酮-丁醇-乙醇(ABE)梭菌发酵一直是丁醇和丙酮的主要来源。

这一章聚焦在产溶剂梭菌的溶剂形成的调整上，特别关注 *Clostridium acetobutylicum* ATCC 824，这是在基因层面上研究得最广泛，也是到近期为止唯一完成测序的产溶剂性梭菌。*Clostridium beijerinckii* 是最近才被定序的，但它的序列和注解并未正式公开。*C. beijerinckii* 和其他梭菌的相关已知具体分子细节也会在本章加以讨论。连续的 C. acetobutylicum 的培养显示三种能够以不同比例分别产出酸类、醇类和酮类的代谢状态。导致这些状态的转换的关键是培养基的 pH 和 NAD(P)H 可用度。当 C. acetobutylicum 生长在葡萄糖和中性 pH 环境中时，其结果是形成一个产酸的培养菌，产出醋酸和丁酸。如生长在低 pH 的葡萄糖上，培养菌就变成产溶剂性，产出丙酮、丁醇和乙醇。最后，在中性 pH 和大量 NAD(P)H 存在的情况下，培养菌被认为是产醇性，因其只产出丁醇和乙醇。

在过去二三十年中，在由产溶剂梭菌生产的溶剂的基础上，科学家们深入研究了一些相关问题，其中包括酸生成作用和孢子形成间的联系、溶剂形成的分子基础、耐酸性和耐溶剂性的原理、如何控制不同发酵产物的比例，以及不同的底物如何影响产出的醇类、酮类和酸类。这一章结合了最新的研究发现来解答这些问题，尤其是在往年未被重视的基因组角度上重新诠释了这些问题。

第二十六章 丁醇的工业生产

丁醇是一个四碳醇，可作为生产可塑剂、醋酸丁酯、丁胺、氨基树脂、乙二醇醚、丙烯酸丁酯和二丁稀酸盐等的直接溶剂。生产生物丁醇是一个相对复杂的过程，因为产溶剂梭菌其实是双相性的发酵。丁醇发酵过程存在好几种限制，如高成本底物、丁醇对发酵微生物的毒性/抑制作用造成发酵液中的丁醇浓度低、产量低、生产力低等。虽然目前面临油价高涨、对生物丁醇的兴趣重燃的大趋势，但如使用石油化工时代之前的传统分批发酵工艺，则生物丁醇将不具有经济竞争力。由纤维素和非纤维素原料组成的生物质含有可发酵的碳水化合物的最丰富的来源，可被发酵成丁醇。所以，提高产溶剂梭菌的性能和利用其他低廉的原料，是促进生物丁醇发酵研究实现有经济竞争力的工业化的策略。为解决产物毒性，提高丁醇生产力，科学家深入研究了各种可选的在线丁醇回收系统，如全蒸发法、渗透萃取(perstraction)、反向渗透、吸附作用、液液萃取和气提工艺等，并结合了同步丁醇发酵。这一章回顾了生物丁醇生产的实践方法，包括从培养基减少或除去丁醇毒性，以及了解发酵微生物的生理学和营养学，以求达到更高的产品专属性和产量。

第二十七章　生物溶剂炼制厂：MixAlco工艺（混合醇类工艺）

现代农业和生物化工的进展，以及高油价的形势，使得生物基溶剂变得切实可行且具经济意义。因为石油化工的发展促进了化学品灵活性更强的新溶剂出现，生物工艺必须能够比之前生产更多品种的溶剂，满足目前的需求量。除 *Clostridium acetobutylicum* 能通过ABE发酵生产好几种不同溶剂之外，其他大多数过程都设计成只生产单一产品。另外发酵需要自由糖分子，必须直接从甘蔗或甜菜等含糖庄稼，或从含淀粉庄稼通过酶解或酸解，或从木质纤维素产出自由糖。尽管面临着更大的难度和复杂性，但因为来源丰富、生产力大，木质纤维素成为最佳原料选择。

目前流行的两大方法——生物发酵和热化学工艺各自存在着缺点。当前，生物发酵的预处理和酶解成本过高，尤其是在维持培养菌的纯度所要求的设备上，成本高居不下，造成产品价格过高。而热化学平台的气化炉成本高，难于操作，防止催化剂中毒的气体清理程序费用也很高。

除生物发酵和热化学平台外，人们还可以利用一种相当有前景的技术生产化学品（如燃料和溶剂等），那就是MixAlco工艺（混合醇类工艺）。该技术由德克萨斯A&M大学的Mark Holtzapple及其研究组发明并申请了专利，其利用某些厌氧微生物的混合培养，因此并不要求无菌性。这些微生物产出一系列从乙酸（C_2）到庚酸（C_7）的羧酸。这些羧酸通过缓冲剂的中和来控制pH，如碳酸钙或碳酸氢铵等，产出碳酸盐。从这些碳酸盐可进行进一步的化学反应，以获得各种化合物，包括100多种燃料和溶剂等，如羧酸、酮类、醇类、酯类、醛类、醚类，还有伯醇和仲醇等。这些化学品中的许多，如醋酸、丙酮、甲基乙基酮、二乙基甲酮、乙醇、丙醇、丁醇、异丙醇、仲丁醇等都是重要的工业溶剂。这些反应的成本合理，原因是碳酸盐的生产成本较低。这种低廉的工艺的核心是发酵，与常规的厌氧产沼气发酵并无差异。这篇论述回顾了发酵下游各种能生成上述化学品的主要反应。

第二十八章　丙酮丁醇梭菌（*Clostridium acetobutylicum*）代谢工程以提高生物燃料生产

为了增加微生物丁醇产出和对高丁醇浓度的耐受性，宿主微生物的代谢途径需接受全局操控。微生物代谢工程是能够从不同的生物体生产许多对化学工业有用的产品的成功策略，如氨基酸、维生素及其前体、类黄酮、化学物前体、辅因子、生物表面活性剂、抗生素、聚合物以及与人体免疫缺陷病毒或抗疟疾药物有关的化疗药物。

代谢工程进行系统性生物合成途径的更改，使代谢物生产水平与宿主生物体原来的产量相异。这些细胞内变化不应比正常特性减损得太厉害。某代谢物的过量生产不应以微生物的正常生长为代价。保留完整的生长档案非常重要，因为某些代谢物只会在生长的某特定阶段中合成，依赖于生长阶段的代谢物生产在丙酮丁醇梭菌发酵中尤其是如此。C. acetobutylicum在指数生长期内发酵糖分至有机酸醋酸盐和丁酸盐，在过渡到平稳期时，酸类开始再同化到有机溶剂丙酮和丁醇，然后在平稳期内才真正产出溶剂。

传统办法包括核心代谢基因敲除或过表达，可实现溶液生产的表现型改进。但该综述中的研究反映了代谢工程过程中的复杂性增加，并确认如要实现菌种工程，意图改良到理想的表现型，必须伴随着对生物体作为一个整体的更深层了解。对关键转录因子的研究发现，跟孢子形成有关的 *Spo0A*，热激蛋白，与脂肪酸合成、代谢通量有关的因子，甚至是用 SolR 研究例证的潜行因子，全部都能影响溶剂产量。

第七卷　微生物采油应用

第二十九章　新兴的油田生物技术:用硝酸盐注射防止油田酸化

传统石油和天然气工业呼唤新技术以减少生产和使用过程对环境的影响,延长使用期限,改进重油和沥青的回收。微生物学是目前可提供新解决方案的学科领域之一。虽然半个多世纪之前人们已经承认了它的潜力,但微生物在调整化石燃料能源资源中油成分的作用却直到现在才被深入了解,而可靠的微生物技术也是近期才得以实施。

世界上大部分的石油开采是通过水注射。在油井中注入水分,能够保持储油区的压力,将密度较小的石油逼浮上来。但这个技术有个缺陷,即可造成硫化物水平升高。因为硫还原细菌(SRB)伴随的可降解石油有机成分的氧化作用,会将硫酸盐还原成硫化物。大量的硫化物预示着毒性提高和腐蚀的风险,所以缓解措施是很有必要的。硝酸盐注射改变了油井内的微生物菌落,令异养硝酸盐还原细菌和某些情况下硝酸盐还原硫化物氧化细菌的活动更为突出,破坏了原来 SRB 的硫代谢反应周期。

为达到防止腐蚀作用的目的,有必要进行更深一层的研究,确定应用硝酸盐注入的科学实验计划。由研发活动带起的这些生物技术应用,将减少与化石燃料相关的水和能源的输入,与其他新兴技术(如碳捕获和地下储存)一起对化石燃料的环保使用作出贡献,同时这也是一个解决今后能源供应问题的对策。

第三十章　老化储区内的甲烷与石油的微生物法回收促进

在可选的能源种类中,甲烷燃烧更清洁,储量丰富且可持续,因此天然气的使用会变得越来越重要。美国境内依然拥有大量的石油储备,在技术可用且必要时可开采或转化这种重要能源。其中一个方法就是将截留石油原地生物降解成更环保的甲烷。各种烃类化合物可被无氧分解,包括链烷和环烷,以及单核或多核芳香烃。这些都是石油中相当重要的烃类组分。

显而易见,许多烃类化合物的氧化反应可协同各种电子受体的还原反应一同进行,因为烃类化合物实质上是远古时期的有机物主要形式,微生物关于烃类的催化反应通常在进化中都会偏向于保留下来。一些分解烃类的微生物可结合代谢途径和互养共营模式,最终将石油中的成分转化成最简单的烃,即甲烷。因此,厌氧代谢在制造形成纯度较高的甲烷流方面有一定的前景。甲烷可作为一种运输燃料是毫无疑问的,其亦可转化成别的燃料。

微生物还能用于增加储油区的石油产量。一些被证实有效的微生物能直接提高油井的产量,或者是提高储油区总体产量。值得指出的是,大多数微生物采油技术都涉及多种活动和产物,很有可能是协同作用。烃类的不完全氧化可形成醇类或脂肪酸,或从微生物细胞诱发生物表面活性剂或生物乳化剂的生产。这些微生物产物的好处是减少黏度并增加石油在储区中的能动性。

第八卷　为生产能源目的探索微生物基因组

第三十一章　探究微生物基因组产生物柴油

生物柴油被定为长链脂肪酸单烷基酯类，可作为压燃式发动机的替代燃料。当下所用的最常见的生物柴油即脂肪酸甲酯。目前，脂肪酸来自从菜油和动物脂肪回收的甘油三酯(TAG)。生物柴油是一种无毒性、硫化物排放低、可完全降解的燃料，因而被许多国家大量生产。在美国，生物柴油的主要来源是大豆油和黄脂膏；而在欧州和亚州地区，其分别主要源自油菜籽和棕榈油。在所有的情况下，生物柴油目前和将来预计的生产成本仍然是石油基产品的2～3倍。基于这些预计，生物柴油无法成为石油基燃料的有力竞争对手。尽管如此，微生物代谢工程还是为降低与生物柴油原料有关的成本提供了一个难得的机会。

虽仍未进行到工业测试阶段，但微生物有望使油菜籽等的产量供应满足生物柴油的需求，并且可设计成利用各种碳源作为产油原料，例如垃圾或者农业副产物等。这一章聚焦于不同微生物(藻类、细菌、酵母等)代谢途径的构造和调节，以生产脂肪酸和TAG。

我们对组成和调节脂肪代谢及油脂积累的生化和遗传学的探究进程在近几年来一直加速发展，负责脂肪酸和TAG生物合成的基因和代谢途径在多种模式生物中被鉴别出来。虽然脂肪生物合成中运用的基本生化反应在生物界中大致相同，但参与其中的不同版本的酶、途径及调控网络正是为了适应环境的精确调整。新兴的基因组对比课题强调的是各种种系发育群中不同调控机理的进化过程；对于一系列无氧、催化和调控步骤的基因转变可能会达到脂肪累积的明显增加。

Contents

Contributors

Peter Aelterman
Laboratory of Microbial Ecology and Technology (LabMET), Ghent University, Coupure Links 653, B-9000 Ghent, Belgium

Ross Anderson
Centre for Gas Hydrate Research, Institute of Petroleum Engineering, Heriot-Watt University, Edinburgh EH14 4AS, United Kingdom

Largus T. Angenent
Dept. of Energy, Environmental & Chemical Engineering, Washington University in St. Louis, St. Louis, MO 63130

Marco A. Báez-Vásquez
396 Mallard Point, Jupiter, FL 33458-8353

Edward A. Bayer
The Maynard I. and Elaine Wishner Chair of Bio-organic Chemistry, Dept. of Biological Chemistry, Weizmann Institute of Science, Rehovot 76100, Israel

George N. Bennett
Dept. of Biochemistry & Cell Biology, Rice University, Houston, TX 77251-1892

Hans P. Blaschek
Center for Advanced Bioenergy Research and Dept. of Food Science & Human Nutrition, University of Illinois, 34 Animal Sciences Laboratory, 1207 W. Gregory Dr., Urbana, IL 61801

Nico Boon
Laboratory of Microbial Ecology and Technology (LabMET), Ghent University, Coupure Links 653, B-9000 Ghent, Belgium

Jacob R. Borden
Dept. of Chemical and Biological Engineering, Northwestern University, Evanston, IL 60208

Rodney J. Bothast
National Corn-to-Ethanol Research Center, Southern Illinois University-Edwardsville, 400 University Park Dr., Edwardsville, IL 62025

Nicky Ciazza
Mascoma Corp., Lebanon, NH 03766

Peter Clauwaert
Laboratory of Microbial Ecology and Technology (LabMET), Ghent University, Coupure Links 653, B-9000 Ghent, Belgium

Laurent Cournac
CEA-Cadarache, DSV, IBEB, SBVME, LB3M, UMR 6191 CNRS CEA Université de la Méditerranée, Bt 161, 13108 Saint Paul Lez Durance Cedex, France

Devin Currie
Thayer School of Engineering, Dartmouth College, Hanover, NH 03755

Bakul C. Dave
Dept. of Chemistry and Biochemistry, Southern Illinois University-Carbondale, Carbondale, IL 62901

Liesje De Schamphelaire
Laboratory of Microbial Ecology and Technology (LabMET), Ghent University, Coupure Links 653, B-9000 Ghent, Belgium

Arnold L. Demain
The Charles A. Dana Research Institute for Scientists Emeriti (R.I.S.E.), HS-330, Drew University, Madison, NJ 07940

Bruce S. Dien
Fermentation Biotechnology Research Unit, National Center for Agricultural Utilization Research, Agricultural Research Service, U.S. Department of Agriculture, 1815 N. University St., Peoria, IL 61604

Roy H. Doi
Section of Molecular & Cellular Biology, University of California, Davis, CA 95616

Thaddeus C. Ezeji
Dept. of Animal Sciences, The Ohio State University, 1680 Madison Ave., Room 305 Gerlaugh Hall, Wooster, OH 44691

James G. Ferry
Dept. of Biochemistry and Molecular Biology, The Pennsylvania State University, University Park, PA 16802

Matthias Gerhardt
BIOPRACT GmbH, Magnusstraße 11, 12489 Berlin, Germany

Maria L. Ghirardi
Chemical and Bioscience Center, National Renewable Energy Laboratory, Golden, CO 80401

Cesar B. Granda
Dept. of Chemical Engineering, Texas A&M University, College Station, TX 77843-3122

Caroline S. Harwood
Dept. of Microbiology, University of Washington, Seattle, WA 98195-7242

Chris Herring
Mascoma Corp., Lebanon, NH 03766

Mark T. Holtzapple
Dept. of Chemical Engineering, Texas A&M University, College Station, TX 77843-3122

Lonnie O. Ingram
Dept. of Microbiology and Cell Science, University of Florida, Gainesville, FL 32611

Natalia Ivanova
Genome Biology Program, DOE-Joint Genome Institute, 2800 Mitchell Dr., Walnut Creek, CA 94598

Laura Jarboe
Dept. of Microbiology and Cell Science, University of Florida, Gainesville, FL 32611

Thomas W. Jeffries
USDA Forest Service, Forest Products Laboratory, and Dept. of Bacteriology, University of Wisconsin, Madison, WI 53726

Shawn W. Jones
Dept. of Chemical and Biological Engineering, Northwestern University, Evanston, IL 60208

Irina A. Kataeva
Dept. of Biochemistry and Molecular Biology, B222A Fred Davison Life Sciences Complex, University of Georgia, Athens, GA 30602

Paul W. King
Chemical and Bioscience Center, National Renewable Energy Laboratory, Golden, CO 80401

Raphael Lamed
Dept. of Molecular Microbiology and Biotechnology, Tel Aviv University, Ramat Aviv 69978, Israel

Z. Lewis Liu
National Center for Agricultural Utilization Research, Agricultural Research Service, U.S. Department of Agriculture, Peoria, IL 61604

Lars G. Ljungdahl
Dept. of Biochemistry and Molecular Biology, A216 Fred Davison Life Sciences Complex, University of Georgia, Athens, GA 30602

Derek R. Lovley
Dept. of Microbiology, University of Massachusetts, Amherst, MA 01003

Gregory W. Luli
Verenium Corp., 4955 Directors Place, San Diego, CA 92121

Athanasios Lykidis
Genome Biology Program, DOE-Joint Genome Institute, 2800 Mitchell Dr., Walnut Creek, CA 94598

Lee R. Lynd
Thayer School of Engineering, Dartmouth College, Hanover, NH 03755, and Mascoma Corp., Lebanon, NH 03766

Michael J. McInerney
Dept. of Botany and Microbiology, 770 Van Vleet Oval, University of Oklahoma, Norman, OK 73019

Anastasios Melis
Dept. of Plant and Microbial Biology, University of California, Berkeley, Berkeley, CA 94720-3102

Dale A. Monceaux
AdvanceBio, LLC, 11427 Reed Hartman Highway, Suite 220, Cincinnati, OH 45241

Kelly P. Nevin
Dept. of Microbiology, University of Massachusetts, Amherst, MA 01003

Michael Newcomb
Dept. of Chemical Engineering, University of Rochester, Rochester, NY 14627

Nancy N. Nichols
Fermentation Biotechnology Research Unit, National Center for Agricultural Utilization Research, Agricultural Research Service, U.S. Department of Agriculture, 1815 N. University St., Peoria, IL 61604

Nick Orem
Mascoma Corp., Lebanon, NH 03766

Eleftherios T. Papoutsakis
Dept. of Chemical and Biological Engineering, Northwestern University, Evanston, IL 60208

Carlos J. Paredes
Dept. of Chemical and Biological Engineering, Northwestern University, Evanston, IL 60208

Matthew C. Posewitz
Environmental Science and Engineering Dept., Colorado School of Mines, Golden, CO 80401

Korneel Rabaey
Advanced Water Management Centre, University of Queensland, Brisbane, QLD 4072, Australia

Manfred Ringpfeil
BIOPRACT GmbH, Magnusstraße 11, 12489 Berlin, Germany

Marc Rousset
CNRS, BIP, 31 chemin Joseph Aiguier, 13402 Marseille Cedex 20, France

Badal C. Saha
National Center for Agricultural Utilization Research, Agricultural Research Service, U.S. Department of Agriculture, Peoria, IL 61604

Kazuo Sakka
Graduate School of Bioresources, Mie University, Tsu 514-8507, Japan

Hans-Joachim Sander
Goldlackweg 5, 06118 Halle (Saale), Germany

Bernhard Schink
Dept. of Biology, University of Konstanz, 78457 Konstanz, Germany

Miles C. Scotcher
Foodborne Contaminants Research Unit, Western Regional Research Center, USDA-ARS, 800 Buchanan St., Albany, CA 94710

Michael Seibert
Chemical and Bioscience Center, National Renewable Energy Laboratory, Golden, CO 80401

Ryan S. Senger
Dept. of Chemical and Biological Engineering, Northwestern University, Evanston, IL 60208

Yuval Shoham
Dept. of Biotechnology and Food Engineering and Institute of Catalysis Science and Technology, Technion—Israel Institute of Technology, Haifa 32000, Israel

Ryan Sillers
Dept. of Chemical and Biological Engineering, Northwestern University, Evanston, IL 60208

Arkady P. Sinitsyn
Dept. of Chemical Enzymology, Moscow State University, Moscow, Russia 119899

Patricia J. Slininger
National Center for Agricultural Utilization Research, Agricultural Research Service, U.S. Department of Agriculture, Peoria, IL 61604

Joseph M. Suflita
Institute for Energy and the Environment and Dept. of Botany and Microbiology, Sarkeys Energy Center, Boyd Street, University of Oklahoma, Norman, OK 73019

Leighann Sullivan
Dept. of Biochemistry & Cell Biology, Rice University, Houston, TX 77251-1892

Ralph S. Tanner
Dept. of Botany and Microbiology, University of Oklahoma, 770 Van Vleet Oval, Norman, OK 73019-6131

Bahman Tohidi
Centre for Gas Hydrate Research, Institute of Petroleum Engineering, Heriot-Watt University, Edinburgh EH14 4AS, United Kingdom

Vladimir N. Uversky
Dept. of Biochemistry and Molecular Biology, Center for Computational Biology and Bioinformatics, Indiana University School of Medicine, 635 Narnhill Dr., MS 4021, Indianapolis, IN 46202-3763

Willy Verstraete
Laboratory of Microbial Ecology and Technology (LabMET), Ghent University, Coupure Links 653, B-9000 Ghent, Belgium

Gerrit Voordouw
Dept. of Biological Sciences, University of Calgary, Calgary, Alberta, T2N 1N4, Canada

Ann C. Wilkie
Soil and Water Science Department, University of Florida—Institute of Food and Agricultural Sciences, Gainesville, FL 32611

Monika Wolf
BIOPRACT GmbH, Magnusstraße 11, 12489 Berlin, Germany

Brian A. Wrenn
National Corn-to-Ethanol Research Center, Southern Illinois University Edwardsville, Edwardsville, IL 62025

J. H. David Wu
Dept. of Chemical Engineering, University of Rochester, Rochester, NY 14627

Preface

This volume, *Bioenergy*, provides concise views of microbial conversions that lead to renewable fuel sources, and microbial activities sparing the use of nonrenewable resources. Each chapter is written by experts currently engaged in the research, who not only present the current status but also lay the foundations for future research. There is no claim of comprehensive coverage; in fact, no boundaries should be implied that might limit the creativity of the readers.

This project grew out of awareness that we as professional microbiologists have an obligation to take the lead in developing renewable sources of fuels that are more nearly carbon-neutral and can replace the demand for dwindling fossil fuels. Many of us can recall the oil crisis of 1973–1974 here in the U.S., when gas prices soared, gas stations had endless lines, and the locking gas-cap was invented. Once again, fuel prices are rising, but this time the cause is increased demand and limited supply rather than politics. If this trend continues unchecked, the economic impact could lead to a breakdown in social structures and perhaps to chaos, not just in the U.S. but across the globe. It is imperative that we exponentially increase the exploration for alternative fuels so that a transition phase can be implemented and the instability that will surely result from severe shortages can be avoided.

No doubt, a single source of alternative energy is unlikely to meet all needs; thus, none can be ignored. Through evolutionary time, microbes have explored almost every chemical reaction to find sustenance. For now, we must focus on sources of raw materials that are sufficiently abundant to make a significant contribution to the voracious demand for energy. Biomass derived from photosynthesis is one obvious source, and microbial conversions of biomass are already proving fruitful.

Thus this book begins with 10 chapters on ethanol production from cellulosic feedstocks, which is more economically feasible and sustainable than ethanol production from corn. These chapters are followed by discussions of the status of energy sources that are in various stages of development or imagination, including methane, methanol, hydrogen, electricity, butanol, and others. In these chapters, a recurring theme is the desperate need for further exploration. We hope that this compilation of these accounts of ongoing research will stimulate and inspire bright scientists who will, in turn, contribute the major breakthroughs needed to make an impact. A brief glance at the Contents will reveal to the interested lay person the remarkable variety of microbial activities that have the potential to contribute to biofuel production.

We would like to thank the American Academy for Microbiology for sponsoring a critical issues colloquium in March 2006 titled "Microbial Energy Conversion," in which all three of us participated. Especially, we thank Carol Colgan, Director of the Academy, whose organizational skills ensured the opportunity for rich discussion. This colloquium clarified the need to generate a book that could summarize the many facets of microbial biofuel involvement and point the way to further advances. *Bioenergy* is the result of that initial discussion.

With some irony, the events of 1973–1974 stimulated the organization of a seminar attended by leading microbiologists and engineers and sponsored by the UN Institute for Training and Research and the Ministry for Research and Technology of the Federal Republic of Germany to review biomass production and the microbial processes involved in the conversion of biomass into energy sources. That seminar resulted in a book titled *Microbial Energy Conversion*, edited by H. G. Schlegel and J. Barnea and published by Gamon Press in 1977. It too is recommended for reading. The similarity in topics underscores the lapse in attention to funding for biofuel development and the unfortunate loss of the sense of urgency that resulted from temporarily plentiful fuel.

Such inattention to alternative fuels is a luxury of the past.

Judy D. Wall
University of Missouri-Columbia

Caroline S. Harwood
University of Washington, Seattle

Arnold Demain
Drew University, Madison, NJ

1. BIOETHANOL

Bioenergy
Edited by J. Wall et al.

Chapter 1

Production of Ethanol from Corn and Sugarcane

NANCY N. NICHOLS, DALE A. MONCEAUX, BRUCE S. DIEN, AND RODNEY J. BOTHAST

The production and use of ethanol for fuel are increasing worldwide in response to economic, security, and environmental concerns. Ethanol is used in the United States as an oxygenate to reduce automotive emissions, to reduce reliance on imported oil, and to reduce the risk of disruption of domestic oil production. Other grounds for use of ethanol include the benefits of producing ethanol for farmers and rural economies. Brazil and the United States are the world's largest ethanol producers and consumers. Ethanol is compatible with conventional automobile engines in blends containing 10 to 22% ethanol with gasoline, and at higher blends in "flexible fuel" vehicles (E85 in the United States and "hydrous" or 95% ethanol in Brazil). Issues related to the use and production of fuel ethanol are discussed in chapter 4 of this book.

In the United States, production of fuel ethanol from grains—primarily corn—at least tripled between 2000 and 2006 and continues to increase (according to the Renewable Fuels Association [www.ethanolrfa.org]). Annual production capacity reached 7.9 billion gallons by 2008, with an additional 5.5 billion gallons of capacity under construction. In 2006, slightly more than 2% (by energy content) of the liquid transportation fuel sold in the United States was ethanol (Collins, 2007). Approximately 20%, or 2.1 billion bushels, of the U.S. corn crop was used for ethanol production in 2006 (U.S. Department of Agriculture, 2007). Until recently, ethanol was produced and used primarily in the midwestern United States, mostly as a 10% blend with gasoline, as a fuel extender and oxygenate to reduce automotive emissions. The phase-out in recent years of methyl tertiary-butyl ether, a petroleum-derived oxygenate, along with the desire to reduce petroleum use, is driving the current expansion of ethanol production and consumption outside the U.S. corn belt.

Brazil has a long history of producing ethanol from domestically grown sugarcane. In 2005, 4 billion gallons of ethanol were produced in Brazil, which accounted for 40% of the transportation fuel sold. Ethanol production consumed 51% of Brazil's sugarcane crop in 2004. Demand for fuel ethanol in Brazil continues to rise with increased blending of gasoline with ethanol, and increasing numbers of flexible fuel vehicles are being produced and sold.

BIOLOGICAL BASIS OF ETHANOL FERMENTATION

Conversion of Corn Starch to Fermentable Sugars

Before fermentation, starch must be extracted from corn kernels and converted enzymatically to glucose. The corn kernel (Fig. 1) usually contains 70% starch along with 9% protein, 4% fat and oil, and 9% fiber on a dry weight basis (Watson, 2003). Starch is present in the endosperm, where it is chemically bound to gluten protein. The oil is found in the germ, which also contains arabinoxylan fiber. The kernel is protected by the pericarp, a fibrous layer with a waxy coating. At the end of the kernel, at the site of attachment to the cob, is the tip cap, which is also formed from arabinoxylan fiber.

Cornstarch is processed to ethanol by either the dry-grind or wet-milling process. The two processes differ in how the kernel is initially treated to access starch for enzymatic hydrolysis. In the dry-grind process, as the name implies, whole corn kernels are ground to coarse flour and carried through the fermentation process. In wet milling, kernels are fractionated into separate components, and only the starch fraction is fermented. The dry-grind and wet-milling processes share common chemical and biological features related to saccharification of starch to glucose and fermentation of glucose to ethanol. Therefore, the underlying

Nancy N. Nichols and Bruce S. Dien • Fermentation Biotechnology Research Unit, National Center for Agricultural Utilization Research, Agricultural Research Service, U.S. Department of Agriculture, 1815 N. University St., Peoria, IL 61604. **Dale A. Monceaux** • AdvanceBio, LLC, 11427 Reed Hartman Highway, Suite 220, Cincinnati, OH 45241. **Rodney J. Bothast** • National Corn-to-Ethanol Research Center, Southern Illinois University-Edwardsville, 400 University Park Drive, Edwardsville, IL 62025.

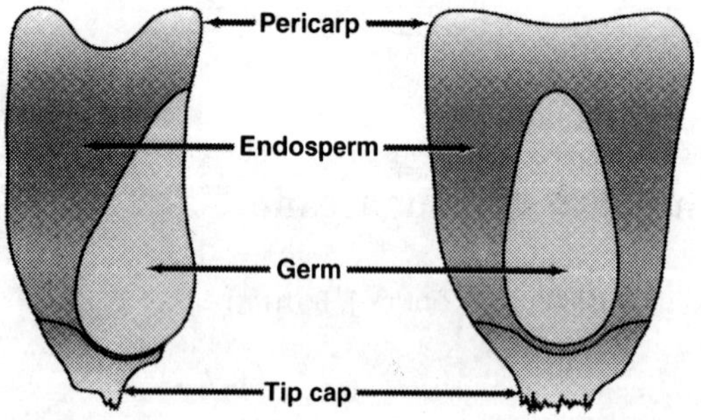

Figure 1. Structure of the corn kernel. Reproduced from Nichols et al., 2006.

basis of starch conversion for both methods will be described, followed by description of each process.

Corn starch occurs as crystalline granules (Fig. 2). The granule is water insoluble, with a hollow hydrophobic core and pores extending from its surface to the hollow center. Starch is present as amylose and amylopectin. Both are polymers of glucose, but amylose is a linear molecule with α-1-4 linkages, while amylopectin contains α-1-6 branch points in addition to α-1-4 linkages. The physical properties of the granule, including melting temperature, are determined to a large extent by the ratio of amylose to amylopectin. Starch is a remarkable natural product, and not surprisingly, the differing properties of starches from various sources direct their end use in food and commercial applications (Tester et al., 2004).

Obtaining soluble sugars from starch involves physical, thermal, and biochemical treatments. First, gelatinization of starch granules is accomplished by heating the starch suspension. Inter- and intramolecular hydrogen bonds are weakened, granules absorb water and swell, and the viscosity of the starch solution increases. The starch solution is thinned (liquefied) by α-amylase (EC 3.2.1.1), an endohydrolyase that breaks internal α-D-(1,4)-glucosidic linkages. The resulting shorter chains, referred to as dextrins, are subsequently saccharified to glucose and the dimer maltose by glucoamylase (EC 3.2.1.3), which hydrolyzes terminal α-D-1,4)-glycosidic linkages at nonreducing ends of chains and can also hydrolyze α-D-(1,6) linkages at branch points.

Amylases play a fundamental role in carbohydrate metabolism and are ubiquitous in plants, animals, and microbes. Major sources of commercial amylases are filamentous fungi, particularly *Aspergillus* species, and thermotolerant *Bacillus* species (Pandey et al., 2000). Historically, alpha-amylase produced from *Bacillus licheniformis* required 50 to 100 ppm of Ca^{2+} and pH 6.2 or higher for optimal stability. Newer versions have been protein engineered to require no added calcium and operate at or below pH 5.6, resulting in greater product yield (Crabb and Shetty, 2003). Amylases are secreted enzymes, allowing enzyme preparations to be obtained as culture supernatants. Other enzymes, in addition to gluco- and alpha-amylases, are sometimes used to aid starch conversion. Pullulanase, a debranching enzyme, may be used to hydrolyze branch points and decrease isomaltose formation, and protease may be added to enhance starch accessibility. Xylanases and cellulases degrade nonstarch plant polymers and are used to decrease viscosity, particularly for fermentation of whole grains such as wheat or barley.

Fermentation of Glucose and Sucrose to Ethanol

The yeast *Saccharomyces cerevisiae* ferments glucose and maltose, derived from starch as described above, by the same fermentation pathway (Fig. 3) used to make beverage alcohol (Russell, 2003). Sucrose, a disaccharide obtained from sugarcane, can be directly transported into yeast cells and converted by intracellular invertase to its glucose and fructose constituents. However, the bulk of sucrose metabolism is thought to initiate with extracellular hydrolysis by periplasmic invertase (Batista et al., 2004). Glucose, fructose, and maltose (a glucose dimer) are transported into yeast cells, and maltose is hydrolyzed intracellularly to glucose. Phosphorylated glucose and fructose are metabolized through glycolysis to pyruvate, which is decarboxylated to acetaldehyde. Ethanol is formed by reduction of acetaldehyde. One glucose molecule is therefore converted to two ethanol and two CO_2 molecules.

With conversion of acetaldehyde to ethanol, NADH is oxidized to NAD^+, which serves to balance the reducing equivalents produced in glycolysis (Fig. 2). However, yeast cells actually gain little energy benefit from fermenting glucose to ethanol, because most of the energy from glucose (heat of combustion [$\Delta H°_c$] of glucose is 2,807 kJ/mol) is retained in the fermentation product ($\Delta H°_c$ is 1,369 kJ/mol for ethanol and 0 for CO_2) (Roels, 1983). The thermodynamic yield for fermentation of glucose to ethanol is 97%. On a mass basis, the theoretical yield is 0.51 g of ethanol and 0.49 g of CO_2 per g of fermented glucose. The actual yield is about 90 to 93% of the theoretical; some loss is due to production of yeast cell mass and side products such as glycerol, citric acid cycle intermediates, and higher alcohols (Ingledew, 1999b).

Yield can be further reduced by contaminating microorganisms, predominantly lactic acid bacteria, which divert a portion of the glucose to alternate fermentation products (Ingledew, 1999b; Connelly, 1999). Contaminants also can disrupt the fermentation by competing for trace nutrients (Bayrock and Ingledew,

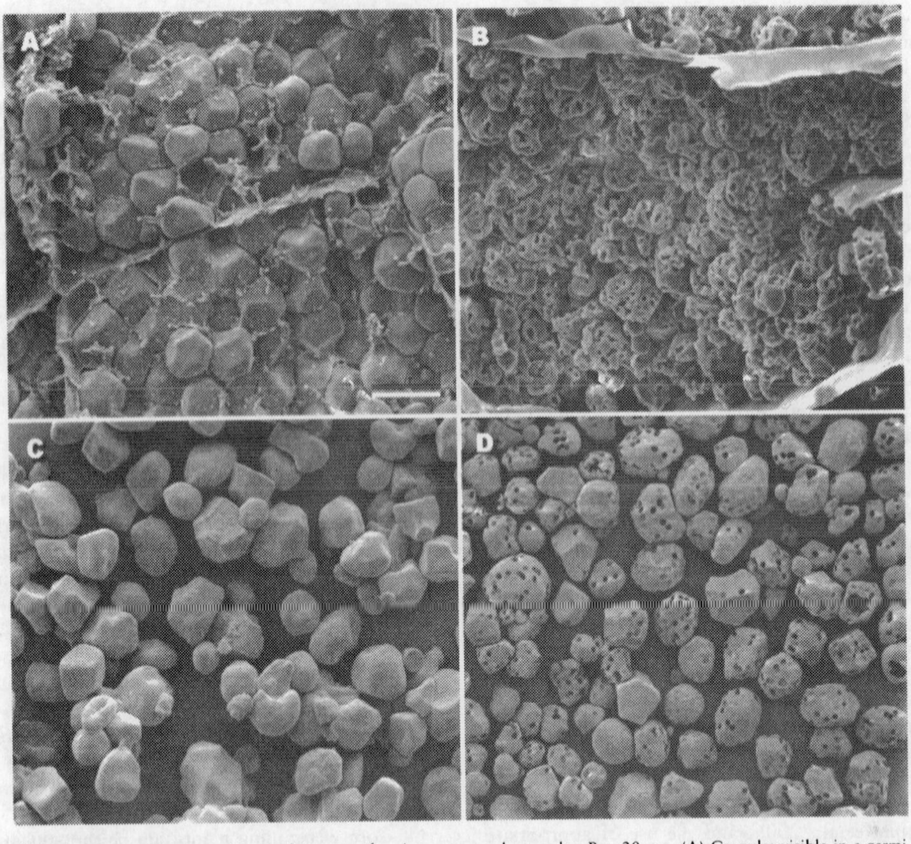

Figure 2. Scanning electron micrograph images of native corn starch granules. Bar, 20 μm. (A) Granules visible in a germinating corn kernel; (B) starch has undergone partial hydrolysis by day 7 of germination; (C) untreated purified starch granules; (D) purified granules treated with *Aspergillus kawachi* alpha-amylase and *Aspergillus niger* glucoamylase (Shetty et al., 2005). Images courtesy of David Johnston and Peter Cooke, U.S. Department of Agriculture, Agricultural Research Service; and Jay Shetty, Genencor International, Inc.

2004). Also, lactic acid and acetic acid produced by contaminants may inhibit yeast growth and metabolism (Connelly, 1999; Narendranath et al., 1997, 2001). In cases of acute "infection," fermentations stall and equipment must be cleaned to remove contaminating organisms. Consequently, contamination control is an important concern. Many bacteria do not grow below pH 4, and the ability of yeast to do so provides a natural method of suppressing the growth of contaminants. Ethanol, produced during the fermentation, also prevents growth of many microbes. Fermentations are frequently operated in "batches" rather than in continuous mode, to prevent well-adapted and faster-growing bacteria from spoiling the fermentation. In addition,

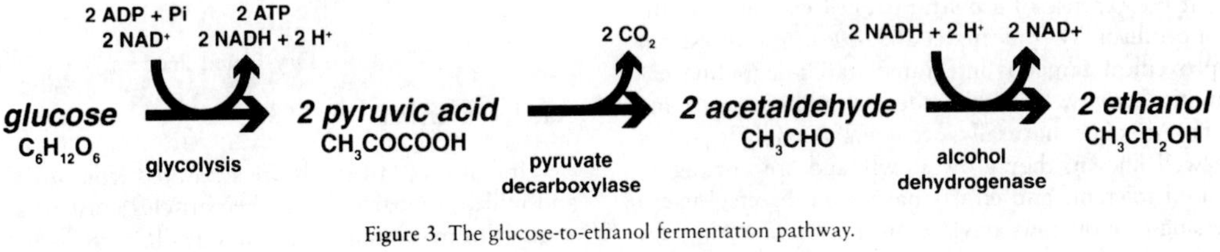

Figure 3. The glucose-to-ethanol fermentation pathway.

some facilities use penicillin or virginiamycin periodically or prophylactically to control contamination (Bayrock et al., 2003). Even so, bacterial titers in fermentation tanks can reach 10^6 to 10^8 organisms/ml (Narendranath et al., 1997; Skinner and Leathers, 2004). Note that yeasts are inoculated at the start of fermentation to an initial viable cell population of approximately 10^7/ml and reach 2.5×10^8 to 3×10^8/ml of fermentation mash (Ingledew, 1999a). So, the contaminant load, on a cell basis, can equal or surpass the yeast titer! Contamination of fermentations has been associated with biofilm microbes. A longitudinal study of dry-grind ethanol plants showed that individual facilities maintained characteristic population profiles whether or not antibiotics were used, and isolates from dry-grind plants were shown to form biofilms under laboratory conditions (Skinner-Nemec et al., 2007).

Ethanol-Fermenting Yeasts

Yeasts are well suited for the massive-scale fermentations typical of fuel ethanol production, because they are robust enough to withstand variations in pH, temperature, and osmotic conditions (Ingledew, 1999b). Nevertheless, research to improve fermenting yeast strains is an area of active investigation. Commercial ethanol fermentations present a stressful environment for yeast. The stresses, which can be synergistic, include high solute and ethanol concentrations, low pH, and the presence of acetic and lactic acids and salts. Adapted or engineered yeast strains (Alper et al., 2006; You et al., 2003; Fujita et al., 2006) and the use of appropriate fermentation conditions (Jones and Ingledew, 1994; Thomas and Ingledew, 1990; Thomas et al., 1994) could allow fermentation of higher substrate concentrations and production of correspondingly increased ethanol concentrations (Casey and Ingledew, 1986; D'Amore et al., 1990). Yeasts with higher ethanol tolerance would allow the use of smaller, less expensive equipment, decrease energy consumption associated with distillation, and reduce the amount of water that must be removed during product recovery.

Strains and fermentation conditions that reduce glycerol production would increase ethanol yield. Glycerol production balances cellular redox and alleviates osmotic stress. So, while shifting carbon from glycerol to ethanol is desirable to increase yield, it must not occur at the expense of growth rate, cell viability, or ethanol productivity (Bideaux et al., 2006). Another strain improvement target is integrated starch-degrading enzymatic activity, with the idea of reducing enzyme costs. Although naturally occurring amylolytic yeasts are well known, they grow slowly and are not highly ethanol tolerant, and efforts have been directed at expressing heterologous amylases in *S. cerevisiae* (Shigechi et al., 2004; Khaw et al., 2006). For sugarcane fermentations, improved ability to ferment high sucrose concentrations is desirable. One approach is to increase sucrose permease activity in cells lacking periplasmic invertase. This alteration could reduce osmotic stress by improving the ability of cells to ferment the disaccharide directly, without producing high external concentrations of glucose and fructose (Badotti et al., 2005). Finally, improved strains for conversion of biomass to ethanol are needed. Work to engineer strains that efficiently metabolize cell wall-related polymers and sugars has been well reviewed (Dien et al., 2003; Hahn-Hägerdal et al., 2006; Jeffries, 2006; van Maris et al., 2006; see also chapter 2 in this volume).

CORN TO ETHANOL PROCESSES

Ethanol is produced from corn by two methods (Fig. 4), dry grind and wet mill. A similar term, corn dry milling, is used to describe the dry fractionation of corn components used primarily for the food industry. In 2006, 82% of the U.S. ethanol-producing capacity was in dry-grind facilities and 18% was at wet-mill plants (Renewable Fuels Association). Dry-grind facilities (Kelsall and Lyons, 2003) produce 2.7 to 2.8 gallons (10.4 liters) of ethanol per bushel (25.5 kg) of corn, as well as animal feed termed distillers' dried grains with solubles (DDGS). DDGS consists of the fermentation residuals, including yeast and nonfermentable parts of the corn.

Corn-wet milling plants can be thought of as biorefineries, because the kernel components are processed into a number of valuable products (Johnson and May, 2003). Wet mills produce approximately 9.5 liters (2.5 gallons) of ethanol per bushel of corn, as well as corn oil, gluten meal, and corn gluten feed. The last two are sold as animal feed, gluten meal as a high-protein feed largely for poultry and corn gluten feed as a low-protein feed for ruminants. In some cases, CO_2 is also captured for sale. The additional coproducts from wet milling add value to the process. However, because wet mills require higher capital and operating expenses, most of the expansion of the corn ethanol industry is in new and larger dry-grind ethanol plants, where future developments will likely include "dry" methods to capture "wet" biorefinery capabilities.

The Corn Dry-Grind Process

Starch conversion

Initially, corn kernels are separated from the chaff and milled to coarse flour. The particles must be small enough to make starch granules available for swelling

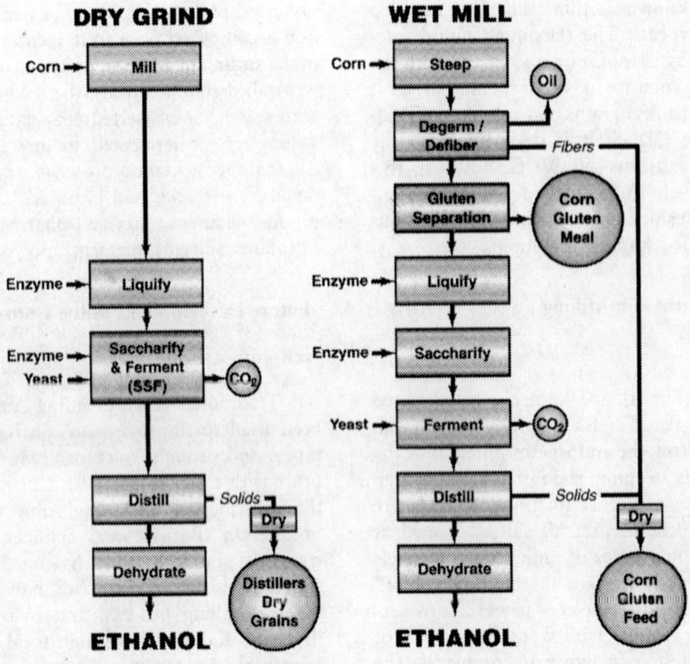

Figure 4. Comparison of the dry-grind and wet-mill processes for production of ethanol. Courtesy of National Corn-to-Ethanol Research Center. Reproduced from Nichols et al., 2006.

and enzymatic hydrolysis, but large enough that residual solids can be physically separated from liquid at the end of the fermentation and distillation. Hydrolysis of starch to fermentable sugars takes place in three stages: (i) gelatinization, the formation of a viscous suspension of starch granules in water; (ii) liquefaction, in which the starch is partially hydrolyzed, polymer size is reduced, and the suspension becomes less viscous; and (iii) saccharification, in which the starch is more completely hydrolyzed to glucose and maltose (Power, 2003). To begin starch conversion, a mash of corn and water is adjusted to pH 6, mixed with alpha-amylase, and processed with high-temperature steam, 110°C or higher, in a jet cooker. Next, additional alpha-amylase is added and the mash is liquefied at 80 to 90°C. The liquefied mash, containing dextrins (starch fragments), is cooled to 32°C, and the pH is reduced to 4.5 to 5.0 in preparation for saccharification and fermentation.

Saccharification and fermentation

The remainder of the dry-grind process utilizes simultaneous saccharification and fermentation. Glucoamylase is added to the liquefied mash, and at the same time, the yeast inoculum (hydrated and propagated in a separate vessel) is added. Glucoamylase cleaves dextrins at α-1,4-glucosidic linkages, releasing glucose and maltose for yeast fermentation in the same tank. Simultaneous saccharification and fermentation limits contamination and reduces osmotic stress, because glucose is consumed as it is formed.

Distillation and dehydration

At the end of fermentation, the ethanol, 12 to 14% or more by volume, is separated from "stillage" containing unfermented residuals (along with yeast cells and fermentation by-products) in aqueous suspension and solution. Conventional distillation of the fermented mash (beer) yields near-azeotropic (96%) ethanol, which is dehydrated using molecular sieves to near-anhydrous (100%) ethanol. To exempt fuel ethanol from beverage alcohol taxes, anhydrous ethanol is blended with denaturant. Typically, 3 to 5% gasoline is used to denature ethanol for use in motor fuels. The dry-grind process lasts 40 to 60 h from start to finish.

Stillage processing and production of animal feed

The stillage remaining after ethanol distillation is separated into predominantly soluble and suspended solid (cake) fractions by centrifugation. A portion of the liquid fraction, known as thin stillage, is recycled (backset) into the process. The remaining liquid fraction is concentrated by evaporation and mixed with the cake. The resulting mixture is used as animal feed, either directly (wet distillers grains and solubles [WDG]) or dried to produce DDGS. It is desirable from the standpoint of energy use to sell WDG; however, this product has a relatively short shelf life. WDG is typically sold to local livestock operations, and the remaining product is dried for storage or shipping.

Corn Wet Milling

Steeping

In wet milling (Fig. 4), the kernel is fractionated into separate components, and a relatively pure starch stream is recovered from the endosperm. Integral to the wet-milling process is steeping, the initial soaking step that enables isolation of kernel fractions. Kernels are hydrated in dilute (0.12 to 0.20%) sulfurous acid at 52°C for 24 to 48 h in a series of tanks. Steep water is recycled countercurrently from tanks holding "older" corn through tanks of progressively "newer" corn kernels. Sulfur dioxide is added with water at the end of the series, to the oldest corn which is completing the steeping process and ready to be milled. Lactic acid-producing bacteria grow in tanks holding the newer corn. The effects of bacteria and their metabolites early in the steep and the increasing sulfur dioxide concentration later in the steep soften kernels and cause them to swell (Hull et al., 1996). Lactic acid has a solubilizing effect on the corn protein matrix, and starch granules are released (Dailey et al., 2000).

Oil, fiber, and gluten separation

After steeping, the germ, fiber, and gluten protein are separated from the kernel by physical methods. The germ fraction, containing most of the corn oil, is separated based on density and washed to remove residual starch and gluten. Fiber is removed from the remaining slurry by screening. Gluten is separated from starch by centrifugation and dried to produce corn gluten meal, a high-protein feed suitable for poultry. A lower-protein mix termed corn gluten feed is made by combining washed fiber with residuals from steep water, germ extraction, and distillation columns. The steep liquid, termed corn steep liquor, can also be sold separately as feed or as a fermentation ingredient.

Starch conversion

Saccharification with alpha- and glucoamylase in wet milling is the same in principle as described for the dry-grind process. However, in wet milling, saccharification is completed prior to fermentation or other end uses of the sugars. In addition to its use in ethanol production, wet-milled starch can also be sold as dried or modified corn starch, or converted enzymatically to dextrins and sweeteners, or fermented to any number of products. Alternate fermentation products include amino acids, vitamins, citric acid, and lactic acid. Some wet mills also produce vitamins, enzymes, pharmaceuticals, nutraceuticals, films, solvents, pigments, polyols, or fibers.

Future Developments in the Corn-to-Ethanol Process

New corn hybrids

Traditional corn-breeding programs have long been used to improve traits such as yield, pest resistance, and drought tolerance. More recently, corn hybrids with traits beneficial for processing have entered the seed market, including some tailored for ethanol production (Bothast and Schlicher, 2005). High extractable starch hybrids having a larger portion of starch accessible to extraction have been developed for the wet-milling process, and hybrids targeted to the dry-grind industry have high total fermentables. Total fermentables refers to all of the starch and sugar fermented in the dry-grind process, including the small percentage of free glucose, fructose, maltose, and sucrose in corn kernels. These hybrids yield 2 to 5% more ethanol than bulk commodity corn.

Transgenic approaches to crop improvement are also being pursued. Alterations in starch biosynthetic pathways are sought that alter starch characteristics—gelling properties, viscosity, reaction efficiencies, etc.—without having negative effects on yield or other kernel characteristics. Physical interactions between various starch biosynthetic isoforms have been disrupted by genetic methods, causing a shift in amylopectin chain length (Myers, 2006). Other hybrids have been developed that accumulate starch-hydrolyzing enzymes in the endosperm and so have "built-in" processing capability. The amylase, encoded by a modified high-temperature archaeal gene, would allow fermentation at higher temperature and lower pH (Craig et al., 2004). Transgenic corn with high free lysine content, originally developed for the animal feed industry, is being introduced at pilot plant scale into a modified dry-grind ethanol process. In this process, corn with high oil content also carries a trait for high free lysine, resulting in DDGS with higher lysine content and potentially higher feed value (Jessen, 2006a).

Low-temperature starch hydrolysis

"Cold hydrolysis" refers to direct enzymatic release of glucose from starch granules. In cold hydrolysis, the traditional model of liquid-phase, enzymatic digestion of liquefied starch at high temperature is replaced by solid-phase reaction with native starch at lower temperature. Starch in its raw, granular form is converted to glucose at the optimum temperature for fermentation, and the need for liquefaction with steam (jet-cooking) is eliminated. The process offers potential energy savings and reduces equipment and maintenance costs related to heat recovery. Early work with raw starch hydrolysis was done with rice (Yamasaki et al., 1963; Matsumoto et al., 1982), and recently developed for use in fermentation of corn starch (Lewis, 2006; Robertson et al., 2006; Shetty et al., 2005; Williams, 2006).

Raw starch hydrolysis in effect mimics the natural process that occurs during germination, in which hydrolyzing enzymes are present in the kernel, in the correct ratio and with the necessary activities, to provide energy for the germinating seed. Enzymes that hydrolyze native starch release free glucose by "drilling" into the starch granules (Fig. 2). Amylases that efficiently hydrolyze native starch have a starch binding domain that binds to granules (Machovič and Janeček, 2006), effectively increasing the concentration of substrate at the enzyme's active site. In general, exo- and endoamylolytic enzymes act synergistically; however, slow reaction rates and incomplete conversion of granules are still problematic in some cases, and larger amounts of enzyme are required compared to conventional liquefaction with steam and enzymes.

Strategies to increase granular starch hydrolyzing activity include increasing the amount of enzyme produced, optimizing synergy between gluco- and alpha-amylases, improving the ability of enzymes to bind and access granular starch, and reducing the particle size of milled corn. Inclusion of acid protease has been shown to improve fermentations (Strohm et al., 2006). Protein engineering has also been undertaken to improve raw starch hydrolysis. Barley alpha-amylase was subjected to directed evolution and yielded mutants with either or both higher specific activity in raw starch hydrolysis and higher concentration in culture supernatant (Wong et al., 2004).

Raw starch hydrolysis may also incorporate very high gravity fermentation (with greater than 30% solids). The low-temperature systems reportedly offer, in addition to energy savings, higher final ethanol yields and concentrations. However, high enzyme loading and cost may present a barrier to wide-scale adoption of no-cook processes. And because heating in the traditional process serves the additional function of partially pasteurizing the feedstock, low-temperature hydrolysis may risk microbial contamination and corresponding loss of yield (Galvez, 2005). Enzyme and process development for cold starch hydrolysis is an area of active interest, and further improvements are likely.

Milling and fractionation technologies

Although a dry-grind facility has lower capital costs than a wet mill, it is limited in its ability to market coproducts. DDGS is typically the sole coproduct of dry-grind ethanol facilities, and consequently, there is potential to increase profitability in the dry-grind facility by capturing some of the biorefinery capacity of the wet mill. The dry-grind industry is pursuing new technologies to fractionate the corn kernel and produce marketable coproducts without using the conventional wet-mill steeping process (Rausch and Belyea, 2006).

Modified corn-processing systems use specially developed milling and separation technologies at the front end of the dry-grind process to separate softer portions of the kernel (germ and fiber) from the harder endosperm (Jessen, 2006a, Jessen, 2006b; Lewis, 2006; Lohrmann, 2006; Singh et al., 2005). In the quick germ and quick germ-quick fiber processes, kernels are soaked for 3 to 12 h in water, followed by coarse grinding and incubation for 2 to 4 h with starch-degrading enzymes (Rausch and Belyea, 2006). Release of soluble carbohydrates increases the specific gravity of the solution, allowing density separation of germ and/or pericarp fiber. In enzymatic milling, proteases are added with the amylases, and fine fiber is also recovered from the endosperm. Modified dry-grind methodologies have been introduced in some new ethanol production facilities and have the potential to be retrofitted at existing facilities.

There are several potential benefits to modifying the dry-grind process (Rausch and Belyea, 2006). Marketable coproducts produced in a modified process would decrease the dependence of dry-grind ethanol facilities on the sale of ethanol and DDGS. For example, enzymatic milling yields (in addition to ethanol) corn oil, purified fiber, and DDGS with 58% protein and 2% fiber content, compared to traditional DDGS having approximately 28% protein and 11% fiber. Higher ethanol concentrations and fermentation rates are also achieved (Singh et al., 2005). And the use of a fractionated starch stream reduces the volume of nonfermentable materials entering the process, which effectively increases processing capacity and efficiency. The facility would also realize energy cost savings because there is less residual material (DDGS) to recover and dry at the end of the process. Although conventional DDGS is marketed as feed primarily for ruminant animals, the new DDGS product could potentially also be fed to nonruminants, for which the amount of fiber that can be included in rations is limited. Expansion of markets for DDGS is desirable because increased ethanol production has resulted in increased availability of DDGS.

Feed coproduct quality and utilization

In addition to increased protein and decreased fiber content of DDGS, other quality issues for DDGS are also being addressed. Due to the decentralized nature of ethanol production, there can be considerable variability in DDGS composition, and standardized analysis and labeling of nutritional content are needed. Consistency of DDGS composition is a significant concern that has resulted in the marketing of "branded" DDGS formulations having guaranteed nutritional qualities. Flowability is another quality issue for DDGS. In some cases DDGS compacts and is difficult to transport and handle, due to factors including moisture content, particle size, and storage and transportation conditions; high priority has been assigned to understanding and preventing DDGS flowability problems (Rosentrater and Muthukumarappan, 2006). Elements, particularly phosphorus, present in DDGS are a potential problem with respect to disposal of manure from animals fed DDGS diets. Microfiltration has been shown to remove significant amounts of phosphorus and other elements from DDGS; adoption of this type of technology would have the additional benefit of removing water from processing streams and reducing the amount of energy needed to dry coproducts (Rausch and Belyea, 2006).

Along with efforts to improve the feed quality of DDGS, new applications for DDGS are being pursued. Nonfeed uses such as a biobased de-icer and weed-suppressing mulch could provide additional niche markets. DDGS is also being explored as a feedstock for ethanol fermentation (Bals et al., 2006; Cotta et al., 2006). Use of the cellulose portion of DDGS for conversion to ethanol would also result in DDGS with higher protein content, which, as stated previously, would be desirable from a feed perspective. Wet-milled corn fiber can also be converted to ethanol (Dien et al., 2006). Both of these potential ethanol feedstocks, DDGS at the dry-grind facility and fiber at the wet mill, are generated on site and require no additional cost for collection and transportation. The fiber component of corn is therefore an attractive starting point to integrate lignocellulosic ethanol production with a conventional starch-to-ethanol process.

CANE TO ETHANOL PROCESSES

As with corn, ethanol is produced from cane (nonstarch sugars) in two primary types of facilities, integrated sugar/ethanol production facilities and autonomous distilleries. Integrated sugar/ethanol production facilities are capable of switching cane processing between sugar and ethanol depending on economic driving forces, whereas autonomous plants are designed to process sugarcane solely for the production of ethanol. The term "autonomous" is a product of the development of the Brazilian ethanol program. Initial development, beginning in the late 1970s, occurred with the addition of ethanol production capacity to existing sugar mills. This was followed by a second wave of stand-alone or autonomous plants dedicated to ethanol production only. This strategy provided for fast construction and lower capital investment during a period of low world sugar pricing. As crude oil prices dipped during the end of the last century, most of the autonomous distilleries were retrofitted with sugar production capacity. Recent escalation in world energy and sugar pricing has spurred rapid growth in both types of facilities, limited mostly by Brazil's ability to bring new land into cane production.

In addition, there are integrated and independent (not associated with sugar mill operations) ethanol production facilities based on by-products of sugar production such as molasses. These facilities have historically been dedicated to the production of beverage and industrial-grade ethanol, with cumulative capacities significantly below the global fuel ethanol market.

One major issue in cane-based ethanol production is the relationship between crop harvest and processing. Unlike corn, which can be transported and stored with no significant change in composition, cane deteriorates rapidly after harvest. Therefore, mills are generally located in close proximity to the cane fields and operate only during harvest season, which occurs from May to November in Brazil's South and East regions, and from September to February in the Northeast (Rípoli et al., 2000). Globally, cane harvest and processing range from as short as 3 months in temperate climates (such as the Southeastern United States) to as long as 11 months in tropical areas with stable climates (Hawaii). The seasonal nature of the cane sugar industry results in reduced return on capital.

Cane Processing

Cane-processing systems for both integrated sugar/ethanol and autonomous production facilities are essentially identical through the front-end process systems. Therefore, this report discusses the integrated sugar/ethanol process, noting differences where pertinent. Sugar and ethanol production processes are shown in Fig. 5 and 6.

Cane is trucked or railed to mills, where it is washed upon receipt to remove dirt, as well as ash and soot in areas where cane fields are burned prior to har-

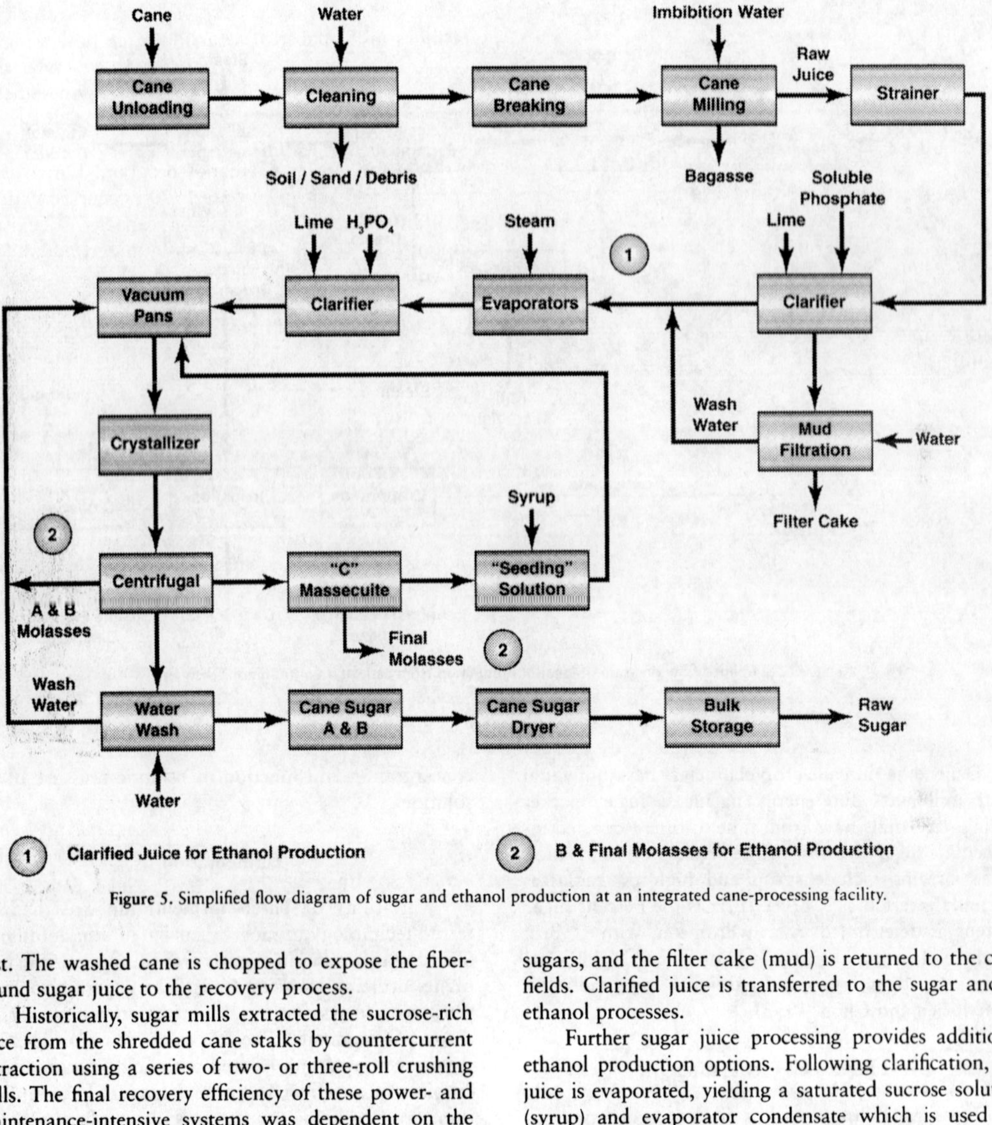

Figure 5. Simplified flow diagram of sugar and ethanol production at an integrated cane-processing facility.

vest. The washed cane is chopped to expose the fiber-bound sugar juice to the recovery process.

Historically, sugar mills extracted the sucrose-rich juice from the shredded cane stalks by countercurrent extraction using a series of two- or three-roll crushing mills. The final recovery efficiency of these power- and maintenance-intensive systems was dependent on the number of extraction stages used as well as the amount of imbibition water. Recently, milling systems have been in competition with diffusion technology, which offers increased sugar recovery and reduced investment, maintenance, and power consumption. The products of cane milling are sugar juice, which is a nominal 15% wt/wt dissolved solids solution at approximately 80 to 85% purity, along with residual fiber (bagasse) at approximately 50% wt/wt dissolved solids and spent wash water. Hydrated lime slurry is added to the heated juice and recycle wash streams prior to gravity clarification. Clarifier sludges are washed in vacuum filters to recover sugars, and the filter cake (mud) is returned to the cane fields. Clarified juice is transferred to the sugar and/or ethanol processes.

Further sugar juice processing provides additional ethanol production options. Following clarification, the juice is evaporated, yielding a saturated sucrose solution (syrup) and evaporator condensate which is used for process and utility water makeup. The syrup is further concentrated in a series of vacuum pans, increasing the sucrose content past saturation and producing a crystalline sucrose-rich solution referred to as magma. The crystalline sucrose/syrup mixture is centrifuged to recover the "A" sugar crystals, leaving behind an A molasses. Concentration and centrifugation processes are typically repeated, producing subsequent "B" sugar and molasses and "C" sugar and molasses, also referred to as "final" or "blackstrap" molasses. Each succeeding step results in improvements in sucrose yield at the expense of lower-grade sugar and molasses.

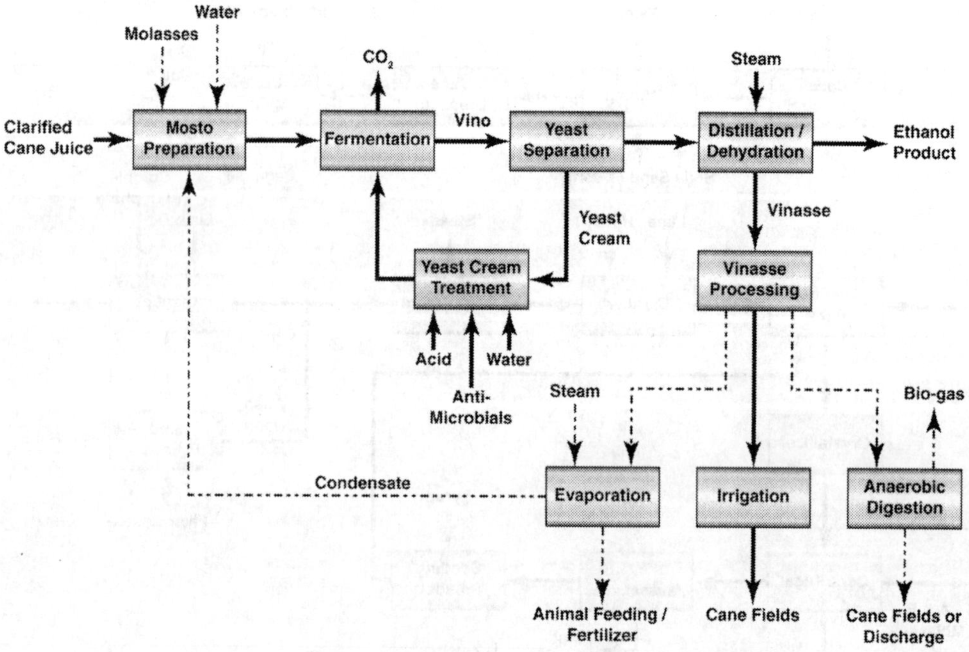

Figure 6. Simplified flow diagram of ethanol production from cane at an autonomous distillery.

Depending on cane crop characteristics, individual sugar mill mass and energy balances, sugar market pricing, etc., mills have from time to time extracted intermediate sugar streams for the production of ethanol. These streams include syrup and high-test molasses (partially inverted, 75° Brix). (Brix is the percent sugar content as determined by a hydrometer, with 1° Brix equal to 1 g of sugar/100 g of juice.) Additional streams are B molasses and excess or low-grade raw sugar (Chen and Chou, 1993).

Feedstock Preparation

The sugar-rich feedstock is diluted, as necessary, to the desired sugar concentration prior to fermentation, producing a "mosto." The desired mosto fermentable sugar concentration varies, depending on feedstock characteristics and the fermentation process employed. Often, juice is fermented "as is" without dilution or concentration. In the case of syrups and high-test molasses, the sugar concentration and corresponding beer ethanol concentration are controlled by the fermentation process employed. With B and C molasses, salt (primarily potassium and magnesium) concentrations, in conjunction with the mosto sugar concentration, must be considered since the final achievable ethanol concentration is a function of osmotic pressure of the solution.

Fermentation

Variations of batch fermentation are the preferred technology in most sugar-based ethanol plants. Since the substrates are nearly free of suspended matter, plants utilizing low-salt feedstocks generally operate yeast recycle systems employing nozzle separators to recover a nominal 15 to 20 wt % yeast cream. The cream is diluted and treated with acid and/or antimicrobials in an effort to control bacterial contamination, prior to being used to inoculate subsequent fermenters. Recycling yeast results in a 2 to 5% increase in ethanol yield, because minimal sugar is diverted to cell growth. Juice-based production facilities typically produce an 8 to 10 vol % (approximately equal to 6.5 to 8% wt/vol) ethanol solution, with fermentation times ranging from 8 to 12 h. Concentrated juice, syrup, and high-test molasses facilities have produced greater than 15 vol % ethanol (approximately 12% wt/vol) from 24-h fermentations. Ethanol concentrations and fermentation times for B and C molasses-based plants vary with feedstock quality but are often limited to less than 8.0 vol % ethanol (approximately 6.5% wt/vol) from ≥30-h fermentations.

Significant attention is paid to the fermenter feed program. Because the primary substrates are readily fermentable sugars, controlling the sugar and salt concentrations and associated osmotic pressure is critical for maximum productivity and efficiency.

Distillation and Dehydration

As with corn-based ethanol production facilities, sugar plants employ conventional distillation to concentrate ethanol to near-azeotropic (96%) concentrations for subsequent dehydration. In most regions, pressure swing adsorption molecular sieve dehydration using Zeolite-based media is the technology of choice. Through the mid-1990s Brazilian plants employed benzene azeotropic distillation technology to dehydrate ethanol. Recent legislation banning the use of carcinogenic benzene resulted in many of the plants converting to cyclohexane-based azeotropic distillation or glycol-based extractive distillation.

In the case of B and C molasses substrates, consideration must be given to the fouling characteristics of the feedstocks. These molasses streams frequently contain high concentrations of calcium salts; therefore, pressure and temperature profiles in the beer still (wash column) and associated heat exchange systems warrant attention to minimize fouling and associated maintenance.

Stillage Processing

Unlike corn alcohol processes, the stillage (vinasse) produced from the fermentation of cane juice, syrups, and molasses contains only low concentrations of protein or lipids. The organic fraction includes nonfermentable sugars, waxes, gums, organic acids, and fiber fines (bagasse) as well as by-products of yeast and bacterial metabolism such as glycerol, lactic acid, and acetic acid. The organic composition, in conjunction with the high potassium and magnesium concentration, restricts the value and use in animal feed rations. Historically, three options have been employed in managing vinasse.

The majority of Brazilian cane juice-based facilities direct the low-solids, low-BOD (biological oxygen demand) vinasse to cane fields as irrigation water, effectively returning the nutrients and organics to the soils. The application of vinasse is typically restricted by potassium content.

In molasses-based facilities producing vinasse with higher dissolved solids content, anaerobic digestion is a viable option. Properly managed, anaerobic digestion can reduce the BOD content by approximately 90% and produce sufficient methane to supply the plant's boiler fuel requirements. Still, the digester effluent BOD value is in excess of allowable discharge limits in many countries. Therefore, as in the case of cane juice plants, the treated vinasse, which still retains the majority of the original mineral and nitrogen content, is applied as irrigation water. Use of vinasse as irrigation water is restricted by the seasonally heavy rainfall in many tropical and subtropical regions. Heavy rainfall results in runoff of organic-laden water into adjacent waterways.

Vinasse evaporation has been implemented with limited success in some factories. Mostly used in molasses-based facilities, the vinasse is concentrated from approximately 10° Brix to a final concentration above 50° Brix, at which point the fluid begins to display thixotropic properties. Due to the high mineral (calcium) content of vinasse, frequent cleaning of evaporator heat transfer surfaces is required.

Bagasse Handling

One critical economic issue differentiating the production of ethanol from corn and that of ethanol from cane resides in the residuals. Whereas the residuals from corn ethanol processing are typically sold as animal feed, the residuals of cane processing are frequently burned for power. A typical mixed juice from extraction contains approximately 15% sugar. The residual fiber, called bagasse, comprises 12 to 14% of the cane and contains 1 to 2% sugar, approximately 50% moisture, and quantities of sand and grit from the field as "ash," yielding about 25 to 30 tons of bagasse per 100 tons of cane or 10 tons of sugar. The fiber-rich portion of the cane remaining after juice extraction releases sufficient energy upon combustion to supply the steam demand of the sugar mill and distillery with residuals for export power generation. Researchers are also currently investigating the use of bagasse as a lignocellulosic-rich feedstock for ethanol production.

CONCLUSION

Production and use of fuel ethanol will continue to increase because of environmental and energy security concerns. Improved strains, enzymes, and feedstocks will increase productivity of the ethanol process, and process improvements will also add value. Future biorefineries that process the whole corn plant (starch, fiber, and stover) or the whole sugarcane plant (sugars and bagasse) could produce liquid fuel, edible oil or sugars, animal feed, power, and polymers or chemical intermediates. Starch and sugars will continue to play an important role in ethanol production, even as lignocellulosic feedstocks come into production.

Acknowledgment. Mention of trade names or commercial products in this article is solely for the purpose of providing specific information and does not imply recommendation or endorsement by the U.S. Department of Agriculture.

REFERENCES

Alper, H., J. Moxley, E. Nevoigt, G. R. Fink, and G. Stephanopoulos. 2006. Engineering yeast transcription machinery for improved ethanol tolerance and production. *Science* **314:**1565–1568.

Badotti, F., A. S. Batista, and B. U. Stambuk. 2005. Sucrose active transport and fermentation by *Saccharomyces cerevisiae. Braz. Arch. Biol. Technol.* **48:**119–127.

Bals, B., B. Dale, and V. Balan. 2006. Enzymatic hydrolysis of distiller's dry grain and solubles using ammonia fiber expansion pretreatment. *Energy Fuels* **20:**2732–2736.

Batista, A. S., L. C. Miletti, and B. U. Stambuk. 2004. Sucrose fermentation by *Saccharomyces cerevisiae* lacking hexose transport. *J. Mol. Microbiol. Biotechnol.* **8:**26–33.

Bayrock, D. P., and W. M. Ingledew. 2004. Inhibition of yeast by lactic acid bacteria in continuous culture: nutrient depletion and/or acid toxicity? *J. Ind. Microbiol. Biotechnol.* **31:**362–368.

Bayrock, D. P., K. C. Thomas, and W. M. Ingledew. 2003. Control of *Lactobacillus* contaminants in continuous fuel ethanol fermentations by constant or pulsed addition of penicillin G. *Appl. Microbiol. Biotechnol.* **62:**498–502.

Bideaux, C., S. Alfenore, X. Cameleyre, C. Molina-Jouve, J.-L. Uribelarrea, and S. E. Guillouet. 2006. Minimization of glycerol production during the high-performance fed-batch ethanolic fermentation process in *Saccharomyces cerevisiae*, using a metabolic model as a prediction tool. *Appl. Environ. Microbiol.* **72:**2134–2140.

Bothast, R. J., and M. A. Schlicher. 2005. Biotechnological processes for conversion of corn into ethanol. *Appl. Microbiol. Biotechnol.* **67:**19–25.

Casey, G. P., and W. M. Ingledew. 1986. Ethanol tolerance in yeasts. *Crit. Rev. Microbiol.* **13:**219–280.

Chen, J. C. P., and C. Chou. 1993. *Cane Sugar Handbook*, 12th ed. John Wiley and Sons, Inc., New York, NY.

Collins, K. 2007. *Statement of Keith Collins, Chief Economist, U.S. Department of Agriculture, Before the U.S. Senate Committee on Agriculture, Nutrition, and Forestry, January 10, 2007.* U.S. GPO, Washington, DC.

Connelly, C. 1999. Bacterial contaminants and their effects on alcohol production, p. 317–334. *In* K. A. Jacques, T. P. Lyons, and D. R. Kelsall (ed.), *The Alcohol Textbook*, 3rd ed. Nottingham University Press, Nottingham, United Kingdom.

Cotta, M. A., B. S. Dien, X.-L. Li, M. Ladisch, N. Mosier, W. Tyner, R. Woodson, H. Blaschek, B. Dale, B. Shanks, and J. Verkade. 2006. Pretreatment and hydrolysis of distiller's grains to fermentable sugars: an integrated approach by the Midwest Consortium for Sustainable Biobased Products and Energy. *In* M. Tumbleson (ed.), *Corn: Nature's Sustainable Resource. Proceedings of the Corn Utilization and Technology Conference, Dallas, TX.* National Corn Growers Association, Chesterfield, MO.

Crabb, D. W., and J. Shetty. 2003. Improving the properties of amylolytic enzymes by protein engineering. *Trends Glycosci. Glycotechnol.* **15:**115–126.

Craig, J. A., J. Batie, W. Chen, S. B. Freeland, M. Kinkima, and M. B. Lanahan. 2004. Expression of starch hydrolyzing enzymes in corn. *In* M. Tumbleson (ed.), *Corn: Feedstock of the Future. Proceedings of the Corn Utilization and Technology Conference, Indianapolis, IN.* National Corn Growers Association and Corn Refiners Association, Washington, DC.

Dailey, O. D., Jr., M. K. Dowd, and J. C. Mayorga. 2000. Influence of lactic acid on the solubilization of protein during corn steeping. *J. Agric. Food Chem.* **48:**1352–1357.

D'Amore, T., C. J. Panchal, I. Russell, and G. G. Stewart. 1990. A study of ethanol tolerance in yeast. *Crit. Rev. Biotechnol.* **9:**287–304.

Dien, B. S., M. A. Cotta, and T. W. Jeffries. 2003. Bacteria engineered for fuel ethanol production: current status. *Appl. Microbiol. Biotechnol.* **63:**258–266.

Dien, B. S., X.-L. Li, L. B. Iten, D. B. Jordan, N. N. Nichols, P. J. O'Bryan, and M. A. Cotta. 2006. Enzymatic saccharification of hot-water pretreated corn fiber for production of monosaccharides. *Enzyme Microb. Technol.* **39:**1137–1144.

Fujita, K., A. Matsuyama, Y. Kobayashi, and H. Iwahashi. 2006. The genome-wide screening of yeast deletion mutants to identify the genes required for tolerance to ethanol and other alcohols. *FEMS Yeast Res.* **6:**744–750.

Galvez, A. 2005. Analyzing cold enzyme starch hydrolysis technology in new ethanol plant design. *Ethanol Producer Magazine* **2005** (January).

Hahn-Hägerdal, B., M. Galbe, M. F. Gorwa-Grauslund, G. Lidén, and G. Zacchi. 2006. Bio-ethanol—the fuel of tomorrow from the residues of today. *Trends Biotechnol.* **24:**549–556.

Hull, S. R., E. Peters, C. Cox, and R. Montgomery. 1996. Composition of corn steep water during experimental steeping. *J. Agric. Food Chem.* **44:**3521–3527.

Ingledew, W. M. 1999a. Yeast—could you build a business on this bug? p. 27–50. *In* T. P. Lyons and K. A. Jacques (ed.), *Under the Microscope. Focal Points for the New Millennium. Biotechnology in the Feed Industry.* Nottingham University Press, Thrompton, United Kingdom.

Ingledew, W. M. 1999b. Alcohol production by *Saccharomyces cerevisiae:* a yeast primer, p. 49–87. *In* K. A. Jacques, T. P. Lyons, and D. R. Kelsall (ed.), *The Alcohol Textbook,* 3rd ed. Nottingham University Press, Nottingham, United Kingdom.

Jeffries, T. W. 2006. Engineering yeasts for xylose metabolism. *Curr. Opin. Biotechnol.* **17:**320–326.

Jessen, H. 2006a. Quality traits: linking the chain from farmer to producer. *Ethanol Producer Magazine* **2006**(October).

Jessen, H. 2006b. Expanding fractionation horizons from food to fuel. *Ethanol Producer Magazine* **2006**(October)**:**70–75.

Johnson, L. A., and J. B. May. 2003. Wet milling: the basis for corn biorefineries, p. 449–494. *In* P. J. White and L. A. Johnson (ed.), *Corn: Chemistry and Technology*, 2nd ed. American Association of Cereal Chemists, St. Paul, MN.

Jones, A. M., and W. M. Ingledew. 1994. Fuel alcohol production: optimization of temperature for efficient very-high-gravity fermentation. *Appl. Environ. Microbiol.* **60:**1048–1051.

Kelsall, D. R., and T. P. Lyons. 2003. Grain dry milling and cooking procedures: extracting sugars in preparation for fermentation, p. 9–21. *In* P. J. White and L. A. Johnson (ed.), *The Alcohol Textbook*, 4th ed. Nottingham University Press, Nottingham, United Kingdom.

Khaw, T. S., Y. Katakura, J. Koh, A. Kondo, M. Ueda, and S. Shioya. 2006. Evaluation of performance of different surface-engineered yeast strains for direct ethanol production from raw starch. *Appl. Microbiol. Biotechnol.* **70:**573–579.

Lewis, S. M. 2006. BPX™ and BFRAC™: innovation in biorefining. *In* M. Tumbleson (ed.), *Corn: Nature's Sustainable Resource. Proceedings of the Corn Utilization and Technology Conference, Dallas, TX.* National Corn Growers Association, Indianapolis, IN.

Lohrmann, T. 2006. Rethinking ethanol coproducts. *Distillers Grains Q.* **Fourth Quarter:**23–24.

Machovič, M., and S. Janeček. 2006. Starch-binding domains in the post-genome era. *Cell. Mol. Life Sci.* **63:**2710–2724.

Matsumoto, N., O. Fukushi, M. Miyanaga, K. Kakihara, E. Nakajima, and H. Yoshizumi. 1982. Industrialization of a noncooking system for alcoholic fermentation from grains. *Agric. Biol. Chem.* **46:**1549–1558.

Myers, A. M. 2006. Genetic modifications of maize starch structure and properties: opportunity for improved corn utilization. *In* M. Tumbleson (ed.), *Corn: Nature's Sustainable Resource. Proceedings of the Corn Utilization and Technology Conference, Dallas, TX.* National Corn Growers Association, Chesterfield, MO.

Narendranath, N. V., S. H. Hynes, K. C. Thomas, and W. M. Ingledew. 1997. Effects of lactobacilli on yeast-catalyzed ethanol fermentations. *Appl. Environ. Microbiol.* **63:**4158–4163.

Narendranath, N. V., K. C. Thomas, and W. M. Ingledew. 2001. Effects of acetic acid and lactic acid on the growth of *Saccharomyces cerevisiae* in a minimal medium. *J. Ind. Microbiol. Biotechnol.* **26:**171–177.

Nichols, N. N., B. S. Dien, R. J. Bothast, and M. A. Cotta. 2006. The corn ethanol industry, p. 59–78. *In* S. Minteer (ed.), *Alcoholic Fuels.* CRC Press, Boca Raton, FL.

Pandey, A., P. Nigam, C. R. Soccol, V. T. Soccol, D. Singh, and R. Mohan. 2000. Advances in microbial amylases. *Biotechnol. Appl. Biochem.* **31:**135–152.

Power, R. F. 2003. Enzymatic conversion of starch to fermentable sugars, p. 23–32. *In* K. A. Jacques, T. P. Lyons, and D. R. Kelsall (ed.), *The Alcohol Textbook*, 4th ed. Nottingham University Press, Nottingham, United Kingdom.

Rausch, K. D., and R. L. Belyea. 2006. The future of coproducts from corn processing. *Appl. Biochem. Biotechnol.* **128:**47–86.

Rípoli, T. C. C., W. F. Molina, and M. L. C. Rípoli. 2000. Energy potential of sugar cane biomass in Brazil. *Scientia Agricola* **57:**677–681.

Robertson, G. H., D. W. S. Wong, C. C. Lee, K. Wagschal, M. R. Smith, and W. J. Orts. 2006. Native or raw starch digestion: a key step in energy efficient biorefining of grain. *J. Agric. Food Chem.* **54:**353–365.

Roels, J. A. 1983. *Energetics and Kinetics in Biotechnology*, p. 23–73. Elsevier Biomedical Press, Amsterdam, The Netherlands.

Rosentrater, K. A., and K. Muthukumarappan. 2006. Corn ethanol coproducts: generation, properties, and future prospects. *Int. Sugar J.* **108:**648–657.

Russell, I. 2003. Understanding yeast fundamentals, p. 85–119. *In* K. A. Jacques, T. P. Lyons, and D. R. Kelsall (ed.), *The Alcohol Textbook*, 4th ed. Nottingham University Press, Nottingham, United Kingdom.

Shetty, J. K., O. J. Lantero, and N. Dunn-Coleman. 2005. Technological advances in ethanol production. *Int. Sugar J.* **107:**605–610.

Shigechi, H., J. Koh, Y. Fujita, T. Matsumoto, Y. Bito, M. Ueda, E. Satoh, H. Fukuda, and A. Kondo. 2004. Direct production of ethanol from raw corn starch via fermentation by use of a novel surface-engineered yeast strain codisplaying glucoamylase and α-amylase. *Appl. Environ. Microbiol.* **70:**5037–5040.

Singh, V., D. B. Johnston, K. Naidu, K. D. Rausch, R. L. Belyea, and M. E. Tumbleson. 2005. Comparison of modified dry-grind processes for fermentation characteristics and DDGS composition. *Cereal Chem.* **82:**187–190.

Skinner, K. A., and T. D. Leathers. 2004. Bacterial contaminants of fuel ethanol production. *J. Ind. Microbiol. Biotechnol.* **31:**401–408.

Skinner-Nemec, K. A., N. N. Nichols, and T. D. Leathers. 2007. Biofilm formation by bacterial contaminants of fuel ethanol production. *Biotechnol. Lett.* **29:**379–383.

Strohm, B., J. Shetty, O. J. Lantero, and C. Pilgrim. 2006. Use of protease with raw starch hydrolyzing enzymes for dry grind corn processes to ethanol. *In* M. Tumbleson (ed.), *Corn: Nature's Sustainable Resource. Proceedings of the Corn Utilization and Technology Conference, Dallas, TX.* National Corn Growers Association, Chesterfield, MO.

Tester, R. F., J. Karkalas, and X. Qi. 2004. Starch—composition, fine structure and architecture. *J. Cereal Sci.* **39:**151–165.

Thomas, K. C., S. H. Hynes, and W. M. Ingledew. 1994. Effects of particulate materials and osmoprotectants on very-high-gravity ethanolic fermentation by *Saccharomyces cerevisiae. Appl. Environ. Microbiol.* **60:**1519–1524.

Thomas, K. C., and W. M. Ingledew. 1990. Fuel alcohol production: effects of free amino nitrogen on fermentation of very-high-gravity wheat mashes. *Appl. Environ. Microbiol.* **56:**2046–2050.

U.S. Department of Agriculture. 2007. *Crop Production 2006 Summary.* U.S. Department of Agriculture National Agricultural Statistics Service, Washington, DC.

van Maris, A. J. A., D. A. Abbott, E. Bellissimi, J. van den Brink, M. Kuyper, M. A. H. Luttik, H. W. Wisselink, W. A. Scheffers, J. P. van Dijken, and J. T. Pronk. 2006. Alcoholic fermentation of carbon sources in biomass hydrolysates by *Saccharomyces cerevisiae*: current status. *Antonie Leeuwenhoek* **90:**391–418.

Watson, S. A. 2003. Description, development, structure, and composition of the corn kernel, p. 69–106. *In* P. J. White and L. A. Johnson (ed.), *Corn: Chemistry and Technology*, 2nd ed. American Association of Cereal Chemists, St. Paul, MN.

Williams, J. 2006. Break it down now. *Ethanol Producer Magazine* **2006**(January).

Wong, D. W. S., S. B. Batt, C. C. Lee, and G. H. Robertson. 2004. High-activity barley α-amylase by directed evolution. *Protein J.* **23:**453–460.

Yamasaki, I., S. Ueda, and T. Shimada. 1963. Alcoholic fermentation of rice without previous cooking by using black-koji amylase. *J. Ferm. Assoc. Jpn.* **21:**83–86.

You, K. M., C.-L. Rosenfield, and D. C. Knipple. 2003. Ethanol tolerance in the yeast *Saccharomyces cerevisiae* is dependent on cellular oleic acid content. *Appl. Environ. Microbiol.* **69:**1499–1503.

Bioenergy
Edited by J. Wall et al.

Chapter 2

Lignocellulosic Biomass Conversion to Ethanol by *Saccharomyces*

Z. LEWIS LIU, BADAL C. SAHA, AND PATRICIA J. SLININGER

As interest in alternative energy sources rises, the concept of agriculture as an energy producer has become increasingly attractive (Outlaw et al., 2005). Renewable biomass, including lignocellulosic materials and agricultural residues, are low-cost materials for bioethanol production (Bothast and Saha, 1997; Wheals et al., 1999; Zaldivar et al., 2001). In the United States, the production of corn grain-based ethanol reached 5 billion gallons in 2006, a fraction of the 140 billion gallons of transportation fuel used annually. The goal is to displace 30% of the nation's 2004 motor gasoline use with ethanol by 2030, and this will require production levels equal to roughly 60 billion gallons a year. If all corn grain now grown in the United States is converted to ethanol, it can satisfy approximately 15% of current gasoline needs. Thus, developing ethanol as fuel, beyond its current role as fuel oxygenate, will require developing lignocellulose as feedstock because of its abundance. In particular, various agricultural residues (corn stover, wheat straw, and rice straw), agricultural processing by-products (corn fiber, rice hulls, and sugarcane bagasse), and energy crops (switchgrass) can be used as low-cost sources of sugars for biofuel production. At present, conversion of lignocellulosic biomass to fermentable sugars presents significant technical and economic challenges, and its success depends largely on the development of effective pretreatment, efficient enzyme conversion of pretreated lignocellulosic substrates to fermentable sugars, and stress-tolerant microbial biocatalysts. Lignocellulosic biomass generates a mixture of hexose and pentose sugars upon pretreatment of itself, or in combination with enzymatic hydrolysis. Traditional *Saccharomyces cerevisiae* ferments glucose to ethanol rapidly and efficiently, but it is limited in its fermentation of pentose sugars (xylose and arabinose) to ethanol. Although some yeasts (*Pachysolen tannophilus*, *Pichia stipitis*, and *Candida shehatae*) have the capability to ferment xylose to ethanol (Du Preez, 1994; Hahn-Hägerdal et al., 1994; Jefferies and Jin, 2004; Prior et al., 1989; Slininger et al., 1987), most strains have low tolerance and slow rates of ethanol fermentation. Genetically modified *S. cerevisiae* has been demonstrated to be able to ferment xylose to ethanol (Kötter et al., 1990). It has been characterized as the best yeast for fermentation of hexose sugars present in lignocellulose-derived hydrolysates due to its ethanol-producing capacity and high inhibitor tolerance. Strains of *S. cerevisiae* have been used extensively to test the fermentability of hydrolysates. For future sustainable and cost efficient lignocellulosic biomass conversion to ethanol, there exist two major challenges: heterogeneous sugar utilization and stress tolerance in engineering microbial catalytic fermentors for bioethanol production. In this chapter, we review the current knowledge on the composition and structure of lignocellulosic biomass, its pretreatment and enzymatic saccharification to simple sugars, and strain development of *S. cerevisiae* for efficient fermentation of the biomass-derived sugars to ethanol.

COMPOSITION AND STRUCTURE OF LIGNOCELLULOSIC BIOMASS

Lignocellulosic biomass includes various agricultural residues (straws, hulls, stems, and stalks), deciduous and coniferous woods, municipal solid wastes (paper, cardboard, yard trash, and wood products), waste from pulp and paper industry, and energy crops (switchgrass, Bermuda grass, Miscanthus). The compositions of these materials vary. The major component is cellulose (35 to 50%), followed by hemicellulose (20 to 35%) and lignin (10 to 25%). Proteins, oils, and ash make up the remaining fraction of lignocellulosic biomass (Wyman, 1994). The structures of these materials are complex, with recalcitrant and heterogeneous characteristics, and native lignocellulose is resistant to enzymatic hydrolysis. In the current model of the structure

Z. Lewis Liu, Badal C. Saha, and Patricia J. Slininger • U.S. Department of Agriculture, Agricultural Research Service, National Center for Agricultural Utilization Research, Peoria, IL 61604.

of lignocellulose, cellulose fibers are embedded in a lignin-polysaccharide matrix. Xylan may play a significant role in the structural integrity of cell walls by both covalent and noncovalent associations (Thomson, 1993).

Cellulose is a linear homopolymer of D-glucose units linked by 1,4-β-D-glucosidic bonds. Hemicelluloses are heteropolymers of pentoses (xylose and arabinose), hexoses (mannose, glucose, and galactose), and sugar acids. Hardwood hemicelluloses contain mostly xylans, whereas softwood hemicelluloses contain mostly glucomannans (McMillan, 1993). Xylans of many plant materials are heteropolysaccharides with homopolymeric backbone chains of 1,4-linked β-D-xylopyranose units. Besides xylose, xylans may contain arabinose, glucuronic acid or its 4-O-methyl ether, and acetic, ferulic, and *p*-coumaric acids. The frequency and composition of branches are dependent on the source of xylan (Aspinall, 1980). The backbone consists of O-acetyl, α-L-arabinofuranosyl, α-1,2-linked glucuronic, or 4-O-methylglucuronic acid substituents. Corn fiber xylan contains 48 to 54% xylose, 33 to 35% arabinose, 5 to 11% galactose, and 3 to 6% glucuronic acid (Doner and Hicks, 1997). About 80% of the xylan backbone in corn fiber contains numerous substitutions of monomeric side chains of arabinose or glucuronic acid linked to O-2 and/or O-3 of xylose residues, as well as substitutions of oligomeric side chains containing arabinose, xylose, and sometimes galactose residues (Saulnier et al., 1995). The heteroxylans in corn fiber are highly cross-linked by diferulic bridges and constitute a network in which the cellulose microfibrils may be embedded (Saulnier and Thibault, 1999). Structural wall proteins might be cross-linked together by isodityrosine bridges and with feruloylated heteroxylans, thus forming an insoluble network (Hood et al., 1991). Ferulic acid is covalently cross-linked to polysaccharides by ester bonds and to components of lignin mainly by ether bonds (Scalbert et al., 1985). In softwood heteroxylans, arabinofuranosyl residues are esterified with *p*-coumaric acids and ferulic acids (Timell, 1967). In hardwood xylans, 60 to 70% of the xylose residues are acetylated. The degree of polymerization of hardwood xylans (150 to 200) is higher than that of softwoods (70 to 130).

PRETREATMENT OF LIGNOCELLULOSIC BIOMASS

The pretreatment of any lignocellulosic biomass is crucial before enzymatic hydrolysis. One objective of pretreatment is to decrease the crystallinity of cellulose, which enhances the hydrolysis of cellulose by cellulases (Focher et al., 1981). Some pretreatments such as alkaline peroxidase improve digestibility but leave cellulose crystallinity intact (Martel and Gould, 1990). The main goals of pretreatment are to remove hemicellulose, to delignify, and to increase available cellulose surface (Dien et al., 2005a). Various pretreatment options are available to fractionate, solubilize, hydrolyze, and separate cellulose, hemicellulose, and lignin components (Wyman, 1994). These include concentrated acid, dilute acid, SO_2, alkali, alkaline peroxide, wet-oxidation, steam explosion (autohydrolysis), ammonia fiber explosion, CO_2 explosion, liquid hot water, and organic solvent treatments (Saha, 2004a). In each option, the biomass is reduced in size, and its physical structure is opened. The effectiveness of dilute acids to catalyze the hydrolysis of hemicellulose to its sugar components is well known. Two categories of dilute acid pretreatment are used: high-temperature (>160°C) continuous flow for low solids loading (5 to 10% [wt/wt]) and low-temperature (<160°C) batch process for high solids loading (10 to 40% [wt/wt]) (Sun and Cheng, 2002). Dilute acid pretreatment at high temperature usually hydrolyzes hemicellulose to its component sugars that are water soluble (Saha et al., 2005). The residue contains cellulose and often much of the lignin. A major problem associated with the dilute acid hydrolysis of lignocellulosic biomass is the poor fermentability of the hydrolysates.

Steam explosion provides effective fractionation of lignocellulosic components at relatively low costs (Nguyen and Saddler, 1991). Optimal solubilization and degradation of hemicellulose are generally achieved by either high-temperature and short-residence-time (270°C, 1 min) or lower-temperature and longer-residence-time (190°C, 10 min) steam explosion (Duff and Murray, 1996). The use of SO_2 as a catalyst during steam pretreatment results in the enzymatic accessibility of cellulose and enhanced recovery of the hemicellulose-derived sugars (Brownell and Saddler, 1984). Steam explosion can induce hemicellulose degradation to furfural and its derivatives and modification of the lignin-related chemicals under high-severity treatment (>200°C, 3 to 5 min, 2 to 3% SO_2) (Ando et al., 1986). Boussaid et al. (1999) recovered around 87% of the original hemicellulose component in the water-soluble fraction by steam explosion of Douglas fir softwood under low-severity conditions (175°C, 7.5 min, 4.5% SO_2). More than 80% of the recovered hemicellulose was in monomeric form. Enzymatic digestibility of the steam-exploded Douglas fir wood chips (105°C, 4.5 min, 4.5% SO_2) was significantly improved using an optimized alkaline peroxide treatment (1% H_2O_2, pH 11.5, 80°C, 45 min) (Yang et al., 2002). About 90% of the lignin in the original wood was solubilized by this procedure, leaving a cellulose-rich residue that was completely hydrolyzed within 48 h, using an enzyme (cellulase) loading of 10 filter paper units/g cellulose.

Steeping of the lignocellulosic biomass in dilute NH_4OH at ambient temperature can be used to remove lignin, acetate, and extractives (Cao et al., 1996). This is followed by dilute acid treatment that readily hydrolyzes the hemicellulose fraction to simple sugars, primarily xylose and glucose. The residual cellulose fraction of biomass can then be enzymatically hydrolyzed to glucose. Garrote et al. (2001) treated *Eucalyptus* wood substrates with water under selected operational conditions (autohydrolysis reaction) to obtain a liquid phase containing hemicellulose decomposition products (mainly acetylated xylooligosaccharides, xylose, and acetic acid). In a further acid-catalyzed step (posthydrolysis reaction), xylooligosaccharides were converted into xylose. The wet-oxidation method can be used for fractionation of lignocellulosics into a solubilized hemicellulose fraction and a solid cellulose fraction susceptible to enzymatic saccharification. Bjerre et al. (1996) found that a combination of alkali and wet oxidation did not generate furfural and 5-hydroxymethylfurfural (HMF). Klinke et al. (2002) characterized the degradation products from alkaline wet oxidation (water, sodium carbonate, oxygen, high temperature, and pressure) of wheat straw. Apart from CO_2 and water, carboxylic acids were the main degradation products from hemicellulose and lignin. Aromatic aldehyde formation was minimized by the addition of alkali and temperature control. Oxygen delignification of kraft pulp removed up to 67% of the lignin from softwood pulp and improved the rate and yield from enzymatic hydrolysis by up to 111 and 174%, respectively (Draude et al., 2001). Supercritical CO_2 explosion was found to be effective for pretreatment of cellulosic materials before enzymatic hydrolysis. Zheng et al. (1995) compared CO_2 explosion with steam and ammonia explosion for pretreatment of sugarcane bagasse. They found that CO_2 explosion was more cost-effective than ammonia explosion, and it did not cause the formation of inhibitory compounds.

ENZYMATIC SACCHARIFICATION OF CELLULOSE AND HEMICELLULOSE

Effective hydrolysis of cellulose to glucose requires the cooperative action of three enzymes: endo-1,4-β-glucanase, exo-1,4-β-glucanase, and β-glucosidase. Endoglucanase acts in a random fashion on the regions of low crystallinity on the cellulosic fiber, whereas exoglucanase removes cellobiose from the nonreducing ends of cellulose chains. Synergism between these two enzymes is attributed to the *endo-exo* form of cooperativeness and has been studied extensively between cellulases in *Trichoderma reesei* in the degradation of cellulose (Henrissat et al., 1985). Besides synergism, the adsorption of the cellulases on the insoluble substrates is a necessary step prior to hydrolysis. In most organisms, cellulases are modular enzymes that consist of a catalytic core connected to a cellulose-binding domain through a flexible and heavily glycosylated linker region (Gilkes et al., 1991). The cellulose binding domain is responsible for bringing the catalytic domain in an appropriate position for the breakdown of cellulose. Binding of cellulases and the formation of cellulose-cellulase complexes are considered critical steps in the hydrolysis of insoluble cellulose (Grethlein, 1985). β-Glucosidase hydrolyzes cellobiose and in some cases cellooligosaccharides to glucose. The enzyme is generally responsible for the regulation of the whole cellulolytic process and is a rate-limiting factor during enzymatic hydrolysis of cellulose, as both endoglucanase and cellobiohydrolase activities are often inhibited by cellobiose (Saha et al., 1995). Thus, β-glucosidase not only produces glucose from cellobiose, but also reduces cellobiose inhibition, allowing the cellulolytic enzymes to function more efficiently. However, most β-glucosidases are subject to end product (glucose) inhibition.

Product inhibition, thermal inactivation, substrate inhibition, low product yield, and high cost of cellulases are some barriers to the commercial development of the enzymatic hydrolysis of cellulose. Many microorganisms are cellulolytic. However, only strains of *Trichoderma* and *Aspergillus* have been studied extensively for cellulase. A newly isolated *Mucor circinelloides* strain produces a complete cellulase enzyme system (Saha, 2004b). The endoglucanase from this strain was found to have a wide pH stability and activity. There is an increasing demand for the development of thermostable, environmentally compatible products and for substrate-tolerant cellulases with increased specificity and activity for the application of converting cellulose to glucose for the fuel ethanol industry.

The cellulose hydrolysis step is a significant component of the total production cost of ethanol from wood (Nguyen and Saddler, 1991). Achieving a high glucose yield is necessary (>85% theoretical) at high substrate loading (>10% [wt/vol]) over short residence times (<4 days). Addition of a surfactant to enzymatic hydrolysis of lignocellulose increases the conversion of cellulose to soluble sugars (Eriksson et al., 2002). Simultaneous saccharification (hydrolysis) of cellulose to glucose and fermentation of glucose to ethanol (SSF) improve the kinetics and economics of biomass conversion by reducing accumulation of hydrolysis products that are inhibitory to cellulase and β-glucosidase. This reduces the contamination risk because of the presence of ethanol and reduces the capital equipment requirements (Philippidis et al., 1993). An important drawback of SSF is that the reaction has to operate at a compromise temperature of around 30°C instead of the optimum temperature for the enzymes, which is between 45 and 50°C. Enzyme recycling, by ultrafiltration of the hydrolysate, can reduce the net enzyme requirement and thus lower costs (Tan et al., 1987). Poor recovery of cellulase was achieved in the case of substrates containing a high proportion of lignin (Tanaka et al., 1988). A preliminary estimate of the cost of ethanol production from wood by SSF technology is $1.22/gal, of which the wood cost is $0.46/gal (Hinman et al., 1992).

Hemicellulases are either glycosyl hydrolases or carbohydrate esterases. The total biodegradation of xylan requires endo-β-1,4-xylanase, β-xylosidase, and several accessory enzymes, such as α-L-arabinofuranosidase, α-glucuronidase, acetylxylan esterase, ferulic acid esterase, and *p*-coumaric acid esterase, which are necessary for hydrolyzing various substituted xylans (Saha and Bothast, 1999b). The endoxylanase attacks the main chains of xylans, and β-xylosidase hydrolyzes xylooligosaccharides to xylose. The α-arabinofuranosidase and α-glucuronidase remove the arabinose and 4-O-methyl glucuronic acid substituents, respectively, from the xylan backbone. The esterases hydrolyze the ester linkages between xylose units of the xylan and acetic acid (acetylxylan esterase) or between arabinose side chain residues and phenolic acids, such as ferulic acid (ferulic acid esterase) and *p*-coumaric acid (*p*-coumaric acid esterase). It is stated that hindrance of lignocellulose biodegradation is associated with phenolic compounds (Hartley and Ford, 1989). The phenolic acids act as a cross-linking agent between lignin and carbohydrates or just between carbohydrates.

Many microorganisms, such as *Penicillium capsulatum* and *Talaromyces emersonii*, possess complete xylan-degrading enzyme systems (Filho et al., 1991). Significant synergistic interactions were observed among endoxylanase, β-xylosidase, α-arabinofuranosidase, and acetylxylan esterase of the thermophilic actinomycete *Thermomonospora fusca* (Bachmann and McCarthy, 1991). Many xylanases do not cleave glycosidic bonds between xylose units which are substituted. The side chains must be cleaved before the xylan backbone can be completely hydrolyzed (Lee and Forsberg, 1987). On the other hand, several accessory enzymes remove side chains only from xylooligosaccharides. These enzymes require a partial hydrolysis of xylan before the side chains can be cleaved (Poutanen et al., 1991). Although the structure of xylan is more complex than cellulose and requires several different enzymes with different specificities for complete hydrolysis, the polysaccharide does not form tightly packed crystalline structures like cellulose and is thus more accessible to enzymatic hydrolysis (Gilbert and Hazlewood, 1993). For effective hydrolysis of xylan, a proper mix of endoxylanase with several accessory enzymes is essential (Saha, 2003). Common enzymes involved in the conversion of cellulose and hemicellulose to simple sugars are presented in Table 1.

At present, the conversion of lignocellulosic biomass to fermentable sugars is not cost-effective. Some of the emerging pretreatment methods such as alkaline peroxide and ammonia fiber explosion generate solubilized and partially degraded hemicellulosic biomass that need to be treated further with enzymes or other means to produce fermentable sugars from them (Saha and Cotta, 2006). With the development of a suitable pretreatment method minimizing the formation of inhibitory compounds for fermentative organisms and use of a proper mixture of highly efficient cellulases and hemicellulases (enzyme cocktail) tailored for each lignocellulosic feedstock, this vast renewable resource can be utilized for production of fuel ethanol by fermentation.

Table 1. Enzymes involved in cellulose and hemicellulose degradation

Enzyme	Systematic name	EC no.	Mode of action
Cellulose			
Endo-1,4-β-glucanase	1,4-β-D-Glucan-4-glucanohydrolase	3.2.1.4	Endohydrolysis of 1,4-β-D-glucosidic linkages
Exo-1,4-β-glucanase releasing cellobiose	1,4-β-D-Glucan cellobiohydrolase	3.2.1.91	Hydrolysis of 1,4-β-D-glucosidic linkages
β-Glucosidase	β-D-Glucoside glucohydrolase	3.2.1.21	Hydrolyzes cellobiose and short-chain cello-oligosaccharides to glucose
Hemicellulose			
Endo-1,4-β-xylanase	1,4-β-D-Xylan xylanohydrolase	3.2.1.8	Hydrolyzes mainly interior β-1,4-xylose linkages of the xylan backbone
α-L-Arabinofuranosidase	α-L-Arabinofuranoside arabinofuranohydrolase	3.2.1.55	Hydrolyzes terminal nonreducing α-arabinofuranose from arabinoxylans
α-Glucuronidase	α-Glucuronoside glucanohydrolase	3.2.1.31	Releases glucuronic acid from glucuronoxylans
Acetylxylan esterase	Acetyl-ester acetylhydrolase	3.1.1.6	Hydrolyzes acetylester bonds in acetylxylans
Ferulic acid esterase	Carboxylic ester hydrolase	3.1.1.1	Hydrolyzes feruloylester bonds in xylans

CONVERSION OF CORN FIBER TO ETHANOL

Corn fiber is a likely first target as a gathered biomass source to incorporate into commercial fuel ethanol production. Corn fiber is available in sufficient quantities from corn milling processes to serve as a low-cost feedstock (Saha and Bothast, 1999a) (see chapter 1 by Nichols et al. for more detail on ethanol production from corn). The production of ethanol from pericarp fiber has the potential to increase ethanol yields by a maximum of 0.3 gal/bushel in a wet-milling process (Gulati et al., 1997). Although wet milling has been the main source, corn fiber is now available as quick fiber (QF) from modified dry-grind ethanol facilities equipped to soak and grind the kernel and then separate both germ and fiber from the endosperm by density fractionation before saccharification (Singh and Eckhoff, 1997; Wahjudi et al., 2000). Pretreatment and enzymatic processes can potentially convert corn fiber or QF to its component fermentable sugars: 37% glucose from starch/cellulose (32% in QF), 18% xylose (22%), 11% arabinose (11%), and 4% galactose (Grohmann and Bothast, 1997; Dien et al., 2004). Pretreatment of QF with dilute sulfuric acid, neutralization with $Ca(OH)_2$ to pH 4.5 followed by SSF at 14.1% (wt/wt) biomass with added cellulase, beta-glucosidase, glucoamylase, and *S. cerevisiae* NRRL Y-2034 (optical density at 600 nm, 0.5, ~0.1 g [dry weight] of cells/liter) yielded 0.096 gal of ethanol/bushel of corn (3.04 lb of QF/bushel), or 0.153 g of ethanol/g of QF by fermentation of the glucose (at 85% of theoretical yield). The ethanol accumulated in the batch SSF process was 23.4 g/liter in 72 h (a volumetric productivity of 0.33 g/liter/h). The volumetric productivity was noted as being about 64% of that on a synthetic medium with comparable sugar concentrations to the hydrolysate, indicating an inhibited rate, likely due to inhibitors such as acetic acid formed during pretreatment and hydrolysis of the QF. A modified fermentation process carried out with the pentose-fermenting recombinant *Escherichia coli* FBR5 yielded 0.116 g of ethanol/g of QF. Since Y-2034 was not able to utilize the pentose sugars, only about two-thirds of the potential ethanol yield was realized. However, much higher yields per gram of QF should be possible if the xylose-fermenting yeast strains are incorporated into the process (Table 2). A recent further adjustment to the modified dry-grind process, enzymatic milling, incorporates additional steps employing enzymes to enhance fractionation and recovery of the endosperm fiber: a starch saccharification step after kernel soaking and grinding prior to density-based removal of the germ and pericarp, followed by a screening step to remove the endosperm fiber from the endosperm starch slurry. The endosperm fiber contains 70% carbohydrates, most of which are fiber-bound starch (57% starch, 5% cellulose, 5% xylose/galactose, and 3% arabinose) (Dien et al., 2005b). These studies suggest that the separated fiber fractions and fiber-free corn mash could be processed in separate batch SSF processes and that the ethanol streams from each process could then be combined for distillation. The fermentation of the separated fiber fractions will further boost the ethanol yield per bushel of corn by 13.3% over that obtained from the starch.

XYLOSE UTILIZATION BY *SACCHAROMYCES*

S. cerevisiae is the classic ethanologenic yeast. It is among the preferred microbial biocatalysts due to its more adaptive and tolerant characteristics compared with other yeasts and engineered bacteria. It can efficiently utilize hexose sugars such as glucose but is limited for pentose sugar utilization. Traditional *S. cerevisiae* strains cannot use xylose as a carbon source to produce ethanol. As mentioned previously, xylose is the predominant pentose sugar derived from hemicelluloses. The inability of the yeast to utilize xylose has limited its use in bioethanol applications, and the development of new strains is needed for this emerging industry. This poses a significant challenge that has been one of the bottlenecks preventing the use of biomass as a low-cost feedstock for ethanol production.

In the past 2 decades, significant advances have been made to improve xylose utilization by *S. cerevisiae*. Although *S. cerevisiae* does not naturally utilize xylose and is considered to dissimilate, rather than assimilate, most xylose, the yeast does possess a pathway that is able to completely oxidize xylose in the presence of D-ribose and other substrates (Gong et al., 1981; van Zyl et al., 1989). It has genetic potential to consume xylose (also see later sections for more yeast genetic potentials). However, a problem is that the majority of xylose is metabolized into xylitol instead of ethanol. In eukaryotes, including natural pentose-fermenting yeast and fungi such as *P. stipitis* (see chapter 3 by Jeffries for more detail on this species), xylose conversion and utilization occur via two oxidoreductase reactions, xylose reduction to xylitol by xylose reductase (XR, *XYL1*, EC 1.1.1.21) and xylitol reduction to xylulose by xylitol dehydrogenase (XDH, *XYL2*, EC 1.1.1.9). Xylulose is then phosphorylated to xylulose-5-phosphate by xylulokinase (*XKS1*/*XYL3*, EC 2.7.1.17) and enters the pentose phosphate pathway. Enzymatic activities of XR and XDH were observed in *S. cerevisiae* about 2 decades ago (Gong et al., 1981; Batt et al., 1986). The first recombinant strain of *S. cerevisiae* incorporated xylose utilization genes *XYL1* and *XYL2* from *P. stipitis*, which allowed it to grow on xylose (Amore et al., 1991; Kötter et al., 1990). The limited xylose utilization was attributed

Table 2. Ethanol fermentation kinetics of selective representative yeast strains

Strain	Conditions[a]	Maximum sp. growth (h^{-1})	Maximum biomass (g/liter)	Volumetric ethanol productivity (g/liter/h)	Maximum ethanol (g/liter)	Biomass yield (g/g)	Ethanol yield (g/g)	By-product (g/liter)	Reference
P. stipitis CSIR-Y633 (CBS 7126)	90 g/liter of xylose. Defined medium (amino acids, NH_4Cl, min. vit), init. 0.2–.4 g dry cells/liter, pH 4, 30°C, 200 ml air/min, 300 rpm, 1.5 liters	0.13	9.2[c]	0.6	33.2	0.1	0.37	0	DuPreez et al., 1986
Saccharomyces 1400 (pLNH33)	53 g/liter of glucose + 56 g/liter of xylose, Bactol peptone, yeast ext., init. 2.3 g dry cells/liter, 30°C, 100 ml/250-ml flask, 150–200 rpm	0.66(glu); 0.19(xyl)[d]	11.5	1.39	47.9	0.084	0.46	5.6 glycerol; 3.2 (xylitol)	Krishnan et al., 1999
S. cerevisiae TMB3001	50 g/liter of glucose + 50 g/liter of xylose, Verduyn's defined minimal medium [$(NH_4)_2SO_4$, min., vit, ergosterol], init. 0.002 g dry cells/liter, pH 5 (controlled), 30°C, 4 liter/5 liter fermentor under N_2 at 600 rpm	0.26	3.5	0.64	23.3	0.034	0.23	8.7 (glycerol); 4.1 (xylitol)	Zaldivar et al., 2002
S. cerevisiae A6 (transformed industrial)	50 g/liter of glucose + 50 g/liter of xylose	0.30	4.6	0.70	25.2	0.046	0.27	8.3 (glycerol); 15.9 (xylitol)	Zaldivar et al., 2002
S. cerevisiae Red Star Ethanol Red	SSF corn mash, init. 250 g/kg total saccharides, 16 mM urea, pH 4.5, 32°C, 4-liter/5-liter fermentor at 300 rpm (N_2 sparge), 10^7 cells/g	0.25	5.3×10^8 cells/g	1.9 g/kg/h	130 g/kg	0.048 (0.055)[b]	0.47 (0.61)[b]		Devantier et al., 2005
CEN.PK.113-7D (MATα)	SSF corn mash, init. 250 g/kg total saccharides, 16 mM urea, pH 4.5, 32°C, 4-liter/5-liter fermentor at 300 rpm (N_2 sparge), 10^7 cells/g	0.23	3.7×10^8 cells/g	1.8 g/kg/h	125 g/kg	0.033 (0.038)[b]	0.47 (0.61)[b]		Devantier et al., 2005
S. cerevisiae RWB 218	100 g/liter of glucose + 25 g/liter of xylose, Verduyn's defined minimal medium [$(NH_4)_2SO_4$, min., vit, ergosterol]), init. 1.1 g of dry cells/liter, pH 5 (controlled), 30°C, 1.5-liter/2-liter fermentor under N_2 at 800 rpm	0.22 (glucose); 0.1 (overall)	10.4	2.17[e]	47.9	0.084	0.38	10.3 glycerol; minor xylitol, acetate, succinate, D-lactate	Kuyper et al., 2005b
P. stipitis NRRL Y-7124 (CBS 5773)	150 g/liter of xylose, optimized defined medium (urea, amino, acids, vit, min.), init. 0.017 g of dry cells/liter, pH 6.5, 25°C, O_2 limited 150-rpm flask	0.13 ± 0.01[f]	4.2 ± 1	0.38 + 0.06	61 ± 9	0.028	0.41 ± 0.06	<5 xylitol, ribitol	Slininger et al., 2006

[a]Abbreviations: min., minerals; vit, vitamins; ext., extract; init., initially; g dry cells, g (dry weight) of cells.
[b]Yields originally expressed as C mol/C mol in the source are given in parentheses.
[c]Values calculated from initial concentrations and yield data, given no residual xylose, as stated in the reference source.
[d]Values obtained by parameter estimation for the proposed model given in the source.
[e]Rate calculated from the maximum ethanol concentration/time at maximum.
[f]Growth rate average over the first 24 h of growth subject to substrate inhibition and oxygen limitation.

to dual cofactor dependence of the XR, the generation of NADPH by the pentose phosphate pathway, and insufficient capacity of the nonoxidative pentose phosphate pathway in *S. cerevisiae* (Kötter and Ciriacy, 1993). Further research showed that a DNA-fusioned *Saccharomyces* strain with xylulokinase gene *XKS1* significantly improved ethanol yield and ability to grow on a mixture of glucose and xylose (Ho et al., 1998). All improved *S. cerevisiae* strains using the three genes were able to produce ethanol at varied yield levels but had a common problem of xylitol production when grown on xylose (Table 3). Xylitol is undesirable in the ethanol production pathway (Fig. 1). Attempts have been made to improve the efficiency of the xylose utilization pathway by selection through directed evolutionary engineering and site-specific chemical mutagenesis (Sonderegger and Sauer, 2003; Wahlbom et al., 2002). Efficiency of xylose utilization and yeast growth rate were improved under anaerobic conditions by different mechanisms. However, xylitol production still occurred at a significant level.

The redox imbalance due to the different cofactor preferences associated with enzyme activities of the two oxidoreductases, XR and XDH, appeared to limit efficient xylose utilization by the yeast in this eukaryotic system. Xylose conversion into xylitol releases one $NADP^+$, and xylitol to xylulose yields one NADH; thus, regeneration of NADPH and NAD^+ is needed to maintain redox balance. In case no electron acceptor is available, such as under anaerobic conditions, yeast cells cannot maintain the redox balance and therefore xylose utilization is limited in this pathway (Fig. 1). Some yeasts appear to have efficient alternative pathways for conversion of xylose, such as *P. stipitis*. In such a route, xylose conversion to xylitol is coupled with NADH rather than NADPH. The released NAD^+ catalyzes the xylitol to D-xylulose reaction to enable

Table 3. A survey of representative *Saccharomyces* strains for improved xylose utilization by genetic engineering and directed adaptation selection

Strain	Background	Condition	Sugar (g liter^{-1})	Medium[a]	Ethanol yield (g g^{-1})	Xylitol production (g g^{-1})	Reference
pRD1	*XYL1, XYL2*	Aerobic batch	Xylose 20	sm	0.13	0.15	Kötter et al., 1990
1400 (pLNH32)	*XYL1, XYL2, XKS1*	Fermentative batch	Glucose 90, xylose 40	cm	0.46	0.17	Ho et al., 1998
H158	*XYL1, XYL2, XKS1*	Anaerobic batch	Glucose 20, xylose 50	sm	0.1	0.61	van Zyl et al., 1999
TMB3001	*XYL1, XYL2, XKS1*	Anaerobic continuous	Glucose 15, xylose 50	sm	0.3	0.03	Eliasson et al., 2000
H158-pXks	*XYL1, XYL2, XKS1*	Fermentative batch	Xylose 80	cm	0.2	0.03	Johansson et al., 2001
H1691	*XYL1, XYL2,XKS1*	Anaerobic batch	Xylose 50	sm	0.09	0.41	Toivari et al., 2001
TMB3102	*xylA*, Δ*gre3*	Anaerobic batch	Xylose 50	sm	0.22	0.31	Traff et al., 2001
TMB3225	*XYL1, XYL2, XKS1*, Δ*zwf1*	Anaerobic batch	Xylose 50	sm	0.41	0.05	Jeppsson et al., 2002
FPL-YSX3	Respiration-deficient *XYL1, XYL2, XYL3*	Aerobic fermentation	Xylose 40	mm	0.09	0.33	Jin et al., 2003
A4	*XYL1, XYL2, XKS1*	Aerobic batch	Glucose 50, xylose 50	sm	0.27	0.27	Zaldivar et al., 2002
TMB3001C1	Adaptation select of TMB3001 (heterozygous populations)	Anaerobic batch	Xylose 20	sm	0.24	0.32	Sonderegger and Sauer, 2003
H2684	*XYL1, XYL2, XKS*, Δ*zwf1, GDP1*	Anaerobic batch	Glucose 24, xylose 24	sm	0.34	0.34	Verho et al., 2003
424A(LNH-ST)	*XYL1, XYL2, XKS1*	Fermentation batch	Xylose 20	cm	0.43	0.1	Sedlak and Ho, 2004
RW 202-AFX	Adaptation select of RWB202 (*xylA*)	Anaerobic batch	Xylose 20	sm	0.42	0.021	Kuyper et al., 2004
TMB3050	Adaptation select of TMB3045 (Δ*gre3, XKS1, TAL1, RKI1, TKL1, RPE1, xylA*)	Oxygen-limited fermentation	Xylose 50	sm	0.29	0.23	Traff et al., 2001; Karhumaa et al., 2005
RWB 218	Adaptation select of RWB 217 (*xylA, XKS1, TAL1, TKL1, RPE1, RKI1*, Δ*gre3)*	Anaerobic batch	Xylose 20	sm	0.41	0.001	Kuyper et al., 2005b

[a]cm, complex medium; sm, synthetic medium; mm, minimum medium.

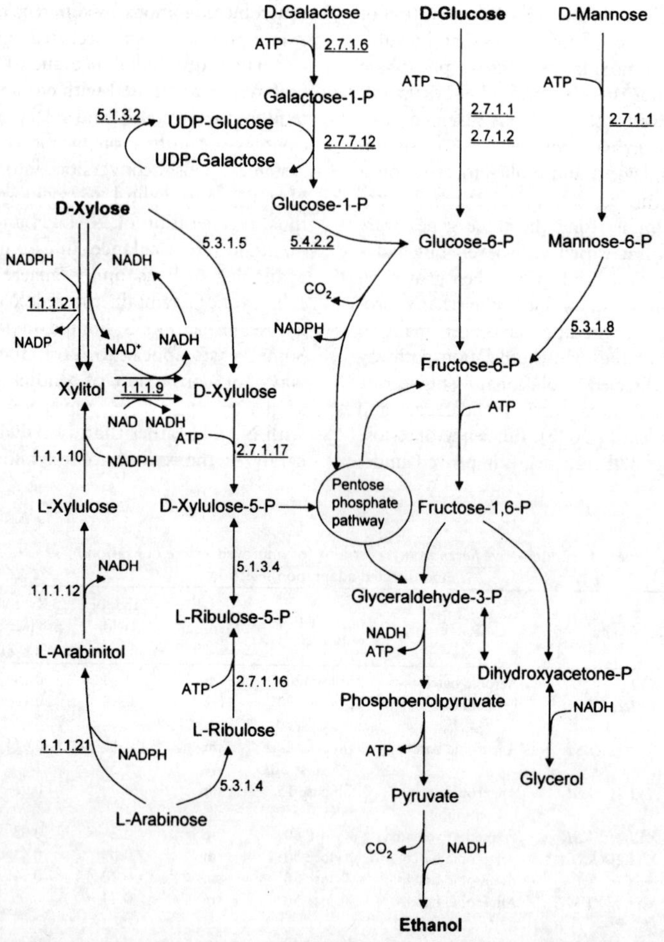

Figure 1. Hexose and pentose catabolism pathways of *Saccharomyces cerevisiae*. Shown are catabolic pathways of yeast in utilization of major hexoses including glucose, galactose, and mannose; and of pentoses including xylose and arabinose for ethanol production. In the diagram, underlined EC numbers represent endogenous enzymes, and those not underlined indicate an exogenous origin (i.e., introduced to the yeast). Enzyme-encoding genes and EC numbers are presented in parentheses as follows: hexokinase (*HXK1/HXK2*, 2.7.1.1); glucokinase (*GLK1*, 2.7.1.2); galactokinase (*GAL1*, 2.7.1.6); galactose-1-phosphate uridylyltransferase (*GAL7*, 2.7.7.12); UDP-glucose 4-epimerase (*GAL10*, 5.1.3.2); phosphoglucomutase (*GAL5/PGM2*, 5.4.2.2); hexokinase 1 (*HXK1*, 2.7.1.1); mannose-6-phosphate isomerase (*PMI40*, 5.3.1.8); xylose reductase/aldose (*GRE3/xyl1*, 1.1.1.21); xylitol dehydrogenase (*XYL2/xyl2*, 1.1.1.9); xylulokinase (*XKS1/xyl3*, 2.7.1.17); xylose isomerase (*xylA*, 5.3.1.5); arabinitol 4-dehydrogenase (*lad1*, 1.1.1.12); L-xylulose reductase (*lxr1*, 1.1.1.10); L-arabinose isomerase (*araA*, 5.3.1.4); L-ribulokinase (*araB*, 2.7.1.16); and L-ribulose-5-phosphate 4-epimerase (*araD*, 5.1.3.4) (adapted from Sonderegger et al., 2004b; Grotkjær et al., 2005; Karhumaa et al., 2005; Jeffries, 2006; and van Maris et al., 2006).

balanced cofactor cycling (Fig. 1) (Bruinenberg et al., 1983, 1984). However, in most cases, including under anaerobic conditions, this pathway is not dominant, and the higher ratios of NADPH/NADH lead to more xylitol production (Bruinenberg et al., 1984; Kuyper et al., 2004; van Maris et al., 2006). Improved pentose fermentation by engineering NADPH regeneration was recently achieved (Verho et al., 2003). Although there is a concern that the anaerobic redox imbalance restricts xylose metabolism by the two-step xylose-xylulose reductions, metabolic flux analysis studies clearly demonstrated that xylose uptake is not the limiting factor for anaerobic growth on xylose (Sonderegger et al., 2004a). Ultimately, restricted xylose metabolism is limited by the rate of ATP production but not only by the redox balance alone. It is known that redox imbalance negatively impacts ATP production. A survey of representative *Saccharomyces* strain improvements is presented in Table 3. Most strains were improved using these three xylose utilization genes. Among these, strain 424A (LNH-ST), derived from chromosome integration of multiple copies of the three genes, appeared to have acceptable performance (Sedlak and Ho, 2004). However, numerous issues still remain to be resolved such as ethanol production rate and yield affected by restricted industrial anaerobic growth conditions, xylose fermentation efficiency, inoculum density, and ethanol yield calculation methods.

In contrast to the multistep pathways demonstrated in eukaryotes, most bacteria initiate xylose metabolism by its isomerization into xylulose directly by xylose isomerase (XI, *xylA*, EC 5.3.1.5) (Fig. 1). Recombinant yeast that expresses XI can be expected to avoid cofactor imbalance associated with reduction to xylitol. Although it was recognized 2 decades ago (Batt et al., 1986), heterogeneous XI expression has not been successfully accomplished in yeast, possibly due to numerous factors including improper protein folding, posttranslational modifications, disulfide bridge formation, and the internal pH of yeast (Ho et al., 1984; Moes et al., 1996; Sarthy et al., 1987; Walfridsson et al., 1996; Karhumaa et al., 2005; Gardonyi and Hahn-Hägerdal, 2003). Utilization of xylose isomerase did not become more significant until the discovery of the anaerobic cellulolytic fungus *Piromyces* sp. E2 from elephant feces (Harhangi et al., 2003; Kuyper et al., 2003). Unlike most fungi, xylose-grown *Piromyces* sp E2 demonstrated a clear bacterial xylose isomerase pathway without typical enzymatic activities of XR and XDH as shown for eukaryotics (Kuyper et al., 2003). The *xylA* gene encoding the *Piromyces* XI was functionally expressed in *S. cerevisiae*. This work conclusively confirmed the yeast's genetic potential to transport and utilize xylose through its endogenous xylulokinase, as previously suggested, regardless of the slow rate of growth and incomplete xylose consumption (Eliasson et al., 2000; Richard et al., 2000). These studies have opened options and marked a new phase for improvement of xylose utilization by *S. cerevisiae*. Demonstration of yeast genetic potential is significant for strain improvement and will be discussed in more detail in the inhibitor stress response sections of this chapter. Selection through a directed evolution method resulted in an improved strain capable of anaerobic growth on xylose with little xylitol produced (Kuyper et al., 2004). Continued efforts in new strain construction focusing on metabolic fluxes further improved anaerobic growth rate using xylose as a sole carbon source (Kuyper et al., 2005a). The new strain RWB 217 has the XI background and overexpression of the following enzymes: xylulokinase (*XKS1*, EC 2.7.1.17), ribose-5-phosphate ketol-isomerase (*RKI1*, EC 5.3.1.6.), D-ribulose-5-phosphate 3-epimerase (*RPE1*, EC 5.1.3.1), transketolase 1 (*TKL1*, EC 2.2.1.1), and transaldolase (*TAL1*, EC 2.2.1.2). Aldose reductase (*GRE3*, EC 1.1.1.21) was deleted for this strain with the intention of limiting xylitol formation (Table 3). However, when grown on glucose-xylose mixtures, this strain preferred glucose, and the xylose consumption was substantially slow, which suggested the low affinity of xylose transportation kinetics. Once again, selection under pressure using directed evolution overcame the low affinity of xylose uptake and resulted in the best current performance of a genetically modified *S. cerevisiae* strain for biomass conversion to ethanol (Kuyper et al., 2005b). It should be pointed out that the complete genetic background and genomic mechanisms of the yeast in xylose fermentation are not clear.

FERMENTATION ON GLUCOSE AND HETEROGENEOUS SUGARS

Ethanol production of yeast on glucose is a classic fermentation example. Strains of *Saccharomyces* perform extremely well in utilization and fermentation using this hexose under both aerobic and anaerobic conditions. Well-functioning hexose transporter family members maintain an easy flow of glucose uptake for the yeast and produce ethanol through glycolysis (Fig. 1). However, diauxic lag appeared to be a problem for native yeast involved in catabolic repression by glucose in the presence of heterogeneous hexoses and pentoses (Slininger et al., 1987). The diauxy has been similarly observed in engineered strains of *S. cerevisiae* (Krishnan et al., 1999; Kuyper et al., 2005b; Zaldivar et al., 2002). In order to effectively use varied mixed sugars obtained from lignocellulosic biomass, it is important for the yeast to have corresponding efficient sugar transporters. Research in this area has been limited but is needed. Corresponding functional enzymes and a well-maintained redox balance are critical. Pathway-based studies in conjunction with these functions are expected to have significant input for future improvement. An illustration of general metabolic pathways of main sugars including hexoses (glucose, mannose, and galactose) and pentoses (arabinose and xylose) is outlined (Fig. 1). See van Maris et al. (2006) for a comprehensive review of fermentations of these sugars and other related compounds including galacturonate and rhamnose.

BIOMASS CONVERSION INHIBITORS

Another technical barrier of biomass conversion to ethanol is the stress conditions involved in the biomass process procedures which interfere with microbial growth and subsequent fermentation. Economic biomass pretreatment generates harsh conditions including high temperature, extreme pH, high substrate concentration, osmotic shifts, and toxic compounds that prohibit yeast growth and fermentation. For economic reasons, dilute acid hydrolysis is commonly used in biomass degradation for hydrolysis of the hemicellulose fraction and increase in fiber porosity to allow enzymatic saccharification and fermentation of the cellulose fraction (Bothast and Saha, 1997; Saha, 2003). However, a major limitation of this method is the generation of numerous by-products and compounds that inhibit microbial growth and metabolism. Some common inhibitors have been identified, including acetic acid, terpenes, alcohols, tannins, and other aromatic compounds released upon hydrolysis of lignocelluloses, i.e., furfural, hydroxymethylfurfural, levulinic acid, and formic acid derived from sugar degradation; phenolics such as 4-hydroxybenzoic acid, vanillin, and catechol from lignin degradation; metals and SO_2 inhibitors resulting from hydrolytic equipment and additives; and ethanol, glycerol, acetic acid, and lactic acid, typically formed during the fermentation (Palmqvist and Hahn-Hägerdal, 2000a, 2000b). More than 100 compounds were detected to have potential inhibitory effects on microbial fermentation (Luo et al., 2002). See Klinke et al. (2004) for a comprehensive review. Synergistic effects of inhibitory compounds in mixture enhance inhibition beyond that expected for a simple sum of the individual compound effects. In addition, the osmolarity of hydrolysates could be a significant factor inhibiting yeast performance (Olsson and Hahn-Hägerdal, 1996). Furfural and HMF are considered to be the most potent and representative inhibitors of yeast growth and fermentation (Chung and Lee, 1985; Olsson and Hahn-Hägerdal, 1996; Taherzadeh et al., 2000a). During biomass degradation by dilute acid treatment, furfural and HMF are derived from dehydration of pentoses and hexoses, respectively (Dunlop, 1948; Antal et al., 1990, 1991; Larsson et al., 1999; Lewkowski, 2001). These compounds reduce enzymatic and biological activities, break down DNA, inhibit protein and RNA synthesis (Sanchez and Bautista, 1988; Khan and Hadi, 1994; Modig et al., 2002), and damage the cell wall of yeast (S. Gorsich and Z. L. Liu, unpublished data). Yeasts can be repressed by the inhibitory complex even at low concentrations (Liu et al., 2004). Most yeasts, including industrial strains, are susceptible to the complexes associated with dilute acid hydrolysis pretreatment (Palmqvist et al., 1999; Taherzadeh et al., 2000; Martin and Jonsson, 2003; Liu et al., 2004). To facilitate fermentation processes, additional remediation treatments, including physical, chemical, or biochemical detoxification procedures, are often required to remove these inhibitory compounds. However, these additional steps add cost and complexity to the process and generate extra waste products (Martinez et al., 2000; Mussatto and Roberto, 2004).

Although *S. cerevisiae* is an ethanol-producing yeast, it is limited in its ability to survive high ethanol concentrations (Ingram and Buttke, 1984). Ethanol stress inhibits enzymes such as the glycolytic enzymes and hexokinase (Augustin et al., 1965; Dombek and Ingram, 1987), causes yeast cells to lose mitochondrial DNA, which is necessary for cellular respiration (Ibeas and Jimenez, 1997), decreases numbers of plasma membrane H(+)-ATPases (Meaden et al., 1999; Monteiro et al., 1994), inhibits endocytosis, alters vacuole morphology (Lucero et al., 1997; Meaden et al., 1999), and changes plasma membrane fluidity (Alexandre et al., 1994). A strong relationship between membrane lipid structure and stress resistance has been documented for ethanol and other stress factors in both prokaryotes and eukaryotes (Ingram, 1976, 1990; Sajbidor et al., 1995; You et al., 2003). With solid loadings for dilute sulfuric acid pretreatment as high as 47% (Dien et al., 2005a) and fermentable glucose and xylose as high as 60% (40:20) by weight, potential ethanol accumulations could approach 10% depending on water addition and ethanol yield. Ethanologenic yeast strains tolerant to high levels of ethanol are needed. Research in this area is limited, and more studies are expected. On the other hand, significant progress has been made recently for inhibitors generated from biomass pretreatment. This chapter focuses on the representative inhibitors furfural and HMF.

Furfural conversion to furan methanol (FM, furfuryl alcohol) by yeasts has been established (Morimoto and Murakami, 1967; Villa et al., 1992; Mohsenzadeh et al., 1998; Liu, 2006) (Fig. 2). It was suggested that furfural is first converted to furan methanol and further reduced to pyromucic acid (Nemirovskii and Kostenko, 1991). Furfural can also break down to form formic acid (Palmqvist and Hahn-Hägerdal, 2000b). Biotransformation of furfural and HMF is due to NADH- and NADPH-coupled enzymes by yeasts (Palmqvist et al., 1999; Larroy et al., 2002; Petersson et al., 2006; Liu, 2006; Z. L. Liu, J. Moon, B. J. Andersh, P. J. Slininger, and S. Weber, submitted for publication). In the presence of furfural, the ATP level is low and cell replication is limited. Glycerol formation is reduced. Furfural has been characterized as an electron acceptor (Wahlbom and Hahn-Hägerdal, 2002). Shortage of NADH was observed in the pres-

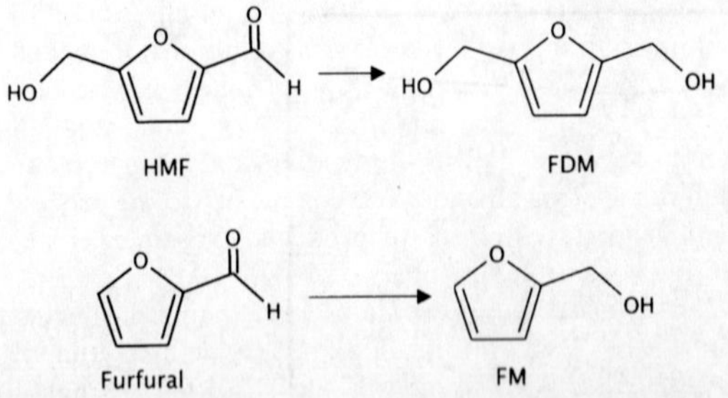

Figure 2. Metabolic conversion products of inhibitors. Furfural is converted to 2-furanmethanol (FM, furfuryl alcohol) and 5-hydroxymethylfurfural (HMF) converted to 2,5-furan-dimethanol (FDM; 2,5-bis-hydroxymethylfuran) (Liu et al., submitted).

ence of furfural. It appears that furfural reduction competes for NADH and interferes with cell glycolysis during regeneration of NAD^+. As a result, furfural can cause accumulation of acetaldehyde, resulting in a delay of acetate and ethanol production. Xylitol excretion is reduced during xylose fermentation when furfural is added into the medium (Wahlbom and Hahn-Hägerdal, 2002). Reduced furfural tolerance was observed for selective deletion mutants of genes in the pentose phosphate pathway (Gorsich et al., 2005, 2006). These observations suggest indirect evidence of potential NADPH-dependent reactions involved in pentose phosphate pathways.

Unlike the well-studied furfural, knowledge on HMF conversion has been limited because there was not a readily available commercial source for an HMF conversion product. Based on the furfural conversion route, HMF has been assumed to convert into HMF alcohol (Nemirovskii et al., 1989). Recently, an HMF metabolic conversion product was isolated and identified as furan-2,5-dimethanol (FDM), also termed 2,5-bis-hydroxymethylfuran (Liu et al., 2004; Liu 2006; Liu et al., submitted) (Fig. 2). HMF has a maximum absorbance at 282 nm, and FDM at 222 nm. FDM was further isolated from cell-free cultural supernatant, purified, and characterized using mass and nuclear magnetic resonance (NMR) spectra analysis (Liu et al., 2004). The signals for the aldehyde proton and the asymmetric spectra of HMF were absent when the purified HMF conversion product was analyzed using NMR. The NMR spectra are consistent with that of a symmetrical molecule with a furan ring. The chemical structure of the metabolite was identified as a compound with a composition of $C_6H_8O_3$ and a molecular weight of 128 (Fig. 2). The identification of FDM clarified the existing literature and provided a basis for later studies on mechanisms of HMF inhibitor detoxification.

MECHANISMS OF IN SITU DETOXIFICATION

On a defined medium, strain Y-12632 showed clear dose-dependent cell growth and metabolic conversion activities in response to varied doses of HMF and/or furfural under controlled conditions. Cell growth was delayed at tolerable concentrations of 10, 30, and 60 mM (Liu et al., 2004; Liu et al., submitted). Metabolic conversion activities in transformation of HMF to FDM, furfural to FM, and glucose to ethanol showed parallel delays with the increase of inhibitor doses compared with a control (Liu et al., 2004). However, this lag phase was not observed at the higher concentration of 120 mM (Liu et al., submitted). Cells were completely repressed at nontolerable concentrations, and no biological activity or HMF transformation was observed. This indicated a biological activity for the transformation of the inhibitors by the yeast. Whereas a wild type was unable to establish a culture in the presence of the HMF and furfural complex, tolerant strain 12HF10 showed a nearly normal cell growth and produced a normal yield of ethanol (Fig. 3 and 4). This suggests that it is possible to in situ detoxify furfural and HMF by ethanologenic *S. cerevisiae*.

A recently described FDM preparation procedure can be used as a standard for metabolic profiling analysis involving HMF using high-pressure liquid chromatography (HPLC) (Liu et al., submitted). This development allowed studies on mechanisms of HMF detoxification. When the tolerant strain was used, furfural and HMF conversion products FM and FDM were found to be accumulated in the medium as yeast growth and fermentation ensued. Furfural and HMF are furan derivatives having a furan ring and a composition of $C_5H_4O_2$ and $C_6H_6O_3$, respectively. Their conversion products FM and FDM contain the furan rings with a composition of $C_5H_6O_2$ and $C_6H_8O_3$, respectively (Fig. 2). As detected by HPLC analysis, during the inhibitor conversion process, these furan elements were intact and persistently existed in the medium at the end of the fermentation. The presence of FM and FDM did not affect yeast growth or ethanol yield. Apparently, the aldehyde functional group in furfural and HMF is toxic to yeast but not the furan ring or associated alcohol functional groups. Clearly, aldehyde reduction is a mechanism involved in the in situ detoxification of furfural and HMF (Liu et al., submitted). Although the toxicity of the aldehyde had been recognized (Leonard and Hajny, 1945; Zaldivar et al., 1999), the mechanism of the detoxification of furfural and HMF had not been convincingly demonstrated until now (Liu et al., submitted). Any potential further reduction or degradation of the furan ring or alcohol groups may not play a significant role for the in situ detoxification of furfural and HMF by the yeast.

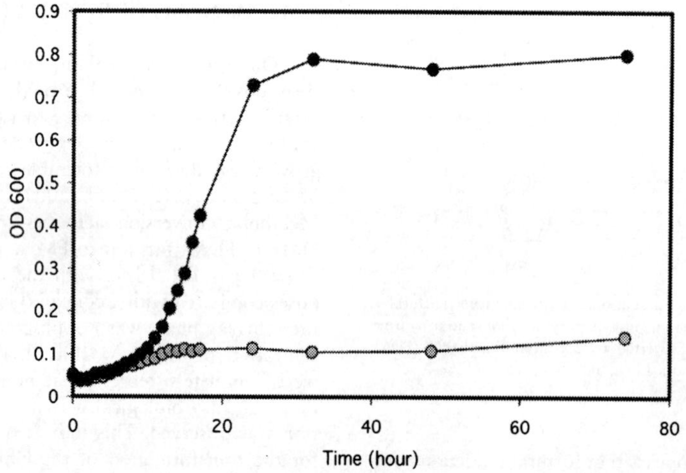

Figure 3. Cell growth under inhibitor stress. Cell growth of *S. cerevisiae* NRRL Y-12632 (gray circles) and strain 12HF10 (black circles) in response to furfural and HMF at 12 mM each on a defined medium (Liu et al., submitted).

Mechanisms of detoxification of the inhibitors by yeast did not involve utilization or degradation of the furan compounds. Therefore, the term furan conversion or furan reduction appeared in the literature but should not be used in the context of these inhibitor conversions (Liu et al., submitted). Furthermore, the use of "furan derivatives" as a general term for "inhibitors" such as furfural and HMF should be avoided since FM and FDM are also furan derivatives, which appear to be less toxic to the yeast. Instead, a category of aldehyde inhibitors, represented by furfural and HMF, should be used, including other aldehydes detected in biomass hydrolysis, such as cinnamaldehyde and veratraldehyde.

NAD(P)H-dependent enzymatic activities were observed under the inhibitor stress conditions (Larroy et al., 2002; Liu, 2006; Nilsson et al., 2005; Petersson et al., 2006; Liu et al., submitted). HMF reduction was earlier reported to have a cofactor preference for NADPH (Wahlbom and Hahn-Hägerdal, 2002). A later study found that a different strain of *S. cerevisiae* showed preference for NADH rather than for NADPH (Nilsson et al., 2005). Recently, enzyme assays of individual functional gene clones showed ADH6 and ADH7 to have stronger reduction activity with NADPH on substrates furfural and HMF, whereas ALD4 and GRE3 displayed stronger activities with NADH (Liu et al., submitted). Although some genes encoding proteins showed dual cofactor reduction activities, the cofactor preference can be easily recognized by the levels of the reduction activity coupled with either cofactor for individual gene clones. However, the whole-cell protein extract of a tolerant strain showed reduction activities on substrate furfural or HMF coupled with either NADH or NADPH (Liu et al., submitted). This acquired reduction activity was particularly strong when it was induced by the inhibitor complex furfural plus HMF. It seems that enzymatic activities of the whole-cell extract reflected a collective activity of numerous functional enzyme-encoding genes, including but not limited to the above tested genes (Liu et al., submitted). More than 400 genes were identified as being differentially expressed significantly under HMF stress conditions (Liu, 2006). Among these, numerous genes were identified as functional enzyme-coding genes, including not-well-characterized genes. Under the inhibitor stress conditions, the in situ detoxification of furfural and HMF by yeast faced much more complicated situations including cell wall and membrane damage and repair. It was suggested that members of the PDR gene family may play a significant role in coping with the inhibitor stress in order for the cell to survive (Liu et al., 2006). Other significantly induced and expressed genes could also be attributed in gene regulatory networks and gene interactions during the inhibitor detoxifications (Liu, 2006; Song and Liu, 2007). Single gene deletion mutations of *ADH6*, *ADH7*, *ALD4*, and *GRE3* showed no detectable growth defect or susceptibility to either furfural or HMF (Liu et al., submitted). This further indicated that a single gene deletion of the above genes does not affect cell growth and tolerance to the inhibitors. Therefore, it is unlikely that a single gene would play a decisive role in furfural and HMF detoxification. Instead, the in situ detoxification of furfural and HMF involves multiple genes, including functional

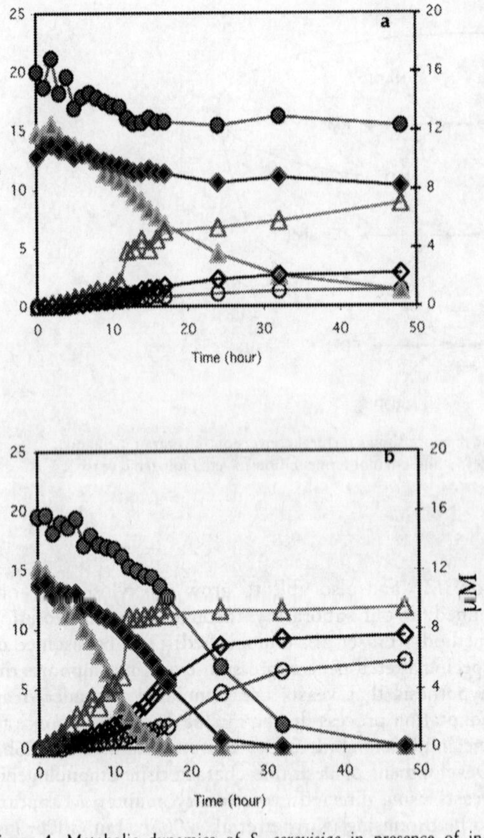

Figure 4. Metabolic dynamics of *S. cerevisiae* in presence of inhibitors. Major metabolic conversion dynamics of NRRL Y-12632 (a) and strain 12HF10 (b) including glucose (●), ethanol (○), HMF (♦), FDM (◇), furfural (▲), and FM (△) in the presence of furfural and HMF at 12 mM each on a defined medium as measured by HPLC analysis. Glucose and ethanol were estimated in grams/liter (left axis), and the remaining values are presented in millimolar (right axis) (Liu et al., submitted).

genes and regulatory genes as well as interactions among these genes and others.

In initial culture, not all genes showing in vitro aldehyde reduction activities are able to function under challenges of furfural and HMF. Absolute quantitative mRNA expression under inhibitor stress revealed that *ADH6*, *ADH7*, *ALD5*, and *ALD6* show varied levels of enhanced expression (Liu et al., submitted). Since they showed aldehyde reduction activities in vitro, these genes are likely to be attributed to the detoxification of furfural and HMF in vivo in the earlier stages, when concentrations of the inhibitors are high. On the other hand, *ALD3*, *ALD4*, and *GRE3* were repressed at varied degrees encountered by the inhibitor challenge. Although they displayed in vitro reduction activities, these genes are unlikely to play a significant role in the detoxification of furfural and HMF in vivo during the earlier stage of the inhibitor challenge. However, these genes could very well contribute to detoxification pathways at a later stage as supported by the strong reduction activities demonstrated with NADH and NADPH by the whole-cell protein extract. This further suggests that complicated gene interactions exist and multiple genes are involved for the in situ detoxification of furfural and HMF.

A refined aldehyde inhibitor conversion pathway for furfural and HMF relevant to glycolysis and ethanol production is presented (Liu et al., submitted) (Fig. 5). The reduction of furfural and HMF competes for cofactor NADH and inhibits glycolysis. It was observed that in the presence of the inhibitors, glucose was not consumed until adequate furfural and/or HMF reduction levels were reached (Liu et al., 2004, 2005). Synergistic inhibition by furfural and HMF has been well recognized as an extended lag phase of cell growth (Larsson et al., 1999; Taherzadeh et al., 2000; Wahlbom and Hahn-Hägerdal, 2002; Liu et al., 2004). These phenomena can be explained by the suggested pathways (Liu et al., submitted). For normal cell growth, NAD^+ needs to be regenerated from NADH to enable continued glycolysis. Furfural and/or HMF can dominate the competition for NADH when they are at higher concentrations; thus, a delayed glycolysis occurs. Once the inhibitors are converted, glucose consumption is observed at a faster rate than the control (Liu et al., 2004). It is likely, with the conversion of furfural to FM and HMF to FDM, that NAD^+ regeneration is easily available and favorable to enhance glucose oxidation in glycolysis. Synergistic competition of NADPH also affects biosynthesis pathways. As a result, the metabolic process can be significantly altered and delayed in the presence of the inhibitors. In addition to the toxicity of the inhibitors causing cell damage, furfural and HMF indeed affect cellular redox balance.

STRAIN IMPROVEMENT BY DIRECTED EVOLUTIONARY ADAPTATION

The economics of fermentation-based bioprocesses rely extensively on the performance of microbial biocatalysts in industrial application. Development of yeast strains that can efficiently utilize heterogeneous sugars and withstand stress conditions in the bioethanol process is key for sustainable, economic and cost-competitive industry dealing with lignocellulosic biomass conversion to ethanol. However, many of the industrially interesting microorganisms obtained thus far are not robust. Recent developments showed that inhibitor-tolerant strains of ethanologenic *S. cerevisiae* with

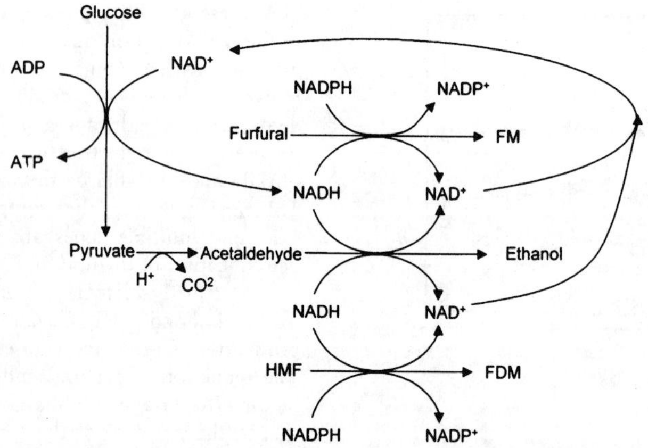

Figure 5. The furfural and HMF conversion pathways. A schematic diagram shows furfural conversion into furan methanol (FM) and HMF into 2,5-furan-dimethanol (FDM) relative to glycolysis and ethanol fermentation for ethanologenic yeast *S. cerevisiae* (adapted from Liu, 2006, and Liu et al., submitted).

enhanced ability to biotransform the inhibitors can be obtained by directed evolutionary adaptation (Liu and Slininger, 2006a; Liu et al., 2004, 2005; Liu et al., submitted). A further-improved strain obtained by this method, 12HF10, detoxified both furfural and HMF and completed ethanol fermentation in 48 h (Fig. 4). This strain did not require acclimation to the inhibitors but functioned as an initial inoculum to establish a culture and complete the fermentation. It significantly enhanced biotransformation of furfural and HMF to alcohol forms, while producing a normal yield of ethanol. This indicated a qualitative change derived from the directed evolutionary adaptation and a genetic alteration involving many genes in cell response to the inhibitors (Color Plate 1). A preliminary improvement of inhibitor tolerance to a sugarcane bagasse hydrolysate was recently reported by adaptation of an engineered xylose-utilizing strain (Martin et al., 2006). Multiple types of mutations that are caused by selection under pressure exist (Z. L. Liu and J. Moon, unpublished). Evidence showed that at least two populations with different phenotypes were recovered from a recombinant strain under selection pressure using an evolutionary method (Sonderegger and Sauer, 2003). Under such conditions, multiple mutations seemed necessary for more integrated functional adaptations. However, the yeast has endogenous genetic potential for further improvement to tolerate inhibitors such as furfural and HMF, as well as for xylose utilization. With minimum recombinant engineering of exogenous xylose isomerase, efficient xylose-utilizing strains of *S. cerevisiae* were obtained through directed evolution (Kuyper et al., 2005b). Strains of *S. cerevisiae* tolerant to HMF and also able to grow on xylose were obtained in our laboratory through directed evolution methods (Liu et al., unpublished). The persistence of specific altered gene expression over time supports the hypothesis that yeasts are stimulated to undergo an adaptation process during the lag phase in response to the inhibitors (Liu, 2006; Liu and Slininger, 2006b). Development of desirable characteristic ethanologenic yeasts using directed evolutionary adaptation appears to be promising (Kuyper et al., 2005b; Liu, 2006; Liu et al., 2005; Liu et al., submitted). The adaptation approach can be an alternative means to improve microbial strain performance. Such adapted strains could be efficiently used for further genetic manipulation. Additional studies in this area are expected for new strain development.

Adaptation is nothing new in the history of yeast utilization. The basis of success depends on the genetic potential of the yeast. As discussed earlier, the genetic potential and the ability of the yeast to withstand and transform furfural and HMF and to utilize xylose are clearly demonstrated. Not all genetic potentials for the stress conditions encountered in the bioethanol conversion process have been experimentally tested. Enrichment of genetic background of the ethanologenic yeast by introducing exogenous gene functions is needed. Thereafter, efficient genomic manipulation can be accomplished. Development of more tolerant strains with enhanced detoxification using directed enzyme evolution has shown promising results (Liu et al., unpublished). The enhanced laboratory procedures significantly speed up biological evolutionary adaptation to the stress conditions encountered by the yeast.

FUNCTIONAL GENOMICS OF ETHANOLOGENIC YEAST

We can assume that desirable yeast strains selected by directed evolutionary adaptation went through systematically integrated adaptation. Thus, the obtained yeasts appear to have balanced functions to withstand inhibitors and utilize xylose for ethanol production. However, the genomic mechanisms of such improved integration are not known. These strains provided excellent materials to study genetic mechanisms, gene interactions, and altered regulatory networks involved in the ethanologenic yeast evolution at the genome level.

Unlike the transient responses of laboratory strains, industrial strains showed more persistent expression patterns (Erasmus et al., 2003; Liu, 2005; Liu and Slininger, 2006b). Yeast genes responded immediately by showing differential expressions soon after exposure to the inhibitor, no later than 10 min (Liu, 2006; Liu and Slininger, 2004b, 2006b). The expression levels of several hundred genes were significantly different for the yeasts under HMF-challenged conditions compared with those of an untreated normal control. These genes demonstrated significant differential expression patterns during the lag phase under HMF stress and are more likely to be responsible for HMF stress tolerance (Color Plate 1). Thus, stress responses are not single-gene-controlled events, but rather an organized global expression program in response to a stress condition. Among those significantly induced genes, members of the pleiotropic drug resistance gene family were suggested to play an important role in coping with inhibitor stress for cell survival (Liu et al., 2006). During the lag phase, constant functional mRNA expression was observed for the ethanologenic yeast *S. cerevisiae*. Some genes showed continued enhanced or repressed expression, while others demonstrated significant dynamics of reversed expression (Liu and Slininger, 2006b). Genes were identified as belonging to categories of biological processes, cellular components, and molecular function. Among these, some genes appeared to be HMF- and/or furfural-specific while others shared functions as a core set of common stress genes. Regulatory elements and transcription factors significant for inhibitor tolerance have been identified (Liu and Sinha, 2006). Computational modeling to infer a genomic regulatory network recently revealed potential interesting gene interactions involved in inhibitor detoxification (Song and Liu, 2007). Genomic DNA sequence variations of the adapted strains were observed for some significant regions both in protein coding and intergenetic regions (Liu et al., unpublished). However, interpretation of some genes was limited by incomplete annotations or lack of known functions.

Fortunately, we have numerous tools available for yeast genomic studies. Among many established resources commonly used are Saccharomyces Genome Database (Fisk et al., 2006), YEASTRACT (Teixeira et al., 2006), Kyoto Encyclopedia of Genes and Genomics (Kanehisa et al., 2006), and Gene Ontology (Ashburner et al., 2000). Recently developed universal external RNA controls provided necessary guidance to improve reliability and reproducibility of expression data (Liu and Slininger, 2007). These controls can be applied to different platforms of microarray and real-time quantitative reverse transcriptase-PCR, including SYBR Green and TaqMan probe-based chemistry. The universal-control concept changed the conventional practice of quantitative reverse transcriptase-PCR and made it possible to use this platform as a high-throughput assay using a robust standard for absolute mRNA quantification (Liu and Palmquist, 2007). Metabolic engineering can improve xylose utilization in direct carbon flux to ethanol production (Ostergaard et al., 2000). A recent comparative metabolic-flux profiling analysis (Fiaux et al., 2003) provides an opportunity to incorporate such information for genetic manipulation to improve nonoxidative pentose phosphate pathway functions for heterogeneous sugar utilization.

Single-gene studies have contributed significantly to our knowledge of gene functions in the past 50 years. They will continue to do so in the future. However, investigations and advances in life science at the genomic level have revolutionized our understanding and changed our view on yeast processing events. Significant gene interactions and genomic regulatory networks need to be considered for efficient genetic manipulation for strain development (Liu and Slininger, 2004a, 2005). A few enhanced functional genes are unlikely to address the challenges encountered in bioethanol conversion. Understanding of genomic mechanisms underlying adapted functions is needed. Such knowledge provides fundamental insight into the integrated alterations in genome architecture, transcriptional profiling, and gene regulatory networks that underline heterogeneous sugar utilization and stress tolerance. Thus, obtained technologies will allow us greater flexibility and power to design and develop more desirable and robust biocatalysts for cost-effective and highly productive lignocellulosic conversion to ethanol for the next decade. Using functional genomics, we imply a broad sense of genomics including integrated approaches and analyses of gene expression and its encoded enzyme activity and protein expression, as well as metabolic outcomes in conjunction with targeted gene functions (Liu, 2007). This should include proteomics and metabolomics, which are parts of the integrated functional genomic approach.

SUMMARY

Given current advances and our understanding in science and technology for ethanologenic yeast, we are optimistic for the future of sustainable economic lignocellulosic biomass conversion to ethanol by *Saccharomyces*. However, to meet challenges of a stable and cost-competitive biomass-to-ethanol industry, research and development by multidisciplinary team efforts are needed; efficient funding and resources are necessary. There are critical issues in agricultural biomass production and industrial processing that need to be addressed using appropriate engineering approaches. From the microbial biocatalyst development point of view, genetic engineering is critical to enriching the comprehensive genetic background of the ethanologenic yeast *S. cerevisiae*. Studies on genomic mechanisms of stress tolerance and efficient heterogeneous sugar utilization are necessary to build our knowledge base for a long-term and efficient genetic manipulation of *S. cerevisiae*. Directed evolutionary genomic adaptation, focused on improvement of specific molecular functions and metabolic dynamics performed under laboratory settings, is a powerful means for improvement and development of desirable strains. Such technology combined with traditional genetic studies will bring us to a new horizon for understanding ethanologenic yeast *S. cerevisiae*. A comprehensive genomic engineering approach will allow us to meet the challenges for efficient lignocellulosic biomass conversion to ethanol in the next decade and beyond.

Acknowledgments. This study was supported by U.S. Department of Agriculture ARS National Program 307/306 and in part supported by National Research Initiative of the USDA Cooperative State Research Education and Extension Service grant no. 2006-35504-17359 to Z.L.L.

REFERENCES

Alexandre, H., I. Rousseaux, and C. Charpentier. 1994. Relationship between ethanol tolerance, lipid composition and plasma membrane fluidity in *Saccharomyces cerevisiae* and *Kloeckera apiculata*. *FEMS Microbiol. Lett.* **124**:17–22.

Amore, R., P. Kotter, C. Kuster, M. Ciriacy, and C. P. Hollenberg. 1991. Cloning and expression in Saccharomyces cerevisiae of the NAD(P)H-dependent xylose reductase-encoding gene (*xyl1*) for the xylose-assimilating yeast *Pichia stipitis*. *Gene* **109**:89–97.

Ando, S., I. Arai, K. Kiyoto, and S. Hanai. 1986. Identification of aromatic monomers in steam-exploded poplar and their influence on ethanol production by *Saccharomyces cerevisiae*. *J. Ferment. Technol.* **64**:567–570.

Antal, M. J., W. S. L. Mok, and G. N. Richards. 1990. Mechanism of formation of 5-(hydroxymethyl)-2-furaldehyde from D-fructose and sucrose. *Carbohydr. Res.* **199**:91–109.

Antal, M. J., T. Leesomboon, W. S. Mok, and G. N. Richards. 1991. Mechanism of formation of 2-furaldehyde from D-xylose. *Carbohydr. Res.* **217**:71–85.

Ashburner, M., C. A. Ball, J. A. Blake, D. Botstein, H. Butler, J. M. Cherry, A. P. Davis, K. Dolinski, S. S. Dwight, J. T. Eppig, M. A. Harris, D. P. Hill, L. Issel-Tarver, A. Kasarskis, S. Lewis, J. C. Matese, J. E. Richardson, M. Ringwald, G. M. Rubin, and G. Sherlock. 2000. Gene ontology: tool for the unification of biology. The Gene Ontology Consortium. *Nat. Genet.* **25**:25–29.

Aspinall, G. O. 1980. Chemistry of cell wall polysaccharides, p. 473–500. *In* J. Preiss (ed.), *The Biochemistry of Plants (A Comprehensive Treatise), vol. 3. Carbohydrates: Structure and Function.* Academic Press, New York, NY.

Augustin, H. W., G. Kopperschlager, H. Steffen, and E. Hofmann. 1965. Hexokinase as limiting factor of anaerobic glucose consumption of *Saccharomyces carlsbergensis* NCYC74. *Biochim. Biophys. Acta* **110**:437–439.

Bachmann, S. L., and A. J. McCarthy. 1991. Purification and cooperative activity of enzymes constituting the xylan-degrading system of *Thermomonospora fusca*. *Appl. Environ. Microbiol.* **57**:2121–2130.

Batt, C. A., S. Carvallo, D. Easson, M. Akedo, and A. J. Sinskey. 1986. Direct evidence for a xylose metabolic pathway in *Saccharomyces cerevisiae*. *Biotechnol. Bioeng.* **67**:549–553.

Bjerre, A. B., A. B. Olesen, T. Fernqvist, A. Ploger, and A. S. Schmidt. 1996. Pretreatment of wheat straw using combined wet oxidation and alkaline hydrolysis resulting in convertible cellulose and hemicellulose. *Bioresour. Technol.* **49**:568–577.

Bothast, R., and B. Saha. 1997. Ethanol production from agricultural biomass substrate. *Adv. Appl. Microbiol.* **44**:261–286.

Boussaid, A., J. Robinson, Y.-J. Cal, D. J. Gregg, and J. N. Saddler. 1999. Fermentability of the hemicellulose-derived sugars from steam-exploded softwood (Douglas-fir). *Biotechnol. Bioeng.* **64**:284–289.

Brownell, H. H., and J. N. Saddler. 1984. Steam explosion pretreatment for enzymatic hydrolysis. *Biotechnol. Bioeng. Symp.* **14**:55–68.

Bruinenberg, P. M., P. H. M. de Bot, J. P. van Dijken, and W. A. Scheffers. 1983. The role of the redox balance in the anaerobic fermentation of xylose by yeasts. *Eur. J. Appl. Microbiol. Biotechnol.* **18**:287–292.

Bruinenberg, P. M., P. H. M. de Bot, J. P. van Dijken, and W. A. Scheffers. 1984. NADH-linked aldose reductase: the key to ethanolic fermentation of xylose by yeasts. *Appl. Microbiol. Biotechnol.* **19**:256–260.

Cao, N. J., M. S. Krishnan, J. X. Du, C. S. Gong, N. W. Y. Ho, Z. D. Chen, and G. T. Tsao. 1996. Ethanol production from corn cob pretreated by the ammonia steeping process using genetically engineered yeast. *Biotechnol. Lett.* **18**:1013–1018.

Chung, I. S., and Y. Y. Lee. 1985. Ethanol fermentation of crude acid hydrolyzate of cellulose using high-level yeast inocula. *Biotechnol. Bioeng.* **27**:308–315.

Devantier, R., S. Pedersen, and L. Olsson. 2005. Transcription analysis of *S. cerevisiae* in VHG fermentation. *Ind. Biotechnol.* **1**:51–63.

Dien, B. S., L. B. Iten, and C. D. Skory. 2005a. Converting herbaceous energy crops to bioethanol; a review with emphasis on pretreatment processes, chapter 23, p. 1–11. *In* C. T. Hou (ed.), *Handbook of Industrial Biocatalysis.* Taylor and Francis, Boca Raton, FL.

Dien, B. S., D. B. Johnston, K. B. Hicks, M. A. Cotta, and V. Singh. 2005b. Hydrolysis and fermentation of pericarp and endosperm fibers recovered from enzymatic corn dry-grind process. *Cereal Chem.* **82**:616–620.

Dien, B. S., N. Nagle, K. B. Hicks, V. Singh, R. A. Moreau, M. P. Tucker, N. N. Nichols, D. B. Johnston, M. A. Cotta, Q. Nguyen, and R. J. Bothast. 2004. Fermentation of "Quick Fiber" produced from a modified corn-milling process into ethanol and recovery of corn fiber oil. *Appl. Biochem. Biotechnol.* **113–116**:937–949.

Dombek, K. M., and L. O. Ingram. 1987. Ethanol production during batch fermentation with *Saccharomyces cerevisiae*: changes in glycolytic enzymes and internal pH. *Appl. Environ. Microbiol.* **53**:1286–1291.

Doner, L. W., and K. B. Hicks. 1997. Isolation of hemicellulose from corn fiber by alkaline hydrogen peroxide extraction. *Cereal Chem.* **74**:176–181.

Draude, K. M., C. B. Kurniawan, and S. J. B. Duff. 2001. Effect of oxygen delignification on the rate and extent of enzymatic hydrolysis of lignocellulosic material. *Bioresour. Technol.* 79:113–120.

Duff, S. J. B., and W. D. Murray. 1996. Bioconversion of forest products industry waste cellulosics to fuel ethanol: a review. *Bioresour. Technol.* 55:1–33.

Dunlop, A. P. 1948. Furfural formation and behavior. *Ind. Eng. Chem.* 40:204–209.

Du Preez, J. C. 1994. Process parameters and environmental factors affecting D-xylose fermentation by yeasts. *Enzyme Microb. Technol.* 16:944–956.

Du Preez, J. C., M. Bosch, and B. A. Prior. 1986. Xylose fermentation by *Candida shehatae* and *Pichia stipitis*: effects of pH, temperature and substrate concentration. *Enzyme Microb. Technol.* 8:360–364.

Eliasson, A., C. Christensson, C. F. Wahlbom, and B. Hahn-Hägerdal. 2000. Anaerobic xylose fermentation by recombinant *Saccharomyces cerevisiae* carrying *XYL1*, *XYL2*, and *XKS1* in mineral medium chemostat cultures. *Appl. Environ. Microbiol.* 66:3381–3386.

Erasmus, D., G. van der Merwe, and H. van Vuure. 2003. Genome-wide expression analysis: metabolic adaptation of *Saccharomyces cerevisiae* to high sugar stress. *Yeast Res.* 3:375–399.

Eriksson, T., J. Borjesson, and F. Tjerneld. 2002. Mechanism of surfactant effect in enzymatic hydrolysis of lignocellulose. *Enzyme Microb. Technol.* 31:353–364.

Fiaux, J., Z. P. Cakar, M. Sonderegger, K. Wuthrich, T. Szyperski, and U. Sauer. 2003. Metabolic-flux profiling of the yeast *Saccharomyces cerevisiae* and *Pichia stipitis*. *Eukaryot. Cell* 2:170–180.

Filho, E. X. F., M. G. Touhy, J. Puls, and M. P. Coughlan. 1991. The xylan-degrading enzyme systems of *Penicillium capsulatum* and *Talaromyces emersonii*. *Biochem. Soc. Trans.* 19:25S.

Fisk, D. G., C. A. Ball, K. Dolinski, S. R. Engel, E. L. Hong, L. Issel-Tarver, K. Schwartz, A. Sethuraman, D. Botstein, and J. M. Cherry. 2006. *Saccharomyces cerevisiae* S288C genome annotation: a working hypothesis. *Yeast* 23:857–865.

Focher, B., A. Marzetti, M. Cattaneo, P. I. Beltrame, and P. J. Carniti. 1981. Effects of structural features of cotton cellulose on enzymatic hydrolysis. *Appl. Polym. Sci.* 26:1989–1999.

Gardonyi, M., and B. Hahn-Hägerdal. 2003. The *Streptomyces rubiginosus* xylose isomerase is misfolded when expressed in *Saccharomyces cerevisiae*. *Enzyme Microb. Technol.* 32:252–259.

Garrote, G., H. Dominguez, and J. C. Parajo. 2001. Generation of xylose solutions from *Eucalyptus globulus* wood by autohydrolysis-posthydrolysis processes: posthydrolysis kinetics. *Bioresour. Technol.* 79:155–164.

Gilbert, H. J., and G. P. Hazlewood. 1993. Bacterial cellulases and xylanases. *J. Gen. Microbiol.* 139:187–194.

Gilkes, N. R., B. Henrissat, D. C. Kilburn, R. C. Miller Jr., and R. A. Warren. 1991. Domains in microbial β-1,4-glycanases: sequence, conservation, function, and enzyme families. *Microbiol. Rev.* 55: 303–315.

Gong, C. S., L. F. Chen, M. C. Flickinger, L. C. Chiang, and G. T. Tsao. 1981. Production of ethanol from D-xylose by using D-xylose isomerase and yeasts. *Appl. Environ. Microbiol.* 41:430–436.

Gorsich, S. W., Z. L. Liu, and P. J. Slininger. 2005. The role of the pentose phosphate pathway in fermentation inhibitor tolerance. *Abstr.* 5-33. *27th Symp. Biotechnol. Fuels Chemicals*, Denver, CO.

Gorsich, S. W., B. S. Dien, N. N. Nichols, P. J. Slininger, Z. L. Liu, and C. Skory. 2006. Tolerance to furfural-induced stress is associated with pentose phosphate pathway genes *ZWF1*, *GND1*, *RPE1*, and *TKL1* in *Saccharomyces cerevisiae*. *Appl. Microbiol. Biotechnol.* 71:339–349.

Grethlein, H. E. 1985. The effect of pore size distribution on the rate of enzymatic hydrolysis of cellulosic substrates. *Bio/Technology* 3:155–160.

Grohmann, K., and R. J. Bothast. 1997. Saccharification of corn fibre by combined treatment with dilute sulphuric acid and enzymes. *Process Biochem.* 32:405–415.

Grotkjær, T., P. Christakopoulos, J. Nielsen, and L. Olsson. 2005. Comparative metabolic network analysis of two xylose fermenting recombinant *Saccharomyces cerevisiae* strains. *Metab. Eng.* 7:437–444.

Gulati, M., K. Kohlmann, M. Ladisch, R. B. Hespell, and R. J. Bothast. 1996. Assessment of ethanol production options from corn products. *Bioresour. Technol.* 58:253–264.

Hahn-Hägerdal, B., H. Jeppsson, K. Skoog, and B. A. Prior. 1994. Biochemistry and physiology of xylose fermentation by yeasts. *Enzyme Microb. Technol.* 16:933–943.

Harhangi, H. R., A. S. Akhmanova, R. Emmens, C. van derDrift, W. T. de Laat, J. P. van Dijken, M. S. Jetten, J. T. Pronk, and H. J. Op den Camp. 2003. Xylose metabolism in the anaerobic fungus *Piromyces* sp. E2 follows the bacterial pathway. *Arch. Microbiol.* 180:134–141.

Hartley, R. D., and C. W. Ford. 1989. Phenolic constituents of plant cell walls and wall biodegradability, p. 135–145. *In* L. G. Lewis and M. G. Paice (ed.), *Plant Cell Wall Polymers, Biogenesis, and Biodegradation*. American Chemical Society, Washington, DC.

Henrissat, B., H. Driguez, C. Viet, and M. Schulein. 1985. Synergism of cellulases from *Trichoderma reesei* in the degradation of cellulose. *Bio/Technology* 3:722–726.

Hinman, N. D., D. J. Schell, C. J. Rieley, P. W. Bergeron, and P. J. Walter. 1992. Preliminary estimate of the cost of ethanol production for SSF technology. *Appl. Biochem. Biotechnol.* 34/35:639–649.

Ho, N. W. Y., Z. Chen, and A. P. Brainard. 1998. Genetically engineered *Saccharomyces* yeast capable of effective cofermentation of glucose and xylose. *Appl. Environ. Microbiol.* 64:1852–1859.

Ho, N. W. Y., P. Stevis, S. Rosenfeld, J. J. Huang, and G. T. Tsao. 1984. Expression of the *E. coli* xylose isomerase gene by a yeast promoter. *Biotechnol. Bioeng. Symp.* 13:245–250.

Hood, E. E., K. R. Hood, and S. E. Fritz. 1991. Hydroxyproline-rich glycoproteins in cell walls of pericarp from maize. *Plant Sci.* 79:13–22.

Ibeas, J. I., and J. Jimenez. 1997. Mitochondrial DNA loss caused by ethanol in *Saccharomyces flor* yeasts. *Appl. Environ. Microbiol.* 63:7–12.

Ingram, L. O. 1976. Adaptation of membrane lipids to alcohols. *J. Bacteriol.* 125:670–678.

Ingram, L. O. 1990. Ethanol tolerance in bacteria. *Crit. Rev. Biotechnol.* 9:305–319.

Ingram, L. O., and T. M. Buttke. 1984. Effects of alcohols on micro-organisms. *Adv. Microb. Physiol.* 25:253–300.

Jeffries, T. W. 2006. Engineering yeasts for xylose metabolism. *Curr. Opin. Biotechnol.* 17:320–326.

Jeffries, T. W., and Y.-S. Jin. 2004. Metabolic engineering for improved fermentation of pentose by yeasts. *Appl. Microbiol. Biotechnol.* 63:495–509.

Jeppsson, M., B. Johansson, B. Hahn-Hägerdal, and M. F. Gorwa-Grauslund. 2002. Reduced oxidative pentose phosphate pathway flux in recombinant xylose-utilizing *Saccharomyces cerevisiae* strains improves the ethanol yield from xylose. *Appl. Environ. Microbiol.* 68:1604–1609.

Jin, Y. S., H. Ni, J. M. Laplaza, and T. W. Jeffries. 2003. Optimal growth and ethanol production from xylose by recombinant *Saccharomyces cerevisiae* require moderate D-xylulokinase activity. *Appl. Environ. Microbiol.* 69:495–503.

Johansson, B., C. Christensson, T. Hobley, and B. Hahn-Hägerdal. 2001. Xylulokinase overexpression in two strains of *Saccharomyces cerevisiae* also expressing xylose reductase and xylitol dehydrogenase and its effect on fermentation of xylose and lignocellulosic hydrolysate. *Appl. Environ. Microbiol.* 67:4249–4255.

Kanehisa, M., S. Goto, M. Hattori, K. F. Aoki-Kinoshita, M. Itoh, S. Kawashima, T. Katayama, M. Araki, and M. Hirakawa. 2006.

From genomics to chemical genomics: new developments in KEGG. *Nucleic Acids Res.* 34:D354–D357.

Karhumaa, K., B. Hahn-Hägerdal, and M. F. Gorwa-Grauslund. 2005. Investigation of limiting metabolic steps in the utilization of xylose by recombinant *Saccharomyces cerevisiae* using metabolic engineering. *Yeast* 22:359–368.

Khan, Q., and S. Hadi. 1994. Inactivation and repair of bacteriophage lambda by furfural. *Biochem. Mol. Biol. Int.* 32:379–385.

Klinke, H. B., B. K. Ahring, A. S. Schmidt, and A. B. Thomson. 2002. Characterization of degradation products from alkaline wet oxidation of wheat straw. *Bioresour. Technol.* 82:15–26.

Klinke, H. B., A. B. Thomsen, and B. K. Ahring. 2004. Inhibition of ethanol-producing yeast and bacteria by degradation products produced during pre-treatment of biomass. *Appl. Microbiol. Biotechnol.* 66:10–26.

Kötter, P., and M. Ciriacy. 1993. Xylose fermentation by *Saccharomyces cerevisiae. Appl. Microbiol. Biotechnol.* 38:776–783.

Kötter P., R. Amore, C. P. Hollenberg, and M. Ciriacy. 1990. Isolation and characterization of the *Pichia stipitis* xylitol dehydrogenase gene, *XYL2*, and construction of a xylose-utilizing *Saccharomyces cerevisiae* transformant. *Curr. Genet.* 18:493–500.

Krishnan, M. S., N. W. Y. Ho, and G. T. Tsao. 1999. Fermentation kinetics of ethanol production from glucose and xylose by recombinant *Saccharomyces* 1400(pLNH33). *Appl. Biochem. Biotechnol.* 77–79:373–388.

Kuyper, M., A. A. Winkler, J. P. van Dijken, and J. T. Pronk. 2004. Minimal metabolic engineering of *Saccharomyces cerevisiae* for efficient anaerobic xylose fermentation: a proof of principle. *FEMS Yeast Res.* 4:655–664.

Kuyper, M., H. R. Harhangi, A. K. Stave, A. A. Winkler, M. S. Jetten, W. T. de Laat, J. J. de Ridder, H. J. Op den Camp, J. P. van Dijken, and J. T. Pronk. 2003. High-level functional expression of a fungal xylose isomerase: the key to efficient ethanolic fermentation of xylose by *Saccharomyces cerevisiae? FEMS Yeast Res.* 4:69–78.

Kuyper, M., M. M. Hartog, M. J. Toirkens, M. J. Almering, A. A. Winkler, J. P. van Dijken, and J. T. Pronk. 2005a. Metabolic engineering of a xylose-isomerase-expressing *Saccharomyces cerevisiae* strain for rapid anaerobic xylose fermentation. *FEMS Yeast Res.* 5:399–409.

Kuyper, M., M. J. Toirkens, J. A. Diderich, A. A. Winkler, J. P. van Dijken, and J. T. Pronk. 2005b. Evolutionary engineering of mixed-sugar utilization by a xylose-fermenting *Saccharomyces cerevisiae* strain. *FEMS Yeast Res.* 5:925–934.

Larroy, C., M. R. Fernadez, E. Gonzalez, X. Pares, and J. A. Biosca. 2002. Characterization of the *Saccharomyces cerevisiae YMR318C (ADH6)* gene product as a broad specificity NADPH-dependent alcohol dehydrogenase: relevance in aldehyde reduction. *Biochem. J.* 361:163–172.

Larsson, S., E. Palmqvist, B. Hahn-Hägerdal, C. Tengborg, K. Stenberg, G. Zacchi, and N. Nilvebrant. 1999. The generation of inhibitors during dilute acid hydrolysis of softwood. *Enzyme Microb. Technol.* 24:151–159.

Lee, S. F., and C. W. Forsberg. 1987. Purification and characterization of an α-L-arabinofuranosidase from *Clostridium acetobutylicum* ATCC 824. *Can. J. Microbiol.* 33:1011–1016.

Leonard, R. H., and G. J. Hajny. 1945. Fermentation of wood sugars to ethyl alcohol. *Ind. Eng. Chem.* 37:390–395.

Lewkowski, J. 2001. Synthesis, chemistry and applications of 5-hydroxymethylfurfural and its derivatives. *Arkivoc* 1:17–54.

Liu, Z. L. 2005. Genomic adaptive response of yeast to biofuel fermentation inhibitors. *Soc. Ind. Microbiol. Annu. Meet.* 2005(s):37.

Liu, Z. L. 2006. Genomic adaptation of ethanologenic yeast to biomass conversion inhibitors. *Appl. Microbiol. Biotechnol.* 73:27–36.

Liu, Z. L. 2007. Genomic engineering of *Saccharomyces cerevisiae* for biomass conversion to ethanol, p. 97. *Abstr. Soc. Ind. Microbiol. Annu. Meet.*

Liu, Z. L., and D. Palmquist. 2007. A robust standard for absolute mRNA quantification of *Saccharomyces cerevisiae* by qRT-PCR using the universal RNA controls. *XXIII Int. Conf. Yeast Gen. Mol. Biol.*, abstr. 0064.

Liu, Z. L., and S. Sinha. 2006. Transcriptional regulatory analysis reveals PDR3 and GCR1 as regulators of significantly induced genes by 5-hydroxymethylfurfural stress involved in bioethanol conversion for ethanologenic yeast *Saccharomyces cerevisiae. Microarray Gene Expr. Data Soc. Meet.* 9:119.

Liu, Z. L., and P. J. Slininger. 2004a. Functional genomic studies of *in situ* detoxification of bioethanol fermentation inhibitors using ethanologenic yeast. *Soc. Ind. Microbiol. Annu. Meet.* S146:85.

Liu, Z. L., and P. J. Slininger. 2004b. Global gene expression analysis of ethanologenic yeasts in adaptation to bioethanol fermentation inhibitory stress. *Genomes* 2004:61.

Liu, Z. L., and P. J. Slininger. 2005. Development of genetically engineered stress tolerant ethanologenic yeasts using integrated functional genomics for effective biomass conversion to ethanol, p. 283–294. *In* J. Outlaw, K. Collins, and J. Duffield (ed.), *Agriculture as a Producer and Consumer of Energy.* CAB International, Wallingford, United Kingdom.

Liu, Z. L., and P. J. Slininger. 2006a. *In situ* detoxification of inhibitors by ethanologenic yeast for biomass conversion to ethanol. *Proc. World BioEnergy* 2006:121.

Liu, Z. L., and P. J. Slininger. 2006b. Transcriptome dynamics of ethanologenic yeast in response to 5-hydroxymethylfurfural stress related to biomass conversion to ethanol, p. 679–684. *In* A. Mendez-Vilas (ed.), *Modern Multidisciplinary Applied Microbiology: Exploiting Microbes and Their Interactions.* Wiley-VCH, Weinheim, Germany.

Liu, Z. L., and P. J. Slininger. 2007. Universal external RNA controls for microbial gene expression analysis using microarray and qRT-PCR. *J. Microbiol. Methods* 68:486–496.

Liu, Z. L., P. J. Slininger, and S. W. Gorsich. 2005. Enhanced biotransformation of furfural and 5-hydroxymethylfurfural by newly developed ethanologenic yeast strains. *Appl. Biochem. Biotechnol.* 121–124:451–460.

Liu, Z. L., P. J. Slininger, and B. J. Andersh. 2006. Induction of pleiotropic drug resistance gene expression indicates important roles of PDR to cope with furfural and 5-hydroxymethylfurfural stress in ethanologenic yeast, p. 169. *Abstr. 27th Symp. Biotechnol. Fuels Chem.*

Liu, Z. L., P. J. Slininger, B. S. Dien, M. A. Berhow, C. P. Kurtzman, and S. W. Gorsich. 2004. Adaptive response of yeasts to furfural and 5-hydroxymethylfurfural and new chemical evidence for HMF conversion to 2,5-bis-hydroxymethylfuran. *J. Ind. Microbiol. Biotechnol.* 31:345–352.

Lucero, P., E. Penalver, E. Moreno, and R. Lagunas. 1997. Moderate concentrations of ethanol inhibit endocytosis of the yeast maltose transporter. *Appl. Environ. Microbiol.* 63:3831–3836.

Luo, C., D. Brink, and H. Blanch. 2002. Identification of potential fermentation inhibitors in conversion of hybrid poplar hydrolyzate to ethanol. *Biomass Bioenergy* 22:125–138.

Martel, P., and J. M. Gould. 1990. Cellulose stability and delignification after alkaline hydrogen peroxide treatment of straw. *J. Appl. Polymer Sci.* 39:707–714.

Martin, C., and L. Jonsson. 2003. Comparison of the resistance of industrial and laboratory strains of *Saccharomyces* and *Zygosaccharomyces* to lignocellulose-derived fermentation inhibitors. *Enzyme Microb. Technol.* 32:386–395.

Martin, C., M. Marcet, O. Almazan, and L. J. Jonsson. 2007. Adaptation of a recombinant xylose-utilizing *Saccharomyces cerevisiae* strain to a sugarcane bagasse hydrolysate with high content of fermentation inhibitors. *Bioresour. Technol.* 98:1767–1773.

Martinez, A., M. E. Rodriguez, S. W. York, J. F. Preston, and L. O. Ingram. 2000. Effect of $Ca(OH)_2$ treatments ("overliming") on the composition and toxicity of bagasse hemicellulose hydrolyzates. *Biotechnol. Bioeng.* 69:526–536.

McMillan, J. D. 1993. Pretreatment of lignocellulosic biomass, p. 292–323. *In* M. E. Himmel, J. O. Baker, and R. P. Overend (ed.), *Enzymatic Conversion of Biomass for Fuel Production.* American Chemical Society, Washington, DC.

Meaden, P. G., N. Arneborg, L. U. Guldfeldt, H. Siegumfeldt, and M. Jakobsen. 1999. Endocytosis and vacuolar morphology in *Saccharomyces cerevisiae* are altered in response to ethanol stress or heat shock. *Yeast* **15**:1211–1222.

Modig, T., G. Liden, and M. Taherzadeh. 2002. Inhibition effects of furfural on alcohol dehydrogenase, aldehyde dehydrogenase and pyruvate dehydrogenase. *Biochem. J.* **363**:769–776.

Moes, C. J., I. S. Pretorius, and W. H. van Zyl. 1996. Cloning and expression of the *Clostridium thermosulfurogenes* D-xylose isomerase gene (*xylA*) in *Saccharomyces cerevisiae. Biotechnol. Lett.* **18**:269–274.

Mohsenzadeh, M., W. Saupe-Thies, G. Sterier, T. Schroeder, F. Francella, P. Ruoff, and L. Rensing. 1998. Temperature adaptation of house keeping and heat shock gene expression in *Neurospora crassa. Fungal Genet. Biol.* 25:31–43.

Monteiro, G. A., P. Supply, A. Goffeau, and I. Sa-Correia. 1994. The *in vivo* activation of *Saccharomyces cerevisiae* plasma membrane H(+)-ATPase by ethanol depends on the expression of the PMA1 gene, but not of the PMA2 gene. *Yeast* **10**:1439–1446.

Morimoto, S., and M. Murakami. 1967. Studies on fermentation products from aldehyde by microorganisms: the fermentative production of furfural alcohol from furfural by yeasts (part I). *J. Ferment. Technol.* 45:442–446.

Mussatto, S. I., and I. C. Roberto. 2004. Alternatives for detoxification of dilute-acid lignocellulosic hydrolyzates for use in fermentative processes: a review. *Bioresour. Technol.* **93**:1–10.

Nemirovskii, V., and V. Kostenko. 1991. Transformation of yeast growth inhibitors which occurs during biochemical processing of wood hydrolysates. *Gidroliz Lesokhimm Prom-st* **1**:16–17.

Nemirovskii, V., L. Gusarova, Y. Rakhmilevich, A. Sizov, and V. Kostenko. 1989. Pathways of furfurol and oxymethyl furfurol conversion in the process of fodder yeast cultivation. *Biotekhnologiya* 5:285–289.

Nguyen, Q. A., and J. N. Saddler. 1991. An integrated model for the technical and economic evaluation of an enzymatic biomass conversion process. *Bioresour. Technol.* 35:275–282.

Nilsson, A., M. F. Gorwa-Grauslund, B. Hahn-Hägerdal, and G. Liden. 2005. Cofactor dependence in furan reduction by *Saccharomyces cerevisiae* in fermentation of acid-hydrolyzed lignocellulose. *Appl. Environ. Microbiol.* 71:7866–7871.

Olsson, L., and B. Hahn-Hägerdal. 1996. Fermentation of lignocellulosic hydrolysates for ethanol production. *Enzyme Microb. Technol.* 18:312–331.

Ostergaard, S., L. Olsson, and J. Nielsen. 2000. Metabolic engineering of *Saccharomyces cerevisiae. Microbiol. Mol. Biol. Rev.* **64**:34–50.

Outlaw, J., K. J. Collins, and J. A. Duffield. 2005. *Agriculture as a Producer and Consumer of Energy.* CABI Publishing, Oxfordshire, Oxford, United Kingdom.

Palmqvist, E., and B. Hahn-Hägerdal. 2000a. Fermentation of lignocellulosic hydrolysates. I. Inhibition and detoxification. *Bioresour. Technol.* 74:17–24.

Palmqvist, E., and B. Hahn-Hägerdal. 2000b. Fermentation of lignocellulosic hydrolysates. II. Inhibitors and mechanisms of inhibition. *Bioresour. Technol.* 74:25–33.

Palmqvist, E., J. Almeida, B. Hahn-Hägerdal. 1999. Influence of furfural on anaerobic glycolytic kinetics of *Saccharomyces cerevisiae* in batch culture. *Biotechnol. Bioeng.* 62:447–454.

Petersson, A., J. R. M. Almeida, T. Modig, K. Karhumaa, B. Hahn-Hägerdal, M. F. Gorwa-Grauslund, and G. Liden. 2006. A 5-hydroxymethyl furfural reducing enzyme encoded by the *Saccharomyces cerevisiae ADH6* gene conveys HMF tolerance. *Yeast* **23**:455–464.

Philippidis, G. P., T. K. Smith, and C. E. Wyman. 1993. Study of the enzymatic hydrolysis of cellulose for production of fuel ethanol by the simultaneous saccharification and fermentation process. *Biotechnol. Bioeng.* 41:846–853.

Poutanen, K., M. Tenkanen, H. Korte, and J. Puls. 1991. Accessory enzymes involved in the hydrolysis of xylans, p. 426–436. *In* G. F. Leatham and M. E. Himmel (ed.), *Enzymes in Biomass Conversion.* American Chemical Society, Washington, DC.

Prior, B. A., S. G. Kilian, and J. C. duPreez. 1989. Fermentation of D-xylose by the yeasts *Candida shehatae* and *Pichia stipitis*: prospects and problems. *Process Biochem.* **Feb**:21–32.

Richard, P., M. H. Toivari, and M. Penttila. 2000. The role of xylulokinase in *Saccharomyces cerevisiae* xylulose catabolism. *FEMS Microbiol. Lett.* **190**:39–43.

Saha, B. C. 2003. Hemicellulose bioconversion. *J. Ind. Microbiol. Biotechnol.* 30:279–291.

Saha, B. C. 2004a. Lignocellulose biodegradation and applications in biotechnology, p. 2–34. *In* B. C. Saha and K. Hayaski (ed.), *Lignocellulose Biodegradation.* American Chemical Society, Washington, DC.

Saha, B. C. 2004b. Production, purification, and properties of endoglucanase from a newly isolated strain of *Mucor circinelloides. Process Biochem.* 39:1871–1876.

Saha, B. C., and M. A. Cotta. 2006. Ethanol production from alkaline peroxide pretreated enzymatically saccharified wheat straw. *Biotechnol. Prog.* 22:449–453.

Saha, B. C., and R. J. Bothast. 1999a. Pretreatment and enzymatic saccharification of corn fiber. *Appl. Biochem. Biotechnol.* 76:65–77.

Saha, B. C., and R. J. Bothast. 1999b. Enzymology of xylan degradation, p. 167–194. *In* S. H. Imam, R. V. Greene, and B. R. Zaidi (ed.), *Biopolymers: Utilizing Nature's Advanced Materials.* American Chemical Society, Washington, DC.

Saha, B. C., L. B. Iten, M. A. Cotta, and Y. V. Wu. 2005. Dilute acid pretreatment, enzymatic saccharification, and fermentation of wheat straw to ethanol. *Process Biochem.* 40:3693–3700.

Saha, B. C., S. N. Freer, and R. J. Bothast. 1995. Thermostable β-glucosidases, p. 197–207. *In* J. N. Saddler and M. H. Penner (ed.), *Enzymatic Degradation of Insoluble Carbohydrates.* American Chemical Society, Washington, DC.

Sajbidor J., Z. Ciesarova, and D. Smogrovicova. 1995. Influence of ethanol on the lipid content and fatty acid composition of *Saccharomyces cerevisiae. Folia Microbiol.* (Praha) **40**:508–510.

Sanchez, B., and J. Bautista. 1988. Effects of furfural and 5-hydroxymethylfurfural on the fermentation of *Saccharomyces cerevisiae* and biomass production from *Candida guilliermondii. Enzyme Microb. Technol.* 10:315–318.

Sarthy, A.V., B. L. McConaughy, Z. Lobo, J. A. Sundstrom, C. E. Furlong, and B. D. Hall. 1987. Expression of the *Escherichia coli* xylose isomerase gene in *Saccharomyces cerevisiae. Appl. Environ. Microbiol.* **53**:1996–2000.

Saulnier, L., and J. F. Thibault. 1999. Ferulic acid and diferulic acids as components of sugar-beet pectins and maize bran heteroxylans. *J. Sci. Food Agric.* 79:396–402.

Saulnier, L., C. Marot, E. Chanliaud, and J. F. Thibault. 1995. Cell wall polysaccharide interactions in maize bran. *Carbohydr. Polymers* 26:279–287.

Scalbert, A., B. Monties, J. Y. Lallemand, E. Guittet, and C. Rolando. 1985. Ether linkages between phenolic acids and lignin fractions from wheat straw. *Phytochemistry* **24**:1359–1362.

Sedlak, M., and N. W. Y. Ho. 2004. Production of ethanol from cellulosic biomass hydrolysates using genetically engineered *Saccharomyces* yeast capable of cofermenting glucose and xylose. *Appl. Biochem. Biotechnol.* 114:403–416.

Singh, V., and S. R. Eckhoff. 1997. Economics of germ preseparation for dry grind ethanol facilities. *Cereal Chem.* 74:462–466.

Slininger, P. J., P. L. Bolen, and C. P. Kurtzman. 1987. *Pachysolen tannophilus*: properties and process considerations for ethanol production from D-xylose. *Enzyme Microb. Technol.* 9:5–15.

Slininger, P. J., B. S. Dien, S. W. Gorsich, and Z. L. Liu. 2006. Nitrogen source and mineral optimization enhance D-xylose conversion to ethanol by the yeast *Pichia stipitis* NRRL Y-7124. *Appl. Microbiol. Biotechnol.* 72:1285–1296.

Sonderegger, M., M. Schumperli, and U. Sauer. 2004a. Metabolic engineering of a phosphoketolase pathway for pentose catabolism in *Saccharomyces cerevisiae*. *Appl. Environ. Microbiol.* 70:2892–2897.

Sonderegger, M., M. Jeppsson, C. Larsson, M. F. Gorwa-Grauslund, E. Boles, L. Olsson, I. Spencer-Martins, B. Hahn-Hägerdal, and U. Sauer. 2004b. Fermentation performance of engineered and evolved xylose-fermenting *Saccharomyces cerevisiae* strains. *Biotechnol. Bioeng.* 87:90–98.

Sonderegger, M., and U. Sauer. 2003. Evolutionary engineering of *Saccharomyces cerevisiae* for anaerobic growth on xylose. *Appl. Environ. Microbiol.* 69:1990–1998.

Song, M., and Z. L. Liu. A linear discrete dynamic system model for temporal gene interaction and regulatory network influence in response to bioethanol conversion inhibitor HMF for ethanologenic yeast. *Lecture Notes Bioinformatics*, in press.

Sun, Y., and J. Cheng. 2002. Hydrolysis of lignocellulosic materials for ethanol production: a review. *Bioresour. Technol.* 83:1–11.

Taherzadeh, M., L. Gustafsson, and C. Niklasson. 2000. Physiological effects of 5-hydroxymethylfurfural on *Saccharomyces cerevisiae*. *Appl. Microbiol. Biotechnol.* 53:701–708.

Tan, L. U. L., E. K. C. Yu, P. Mayers, and J. N. Saddler. 1987. Column cellulose hydrolysis reactor: the effect of retention time, temperature, cellulase concentration, and exogeneously added cellobiose on the overall process. *Appl. Microbiol. Biotechnol.* 26: 21–27.

Tanaka, M., M. Fukui, and R. Matsuno. 1988. Effect of pore size in substrate and diffusion of enzyme on hydrolysis of cellulosic materials with cellulases. *Biotechnol. Bioeng.* 32:897–902.

Teixeira, M. C., P. Monteiro, P. Jain, S. Tenreiro, A. R. Fernandes, N. P. Mira, M. Alenquer, A. T. Freitas, A. L. Oliveira, and I. Sá-Correia. 2006. The YEASTRACT database: a tool for the analysis of transcription regulatory associations in *Saccharomyces cerevisiae*. *Nucleic. Acids Res.* 34:D446–D451.

Thomson, J. A. 1993. Molecular biology of xylan degradation. *FEMS Microbiol. Rev.* 104:65–82.

Timell, T. E. 1967. Recent progress in the chemistry of wood hemicelluloses. *Wood Sci. Technol.* 1:45–70.

Toivari, M. H., A. Aristidou, L. Ruohonen, and M. Penttila. 2001. Conversion of xylose to ethanol by recombinant *Saccharomyces cerevisiae:* importance of xylulokinase (*XKSI*) and oxygen availability. *Metab. Eng.* 3:236–249.

Traff, K. L., R. R. Otero Cordero, W. H. van Zyl, and B. Hahn-Hagerdal. 2001. Deletion of the *GRE3* aldose reductase gene and its influence on xylose metabolism in recombinant strains of *Saccharomyces cerevisiae* expression the *xylA* and *XKSI* genes. *Appl. Environ. Microbiol.* 67:5668–5674.

van Maris, A. J. A., D. A. Abbott, E. Bellissimi, J. van den Brink, M. Kuyper, M. A. H. Luttik, H. W. Wisselink, W. A. Scheffers, J. P. van Dijken, and J. T. Pronk. 2006. Alcoholic fermentation of carbon sources in biomass hydrolysates by *Saccharomyces cerevisiae:* current status. *Antonie Leeuwenhoek* 90:391–418.

van Zyl, C., B. A. Prior, S. G. Kilian, and J. L. Kock. 1989. D-xylose utilization by *Saccharomyces cerevisiae*. *J. Gen. Microbiol.* 135: 2791–2798.

van Zyl, W. H., A. Elliasson, T. Hobley, and B. Hahn-Hagerdal. 1999. Xylose utilization by recombinant strains of *Saccharomyces cerevisiae* on different carbon sources. *Appl. Microbiol. Biotechnol.* 52:829–833.

Verho, R., J. Londesborough, M. Penttila, and P. Richard. 2003. Engineering redox cofactor regeneration for improved pentose fermentation in *Saccharomyces cerevisiae*. *Appl. Environ. Microbiol.* 69:5892–5897.

Villa, G. P., R. Bartroli, R. Lopez, M. Guerra, M. Enrique, M. Penas, E. Rodriquez, D. Redondo, I. Iglesias, and M. Diaz. 1992. Microbial transformation of furfural to furfuryl alcohol by *Saccharomyces cerevisiae*. *Acta Biotechnol.* 12:509–512.

Wahjudi, J., L. Xu, P. Yang, V. Singh, P. Buriak, K. D. Rausch, A. J. McAloon, M. E. Tumbleson, and S. R. Eckhoff. 2000. The quick fiber process: effect of mash temperature, dry solids and residual germ on fiber yield and purity. *Cereal Chem.* 77:640–644.

Wahlbom, C. F., and B. Hahn-Hägerdal. 2002. Furfural, 5-hydroxymethylfurfural, and acetone act as external electron acceptors during anaerobic fermentation of xylose in recombinant *Saccharomyces cerevisiae*. *Biotechnol. Bioeng.* 78:172–178.

Wahlbom, C. F., W. H. van Zyl, L. J. Jonsson, B. Hahn-Hagerdal, and R. R. Otero. 2003. Generation of the improved recombinant xylose-utilizing *Saccharomyces cerevisiae* TMB 3400 by random mutagenesis and physiological comparison with *Pichis stipitis* CBS 6054. *FEMS Yeast Res.* 3:319–326.

Walfridsson, M., X. Bao, M. Anderlund, G. Lilius, L. Bulow, and B. Hahn-Hägerdal. 1996. Ethanolic fermentation of xylose with *Saccharomyces cerevisiae* harboring the *Thermus thermophilus xylA* gene, which expresses an active xylose (glucose) isomerase. *Appl. Environ. Microbiol.* 62:4648–4651.

Wheals, A. E., L. C. Basso, D. M. Alves, and H. V. Amorim. 1999. Fuel ethanol after 25 years. *Trends Biotechnol.* 17:482–487.

Wyman, C. E. 1994. Ethanol from lignocellulosic biomass: technology, economics, and opportunities. *Bioresour. Technol.* 50:3–16.

Yang, B., A. Boussaid, S. D. Mansfield, D. J. Gregg, and J. N. Saddler. 2002. Fast and efficient alkaline peroxide treatment to enhance the enzymatic digestibility of steam-exploded softwood substrates. *Biotechnol. Bioeng.* 77:678–684.

You, K. M., C.-L. Rosenfield, and D. C. Knipple. 2003. Ethanol tolerance in the yeast *Saccharomyces cerevisiae* is dependent on cellular oleic acid content. *Appl. Environ. Microbiol.* 69:1499–1503.

Zaldivar, J., A. Martinez, and L. O. Ingram. 1999. Effect of selected aldehydes on the growth and fermentation of ethanologenic *Escherichia coli*. *Biotechnol. Bioeng.* 65:24–33.

Zaldivar, J., J. Nielsen, and L. Olsson. 2001. Fuel ethanol production from lignocellulose: a challenge for metabolic engineering and process integration. *Appl. Microbiol. Biotechnol.* 56:17–34.

Zaldivar, J., A. Borges, B. Johansson, H. P. Smits, S. G. Villas-Boas, J. Nielsen, and L. Olsson. 2002. Fermentation performance and intracellular metabolite patterns in laboratory and industrial xylose-fermenting *Saccharomyces cerevisiae*. *Appl. Microbiol. Biotechnol.* 59:436–442.

Zheng, Y., H. M. Lin, J. Wen, N. Cao, X. Yu, and G. T. Tsao. 1995. Supercritical carbon dioxide explosion as a pretreatment for cellulose hydrolysis. *Biotechnol. Lett.* 17:845–850.

Bioenergy
Edited by J. Wall et al.

Chapter 3

Engineering the *Pichia stipitis* Genome for Fermentation of Hemicellulose Hydrolysates

THOMAS W. JEFFRIES

Pichia stipitis Pignal (1967) is a hemiascomycetous yeast (Kurtzman, 1990; Melake et al., 1996; Vaughan Martini, 1984), closely related to several yeast endosymbionts of passalid beetles (Nardi et al., 2006) that inhabit and degrade white-rotted hardwood (Suh et al., 2003, 2004). It seems to be well adapted to this environment because the *P. stipitis* genomic sequence reveals numerous features that would enable survival and growth in a wood-inhabiting insect's gut (Nardi et al., 2006). It has the capacity to grow on and ferment xylan (Lee et al., 1986; Ozcan et al., 1991) and to use all of the major sugars found in wood. It both assimilates and ferments cellobiose (Parekh and Wayman, 1986; Parekh et al., 1988). In addition, it has been reported to transform low-molecular-weight lignin-related aromatic aldehydes such as 3,4-dimethoxybenzaldehyde and 3-methoxy-4-hydroxybenzylaldehyde to their corresponding alcohols and acids during aerobic culture (Targonski, 1992).

P. stipitis has been studied mainly for its capacity to ferment D-xylose to ethanol. Xylose is a five-carbon sugar that makes up about 15 to 25% of all hardwoods and agricultural residues (Jeffries and Shi, 1999). Its fermentation is therefore essential for the economic conversion of lignocellulose to ethanol (Gulati et al., 1996; Hinman et al., 1989; Saha et al., 1998). *P. stipitis* can produce up to 57 g of ethanol per liter from xylose (Ferrari et al., 1992) when the pH is maintained between 4.5 and 6 (Slininger et al., 1990). The optimal fermentation temperature is 25 to 26°C for xylose and 30°C for glucose. A maximum volumetric ethanol productivity of 0.9 g of ethanol $l^{-1} \cdot h^{-1}$ has been reported (du Preez et al., 1986). The specific ethanol production rate is between 0.15 and 0.17 g of ethanol·g of $cells^{-1} \cdot h^{-1}$ on xylose and up to 0.35 $g \cdot g^{-1} \cdot h^{-1}$ on glucose (Grootjen et al., 1990). The optimal oxygen transfer rate is 3 to 5 mmol of $O_2 \cdot l^{-1} \cdot h^{-1}$ (Guebel et al., 1991). Under oxygen-limited conditions, *P. stipitis* shows a maximum ethanol yield from xylose of 0.47 g/g (Ligthelm et al., 1988). Ethanol production from acid hydrolysates of wheat straw and eucalyptus wood has been reported with *P. stipitis* at yields of 0.35 to 0.41 g/g (Ferrari et al., 1992; Nigam, 2001b). Fermentation of D-xylose by *P. stipitis* depends on the acetic acid concentration and the availability of oxygen. The volumetric rate of ethanol production is inhibited 50% by acetic acid at concentrations of 0.8 and 13.8 g l^{-1} at pH 5.1 and 6.5, respectively, under anaerobic conditions (Van Zyl et al., 1991). *P. stipitis* can be adapted to tolerate higher concentrations of acetic acid (Mohandas et al., 1995) and hardwood hydrolysates (Nigam, 2001a).

P. stipitis has been a source of genes for engineering xylose metabolism in *Saccharomyces cerevisiae*—a task that has been undertaken in numerous laboratories around the world (Jeffries and Jin, 2004). Much less effort has gone into engineering *P. stipitis* for improved xylose metabolism. In addition to genes for xylose assimilation, *P. stipitis* possesses the regulatory machinery that enables the efficient conversion of xylose under oxygen-limited conditions. While such engineering is conceivable, alteration of transcriptional regulators and *cis*-acting factors could be very difficult. *P. stipitis* shunts most of its metabolic flux into ethanol and produces very little xylitol, but its xylose fermentation rate is low relative to that of *S. cerevisiae* on glucose. Strain development for improved fermentation rates and higher ethanol tolerance is still needed. Increasing the capacity of *P. stipitis* for rapid xylose fermentation could therefore greatly improve its usefulness in commercial xylose fermentations.

GENETIC SYSTEM

P. stipitis is homothallic, but its ploidy has been difficult to establish (Gupthar, 1994). The high mutational frequency of wild-type strains suggests that they

Thomas W. Jeffries • USDA, Forest Service, Forest Products Laboratory, and University of Wisconsin, Department of Bacteriology, Madison, WI 53726.

are haploid. Genome sequencing of *P. stipitis* CBS 6054 did not indicate diploid polymorphisms. Since these strains can also form spores, they are homothallic (Melake et al., 1996). Sporulation can be induced by cultivation under nutritionally poor conditions, and it is possible to increase the ploidy through fusion or mating (Gupthar, 1987; Gupthar and Garnett, 1987). Fusions between *P. stipitis* and *Candida shehatae* can lead to incorporation of *C. shehatae* DNA into the *P. stipitis* genome (Gupthar and Garnett, 1987; Selebano et al., 1993), but fusions between *P. stipitis* and *S. cerevisiae* tend to segregate (Gupthar, 1992). Electrophoretic karyotyping indicated that both *P. stipitis* and its presumptive anamorph have six chromosomes (Passoth et al., 1992), but genome sequencing has shown that two pairs of chromosomes are very similar in size and that the actual number of chromosomes is eight (Jeffries et al., 2007).

GENETIC TRANSFORMATION

The first genetic transformation system for *P. stipitis* described by Ho et al. (1991) was based on the kanamycin resistance gene (Km^r); however, the transformation frequency was very low. Much higher transformation frequencies have been achieved using auxotrophs of *ura3* along with the homologous native *URA3* gene and an autonomous replication sequence, *ARS2* (Yang et al., 1994). This system has been further extended by disrupting and recovering the *leu2* auxotrophic marker (Lu et al., 1998). Heterologous complementation of *his3* has also been used to transform *P. stipitis* (Morosoli et al., 1993; Piontek et al., 1998).

Orotidine-5′-P decarboxylase (*ura3*) mutants can be obtained by selecting on minimal medium that contains 50 μg of uracil/ml (or 100 μg of uridine/ml) along with 1 mg of 5′-fluoroorotic acid (5-FOA)/ml (Boeke et al., 1984). This method has been used successfully with *P. stipitis* to obtain spontaneous mutants from cells that have undergone sporulation, but the technique has not been successful in obtaining mutants from fresh or actively growing cells. Three *P. stipitis ura3* mutants have proven useful. These are TJ-26, PSU1, and UC7 (Fig. 1). Of these, TJ26 and PSU1 are relatively stable point mutations; UC7 is an insertional mutant and is completely stable. All three of these mutants grow relatively poorly—even when the medium is supplemented with uracil—but all three show normal growth when complemented with the native *URA3* gene.

The most useful recipient hosts for genetic manipulation and metabolic engineering are FPL-UC7 (*ura3*), FPL-PLU20 (*ura3, leu2Δ*), and FPL-PLU5 (*ura3, leu2Δ*). These are completely stable with respect to their selectable markers. These strains also contain a number of mutations that distinguish them from the parental strain, CBS 6054. FPL-061 was obtained following nitrosoguanidine mutagenesis and selection on several different slowly utilized carbon sources in the presence of respiration inhibitors (Jeffries and Livingston, 1992;

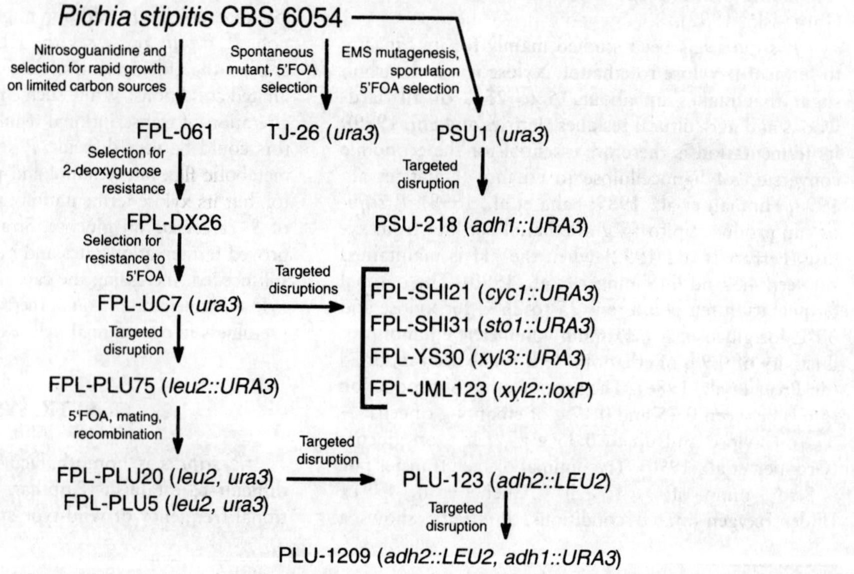

Figure 1. Mutants developed from *P. stipitis* CBS 6054.

Sreenath and Jeffries, 1997). This strain was then selected for resistance to 2-deoxyglucose in order to obtain higher rates of xylose utilization in the presence of glucose (Sreenath and Jeffries, 1999). We then selected for resistance to 5-FOA to obtain the *ura3* strain, UC7, and then used targeted disruption with flanking direct repeats to obtain the PLU20 and PLU5 strains (*ura3*, *leu2*Δ) (Lu et al., 1998).

The report of Ho et al. (1991) notwithstanding, drug resistance markers for genetic transformation of *P. stipitis* have been very difficult to identify and use. This is in part because *P. stipitis* uses the nonconventional yeast codon system 12 in which CUG codes for serine rather than leucine (Santos and Tuite, 1995; Sugita and Nakase, 1999). While CUG is not commonly used by *P. stipitis*, it is frequently used in bacterial genes that comprise many drug resistance markers. Modification of the CUG codons in the *Sh ble* selectable marker into other codons specifying leucine enables use of phleomycin D1 (Zeocin) as an antibiotic in selecting for transformation of nonauxotrophic *P. stipitis* strains (Laplaza et al., 2006). In addition to modifying *Sh ble*, Laplaza et al. (2006) modified the CUG codons in Cre recombinase (Guldener et al., 1996) and developed an improved vector with *URA3* flanked by *loxP* sites to enable easy excision and repeated use of the auxotrophic selectable marker.

Genetic transformation techniques for *P. stipitis* generally are similar to those used for *S. cerevisiae* except that transformation frequencies are usually lower, and targeted disruptions are more difficult. Genetic transformation with *URA3* is more efficient than transformation with the modified *Sh ble* marker. Complementation with autonomous vectors is much more efficient than site-specific disruptions. With an ARS-based plasmid, it is possible to obtain transformants with 10 μg of DNA. When carrying out a site-specific disruption using a linearized fragment, it is generally necessary to use at least 500 bp of flanking sequence on each end of the target gene and to use 10 times as much DNA (i.e., approximately 100 μg per transformation mixture). The number of transformants obtained decreases, but the specificity of the targeted disruption increases with increasing length of flanking DNA. For example, with flanking regions of 500 to 1,000 bp, it is generally necessary to screen ~50 to 100 transformants before obtaining a site-specific disruptant. With 1,500 bp, disruptants can be obtained from as few as 30 transformants.

The success of the transformation depends greatly on the viability of the recipient host. It is very important to have a young, vigorously growing culture. This is best accomplished by transferring the culture from a fresh plate into broth the night before the transformation. At the time of harvest for transformation, the cell density should not exceed an optical density at 600 nm of 1.0.

Table 1. Fermentation characteristics of three strains of *P. stipitis* grown in shake flasks

Fermentation parameters[a]	CBS 6054 (wild type)	UC7 (*ura3*)	Shi21 (*cyc1*)
Y_x ($g \cdot g^{-1}$)	0.16	0.17	0.09
Y_p ($g \cdot g^{-1}$)	0.41	0.38	0.46
Q_{etoh} ($g \cdot g^{-1} \cdot h^{-1}$)	0.04	0.03	0.06
Q_{xo} ($g \cdot g^{-1} \cdot h^{-1}$)	0.11	0.09	0.13

[a] Y_x, cell yield; Y_p, ethanol yield; Q_{etoh}, specific ethanol production rate; Q_{xo}, specific xylose consumption rate.

We have used targeted disruptions to obtain mutants of beta-isopropylmalate dehydrogenase (*LEU2*) (Lu et al., 1998); cytochrome *c*, *CYC1* (Shi et al., 1999); the alternative oxidase, *STO1* (Shi et al., 2002); alcohol dehydrogenase 1 and 2 (*ADH1* and *ADH2*) (Cho and Jeffries, 1998); D-xylulokinase, *XYL3* (Jin et al., 2002); and xylitol dehydrogenase, *XYL2* (Laplaza et al., 2006) (Fig. 1).

Deletion of cytochrome *c* proved to be a particularly effective means of increasing the specific cell yield of ethanol. *P. stipitis* is a petite-negative yeast in that cells completely deficient in mitochondria are not viable (Alexander and Jeffries, 1990). When *CYC1* is disrupted, however, the cells completely lose their terminal cytochrome oxidase complex and produce ethanol with a 50% higher specific fermentation rate (Shi et al., 1999) (Table 1).

METABOLIC REGULATION

Cultivation conditions can strongly affect expression of fermentative enzymes in *P. stipitis*. Unlike *S. cerevisiae*, which regulates fermentation by sensing the presence of glucose, *P. stipitis* induces fermentative activity in response to oxygen limitation (Klinner et al., 2005; Passoth et al., 1998, 2003). Alcohol dehydrogenase and pyruvate decarboxylase are induced in response to oxygen limitation (Cho and Jeffries, 1998, 1999; Klinner et al., 2005; Passoth et al., 1998, 2003), and transcription of a number of other genes is affected as well.

One of the biggest problems that *P. stipitis* must overcome in converting xylose into ethanol is an imbalance of cofactors that arise during xylose assimilation. *P. stipitis* is a bit unusual in that its enzyme for xylose reductase, Xyl1, accepts either NADH or NADPH as a cofactor. The second enzyme in this pathway, xylitol dehydrogenase, Xyl2, accepts only NAD. Xyl1, however, has a higher affinity for NADPH than for NADH and tends to use it preferentially. The coupling of Xyl1 and Xyl2 activities therefore tends to result in the consumption of NADPH and accumulation of NADH. At the same time,

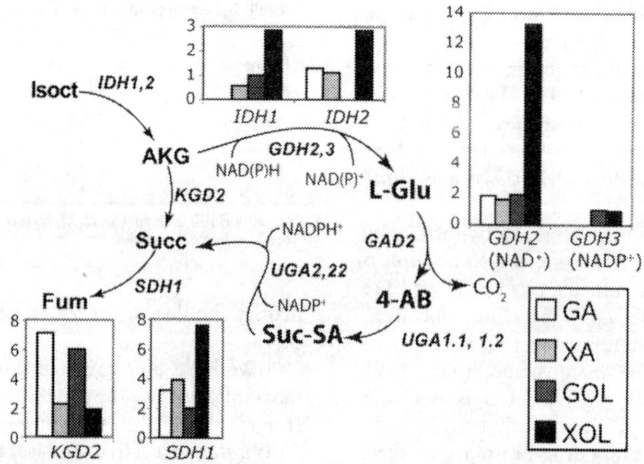

Figure 2. Relative expression of transcripts for the glutamate decarboxylase bypass. Abbreviations: *IDH1* and *IDH2*, isocitrate dehydrogenase 1 and 2; *GDH2*, NAD-specific glutamate dehydrogenase; *GDH3*, NADP-specific glutamate dehydrogenase; *GAD2*, glutamate decarboxylase 2; *UGA1.1* and *UGA1.2*, 4-aminobutyrate aminotransferase; *UGA2* and *UGA22*, succinate semialdehyde dehydrogenase; *KGD2*, 2-ketoglutarate dehydrogenase; Isoct, isocitrate; AKG, 2-keto-glutarate; L-Glu, L-glutamate; 4-AB, 4-aminobutyrate; Suc-SA, succinate semialdehyde; Succ, succinate; Fum, fumarate; GA, glucose aerobic; XA, xylose aerobic; GOL, glucose oxygen limited; XOL, xylose oxygen limited.

excess NADH production (over NAD consumption) arises during oxygen-limited growth because the overall composition of yeast cells is generally more oxidized than the sugar they grow on. Thus, when growing on xylose under oxygen limitation, NADH is present in excess.

P. stipitis compensates for this imbalance in a number of ways. First, it appears to strongly induce expression of NAD-specific glutamate dehydrogenase (*GDH2*) when cells are grown on xylose under oxygen limitation (Fig. 2) (Jeffries et al., 2007). This enzyme consumes NADH to convert 2-keto-glutarate (AKG) into L-glutamate, which can then be decarboxylated by glutamate decarboxylase 2 (*GAD2*) to form 4-aminobutyrate. Transamination by 4-aminobutyrate aminotransferase (*UGA1.1* or *UGA1.2*) converts this into succinate semialdehyd, which is then oxidized by NADP to succinate by succinate semialdehyde dehydrogenase (*UGA2* or *UGA22*). The net effect of this bypass converts NADH into NADPH. 2-Ketoglutarate dehydrogenase (*KGD2*) normally converts AKG into succinate with the generation of NADH, but this enzyme is subject to allosteric inhibition by NADH, so when NADH levels are high during oxygen-limited growth on xylose, *GDH2* is induced and the redox cofactor imbalance can be reduced. The transcript levels of isocitrate dehydrogenase (*IDH1* or *IDH2*) and succinate dehydrogenase (*SDH1*) are higher when cells are growing on xylose under oxygen-limited conditions, but the levels of 2-ketoglutarate are lower, further indicating that the NADH-specific-glutamate dehydrogenase (*GDH2*) is active under these conditions. This bypass has been engineered into *S. cerevisiae*, where it has some of the same effect (Grotkjaer, 2005), but it appears to exist naturally in *P. stipitis*.

Another route by which *P. stipitis* appears to use excess reductant when growing on xylose is through the induction of genes for lipid synthesis. Preliminary data based on expressed sequence tags indicate that transcripts for fatty acid synthase (*FAS2*) stearoyl-coenzyme A desaturase (*OLE1*) are induced under oxygen-limiting conditions (Fig. 3).

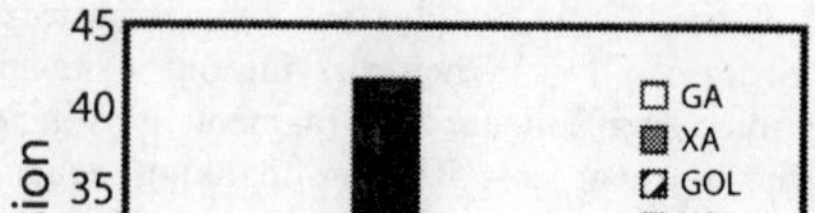

Figure 3. Induction of transcripts for lipid synthesis under oxygen-limiting conditions. Abbreviations: GA, glucose aerobic; XA, xylose aerobic; GOL, glucose oxygen limited; XOL, xylose oxygen limited.

GENOME ORGANIZATION

Xylanase, Cellulase, β-Glucosidase, and Mannanase Utilization

Xylanase production by *P. stipitis* has been recognized since 1991 (Basaran et al., 2000; Ozcan et al., 1991), and the organism has also been transformed with heterologous xylanases to increase xylanase activity (Den Haan and Van Zyl, 2001, 2003; Morosoli et al., 1993; Passoth and Hahn-Hägerdal, 2000). A xylanase gene (*xynA*) cloned from *P. stipitis* NRRL Y-11543 (Basaran et al., 2001) belongs to glycosyl hydrolase family 11. This xylanase gene was not found in the genome of *P. stipitis* CBS 6054, but another xylanase gene (*XYN1*) belonging to glycoside hydrolase family 10 was present (Jeffries et al., 2007). Preliminary expression analyses suggest that the *XYN1* gene is induced when cells are cultivated on xylose.

P. stipitis also appears to possess considerable capacity for assimilation and fermentation of cellulose oligosaccharides. As mentioned earlier, *P. stipitis* can ferment cellobiose (Parekh and Wayman, 1986; Parekh et al., 1988; Sibirny et al., 2003), which is a trait that could be very useful in the simultaneous enzymatic saccharification and fermentation of cellulose, but which is also beneficial in the metabolism of partial cellulose hydrolysates formed by thermochemical pretreatments. The genome contains seven β-glucosidases (*BGL1-7*) and three β-(1→4) endoglucosidases (*BGL1-3*), but what is really striking is the way that they are arranged in the genome. In all but two instances, the β-glucosidase is found immediately adjacent or proximal to one of the three endoglucanases, a sugar transporter, or both (Fig. 4). In each case, the β-glucosidase belongs to glycoside hydrolase family 3 and the endoglucanase to family 5.

Examination of each of these gene clusters shows that the orientation of the β-glucosidase with the endoglucosidase or the sugar transporter is virtually identical in the case of the *BGL1* and *BGL3* clusters on chromosomes 4 and 6, respectively. This indicates that these two clusters probably arose through duplication. Gene orientations in the other clusters are different in each instance, which implies that they did not arise through duplication and that the association of these genes with one another imparts a survival advantage. Further evidence of gene duplication is found in the phylogenetic relationships among the hexose transporters and β-glucosidases in each of these clusters

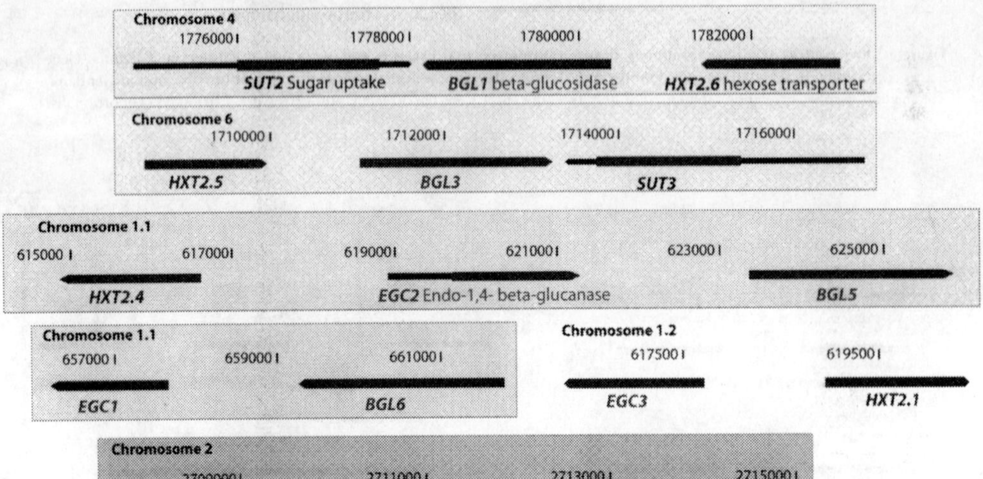

Figure 4. Gene clusters of β-glucosidases and endoglucanases with sugar transporters in the *P. stipitis* genome. Approximate chromosome coordinates are shown. All constructs were derived from the original sequence deposit at the Joint Genome Institute website (http://genome.jgi-psf.org/Picst3/Picst3.home.html). The complete genome is also found in GenBank.

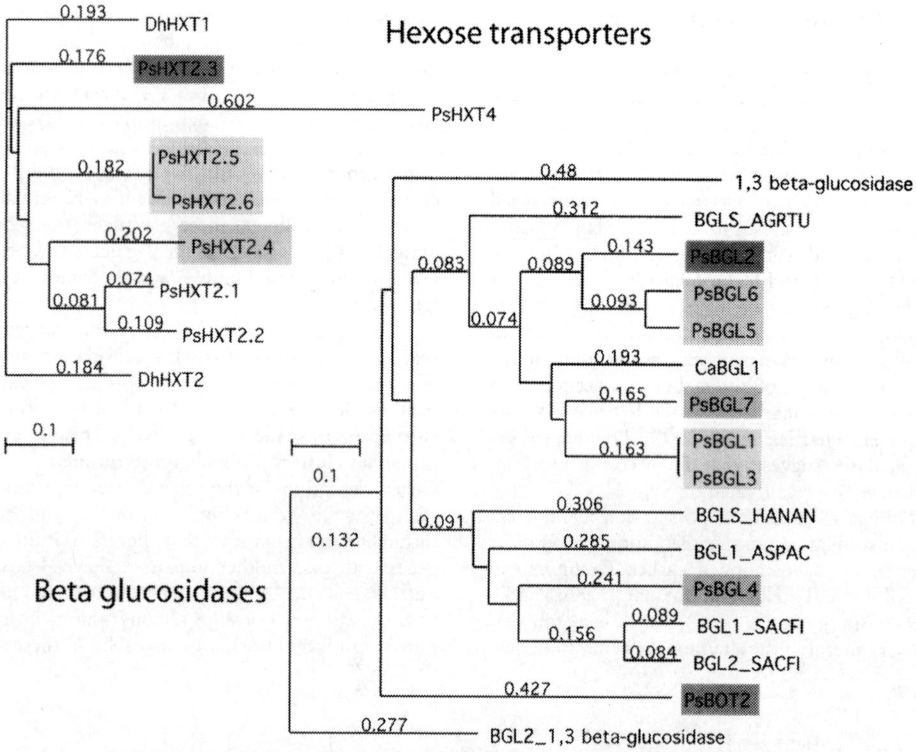

Figure 5. Phylogenetic relationships among hexose transporters and beta-glucosidases found in clusters in *P. stipitis*. Gene names correspond to designations given on the Joint Genome Institute website (see legend to Fig. 4 for URL) and in GenBank. The method used was Neighbor Joining with Best Tree; distances were uncorrected, and gaps were distributed proportionally.

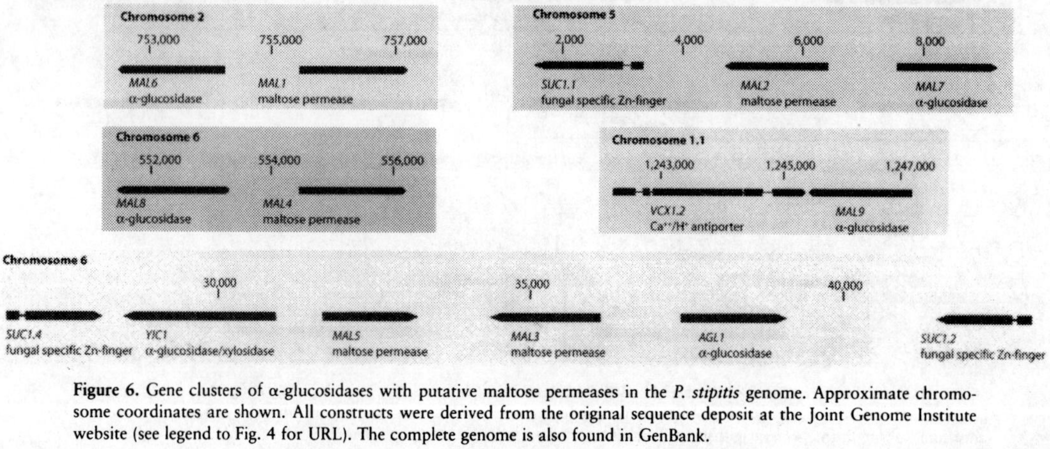

Figure 6. Gene clusters of α-glucosidases with putative maltose permeases in the *P. stipitis* genome. Approximate chromosome coordinates are shown. All constructs were derived from the original sequence deposit at the Joint Genome Institute website (see legend to Fig. 4 for URL). The complete genome is also found in GenBank.

(Fig. 5). *BGL1* and *BGL3*, which are in the paired triplet clusters, are most closely related taxonomically, as are *HXT2.5* and *HXT2.6*. *BGL5* and *BGL6* are closely related as well. They do not share an *HXT* gene in their clusters, but *EGC1* and *EGC2* are more closely related to one another than to *EGC3* (data not shown).

Maltose Utilization

A similar genome organization is observed in the case of α-glucosidases and maltose permeases. In this case, the genome contains five maltose permeases and five α-glucosidases. In four instances, these are arranged in pairs with each pair containing a permease and a glucosidase in a divergent orientation. In one locus, two permeases and two glucosidase/xylosidase genes are found along with putative fungal-specific Zn finger regulatory proteins (Fig. 6). Phylogenetically, the five maltose permeases (*MAL 1-5*) form a relatively close clade (Fig. 7). The one exception is *MAL5*, which is also found associated with *YIC1*, which is the most divergent of all of the α-glucosidases (Fig. 8). In comparing

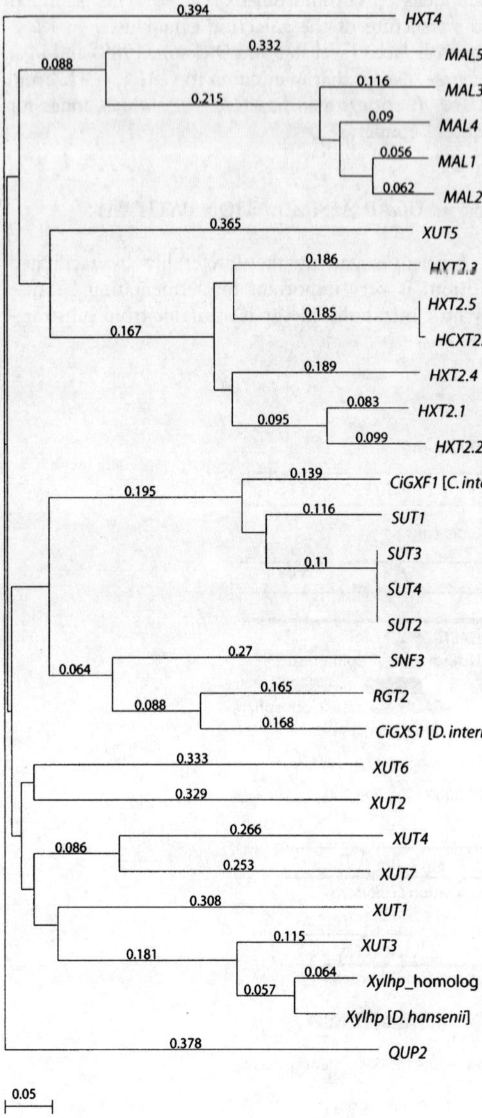

Figure 7. Phylogenetic relationships of putative sugar transporters from *P. stipitis*. Except as noted, all genes are from *P. stipitis*. Gene names correspond to designations given on the Joint Genome Institute website and in GenBank. Sequences for *CiGXF1* and *CiGXS1* are from *C. intermedia*; sequences for *Xylhp* and *Xylhp* homolog are from *D. hansenii*. The method used was Neighbor Joining with Best Tree; distances were uncorrected, and gaps were distributed proportionally.

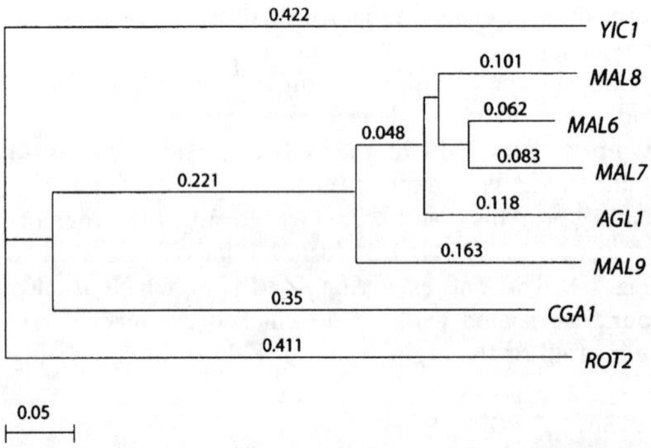

Figure 8. Phylogenetic relationships of α-glucosidases from *P. stipitis*. Gene names correspond to designations given on the Joint Genome Institute website and in GenBank. The method used was Neighbor Joining with Best Tree; distances were uncorrected, and gaps were distributed proportionally.

the relationships among maltose permeases, *MAL1* and *MAL2* are the closest to one another. The α-glucosidases with which they are associated, *MAL6* and *MAL7*, are most closely related (compare Fig. 6, 7, and 8). Likewise, the other paired genes show increasing divergence, suggesting that these could have arisen through successive duplication.

Galactose and Mannose Utilization

The genome contains three family 2 glycoside hydrolases in three different loci. These are comprised of one β-galactosidase (*LAC4*), one β-mannosidase (*MAN2*), and one β-galactosidase (*BMS1*), the latter of which is immediately adjacent to a putative lactose permease (*LAC3*). *P. stipitis* also has a complete set of genes for galactose metabolism, and many of these are found in a single locus on chromosome 3 (Fig. 9). This is similar to the structure of the galactose gene cluster in *Kluyveromyces lactis* (Webster and Dickson, 1988) and in *S. cerevisiae*, except that in addition to *GAL1*, *GAL7*, and *GAL10*, *P. stipitis* also has *GAL102*, which codes for glucose-4-epimerase.

SUGAR ASSIMILATION PATHWAYS

Efficient sugar uptake (preferably by facilitated diffusion) is very important for fermentation because very little metabolic energy is available from substrate-

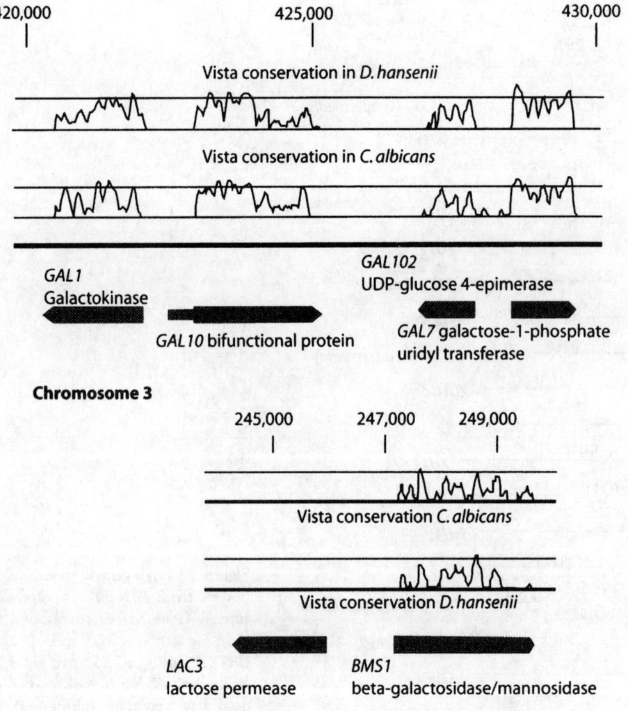

Figure 9. Galactose and lactose gene clusters in *P. stipitis* chromosome 3.

level phosphorylation and the restricted respiration possible under oxygen limitation. Since less metabolic energy is available from the assimilation of xylose than from glucose, this problem is even more critical to xylose fermentation.

Glucose and Xylose Transport

Xylose-specific sugar transporters are critically important for xylose assimilation in the presence of glucose. Glucose and xylose have very similar structures that differ only by the absence of a C6-OH group in xylose. Both sugars are recognized by xylose isomerases, which equilibrate the conversion of xylose or glucose into xylulose or fructose, respectively. In *S. cerevisiae* glucose transporters mediate xylose uptake. Their affinities are, however, much higher for glucose than for xylose, which means that in glucose-xylose mixtures, glucose uptake is almost complete before xylose is assimilated—even in cells that are engineered for xylose metabolism. *P. stipitis* likewise takes up glucose before xylose, but it appears to have several transporters that are relatively specific for xylose.

Thus far, three research groups have described putative xylose transporters from various yeasts. Weierstall et al. (1999) cloned, sequenced, and characterized three sugar uptake genes (*SUT1*, *SUT2*, and *SUT3*) from *P. stipitis* by complementing a glucose transporter-deficient mutant of *S. cerevisiae,* and completion of the *P. stipitis* genome revealed a fourth member of this closely related family (Fig. 7). These four proteins are highly similar in structure. Sut2 and Sut3 differ only in a single amino acid. Disruption of *SUT1* resulted in the loss of low-affinity glucose uptake. Leandro et al. (2006) described two glucose/xylose transporters from *Candida intermedia*, and Nobre and Lucas (A. Nobre and C. Lucas, unpublished data) described a putative xylose permease from *Debaryomyces hansenii*.

When these putative xylose transporters are aligned with putative sugar transporters from *P. stipitis* and other related yeasts, a number of proteins cluster together (Fig. 7). The *C. intermedia* glucose/xylose facilitator (*GXF1*) clusters most closely to the *P. stipitis* *SUT1-4* transporter family. However, the *C. intermedia* putative glucose/xylose symporter (*GXS1*) clusters most closely to a *P. stipitis* protein (*RGT2*) that is similar to a high-affinity glucose sensor. Moreover, the putative xylose transport protein from *D. hansenii*, Xylhp, clusters with the *P. stipitis* transporter *XUT3*. Because glucose and xylose are so similar, relatively few changes are necessary to alter substrate affinities; thus, in the absence of functional or regulatory information we cannot draw too many conclusions from these structural homologies between the proteins.

CONCLUSION

The unconventional yeast *P. stipitis* is potentially well suited for the fermentation of xylose and cellulose oligosaccharides from hydrolysates of lignocellulose. It produces ethanol with a relatively high yield and can metabolize all of the common sugars. A genetic system has been developed that includes the auxotrophic markers *URA3* and *LEU3* along with modified forms of the phleomycin D1 resistance marker, *Sh ble*, and the Cre recombinase. This system has been used to disrupt genes for respiration, which has resulted in higher specific fermentation rates. Expression analyses based on expressed sequence tags and gene chips can guide genetic modifications. Many genes in *P. stipitis* are found in functionally related clusters. Further metabolic engineering and strain selection are needed to increase the overall fermentation rate and ethanol tolerance of *P. stipitis* for the commercial bioconversion of hemicellulose hydrolysates.

REFERENCES

Alexander, M. A., and T. W. Jeffries. 1990. Respiratory efficiency and metabolite partitioning as regulatory phenomena in yeasts. *Enzyme Microb. Technol.* **12:**2–19.

Basaran, P., N. Basaran, and Y. D. Hang. 2000. Isolation and characterization of *Pichia stipitis* mutants with enhanced xylanase activity. *World J. Microbiol. Biotechnol.* **16:**545–550.

Basaran, P., Y. D. Hang, N. Basaran, and R. W. Worobo. 2001. Cloning and heterologous expression of xylanase from *Pichia stipitis* in *Escherichia coli*. *J. Appl. Microbiol.* **90:**248–255.

Boeke, J. D., F. LaCroute, and G. R. Fink. 1984. A positive selection for mutants lacking orotidine-5′-phosphate decarboxylase activity in yeast: 5-fluoro-orotic acid resistance. *Mol. Gen. Genet.* **197:** 345–346.

Cho, J. Y., and T. W. Jeffries. 1998. *Pichia stipitis* genes for alcohol dehydrogenase with fermentative and respiratory functions. *Appl. Environ. Microbiol.* **64:**1350–1358.

Cho, J. Y., and T. W. Jeffries. 1999. Transcriptional control of ADH genes in the xylose-fermenting yeast *Pichia stipitis*. *Appl. Environ. Microbiol.* **65:**2363–2368.

Den Haan, R., and W. H. Van Zyl. 2001. Differential expression of the *Trichoderma reesei* beta-xylanase II (xyn2) gene in the xylose-fermenting yeast *Pichia stipitis*. *Appl. Microbiol. Biotechnol.* **57:** 521–527.

Den Haan, R., and W. H. Van Zyl. 2003. Enhanced xylan degradation and utilisation by *Pichia stipitis* overproducing fungal xylanolytic enzymes. *Enzyme Microb. Technol.* **33:**620–628.

du Preez, J. C., M. Bosch, and B. A. Prior. 1986. Xylose fermentation by *Candida shehatae* and *Pichia stipitis*—effects of pH, temperature and substrate concentration. *Enzyme Microb. Technol.* **8:**360–364.

Ferrari, M. D., E. Neirotti, C. Albornoz, and E. Saucedo. 1992. Ethanol-production from eucalyptus wood hemicellulose hydrolysate by *Pichia stipitis*. *Biotechnol. Bioeng.* **40:**753–759.

Grootjen, D. R. J., R. Vanderlans, and K. Luyben. 1990. Effects of the aeration rate on the fermentation of glucose and xylose by *Pichia stipitis* CBS-5773. *Enzyme Microb. Technol.* **12:**20–23.

Grotkjaer, T., P. Christakopoulos, J. Nielsen, and L. Olsson. 2005. Comparative metabolic network analysis of two xylose fermenting recombinant *Saccharomyces cerevisiae* strains. *Metab. Eng.* **7:**437–444.

Guebel, D. V., A. Cordenons, B. C. Nudel, and A. M. Giulietti. 1991. Influence of oxygen-transfer rate and media composition on fermentation of D-xylose by *Pichia stipitis* NRRL Y-7124. *J. Ind. Microbiol.* **7**:287–291.

Gulati, M., K. Kohlmann, M. R. Ladisch, R. Hespell, and R. J. Bothast. 1996. Assessment of ethanol production options for corn products. *Bioresour. Technol.* **58**:253–264.

Guldener, U., S. Heck, T. Fiedler, J. Beinhauer, and J. H. Hegemann. 1996. A new efficient gene disruption cassette for repeated use in budding yeast. *Nucleic Acids Res.* **24**:2519–2524.

Gupthar, A. S. 1987. Construction of a series of *Pichia stipitis* strains with increased DNA contents. *Curr. Genet.* **12**:605–610.

Gupthar, A. S. 1992. Segregation of altered parental properties in fusions between *Saccharomyces cerevisiae* and the D-xylose fermenting yeasts *Candida shehatae* and *Pichia stipitis*. *Can. J. Microbiol.* **38**:1233–1237.

Gupthar, A. S. 1994. Theoretical and practical aspects of ploidy estimation in *Pichia stipitis*. *Mycol. Res.* **98**:716–718.

Gupthar, A. S., and H. M. Garnett. 1987. Hybridization of *Pichia stipitis* with its presumptive imperfect partner *Candida shehatae*. *Curr. Genet.* **12**:199–204.

Hinman, N. D., J. D. Wright, W. Hoagland, and C. E. Wyman. 1989. Xylose fermentation—an economic analysis. *Appl. Biochem. Biotechnol.* **20–1**:391–401.

Ho, N. W., D. Petros, and X. X. Deng. 1991. Genetic transformation of xylose-fermenting yeast *Pichia stipitis*. *Appl. Biochem. Biotechnol.* **28–29**:369–375.

Jeffries, T. W., I. V. Grigoriev, J. Grimwood, J. M. Laplaza, A. Aerts, A. Salamov, J. Schmutz, E. Lindquist, P. Dehal, H. Shapiro, Y. S. Jin, V. Passoth, and P. M. Richardson. 2007. Genome sequence of the lignocellulose-bioconverting and xylose-fermenting yeast *Pichia stipitis*. *Nat. Biotechnol.* **25**:319–326.

Jeffries, T. W., and Y. S. Jin. 2004. Metabolic engineering for improved fermentation of pentoses by yeasts. *Appl. Microbiol. Biotechnol.* **63**:495–509.

Jeffries, T. W., and P. L. Livingston. June 30, 1992. Xylose fermenting yeast mutants. U.S. patent 5,126,266.

Jeffries, T. W., and N. Q. Shi. 1999. Genetic engineering for improved xylose fermentation by yeasts. *Adv. Biochem. Eng. Biotechnol.* **65**:117–161.

Jin, Y. S., S. Jones, N. Q. Shi, and T. W. Jeffries. 2002. Molecular cloning of *XYL3* (D-xylulokinase) from *Pichia stipitis* and characterization of its physiological function. *Appl. Environ. Microbiol.* **68**:1232–1239.

Klinner, U., S. Fluthgraf, S. Freese, and V. Passoth. 2005. Aerobic induction of respiro-fermentative growth by decreasing oxygen tensions in the respiratory yeast *Pichia stipitis*. *Appl. Microbiol. Biotechnol.* **67**:247–253.

Kurtzman, C. P. 1990. *Candida shehatae*—genetic diversity and phylogenetic relationships with other xylose-fermenting yeasts. *Antonie Leeuwenhoek* **57**:215–222.

Laplaza, J. M., B. R. Torres, Y. S. Jin, and T. W. Jeffries. 2006. Sh ble and Cre adapted for functional genomics and metabolic engineering of *Pichia stipitis*. *Enzyme Microb. Technol.* **38**:741–747.

Leandro, M. J., P. Goncalves, and I. Spencer-Martins. 2006. Two glucose/xylose transporter genes from the yeast *Candida intermedia*: first molecular characterization of a yeast xylose-H^+ symporter. *Biochem. J.* **395**:543–549.

Lee, H., P. Biely, R. K. Latta, M. F. S. Barbosa, and H. Schneider. 1986. Utilization of xylan by yeasts and its conversion to ethanol by *Pichia stipitis* strains. *Appl. Environ. Microbiol.* **52**:320–324.

Ligthelm, M. E., B. A. Prior, and J. C. Dupreez. 1988. The oxygen requirements of yeasts for the fermentation of D-xylose and D-glucose to ethanol. *Appl. Microbiol. Biotechnol.* **28**:63–68.

Lu, P., B. P. Davis, J. Hendrick, and T. W. Jeffries. 1998. Cloning and disruption of the beta-isopropylmalate dehydrogenase gene (*LEU2*) of *Pichia stipitis* with *URA3* and recovery of the double auxotroph. *Appl. Microbiol. Biotechnol.* **49**:141–146.

Melake, T., V. V. Passoth, and U. Klinner. 1996. Characterization of the genetic system of the xylose-fermenting yeast *Pichia stipitis*. *Curr. Microbiol.* **33**:237–242.

Mohandas, D. V., D. R. Whelan, and C. J. Panchal. 1995. Development of xylose-fermenting yeasts for ethanol-production at high acetic-acid concentrations. *Appl. Biochem. Biotechnol.* **51–52**:307–318.

Morosoli, R., E. Zalce, and S. Durand. 1993. Secretion of a *Cryptococcus albidus* xylanase in *Pichia stipitis* resulting in a xylan fermenting transformant. *Curr. Genet.* **24**:94–99.

Nardi, J. B., C. M. Bee, L. A. Miller, N. H. Nguyen, S. O. Suh, and M. Blackwell. 2006. Communities of microbes that inhabit the changing hindgut landscape of a subsocial beetle. *Arthr. Struct. Dev.* **35**:57–68.

Nigam, J. N. 2001a. Development of xylose-fermenting yeast *Pichia stipitis* for ethanol production through adaptation on hardwood hemicellulose acid prehydrolysate. *J. Appl. Microbiol.* **90**:208–215.

Nigam, J. N. 2001b. Ethanol production from wheat straw hemicellulose hydrolysate by *Pichia stipitis*. *J. Biotechnol.* **87**:17–27.

Ozcan, S., P. Kötter, and M. Ciriacy. 1991. Xylan-hydrolyzing enzymes of the yeast *Pichia stipitis*. *Appl. Microbiol. Biotechnol.* **36**:190–195.

Parekh, S., and M. Wayman. 1986. Fermentation of cellobiose and wood sugars to ethanol by *Candida shehatae* and *Pichia stipitis*. *Biotechnol. Lett.* **8**:597–600.

Parekh, S. R., R. S. Parekh, and M. Wayman. 1988. Fermentation of xylose and cellobiose by *Pichia stipitis* and *Brettanomyces clausenii*. *Appl. Biochem. Biotechnol.* **18**:325–338.

Passoth, V., M. Cohn, B. Schafer, B. Hahn-Hägerdal, and U. Klinner. 2003. Analysis of the hypoxia-induced *ADH2* promoter of the respiratory yeast *Pichia stipitis* reveals a new mechanism for sensing of oxygen limitation in yeast. *Yeast* **20**:39–51.

Passoth, V., and B. Hahn-Hägerdal. 2000. Production of a heterologous endo-1,4-beta-xylanase in the yeast *Pichia stipitis* with an O-2-regulated promoter. *Enzyme Microb. Technol.* **26**:781–784.

Passoth, V., M. Hansen, U. Klinner, and C. C. Emeis. 1992. The electrophoretic banding pattern of the chromosomes of *Pichia stipitis* and *Candida shehatae*. *Curr. Genet.* **22**:429–431.

Passoth, V., B. Schafer, B. Liebel, T. Weierstall, and U. Klinner. 1998. Molecular cloning of alcohol dehydrogenase genes of the yeast *Pichia stipitis* and identification of the fermentative ADH. *Yeast* **14**:1311–1325.

Piontek, M., J. Hagedorn, C. P. Hollenberg, G. Gellissen, and A. W. Strasser. 1998. Two novel gene expression systems based on the yeasts *Schwanniomyces occidentalis* and *Pichia stipitis*. *Appl. Microbiol. Biotechnol.* **50**:331–338.

Saha, B. C., B. S. Dien, and R. J. Bothast. 1998. Fuel ethanol production from corn fiber—current status and technical prospects. *Appl. Biochem. Biotechnol.* **70–72**:115–125.

Santos, M. A. S., and M. F. Tuite. 1995. The CUG codon is decoded in-vivo as serine and not leucine in *Candida albicans*. *Nucleic Acids Res.* **23**:1481–1486.

Selebano, E. T., R. Govinden, D. Pillay, B. Pillay, and A. S. Gupthar. 1993. Genomic comparisons among parental and fusant strains of *Candida shehatae* and *Pichia stipitis*. *Curr. Genet.* **23**:468–471.

Shi, N. Q., J. Cruz, F. Sherman, and T. W. Jeffries. 2002. SHAM-sensitive alternative respiration in the xylose-metabolizing yeast *Pichia stipitis*. *Yeast* **19**:1203–1220.

Shi, N. Q., B. Davis, F. Sherman, J. Cruz, and T. W. Jeffries. 1999. Disruption of the cytochrome c gene in xylose-utilizing yeast *Pichia stipitis* leads to higher ethanol production. *Yeast* **15**:1021–1030.

Sibirny, A. A., O. B. Ryabova, O. M. Chmil, V. Sibirny, Z. Kotylak, and D. Grabek. 2003. Xylose and cellobiose fermentation to ethanol by the thermotolerant methylotrophic yeast *Hansenula polymorpha* and by xylose fermenting yeast *Pichia stipitis*. *Yeast* **20**:S219–S219.

Slininger, P. J., R. J. Bothast, M. R. Ladisch, and M. R. Okos. 1990. Optimum pH and temperature conditions for xylose fermentation by *Pichia stipitis*. *Biotechnol. Bioeng.* **35**:727–731.

Sreenath, H. K., and T. W. Jeffries. 1999. 2-Deoxyglucose as a selective agent for derepressed mutants of *Pichia stipitis*. *Appl. Biochem. Biotechnol.* **77–79**:211–222.

Sreenath, H. K., and T. W. Jeffries. 1997. Diminished respirative growth and enhanced assimilative sugar uptake result in higher specific fermentation rates by the mutant *Pichia stipitis* FPL-061. *Appl. Biochem. Biotechnol.* **63–65**:109–116.

Sugita, T., and T. Nakase. 1999. Non-universal usage of the leucine CUG codon and the molecular phylogeny of the genus *Candida*. *Syst. Appl. Microbiol.* **22**:79–86.

Suh, S. O., C. J. Marshall, J. V. McHugh, and M. Blackwell. 2003. Wood ingestion by passalid beetles in the presence of xylose-fermenting gut yeasts. *Mol. Ecol.* **12**:3137–3145.

Suh, S. O., M. M. White, N. H. Nguyen, and M. Blackwell. 2004. The status and characterization of *Enteroramus dimorphus*: a xylose-fermenting yeast attached to the gut of beetles. *Mycologia* **96**:756–760.

Targonski, Z. 1992. Biotransformation of lignin-related aromatic-compounds by *Pichia stipitis* Pignal. *Zentbl. Mikrobiol.* **147**:244–249.

Van Zyl, C., B. A. Prior, and J. C. Du Preez. 1991. Acetic-acid inhibition of D-xylose fermentation by *Pichia stipitis*. *Enzyme Microb. Technol.* **13**:82–86.

Vaughan Martini, A. E. 1984. Comparazione dei genomi del lievito *Pichia stipitis* e de alcune specie imperfette affini. *Ann. Fac. Agr. Univ. Perugia* **38B**:331–335.

Webster, T. D., and R. C. Dickson. 1988. The organization and transcription of the galactose gene cluster of *Kluyveromyces lactis*. *Nucleic Acids Res.* **16**:8011–8028.

Weierstall, T., C. P. Hollenberg, and E. Boles. 1999. Cloning and characterization of three genes (*SUT*1-3) encoding glucose transporters of the yeast *Pichia stipitis*. *Mol. Microbiol.* **31**:871–883.

Yang, V. W., J. A. Marks, B. P. Davis, and T. W. Jeffries. 1994. High-efficiency transformation of *Pichia stipitis* based on its *URA3* gene and a homologous autonomous replication sequence, *ARS2*. *Appl. Environ. Microbiol.* **60**:4245–4254.

Bioenergy
Edited by J. Wall et al.

Chapter 4

Ethanol, Biomass, and Clostridia

Marco A. Báez-Vásquez and Arnold L. Demain

We are living in an unprecedented time in the history of mankind with respect to economic development based on available energy sources. As the debate continues in U.S. political and economic circles, alternative renewable energy sources again become the epicenter of this issue. The history of this paradigm has been up and down for more than 30 years, being subject to oil crisis volatility in a cyclical manner. Economical and technological approaches have been suggested to overcome oil dependence from foreign sources, and strong scientific commitment has been devoted to this paramount challenge. Today, unprecedented global environmental issues strongly related to the social and economic impact of the energy sector are dominating the international agenda. It is clear that the petroleum-based economy is getting closer and closer to the end of its life cycle. Therefore, it is crucial to anticipate and to avoid any shortfall in future supply and to provide access to new bioenergy alternatives for the marketplace.

Bioethanol and other biofuels are very important to foster energy independence and reduce gas emissions. A vigorous debate on gradual substitution of petroleum by use of renewable alternatives such as starch or sugar to ethanol or cellulosic biofuels dominates the political and economic agenda worldwide. For more than 30 years, petroleum and its derivatives have dominated the economy, becoming important commodities in multiple industrial sectors which transform them into products and services to satisfy population needs. In the United States, this dependency has already reached its saturation point, putting at risk our economy, energy security, and homeland security as well as the environment (Greene and Roth, 2006; Sanderson, 2007).

PETROLEUM VERSUS ETHANOL AS FUELS

The economy, energy security, climate protection, and minimization of the threat of global climate change constitute the major factors in favor of an ambitious and stable renewable energy policy in the United States and abroad (U.S. Department of Energy, 2006). The U.S. economy is firmly attached to imported oil and its derivatives at present. Severe economic and social problems could be generated as a consequence of any unexpected disruption in the oil supply. Because the United States is a major oil importer, this represents an unstable situation that becomes more prominent during geopolitical unrest which jeopardizes energy security and the stability of the world's oil supply. In addition, increasing energy demands from China, India, and other nations raise the pressure to compete for oil supplies. It is clear that global oil reserves and new petroleum discoveries will not be sufficient to meet the annual demand worldwide. In addition, the strong variation in market prices makes the whole scenario very unstable.

Petroleum import is the largest single part of the U.S. trade deficit. Our transportation and industries rely heavily on technologies that use fossil energy such as coal, petroleum, and natural gas (Greene et al., 2004; Hess, 2006a, 2006b). The United States consumes 25% of the world's petroleum, most of which is imported. Imports in 2004 were about $90 to $100 billion for 62% of the oil used just for transportation and industry. In 2005, imports increased to $140 billion. The dependence on oil is so strong that we import over 12 million barrels per day. Our transportation needs require 15 million barrels per day. This accelerated consumption is forcing a severe climate change, diminishing the quality of our environment. U.S. transportation alone generates tremendous amounts of "greenhouse gases" (GHG), mainly carbon dioxide. Carbon dioxide is expected to increase from around 1.9 billion metric tons in 2004 to about 2.7 billion metric tons in 2030.

Biofuels could help to eventually yield the equivalent of 8 million barrels per day. Therefore, biomass presents an opportunity to counteract our dependence

Marco A. Báez-Vásquez • 396 Mallard Point, Jupiter, FL 33458-8353. **Arnold L. Demain** • The Charles A. Dana Research Institute for Scientists Emeriti (R.I.S.E.), HS-330, Drew University, Madison, NJ 07940.

on fossil energy sources. As a biofuel, ethanol enhances our energy security and fosters a favorable trade balance, besides being an excellent transportation fuel in comparison to gasoline. Among key attributes of ethanol are cleaner and more efficient burning, a higher octane rating, and replacement of tetraethyl lead as an octane enhancer. It causes a disproportionate increase in octane rating when blended with gasoline. It reduces smog formation due to low volatility. Most importantly, it helps to decrease net carbon dioxide emissions and to produce a lower level of particles and toxic emissions. Air pollution is averted by production of lower levels of ozone precursors. Ethanol is less toxic to humans; it contains no sulfur and produces low levels of particulates and toxic emissions. It is used as an oxygenate to reduce auto emissions, replacing methyl *tert*–butyl ether. Biomass could provide major new crops for a depressed agricultural economy (Wyman and Hinman, 1990; Lynd et al., 1991; Greene et al., 2004).

Historically, Brazil has been a leading nation in renewable liquid fuels. Ethanol has been produced from sugarcane since the 1990s; an aggressive program has achieved the use of neat ethanol in new cars and blends of 20 to 22% for older models. During 2005, the blend was increased to 25%. The renewable fuels program in Brazil has been an excellent example to the rest of the world. Production has increased up to 3.8 billion gal of ethanol/year, which represents 40% of Brazil's domestic fuel consumption.

During the past 15 years, gasoline consumption in the United States increased from 112 billion gal in 1990 to 132 billion gal in 2005. Corn-based ethanol production became commercially viable. Several cooperatives and private ventures fostered corn-to-ethanol production. Since the early 1980s, 1 to 10% ethanol has been used to blend with gasoline. Ethanol production has gradually increased; in 1987, corn was used to produce 850 million gal of anhydrous ethanol. During the last 12 years (Fig. 1), ethanol production from corn increased from 1 billion to 3.8 billion gal (Wyman and Hinman, 1990; Wyman, 1994; Lynd et al., 1991; Lynd, 1996; Meilenz, 2001; Fox, 2002; Wirth et al., 2003), and in 2006, it rose to 4.8 billion gal. This impressive growth is derived from a growing number of production facilities, resulting in an impact on the U.S. economy, agriculture, energy, and the transportation sectors. The rest of the world produces about 8 billion gal per year of ethanol (Gray et al., 2006).

About 17 to 20% of the 2006 corn crop, equivalent to 2.15 billion bushels, was utilized to produce ethanol. Requirements for 2007 and 2008 are estimated to reach 3 billion and 4 billion bushels, respectively. However, such production has supported only 2% of the national transportation requirements. If the

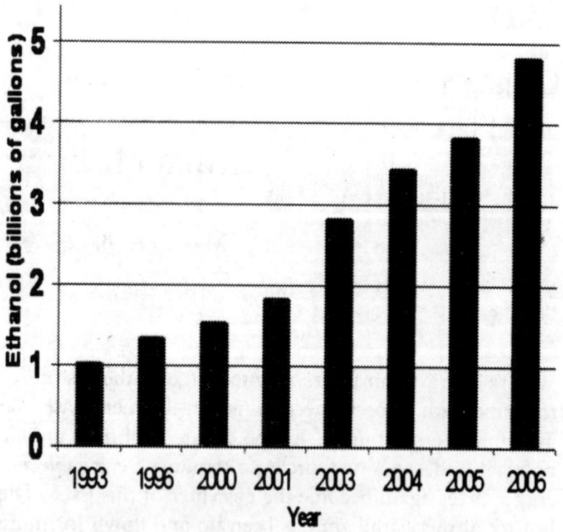

Figure 1. Production of ethanol from U.S. corn.

projected ethanol demand were to be fulfilled only with corn, it would require more corn than the United States currently produces and would compete with food crops. Such a limited availability of corn and a restricted capacity to expand its production due to infrastructure constraints put a ceiling on corn-to-ethanol production goals (Greene et al., 2004; Farrell et al., 2006). Therefore, to continue the plan to meet rapidly growing fuel for transportation demands, it is essential for the biofuels sector to move away from food and grain crops into alternative renewable feedstocks such as nonfood lignocellulosic biomass. The content of cellulose and hemicellulose in these biomass feedstocks represents a sustainable alternative, i.e., a cheap and renewable energy source to increase biofuel production and to improve energy balance with less contribution to greenhouse effects. The cost of straw is much less than that of petroleum, i.e., $27 per ton compared to $340 per ton. It is also much less costly than corn and wheat ($135/ton) or sugar ($340/ton) (Soetoert and Vandamme, 2006). It is important to appreciate, however, that in contrast to the corn-to-ethanol process, biomass feedstocks require more advanced pretreatment to unlock the fermentable sugars from the cellulose and hemicellulose components (Wyman et al., 2005; Allen et al., 2001; Gorhmann et al., 1985; Galbe and Zacchi, 2002). Compelling arguments in support of lignocellulosic ethanol include an increase in our energy security and our trade balance, as well as a reduction of greenhouse emissions and air pollution (U.S. Department of Energy, 2005). This should help in our battle against global warming. The difference between projected ethanol selling prices from lignocellulose and the actual selling price of ethanol from corn is depicted

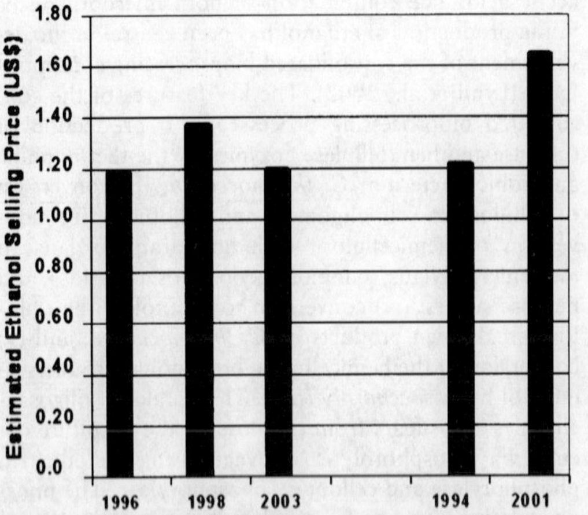

Figure 2. A comparison of estimated selling prices of bioethanol from lignocellulosic biomass and bioethanol from corn.

in Fig. 2, where it can be seen that renewable lignocellulose biomass could compete with corn.

AGRICULTURAL AND FOREST LIGNOCELLULOSIC BIOMASS

There are various reports on lignocellulosic ethanol production using biomass. These indicate that ethanol from lignocellulosic biomass reduces GHG emissions by about 80% compared to gasoline, whereas ethanol from corn reduces them by 20 to 30%. Bioenergy crops are able to balance CO_2 emissions by converting atmospheric CO_2 into organic carbon in biomass (U.S. Department of Energy and Argonne Labs, 2005). Agricultural residues are renewable, chiefly unexploited, and inexpensive. They include leaves, stems, and stalks from sources such as corn fiber, corn stover, sugarcane bagasse, rice hulls, woody crops, and forest residues. Also, there are multiple sources of waste from industrial and agricultural processes, e.g., citrus peel waste, sawdust, paper pulp, industrial waste, municipal solid waste, and paper sludge. In addition, dedicated energy crops for biofuels could include perennial grasses such as switchgrass and other forage feedstocks such as *Miscanthus*, Bermuda grass, elephant grass, etc. These possess desirable environmental qualities, utilize water more efficiently, have high crop yields (above 5 dry tons per acre), prevent soil erosion, contribute to soil fertility, and require minimal supplies of fertilizers and pesticides. Most importantly, perennial grasses can be adapted to various geographical soils and weather conditions across the United States (Greene et al., 2004). If agronomic practices and breeding programs on perennial switchgrass can be improved, yields may increase to more than 14 dry tons per acre. Continuous improvement could yield 114 million acres of switchgrass by 2050, which could generate 165 billion gal of ethanol, equivalent to 108 billion gal of gasoline. However, economics will be the final determinant for commitment to this promising effort by farmers (Greer, 2005).

In addition to cellulose and hemicellulose, another key biomass component is lignin, a complex aromatic polymer which comprises 15 to 25% of plant biomass and contains 40% of the energy. The potential of lignin as an energy source for power generation is considerable. Most importantly, lignin as a by-product from fractionation and conversion operations is a renewable fuel itself with minimal GHG problems mainly because the CO_2 released will be taken up by the plant biomass during growth (Greene et al, 2004; Perlack et al., 2005).

The new "30 × 30" targets proposed by the U.S. Department of Energy (DOE) include satisfying 30% of the fossil fuel requirements by 2030 with 60 billion gal of ethanol per year. Such a goal relies on lignocellulosic biomass. Biomass feedstock availability in 1999 was more than 68 million dry tons comprising agricultural waste, forest residues, and mill waste; this could generate over 7 billion gal of ethanol. By 2005, lignocellulosic biomass had grown to 180 to 200 million dry tons per year, which could yield 16 billion gal of ethanol (Bothast et al., 1999). A recent study indicates a future U.S. availability of 368 million dry tons of biomass from forest lands and 998 million dry tons from agricultural lands, altogether 1.3 billion dry tons. This would be enough biomass to replace about 40% of potential domestic fossil fuel requirements (Perlack et al., 2005). The world content of lignocellulosic biomass is 10 to 50 billion tons (Sticklen, 2006).

DIFFICULTIES WITH HYDROLYSIS OF CELLULOSE TO UNLOCK FERMENTABLE SUGARS

An important requirement for the cost-effective production of liquid biofuels is to unlock the fermentable sugars present in cellulosic and hemicellulosic polysaccharides. These polymers are normally associated with, and surrounded by, a lignin seal. In order to deconstruct highly ordered, tightly packed crystalline and amorphous cellulose regions, pretreatment technologies are used. This helps to disrupt crystallinity, remove the lignin seal, increase pore volume, and solubilize cellulose and hemicellulose, thus making target polymers susceptible to enzymatic attack (Wyman et al., 2005).

Continuous technology optimization through coordinated development efforts on biomass pretreatment technologies is a very active field dedicated to enhance the total yield of fermentable sugars prior to the microbial fermentation stage (Lynd et al., 2002).

THE USE OF ANAEROBIC COCULTURES FOR PRODUCTION OF LIGNOCELLULOSIC ETHANOL

Much of the saccharification work done up until the present has utilized hydrolytic enzymes from the fungus *Trichoderma reesei*, in conjunction with fermentation by the yeast *Saccharomyces cerevisiae* in the simultaneous saccharification and fermentation process (Philippidis et al., 1993). An alternative technology that integrates both saccharification and fermentation into the microbial cell involves the use of a thermophilic, anaerobic, ethanologenic bacterium such as *Clostridium thermocellum* (Demain and Wu, 1989; Beguin and Aubert, 1994; Lynd et al., 2002). The cellulase system in *C. thermocellum* comprises multiple enzyme complexes (Beguin and Lemaire, 1996; Bumazkin et al., 1990; Guimaraes et al., 2002). It contains multiple endo-β-glucanases, four exoglucanases, two β-glucosidases, one cellodextrin phosphorylase, one cellobiose phosphorylase, six xylanases, minor β-xylosidase, minor β-galactosidase, minor β-mannosidase, two lichenases, two laminarinases, pectin lyase, polygalacturonate hydrolase, and pectin methylesterase.

An improved process to maximize the hydrolysis and fermentation of lignocellulosic/hemicellulosic biomass after mild acid treatment is the use of a clostridial coculture as depicted in Fig. 3 (Demain et al., 2005). *C. thermocellum* has the capacity to produce cellobiose and cellodextrins from cellulose, which are transformed into ethanol, lactic acid, and acetic acid. In the case of hemicellulose, enzymatic saccharification by *C. thermocellum* occurs, generating xylose and xylobiose; both are utilized by *Thermoanaerobacterium saccharolyticum* for production of ethanol, lactic acid, and acetic acid. The combination of both microorganisms in the production of ethanol has been central in the development of the consolidated bioprocessing concept of Lynd (Lynd et al., 2002). The key features of the consolidated bioprocessing process are (i) production of cellulase and hemicellulase enzymes by the thermophilic anaerobic bacterium *C. thermocellum*; (ii) conversion of cellulose to cellooligomers and cellobiose; (iii) conversion of hemicellulose including arabinoxylans to monomeric xylans, xylobiose, xylooligosaccharides, and pentose sugars; (iv) conversion to ethanol of the cellulose breakdown products by *C. thermocellum*; and (v) conversion of the hemicellulose breakdown products to ethanol by *T. saccharolyticum*. The uptake of oligosaccharides and sugars from cellulose by the coculture occurs via phosphorolytic cleavage using cellodextrin phosphorylase and cellobiose phosphorylase. The phosphorylative cleavage by cellular phosphorylases seems to be much more active than the exogenous hydrolysis of cellulose to cellobiose due to lower K_m values; thus, the cellodextrins are better utilized than is cellobiose (Zhang and Lynd, 2004). Another key item for cell metabolism is the positive ATP balance generated during phosphorylation (Lynd et al., 2002).

The appeal of an anaerobic thermophilic coculture system lies in (i) the single-step nature of saccharification and ethanol production, thus eliminating the cost of fungal enzyme(s); (ii) the lack of need for expensive oxygen transfer; (iii) fewer contamination problems; (iv) minimal cooling requirement; (v) facilitation of ethanol removal and recovery due to the thermophilic nature of the microbes; (vi) use of robust microorganisms with stable enzymes; (vii) low cell yield allowing more substrate to be converted to ethanol; (viii) high rates of growth and metabolism of cellulose; and (ix) the ability to use cellulose, hemicellulose, glucose, starch, xylan, xylose, mannose, galactose, and arabinose (Lynd et al., 2002; Tyurin et al., 2004, 2006; Desai et al., 2004).

Although the coculture system is very promising, several barriers exist such as end product inhibition by the produced ethanol (Herrero et al., 1985). However, this disadvantage can almost be counterbalanced by the

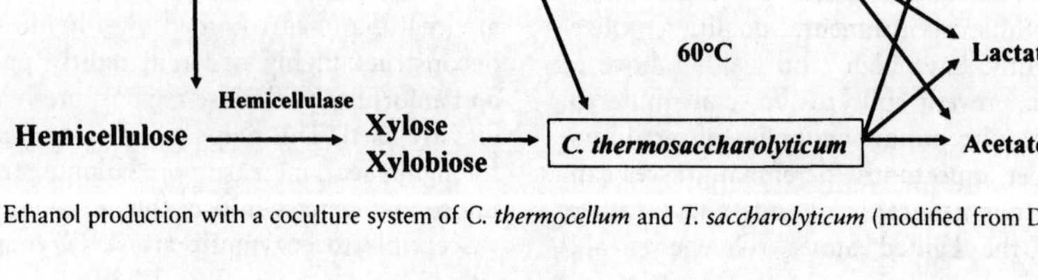

Figure 3. Ethanol production with a coculture system of *C. thermocellum* and *T. saccharolyticum* (modified from Demain et al., 2005).

process of ethanol distillation from dilute broths. Continuous ethanol removal gives a twofold increase in ethanol yield when only 37% of the ethanol has been removed. Furthermore, a mutant of *C. thermocellum* has been obtained that is tolerant to 60 g of ethanol per liter and is capable of producing 26 g of ethanol per liter (Lynd et al., 2005).

Another disadvantage has been a low yield due to the production of side products lactate and acetate. However, high-frequency gene transfer systems including that of electroporation have been used (Tyurin et al., 2006) to eliminate the genes *ldh* and *ack* involved in their production in both members of the coculture. In the case of *T. saccharolyticum* (Desai et al., 2004), single-knockout mutants for lactate and acetate were obtained first and then a double-knockout mutant was generated, which presented a substantial reduction of lactate and acetate production and an ethanol production four times higher than that of the wild strain. In addition, the double-knockout mutant was capable of utilizing xylose more efficiently (Lynd et al, 2005; Tyurin et al., 2006).

CELLULOSOMES

An important hydrolytic complex is the *C. thermocellum* cellulosome. The work by Bayer and Lamed (1986) on the structure of the cell surface multisubunit complex called the cellulosome has been very important for understanding the detailed interactions between the organism and its enzymes and the binding affinity to cellulose through the cellulosic binding domain. The cellulosome is a macromolecular machine dedicated to catalyze concerted, synergistic, and efficient catalysis. It is composed of a large protein called scaffoldin (also known as CipA, SL, or S1), which contains the cellulosic binding domain, and multiple domains called cohesins, which assemble the enzymatic subunits. These cohesins bind the dockerin subunits (type I dockerins) of the relevant enzymes. The scaffoldin also contains dockerins (type II) that anchor the cellulosomes to the cell surface. Much information on the scaffoldin structure was revealed by the genetic sequencing of the gene by Gerngross et al. (1993). Cellulosomes are bound to the cell surface during the early log phase of growth and are released into the medium during the late log phase. During the stationary phase, the cellulosome complex remains attached to cellulose in the medium (Bayer et al., 1998). Extensive work has been dedicated to the understanding of each component, including the purification, cloning, expression, and sequencing of each relevant gene. Besides *C. thermocellum*, there are other microorganisms which contain cellulosome type structures such as *Clostridium cellulovorans*, *Clostridium cellulolyticum*, *Clostridium josui*, *Acetovibrio cellulolyticus*, *Bacteroides cellulosolvens*, *Ruminococcus albus*, *Ruminococcus flavefaciens*, *Vibrio* sp., and some species of the fungal genera *Neocallimastix*, *Piromyces*, and *Orpinomyces* (Shoham et al., 1999).

FINAL COMMENTS

The future of bioethanol looks very bright. Corn-to-ethanol efforts have received a tax credit of $0.51 per gal. In 2005, the U.S. Congress set a goal of increasing ethanol production to 7.5 billion gal by 2012 to provide 6% of U.S. transportation fuel. In the same year, President George W. Bush proposed an increase in ethanol production to 45 billion gal by 2030, of which only 10% would be from corn. In 2006, DOE offered $160 million to industry for the construction of three biorefineries for which 40% of the funds would come from the Federal Government and 60% from industry. Also, the U.S. Department of Agriculture offered $188 million in loan guarantees and grants for renewable energy and energy efficiency efforts (Moriera, 2005). In 2006 and 2007, many biotechnology companies went into biofuels either alone or with companies of the petroleum and chemical industries. It is obvious that for bioethanol, the best is yet to come.

REFERENCES

Allen, S. G., D. Schulman, J. Lichwa, M. J. Antal, E. Jennings, and R. Elander. 2001. A comparison of aqueous and dilute-acid single-temperature pretreatment of yellow poplar sawdust. *Ind. Eng. Chem. Res.* 40:2352–2361.

Bayer, E. A., and R. Lamed. 1986. Ultrastructure of the cell surface cellulosome of *Clostridium thermocellum* and its interaction with cellulose. *J. Bacteriol.* 167:828–836.

Bayer, E. A., L. J. W. Shimon, Y. Shoham, and R. Lamed. 1998. Cellulosomes—structure and ultrastructure. *J. Struct. Biol.* **124:** 221–234.

Beguin, P., and J. P. Aubert. 1994. The biological degradation of cellulose. *FEMS Microbiol. Rev.* **13:**25–58.

Beguin, P., and M. Lemaire. 1996. The cellulosome: an exocellular, multiprotein complex specialized in cellulose degradation. *Crit. Rev. Biochem. Mol. Biol.* **31:**201–236.

Bothast, R., N. N. Nichols, and B. S. Dien. 1999. Fermentations with new recombinant organisms. *Biotechnol. Prog.* **75:**867–875.

Bumazkin, B. K., G. A. Velikodsorsakaya, K. Tuka, M. A. Mogutov, and A. Strongin. 1990. Cloning of *Clostridium thermocellum* endoglucanase genes in *Escherichia coli*. *Biochem. Biophys. Res. Commun.* **168:**1326–1327.

Demain, A. L., and J. H. D. Wu. 1989. The cellulose complex of *Clostridium thermocellum*, p. 68–86. *In* T. K. Ghose (ed.), *First Generation of Bioprocess Engineering*. Ellis Horwood, Chichester, United Kingdom.

Demain, A. L., M. Newcombe, and J. H. D. Wu. 2005. Cellulase, clostridia, and ethanol. *Microbiol. Mol. Biol. Rev.* **69:**124–154.

Desai, S. G., M. L. Guerinot, and L. R. Lynd. 2004. Cloning of the L-lactate dehydrogenase gene and elimination of lactic acid production via gene knockout in *Thermoanaerobacterium saccharolyticum* JW/SL-YS485. *Appl. Microbiol. Biotechnol.* **65:**600–605.

Farrell, A. E., R. J. Plevin, B. T. Turner, A. D. Jones, M. O'Hare, and D. M. Kammen. 2006. Ethanol can contribute to energy and environmental goals. *Science* **311**:506–508.

Fox, J. L. 2002. Legislation, technology boosting renewable fuel, materials uses. *ASM News* **68**:480–481.

Galbe, M., and G. Zacchi. 2002. A review of the production of ethanol from softwood. *Appl. Microbiol. Biotechnol.* **59**:618–628.

Gerngross, U. T., M. P. M. Romaniec, N. S. Huskisson, and A. L. Demain. 1993. Sequencing of a *Clostridium thermocellum* gene (*cipA*) encoding the cellulosomal SL-protein reveals an unusual degree of internal homology. *Mol. Microbiol.* **8**:325–334.

Gorhmann, K., R. Torget, and M. Himmel. 1985. Optimization of dilute acid pretreatment of biomass. *Biotechnol. Bioeng. Symp.* **15**: 59–80.

Gray, K. A., L. Zhao, and M. Emptage. 2006. Bioethanol. *Curr. Opin. Chem. Biol.* **10**:141–146.

Greene, N., et al. 2004. Growing energy. How biofuels can help end America's oil dependence. *Nat. Res. Def. Council Rep.* **2004**:1–86.

Greene, N., and R. Roth. 2006. Ethanol: energy well spent. A review of corn and cellulosic energy balances in the scientific literature to date. *Ind. Biotechnol.* **2**:36–39.

Greer, D. May 2005. Creating cellulosic ethanol: spinning straw into fuel. http//:www.harvestcleanenergy.org/enews/enews_0505/enews_0505_Cellulosic_Ethanol.htm.

Guimaraes, B. G., H. Soucho, B. L. Lytle, J. H. D. Wu, and P. M. Alzari. 2002. The crystal structure and catalytic mechanism of the cellobiohydrolase CelS, the major enzymatic component of the *Clostridium themocellum* cellulosome. *J. Mol. Biol.* **320**:578–596.

Herrero, A. A., R. F. Gomez, and M. F. Roberts. 1985. ^{35}P-NMR studies of *Clostridium thermocellum*. Mechanism of end product inhibition by ethanol. *J. Biol. Chem.* **260**:7442–7451.

Hess, G. 2006a. Bush promotes alternative fuel. *Chem. Eng. News* **84**(10):50–56.

Hess, G. 2006b. Push for biofuels seen in farm bill. *Chem. Eng. News* **84**(21):29–31.

Lynd, L. R., J. H. Cushman, R. J. Nichols, and C. E. Wyman. 1991. Fuel ethanol from cellulosic biomass. *Science* **251**:1318–1323.

Lynd, L. R. 1996. Overview and evaluation of fuel ethanol from cellulosic biomass: technology, economics, the environment and policy. *Annu. Rev. Energy Environ.* **21**:403–465.

Lynd, L. R., P. J. Weimer, W. H. van Zyl, and I. S. Pretorius. 2002. Microbial cellulose utilization: fundamentals and biotechnology. *Microbiol. Mol. Biol. Rev.* **66**:506–577.

Lynd, L. R., J. Jin, J. G. Michels, C. E. Wyman, and B. E. Dale. 2003. Bioenergy: background, potential, and policy. A policy briefing prepared for the Center for Strategic & International Studies, Washington, DC. http://i-farmtools.org/ref/Lynd_et_al_2002.pdf.

Lynd, L. R., W. H. van Zyl, J. E. McBride, and M. Laser. 2005. Consolidated bioprocessing of cellulosic biomass: an update. *Curr. Opin. Biotechnol.* **16**:577–583.

Meilenz, J. R. 2001. Ethanol production from biomass: technology and commercialization status. *Curr. Opin. Microbiol.* **4**:324–329.

Moriera, N. 2005. Growing expectations. New technology could turn fuel into a bumper crop. *Sci. News* **168**:216–218.

Perlack, R. D., L. L. Wright, A. F. Turhollow, R. L. Graham, B. J. Stokes, and D. C. Erbach. 2005. *Biomass as Feedstock for a Bioenergy and Bioproducts Industry: The Technical Feasibility of a Billion-Ton Annual Supply.* Oak Ridge Natl. Lab. 1–78. http://feedstockreview.ornl.gov/pdf/billion_ton_visio.pdf.

Philippidis, G., T. K. Smith, and C. E. Wyman. 1993. Study of the enzymatic hydrolysis of cellulose for production of fuel ethanol by simultaneous saccharification and fermentation process. *Biotechnol. Bioeng.* **41**:846–853.

Sanderson, K. W. 2007. Are ethanol and other biofuels technologies part of the answer for energy independence? *Cereal Food World* **52**:5–7.

Shoham, Y., R. Lamed, and E. A. Bayer. 1999. The cellulosome concept as an efficient microbial strategy for the degradation of insoluble polysaccharides. *Trends Microbiol.* **7**:275–281.

Soetoert, W., and E. Vandamme. 2006. The impact of industrial biotechnology. *Biotechnol. J.* **1**:756–769.

Sticklen, M. 2006. Plant genetic engineering to improve biomass characteristics for biofuels. *Curr. Opin. Biotechnol.* **17**:315–319.

Tyurin, M., S. G. Desai, and L. R. Lynd. 2004. Electrotransformation studies in *Clostridium thermocellum*. *Appl. Environ. Microbiol.* **70**:883–890.

Tyurin, M. V., L. R. Lynd, and J. Weigel. 2006. Gene transfer systems for obligately anaerobic thermophilic bacteria. *Methods Microbiol.* **35**:309–330.

U.S. Department of Energy. 2005. DOE report. *Genomics: GTL Roadmap: Systems Biology for Energy and Environment.* U.S. Department of Energy Office of Science. http://genomicsgtl.energy.gov/roadmap/.

U.S. Department of Energy. 2006. DOE report. *Breaking the Biological Barriers to Cellulosic Ethanol: A Joint Research Agenda. A Research Roadmap Resulting from the Biomass to Biofuel Workshop, Dec. 7–9, 2005 at Rockville, MD.* U.S. Department of Energy. Office of Science and Office of Energy Efficiency and Renewable Energy. http://genomicsgtl.energy.gov/biofuels/b2bworkshop.shtml.

U.S. Department of Energy and Argonne Labs. 2005. Ethanol study. DOE Office of Energy Efficiency and Renewable Energy. http://www.ncga.com/public_policy/PDF/03_26_05ArgonneNatlLabEthanolStudy.pdf.

Wirth, T. E., C. B. Gray, and J. D. Podesta. 2003. The future of energy policy. *Foreign Affairs* **82**:132–155.

Wyman, C. E., and N. D. Hinman. 1990. Ethanol. Fundamentals of production from renewable feedstocks and use as a transportation fuel. *Appl. Biochem. Biotechnol.* **24-25**:735–753.

Wyman, C. E. 1994. Alternative fuels from biomass and their impact on carbon dioxide accumulation. *Appl. Biochem. Biotechnol.* **45-46**:897–915.

Wyman, C. E., B. E. Dale, R. T. Elander, M. Hotzapple, M. R. Ladisch, and Y. Y. Lee. 2005. Coordinated development of leading biomass pretreatment technologies. *Bioresour. Technol.* **96**:1959–1966.

Zhang, Y.-H. P., and L. R. Lynd. 2004. Kinetics and relative importance of phosphorylytic and hydrolytic cleavage of cellodextrins and cellobiose in cell extracts of *Clostridium thermocellum*. *Appl. Environ. Microbiol.* **70**:1563–1569.

Bioenergy
Edited by J. Wall et al.

Chapter 5

Consolidated Bioprocessing of Cellulosic Biomass to Ethanol Using Thermophilic Bacteria

LEE R. LYND, DEVIN CURRIE, NICKY CIAZZA, CHRIS HERRING, AND NICK OREM

At a representative cost of $50/metric ton (dry basis), the purchase price of cellulosic biomass is $3/GJ, the same as if petroleum were about $17/barrel. Thus, cellulosic biomass is a highly competitive raw material for fuel production. The immediate factor impeding the emergence of an industry converting cellulosic biomass into liquid fuels on a large scale is the high cost of processing rather than the cost of feedstock. Within the processing domain, potential research-and-development-driven improvements in converting biomass to sugars offer much larger cost savings than improvements in converting sugars to fuels (Lynd, et al., 2008).

Among options to lower the cost of converting biomass to sugars, one-step microbial conversion in the absence of added saccharolytic enzymes, or consolidated bioprocessing (CBP), is gaining increasing recognition as a potentially transformative breakthrough. A fourfold reduction in the cost of biological processing and a twofold reduction in the cost of processing overall are projected when a mature CBP process is substituted for a hypothetical advanced simultaneous saccharification and cofermentation process featuring cellulase costing US$0.10 per gal of ethanol (Lynd et al., 2007). The U.S. Department of Energy Biomass Program multiyear technical plan states: "Making the leap from technology that can compete in niche or marginal markets for fuels and products also requires expanding the array of possible concepts and strategies for processing biomass. Concepts such as consolidated bioprocessing . . . offer new possibilities for leapfrog improvements in yield and cost"(U.S. Department of Energy, 2004). A detailed analysis of mature biomass conversion processes by Greene et al. (2004) finds CBP to be responsible for the largest cost reduction of all research-and-development-driven improvements incorporated into mature technology scenarios featuring projected ethanol selling prices of less than US$0.70 per gal. A recent report entitled *Breaking the Biological Barriers to Cellulosic Ethanol* states that CBP is "widely considered the ultimate low-cost configuration for cellulose hydrolysis and fermentation . . ." (U.S. Department of Energy, 2005).

Realization of the potential benefits of ethanol production via CBP requires a microbe, or combination of microbes, able to rapidly utilize cellulose and other components of pretreated biomass while at the same time producing ethanol at high yield and titer. Since no known naturally occurring microbes exhibit this combination of properties to the desired extent, it is appropriate to consider prospects for developing CBP-enabling microbes by genetic engineering. Such development can proceed via either of two strategies: (i) improving ethanol formation properties of organisms able to utilize cellulose and other biomass components, or (ii) conferring the ability to utilize cellulose and other biomass components to organisms able to produce ethanol at high yield and titer.

Significant success has been realized via both strategies in the case of engineering ethanol-producing microbes able to utilize soluble nonglucose sugars arising from hemicellulose (xylose, arabinose, galactose, and mannose). Investigators at the University of Florida led by Lonnie Ingram have improved ethanol production by enteric bacteria that are naturally able to utilize nonglucose sugars (Kim et al., 2007; Underwood et al., 2002; Ingram et al., 1998; Lindsay et al., 1995; Wood and Ingram, 1992). The ability to utilize nonglucose sugars has been engineered into a *Saccharomyces* sp. by Ho and coworkers at Purdue University (Sedlak and Ho, 2004; Ho et al., 1999); into *Zymomonas mobilis* by Zhang, Deanda, and coworkers at the National Renewable Energy Laboratory (Deanda et al., 1996; Zhang et al., 1995); and into *Saccharomyces cerevisiae* by Kuyper, Pronk, van Dijken, and coworkers at Delft

Lee R. Lynd • Thayer School of Engineering, Dartmouth College, Hanover, NH 03755, and Mascoma Corp., Lebanon, NH 03766. **Devin Currie** • Thayer School of Engineering, Dartmouth College, Hanover, NH 03755. **Nicky Ciazza, Chris Herring, and Nick Orem** • Mascoma Corp., Lebanon, NH 03766.

Technical University (Kuyper et al., 2004, 2005; Pronk et al., 2005).

Compared to developing organisms able to produce ethanol from nonglucose sugars, developing microbes able to produce ethanol at high yields and titers from cellulose in the absence of saccharolytic enzymes is much less advanced. Thermophiles are of particular interest for CBP because of the positive correlation between temperature and rates of microbially mediated hydrolysis of crystalline cellulose reported in the literature (Lynd et al., 2002). To produce ethanol, or other reduced end products, at high yield, metabolism must be fermentative rather than respiratory. *Clostridium thermocellum* is the most thoroughly characterized cellulolytic, fermentative, thermophilic microbe, and it utilizes cellulose at rates among the highest observed to date. There are, however, described microbes in addition to *C. thermocellum* that are moderately thermophilic (with an optimal growth temperature near 60°C), as well as hyperthermophilic cellulolytic bacteria, that warrant investigation as potential CBP hosts.

This chapter considers use of thermophilic bacteria to produce ethanol in a CBP configuration. Our main focus is cellulose conversion, as this represents the largest technical challenge for development of CBP-enabling microorganisms, with an emphasis on recent developments pertaining to processing of cellulose by cellulolytic microorganisms. Comprehensive reviews of the cellulase enzyme system of *C. thermocellum* may be found elsewhere (Demain et al., 2005; chapters 6 and 9 in this volume). Conversion of nonglucose sugars is also addressed, as such conversion is an essential part of CBP and recent progress in developing thermophiles for conversion of nonglucose sugars is relevant to thermophilic cellulose conversion. Engineering microbes able to utilize cellulose as a result of heterologous expression of cellulase enzymes is a promising strategy in our view but is not considered in this review. This alternative CBP organism development strategy is considered in detail elsewhere for yeasts (van Zyl et al., 2007) as well as other microbes (Lynd et al., 2002).

We first consider thermophilic bacteria of interest for CBP with respect to diversity and ecology, utilization of nonglucose sugars, and ethanol tolerance. Fundamentals of microbial cellulose utilization are then addressed, including substrate assimilation and bioenergetics, enzyme-microbe synergy, substrate capture, control of cellulase synthesis, and rates of cellulose utilization. Next, we review recent progress pursuant to development of CBP-enabling microbes relative to nonspecific mutagenesis, genetic system development, and metabolic engineering. We close by offering an updated perspective on the feasibility and desirability of CBP as well as remaining obstacles and uncertainties.

THERMOPHILIC BACTERIA OF INTEREST FOR CBP

Diversity and Ecology

Cellulose and hemicellulose represent the most abundant, renewable polysaccharides in nature and can be found in a wide array of ecological settings. Thus, the organisms which have evolved to utilize these materials as carbon and energy sources occupy niches that are both numerous and distinct. However, this section focuses only on those organisms that are anaerobic, thermophilic (with optimal growth temperatures above 50°C), and of interest for conversion of cellulose and hemicellulose to ethanol via CBP. For the most part, organisms in this category are low G+C gram-positive bacteria with cell walls that are strictly anaerobic and belong to the phylum *Firmicutes*. These properties are consistent with the placement of these organisms into the class *Clostridia* and distinguish them from the remaining classes of low G+C gram-positive bacteria: the *Mollicutes* (which lack cell walls) and *Bacilli* (which are aerobes or facultative anaerobes). Within the class *Clostridia*, most of the organisms considered for the thermophilic CBP conversion of cellulose to ethanol are found within the orders *Clostridiales* and *Thermoanaerobacteriales*, with a significant proportion of cellulolytic organisms belonging to the former and hemicellulolytic organisms belonging to the latter.

Thermophilic cellulolytic microbes of interest for CBP

A list of the organisms discussed in this section and related physiological characteristics and phenotypes can be found in Table 1.

A majority of the cellulolytic thermophiles found in the order *Clostridiales* that are of interest to CBP conversion of cellulose to ethanol belong to the genus *Clostridium*, of the family *Clostridiaceae*. The most thoroughly studied cellulolytic thermophile is *C. thermocellum*, which exhibits among the highest growth rates of known microbes observed on crystalline cellulose (Lynd et al., 2002). A report in 1954 identified this organism as the first reported thermophilic and cellulolytic species of the genus *Clostridium* (McBee, 1954). The second report of such an organism, in the early 1980s, characterized *Clostridium stercorarium* (Madden, 1983), and since then more thermophilic, cellulolytic subspecies of *C. stercorarium* have been isolated and characterized (Fardeau et al., 2001; Toda et al., 1988). In the late 1980s another thermophilic and cellulolytic species belonging to the genus *Clostridium* was isolated and catalogued as *Clostridium thermocopriae* (Jin et al., 1988). Subsequent phylogenetic analyses showing high 16S sequence similarity and bootstrap

Table 1. Physiological characteristics and phenotypes of discussed species

Strain	Optimal temp (°C) (range)	Optimal pH (range)	GC mol%	Morphology	Cell size μM	Substrates utilized
C. thermocopriae	60 (47–74)	6.5–7.3 (6–8)	37.2	Rods	~0.6 × 2.2–6.0	Arabinose, glucose, fructose, galactose, ribose, cellobiose, lactose, maltose, mannose, sucrose, trehalose, raffinose, sorbitol, salicin, soluble starch, glycogen, amygdalin, cellulose, xylan
C. saccharolyticus	70 (45–80)	7.0 (5.5–8.0)	37.5	Straight rods	0.4–0.6 × 3.0–4.0	Arabinose, amorphous cellulose, Avicell, cellobiose, fructose, galactose, glycogen, gum guar, gum locust bean, lactose, aminarin, lichenin, mannose, maltose, pullulan, pectin, rhamnose, Sigmacell (20, 50, 100), starch, sucrose, xylan, xylose
C. kristjanssonii	78 (50–82)	7.0 (5.8–8.0)	35	Rods	0.7–1.0 × 2.8–9.4	Avicel, cellobiose, dextrin, fructose, galactose, glucose, lactose, maltose, mannose, pectin, salicin, soluble starch, sucrose, trehalose, xylan, xylose
C. owensensis	75 (50–80)	7.5 (5.5–9)	36.6	Straight rods	0.5–0.8 × 2–5	Arabinose, cellulose, cellobiose, dextrin, fructose, galactose, glucose, glycogen, inositol, lactose, mannitol, mannose, maltose, pectin, raffinose, rhamnose, ribose, starch, sucrose, tagatose, xylan, xylose
C. lactoaceticus	68 (50–78)	7.0 (5.8–8.2)	35.2	Rods	0.7 × 1.5–3.5	Avicel, cellobiose, lactose, maltose, pectin, starch, xylan, xylose
C. acetigenus	68 (50–78)	7.0 (5.2–8.5)	35.7	Rods	3.6–5.9 × 0.7–1.0	Arabinose, cellobiose, fructose, galactose, glucose, lactose, maltose, mannose, raffinose, soluble starch, sucrose, trehalose, xylan, xylose
C. stercorarium subsp. stercorarium	65	7.3	39	Straight rods	0.7–0.8 × 2.7–7.7	Arabinose, galactose, glycogen, lactose, maltose, mannose, melibiose, rhamnose, ribose, starch, xylose
C. stercorarium subsp. leptospartum	60 (45–71)	7.5 (6.7–8.9)	43	Rods	0.3 × 6.0	Xylose, galactose, glucose, mannose, cellobiose, lactose, maltose, melibiose, raffinose
C. stercorarium subsp. thermolacticum	60–65 (50–70)	7–7.2 (6.8–7.4)	41.9–42.3	Slightly curved rods	2–4 × 0.3–0.4	Xylan, xylose, glucose, lactose, starch, cellobiose, fructose, melibiose, pectin, ribose, sucrose, raffinose, salicin, mannose, maltose, rhamnose, cellulose
C. straminisolvens	50–55 (45–65)	7.5 (6.0–8.5)	41.3	Straight/slightly curved rods	0.5–1.0 × 3.0–8.0	Cellulose, cellobiose
Moorella sp. F21	60	NR[a]	NR	Rods	NR	NR

[a]NR, not reported.

values of 100 indicate that *C. thermocopriae* should be reclassified in the genus *Thermoanaerobacter* (Collins et al., 1994). More recently, in 2004, a cellulolytic and thermophilic *Clostridium* species, *Clostridium straminisolvens*, was reported (Kato et al., 2004).

The other genus in the order *Clostridiales* containing multiple cellulolytic thermophiles is the *Caldicellulosiruptor* of the family *Syntrophomonadaceae*. Translated from Latin, *Caldicellulosiruptor* means cellulose-breaker living under hot conditions. This is a genus for which many of the representative organisms have not been deposited into more than one international strain collection. The type species of this genus, *Caldicellulosiruptor saccharolyticus*, was identified in 1994 (Rainey et al., 1994). The anaerobic, thermophilic, and saccharolytic nature of this organism narrowed its phylogenetic classification into any of the genera *Clostridium*, *Thermoanaerobacter*, *Thermoanaerobium*, and *Thermoanaerobacterium*. Since it did not produce spores, it could not be classified as a *Clostridium* or *Thermoanaerobacter* (Rainey et al., 1994). Likewise, it could not be part of the genus *Thermoanaerobacterium* because it did not reduce thiosulfate (Rainey et al., 1994). It could be distinguished from the genus *Thermoanaerobium* based on its ability to utilize microcrystalline cellulose. Thus, *C. saccharolyticus* became the type species in a phylogenetically distinct lineage that is now referred to as the *Caldicellulosiruptor*. Three additional cellulolytic organisms belonging to this genus, *Caldicellulosiruptor lactoaceticus*, *Caldicellulosiruptor owensensis*, and *Caldicellulosiruptor kristjanssonii*, have been isolated, characterized, and validated (Bredholt et al., 1999; Huang et al., 1998; Mladenovska and Mathrani, 1995). Furthermore, phylogenetic studies have reclassified existing cellulolytic organisms into the genus *Caldicellulosiruptor*. For example, *Thermoanaerobium acetigenum* has been reclassified to the *Caldicellulosiruptor* genus based on 16S rRNA gene analysis and renamed *Caldicellulosiruptor acetigenus* (Onyenwoke et al., 2006). High intracluster similarity (>95%) and bootstrap calculations support the placement of the anaerobic cellulolytic thermophile *Anaerocellum thermophilum* into this group; however, this strain has yet to be validated (Collins et al., 1994; Zverlov et al., 1998).

Distinct from the order *Clostridiales*, yet still common to the class of *Clostridia*, a cellulolytic thermophile isolated in 2003 was designated *Moorella* sp. strain F21 (Karita et al., 2003). This strain was closely related to *Moorella thermoacetica*, the type strain of this genus (Karita et al., 2003). The genus *Moorella* belongs to the family *Thermoanaerobacteriaceae*, which is in the order of *Thermoanaerobacteriales*. *Moorella* sp. strain F21 was isolated from soil by enrichment on microcrystalline cellulose (Avicel) (Karita et al., 2003). Interestingly, this organism makes no cellulosomes and is the only known cellulolytic organism in the genus *Moorella*. Thus, it is a formal possibility that this organism has acquired a cellulase system through horizontal gene transfer.

The ecological settings within which the cellulolytic and thermophilic *Clostridium* species are found appear to be similar. *C. thermocellum* has been isolated from soils, compost, animal feces, sewage digesters, and hot springs (Bayer et al., 1983; Chuvil'skya et al., 1986; Han et al., 2002; McBee, 1950; Ozkan et al., 2001). Similarly, *C. thermocopriae* has been isolated from compost of cattle feces and grass, camel feces, soil, and hot springs (Jin et al., 1988). The species name "*thermocopriae*" is of Greek origin and translates to "heat compost." The major theme present throughout the above environments is an abundance of cellulosic material and often, but not always, elevated temperatures. This trend is again observed within the isolated subspecies of *C. stercorarium*, which was named using a Latin word for dung, referring to the source of isolation of the type strain of this species, a compost heap. The type strain is now referred to as *C. stercorarium* subsp. *stercorarium* (Madden, 1983). *C. stercorarium* subsp. *leptospartum* (formerly *Thermobacteroides leptospartum*) was isolated from manure, and *C. stercorarium* subsp. *thermolacticum* (formerly *Clostridium thermolacticum*) has been found in sediments, anaerobic digesters, and cattle manure (Fardeau et al., 2001; Le et al., 1985; Toda et al., 1988).

Another common feature of all the *Clostridium* species mentioned above is that they were enriched and isolated under anaerobic conditions in the presence of cellulose. The lone exception is *C. straminisolvens* (Kato et al., 2004). Like the other cellulolytic anaerobic thermophiles of the genus *Clostridium*, this species was present in animal feces and enriched on cellulolytic material, in this case, rice straw (Kato et al., 2004). The species name of this organism is of Latin origin and translates to straw-dissolving. However, the cellulolytic digester used to isolate *C. straminisolvens* was engineered to provide aeration from the bottom of the reactor (Haruta et al., 2002). It is likely that the digester was not completely aerobic and that *C. straminisolvens*, which requires anaerobic conditions for growth, was able to find a suitable microenvironment with low O_2 concentrations. It is unlikely a mere coincidence that *C. straminisolvens* is moderately aerotolerant and can persist in an atmosphere containing up to 4% O_2, and this is reflected in the enrichment of this organism over other *Clostridium* species under the engineered conditions of the digester (Kato et al., 2004).

Like *Clostridium*, *Caldicellulosiruptor* species also inhabit similar ecological settings that feature elevated temperatures and that are rich in cellulosic materials, and they have been isolated under anaerobic conditions in the presence of cellulose. The type strain for this genus, *C. saccharolyticus*, was isolated from a piece of wood in the flow of a geothermal spring in Taupo, New Zealand (Rainey et al., 1994), while *C. kristjanssonii* and *C. lactoaceticus* were both isolated from Icelandic hot springs (Bredholt et al., 1999; Mladenovska and Mathrani, 1995). The nonvalidated strain "*Anaerocellum thermophilum*" was isolated from a hot spring containing residues of higher plants in the Valley of the Geysers on the Kamchatka peninsula in Russia (Zverlov et al., 1998). In comparison to *Clostridium* species, *Caldicellulosiruptor* species have slightly higher optimal growth temperatures, and this may explain why these organisms are found mainly in geothermal environments as opposed to animal feces, compost piles, and soils. This being said, *C. owensensis* was isolated from the sediment of a shallow freshwater pond in the Owens Lake bed area in California, which may not appear to be an obvious ecological setting to find an organism with an optimal growth temperature of 70°C (Huang et al., 1998). However, the authors claim that during the summer months, when the area is subject to dryness, temperatures could increase to the point of favoring thermophilic microbes. Another common feature regarding the isolation of *Caldicellulosiruptor* species pertains to pH. When sampling hot springs in New Zealand, Sissons et al. found that *Caldicellulosiruptor* species were isolated only from springs with relatively neutral pH and none were obtained from acid pools (Sissons et al., 1987).

From the standpoint of industrial hardiness, an acidiphilic thermophile would be advantageous but not necessarily required for the CBP conversion of cellulose to ethanol. In laboratory cultures, there are no reports of an anaerobe that is able to grow below pH 6.0 via a cellulose-based metabolism (Lynd et al., 2002). However, cellulose degradation is observed at pH values as low as 4.5 in environmental settings and among anaerobic mixed cultures (Chyi and Dague, 1994). One interpretation of the above is that a cellulolytic community is more tolerant of acidic pH than the individual, cellulolytic organisms that constitute it. A communal setting could provide buffered microenvironments that protect cellulolytic microbes against low pH. For instance, ruminal bacteria are able to adhere to and hydrolyze cellulose below pH 6.0 but growth occurs only after a protective glycocalyx is synthesized and the pH is raised (Mourino et al., 2001). Hemicellulolytic bacteria (discussed in the following section) have a broader pH range than cellulolytic organisms, and numerous examples of anaerobes growing on hemicellulosic material below pH 6.0 exist. Interestingly, cellulolytic and hemicellulolytic bacteria are commonly cultured from the same environment, which can have a different pH than what is optimal for the individual organisms isolated. Thus, the cellulolytic community is more dynamic than the sum of its parts, a characteristic that might explain how this process has become so widespread in nature. It may also explain why a process that is so widespread has produced only neutrophilic anaerobes.

In addition to low pH, high temperature is also desirable for CBP conversion of cellulose to ethanol from standpoints of process engineering and biocatalysts. A correlation exists whereby cellulose hydrolysis increases as a function of temperature. More specifically, as the optimal growth temperatures of cellulolytic organism are compared, those with higher growth temperatures exhibit higher rates of cellulose hydrolysis. This is true of both aerobic and anaerobic organisms. *A. thermophilum* has an optimal growth temperature of 75°C and has been observed to hydrolyze cellulose above 80°C, making this the highest temperature known for microbial cellulose utilization by an anaerobe (Svetlichny et al., 1990). Furthermore, purified enzymes from *A. thermophilum* have maximal activity at 95 and 85°C on carboxymethyl cellulose and microcrystalline cellulose, respectively (Zverlov et al., 1998). These findings have obvious implications from an industrial perspective, since the rate and conditions that lead to cellulose hydrolysis will directly impact the economics of biomass conversion and ethanol production.

Thermophilic and hemicellulolytic organisms of interest to CBP conversion of cellulose to ethanol

After cellulose, hemicellulolytic material such as xylan represents the second most abundant product of photosynthesis. However, the number of thermophilic organisms isolated and characterized that utilize hemicellulose substrates exceeds the number of cellulose-utilizing species. While the bulk of the hemicellulose-utilizing species belong to the family *Thermoanaerobacteriaceae* in the order *Thermoanaerobacteriales* of the class *Clostridia*, there are reports of hemicellulose-utilizing thermophiles in distinct phylogenetic groups. Phylogenetic diversity is something not observed with the cellulose-utilizing thermophiles mentioned above, which all belong to the family *Clostridiaceae*, with the exceptions being *C. thermocopriae* and *Moorella* sp. strain F21 of the family *Thermoanaerobacteriaceae*. Thus, in addition to number, the diversity of hemicellulose-utilizing organisms exceeds that of the cellulose-utilizing species. The following paragraphs do not attempt to capture the qualities of individual hemicellulose-utilizing thermophiles but instead discuss the diversity and related ecology of these organisms.

Within the order *Thermoanaerobacteriales* of the family *Thermoanaerobacteriaceae* there are 19 genera. Of these, the most dominant with respect to hemicellulose-utilizing thermophiles are *Thermoanaerobacterium* and *Thermoanaerobacter*. Representative organisms of the above genera relevant to the CBP conversion of cellulose to ethanol include, but are not limited to, *Thermoanaerobacterium saccharolyticum*, *Thermoanaerobacterium xylanolyticum*, *Thermoanaerobacterium aotearoense*, *Thermoanaerobacterium zeae*, *Thermoanaerobacter brockii*, *Thermoanaerobacter ethanolicus*, *Thermoanaerobacter thermohydrosulfuricus*, *Thermoanaerobacter acetoethylicus*, and *Thermoanaerobacter mathrannii*.

Many of the above organisms were isolated from surface geothermal features such as hot springs. *T. saccharolyticum* and *T. xylanolyticum* were isolated from sediments from Frying Pan Springs in Yellowstone National Park (Lee et al., 1993). *T. brockii*, first reported in the genus *Thermoanaerobium* in 1979 by Zeikus, was also isolated from hot springs in Yellowstone National Park (Zeikus et al., 1979). *T. ethanolicus*, as indicated by its name, produces mainly ethanol as a fermentation product and is yet another organism isolated from geothermal features within Yellowstone National Park (Wiegel and Ljungdahl, 1981). *T. aotearoense* was isolated from geothermally heated water and sediments from hot pools in New Zealand, and *T. mathranii* was isolated from an Icelandic hot spring (Larsen et al., 1997; Liu et al., 1996).

Unlike the cellulolytic clostridium species, the hemicellulose-utilizing *Thermoanaerobacterium* species are more diverse with respect to pH. At the time of sampling, *T. saccharolyticum* and *T. xylanolyticum* were isolated from acidic hydrothermal areas that had a pH H range of 4.5 to 6.0, and this is reflected in a pH 6.0 growth optimum for these organisms (Lee et al., 1993). In contrast, *T. mathranii* was isolated from 70°C sediment that had a pH of 8.5 (Larsen et al., 1997). In a given study, *T. ethanolicus* strains have been isolated from hydrothermal water and mud samples from a pool with a pH of 8.8 and in a separate pool that has a pH measuring 5.5. As mentioned above, cellulolytic organisms such as the *Caldicellulosiruptor* species were never isolated from acidic hydrothermal pools. *T. aotearoense* is even more acidophilic than the species mentioned above: it has an optimal pH for growth of 5.2 and does not tolerate pH above 7.0 (Liu et al., 1996). The biochemical, physiological, and genetic traits that lend to the acidiphilic nature of some of these organisms are of interest to CBP conversion of biomass to ethanol as an industrial processes, because it is desirable to perform fermentations at low pH in order to minimize microbial contamination and minimize caustic addition to neutralize carbonic acid production.

Thermophilic and hemicellulolytic organisms such as *T. zeae* and *T. thermohydrosulfuricus* were isolated from industrial environments. *T. zeae* originated from organic waste from a vegetable canning facility in Hooperston, IL, and *T. thermohydrosulfuricus* originated from an Austrian sugar factory (Cann et al., 2001; Klaushofer and Parkkinen, 1965). The latter has also been found in mesophilic and thermophilic environments including, not surprisingly, geothermal springs in Yellowstone National Park (Wiegel and Ljungdahl, 1981). Thus, with a proper enrichment procedure hemicellulolytic thermophiles may be obtained from a wide array of environments, and it appears that geothermal areas are particularly promising in this context.

Hemicellulose-utilizing, thermophilic anaerobes have also been found in the genera *Thermotoga* and *Petrotoga*. Both of these genera are in the order *Thermotogales*, of the class *Thermotogae*, which is the sole class in the phylum *Thermotogae*. In addition to the *Aquificales*, the *Thermotogales* represent the deepest phylogenetic branch in the domain *Bacteria* (Winker and Woese, 1991). *Thermotoga maritima* was isolated from a marine geothermal sediment at Vulcano, Italy, and its ability to produce hydrogen from cellulose and xylan has gained attention in the realm of renewable energy (Sleyter and Stetter, 1986). Interestingly, the genomic sequence of this organism has revealed that approximately one-quarter of the genome is archaeal in nature, suggesting rampant horizontal gene transfer. *Thermotoga hypogea* is a xylanolytic member of the same genus and does produce trace amounts of ethanol during xylan fermentations (Fardeau et al., 1997). It was isolated from an oil-producing well in Cameroon (western Africa) that had a temperature of 66°C and contained 12 g of sodium chloride per liter, indicating that marine or saline environments may produce desirable organisms for CBP production of ethanol (Fardeau et al., 1997). Additionally, *Petrotoga mexicana* was isolated from an oil well in the Gulf of Mexico, also a saline environment (Miranda-Tello et al., 2004). However, this organism did not produce ethanol during fermentation of xylan (Miranda-Tello et al., 2004).

Utilization of Nonglucose Sugars

The ability of a CBP organism to utilize not only the cellulosic but also the hemicellulosic components of feedstock biomass is crucially important. While the precise composition of different biomass feedstocks varies, the hemicellulose fraction typically makes up approximately one-third of the available carbohydrate (Lee, 1997). In contrast to cellulose, which is a high-molecular-weight homopolysaccharide consisting of 4,000 to 8,000 glucose subunits, hemicellulose is a relatively short heteropolysaccharide consisting of around 200 subunits. The subunit composition varies among plant species, with xylose as the major component sugar in hemicellulose from angiosperms (including hardwoods and herbaceous plants), whereas gymnosperms contain both xylans and glucomannans (McMillan, 1993; Sjöström, 1993). Hemicellulose from both gymnosperms and angiosperms also contains lesser amounts of galactose and arabinose (Burke et al., 1974; Carpita et al., 1996; Robinson et al., 2002; Sjöström, 1993; Thomas et al., 1987; Jin et al., 2006). In angiosperms, xylan exists as O-acetyl-4-O-methylglucuronoxylan, whereas in gymnosperms xylan occurs as arabino-4-O-methylglucuronoxylan (Sunna and Antranikian, 1997). For hardwood xylan, 70% of the β-D-xylopyranose units are acetylated at either the C-2 or C-3 position or both (Coughlan and Hazlewood, 1993), and one in 10 of these units has a 1,2-linked 4-O-methy-α-D-glucuronic acid residue at the C-2 position. Softwood xylans are not acetylated. However, the 4-O-methylglucuronic acid and the L-arabinofuranose residues are coupled to the C-2 and C-3 positions, respectively, of the relevant xylopyranose backbone units (Coughlan and Hazlewood, 1993). The structure of hemicellulose also differs from that of cellulose, which is crystalline, strong, and resistant to hydrolysis, in that it is random, amorphic, and relatively weak and can be hydrolyzed by a variety of chemical or enzymatic means (Mosier et al., 2005; Saha, 2003; Sun and Cheng, 2002). Additionally, hemicellulose covers the cellulose in the plant cell wall and therefore its solubilization greatly increases the rate of cellulose hydrolysis by increasing the accessibility of the cellulose to cellulase enzymes (Fernandez-Bolanos et al., 2001; Laureano-Perez et al., 2005). Since some pretreatment processes solubilize essentially all of the hemicellulose (Lynd et al., 2002), it may or may not be necessary that insoluble hemicellulose be enzymatically hydrolyzed in the context of industrial processing of cellulosic biomass. Given that pretreatment represents the most expensive single operation in cellulosic ethanol production, removing or reducing the need for this step would have a large beneficial impact on the cost of production (Wyman, 2007). Regardless of the ultimate role pretreatment plays in cellulosic ethanol production, effective fermentation of nonglucose sugars originating from hemicellulose is highly advantageous for all biomass conversion processes.

Hemicellulose hydrolysis and utilization

The major component of angiosperm hemicellulose is xylan, which is a largely linear polysaccharide made of β-1,4-linked xylopyranoside residues but also contains side chains, such as arabinofuranose in arabinoxylan or glucuronic acid in glucuronoxylan, as well as other sugars (Izydorczyk and Biliaderis, 1995). Additionally, xylan can be esterified or acylated by phenolic residues such as ferulic acid and *p*-coumaric acid. Therefore, it is not surprising that hemicellulolytic organisms must express a large set of cooperating enzymes in order to efficiently and completely hydrolyze hemicellulose and assimilate products of this hydrolysis. In the anaerobic thermophile *Clostridium stercorarium*, xylan degradation is accomplished by at least eight different enzymes (Adelsberger et al., 2004).

The first step in the hydrolysis of hemicellulose is the depolymerization of the xylan backbone by endoxylanases. Bacterial xylanases typically belong to two families, Family 10 and Family 11. The Family 10 xylanases have a catalytic domain, which is a cylindrical α/β barrel resembling a salad bowl. The catalytic site is at the narrower end of the bowl near the C terminus of the β-barrel (Derewenda et al., 1994; Harris et al., 1994). These enzymes have five xylopyanose binding sites and belong to a superfamily of enzymes that includes Family A cellulases β-glucosidase, β-galactosidase, β-(1,3)-glucanases, and β-(1,3;1,4)-glucanases (Jenkins et al., 1995). Family 10 xylanases tend to have relatively high molecular weights and typically produce oligosaccharides with a low degree of polymerization (Jeffries, 1996; White et al., 1994). Family 11 xylanases have a catalytic domain that is comprised primarily of β-pleated sheets that form a trough surrounding the catalytic site. Protruding into this trough is a long loop that terminates in an isoleucine residue (Miao et al., 1994; Withers and Aebersold, 1995). The structure of Family 11 xylanases has been likened to the palm and fingers of the cupped right hand with the thumb serving as the loop protruding into the trough (Torronen and Rouvinen, 1995). The resulting xylooligosaccharides are then further degraded by β-xylosidases and side groups are cleaved by specific glycosidases such as β-gluronidases and α-arabinosidases (Bronnenmeier et al., 1995; Donaghy et al., 2000; Kormelink et al., 1993; Kosugi et al., 2002; Nagy et al., 2002; Schwarz et al., 1995). There is a synergistic activity between endoxylanases and glycosidases, whereby the removal of side groups enhances the activity of endoxylanases towards the molecular backbone. This was shown in experiments using biochemically purified fungal and bacterial xylanases, xylosidases, and arabinosidases (Sørensen et al., 2003; Suh et al., 1996). Additionally, the expression of both xylanases and xylosidases seems to be induced by growth on xylan and to a lesser extent growth on xylose (Adelsberger et al., 2004; Baba et al., 1994). The expression of genes encoding xylanolytic enzymes are also subjected to catabolite repression in the presence of favored carbon sources such as glucose.

Once the component sugars have been liberated, they must be taken up by the organism. Although common in mesophilic bacteria, most of the thermophilic saccharolytic bacteria investigated to date do not possess phosphotransferase transport systems, which simultaneously transport and phosphorylate sugars with phosphoenolpyruvate serving as the phosphoryl donor. Instead, it is believed that in thermophilic bacteria, sugar uptake is routed through ATP binding cassette (ABC) transporter systems (Erbeznik et al., 1998b, 2004; Jones et al., 2000, 2002a, 2002b). ABC importers belong to two classes, the sugar transporters or carbohydrate uptake transporter (CUT) family and the di/oligopeptide transport, or Opp family, which differ not only in substrate specificity but also in the architecture of the transporter complex (Schneider, 2001). The CUT transporter family is further broken down into two subfamilies; the CUT1 family, which transports mainly di/oligosaccharides, and the CUT2 family, which is involved in monosaccharide transport. ABC transporters are comprised of four subunits or domains: hydrophobic membrane-spanning domains, which make up the translocation pathway, and two hydrophilic nucleotide binding domains, which provide the energy for active transport. ABC transporters that mediate sugar uptake employ a high-affinity solute binding protein that is tethered to the cell surface in the case of gram-negative bacteria or fused to the transporter itself in the case of gram-positive bacteria (Davidson and Chen, 2004; van der Heide and Poolman, 2002). These binding proteins are generally specific for a single or limited number of sugars or oligosaccharides. In *C. thermocellum* cellulose utilization, it has been demonstrated that cellobiose and cellodextrins are transported as well as glucose (Strobel et al., 1995). Because transport is driven by ATP hydrolysis, it has been demonstrated that from an energetic perspective it is more efficient to uptake an intact oligomer than to cleave it extracellularly and transport the resulting monomer sugars (Muir et al., 1985; Ng and Zeikus, 1982; Strobel, 1995). This mechanism may or may not be important for pentose sugars; however, there is some evidence of xylodextrins being imported in a similar fashion (Tsujibo et al., 2004). It is also possible that xylodextrins are degraded extracellularly, which is thought to relieve product inhibition of xylanases (Adelsberger et al., 2004). Once inside the cell, pentoses are shuttled into the Entner-Doudoroff or Embden-Meyerhoff-Parnas pathway, depending on the organism, through xylulose-5-*p*. Xylose is isomerized into xylulose by xylose isomerase and phosphorylated by xylulose kinase (Erbeznik et al., 1998a). In the case of arabinose it is slightly more complex. Arabinose is isomerized by L-arabinose isomerase to L-ribulose, which is then phosphorylated by L-ribulose kinase. Finally, L-ribulose-5-*p* is converted to xylulose-5-*p* by L-ribulose-5-phosphate-4-epimerase.

Ethanol Tolerance

The ethanol concentrations encountered in lignocellulose processing are likely to be lower than in processing starch and soluble sugars due to a variety of compounding factors. These include difficulties in generating high concentrations of sugars from solids as well as constraints on ethanol production imposed by high temperatures used to reduce cellulase loading and the presence of inhibitors present in pretreated biomass. Consistent with this, ethanol concentrations of about 5% (wt/vol) are assumed in both near-term (Aden et al., 2002) and mature process designs for bioethanol production from lignocellulose (Greene et al., 2004). Operation at this concentration does not preclude highly cost-effective and efficient processing of cellulosic biomass (Greene et al., 2004). However, at ethanol concentrations of <5% (wt/vol) the economic penalty becomes significant and progressively larger (Zacchi and Axelsson, 1989).

Much is known about the ethanol tolerance of the yeast *S. cerevisiae* (D'Amore et al., 1990; Jeffries and Jin, 2000; Zaldivar et al., 2001). Tolerance of ethanol by *S. cerevisiae* has been reported at titers greater than 20% (wt/vol) (Walker, 1998), although growth generally ceases at lower ethanol concentrations, particularly in the presence of inhibitors in pretreated biomass (Walker, 1998; Bazua and Wilke, 1977; Ghose and Tyagi, 1979). The ethanol tolerance of bacteria has also been studied (Zaldivar et al., 2001; Demain et al., 2005; Ingram, 1990). The ethanol-producing bacterium *Z. mobilis* can grow at ethanol concentrations as high as 9% (vol/vol) (Ingram, 1990) but does not natively utilize xylose or arabinose. The ethanol tolerance of the mesophilic facultative anaerobe *Escherichia coli* is lower than that of yeast and *Z. mobilis* but has been increased using both serial transfer (Yomano et al., 1998) and a novel directed mutagenesis strategy focusing on the global transcription factor sigma-70 (Alper and Stephanopoulos, 2007).

Methods for evaluating ethanol tolerance vary (Lynd et al., 1991), which can complicate comparison of data from different studies and different organisms. Variables that affect measurements of tolerance include the strain tested; temperature; medium; amount of inoculum; whether ethanol is added and/or produced endogenously; whether the test is in batch, fed-batch, or continuous mode; whether final optical density (OD), viability, or growth rate is measured; whether complete or partial inhibition is examined; and whether the organism has been preadapted to ethanol. In the discussion that follows, we refer to the ethanol concentration resulting in 50% inhibition (IC_{50}).

Despite variations in methods for measuring ethanol tolerance, some comparison can be made between thermophilic anaerobes that were not preadapted—or were minimally adapted—to high ethanol concentrations (see Lynd et al., 1991, for an earlier comparison). The batch mode IC_{50} of *C. thermocellum* ranges from 0.4 to 2.7% (wt/vol), depending on the isolate (Herrero and Gomez, 1980; Sato et al., 1993). An isolate in which the final OD is reduced by 50% in 5.0% (vol/vol) ethanol has been described (Sudha Rani and Seenayya, 1999). Ethanol production has been described for other cellulolytic organisms, but to our knowledge, ethanol tolerance has not been quantitatively evaluated in the literature. Among noncellulolytic thermophilic anaerobes, the final OD of *Thermoanaerobacterium thermosaccharolyticum* was 50% reduced at 3.2% (wt/vol) ethanol (Wang et al., 1983). Measurements in continuous mode showed an IC_{50} of 4.0% (wt/vol) ethanol (Baskaran et al., 1995). The IC_{50} of *Thermoanaerobacter mathranii* was between 1 and 2% (wt/vol) ethanol at 70°C (Larsen et al., 1997). For *Thermoanaerobacter pseudoethanolicus* strain 39E the IC_{50} was ~1% (vol/vol) ethanol (Lovitt et al., 1984; Burdette et al., 2002), and *Thermoanaerobacter ethanolicus* strain JW200 was slightly inhibited at ethanol concentrations of 2.5% (vol/vol) (Wiegel and Ljungdahl, 1981).

It is clear that high ethanol tolerance is not a constitutive property in thermophilic bacteria, with many reports of improved ethanol tolerance achieved by extended exposure to increased ethanol concentrations, and no reports of failure to increase ethanol tolerance by this approach. Continuous-culture methods have been used to select for ethanol-tolerant strains (Lynd et al., 1991; Klapatch et al., 1994). Serial-transfer methods, in which cells are repeatedly grown in batch culture and transferred to fresh media containing higher and higher initial ethanol levels, have been very successful in improving the ethanol tolerance of *C. thermocellum*. Herrero and Gomez (1980) increased the IC_{50} from <1 to 3.0% (wt/vol) ethanol, while Williams et al. (2007) increased the IC_{50} to >5.0%. Using *Thermoanaerobacter pseudoethanolicus* strain 39E, Lovitt et al. (1984) showed data indicating an increase in the IC_{50} from ~1% (vol/vol) to ~5.5% (vol/vol) ethanol. Burdette et al. (2002) increased the IC_{50} of strain 39E from ~1 to 8% (vol/vol) ethanol. It appears that many and perhaps most strains of thermophilic saccharolytic bacteria—including both cellulose-utilizing and hemicellulose-utilizing strains—can be adapted to grow in the presence of added ethanol concentrations of up to at least 6% (wt/vol). Ethanol-adapted mutants typically show lower cell yield and/or lower growth rates in the absence of ethanol than the wild type (WT), but in some reports they show marked stimulation in low concentrations of ethanol.

Temperature plays a major role in ethanol tolerance. In general, higher levels of ethanol are tolerated at lower temperatures, and larger increases in tolerance can be achieved by conducting adaptations at lower temperatures. Thus, the optimization of temperature for CBP must account for not only enzyme activity and bacterial growth rate but also ethanol tolerance. It is important to also consider that cell growth is generally more sensitive to ethanol than is fermentation and that significant productivity can occur above ethanol levels at which growth is completely halted. Another key consideration is the stability of ethanol tolerance as a trait—i.e., for how many generations an ethanol-tolerant strain can be cultured in low-ethanol medium without losing its elevated tolerance. Finally, there remains a consistent—and persistent—difference between the maximum concentration of added ethanol tolerated and the maximum amount produced by actual cultures (Lynd and Zhang, 2002). Such gaps have been noted and eventually closed for other organisms and products (Lynd et al., 2005), lending support to the notion that this can also be accomplished for thermophilic ethanol production. This, however, remains to be demonstrated.

The basis of ethanol tolerance in thermophilic anaerobes can be studied by determining the mode of action of ethanol inhibition (Ingram, 1990; Jones, 1989) or by identifying the molecular or genetic adaptations that occur in generating ethanol-tolerant mutants. In thermophilic anaerobes, two broad classes of adaptations have been identified: those affecting membrane properties and those affecting metabolic machinery (Larsen et al., 1997). Studies by Herrero et al. with *C. thermocellum* found that cells respond to ethanol treatment in three distinct growth phases (Herrero and Gomez, 1980; Herrero et al., 1982). They also observed an alteration of the membrane fatty acid content that lowers the overall melting point and increases membrane fluidity. The response of an ethanol-tolerant mutant was similar in kind but reduced in degree for any given ethanol concentration, suggesting that the ethanol response of WT cells was maladaptive or too extreme. Williams et al. have found extensive differences in membrane protein content between WT cells and an ethanol-adapted strain (Williams et al., 2007). Burdette et al. found that an ethanol-adapted strain of *T. pseudoethanolicus*, strain 39E, had increased levels of long-chain fatty acids (Burdette et al., 2002).

Evidence for effects of ethanol on metabolism was reported by Herrero et al., concluding that ethanol had extensive effects on glycolysis and that an ethanol-tolerant mutant had increased levels of early glycolytic intermediates, possibly related to protein-membrane interactions (Herrero et al., 1985). Studies with *T. pseudoethanolicus* 39E have shown that ethanol raised the NADH/NAD ratio in WT strains but not in an ethanol-adapted mutant (Lovitt et al., 1988). Another study with an ethanol-adapted derivative of 39E showed that the mutant lacked one of two alcohol dehydrogenases (ADHs) and was no longer stimulated by the addition of propanone (Lovitt et al., 1984). These observations support a model in which cells possess two ADH enzymes, one producing ethanol and the other consuming it and producing NADH. High ethanol concentrations then result in a buildup of NADH, which in turn inhibits glycolysis. The ethanol-adapted mutant avoids this problem by elimination of the ethanol-consuming ADH.

Although the ethanol tolerance of potential CBP organisms is not innately high, adaptation experiments have shown that tolerance can be raised to levels above the 4 to 5 weight percent threshold for economic viability. Considering that most proteins are only appreciably inhibited by ethanol concentrations at substantially higher concentrations than 5% wt/vol (Millar et al., 1982), including cellulases from both mesophilic (Ghose et al., 1982) and thermophilic (Bernardez et al., 1994) organisms, inhibition of cellulose hydrolysis should not prevent further increases. Tolerance to other inhibitors present in pretreated biomass such as acetic acid, furfural, and hydroxymethyl furfural may also be achieved using similar approaches (Klinke et al., 2004). There is no obvious reason why these traits cannot be combined in one organism with other desired CBP properties, given the right genetic tools and increased mechanistic understanding.

FUNDAMENTALS OF MICROBIAL CELLULOSE UTILIZATION

Hydrolysis and Substrate Assimilation

The complexed cellulase enzyme systems, or cellulosomes, of anaerobic microorganisms have a different fundamental architecture relative to the noncomplexed systems of aerobic microorganisms (Lynd et al., 2002). Notwithstanding this deeply rooted difference, cellulose hydrolysis by anaerobes has often been assumed to proceed via cellobiose-liberating exoglucanases as per the accepted mechanism for aerobes. This assumption is understandable in light of the well-established mechanism for aerobic cellulase systems together with the widespread observation that cellobiose accumulation is routinely observed in cell-free experiments involving anaerobic cellulase systems and, under some conditions, whole-cell experiments as well (Lynd and Zhang, 2002).

In the case of *C. thermocellum*, considerable evidence supports the proposition that solubilization in fact does not occur primarily via the action of cellobiohydrolases. Rather, uptake of cellulose hydrolysis products appears to involve cellodextrins of average degree of polymerization equal to about 4, implying that the immediate products of cellulose solubilization have an average degree of polymerization of at least this value. Evidence supporting this conclusion is provided by both bioenergetic modeling and labeling studies as discussed in more detail elsewhere (Lynd et al., 2007; Zhang and Lynd, 2005b).

Uptake of hemicellulose hydrolysis products is considered above (see "Utilization of nonglucose sugars"). Assimilation of cellulose hydrolysis products has been investigated in most detail among thermophilic, saccharolytic bacteria for *C. thermocellum*. Strobel (1995) showed that resting cells of *C. thermocellum* take up [^{14}C]cellodextrins and [^{14}C]cellobiose by a common ATP-dependent mechanism whereas glucose is taken up by a distinct mechanism that is also ATP dependent. Inhibitor studies further confirmed that transport is dependent upon ATP but not the proton motive force. These observations are consistent with cellobiose and cellodextrin transport by an ABC protein. There is no evidence for phosphoenolpyruvate-dependent transport of cellulose hydrolysis products in *C. thermocellum*, or in other cellulolytic bacteria for that matter (Lynd et al., 2002).

Control of Cellulase Synthesis

An important dimension of microbial cellulose utilization is the mechanism(s) by which the cell controls the expression of cellulases and related proteins (i.e., the scaffoldin *cipA*). Reviews have previously been written on the subject, most recently by Demain et al., in 2005. As with most other aspects of cellulose utilization by thermophilic anaerobes, investigation thus far has focused on *C. thermocellum*.

Johnson et al. (1985) were the first to clearly show a relationship between high carbohydrate concentrations and the repression of cellulase production. Cultures grown on cellobiose exhibited a substantial lag phase (110 to 125 h) when transferred to media containing either sorbitol, fructose, or glucose. These strains, with the exception of those grown on glucose, exhibited a five- to sixfold increase in the production of cellulases over cellobiose-grown cultures while acclimatizing to the new carbohydrate. This cellulase production sharply declined once the cells had begun to metabolize the new carbon source. Cultures adapted for growth on sorbitol and fructose had a lower specific cellulase activity than those grown on cellobiose, even though growth was faster on cellobiose. These experiments established a model for *C. thermocellum* ATCC 27405 cellulase regulation based on the concentration of available soluble saccharides but left unanswered the question as to why this was seen.

Further research brought about the discovery of the cellulosome complex as well as specific component enzymes of the cellulosome. Some of the first characterization work of cellulosomal enzyme expression was done by Mishra et al. (1991) on endoglucanases encoded by *celA*, *celC*, *celD*, and *celF* from *C. thermocellum* NCIB 10682. It was observed that *celA*, *celD*, and *celF* were all expressed in late exponential and early stationary phases whereas *celC* was transcribed in early stationary phase when grown in batch culture on cellobiose. The authors concluded that a system similar to catabolite repression may be responsible for the apparent inhibition of cellulase synthesis and proposed that over time as cellobiose was metabolized, its concentration would drop below the point necessary to inhibit cellulase transcription. This work contradicted earlier studies which had concluded that the endoglucanase activity in *C. thermocellum* was constitutive (Shinmyo et al., 1979; Garcia-Martinez et al., 1980).

In contrast to the conclusions drawn by Mishra et al. (1991), Dror et al. (2003a, 2003b) proposed that growth rate, rather than substrate or catabolite concentration, is responsible for cellulase regulation. Using RNase protection assays, the group was able to quantify the relative number of transcripts per cell of *celS* (Dror et al., 2003a) and of *olpB*, *orf2*, *sdbA*, and *cipA* (Dror et al., 2003b) from *C. thermocellum* YS in either batch or continuous culture. For nitrogen-limited continuous cultures with excess cellobiose, as the growth rate decreased (0.14 to 0.07 h^{-1}) the *celS* transcript level increased more than threefold (30 to 100 transcripts/cell). A similar effect was seen for transcript levels of *cipA*, *olpB*, and *orf2* in continuous culture when grown under nitrogen limitation and an excess of cellobiose, but the effect was not seen for *sdbA*. The authors were unable to locate catabolite responsive element (CRE) consensus sequences, regulatory sequences related to catabolite repression, in any of the currently studied cellulosome-related genes other than *orf2*. From these data it was concluded that regulation occurred as a function of growth rate and that no known catabolite repression was active. More recent work, however, has continued to support the idea that cellulase regulation does indeed involve carbon catabolite repression (CCR), substrate presence, or concentration of solubilized carbohydrates at some level.

Zhang and Lynd studied cellulase production in batch and continuous cultures grown on cellulose and cellobiose (Zhang and Lynd, 2005a). The cellulose-grown batch cultures displayed a ninefold increase in cellulase content over cellobiose-grown cells. In continuous cultures at a given dilution rate, cellulase yields were 1.34- to 2.36-fold higher for Avicel-grown cultures than those grown on cellobiose. This increased cellulase yield for growth on crystalline cellulose over cellobiose indicates that cellobiose is not the only breakdown product of cellulose and/or that *C. thermocellum* can sense the presence of cellulose (i.e., via adhesion), triggering upregulation of cellulases. Additionally, it was reported that using a degenerative CRE consensus sequence it was possible to locate over 100 putative CREs in *C. thermocellum* ATCC 27405 (Zhang and Lynd, 2005a). Two of these were located within *cipA* (positions +935 and +5231). Warner and Lolkema (2003) also support the conclusion that CCR is active in *C. thermocellum* by revealing the presence of three different Hpr kinase/phosphatases which are considered to be essential components of gram-positive CCR. If these CRE sequences are confirmed experimentally, it will lend substantial support to the idea of a solubilized sugar regulation mechanism acting on cellulases.

Subsequent work using real-time PCR (Stevenson and Weimer, 2005) quantified the expression of 17 genes from *C. thermocellum* ATCC 27405 as affected by growth rate and substrate type in continuous culture. Continuous cultures were grown under cellulose or cellobiose limitation at various dilution rates (0.013 to 0.16 h^{-1}). The results suggest that it may be the presence of cellulose itself that affects expression of some genes rather than soluble substrates.

Yet another example of catabolite-derived regulation comes from Newcomb et al. (2007). GlyR3, which is cotranscribed with *celC*, *glyR3*, and *licA*, was shown to be a negative regulator of the *celC* operon in *C. thermocellum* ATCC 27405. This repression was relieved by the addition of laminaribiose in a dose-dependent fashion. Laminaribiose is a β-1-3-linked glucose dimer which is a breakdown product of β-1-3-glucan, upon which both *celC* and *licA* are active. GlyR3 binds to a near-perfect 18-bp palindrome within the *celC* promoter region and shares sequence homology to LacI binding sites. Laminaribiose inhibits the binding of GlyR3 to DNA, thus allowing transcription. Finally, as these genes are transcribed in a polycistronic manner, they make up the first confirmed cellulase operon in *C. thermocellum*. Newcomb et al. (2007) are the first to provide concrete evidence of solubilized sugars regulating enzyme expression in *C. thermocellum* ATCC 27405.

Bioenergetics of Cellulose Utilization

In order for CBP to be feasible from a bioenergetic perspective, it is crucial that the ATP demand associated with synthesizing the significant amount of cellulase necessary to mediate rapid cellulose hydrolysis be provided from anaerobic metabolism producing ethanol as an end product. While the existence of cellulolytic microorganisms in nature provides some reassurance on this point, most such organisms benefit from

ATP-generating processes associated with acetate kinase—not available to a hypothetical CBP organism engineered to make only ethanol.

Current understanding of how microbes solve the bioenergetic challenge of cellulose utilization under anaerobic conditions is most advanced in the case of *C. thermocellum*. ATP-generating metabolic processes in *C. thermocellum* include intracellular phosphorolytic cleavage of β-glucosidic bonds by cellodextrin and cellobiose phosphorylase (Strobel et al., 1995; Zhang and Lynd, 2003), glycolysis via the Embden-Meyerhoff pathway, and the action of acetate kinase. ATP-consuming metabolic processes include substrate transport via an adenosine binding cassette system (Mitchell, 1998; Lynd et al., 2002; Zhang and Lynd, 2005b), cell synthesis, cellulase synthesis, and nonbiosynthetic maintenance functions. As presented in more detail elsewhere (Zhang and Lynd, 2005b; Lynd et al., 2007), Zhang and Lynd developed a model for the bioenergetics of cellulase synthesis in *C. thermocellum* in which each of these processes is evaluated quantitatively. This model supports the interpretation that two bioenergetic benefits specific to growth on cellulose are operative: (i) reduced ATP demand for substrate uptake due to assimilation of glucan oligomers with average length of about 4 during growth on cellulose, with the ATP requirement for transport equal to $1/n$, where n is the length of assimilated cellodextrin; and (ii) intracellular phosphorolytic cleavage of cellodextrins of length n to a cellodextrin of $n - 1$ and glucose-1-phosphate.

The magnitude of ATP made available by these processes during growth on cellulose was shown to exceed the cost of cellulase synthesis. It may also be observed that although the rate of cellulose hydrolysis is rightly thought of as being controlled by processes that occur outside the cell, current understanding is consistent with roughly three-quarters of the beta-glucosidic bonds initially present in cellulose cleaved intracellularly during cellulose utilization by *C. thermocellum*. Investigation of the generality of bioenergetic understanding gained with the *C. thermocellum* model system is an important topic for future research.

Enzyme-Microbe Synergy and Natural Selection Involving Cellulose-Adhered Microbes

Cellulose hydrolysis can be mediated by cellulase enzymes acting in the absence of cells, by cellulases acting in the presence of cells but with no cell-enzyme attachment, or by cellulases attached to the cells that produce them. In the last case, hydrolysis is mediated by ternary cellulose-enzyme-microbe (CEM) complexes rather than binary cellulose-enzyme (CE) complexes. For anaerobic cellulolytic bacteria found in nature, CEM complexes are commonly formed and are thought to be the major agent of cellulose hydrolysis (Lynd et al., 2002). Potential benefits of CEM complexes for cellulolytic microorganisms have been suggested, including preferred access to hydrolysis products and local concentration of cellulases (Lynd et al., 2002; Adams et al., 2006; Miron et al., 2001; Schwarz, 2001; Shoham et al., 1999). Microbial adhesion to cellulose surfaces is reviewed elsewhere (Miron et al., 2001).

The specific activity of the *C. thermocellum* cellulase enzyme system has recently been compared for growing cultures relative to controls under which the cellulase acts independently of cells (Lu et al., 2006). Depending on conditions, a 2.7- to 4.7-fold-higher specific activity was observed for growing cultures than for the control, indicative of enzyme-microbe synergy. It may be noted that cellulose hydrolysis by growing cultures of cellulolytic anaerobes, and *C. thermocellum* in particular, is thought to be mediated primarily by CEM complexes, whereas hydrolysis in the control was mediated solely by CE complexes. The mechanistic basis of enzyme-microbe synergy is not inhibition by the bulk phase concentration of hydrolysis products and presumably involves events and conditions operative at the cellulose surface where hydrolysis occurs but is otherwise not known. Enzyme-microbe synergy of the magnitude observed for *C. thermocellum* is a potentially important advantage of microbial hydrolysis as occurs in CBP compared to the purely enzymatic hydrolysis that occurs in other process configurations for cellulose hydrolysis and fermentation. Further investigation of such synergy for organisms in addition to *C. thermocellum*, and under a broader range of conditions and substrates, is of considerable interest.

Natural selection rewards populations that arise from variant cells that are phenotypically different from the parent population and realize a benefit as a result of this difference. In the case of cellulolytic microbes, adherence and preferential access to hydrolysis products would appear to be necessary in order for a variant cell with a more active, more effective, or more plentiful cellulase system to have a selective advantage compared to nonvariant cells (Fan et al., 2005). While conceptually attractive and suggested in the literature (Fan et al., 2005; Weimer, 1996; Lynd et al., 2002), the hypothesis that cellulose hydrolysis products are "captured" by adherent cellulolytic microbes has yet to be experimentally proven. Molecular modeling indicates that the water at the cellulose surface has markedly different properties from bulk water (Matthews et al., 2006). The same can be expected to be true of the surface of cells. It seems reasonable to speculate that the "cell-substrate gap," that is, the space between adhered cellulolytic microbes and cellulose, has been the target of intensive selective pressure with respect to modifying physical chemical properties to maximize the rate of hydrolysis and to capturing hydrolysis products before they enter the bulk solution and are available to nonadhered cells. Conceptual and analytical tools developed for study of microbial biofilms have been applied to microbial cellulose utilization to a limited extent but appear promising (Lynd et al., 2007).

Cellulose Hydrolysis Rates

Rates of cellulose hydrolysis by thermophilic bacteria and their cellulases have been investigated primarily for *C. thermocellum*. As reviewed elsewhere, literature from the 1980s and 1990s provides strong evidence supporting the proposition that the specific activity of crystalline cellulose hydrolysis on a per-unit-cellulase-protein basis is higher for the complexed *C. thermocellum* cellulase system than for the noncomplexed cellulase system of the mesophilic aerobic fungus *Trichoderma reesei*. The number of experimental studies that speak directly to this point is, however, quite small, and extensive data are not available with respect to variables such as substrates, enzyme preparation method, enzyme/substrate ratio, extent of reaction, and comparisons on an absolute-rate basis as well as a specific-activity basis. As a result, the magnitude of the rate advantage of in vitro cellulose hydrolysis by the *C. thermocellum* cellulase system compared to the *T. reesei* system is not definitively known.

Johnson et al. (1982) reported 50-fold-higher specific activity for cellulase present in the culture broth of *C. thermocellum* than for reconstituted cellulase of *T. reesei*. The ratio of crude protein to cellulose used in the experiments upon which this calculation was based was much higher (48-fold) for *T. reesei* than for *C. thermocellum*, raising the possibility that rate saturation may have been operative and that a higher specific activity would have been calculated for *T. reesei* at lower protein loadings. When rates of hydrolysis by broth dilutions from *C. thermocellum* and *T. reesei* having the same amount of crude protein were compared, reported in E. Johnson's Ph.D. thesis (Johnson, 1983), the *C. thermocellum* system exhibited about 15-fold-higher specific activity on Avicel (Lynd and Zhang, 2002). Remarkably, and not followed up, to our knowledge, since Johnson's seminal work a quarter-century ago, the rate of in vitro hydrolysis by the *C. thermocellum* cellulase system increased slightly (~2-fold) for hydrolysis of phosphoric acid-swollen cellulose compared to Avicel, whereas the hydrolysis rate mediated by the *T. reesei* system increased about 50-fold (Johnson, 1983). Wilson and Wood (1992) compared hydrolysis rates of cellulase preparations from *C. thermocellum* and *T. reesei* acting on cotton over a 10-fold range of protein concentration. For all but the lowest protein concentration tested, the specific activity was greater for the *C. thermocellum* system by roughly 1 order of magnitude. It is interesting that the specific activity of the complexed cellulase produced by the mesophilic anaerobic fungus *Neocallimastix frontalis* was found to be comparable to that of *C. thermocellum* in this study.

While much remains to be elucidated regarding relative rates of hydrolysis mediated by cell-free cellulase preparations, the extent of study afforded to kinetics of cellulose hydrolysis by cellulolytic microorganisms, particularly relevant for CBP, is considerably more sparse. The specific growth rate on cellulose, μ (per hour), may readily be defined as the ratio of cell formation to cell concentration and may be arbitrarily set in continuous culture. However, exponential growth is seldom observed in batch cultures of cellulolytic microorganisms due to the limited accessibility of cellulose (Lynd et al., 2002). Available data obtained for various microorganisms grown on crystalline cellulose under similar but not identical conditions showed that μ increases linearly with respect to temperature over the range of 28 to 75°C (Lynd et al., 2002). Similar growth rates are observed for anaerobes and aerobes at a given temperature, indicating that the cell-specific rate of cellulose hydrolysis is severalfold higher for anaerobes (Lynd et al., 2002).

Table 2 presents comparative specific activities for *T. reesei* and *C. thermocellum* cellulases under a variety of conditions. A value of 0.6 U per mg of protein under the conditions specified for the filter paper assay (Ghose, 1987) is representative for *T. reesei* and similar aerobic fungi based on many studies compiled by Esterbauer and others in 1991 (Esterbauer et al., 1991). Although improved fungal cellulase production has been an intensive focus since that time, reports known to us in either the scientific or patent literature have not demonstrated a significant improvement in specific activity measured by the standard filter paper assay. The activity of *T. reesei* cellulase at 37°C, near the upper limit tolerated by established ethanol-producing microorganisms, is much less widely documented than at 50°C. Data obtained in the Lynd lab (K. Podkaminer and S. House, unpublished data) indicate that changing the reaction temperature from 50 to 37°C under controlled conditions results in about a 2.5-fold reduction in the rate of hydrolysis by a commercial *T. reesei* cellulase preparation with supplemental beta-glucosidase. Fungal cellulases can, of course, be used to hydrolyze cellulose at temperatures higher than those tolerated by available fermentative organisms by performing hydrolysis and fermentation in two separate steps. While this approach allows higher cellulase-specific activity to be realized, it entails other compromises as reviewed elsewhere (Lynd et al., 2002). In summary, available information indicates that the specific activity of commercial

Table 2. Comparative rates of cellulose hydrolysis under dilute conditions

Organism and conditions	% Cellulose hydrolyzed (μmol/min/mg)	Excess enzyme?	Sp act
T. reesei cellulase (in vitro, filter paper assay)			
50°C[a]	0.04	No	0.60
37°C[b]	0.04	No	0.24
C. thermocellum (60°C)[c]			
Cellulase (in vitro)	0.05 to 0.1	No	2.5
Culture (total cellulase basis)	0.65	Yes	11.5

[a]Data from Esterbrauer et al., 1991.
[b]Data from the Lynd lab (Podkaminer, unpublished).
[c]Data for continuous cultures from Lu et al., 2006.

T. reesei cellulase preparations on model cellulosic substrates such as filter paper is in the range of 0.24 to 0.6 units/mg of protein. It should be understood that this activity is based on initial rates measured at very low cellulose concentrations and thus substantially exceeds rates anticipated in industrial processes featuring extensive hydrolysis of pretreated feedstocks at high concentrations.

In the hands of multiple investigators from the Lynd lab (Zhang and Lynd, 2003; Lu et al., 2006), we find the activity of the *C. thermocellum* cellulosome measured in vitro to be about 2.5 U/mg. It may be noted that the cellulosome preparations used to generate the data in Table 2 were prepared by affinity digestion (Morag et al., 1992) with high (≥80%) recovery of the activity present in the original broth. In addition, these data are based on initial rates at low cellulose concentrations with careful attention to avoid enzyme saturation—as is the case with the standard filter paper assay. Avicel is used to assay *C. thermocellum* activity instead of filter paper. However, these two substrates exhibit rather similar activities at low cellulase loadings for both the *T. reesei* and *C. thermocellum* systems (Johnson et al., 1982; Zhang and Lynd, 2006). The specific activity of 2.5 U/mg observed in the *C. thermocellum* cellulase system measured in vitro (Table 2) exceeds that representative of *T. reesei* at 50 and 37°C by 4.2-fold and 10.4-fold, respectively. Consistent with the phenomenon of enzyme-microbe synergy discussed above, higher cellulase specific activities are observed in growing cultures than in in vitro assays. In particular, a specific activity of 11.5 U/mg is observed for *C. thermocellum* growing in continuous culture. This value (Table 2) exceeds that achieved by the *T. reesei* system by a substantial margin, 19-fold and 48-fold compared to data at 50 and 37°C, respectively. These ratios are likely a conservative estimate of relative activity since cellulase is in excess for the *C. thermocellum* culture but not the *T. reesei* assay, and also because the percentage of cellulose hydrolyzed is 1 order of magnitude higher for *C. thermocellum* culture than for the *T. reesei* assay.

More measurements and further study will be needed to further bring into focus the matter of the relative activity of thermophilic microbes and cellulases, including organisms other than *C. thermocellum*, and to make informative comparisons with cultures and enzymes from *T. reesei* as well as additional mesophiles and/or aerobes. While acknowledging this, available data support the understanding that (i) the cell-specific rate of cellulose hydrolysis is substantially higher for thermophilic anaerobic microbes than for any described mesophilic bacterium, whether aerobic or anaerobic; and (ii) the protein-specific rate of cellulose hydrolysis exhibited by multicomponent cellulase systems is substantially higher for thermophilic anaerobes than for commercial preparations produced by *T. reesei*, and this difference is substantially enhanced when specific activity for the thermophilic system is measured in growing cultures rather than in vitro.

With respect to the second point above, the question of whether cultures of *T. reesei* and other aerobes exhibit enzyme-microbe synergy as does *C. thermocellum* is of interest from a scientific perspective. From an applied perspective in the context of biofuel production, however, the relevant comparison is to *T. reesei* cellulase acting independently of growing cultures of this organism since such cultures generate ATP from respiratory metabolism and are thus not a practical means of producing reduced end products such as ethanol industrially.

ORGANISM DEVELOPMENT

Availability and Development of Genetic Tools

As reviewed in detail elsewhere (Slapack et al., 1987; Lynd, 1989; Lynd et al., 2002), the 1980s and 1990s saw a considerable effort devoted to improving and developing strains of thermophilic bacteria for biofuel production using classical approaches: screening, selection, nonspecific mutagenesis, and manipulation of culture conditions such as nutrients and end product concentrations. Several more-recent studies have also used strain selection to generate ethanol-tolerant strains, as reviewed above in "Ethanol tolerance." Viewed in aggregate, application of classical techniques has established that fermentation with near-theoretical ethanol yields is possible under some conditions and that strains can be selected to be tolerant to added ethanol at concentrations sufficient to not preclude commercial application. However, this body of work did not achieve two key requirements for use in an industrial process: (i) establishment of robust strains that stably and reliably produce ethanol at high yields over a range of conditions, and (ii) establishment of strains that produce ethanol at the highest concentrations of ethanol that are tolerated. Overcoming these barriers will likely require metabolic engineering using recombinant DNA technology.

Although foreign gene expression was first reported over 30 years ago and has seen tremendous application since that time, genetic tools are well developed for only a few "workhorse" microorganisms. Thermophilic, saccharolytic anaerobes have been particularly slow to move into the circle of organisms amenable to genetic manipulation as reviewed elsewhere (Mai and Wiegel, 2000; Lynd et al., 2002; Tyurin et al., 2006). Soutschek-Bauer et al. achieved transformation of *Clostridium thermohydrosulfuricum* (now *Thermoanaerobacter ethanolicus*) using chemically induced competence in 1985. In 1987, Tsoi and others reported protoplast transformation of *C. thermocellum* (Tsoi et al., 1987). Subsequently, Klapatch and others reported electrotransformation of *Clostridium* (now *Thermoanaerobacterium*) *thermosaccharolyticum* in 1996 (Klapatch et al., 1996), Mai and coworkers reported electrotransformation of *Thermoanaerobacterium saccharolyticum* in 1997 (Mai et al., 1997), and Tyurin and coworkers reported electrotransformation of *C. thermocellum* in 2004 (Tyurin et al., 2004).

Experience both with thermophiles and with mesophilic gram-positive bacteria indicates that there is a difference between an initial report of transformation, typically involving plasmid detection in a presumptive transformant and retransformation using recovered plasmid in an initial investigator's hands, and a robust system that works for a variety of investigators and can actually be used to develop modified organisms. For cellulolytic clostridia, demonstrating targeted gene knockout following an initial report of transformation has been problematic in the case of both *C. thermocellum* (unpublished results from the Lynd group) and *C. cellulolyticum* (C. Tardiff, personal communication). Further effort and progress in the challenging area of development of genetic tools that work for thermophilic anaerobes are clearly warranted.

Metabolic Engineering

In the first example of using genetic tools to modify the metabolism of thermophilic anaerobes known to us, Mai and Wiegel expressed four saccharolytic enzymes, ManA from *Caldicellulosiruptor*, CelD from *C. thermocellum*, CelS from *C. thermocellum*, and CbhA from *C. thermocellum*, in *T. saccharolyticum* (Mai and Wiegel, 2000). Functional activity was detected, but growth of transformed strains on nonnative substrates was not investigated. The heterologous expression of pyruvate decarboxylase and ADH from *Z. mobilis* in *C. cellulolyticum* reported in 2002 by Guedon et al. is a significant milestone, although it does not involve thermophilic microbes (Guedon et al., 2002). These authors demonstrated functional expression and alleviation of growth limitation, which they attribute to diminished accumulation of pyruvate. Interestingly, although an additional route to ethanol formation is provided, ethanol:acetate ratios decreased somewhat in recombinant strains expressing pyruvate decarboxylase and ADH.

Metabolic engineering in a thermophilic host resulting in increased ethanol yields was achieved by Desai et al. in 2004 (Desai et al., 2004). Homologous recombination-mediated gene knockout was used to eliminate lactic acid production during fermentation of glucose and xylose by *T. saccharolyticum*, which was accompanied by increased yields of the remaining fermentation products: ethanol and acetic acid. More recently, gene knockout was used to prepare strains of this organism deficient in acetate production and—significantly—a double-knockout strain (ALK2) producing ethanol as the only organic product (A. J. Shaw, unpublished data). Strain ALK2 grows at rates similar to that of the WT and has shown no sign of instability. Whereas all previously described organisms carrying out a homoethanol fermentation metabolize pyruvate to acetaldehyde using pyruvate dehydrogenase, strain ALK2 converts pyruvate to acetyl-coenzyme A via pyruvate-ferredoxin oxidoreductase, which is reduced to acetaldehyde and then to ethanol.

After decades of anticipation, it seems that the era of being able to apply genetic tools to improve production of ethanol from thermophilic bacteria is upon us. The next few years promise to be exciting.

REFERENCES

Adams, J. J., G. Pal, Z. Jia, and S. P. Smith. 2006. Mechanism of bacterial cell-surface attachment revealed by the structure of cellulosomal type II cohesin-dockerin complex. *Proc. Natl. Acad. Sci. USA* **103**:305–310.

Adelsberger, H., C. Hertel, E. Glawischnig, V. V. Zverlov, and W. H. Schwarz. 2004. Enzyme system of *Clostridium stercorarium* for hydrolysis of arabinoxylan: reconstitution of the in vivo system from recombinant enzymes. *Microbiology* **150**:2257–2266.

Aden, A., M. Ruth, K. Ibsen, J. Jechura, K. Neeves, J. Sheehan, B. Wallace, L. Montague, A. Slayton, and J. Lukas. 2002. *Lignocellulosic*

Biomass to Ethanol Process Design and Economics Utilizing Co-Current Dilute Acid Prehydrolysis and Enzymatic Hydrolysis for Corn Stover. NREL/TP-510-32438. National Renewable Energy Laboratory, Golden, CO.

Alper, H., and G. Stephanopoulos. 2007. Global transcription machinery engineering: a new approach for improving cellular phenotype. *Metab. Eng.* 9:258–267.

Baba, T., R. Shinke, and T. Nanmori. 1994. Identification and characterization of clustered genes for thermostable xylan-degrading enzymes, beta-xylosidase and xylanase, of *Bacillus stearothermophilus* 21. *Appl. Environ. Microbiol.* 60:2252–2258.

Baskaran, S., H. J. Ahn, and L. R. Lynd. 1995. Investigation of the ethanol tolerance of *Clostridium thermosaccharolyticum* in continuous culture. *Biotechnol. Prog.* 11:276–281.

Bayer, E. A., R. Kenig, and R. Lamed. 1983. Adherence of *Clostridium thermocellum* to cellulose. *J. Bacteriol.* 156:818–827.

Bazua, C. D., and C. R. Wilke. 1977. Ethanol effects on the kinetics of a continuous fermentation with *Saccharomyces cerevisiae. Biotechnol. Bioeng. Symp.* 7:105–118.

Bernardez, T. D., K. A. Lyford, and L. R. Lynd. 1994. Kinetics of the extracellular cellulases of Clostridium thermocellum acting on pretreated mixed hardwood and Avicel. *Appl. Microbiol. Biotechnol.* 41:620–625.

Bredholt, S., J. Sonne-Hansen, P. Nielsen, I. Mathrani, and B. K. Ahring. 1999. *Caldicellulosiruptor kristjanssonii* sp. nov., a cellulolytic, extremely thermophilic, anaerobic bacterium. *Int. J. Syst. Bacteriol.* 49:991–996.

Bronnenmeier, K., H. Meissner, S. Stocker, and W. Staudenbauer. 1995. alpha-D-glucuronidases from the xylanolytic thermophiles *Clostridium stercorarium and Thermoanaerobacterium saccharolyticum. Microbiology* 141:2033–2040.

Burdette, D. S., S. H. Jung, G. J. Shen, R. I. Hollingsworth, and J. G. Zeikus. 2002. Physiological function of alcohol dehydrogenases and long-chain (C_{30}) fatty acids in alcohol tolerance of *Thermoanaerobacter ethanolicus. Appl. Environ. Microbiol.* 68:1914–1918.

Burke, D., P. Kaufman, M. McNeil, and P. Albersheim. 1974. The structure of plant cell walls. VI. A survey of the walls of suspension-cultured monocots. *Plant Physiol.* 54:109–115.

Cann, K. O., P. G. Stroot, K. R. Mackie, B. R. White, and R. I. Mackie. 2001. Characterization of two novel saccharolytic anaerobic thermophiles, *Thermoanaerobacterium polysaccharolyticum* sp. nov. and *Thermoanaerobacterium zeae* sp. nov., and emendation of the genus *Thermoanaerobacterium. Int. J. Syst. Bacteriol.* 51:293–302.

Carpita, N., M. McCann, and L. R. Griffing. 1996. The plant extracellular matrix: news from the cell's frontier. *Plant Cell* 8:1451–1463.

Chuvil'skya, N. A., N. P. Golovchenko, B. F. Belokopytov, and V. K. Akimenko. 1986. Isolation, identification and some physiological properties of *Clostridium thermocellum. Prikl. Biokhim. Mikrobiol.* 22:800–805.

Chyi, Y. T., and R. R. Dague. 1994. Effects of particulate size in anaerobic acidogenesis using cellulose as a sole carbon source. *Water Environ. Res.* 66:670–678.

Collins, M. D., P. A. Lawson, A. Willems, J. J. Cordoba, J. Fernandez-Garayzabal, P. Garcia, J. Cai, H. Hippe, and J. A. Farrow. 1994. The phylogeny of the genus *Clostridium*: proposal of five new genera and eleven new species combinations. *Int. J. Syst. Bacteriol.* 44:812–826.

Coughlan, M. P., and G. P. Hazlewood. 1993. beta-1,4-D-xylan-degrading enzyme systems: biochemistry, molecular biology and applications. *Biotechnol. Appl. Biochem.* 17(Pt. 3):259–289.

D'Amore, T., C. J. Panchal, I. Russell, and G. G. Stewart. 1990. A study of ethanol tolerance in yeast. *Crit. Rev. Biotechnol.* 9:287–304.

Davidson, A. L., and J. Chen. 2004. ATP-binding cassette transporters in bacteria. *Annu. Rev. Biochem.* 73:241–268.

Deanda, K., M. Zhang, C. Eddy, and S. Picataggio. 1996. Development of an arabinose-fermenting *Zymomonas mobilis* strain by metabolic pathway engineering. *Appl. Environ. Microbiol.* 62:4465–4470.

Demain, A. L., M. Newcomb, and J. H. Wu. 2005. Cellulase, clostridia, and ethanol. *Microbiol. Mol. Biol. Rev.* 69:124–154.

Derewenda, U., L. Swenson, R. Green, Y. Wei, R. Morosoli, F. Shareck, D. Kluepfel, and Z. S. Derewenda. 1994. Crystal structure, at 2.6-A resolution, of the *Streptomyces lividans* xylanase A, a member of the F family of beta-1,4-D-glycanases. *J. Biol. Chem.* 269:20811–20814.

Desai, S. G., M. L. Guerinot, and L. R. Lynd. 2004. Cloning of L-lactate dehydrogenase and elimination of lactic acid production via gene knockout in *Thermoanaerobacterium saccharolyticum* JW/SL-YS485. *Appl. Microbiol. Biotechnol.* 65:600–605.

Donaghy, J. A., K. Bronnenmeier, P. F. Soto-Kelly, and A. M. McKay. 2000. Purification and characterization of an extracellular feruloyl esterase from the thermophilic anaerobe *Clostridium stercorarium. J. Appl. Microbiol.* 88:458–466.

Dror, T. W., A. Rolider, E. A. Bayer, R. Lamed, and Y. Shoham. 2003a. Regulation of expression of scaffoldin-related genes in *Clostridium thermocellum. J. Bacteriol.* 185:5109–5116.

Dror, T. W., E. Morag, A. Rolider, E. A. Bayer, R. Lamed, and Y. Shoham. 2003b. Regulation of the cellulosomal CelS (*cel48A*) gene of *Clostridium thermocellum* is growth rate dependent. *J. Bacteriol.* 185:3042–3048.

Erbeznik, M., K. A. Dawson, and H. J. Strobel. 1998a. Cloning and characterization of transcription of the *xylAB* operon in *Thermoanaerobacter ethanolicus. J. Bacteriol.* 180:1103–1109.

Erbeznik, M., H. J. Strobel, K. A. Dawson, and C. R. Jones. 1998b. The D-xylose-binding protein, XylF, from *Thermoanaerobacter ethanolicus* 39E: cloning, molecular analysis, and expression of the structural gene. *J. Bacteriol.* 180:3570–3577.

Erbeznik, M., S. E. Hudson, A. B. Herrman, and H. J. Strobel. 2004. Molecular analysis of the xylFGH operon, coding for xylose ABC transport, in *Thermoanaerobacter ethanolicus. Curr. Microbiol.* 48:299.

Esterbauer, H., W. Steiner, I. Labudova, A. Hermann, and M. Hayn. 1991. Production of *Trichoderma* cellulase in laboratory and pilot scale. *Bioresour. Technol.* 36:51–65.

Fan, Z., J. E. McBride, W. H. van Zyl, and L. R. Lynd. 2005. Theoretical analysis of selection-based strain improvement for microorganisms with growth dependent on extracytoplasmic enzymes. *Biotechnol. Bioeng.* 92:35–44.

Fardeau, M. L., B. Ollivier, B. K. C. Patel, M. Magot, P. Thomas, A. Rimbault, F. Rocchiccioli, and J. L. Garcia. 1997. *Thermotoga hypogea* sp. nov., a xylanolytic, thermophilic bacterium from an oil-producing well. *Int. J. Syst. Bacteriol.* 47:1013–1019.

Fardeau, M. L., B. Ollivier, J. L. Garcia, and B. K. Patel. 2001. Transfer of *Thermobacteroides leptospartum* and *Clostridium thermolacticum* as *Clostridium stercorarium* subsp. *leptospartum* subsp. *thermolacticum* subsp. nov., comb. nov. and *C. stercorarium* subsp. *thermolacticum* subsp. nov., comb. nov. *Int. J. Syst. Evol. Microbiol.* 51:1127–1131.

Fernandez-Bolanos, J., B. Felizon, A. Heredia, R. Rodriguez, R. Guillen, and A. Jimenez. 2001. Steam-explosion of olive stones: hemicellulose solubilization and enhancement of enzymatic hydrolysis of cellulose. *Bioresour. Technol.* 79:61.

Garcia-Martinez, D. V., A. Shinmyo, A. Madia, and A. L. Demain. 1980. Studies on cellulase production by *Clostridium thermocellum. Eur. J. Appl. Microbiol. Biotechnol.* 9:189–197.

Ghose, P., N. B. Pamment, and W. R. B. Martin. 1982. Simultaneous saccharification and fermentation of cellulose: effect of β-D-glucose activity and ethanol inhibition of cellulases. *Enzyme Microb. Technol.* 4:425–430.

Ghose, T. K. 1987. Measurement of cellulase activities. *Pure Appl. Chem.* 59:257–268.

Ghose, T. K., and R. D. Tyagi. 1979. Rapid ethanol fermentation of cellulose hydrolysate. 2. Product and substrate inhibition and optimization of fermentor design. *Biotechnol. Bioeng. Symp.* 21:1400–1420.

Greene, N., F. E. Celik, B. Dale, M. Jackson, K. Jayawardhana, H. Jin, E. D. Larson, M. Laser, L. Lynd, and D. MacKenzie. 2004. Growing energy: how biofuels can help end America's oil dependence. National Resources Defense Council, Washington, DC. http://www.nrdc.org/air/energy/biofuels/biofuels.pdf.

Guedon, E., M. Desvaux, and H. Petitdemange. 2002. Improvement of cellulolytic properties of *Clostridium cellulolyticum* by metabolic engineering. *Appl. Environ. Microbiol.* 68:53–58.

Han, R., H. Min, M. Che, and Y. Zhao. 2002. Isolation, identification and phylogenetic analysis of a thermophilic cellulolytic anaerobic bacterium. *Wei Sheng Wu Xue Bao.* 42:138–144. (In Chinese.)

Harris, G. W., J. A. Jenkins, I. Connerton, N. Cummings, L. Lo Leggio, M. Scott, G. P. Hazlewood, J. I. Laurie, H. J. Gilbert, and R. W. Pickersgill. 1994. Structure of the catalytic core of the family F xylanase from *Pseudomonas fluorescens* and identification of the xylopentaose-binding sites. *Structure* 2:1107–1116.

Haruta, S., Z. Cui, Z. Huang, M. Li, M. Ishii, and Y. Igarashi. 2002. Construction of a stable microbial community with high cellulose-degradation ability. *Appl. Microbiol. Biotechnol.* 59:529–534.

Herrero, A. A., and R. F. Gomez. 1980. Development of ethanol tolerance in *Clostridium thermocellum*: effect of growth temperature. *Appl. Environ. Microbiol.* 40:571–577.

Herrero, A. A., R. F. Gomez, and M. F. Roberts. 1982. Ethanol induced changes in the membrane lipid composition of *C. thermocellum*. *Biochim. Biophys. Acta* 693:195–204.

Herrero, A. A., R. F. Gomez, and M. F. Roberts. 1985. 31P NMR studies of *Clostridium thermocellum*. Mechanism of end product inhibition by ethanol. *J. Biol. Chem.* 260:7442–7451.

Ho, N. W. Y., Z. Chen, A. Brainard, and M. Sedlak. 1999. Successful design and development of genetically engineered *Saccharomyces* yeasts for effective cofermentation of glucose and xylose from cellulosic biomass to fuel ethanol. *Adv. Biochem. Eng. Biotechnol.* 65:164–192.

Huang, C., B. K. Patel, R. A. Mah, and L. Baresi. 1998. *Caldicellulosiruptor owensensis* sp. nov., an anaerobic, extremely thermophilic, xylanolytic bacterium. *Int. J. Syst. Bacteriol.* 48:91–97.

Ingram, L. O. 1990. Ethanol tolerance in bacteria. *Crit. Rev. Biotechnol.* 9:305–319.

Ingram, L. O., P. F. Gomez, X. Lai, M. Moniruzzaman, B. E. Wood, L. P. Yomano, and S. W. York. 1998. Metabolic engineering of bacteria for ethanol production. *Biotechnol. Bioeng.* 58:204–214.

Izydorczyk, M. S., and C. G. Biliaderis. 1995. Cereal arabinoxylans: advances in structure and physicochemical properties. *Carbohydr. Polymers* 28:48.

Jeffries, T. W. 1996. Biochemistry and genetics of microbial xylanases. *Curr. Opin. Biotechnol.* 7:337–342.

Jeffries, T. W., and Y. S. Jin. 2000. Ethanol and thermotolerance in the bioconversion of xylose by yeasts. *Adv. Appl. Microbiol.* 47:221–268.

Jenkins, J., L. Lo Leggio, G. Harris, and R. Pickersgill. 1995. [beta]-Glucosidase, [beta]-galactosidase, family A cellulases, family F xylanases and two barley glycanases form a superfamily of enzymes with 8-fold [beta]/[alpha] architecture and with two conserved glutamates near the carboxy-terminal ends of [beta]-strands four and seven. *FEBS Lett.* 362:285.

Jin, F., K. Yamasato, and K. Toda. 1988. *Clostridium thermocopriae* sp. nov., a cellulolytic thermophile from animal feces, compost, soil, and a hot spring in Japan. *Int. J. Syst. Bacteriol.* 38:279–281.

Jin Z., K. S. Katsumata, T. B. Lam, and K. Iiyama. 2006. Covalent linkages between cellulose and lignin in cell walls of coniferous and nonconiferous woods. *Biopolymers* 83:103–110.

Johnson, E. A. 1983. Regulation of cellulase activity and synthesis in *Clostridium thermocellum*. Ph.D. dissertation. Massachusetts Institute of Technology, Cambridge.

Johnson, E. A., F. Bouchot, and A. L. Demain. 1985. Regulation of cellulase formation in *Clostridium thermocellum*. *J. Gen. Microbiol.* 131:2303–2308.

Johnson, E. A., M. Sakajoh, G. Halliwell, A. Madia, and A. L. Demain. 1982. Saccharification of complex cellulosic substrates by the cellulase system from *Clostridium thermocellum*. *Appl. Environ. Microbiol.* 43:1125–1132.

Jones, C., M. Ray, and H. Strobel. 2002a. Cloning and transcriptional analysis of the *Thermoanaerobacter ethanolicus* strain 39E maltose ABC transport system. *Extremophiles* 6:299.

Jones, C. R., M. Ray, and H. J. Strobel. 2002b. Transcriptional analysis of the xylose ABC transport operons in the thermophilic anaerobe *Thermoanaerobacter ethanolicus*. *Curr. Microbiol.* 45:62.

Jones, C. R., M. Ray, K. A. Dawson, and H. J. Strobel. 2000. High-affinity maltose binding and transport by the thermophilic anaerobe *Thermoanaerobacter ethanolicus* 39E. *Appl. Environ. Microbiol.* 66:995-1000.

Jones, R. P. 1989. Biological principles for the effects of ethanol. *Enzyme Microb. Technol.* 11:130–153.

Karita, S., K. Nakayama, M. Goto, K. Sakka, W. J. Kim, and S. Ogawa. 2003. A novel cellulolytic, anaerobic, and thermophilic bacterium, *Moorella* sp. strain F21. *Biosci. Biotechnol. Biochem.* 67:183–185.

Kato, S., S. Haruta, Z. J. Cui, M. Ishii, A. Yokota, and Y. Igarashi. 2004. *Clostridium straminisolvens* sp. nov., a moderately thermophilic, aerotolerant and cellulolytic bacterium isolated from a cellulose-degrading bacterial community. *Int. J. Syst. Evol. Microbiol.* 54:2043–2047.

Kim, Y., L. O. Ingram, and K. T. Shanmugam. 2007. Construction of an *Escherichia coli* K-12 mutant for homoethanologenic fermentation of glucose or xylose without foreign genes. *Appl. Environ. Microbiol.* 73:1766–1771.

Klapatch, T. R., D. A. Hogsett, S. Baskaran, and S. Pal. 1994. Organism development and characterization for ethanol production using thermophilic bacteria. *Appl. Biochem. Biotechnol.* 45/46:209–223.

Klapatch, T. R., M. L. Guerinot, and L. R. Lynd. 1996. Electrotransformation of *Clostridium thermosaccharolyticum*. *J. Ind. Microbiol.* 16:342–347.

Klaushofer, H., and E. Parkkinen. 1965. Zur Frage der Bedeutung aerober und anaerober thermophiler Sporenbildner als Infektionsursache in Rubenzucker-Fabriken. I. *Clostridium thermohydrosulfuricum*, eine neue Art eines saccharoseabbauenden, thermophilen, schwefelwasserstoff bildenden Clostridiums. *Zuckerind* 15:445–449.

Klinke, H. B., A. B. Thomsen, and B. K. Ahring. 2004. Inhibition of ethanol-producing yeast and bacteria by degradation products produced during pre-treatment of biomass. *Appl. Microbiol. Biotechnol.* 66:10–26.

Kormelink, F. J. M., H. Gruppen, and A. G. J. Voragen. 1993. Mode of action of (1-->4)-beta-D-arabinoxylan arabinofuranohydrolase (AXH) and alpha-L-arabinofuranosidases on alkali-extractable wheat-flour arabinoxylan. *Carbohydr. Res.* 249:353.

Kosugi, A., K. Murashima, and R. H. Doi. 2002. Xylanase and acetyl xylan esterase activities of XynA, a key subunit of the *Clostridium cellulovorans* cellulosome for xylan degradation. *Appl. Environ. Microbiol.* 68:6399–6402.

Kuyper, M., A. A. Winkler, J. P. van Dijken, and J. T. Pronk. 2004. Minimal metabolic engineering of *Saccharomyces cerevisiae* for efficient anaerobic xylose fermentation: a proof of principle. *FEMS Yeast Res.* 4:655–664.

Kuyper, M., M. J. Toirkens, J. A. Diderich, A. A. Winkler, J. P. van Dijken, and J. T. Pronk. 2005. Evolutionary engineering of mixed-sugar utilization by a xylose-fermenting *Saccharomyces cerevisiae* strain. *FEMS Yeast Res.* 5:925–934.

Larsen, L., P. Nielsen, and B. K. Ahring. 1997. *Thermoanaerobacter mathranii* sp. nov., an ethanol-producing, extremely thermophilic anaerobic bacterium from a Hot Spring in Iceland. *Arch. Microbiol.* **168:**114–119.

Laureano-Perez, L., F. Teymouri, H. Alizadeh, and B. E. Dale. 2005. Understanding factors that limit enzymatic hydrolysis of biomass: characterization of pretreated corn stover. *Appl. Biochem. Biotechnol.* **121–124:**1081–1099.

Le, R. P., H. C. Dubourguier, G. Albagnac, and G. Prensier. 1985. Characterization of *Clostridium thermolacticum* sp.nov., a hydrolytic thermophilic anaerobe producing high amoounts of lactate. *Syst. Appl. Microbiol.* **6:**196–202.

Lee, J. 1997. Biological conversion of lignocellulosic biomass to ethanol. *J. Biotechnol.* **56:**1–24.

Lee, Y. E., M. K. Jain, C. Lee, S. E. Lowe, and G. J. Zeikus. 1993. Taxonomic distinction of saccharolytic thermophilic anaerobes: description of *Thermoanaerobacterium xylanolyticum* gen. nov., sp. nov., and *Thermoanaerobacterium saccharolyticum* gen. nov., sp. nov.; reclassification of *Thermoanaerobium brockii*, *Clostridium thermosulfurogenes*, and *Clostridium thermohydrosulfuricum* E100–69 as *Thermoanaerobacter brockii* comb. nov., *Thermoanaerobacterium thermosulfurigenes* comb. nov., and *Thermoanaerobacter thermohydrosulfuricus* comb. nov., respectively; and transfer of *Clostridium thermohydrosulfuricum* 39E to *Thermoanaerobacter ethanolicus*. *Int. J. Syst. Bacteriol.* **43:**41–51.

Lindsay, S. E., R. J. Bothast, and L. O. Ingram. 1995. Improved strains of recombinant *Escherichia coli* for ethanol production from sugar mixtures. *Appl. Microbiol. Biotechnol.* **43:**70–75.

Liu, S. Y., F. A. Rainey, H. W. Morgan, F. Mayer, and J. Wiegel. 1996. *Thermoanaerobacterium aotearoense* sp. nov., a slightly acidophilic, anaerobic thermophile isolated from various hot springs in New Zealand, and emendation of the genus *Thermoanaerobacterium*. *Int. J. Syst. Bacteriol.* **46:**388–396.

Lovitt, R. W., G.-J. Shen, and J. G. Zeikus. 1988. Ethanol production by thermophilic bacteria: biochemical basis for ethanol and hydrogen tolerance in *Clostridium thermohydrosulfuricum*. *J. Bacteriol.* **170:**2809–2815.

Lovitt, R. W., R. Longin, and J. G. Zeikus. 1984. Ethanol production by thermophilic bacteria: physiological comparison of solvent effects on parent and alcohol-tolerant strains of *Clostridium thermohydrosulfuricum*. *Appl. Environ. Microbiol.* **48:**171–177.

Lu, Y. P., Y. H. P. Zhang, and L. R. Lynd. 2006. Enzyme-microbe synergy during cellulose hydrolysis by *Clostridium thermocellum*. *Proc. Natl. Acad. Sci. USA* **103:**16165–16169.

Lynd, L. R. 1989. Ethanol production from lignocellulosic substrates using thermophilic bacteria: critical evaluation of potential and review. *Adv. Biochem. Eng. Biotechnol.* **38:**1–52.

Lynd, L. R., H. J. Ahn, G. Anderson, P. Hill, S. D. Kersey, and T. Klapatch. 1991. Thermophilic ethanol production: investigation of ethanol yield and tolerance in continuous culture. *Appl. Biochem. Biotechnol.* **28/29:**549–570.

Lynd, L. R., M. S. Laser, D. Bransby, B. E. Dale, B. Davison, R. Hamilton, M. Himmel, M. Keller, J. D. McMillan, J. Sheehan, and C. E. Wyman. 2008. How biotech can transform biofuels. *Nat. Biotechnol.* **26:**169–172.

Lynd, L. R., W. H. van Zyl, J. E. McBride, and M. Laser. 2005. Consolidated bioprocessing of cellulosic biomass: an update. *Curr. Opin. Biotechnol.* **16:**577–583.

Lynd, L. R., P. J. Weimer, W. H. van Zyl, and I. S. Pretorius. 2002. Microbial cellulose utilization: fundamentals and biotechnology. *Microbiol. Mol. Biol. Rev.* **66:**506–577.

Lynd, L. R., P. J. Weimer, G. Wolfaardt, and Y. H. P. Zhang. 2007. Cellulose hydrolysis by *Clostridium thermocellum*: a microbial perspective, p. 95–117. *In* V. Uversky and I. A. Kataeva (ed.), *Molecular Anatomy and Physiology of Proteins Cellulosome.* Nova Science Publishers Inc., Hauppage, NY.

Lynd, L. R., and Y. H. Zhang. 2002. Quantitative determination of cellulase concentration as distinct from cell concentration in studies of microbial cellulose utilization: analytical framework and methodological approach. *Biotechnol. Bioeng.* 77:467–475.

Madden, R. H. 1983. Isolation and characterization of *Clostridium stercorarium* sp. nov., cellulolytic thermophile. *Int. J. Syst. Bacteriol.* 33:837–840.

Mai, V., and J. Wiegel. 2000. Advances in development of a genetic system for *Thermoanaerobacterium* spp.: expression of genes encoding hydrolytic enzymes, development of a second shuttle vector, and integration of genes into the chromosome. *Appl. Environ. Microbiol.* **66:**4817–4821.

Mai, V., W. W. Lorenz, and J. Wiegel. 1997. Transformation of *Thermoanaerobacterium* sp. strain JW/SL-YS485 with plasmid pIKM1 conferring kanamycin resistance. *FEMS Microbiol. Lett.* **148:**163–167.

Matthews, J. F., C. E. Skopec, P. E. Mason, P. Zuccato, R. W. Torget, J. Sugiyama, M. E. Himmel, and J. W. Brady. 2006. Computer simulation studies of microcrystalline cellulose I beta. *Carbohydr. Res.* **341:**138–152.

McBee, R. H. 1950. The anaerobic thermophilic cellulolytic bacteria. *Bacteriol. Rev.* **14:**51–63.

McBee, R. H. 1954. The characteristics of *Clostridium thermocellum*. *J. Bacteriol.* **67:**505–506.

McMillan, J. D. 1993. Pretreatment of lignocellulosic biomass, p. 292–323. *In* M. E. Himmel, J. O. Baker, and R. P. Overend (ed.), *Enzymatic Conversion of Biomass for Fuel Production.* American Chemical Society, Washington, DC.

Miao, S., L. Ziser, R. Aebersold, and S. G. Withers. 1994. Identification of glutamic acid 78 as the active site nucleophile in *Bacillus subtilis* xylanase using electrospray tandem mass spectrometry. *Biochemistry* **33:**7027–7032.

Millar, D. G., K. Griffiths-Smith, E. Algar, and R. K. Scopes. 1982. Activity and stability of glycolytic enzymes in the presence of ethanol. *Biotechnol. Lett.* **4:**601–606.

Miranda-Tello, E., M. L. Fardeau, P. Thomas, F. Ramirez, L. Cassalot, J. L. Cayol, J. L. Garcia, and B. Ollivier. 2004. *Pertoga mexicana* sp. nov., a novel thermophilic anaerobic and xylanolytic bacterium isolated from an oil-producing well in the Gulf of Mexico. *Int. J. Syst. Evol. Microbiol.* **54:**169–174.

Miron, J., D. Ben-Ghedalia, and M. Morrison. 2001. Adhesion mechanisms of rumen cellulolytic bacteria. *J. Dairy Sci.* **84:**1294–1309.

Mishra, S., P. Beguin, and J. P. Aubert. 1991. Transcription of *Clostridium thermocellum* endoglucanase genes *celF* and *celD*. *J. Bacteriol.* **173:**80–85.

Mitchell, W. J. 1998. Physiology of carbohydrate to solvent conversion by clostridia. *Adv. Microb. Physiol.* **39:**31–130.

Mladenovska, Z., and I. Mathrani. 1995. Isolation and characterization of *Caldicellulosiruptor lactoaceticus* sp. nov., an extremely thermophilic, cellulolytic, anaerobic bacterium. *Arch. Microbiol.* **163:**223–230.

Morag, E., E. A. Bayer, and R. Lamed. 1992. Affinity digestion for the near-total recovery of purified cellulosome from *Clostridium thermocellum*. *Enzyme Microb. Technol.* **14:**289–292.

Mosier, N., C. Wyman, B. Dale, R. Elander, Y. Y. Lee, M. Holtzapple, and M. Ladisch. 2005. Features of promising technologies for pretreatment of lignocellulosic biomass. *Bioresour. Technol.* **96:**686.

Mourino, F. M., R. Akkarawongsa, and P. J. Weimer. 2001. Initial pH as a determinant of cellulose digestion rate by mixed ruminal microorganisms in vitro. *J. Dairy Sci.* **84:**848–859.

Muir, M., L. Williams, and T. Ferenci. 1985. Influence of transport energization on the growth yield of *Escherichia coli*. *J. Bacteriol.* **163:**1237–1242.

Nagy, T., K. Emami, C. M. G. A. Fontes, L. M. A. Ferreira, D. R. Humphry, and H. J. Gilbert. 2002. The membrane-bound {alpha}-

glucuronidase from *Pseudomonas cellulosa* hydrolyzes 4-O-methyl-D-glucuronoxylooligosaccharides but not 4-O-methyl-D-glucuronoxylan. *J. Bacteriol.* 184:4925–4929.

Newcomb, M., C. Y. Chen, and J. H. D. Wu. 2007. Induction of the celC operon of *Clostridium thermocellum* by laminaribiose. *Proc. Natl. Acad. Sci. USA* 104:3747–3752.

Ng, T. K., and J. G. Zeikus. 1982. Differential metabolism of cellobiose and glucose by *Clostridium thermocellum* and *Clostridium thermohydrosulfuricum. J. Bacteriol.* 150:1391–1399.

Onyenwoke, R., Y. Lee, S. Dabrowski, B. K. Ahring, and J. Wiegel. 2006. Reclassification of *Thermoanaerobium acetigenum* as *Caldicellulosiruptor acetigenus* comb. nov. and emendation of the genus description. *Int. J. Syst. Bacteriol.* 56:1391–1395.

Ozkan, M., S. G. Desai, Y. Zhang, D. M. Stevenson, J. Beane, E. A. White, M. L. Guerinot, and L. R. Lynd. 2001. Characterization of 13 newly isolated strains of anaerobic, cellulolytic, thermophilic bacteria. *J. Ind. Microbiol. Biotechnol.* 27:275–280.

Pronk, J., M. Kuyper, M. Toirkens, R. Winkler, H. van Dijken, and W. de Laat. 2005. Engineering *Saccharomyces cerevisiae* for xylose utilization. *J. Biotechnol.* 118:S86–S87.

Rainey, F. A., A. M. Donnison, P. H. Janssen, D. Saul, A. Rodrigo, P. L. Bergquist, R. M. Daniel, E. Stackebrandt, and H. W. Morgan. 1994. Description of *Caldicellulosiruptor saccharolyticus* gen. nov., sp. nov: an obligately anaerobic, extremely thermophilic, cellulolytic bacterium. *FEMS Microbiol. Lett.* 120:263–266.

Robinson, J., J. Keating, A. Boussaid, S. Mansfield, and J. Saddler. 2002. The influence of bark on the fermentation of Douglas-fir whitewood pre-hydrolysates. *Appl. Microbiol. Biotechnol.* 59:448.

Saha, B. C. 2003. Hemicellulose bioconversion. *J. Ind. Microbiol. Biotechnol.* 30:291.

Sato, K., M. Tomita, S. Yonemura, S. Goto, K. Sekine, E. Okuma, Y. Takagi, K. Hon-nami, and T. Saiki. 1993. Characterization of and ethanol hyper-production by *Clostridium thermocellum* I-1-B. *Biosci. Biotechnol. Biochem.* 57:2116–2121.

Schneider, E. 2001. ABC transporters catalyzing carbohydrate uptake. *Res. Microbiol.* 152:310.

Schwarz, W. H., K. Bronnenmeier, B. Krause, F. Lottspeich, and W. L. Staudenbauer. 1995. Debranching of arabinoxylan: properties of the thermoactive recombinant a-L-arabinofuranosidase from *Clostridium stercorarium* (ArfB). *Appl. Microbiol. Biotechnol.* 43:860.

Schwarz, W. H. 2001. The cellulosome and cellulose degradation by anaerobic bacteria. *Appl. Microbiol. Biotechnol.* 56:634–649.

Sedlak, M., and N. W. Y. Ho. 2004. Production of ethanol from cellulosic biomass hydrolysates using genetically engineered *Saccharomyces* yeast capable of cofermenting glucose and xylose. *Appl. Biochem. Biotechnol.* 113–116:403–416.

Shinmyo, A., D. V. Garcia-Martinez, and A. L. Demain. 1979. Studies on the extracellular cellulolytic enzyme complex produced by *Clostridium thermocellum. J. Appl. Biochem.* 1:202–209.

Shoham, Y., R. Lamed, and E. A. Bayer. 1999. The cellulosome concept as an efficient microbial strategy for the degradation of insoluble polysaccharides. *Trends Microbiol.* 7:275–281.

Sissons, C. H., K. R. Sharrock, R. M. Daniel, and H. W. Morgan. 1987. Isolation of cellulolytic anaerobic extreme thermophiles from New Zealand thermal sites. *Appl. Environ. Microbiol.* 53:832–838.

Sjöström, E. 1993. *Wood Chemistry: Fundamentals and Applications.* Academic Press, San Diego, CA.

Slapack, G. E., I. Russell, and G. G. Stewart. 1987. Thermophilic microbes in ethanol production. CRC Press, Inc., Boca Raton, FL.

Sleyter, U. B., and K. O. Stetter. 1986. *Thermotoga maritima* sp. nov. represents a new genus of unique extremely thermophilic eubacteria growing up to 90°C. *Arch. Microbiol.* 144:324–333.

Sørensen, H. R., A. S. Meyer, and S. Pedersen. 2003. Enzymatic hydrolysis of water-soluble wheat arabinoxylan. 1. Synergy between alpha-L-arabinofuranosidases, endo-1,4-beta-xylanases, and beta-xylosidase activities. *Biotechnol. Bioeng.* 81:726–731.

Soutschek-Bauer, E., L. Hartl, and W. L. Staudenbauer. 1985. Transformation of *Clostridium thermohydrosulfuricum* DSM 586 with plasmid DNA. *Biotechnol. Lett.* 7:705–710.

Stevenson, D. M., and P. J. Weimer. 2005. Expression of 17 genes in *Clostridium thermocellum* ATCC 27405 during fermentation of cellulose or cellobiose in continuous culture. *Appl. Environ. Microbiol.* 71:4672–4678.

Strobel, H. J. 1995. Growth of the thermophilic bacterium *Clostridium thermocellum* in continuous culture. *Curr. Microbiol.* 31:214.

Strobel, H. J., F. C. Caldwell, and K. A. Dawson. 1995. Carbohydrate transport by the anaerobic thermophile *Clostridium thermocellum* LQRI. *Appl. Environ. Microbiol.* 61:4012–4015.

Sudha Rani, K., and G. Seenayya. 1999. High ethanol tolerance of new isolates of *Clostridium thermocellum* strains SS21 and SS22. *World J. Microbiol. Biotechnol.* 15:173–178.

Suh, J. H., S. G. Cho, and Y. J. Choi. 1996. Synergistic effects among endo-xylanase, {beta}-xylosidase, and {alpha}-L-arabinofuranosidase from *Bacillus stearothermophilus. J. Microbiol. Biotechnol.* 6:179–183.

Sun, Y., and J. Cheng. 2002. Hydrolysis of lignocellulosic materials for ethanol production: a review. *Bioresour. Technol.* 83:1–11.

Sunna, A., and G. Antranikian. 1997. Xylanolytic enzymes from fungi and bacteria. *Crit. Rev. Biotechnol.* 17:39–67.

Svetlichny, V. A., T. P. Svetlichnaya, N. A. Chernykh, and G. A. Zavarzin. 1990. *Anaerocellum thermophilum* gen. nov., sp. nov., an extreme thermophilic cellulolytic eubacterium isolated from the hot springs in the Valley of Geysers. *Mikrobiologiya* 59:871–879.

Thomas, J. R., M. McNeil, A. G. Darvill, and P. Albersheim. 1987. Structure of plant cell walls. XIX. Isolation and characterization of wall polysaccharides from suspension-cultured Douglas Fir cells. *Plant Physiol.* 83:659–671.

Toda, Y., T. Saiki, T. Uozumi, and T. Beppu. 1988. Isolation and characterization of a protease-producing thermophilic, anaerobic bacterium, *Thermobacteroides leptospartum* sp. nov. *Agric. Biol. Chem.* 52:1339–1344.

Torronen, A., and J. Rouvinen. 1995. Structural comparison of two major endo-1,4-xylanases from *Trichoderma reesei. Biochemistry* 34:847–856.

Tsoi, T. V., N. A. Chuvil'skaia, I. Atakishieva, T. Dzhavakhishvili, and V. K. Akimenko. 1987. *Clostridium thermocellum*—a new object of genetic studies. *Mol. Gen. Mikrobiol. Virusol.* 1987:18–23. (In Russian.)

Tsujibo, H., M. Kosaka, S. Ikenishi, T. Sato, K. Miyamoto, and Y. Inamori. 2004. Molecular characterization of a high-affinity xylobiose transporter of *Streptomyces thermoviolaceus* OPC-520 and its transcriptional regulation. *J. Bacteriol.* 186:1029–1037.

Tyurin, M. V., L. R. Lynd, and J. Wiegel. 2006. Gene transfer systems for obligately anaerobic thermophilic bacteria. *Extremophiles* 35:309–330.

Tyurin, M. V., S. G. Desai, and L. R. Lynd. 2004. Electrotransformation of *Clostridium thermocellum. Appl. Environ. Microbiol.* 70:883–890.

Underwood, S. A., S. Zhou, T. B. Causey, L. P. Yomano, K. T. Shanmugam, and L. O. Ingram. 2002. Genetic changes to optimize carbon partitioning between ethanol and biosynthesis in ethanologenic *Escherichia coli. Appl. Environ. Microbiol.* 68:6263–6272.

U.S. Department of Energy. 2004. *Biomass Program Multi-Year Technical Plan.* Biomass Program, U.S. Department of Energy, Washington, DC.

U.S. Department of Energy. 2005. *Breaking the Biological Barriers to Cellulosic Ethanol: a Joint Research Agenda.* U.S. Department of Energy, Rockville, MD. http://genomicsgtl.energy.gov/biofuels/.

van der Heide, T., and B. Poolman. 2002. ABC transporters: one, two or four extracytoplasmic substrate-binding sites? *EMBO Rep.* 3:938–943.

van Zyl, W. H., L. R. Lynd, R. den Haan, and J. E. McBride. 2007. Consolidated bioprocessing for bioethanol production using *Saccharomyces cerevisiae*. *Adv. Biochem. Eng. Biotechnol.* 108:205–235.

Walker, G. 1998. Yeast growth, p. 101–202. *In* G. Walker (ed.), *Yeast: Physiology and Biotechnology.* Wiley, New York, NY.

Wang, D. I. C., G. C. Avgerinos. I. Biocic, S. D. Fang, and H. Y. Fang. 1983. Ethanol from cellulosic biomass. *Philos. Trans. R. Soc. London* B300:323–333.

Warner, J. B., and J. S. Lolkema. 2003. CcpA-dependent carbon catabolite repression in bacteria. *Microbiol. Mol. Biol. Rev.* 67: 475–490.

Weimer, P. J. 1996. Why don't ruminal bacteria digest cellulose faster? *J. Dairy Sci.* 79:1496–1502.

White, A., S. G. Withers, N. R. Gilkes, and D. R. Rose. 1994. Crystal structure of the catalytic domain of the beta-1,4-glycanase cex from *Cellulomonas fimi*. *Biochemistry* 33:12546–12552.

Wiegel, J., and L. G. Ljungdahl. 1981. *Thermoanaerobacter ethanolicus* gen nov., spec. nov., a new, extreme thermophilic anaerobic bacterium. *Arch. Microbiol.* 128:343–348.

Williams, T. I., J. C. Combs, B. C. Lynn, and H. J. Strobel. 2007. Proteomic profile changes in membranes of ethanol-tolerant *Clostridium thermocellum*. *Appl. Microbiol. Biotechnol.* 74:422–432.

Wilson, C. A., and T. M. Wood. 1992. The anaerobic fungus *Neocallimastix frontalis*: isolation and properties of a cellulosome-type enzyme fraction with the capacity to solubilize hydrogen-bond-ordered cellulose. *Appl. Microbiol. Biotechnol.* 37:125–129.

Winker, S., and C. R. Woese. 1991. A definition of the domains *Archaea*, *Bacteria* and *Eucarya* in terms of small subunit ribosomal RNA characteristics. *Syst. Appl. Microbiol.* 14:305–310.

Withers, S. G., and R. Aebersold. 1995. Approaches to labeling and identification of active site residues in glycosidases. *Protein Sci.* 4:361–372.

Wood, B. E., and L. O. Ingram. 1992. Ethanol production from cellobiose, amorphous cellulose, and crystalline cellulose by recombinant *Klebsiella oxytoca* containing chromosomally integrated *Zymomonas mobilis* genes for ethanol production and plasmids expressing thermostable cellulase genes from *Clostridium thermocellum*. *Appl. Environ. Microbiol.* 58:2103–2110.

Wyman, C. E. 2007. What is (and is not) vital to advancing cellulosic ethanol. *Trends Biotechnol.* 25:157.

Yomano, L. P., S. W. York, and L. O. Ingram. 1998. Isolation and characterization of ethanol-tolerant mutants of *E. coli* KO11 for fuel ethanol production. *J. Ind. Microbiol. Biotechnol.* 20:132–138.

Zacchi, G., and A. Axelsson. 1989. Economic evaluation of preconcentration in production of ethanol from dilute sugar solutions. *Biotechnol. Bioeng.* 34:223–233.

Zaldivar, J., J. Nielsen, and L. Olsson. 2001. Fuel ethanol production from lignocellulose: a challenge for metabolic engineering and process integration. *Appl. Microbiol. Biotechnol.* 56:17–34.

Zeikus, G. J., P. W. Hegge, and M. A. Anderson. 1979. *Thermoanaerobium brockii* gen. nov. and sp. nov., a new chemoorganotrophic, caldoactive, anaerobic bacterium. *Arch. Microbiol.* 122:41–48.

Zhang, M., C. Eddy, K. Deanda, M. Finkelstein, and S. Picataggio. 1995. Metabolic engineering of a pentose metabolism pathway in ethanologenic *Zymomonas mobilis*. *Science* 267:240–243.

Zhang, Y. H., and L. R. Lynd. 2005a. Regulation of cellulase synthesis in batch and continuous cultures of *Clostridium thermocellum*. *J. Bacteriol.* 187:99–106.

Zhang, Y. H., and L. R. Lynd. 2005b. Cellulose utilization by *Clostridium thermocellum*: bioenergetics and hydrolysis product assimilation. *Proc. Natl. Acad. Sci. USA* 102:7321–7325.

Zhang, Y. H., and L. R. Lynd. 2006. A functionally based model for hydrolysis of cellulose by fungal cellulase. *Biotechnol. Bioeng.* 94:888–898.

Zhang, Y. H., and L. R. Lynd. 2003. Quantification of cell and cellulase mass concentrations during anaerobic cellulose fermentation: development of an enzyme-linked immunosorbent assay-based method with application to *Clostridium thermocellum* batch cultures. *Anal. Chem.* 75:219–227.

Zverlov, V., S. Mahr, K. Riedel, and K. Bronnenmeier. 1998. Properties and gene structure of a bifunctional cellulolytic enzyme (CelA) from the extreme thermophile *'Anaerocellum thermophilum'* with separate glycosyl hydrolase family 9 and 48 catalytic domains. *Microbiology* 144(Pt. 2):457–465.

Bioenergy
Edited by J. Wall et al.

Chapter 6

Cellulosome-Enhanced Conversion of Biomass: On the Road to Bioethanol

EDWARD A. BAYER, YUVAL SHOHAM, AND RAPHAEL LAMED

In early studies (Leatherwood, 1969), it was reported that effective cellulose digestion by colonies of *Ruminococcus albus* was dependent on a combination of a separate "affinity factor" with hydrolytic components. A decade later, it was proposed that components of the cellulase system in *Clostridium thermocellum* may form aggregates which interfere with the purification of individual enzymes (Ait et al., 1979). In contrast to the fungal cellulases, the isolation of individual cellulases from this organism and related anaerobic bacteria proved problematic. *C. thermocellum* had indeed been recognized as a particularly efficient cellulose-degrading bacterium, but its cellulolytic enzymes remained a mystery. Consequently, the study of fungal cellulase systems, such as *Trichoderma reesei*, prevailed (Reese et al., 1950; Mandels and Reese, 1957; Reese, 1975, 1976a, 1976b; Reese and Mandels, 1980).

The efficient degradation of cellulose by *C. thermocellum* was brought to the attention of the scientific community through a series of pivotal articles by Demain and coworkers (Shinmyo et al., 1979; Garcia-Martinez et al., 1980; Johnson et al., 1982a, 1982b; Duong et al., 1983; Johnson and Demain, 1984), who compared the unique specific activity of its cellulase system with the best-known enzyme system of the fungus, *Trichoderma viride* (later renamed *T. reesei* after Elwyn T. Reese). These authors reported much lower (~50-fold) levels of secreted proteins in *C. thermocellum* that generated levels of activity on microcrystalline cellulose (Avicel) similar to those of the fungal system (Johnson et al., 1982b). The *C. thermocellum* cellulase system was highly active on recalcitrant forms of cellulose compared to the amorphous forms, and the authors introduced the term "true cellulase activity" to distinguish between the two (Johnson et al., 1982b). Several other fundamental properties of the *C. thermocellum* cellulase system were established in early work by Demain and colleagues, including the dependence on calcium ions and thiols (Johnson and Demain, 1984) and the strong inhibition by cellobiose when acting on microcrystalline cellulose (Shinmyo et al., 1979; Johnson et al., 1982a). These studies were corroborated by the research of Zeikus and colleagues (Ng et al., 1977; Weimer and Zeikus, 1977; Lamed and Zeikus, 1980; Ng and Zeikus, 1981, 1982). The advantages of the cellulase system of *C. thermocellum* have been recently documented in a review on the enhanced potential of cellulosomes to counteract the inherent barriers of cellulose recalcitrance (Bayer et al., in press). Numerous reviews on various aspects of the cellulosome have been authored by many different laboratories during the past 2 decades (Lamed et al., 1983a, 1983b; Lamed and Bayer, 1988a, 1988b, 1993; Bayer and Lamed, 1992, 2006; Felix and Ljungdahl, 1993; Wu, 1993; Bayer et al., 1994, 1995, 1996, 1998a, 1998b, 1998c, 1999, 2000, 2001, 2007; Béguin and Lemaire, 1996; Karita et al., 1997; Béguin and Alzari, 1998; Béguin et al., 1999; Shoham et al., 1999; Morrison and Miron, 2000; Doi and Tamura, 2001; Schwarz, 2001; Doi et al., 2003, 2004; Belaich et al., 2004; Doi and Kosugi, 2004; Demain et al., 2005; Desvaux, 2005; Gilbert, 2007).

THE *CLOSTRIDIUM THERMOCELLUM* PROTOTYPE

The cellulosome was first isolated on the basis of the cellulose-binding function of the anaerobic thermophilic bacterium *C. thermocellum* (Bayer et al., 1983). A very large multisubunit supramolecular complex was isolated, rather than a small protein (Lamed et al.,

Edward A. Bayer • The Maynard I. and Elaine Wishner Chair of Bio-organic Chemistry, Department of Biological Chemistry, Weizmann Institute of Science, Rehovot 76100, Israel. **Yuval Shoham** • Department of Biotechnology and Food Engineering, and Institute of Catalysis Science and Technology, Technion—Israel Institute of Technology, Haifa 32000 Israel. **Raphael Lamed** • Department of Molecular Microbiology and Biotechnology, Tel Aviv University, Ramat Aviv 69978, Israel.

1983a, 1983b), and a combination of biochemical, biophysical, immunochemical, and ultrastructural techniques led to the definition of the cellulosome concept. The discrete, multienzyme cellulosome complex was thus defined as a highly organized multicellulase complex, which appeared to be responsible both for cell adhesion to the substrate and for economy in enzyme secretion (Lamed et al., 1983a, 1983b). The definition was eventually resolved by the advent of molecular biology techniques and the accumulated sequences of various cellulosomal components (Hazlewood et al., 1988; Tokatlidis et al., 1991; Salamitou et al., 1992; Shoseyov et al., 1992; Fujino et al., 1993; Gerngross et al., 1993).

Early on, we compared the properties of the purified cellulosome to those reported for the crude cellulase system of *C. thermocellum* (Lamed et al., 1985). Like the crude enzyme system, the purified cellulosome exhibited the decisive "true cellulase activity" and caused extensive solubilization of crystalline cellulose substrates. This was the first instance that a single entity was shown to effect such activity. Like the crude cellulase system, cellulosome-induced degradation of cellulose was enhanced by both calcium ions and thiols and was subject to feedback inhibition by cellobiose. In short, the data suggested that the efficient degradation of crystalline cellulose by *C. thermocellum* essentially reflected the major contribution of its cellulosome.

Cellulosome architecture is defined by the selective complementary intermodular interactions among its components (Bayer and Lamed, 1992; Wu, 1993; Bayer et al., 1994). Two different families of complementary modules were discovered from the sequencing of recombinant cellulosome subunits and subsequent bioinformatics analyses. One of these types of subunits was first sequenced in *C. thermocellum* and *C. cellulovorans* (Shoseyov et al., 1992; Gerngross et al., 1993) and initially referred to as CipA (cellulosome-integrating protein A) and CbpA (cellulose-binding protein A), respectively. Eventually, this type of subunit was termed scaffoldin (Bayer et al., 1994). The scaffoldin subunit includes a series of reiterated segments, later called cohesins (Bayer et al., 1994). In parallel, various cellulosomal enzymes were sequenced that could also be divided into modules—a catalytic module and a dockerin, among others—frequently separated by distinctive linker segments. The real breakthrough came with the experimental demonstration that the cohesins bound strongly to the dockerins (Tokatlidis et al., 1991, 1993; Salamitou et al., 1992, 1994). Surprisingly, the cohesins of a given primary scaffoldin seemed to recognize the enzyme-bearing dockerins without apparent discrimination (Yaron et al., 1995; Lytle et al., 1996; Pagès et al., 1997a). The close proximity of the cellulosomal enzymatic components is thought to be one of the major factors that account for their enhanced synergistic action (Lamed et al., 1983a, 1983b; Lamed and Bayer, 1988a, 1988b; Shoham et al., 1999; Fierobe et al., 2002; Bayer et al., 2004). In contrast to the enzyme-borne dockerins, the C-terminal dockerin of the primary *C. thermocellum* scaffoldin clearly distinguishes between its own cohesins (which it failed to bind) and those of several surface-bound anchoring scaffoldins (to which it bound strongly) (Salamitou et al., 1994; Leibovitz and Béguin, 1996). The cohesin-dockerin interaction between the enzymes and the primary scaffoldin was defined as type I, and that between the C-terminal scaffoldin-borne dockerin and the cohesin(s) of the anchoring scaffoldins was termed type II. By understanding the arrangement of the different modules of the cellulosomal subunits and their various specificities, the architecture of a given cellulosome in a given organism can be determined.

In addition to the cohesin-dockerin interaction, another equally important module is included in the primary scaffoldin of *C. thermocellum* as well as in that of most other known cellulosome-producing bacteria. The cellulose-specific carbohydrate-binding module (CBM) accounts for the attachment of the cellulosome to the cellulose substrate (Tormo et al., 1996; Shimon et al., 2000). Since the cellulosome is attached to the cell surface, the CBM also serves as the molecular basis for adhesion of the entire cell to the cellulose substrate (Bayer et al., 1996). It is intriguing that such a small protein module of about 150 amino acid residues can account for such a strong attachment to the substrate. The probable explanation of course is that it is present on the cell surface in multiple copies, which together provide a very strong adhesion of the cell to its insoluble substrate.

The architecture of the *C. thermocellum* cellulosome is presented in Fig. 1. The genome of *C. thermocellum* has recently been sequenced to completion (http://genome.jgi-psf.org/draft_microbes/cloth/cloth.home.html), which now aids in the analysis of its cellulosomal components of over 70 different dockerin-containing proteins (Schwarz et al., 2004)—most, but not all, are enzymes that function in dismantling plant cell wall polysaccharides. These have been catalogued from this bacterium (Table 1). The primary CipA scaffoldin bears nine cohesins to which any of these proteins (notably enzymes) can attach (Gerngross et al., 1993). At the C terminus of the primary scaffoldin is a divergent type II dockerin (Salamitou et al., 1994) that binds selectively to one of three known anchoring scaffoldins (SdbA, Orf2p, and OlpB), each of which bears one or more type II cohesins at the N terminus and an S-layer homology (SLH) module at the C terminus (Lemaire et al., 1995, 1998; Leibovitz et al., 1997). The interconnecting linker regions of the anchoring scaffoldins are

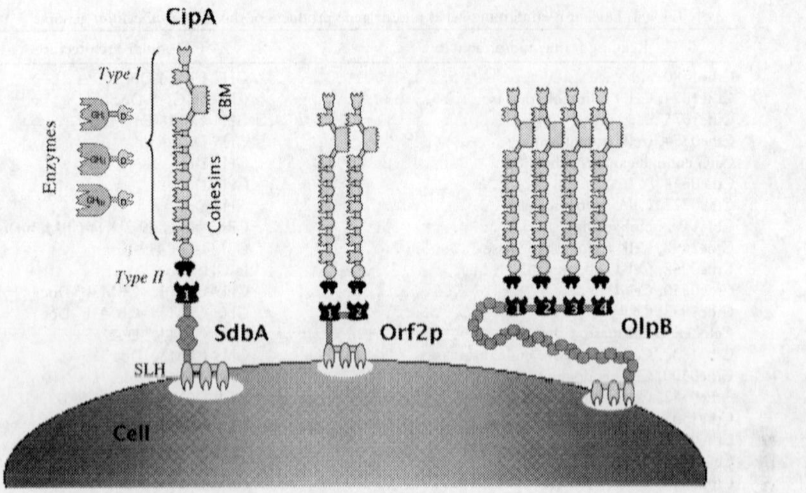

Figure 1. Cellulosome architecture of *C. thermocellum*. The cellulosomal enzymes are attached via their type I dockerin modules to the type I cohesins of the CipA scaffoldin. In turn, the CipA scaffoldin is attached, via its divergent type II dockerin, to a series of anchoring scaffoldins—SdbA, Orf2p, and OlpB—which bear one, two, and four type II cohesins, respectively. The anchoring scaffoldins carry at their C terminus an SLH module that anchors the cellulosome to the bacterial cell surface. A single CBM on the CipA scaffoldin binds the cellulosome and the entire cell to the cellulose substrate.

very different in length and amino acid content, which appear to reflect different properties and, perhaps, function (Bayer et al., 1998b). The SLH modules have been found in S-layer proteins of different types and various cell surface enzymes such as glycoside hydrolases, and they serve to anchor secreted proteins onto the bacterial cell surface (Lupas et al., 1994). They are composed of either a single or repeating subdomains, which contain highly conserved sequence patterns that may contribute to SLH function. The SLH modules may bind to the cell surface via the peptidoglycan or secondary cell wall polymers. Although biochemical information of SLH binding is available (Engelhardt and Peters, 1998; Mesnage et al., 1999, 2000; Zhao et al., 2006), its structure is still unknown.

CELLULOSOME DIVERGENCE

Over the years, many different research groups have contributed to the overall knowledge accumulated for the *C. thermocellum* cellulosome. This bacterium has thus served as the standard reference for a cellulosome-producing microorganism. An early report suggested the presence of cellulosomes in a number of different bacterial species (Lamed et al., 1987). Eventually, various research groups began studying additional cellulosome-producing bacteria, which proved to exhibit striking properties that clearly diverged from the *C. thermocellum* paradigm. Cellulosomes have been discovered and verified in *Clostridium cellulolyticum* (Belaich et al., 1997), *Clostridium cellulovorans* (Lamed et al., 1987; Doi et al., 1994), *Acetivibrio cellulolyticus* (Lamed et al., 1987; Ding et al., 1999), *Bacteroides cellulosolvens* (Lamed et al., 1991), *Ruminococcus albus* (Lamed et al., 1987; Morrison and Miron, 2000) and *Ruminococcus flavefaciens* (Kirby et al., 1997). Verification of a probable cellulosome producer is inherent in the discovery of cohesin and dockerin sequences in a cellulolytic bacterium (Bayer et al., 1998a). A compendium of the currently known scaffoldins appears in Fig. 2.

Several clostridial strains, for example, produce a scaffoldin subunit that lacks a C-terminal dockerin. It is currently unclear how the cellulosomes are connected, if at all, to the cell surface in such bacteria. Thus, *C. cellulovorans* (Shoseyov et al., 1992), *C. cellulolyticum* (Pagès et al., 1999), and *Clostridium josui* (Kakiuchi et al., 1998) all produce similar types of scaffoldins that lack the type II dockerin but exhibit different numbers of cohesins and an X module of currently undefined function (Fig. 2). A fourth scaffoldin of this type has been discovered by surprise in the genome of *Clostridium acetobutylicum* (Nolling et al., 2001). Its scaffoldin is characterized by five cohesins and numerous X modules. All four of these bacteria include numerous dockerin-containing cellulase genes which occur downstream of the scaffoldin gene, thereby comprising a cellulosome gene cluster. The cluster of *C. acetobutylicum* is especially intriguing since this bacterium has never been considered to display cellulose-degrading activity.

Table 1. Dockerin-containing genes and/or gene products of the *C. thermocellum* genome[a]

Reading frame and/or activity	Modular architecture
Cthe1580	GH2-CBM6-Doc
Cthe1674, CelO cellobiohydrolase	CBM3b-GH5-Doc
Cthe1575	GH5-CBM6-Fn3–Doc
Cthe0374, CelB endoglucanase	GH5-Doc
CelG endoglucanase, Cthe0885	GH5-Doc
Cthe0444	GH5-Doc
Cthe0722, CelA endoglucanase	GH8-Doc
CbhA cellobiohydrolase	CBM4-Ig-GH9-2(X1)-CBM3b-Doc
Cthe2598, CelK cellobiohydrolase	CBM4-Ig-GH9-Doc
Cthe0968, CelD endoglucanase	Ig-GH9-Doc
Cthe0850, Cel9U	GH9-CBM3c′-CBM3b′-Doc
Cthe1953, Cel9V	GH9-CBM3c′-CBM3b′-Doc
CelN endoglucanase, Cthe1222	GH9-CBM3c″-Doc
Cthe1837, CelR processive endoglucanase	GH9-CBM3c–Doc
Cthe0300, CelQ endoglucanase	GH9-CBM3c-Doc
Cthe0382, CelF endoglucanase	GH9-CBM3c-Doc
Cthe1308	GH9-CBM3c–Doc
Cthe0727	GH9-Doc
CelT endoglucanase	GH9-Doc
Cthe0688, XynD xylanase	CBM22-GH10–Doc
Cthe0626, XynC xylanase	CBM22-GH10-Doc
Cthe1161, XynA or XynU xylanase	GH11-CBM4-Doc-CE4
XynB, XynV xylanase	GH11-CBM4-Doc
LicB lichenase	GH16-Doc
ChiA chitinase	GH18-Doc
Cthe0533, ManA mannanase	CBM-GH26-Doc
Cthe2142	GH26-Doc
Cthe1127	GH30-CBM6-Doc
Cthe2333	GH53-Doc
Cthe0269, laminarinase	GH81-Doc
Cthe1665	GH39-2(CBM6)-Doc
Cthe1579	GH43-CBM6-Doc
Cthe0268	GH43-CBM13-Doc
Cthe0484	GH43-2(CBM6)-Doc
Cthe0939, CelS exoglucanase	GH48-Doc
Cthe2335, XghA xyloglucanhydrolase	GH74-CBM2-Doc
Cthe0066, carbohydrate esterase	Fn3-CE12-Doc-CBM6-CE12
Cthe1577, carbohydrate esterase	CE1-CBM6-Doc
Cthe2008	GH28-Doc
Cthe2236, pectate lyase	PL1-Doc-CBM6
Cthe1810, pectate lyase	Doc-CBM6-PL9
Cthe2234, pectate lyase	PL10-UNK-Doc
Cthe0702, pectate lyase	Doc-CBM6-PL11
Cthe0301, CelJ cellulase	CBM30-Ig-GH9-GH44-Doc-CBM44
Cthe0837, CelH endoglucanase	GH26-GH5-CBM9-Doc
Cthe1667	GH30-GH54-GH43-Doc
Cthe1211	GH54-Doc-GH43
Cthe1666	GH54-GH43-Doc
XynZ xylanase, Cthe1691	CE1-CBM6-Doc-GH10
XynY xylanase, Cthe2036	CBM22-GH10-CBM22-Doc-CE1
Cthe0940, Cthe2702, Cthe2514, CelE endoglucanase	GH5-Doc-CE2
Cthe1412	Fn3-Doc-serpin
Cthe1413	Fn3-Doc-serpin
Cthe0694	2(UNK)-UNK-UNK(CelP)-Doc
Cthe1578	UNK-CBM6-Doc
CseP, Cthe1223	UNK-Doc
Cthe1474	Doc-UNK
Cthe0287	UNK1-UNK2-Doc
Cthe0416	Doc-UNK
Cthe0073	UNK-Doc
Cthe0649	UNK-Doc

[a]Abbreviations: Doc, dockerin; GH*n*, family-*n* glycoside hydrolase; CE*n*, family-*n* carbohydrate esterase; PL*n*, family-*n* pectate lyase; CBM*n*, family-*n* carbohydrate-binding module; Fn3, fibronectin-3-like module; UNK, module of unknown function.

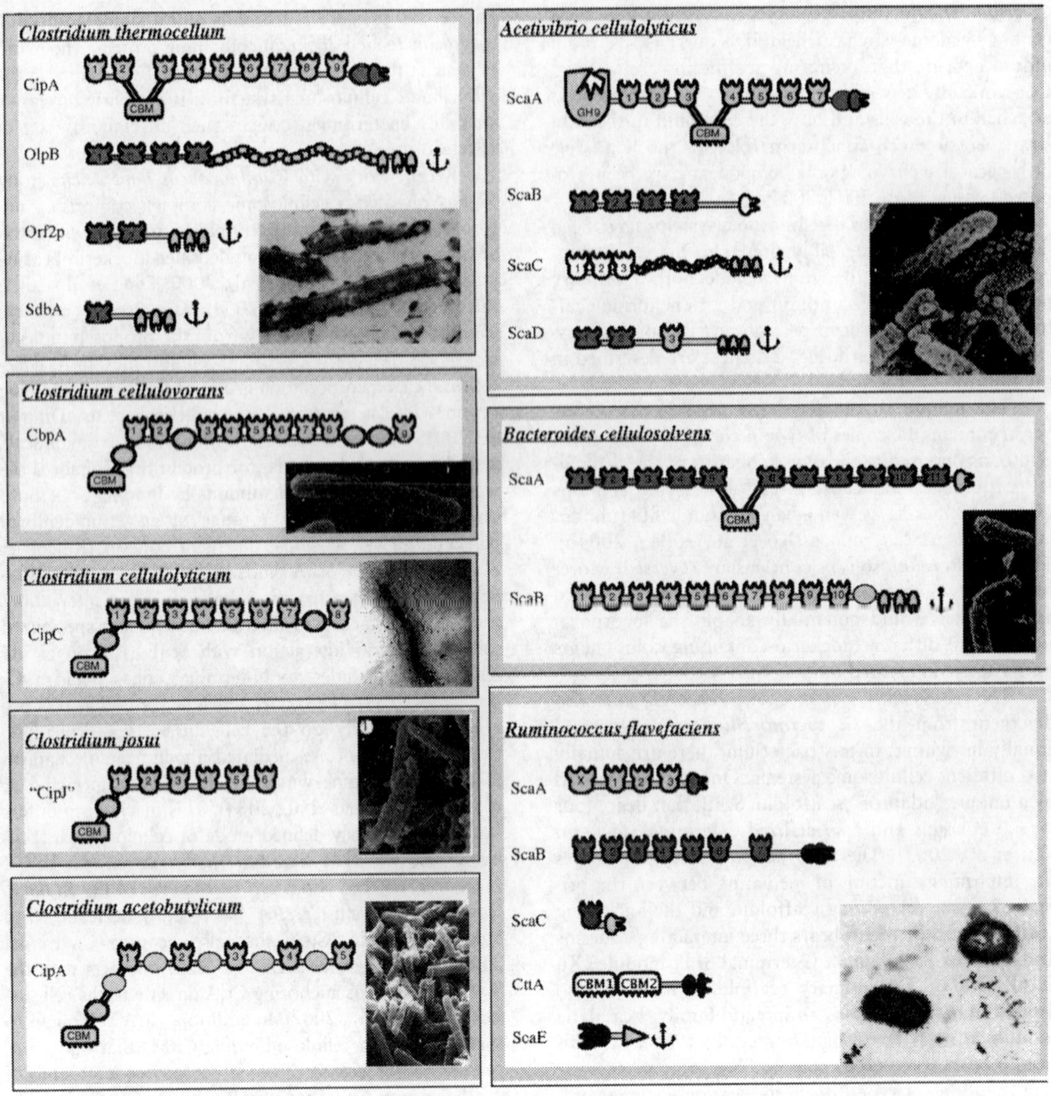

Figure 2. Scaffoldins of different cellulosome-producing bacteria. Micrographs of the bacteria are also included. The four scaffoldins of the *C. thermocellum* paradigm are shown. The type II cohesin-dockerin pairs are shown in a darker shade of gray, and the anchoring component, SLH module or sortase signal, is designated by the adjacent symbol of an anchor. The other clostridial species are characterized with a single scaffoldin. The four scaffoldins of the *A. cellulolyticus* system are more cross-interactive than that of the *C. thermocellum* paradigm (see the text for details). The reversed types of cohesin-dockerin pairings are evident in the *B. cellulosolvens* system, as are the two exceptionally large scaffoldins. The *R. flavefaciens* system is especially elaborate, with the single-cohesin ScaC "adaptor" scaffoldin providing the means with which to modify the repertoire of cellulosomal components.

Indeed, unlike *C. cellulovorans, C. cellulolyticum*, and *C. josui*, the cellulosome of *C. acetobutylicum* appears to be functionally crippled, both in the activity of its enzymes and the functionality of its cohesin-dockerin interaction (Sabathe et al., 2002; Lopez-Contreras et al., 2003, 2004; Sabathe and Soucaille, 2003). The presence of this well-defined cellulosome cluster in this bacterium remains a mystery. Defined cohesin-containing anchoring scaffoldins have yet to be described for any of these species. A draft genome has been reported for

C. cellulolyticum (http://genamics.com/cgi-bin/genamics/genomes/genomesearch.cgi?field=ID&query=437), and it indeed appears that anchoring scaffoldins are lacking. Consequently, it is still unclear how the cellulosome is attached to the cells and how the cells bind to the substrate. Possible cell-attachment roles for the X module or a particular enzyme-borne component have been proposed (Kosugi et al., 2002, 2004).

Two other bacterial cellulosome systems have been described that are ostensibly similar to the *C. thermocellum* paradigm in that they include both a primary (enzyme-integrating) scaffoldin and an anchoring scaffoldin. However, closer inspection reveals novel digressions from the standard (Fig. 2). These are described in the following paragraphs.

The unique *B. cellulosolvens* primary scaffoldin, ScaA, contains 11 copies of type II cohesins (Ding et al., 2000), as opposed to the type I cohesins of the clostridial scaffoldins. Its anchoring scaffoldin, ScaB, contains 10 type I cohesins as well as an X-module/SLH modular dyad at its C terminus (Xu et al., 2004a, 2004b). Thus, the *B. cellulosolvens* cohesins are reversed! Moreover, the two scaffoldins are the largest yet described and together would potentially enable the incorporation of 110 different dockerin-containing components into a single integrated unit.

The cellulosome system of *A. cellulolyticus* is also divergent from the *C. thermocellum* paradigm and equally intriguing. In this bacterium, there are actually two different cellulosome systems. One is characterized by a unique "adaptor" scaffoldin, ScaB, that bears four type II cohesins and a specialized C-terminal dockerin (Xu et al., 2003). The ScaB adaptor scaffoldin serves the interesting function of mediating between the primary, enzyme-integrating scaffoldin and the anchoring scaffoldin, ScaB, which bears three interacting cohesins and, like the paradigm, a C-terminal SLH module (Xu et al., 2003). The primary scaffoldin, ScaA, is also unique in that it contains an integral family 9 catalytic module at its N terminus (Ding et al., 1999). In addition, it bears seven type I cohesins that serve to incorporate the different type I dockerin-bearing enzyme components. It also carries a C-terminal type II dockerin. Thus, this intricate tri-scaffoldin system serves to further amplify the number of components that can be incorporated into a cellulosome and has the potential to incorporate 96 enzymes and related components onto the *A. cellulolyticus* surface. As mentioned above, the bacterium is characterized by a second cellulosome system, which includes a fascinating anchoring scaffoldin, ScaD, that bears, in addition to the defining SLH module at its C terminus, two different types of cohesins: two type II cohesins and a single type I cohesin (Xu et al., 2004a, 2004b). This arrangement suggests that the second *A. cellulolyticus* cellulosome system can incorporate two primary scaffoldins and a single additional enzyme or related dockerin-containing protein. The mathematics in this case indicate a total of 17 enzymes in the subordinate cellulosome system. It is currently unknown why this bacterium produces such an elaborate set of cellulosomes.

Recent work with *Ruminococcus flavefaciens* strain 17 has revealed a cellulosome complex comprising numerous cohesin-containing scaffoldins, together with interacting enzymes and other unidentified dockerin-bearing proteins (Fig. 2) (Aurilia et al., 2000; Ding et al., 2001; Rincon et al., 2001, 2003, 2004). The assembly of these components differs markedly from the proposed molecular architecture in the clostridial cellulosomes. In *R. flavefaciens,* ScaA incorporates a group of dockerin-containing enzymes into its three resident cohesin repeats (Ding et al., 2001; Rincon et al., 2003). In addition, a small ScaC scaffoldin serves as an adaptor protein that enhances the repertoire of cellulosomal subunits by binding ScaA both via its dockerin and to a range of as-yet-unidentified polypeptides via its single divergent cohesin (Rincon et al., 2004). In turn, ScaA binds to any of seven cohesin repeats of ScaB via a specific cohesin-dockerin interaction, and ScaB is attached to the cell surface via a specialized cohesin-dockerin interaction with ScaE (Rincon et al., 2005). ScaE includes an N-terminal cohesin and a C-terminal LPXTG-like motif, which suggests that it is positioned covalently on the cell surface via proteolytic cleavage and a sortase-mediated attachment mechanism (Navarre and Schneewind, 1994, 1999; Schneewind et al., 1995; Ton-That et al., 2004). This mechanism differs from the previously defined mode of cellulosome attachment, i.e., via SLH modules (Leibovitz et al., 1997). Moreover, another important component of the *R. flavefaciens sca* gene cluster, *cttA*, has recently been shown to encode a protein, CttA, that also bears a C-terminal XDoc, which, like the XDoc of ScaB, interacts with the ScaE cohesin, thus anchoring CttA directly to the cell surface (Rincon et al., 2007). In addition, CttA is thought to contain a pair of cellulose-binding CBMs that may compensate for the lack of CBMs in the known scaffoldins of this bacterium (i.e., ScaA, ScaB, and ScaC). CttA might therefore be one of the major, if not *the* major factor for binding of the bacterium to fibrous fraction in the rumen environment. The system in *R. flavefaciens*, therefore, appears much more intricate and divergent than those in the other cellulosome-producing species (Bayer et al., 1998a, 1998c, 2004).

CELLULOSOME COMPONENTS

Cellulosomes were originally thought to contain essentially cellulose-degrading enzymes, and the cellulosome was initially defined as "a discrete cell surface organelle that exhibits cellulose-binding and various cellulolytic activities." Eventually, it became clear that cellulosomes also contained noncellulolytic enzymes (Grépinet et al., 1988; Hazlewood et al., 1988; Morag et al., 1990; Zverlov et al., 1994, 2002; Halstead et al., 1999; Blum et al., 2000; Tamaru and Doi, 2001; Pagès et al., 2003). The first organism with cellulosomal components whose genome was completely sequenced was *C. acetobutylicum* (Nolling et al., 2001). However, the *C. acetobutylicum* genome provided little insight into cellulosome complexity, as its cellulosome is most likely ineffective or crippled (Sabathe et al., 2002; Sabathe and Soucaille, 2003). However, the recent sequencing of three true cellulosome-producing bacteria, *C. thermocellum* (http://genome.jgi-psf.org/draft_microbes/cloth/cloth.home.html), *R. flavefaciens* (Antonopoulos et al., 2004), and *C. cellulolyticum* (http://genome.jgi-psf.org/draft_microbes/cloce/cloce.home.html), furnished a glimpse into the utter complexity of cellulosomes from different species. The number of dockerin-containing proteins encoded by these genomes (between ~60 and ~170 depending on the bacterium) is clearly much larger than the cohesin sites of a single scaffoldin. Moreover, most of the enzymes are actually noncellulolytic and appear to specialize in the degradation of hemicelluloses and pectin. Each genome also reveals the presence of dockerin modules attached to proteins of completely unknown function.

The polysaccharide-degrading enzymes (glycoside hydrolases and polysaccharide lyases) and associated carbohydrate esterases are classified in numerous families based on amino acid sequence similarities (Henrissat, 1991; Henrissat and Bairoch, 1993; Henrissat and Romeu, 1995; Coutinho and Henrissat, 1999a, 1999b; Henrissat and Davies, 2000; Bourne and Henrissat, 2001; Coutinho et al., 2003; Henrissat et al., 2003). These are accessible online via the Carbohydrate-active enzyme database (http://afmb.cnrs-mrs.fr/CAZY/), which provides an essential database for the entire field. Preliminary analysis indicates that each of the abovementioned genomes contains only one critical cellulosomal GH48 cellobiohydrolase, with the exception of *C. thermocellum*, which contains two: one cellulosomal (dockerin bearing) and one free (CBM bearing). On the other side of the scale, we have the GH5 and GH9 endoglucanases, which appear in multiple copies. In *C. thermocellum*, the dockerin-bearing GH48 enzyme is believed to be a major and decisive component of the cellulosome, when grown on microcrystalline cellulose (Lamed et al., 1983a, 1983b; Wu et al., 1988; Morag et al., 1991; Wang et al., 1993; Ali et al., 1995; Dror et al., 2003a, 2003b; Zverlov et al., 2005); growth of the bacterium on cellobiose results in reduced amounts of this component in the *C. thermocellum* cellulosome (Bayer et al., 1985; Dror et al., 2003a, 2003b). The GH9 enzymes include several different modular themes and consequent alterations in activity patterns (Gilad et al., 2003; Arai et al., 2006; Jindou et al., 2006), among which is the capacity of the GH9-CBM3c pair to exhibit processive endoglucanase activity. The versatility in the repertoire of the cellulosomal GH9 enzymes may be a necessary and advantageous adaptation of cellulosomes. The various types of GH9 enzymes may undergo extensive regulation in response to changing product profiles, during the course of degradation of the plant cell wall. Interestingly, no member of family GH6 (candidate cellobiohydrolases) has been found in cellulosomal organisms so far, although they are common components of some aerobic cellulolytic bacteria and fungi.

A cursory look at the genes contained in the *C. thermocellum* genome that encode dockerin-containing proteins reveals a tremendously diverse set of cellulosome components that can be produced by this organism (Table 1). In *C. thermocellum*, the intricacy of cellulosome assembly is further augmented by the complexity of some of its enzymes. Some of the cellulosomal enzymes are unusually large and contain numerous modules, and several contain more than one catalytic module (Bayer et al., 1998c, 2000, 2001; Schwarz et al., 2004; Zverlov et al., 2005). The ~180-kDa *C. thermocellum* CelJ, for example, contains seven modules, including a GH9 and GH44 (Ahsan et al., 1996, 1997); and CelH (~100 kDa) bears a GH5 and a GH26. The ~140-kDa family 9 CbhA contains six modules (Zverlov et al., 1998). XynY and XynZ both contain a GH10 together with a family 1 carbohydrate esterase (a feruloyl esterase) (Fernandes et al., 1999; Blum et al., 2000; Prates et al., 2001; Tarbouriech et al., 2005).

Such bifunctional cellulosomal enzymes appear to be highly specialized for concerted cleavage of critical bonds of the native substrate (Bayer et al., in press). For example, the combination of a GH10 and feruloyl esterase in the xylanases would appear to be particularly effective in severing the covalent linkage between xylan and lignin in the plant cell wall. Why would the bacterium "choose" the solution of placing complementary activities here on the same polypeptide when it could have just incorporated them in the cellulosome? The answer to this may shed some light on the "proximity" effect in the cellulosome: if complementary enzymatic actions must occur on the *same* polymer chain, then the optimal solution is to place the enzymes on the same polypeptide. The other obvious alternative would have been to just anchor the complementary activities on the same cellulosome. If the enzyme activities have to be in the same *microscopic* space, however, then the resultant cellulosome would probably be superior for degrading that particular microenvironment of the substrate (i.e., the plant cell wall). In addition, cellulosome composition can be further adjusted and refined by the bacterium, whereas the ratio between the catalytic activities in a bifunctional polypeptide cannot. The cellulosome can thus assemble, in a much more versatile manner, a larger number of catalytic subunits than that achieved by a single polypeptide. In many cases, however, the cellulosome-producing bacterium shows an even greater degree of versatility. For example, the *C. thermocellum* genome contains three different dockerin-bearing enzymes that include a single family 43 catalytic module with one or more accessory CBMs. They also contain three other bifunctional enzymes that also bear a family 43 glycoside hydrolase, but in conjunction with a family 54 catalytic module. These three enzymes contain dockerins in different positions, and one also includes a family 30 glycoside hydrolase, thus forming a trifunctional cellulosomal enzyme!

Polysaccharide-degrading enzymes, whether hydrolases, lyases, or accessory esterases, frequently contain appended CBMs, the function of which is to target the enzyme to plant cell wall components (Tomme et al., 1995; Linder and Teeri, 1997; Boraston et al., 2004). It has been suggested that bacteria have evolved different CBMs to target different polysaccharides or even different regions of a given polysaccharide (Carrard et al., 2000). Cellulose-binding domains promote hydrolysis of different sites on crystalline cellulose (Blake et al., 2006). The number and the family membership of CBM-containing proteins in *C. thermocellum*, *R. flavefaciens*, and *C. cellulolyticum* are highly variable. Nevertheless, it is interesting that although some CBMs are common to cellulosomes and free bacterial systems (CBM3 and CBM6 for instance), the available evidence (sparse though it may be) suggests that cellulosomes appear to be completely devoid of CBM2 and CBM10 members, which are particularly abundant in "free" cellulase-producing bacteria.

REGULATION AND EXPRESSION OF CELLULOSOME COMPONENTS

The cellulose utilization systems in cellulosome-producing bacteria include over 100 different genes that must be orchestrated and timely expressed. Many lines of evidence indicate that the composition of the cellulosomes varies when the bacteria are grown on different substrates and under different conditions. However, relatively little is known about the exact mechanism and regulatory components involved in the regulation of the cellulose-utilization genes, and even less is known about how the different compositions of the cellulosome affect its ability to degrade the corresponding substrates.

Several intrinsic challenges exist when studying gene regulation in cellulosome-producing bacteria. First, until very recently, there has been a lack of the relevant genetic tools (transformation procedures, plasmids, etc.) for these bacteria. Secondly, many of the gene products cross-react immunogenically and share similar enzymatic activities, rendering expression of individual genes difficult to follow on the protein level. In addition, the various cellulolytic components can either be free in solution, attached to the cell, or attached to the insoluble substrate. Thus, quantitative assessment of the various components requires careful differentiation between the molecular species. Lastly, many of the powerful experimental approaches that are based on growing cultures in continuous or batch cultures probably do not emulate the circumstances found in nature, whereby the cells are closely associated to the insoluble cellulose matrix under conditions of passive or limited diffusion. These conditions are difficult to mimic in a normal laboratory setup. Nevertheless, much progress has been made in recent years, with the development of novel experimental approaches that aim to quantify gene expression in cellulosome-producing bacteria. Taken together with the elucidation of the genome sequences of several cellulosome-producing strains, it is likely that our understanding will advance significantly in the coming years. Recent reviews that cover some of the regulation aspects of cellulosome expression have been published (Shoham et al., 1999; Bayer et al., 2004; Doi and Kosugi, 2004; Demain et al., 2005; Desvaux, 2005; Gilbert, 2007). To date, the most studied cellulosome systems are from *C. cellulolyticum*, *C. cellulovorans*, and *C. thermocellum*.

Regulation in *Clostridium cellulolyticum*

C. cellulolyticum is a mesophilic cellulosome-producing bacterium, capable of utilizing crystalline cellulose as well as glucose and cellobiose. The major cellulosomal genes are located in a large 26-kb cluster. The scaffoldin-coding gene, *cipC*, is the first gene in the cluster, followed downstream by 11 genes coding for eight cellulases, a mannanase, and a pectinase (Pagès et al., 1996, 1997b, 1999; Belaich et al., 1997, 1999; Bayer et al., 2004). Transformation procedures have recently been worked out for this strain (Tardif et al., 2001) and allowed the use of insertion elements, complementation (Maamar et al., 2003, 2004), and antisense RNA technology (Perret et al., 2004c) for studying some regulatory aspects. Transcriptional linkage between all of the open reading frames in the cluster was demonstrated by reverse transcription-PCR; however, Northern hybridization revealed only a 14-kb messenger, starting from the 5′-end of the cluster and including the first five genes (Maamar et al., 2006). It was suggested that the primary transcript can be processed in such a way as to give several secondary messengers displaying different stabilities. When the *cipC* gene was disrupted, the ability to hydrolyze crystalline cellulose was severely impaired (Maamar et al., 2004). Moreover, mutations in the *cipC* gene appear to have strong polar effects. Recently, antisense RNA methodology was used to control the expression of the major cellulosomal cellulase, Cel48F (Perret et al., 2004c). Cultures harboring the Cel48F antisense construct exhibited lower expression of Cel48F and produced fewer cellulosomes, and their activity towards Avicel was 30% lower. Interestingly, these cultures overproduced two proteins (P105 and P98) of unknown function that are not part of the cellulosome complex but presumably represent free glycoside hydrolases. *C. cellulolyticum* was also a subject of in-depth studies of its carbon metabolism (Desvaux et al., 2000, 2001a, 2001b, 2001c, 2005; Desvaux and Petitdemange, 2001; Guedon et al., 2002; Desvaux, 2005). By combining metabolic flux analyses and continuous cultures, key steps in carbon metabolism were identified.

Regulation in *Clostridium cellulovorans*

C. cellulovorans is a cellulosome-producing anaerobic mesophilic bacterium that utilizes not only cellulose and cellodextrins but also hemicelluloses and other carbon sources (Doi et al., 2003; Doi and Kosugi, 2004). As in *C. cellulolyticum*, many of the cellulose-utilization-related genes are clustered on the chromosome. Regulation studies with *C. cellulovorans* included measurements of mRNA levels, enzymatic activities, and protein levels (Han et al., 2003a, 2003b, 2004, 2005a, 2005b). Growth on natural polymers such as cellulose, xylan, or pectin induced high expression of most genes including the cellulosome scaffolding gene, *cbpA*. The expression of the key genes was moderate when cellobiose or fructose served as the carbon source; poor with lactose, mannose, or locust bean gum; and minimal or nonexistent with glucose, galactose, maltose, or sucrose. The promoter region of several genes was analyzed by RNA ligase-mediated rapid amplification of 5′ cDNA ends by PCR and revealed strong similarity to the σ^A consensus sequences of gram-positive bacteria. The regulation of the cellulosomal genes in this bacterium appears to be coordinated, induced by some sugars, and probably under some type of carbon catabolite repression.

Regulation in *Clostridium thermocellum*

In *C. thermocellum*, there are over 100 genes that are related to cellulose utilization and the cellulosome. Based on the recently published genome sequence (http://genome.jgi-psf.org/finished_microbes/cloth/cloth.home.html), the genes appear to be scattered on the chromosome and no large clusters can be identified (Guglielmi and Béguin, 1998). We can divide the various genes into functional groups including the major cellulosomal scaffoldin, the different dockerin-bearing catalytic components, the cell surface-anchoring proteins, and the ABC sugar transporters that mediate the transport of cellodextrins into the cell.

Quite surprisingly, *C. thermocellum* appears to be able to utilize only glucose-based oligosaccharides, comprising either β-1-4-linked cellodextrins down to cellobiose or β-1-3-glucan. Although the bacterium is capable of adapting to grow on glucose, fructose, or sorbitol, this adaptation is probably a result of mutations in the ABC sugar transport systems that allow the nonspecific uptake of these sugars. Since *C. thermocellum* grows only on glucose-based compounds, it is not clear to what extent induction systems are involved in regulating the cellulose-utilization genes.

Recent regulation studies with *C. thermocellum* included the measurement of the transcriptional level of cellulose utilization genes under different growth conditions (Dror et al., 2003a, 2003b, 2005), as well as analysis of protein level by two-dimensional gel electrophoresis (Zverlov et al., 2005) or enzyme-linked immunosorbent assay-based methods (Zhang and Lynd, 2005a, 2005b). Using RNase protection assays, Dror et al. determined the number of mRNA transcripts per cell for the major family 48 cellulosomal cellobiohydrolase CelS (Cel48A) (Dror et al., 2003a, 2003b); the scaffoldin gene *cipA*; three cell surface anchoring proteins (scaffoldins) that interact with the CipA scaffoldin, OlpB, Orf2p, and SdbA (Dror et al., 2003b); three endoglucanases, family 5 CelB, family 5 CelG, and family 9 CelD; and the family 10 xylanase, XynC (Dror et al., 2005). Growth conditions used in these works included batch growth on cellulose or cellobiose and continuous culture under carbon (cellobiose) or nitrogen limitation.

In general, the transcript level of most of both the structural and enzyme-encoding genes was influenced by the growth rate (controlled by carbon or nitrogen limitation), whereas higher levels of expression were obtained at low growth rates. Independent primer extension analyses were in good agreement with RNase protection assays. For the *cipA* gene, two alternative transcriptional start sites were identified at −81 and −50 bp, upstream of the translational start site. The potential promoters displayed homology to the known sigma factors σ^A and σ^L (σ^{54}) of *Bacillus subtilis*.

In a different study, Stevenson and Weimer looked at the expression level in *C. thermocellum* of 17 genes, part of which are involved in cellulose utilization

(Stevenson and Weimer, 2005). The cultures were grown on Avicel or cellobiose in continuous culture, and the transcriptional level of these genes was quantified by real-time PCR with normalization to two reference genes, *recA* and the 16S rRNA. When growth was carried out on cellobiose-limited continuous culture, the genes *celS* and *cipA*, encoding the two major cellulosome components, exhibited a pattern similar to that observed by Dror et al. (Dror et al., 2003a, 2003b), i.e., higher expression at low growth rates. However, this trend was not observed in continuous cultures operating with cellulose as the carbon source. In this context, Zhang and Lynd (Zhang and Lynd, 2005a, 2005b) took a somewhat different approach for studying the regulation of cellulase synthesis, using an enzyme-linked immunosorbent assay, based on antibodies raised against a peptide sequence for the scaffoldin protein (CipA). In the latter work, batch and continuous cultures were grown on either Avicel or cellobiose, and the cell-specific cellulase synthesis (expressed in milligrams of cellulase per gram [dry weight] of cells) was determined. Avicel-grown batch cultures exhibited ninefold-higher cellulase synthesis than cellobiose-grown cultures. However, in continuous cultures growing on either cellobiose or Avicel, the differences in cellulase synthesis between the two substrates were much smaller. Interestingly, in the studies of both Zhang and Lynd (2005a, 2005b) and Stevenson and Weimer (2005), cellulase-related gene expression was not influenced by the dilution rate in continuous cultures operating with cellulose, compared to cellobiose-limited continuous cultures. The reason for these inconsistencies probably stems from the fact that continuous cultures operating with an insoluble substrate such as Avicel do not follow the normal steady-state equation for continuous culture. With this system, the cells are physically attached to the cellulose particles (at least in part) and therefore are probably limited by the available surface area of the cellulose particles and not really limited for the carbon source per se. Thus, their specific growth rate is not proportional or equal to the dilution rate as in the case of cellobiose-limited continuous cultures.

Very recently, Newcomb et al. (2007) identified the first cellulase gene transcriptional regulatory protein, GlyR3, of *C. thermocellum*. This protein binds specifically to an 18-bp palindrome sequence, located in the operator region of an operon encoding for *celC* and *licA*. The gene products of these genes hydrolyze β-1,3-glucan such as lichenan and laminarin. Laminaribiose appears to be the molecular inducer of this operon, as it inhibits GlyR3 binding to the promoter region as demonstrated both in electrophoretic mobility shift assay and in vitro transcription assay. It is also interesting that both CelC and LicA are noncellulosomal enzymes. The family 5 CelC represents a simple catalytic module without any ancillary modules, such as a CBM or (by definition a noncellulosomal enzyme) a dockerin. The modular architecture of CelC indicates that it acts as a free enzyme. The family 16 LicA, on the other hand, includes an N-terminal SLH module, indicating its association with the bacterial cell surface. In addition, it bears four family 4 CBMs, which presumably serve to help process β-1,3 glucan substrates in concert with the active site of the family 16 catalytic module.

All of the regulatory studies in *C. thermocellum* indicate that at high growth rates the expression of most of the studied cellulosomal genes is down-regulated. The mechanism by which this effect takes place is not known. One possible mechanism is carbon catabolite repression, CCR. In gram-positive bacteria CCR is mediated by the catabolite control protein, CcpA, which serves as a negative regulator belonging to the GalR-LacI family of repressors (Warner and Lolkema, 2003). CcpA employs as an allosteric corepressor the phosphoprotein HPr-Ser-P, which is formed upon growth of cells in the presence of glucose. The CcpA-HPr complex binds specific catabolite responsive elements, *cre*, located usually at the operator and the upstream regions of catabolite-repressed genes. Cells harboring the CcpA, HPr, and HPr kinase proteins are likely to utilize a CcpA-mediated catabolite repression mechanism. Based on the observations that at high concentrations of cellobiose specific cellulase synthesis is reduced, it has been suggested that cellulase synthesis is regulated by CCR (Zhang and Lynd, 2005a, 2005b). Support for such a mechanism comes from the finding that the genome of *C. thermocellum* contains three putative LacI/GalR transcriptional regulatory genes (GI 125972734, 125975290, and 125974990) with some homology to the *Bacillus subtilis* CcpA, an HPr protein with the conserved Ser and His residues (GI 125975217) and an HPr-kinase gene (GI 125972634). In addition, over 100 putative *cre* sequences can be found in the genome by using a degenerate *cre* consensus sequence (WGW-NANCGNTNNCW) (Zhang and Lynd, 2005a, 2005b).

In order to further study the possibility that CCR exists in *C. thermocellum*, we have recently cloned the appropriate genes and tested in vitro their functionality (Nechooshtan, 2007). The three putative key proteins, CcpA (GI 125972734), HPr serine kinase, and HPr, were prepared as His-tagged fusion proteins and were purified to homogeneity. We were able to obtain phosphorylation of HPr by the HPr serine kinase in vitro in the presence of ATP and fructose-1,6-biphosphate. However, we failed to demonstrate any binding of the putative CcpA (with or without phosphorylated Hpr) to *cre* elements (both from *C. thermocellum* and *B. subtilis*) using electrophoretic shift mobility assays. To reconcile these results, we further examined the sequence of the putative *C. thermocellum* CcpA protein using the recently published crystal structure of *Bacillus megaterium* CcpA, complexed with HPr-Ser-P and a *cre* element (Schumacher et al., 2004, 2007). The structure allowed us not only to analyze the overall homology of the *C. thermocellum* putative CcpA to known CcpAs, but also to examine crucial conserved residues that directly participate in the binding of CcpA to Hpr-S-P and the *cre* DNA. Based on the crystal structure of *B. megaterium* CcpA, there are 12 residues that closely interact with the HPr-Ser-P corepressor and 19 residues that interact with the *cre* DNA. All of these residues are almost completely conserved in all known CcpA proteins. However, the *C. thermocellum* LacI transcriptional regulator (GI 125972734, the CcpA homolog) contains only 12 of the 19 DNA-binding residues and only 2 of 12 residues that would interact with HPr-Ser-P. These results suggest that *C. thermocellum* in fact lacks a functional CcpA, and if CCR indeed exists in this bacterium, it would not operate according to the classic mechanism, i.e., via CcpA and the HPr-Ser-P corepressor, found in gram-positive bacteria. Further research into the mechanism of regulation and expression of cellulosome components in *C. thermocellum* is therefore warranted.

THE POTENTIAL OF COCULTURE FERMENTATION

C. thermocellum has the metabolic capacity of producing ethanol, molecular hydrogen, carbon dioxide, and lactic and acetic acids as final end products. Unfortunately, the bacterium is strongly inhibited at relatively low concentrations (<2% vol/vol) of ethanol (Herrero and Gomez, 1980). Nevertheless, strains that tolerate up to 8% have been selected (Herrero and Gomez, 1980; Lamed and Zeikus, 1980; Herrero et al., 1982; Jones, 1989; Rani and Seenayya, 1999). The establishment of ethanol-tolerant strains appears to have a pronounced effect on membrane composition (Williams et al., 2007) and may in fact have a deleterious effect on the production of cellulosomes and consequent degradation of cellulose.

Another problem with *C. thermocellum* in monoculture concerns its capacity via cellulosome action to produce copious quantities of cellobiose and other cellodextrins. The cellulosomal enzymes are characteristically subject to product inhibition, whereby the soluble sugars occupy the active site, thus blocking entry and hydrolysis of the intact cellulose chain. The inhibitory action is partially or tentatively alleviated by cellular uptake and assimilation of the cellulose degradation products, but when excess soluble sugars accumulate, cellulose digestion comes to a halt.

An additional consideration is the product pattern of *C. thermocellum* when grown on more complex substrates, such as the plant cell wall (its natural substrate). In addition to cellulose, this bacterium has the capacity to degrade most or all of the other plant cell wall polysaccharides, i.e., the hemicelluloses and pectins. Previously sequenced dockerin-containing genes and the now-sequenced genome of this bacterium have revealed the presence of numerous different xylanases, mannanases, lichenases, and laminarinases, a xyloglucan hydrolase, arabinofuranosidases, several carbohydrate esterases, and pectate lyases. But the bacterium itself cannot utilize or assimilate the resultant end products, since it lacks key enzymes in the appropriate metabolic pathways.

Another critical problem is the accumulation of the acid products, i.e., acetate and lactate. If our final goal is to produce products appropriate for the biofuels initiative, then clearly acetate and lactate are useless for our needs.

The concept of directly converting biomass to ethanol by a mixed clostridial fermentation was fashionable some 30 years ago when it was found that the product pattern of *C. thermocellum* in favor of ethanol could become almost quantitative in stable coculture with another ethanol-producing anaerobe. It was speculated that *C. thermocellum* was sharing the cellobiose or other cellooligodextrins with the other saccharolytic-ethanologenic microbe, which resulted in a faster growth rate and productivity. *Clostridium thermosaccharolyticum* or *Clostridium thermohydrosulfuricum* strains were the initial chosen partners in different research institutes (Dartmouth, MIT, University of Wisconsin, University of Georgia, etc.). Moreover, the saccharolytic strains would be capable of utilizing the hemicellulose degradation products.

The use of thermophilic microorganisms for biotechnological purposes has been recognized for decades (Lamed et al., 1988; Lamed and Bayer, 1991). The concept of thermophilic ethanol fermentation is very simple: fermentation under conditions of high temperature (60 to 70°C) would eliminate or reduce the need for power-consuming refrigeration and aeration of large reactor vessels. The same approach would facilitate concomitant removal of the volatile ethanol by evaporation or distillation. Low ethanol concentrations would thus ensue, thereby obviating the need for an organism like yeast that would be resistant to very high ethanol concentrations. In this context, the cellulosome-producing, thermophilic anaerobe *C. thermocellum* would seem to present an ideal practical means of directly converting cellulose and hemicellulose to fuel ethanol.

Symbiotic interactions among microorganisms are widespread natural phenomena, which constitute a primary ecological axiom. In fact, it is frequently difficult to prepare a monoculture of cellulolytic bacteria from native sources (Lamed et al., 1991; Erbeznik et al., 1997). In this context, the establishment of stable, mixed cocultures among a variety of cellulolytic and noncellulolytic (saccharolytic) bacteria is a general phenomenon that readily occurs in nature. Several different factors may, in theory, dictate the mode of interaction between different bacterial strains that leads to the formation of such cocultures. It is easier, however, to postulate logical alternatives that may account for this phenomenon than to provide concrete evidence in their support.

For example, noncellulolytic bacteria, i.e., *C. thermosaccharolyticum* (Wang et al., 1983; Saddler and Chan, 1984; Venkateswaren and Demain, 1986), *C. thermohydrosulfuricum* (Ng and Zeikus, 1981; Saddler and Chan, 1984; Germain et al., 1986), *Thermoanaerobacter ethanolicus* (Wiegel et al., 1985), and *Thermoanaerobium brockii* (Lamed and Zeikus, 1980), are incapable of growing on cellulosic substrates in monoculture. In the presence of a cellulolytic bacterium, however, a competition for the hydrolysis products takes place, which leads to the coexistence of the two (or more) strains. With regards to hemicellulose degradation, the situation is even more intriguing. Although, as mentioned above, *C. thermocellum* is capable of degrading xylan, presumably to xylose and xylobiose (Zeikus et al., 1981; Slapack et al., 1987), it is incapable of metabolizing these saccharides. The fact that *C. thermocellum* can also produce other hemicellulases in addition to the numerous xylanases, e.g., mannanases, lichenases, laminarinases, arabinofuranosidases, etc., renders this bacterium an ideal choice as a general polymer degrader in anaerobic thermophilic fermentations. Although *C. thermocellum* is entirely incapable of utilizing any of the resultant sugar products other than those of cellulose, they can be readily utilized by the above-mentioned saccharolytic strains. Moreover, the same saccharolytic partner, in coculture with *C. thermocellum*, serves to purge the extracellular medium from the soluble cellulose degradation products (i.e., cellobiose and other cellodextrins), thus relieving its enzymes (notably those of the cellulosome) from inhibitory action of the cellulosic products. The noncellulolytic ethanologenic strains can be used to produce ethanol from any of the accumulated soluble sugar products produced by *C. thermocellum*.

Under laboratory conditions involving degradation of cellulose by multispecies, it is clear that the secondary saccharolytic strain is dependent in absolute terms upon the cellulolytic strain. Much less evident is the opposite question, i.e., in what sense does the cellulolytic organism benefit from the presence of the accompanying strain? In fact, starvation and death of the cellulolytic bacterium may result from the more rapid assimilation of soluble sugars by the saccharolytic strain. On the other hand, nutritional requirements of the cellulolytic bacterium may be satisfied by substances (e.g., metabolic intermediates, vitamins, cofactors, and the like), which are secreted by the noncellulolytic strain. The latter may also serve to sequester materials toxic to the polymer-degrading strain, as demonstrated for a defined mesophilic coculture (Murray et al., 1986).

The action of the exocellular protuberance-bound cellulosome may serve to delay or limit diffusional loss of the hydrolyzed sugar to the environment and/or competing bacteria. Together with the adherence and transport phenomena, this may provide *C. thermocellum* with a clear ecological advantage in initial stages of coculture fermentation. As fermentation proceeds, the extracellular enzymes would enrich the environment with diffusible sugars that would then be readily available for competing saccharolytic organisms. The accumulation of fermentation products would eventually inhibit the metabolic activity of the cellulolytic organism (Carreira et al., 1983), and the character of the fermentation end products would mainly reflect the enzymatic apparatus of the noncellulolytic organism.

In an applicative sense, the purpose of coculture fermentation is essentially to "correct" or to improve deficient or undesirable characteristics in the cellulolytic strain when grown in monoculture. Such defects include poor product pattern (low ethanol yields), insufficient utilization of cellulosic substrates, inability to metabolize hemicellulose-derived pentoses, and unacceptable nutritional requirements.

Thus, the coculture approach potentially enables the resolution of a desired product pattern by appropriate selection of a secondary bacterial strain, which exhibits metabolic pathways that complement the primary cellulolytic strain. This approach may in the long run be advantageous over direct genetic manipulations of the cellulolytic bacterial strain. Of course, a mixed approach, employing cocultures of genetically improved primary and/or secondary strains, may eventually prove to be the preferred route.

It is clear, however, that more research using additional combinations of potentially attractive cellulolytic and saccharolytic strains would be desirable. The major problem in conducting such studies is their laborious, and often unproductive, nature, which has intimidated the majority of researchers in the past several decades.

FUTURE APPLICATIONS OF CELLULOSOMES FOR BIOMASS CONVERSION

Designer Cellulosomes

Designer cellulosomes are artificial complexes, composed of different cellulases and related enzymes, fabricated from recombinant cellulosome components

(Fig. 3). In this approach, cohesin-containing scaffoldins are mixed with dockerin-containing enzyme, thus forming artificial cellulosome complexes. By adjusting the concentration of enzyme components with respect to the scaffoldin, the composition of the resultant cellulosomes can be controlled. If the same species (i.e., binding specificity) of cohesin and dockerin is used to produce the scaffoldin and enzyme components, the cellulosomes produced would presumably be heterogeneous in enzyme content. This problem can be circumvented by using many different types of cohesins and dockerins, derived from many different bacterial sources. Thus, a chimeric scaffoldin containing different species of cohesins can be prepared, and enzymes can be designed to contain the complementary dockerins. Precise incorporation of the desired enzymes can then be achieved. The designer cellulosomes can be employed as a tool both for understanding cellulosome action and for future biotechnological application, e.g., for waste management (Bayer et al., 2007) and for production of biofuels (U.S. Department of Energy, 2006; Bayer et al., in press). Such an approach was first proposed over 1 decade ago (Bayer and Lamed, 1992; Bayer et al., 1994; Ohmiya et al., 2003).

The initial experimental feasibility study of the designer cellulosome concept involved the preparation of small artificial cellulosomes, the activity of which was examined for efficient degradation of specific substrates (Fierobe et al., 2001). In that study, small chimeric scaffoldins that contained two divergent cohesins and matching dockerin-bearing enzymes were prepared. The composition and spatial arrangement of the resultant designer cellulosomes were shown to be stably maintained, and their synergistic action on recalcitrant cellulosic substrates was demonstrated.

The controlled incorporation of the enzyme components allowed us to examine more systematically the factors important for efficient cellulosome action. In the future, designer cellulosomes may eventually find use in the efficient degradation of large quantities of cellulosic substrates. More extensive studies using the designer cellulosome approach (Fierobe et al., 2002) have served to define two factors that were originally proposed to enhance deconstruction of recalcitrant forms of cellulose (Lamed et al., 1983a, 1983b). One factor involves the targeting of cellulosomes to the substrate surface by the resident scaffoldin-bearing CBM. The second factor reflects the proximity of the enzyme components within the cellulosome complex. Enhanced deconstruction of pure forms of crystalline cellulose has been shown to reflect the extent of recalcitrance of the cellulosic substrate. In the case of crude native substrates, such as wheat straw, the inclusion of a xylanase into designer cellulosomes, together with potent cellulases, has

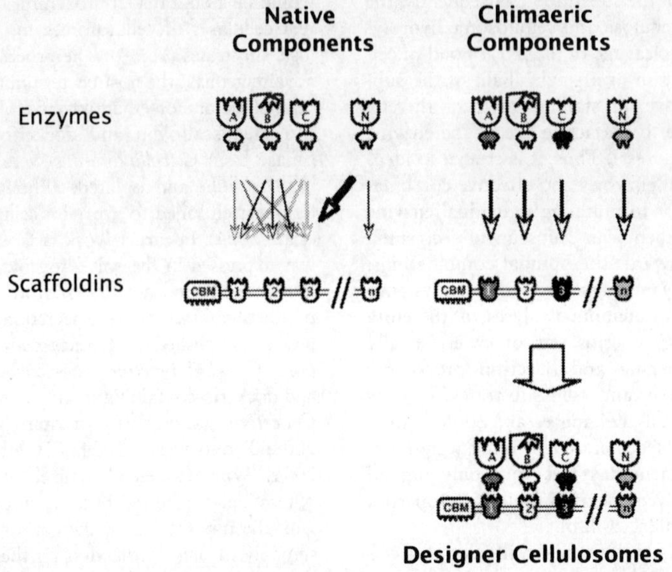

Figure 3. Schematic representation of the designer cellulosome concept. In the native complex (left) the specificity of the cohesin-dockerin interaction is usually identical. If the native components are used, we would expect random incorporation of the enzymes, resulting in a large collection of artificial cellulosomes heterogeneous in their content. In designer cellulosomes (right) composed of matching chimeric components, the specificity of the different cohesin-dockerin pairs is divergent, thus facilitating the controlled incorporation of enzyme components.

been demonstrated to dramatically improve the degradation of complex lignocellulosic substrates (Fierobe et al., 2005).

Mixed minicellulosomes have been reconstituted from recombinant cellulosomal enzymes and truncated scaffoldin components of *C. cellulovorans* (Murashima et al., 2002a, 2002b). In this case, the nonspecificity of the native cohesin-dockerin interaction prevented the controlled incorporation of the enzymes, and a heterogeneous mixture likely resulted. Nevertheless, the reconstituted cellulosomes exhibited synergistic activity on cellulosic substrates. Although the cohesin-dockerin interaction in a single bacterial scaffoldin is presumably nonspecific, the authors implied that the action of minicellulosomes on different types of celluloses and hemicelluloses indicated that the *C. cellulovorans* cohesin-dockerin interaction may be more selective than originally believed (Koukiekolo et al., 2005).

In recent studies (Hughes et al., 2006), combinatorial and robotic handling methods have been employed to try to further improve cellulase activity in individual cellulase enzymes. A similar approach can also be applied to improve the synergistic action of cellulosomal enzymatic components. However, it should be constantly borne in mind that the defining characteristic of cellulase and cellulosome action is *not* the improvement of individual cellulolytic enzymes, but how the different cellulases work together to counteract the recalcitrant properties of the substrate. Consequently, the rate-limiting step in cellulose and cellulosome hydrolysis is *not* the catalytic cleavage of the β-1,4 bond of cellulose, but the disruption of a single chain of the substrate from its native crystalline matrix, thereby rendering it accessible to the active site of the enzyme (Bayer et al., 2007, in press). Thus, it is crucial to identify efficient individual enzymes and effective combinations of enzymes, prior to mutating individual enzyme components. Only then can subsequent screening rounds be directed towards the optimal combinations. The evaluation of relevant substrates and assay systems, appropriate for high-throughput analyses, is therefore crucial for identifying superior sets of synergistically acting enzymes. Screening and selection procedures should be based on relevant solid substrates, such as paper or plant cell walls (Zhang et al., 2006). However, the application of protein engineering techniques for enzyme improvement has met with only limited success, e.g., improved thermostability properties (Murashima et al., 2002a, 2002b).

In the future, the development of the designer cellulosome approach for decomposition of recalcitrant cellulosic substrates will likely concentrate on several fronts (Bayer et al., in press), including (i) incorporation of currently available enzymes into designer cellulosomes; (ii) development of novel or improved cellulosomal components, i.e., scaffoldins, cohesins, dockerins, and CBMs; and (iii) rational design and/or directed evolution of superior types of hybrid enzymes and the evaluation of their enhanced synergistic action within designer cellulosomes. The desired result will be improved biomass degradation with concurrent increased understanding of the structure-function relationship of cellulosome components.

Cell Surface Engineering of Cellulosomes

Historically, numerous strategies have been considered for engineering microbes for efficient degradation of plant cell wall polysaccharides. The biomass-to-bioenergy initiative has prompted renewed interest and new approaches in this endeavor (U.S. Department of Energy, 2006; Ragauskas et al., 2006; Schubert, 2006; Himmel et al., 2007). In this context, naturally occurring cellulose-degrading bacteria and fungi may be engineered further to improve the profile of useful products (e.g., ethanol). In addition, components of a cellulase system can be introduced by genetic means into noncellulolytic microorganisms that produce a desired product. A preferred approach would be to combine both strategies such that both cellulolytic enzymes and a suitable product profile can be achieved in a single cell.

Heterologous expression of designer cellulosome components in a suitable industrial host cell system would be a desirable approach for overproduction of potent cellulases or cellulosomes for degradation of cellulosic substrates (Fig. 4). The process will be additionally advantageous if the host bacterium is, or can be rendered, cellulolytic and/or ethanologenic. Using this approach, a truncated scaffoldin and dockerin-containing endoglucanase from *C. cellulovorans* were recently coexpressed in *B. subtilis* and isolated, although the resultant bacterium itself failed to grow on cellulosic substrates (Cho et al., 2004). In earlier work, a *C. cellulovorans* cellulase was expressed in the solventogenic bacterium, *C. acetobutylicum* (Kim et al., 1994). More recently, heterologous production, assembly, and secretion of a minicellulosome was accomplished in *C. acetobutylicum*. For this purpose, *C. cellulolyticum* genes encoding a miniscaffoldin and dockerin-containing mannanase were cloned into the *C. acetobutylicum* host bacterium (Sabathe and Soucaille, 2003; Perret et al., 2004a, 2004b; Mingardon et al., 2005). When expressed by itself, the mannanase lost its N-terminal dockerin, but upon coexpression together with the miniscaffoldin, the enzyme was stabilized, presumably through protection by the scaffoldin-borne cohesin. Interestingly, the *C. acetobutylicum* genome itself includes a complete, but essentially inactive, cellulosome gene cluster of its own (Sabathe et al., 2002).

Other types of cellulosomal and noncellulosomal carbohydrate-active enzymes have been expressed in

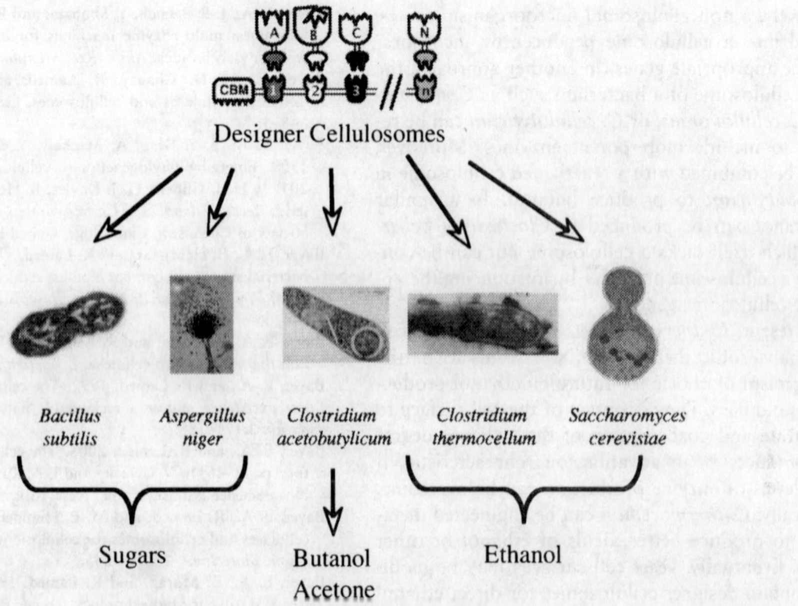

Figure 4. Engineering of potent cellulolytic microorganisms for production of biofuels. Bacterial, mold, or yeast host cell systems can be converted into a cellulosome producer by cloning appropriate genes that code for cellulases and/or designer cellulosome components (i.e., a chimeric scaffoldin and desired dockerin-containing hybrid enzymes). The resultant cellulolytic cells can be grown directly on cellulosic biomass to produce the desired end product, e.g., sugar intermediates or final biofuel.

different host cell systems. For example, *Aspergillus niger* was used as a host cell system for expression of a hybrid dockerin-containing feruloyl esterase (Levasseur et al., 2004). Similarly (Levasseur et al., 2005), a bifunctional noncellulosomal chimeric enzyme, comprising a feruloyl esterase fused to a xylanase and CBM, was also overexpressed. The expressed chimeric enzyme exhibited both xylanase and esterase activities and bound to cellulose by virtue of the CBM. The catalytic components of the bifunctional enzyme appeared to act synergistically in degrading complex substrates.

Efforts have been recently devoted to construct a whole-cell biocatalyst with multiple enzymes for synergistic and sequential degradation of cellulose. Three types of cellulolytic enzymes were thus codisplayed on the cell surface of the yeast *Saccharomyces cerevisiae* (Fujita et al., 2004). In another work (Ito et al., 2004), an endoglucanase fused with several CBMs was incorporated into yeast, which displayed enhanced binding affinity and hydrolytic activity. In addition, a xylan-fermenting yeast strain was constructed by combined surface and metabolic engineering, and the direct conversion of xylan to ethanol was demonstrated (Katahira et al., 2004). More recently, the same authors (Katahira et al., 2006) employed a recombinant xylose- and cellooligosaccharide-assimilating yeast strain to produce ethanol from acid-treated wood chip hydrolysate.

In order to streamline and economize processing strategies, future trends include the combination of key production steps, which should reduce overall process complexity and cost. Combined biomass processing technology, termed consolidated bioprocessing (CBP), has thus been proposed by Lynd and colleagues (Lynd et al., 2005), initially based on *C. thermocellum*. In this regard, efficient oligosaccharide uptake combined with intracellular phosphorolytic cleavage of β-glucosidic bonds is considered to provide bioenergetic benefits specific to growth on cellulose (Zhang and Lynd, 2005a, 2005b). These advantages are predicted to exceed the bioenergetic cost of cellulase synthesis, which would render anaerobic processing of cellulosic biomass feasible without added saccharolytic enzymes. CBP was recently extended for direct production of ethanol in yeast by cloning an endoglucanase and a β-glucosidase in *S. cerevisiae*. The recombinant strain was capable of growing on phosphoric acid-swollen cellulose (Den Haan et al., 2007).

Designer cellulosomes can be combined with any of these strategies. For example, a host cell can be altered to include separate enzymatic steps for sugar production.

Alternatively, a noncellulosomal microorganism can be converted into a cellulosome producer by incorporation of the appropriate genes. In another approach, the resident cellulosome of a bacterium such as *C. thermocellum*, *C. cellulovorans*, or *C. cellulolyticum* can be redesigned to include more-potent enzymes. Moreover, CBP can be combined with a reactivated cellulosome in *C. acetobutylicum* to produce butanol. In a similar vein, butanol can be produced in *Clostridium beijerinckii,* which itself lacks a cellulosome but can be converted to a cellulosome producer by introducing the appropriate cellulosome genes.

At present, *C. thermocellum,* as a very potent cellulolytic, anaerobic thermophile, still seems to be the microorganism of choice for future bioethanol production from biomass. Direct contact of the cell surface to the substrate and coabsorption of the cellulose degradation products is an advantageous characteristic, if the low levels of enzyme production can be overcome. Theoretically, *C. thermocellum* can be engineered metabolically to produce better yields of ethanol or other products. Eventually, yeast cell surfaces may be modified to contain designer cellulosomes for direct ethanol conversion. The combination of CBP with the designer cellulosome concept may ultimately provide optimized degradation of specific cellulosic feedstocks for bioethanol production. In any case, the study of the designer cellulosomes and genetic engineering of *C. thermocellum* are important steps necessary for evaluation and development of this approach.

REFERENCES

Ahsan, M. M., T. Kimura, S. Karita, K. Sakka, and K. Ohmiya. 1996. Cloning, DNA sequencing, and expression of the gene encoding *Clostridium thermocellum* cellulase CelJ, the largest catalytic component of the cellulosome. *J. Bacteriol.* **178:**5732–5740.

Ahsan, M. M., M. Matsumoto, S. Karita, T. Kimura, K. Sakka, and K. Ohmiya. 1997. Purification and characterization of the family J catalytic domain derived from the *Clostridium thermocellum* endoglucanase CelJ. *Biosci. Biotechnol. Biochem.* **61:**427–431.

Ait, N., N. Creuzet, and P. Forget. 1979. Partial purification of cellulase from *Clostridium thermocellum. J. Gen. Microbiol.* **113:**399–402.

Ali, B. R., M. P. Romaniec, G. P. Hazlewood, and R. B. Freedman. 1995. Characterization of the subunits in an apparently homogeneous subpopulation of *Clostridium thermocellum* cellulosomes. *Enzyme Microb. Technol.* **17:**705–711.

Antonopoulos, D. A., K. E. Nelson, M. Morrison, and B. A. White. 2004. Strain-specific genomic regions of *Ruminococcus flavefaciens* FD-1 as revealed by combinatorial random-phase genome sequencing and suppressive subtractive hybridization. *Environ. Microbiol.* **6:**335–346.

Arai, T., A. Kosugi, H. Chan, R. Koukiekolo, H. Yukawa, M. Inui, and R. H. Doi. 2006. Properties of cellulosomal family 9 cellulases from *Clostridium cellulovorans. Appl. Microbiol. Biotechnol.* **71:** 654–660.

Aurilia, V., J. C. Martin, S. I. McCrae, K. P. Scott, M. T. Rincon, and H. J. Flint. 2000. Three multidomain esterases from the cellulolytic rumen anaerobe *Ruminococcus flavefaciens* 17 that carry divergent dockerin sequences. *Microbiology* **146:**1391–1397.

Bayer, E. A., J.-P. Belaich, Y. Shoham, and R. Lamed. 2004. The cellulosomes: multi-enzyme machines for degradation of plant cell wall polysaccharides. *Annu. Rev. Microbiol.* **58:**521–554.

Bayer, E. A., H. Chanzy, R. Lamed, and Y. Shoham. 1998a. Cellulose, cellulases and cellulosomes. *Curr. Opin. Struct. Biol.* **8:** 548–557.

Bayer, E. A., S.-Y. Ding, A. Mechaly, Y. Shoham, and R. Lamed. 1999. Emerging phylogenetics of cellulosome structure, p. 189–201. *In* H. J. Gilbert, G. J. Davies, B. Henrissat, and B. Svensson (ed.), *Recent Advances in Carbohydrate Bioengineering.* The Royal Society of Chemistry, Cambridge, United Kingdom.

Bayer, E. A., B. Henrissat, and R. Lamed. The cellulosome: a natural bacterial strategy to combat biomass recalcitrance. *In* M. E. Himmel (ed.), *Biomass Recalcitrance*, in press. Blackwell, London, United Kingdom.

Bayer, E. A., R. Kenig, and R. Lamed. 1983. Adherence of *Clostridium thermocellum* to cellulose. *J. Bacteriol.* **156:**818–827.

Bayer, E. A., and R. Lamed. 1992. The cellulose paradox: pollutant *par excellence* and/or a reclaimable natural resource? *Biodegradation* **3:**171–188.

Bayer, E. A., and R. Lamed. 2006. The cellulosome saga: early history, p. 11–46. *In* V. Uversky and I. A. Kataeva (ed.), *Cellulosome.* Nova Science Publishers, Inc, New York, NY.

Bayer, E. A., R. Lamed, and M. E. Himmel. 2007. The potential of cellulases and cellulosomes for cellulosic waste management. *Curr. Opin. Biotechnol.* **18:**237–245.

Bayer, E. A., E. Morag, and R. Lamed. 1994. The cellulosome—a treasure-trove for biotechnology. *Trends Biotechnol.* **12:**378–386.

Bayer, E. A., E. Morag, R. Lamed, S. Yaron, and Y. Shoham. 1998b. Cellulosome structure: four-pronged attack using biochemistry, molecular biology, crystallography and bioinformatics, p. 39–65. *In* M. Claeyssens, W. Nerinckx, and K. Piens (ed.), *Carbohydrases from* Trichoderma reesei *and Other Microorganisms.* The Royal Society of Chemistry, London, United Kingdom.

Bayer, E. A., E. Morag, Y. Shoham, J. Tormo, and R. Lamed. 1996. The cellulosome: a cell-surface organelle for the adhesion to and degradation of cellulose, p. 155–182. *In* M. Fletcher (ed.), *Bacterial Adhesion: Molecular and Ecological Diversity.* Wiley-Liss, Inc., New York, NY.

Bayer, E. A., E. Morag, M. Wilchek, R. Lamed, S. Yaron, and Y. Shoham. 1995. Cellulosome domains for novel biotechnological application, p. 251–259. *In* S. B. Petersen, B. Svensson, and S. Pedersen (ed.), *Carbohydrate Bioengineering.* Elsevier Science B.V., Amsterdam, The Netherlands.

Bayer, E. A., E. Setter, and R. Lamed. 1985. Organization and distribution of the cellulosome in *Clostridium thermocellum. J. Bacteriol.* **163:**552–559.

Bayer, E. A., L. J. W. Shimon, R. Lamed, and Y. Shoham. 1998c. Cellulosomes: structure and ultrastructure. *J. Struct. Biol.* **124:** 221–234.

Bayer, E. A., Y. Shoham, and R. Lamed. September 2001, posting date. Cellulose-decomposing prokaryotes and their enzyme systems. *In* M. Dworkin, S. Falkow, E. Rosenberg, K.-H. Schleifer, and E. Stackebrandt (ed.), *The Prokaryotes: An Evolving Electronic Resource for the Microbiological Community*, 3rd ed. Springer-Verlag, New York, NY. http://link.springer.de/link/service/books/10125/index.htm.

Bayer, E. A., Y. Shoham, and R. Lamed. 2000. The cellulosome—an exocellular organelle for degrading plant cell wall polysaccharides, p. 387–439. *In* R. J. Doyle (ed.), *Glycomicrobiology.* Kluwer Academic/Plenum Publishers, New York, NY.

Béguin, P., and P. M. Alzari. 1998. The cellulosome of *Clostridium thermocellum. Biochem. Soc. Trans.* **26:**178–185.

Béguin, P., S. Chauvaux, G. Guglielmi, M. Matuschek, E. Leibovitz, M.-K. Chaveroche, I. Miras, P. Alzari, and P. Gounon. 1999. The *Clostridium thermocellum* cellulosome: organization and mode of attachment to the cell, p. 437-443. *In* K. Ohmiya, K. Hayashi, K. Sakka, Y. Kobayashi, S. Karita, and T. Kimura (ed.), *Genetics, Biochemistry and Ecology of Cellulose Degradation*. Uni Publishers Co., Tokyo, Japan.

Béguin, P., and M. Lemaire. 1996. The cellulosome: an exocellular, multiprotein complex specialized in cellulose degradation. *Crit. Rev. Biochem. Mol. Biol.* 31:201–236.

Belaich, J.-P., A. Belaich, H.-P. Fierobe, L. Gal, C. Gaudin, S. Pagès, C. Reverbel-Leroy, and C. Tardif. 1999. The cellulolytic system of *Clostridium cellulolyticum*, p. 479–487. *In* K. Ohmiya, K. Hayashi, K. Sakka, Y. Kobayashi, S. Karita, and T. Kimura (ed.), *Genetics, Biochemistry and Ecology of Cellulose Degradation*. Uni Publishers Co., Tokyo, Japan.

Belaich, J.-P., C. Tardif, A. Belaich, and C. Gaudin. 1997. The cellulolytic system of *Clostridium cellulolyticum*. *J. Biotechnol.* 57:3–14.

Belaich, J. P., H.-P. Fierobe, S. Pagès, A. Belaich, C. Tardif, E. A. Bayer, A. Mechaly, and O. Valette. 2004. From native to engineered cellulosomes, p. 167–174. *In* K. Ohmiya, K. Sakka, S. Karita, T. Kimura, and M. Sakka (ed.), *Genetics, Biotechnology of Lignocellulose Degradation and Biomass Utilization*. Uni Publishers Co., Ltd., Tokyo, Japan.

Blake, A. W., L. McCartney, J. E. Flint, D. N. Bolam, A. B. Boraston, H. J. Gilbert, and J. P. Knox. 2006. Understanding the biological rationale for the diversity of cellulose-directed carbohydrate-binding modules in prokaryotic enzymes. *J. Biol. Chem.* 281:29321–29329.

Blum, D. L., I. A. Kataeva, X. L. Li, and L. G. Ljungdahl. 2000. Feruloyl esterase activity of the *Clostridium thermocellum* cellulosome can be attributed to previously unknown domains of XynY and XynZ. *J. Bacteriol.* 182:1346–1351.

Boraston, A. B., D. N. Bolam, H. J. Gilbert, and G. J. Davies. 2004. Carbohydrate-binding modules: fine-tuning polysaccharide recognition. *Biochem. J.* 382:769–781.

Bourne, Y., and B. Henrissat. 2001. Glycoside hydrolases and glycosyltransferases: families and functional modules. *Curr. Opin. Struct. Biol.* 11:593–600.

Carrard, G., A. Koivula, H. Soderlund, and P. Béguin. 2000. Cellulose-binding domains promote hydrolysis of different sites on crystalline cellulose. *Proc. Natl. Acad. Sci. USA* 97:10342–10347.

Carreira, L. H., and L. G. Ljungdahl. 1983. Production of ethanol from biomass using anaerobic thermophilic bacteria, p. 1–29. *In* D. L. Wise, L. H. Carreira, and L. G. Ljungdahl (ed.), *Liquid Fuel Developments*. CRC Press, Inc., Boca Raton, FL.

Cho, H. Y., H. Yukawa, M. Inui, R. H. Doi, and S. L. Wong. 2004. Production of minicellulosomes from *Clostridium cellulovorans* in *Bacillus subtilis* WB800. *Appl. Environ. Microbiol.* 70:5704–5707.

Coutinho, P. M., E. Deleury, and B. Henrissat. 2003. The families of carbohydrate-active enzymes in the genomic era. *J. Appl. Glycosci.* 50:241–244.

Coutinho, P. M., and B. Henrissat. 1999a. Carbohydrate-active enzymes: an integrated database approach, p. 3–12. *In* H. J. Gilbert, G. J. Davies, B. Henrissat, and B. Svensson (ed.), *Recent Advances in Carbohydrate Bioengineering*. The Royal Society of Chemistry, Cambridge, United Kingdom.

Coutinho, P. M., and B. Henrissat. 1999b. The modular structure of cellulases and other carbohydrate-active enzymes: an integrated database approach, p. 15–23. *In* K. Ohmiya, K. Hayashi, K. Sakka, Y. Kobayashi, S. Karita, and T. Kimura (ed.), *Genetics, Biochemistry and Ecology of Cellulose Degradation*. Uni Publishers Co., Tokyo, Japan.

Demain, A. L., M. Newcomb, and J. H. Wu. 2005. Cellulase, clostridia, and ethanol. *Microbiol. Mol. Biol. Rev.* 69:124–154.

Den Haan, R., S. H. Rose, L. R. Lynd, and W. H. van Zyl. 2007. Hydrolysis and fermentation of amorphous cellulose by recombinant *Saccharomyces cerevisiae*. *Metab. Eng.* 9:87–94.

Desvaux, M. 2005. *Clostridium cellulolyticum*: model organism of mesophilic cellulolytic clostridia. *FEMS Microbiol. Rev.* 29:741–764.

Desvaux, M., E. Guedon, and H. Petitdemange. 2001a. Carbon flux distribution and kinetics of cellulose fermentation in steady-state continuous cultures of *Clostridium cellulolyticum* on a chemically defined medium. *J. Bacteriol.* 183:119–130.

Desvaux, M., E. Guedon, and H. Petitdemange. 2000. Cellulose catabolism by *Clostridium cellulolyticum* growing in batch culture on defined medium. *Appl. Environ. Microbiol.* 66:2461–2470.

Desvaux, M., E. Guedon, and H. Petitdemange. 2001b. Kinetics and metabolism of cellulose degradation at high substrate concentrations in steady-state continuous cultures of *Clostridium cellulolyticum* on a chemically defined medium. *Appl. Environ. Microbiol.* 67:3837–3845.

Desvaux, M., E. Guedon, and H. Petitdemange. 2001c. Metabolic flux in cellulose batch and cellulose-fed continuous cultures of *Clostridium cellulolyticum* in response to acidic environment. *Microbiology* 147:1461–1471.

Desvaux, M., A. Khan, A. Scott-Tucker, R. R. Chaudhuri, M. J. Pallen, and I. R. Henderson. 2005. Genomic analysis of the protein secretion systems in *Clostridium acetobutylicum* ATCC 824. *Biochim. Biophys. Acta* 1745:223–253.

Desvaux, M., and H. Petitdemange. 2001. Flux analysis of the metabolism of *Clostridium cellulolyticum* grown in cellulose-fed continuous culture on a chemically defined medium under ammonium-limited conditions. *Appl. Environ. Microbiol.* 67:3846–3851.

Ding, S.-Y., E. A. Bayer, D. Steiner, Y. Shoham, and R. Lamed. 1999. A novel cellulosomal scaffoldin from *Acetivibrio cellulolyticus* that contains a family-9 glycosyl hydrolase. *J. Bacteriol.* 181:6720–6729.

Ding, S.-Y., E. A. Bayer, D. Steiner, Y. Shoham, and R. Lamed. 2000. A scaffoldin of the *Bacteroides cellulosolvens* cellulosome that contains 11 type II cohesins. *J. Bacteriol.* 182:4915–4925.

Ding, S.-Y., M. T. Rincon, R. Lamed, J. C. Martin, S. I. McCrae, V. Aurilia, Y. Shoham, E. A. Bayer, and H. J. Flint. 2001. Cellulosomal scaffoldin-like proteins from *Ruminococcus flavefaciens*. *J. Bacteriol.* 183:1945–1953.

Doi, R. H., M. Goldstein, S. Hashida, J. S. Park, and M. Takagi. 1994. The *Clostridium cellulovorans* cellulosome. *Crit. Rev. Microbiol.* 20:87–93.

Doi, R. H., and A. Kosugi. 2004. Cellulosomes: plant-cell-wall-degrading enzyme complexes. *Nat. Rev. Microbiol.* 2:541–551.

Doi, R. H., A. Kosugi, K. Murashima, Y. Tamaru, and S. O. Han. 2003. Cellulosomes from mesophilic bacteria. *J. Bacteriol.* 185:5907–5914.

Doi, R. H., and Y. Tamura. 2001. The *Clostridium cellulovorans* cellulosome: an enzyme complex with plant cell wall degrading activity. *Chem. Rec.* 1:24–32.

Dror, T. W., E. Morag, A. Rolider, E. A. Bayer, R. Lamed, and Y. Shoham. 2003a. Regulation of the cellulosomal *celS* (*cel48A*) gene of *Clostridium thermocellum* is growth-rate dependent. *J. Bacteriol.* 185:3042–3048.

Dror, T. W., A. Rolider, E. A. Bayer, R. Lamed, and Y. Shoham. 2003b. Regulation of expression of scaffoldin-related genes in *Clostridium thermocellum*. *J. Bacteriol.* 185:5109–5116.

Dror, T. W., A. Rolider, E. A. Bayer, R. Lamed, and Y. Shoham. 2005. Regulation of major cellulosomal endoglucanases of *Clostridium thermocellum* differs from that of a prominent cellulosomal xylanase. *J. Bacteriol.* 187:2261–2266.

Duong, C. T. V., E. A. Johnson, and A. L. Demain. 1983. Thermophilic, anaerobic and cellulolytic bacteria. *Enzyme Ferment. Biotechnol.* 7:156–195.

Engelhardt, H., and J. Peters. 1998. Structural research on surface layers: a focus on stability, surface layer homology domains, and surface layer-cell wall interactions. *J. Struct. Biol.* 124:276–302.

Erbeznik, M., C. R. Jones, K. A. Dawson, and H. J. Strobel. 1997. *Clostridium thermocellum* JW20 (ATCC 31549) is a coculture with *Thermoanaerobacter ethanolicus*. *Appl. Environ. Microbiol.* 63:2949–2951.

Felix, C. R., and L. G. Ljungdahl. 1993. The cellulosome—the exocellular organelle of *Clostridium*. *Annu. Rev. Microbiol.* 47:791–819.

Fernandes, A. C., C. M. Fontes, H. J. Gilbert, G. P. Hazlewood, T. H. Fernandes, and L. M. A. Ferreira. 1999. Homologous xylanases from *Clostridium thermocellum:* evidence for bi-functional activity, synergism between xylanase catalytic modules and the presence of xylan-binding domains in enzyme complexes. *Biochem. J.* 342:105–110.

Fierobe, H.-P., E. A. Bayer, C. Tardif, M. Czjzek, A. Mechaly, A. Belaich, R. Lamed, Y. Shoham, and J.-P. Belaich. 2002. Degradation of cellulose substrates by cellulosome chimeras: substrate targeting versus proximity of enzyme components. *J. Biol. Chem.* 277:49621–49630.

Fierobe, H.-P., A. Mechaly, C. Tardif, A. Belaich, R. Lamed, Y. Shoham, J.-P. Belaich, and E. A. Bayer. 2001. Design and production of active cellulosome chimeras: selective incorporation of dockerin-containing enzymes into defined functional complexes. *J. Biol. Chem.* 276:21257–21261.

Fierobe, H.-P., F. Mingardon, A. Mechaly, A. Belaich, M. T. Rincon, R. Lamed, C. Tardif, J.-P. Belaich, and E. A. Bayer. 2005. Action of designer cellulosomes on homogeneous versus complex substrates: controlled incorporation of three distinct enzymes into a defined tri-functional scaffoldin. *J. Biol. Chem.* 280:16325–16334.

Fujino, T., P. Béguin, and J.-P. Aubert. 1993. Organization of a *Clostridium thermocellum* gene cluster encoding the cellulosomal scaffolding protein CipA and a protein possibly involved in attachment of the cellulosome to the cell surface. *J. Bacteriol.* 175:1891–1899.

Fujita, Y., J. Ito, M. Ueda, H. Fukuda, and A. Kondo. 2004. Synergistic saccharification, and direct fermentation to ethanol, of amorphous cellulose by use of an engineered yeast strain codisplaying three types of cellulolytic enzyme. *Appl. Environ. Microbiol.* 70: 1207–1212.

Garcia-Martinez, D. V., A. Shinmyo, A. Madia, and A. L. Demain. 1980. Studies on cellulase production by *Clostridium thermocellum*. *Eur. J. Appl. Microbiol. Biotechnol.* 9:189–197.

Germain, P., F. Toukourou, and L. Donaduzzi. 1986. Ethanol production by anaerobic thermophilic bacteria: regulation of lactate dehydrogenase activity in *Clostridium thermohydrosulfuricum*. *Appl. Microbiol. Biotechnol.* 24:300–305.

Gerngross, U. T., M. P. M. Romaniec, T. Kobayashi, N. S. Huskisson, and A. L. Demain. 1993. Sequencing of a *Clostridium thermocellum* gene (cipA) encoding the cellulosomal S_L-protein reveals an unusual degree of internal homology. *Mol. Microbiol.* 8:325–334.

Gilad, R., L. Rabinovich, S. Yaron, E. A. Bayer, R. Lamed, H. J. Gilbert, and Y. Shoham. 2003. CelI, a non-cellulosomal family-9 enzyme from *Clostridium thermocellum*, is a processive endoglucanase that degrades crystalline cellulose. *J. Bacteriol.* 185:391–398.

Gilbert, H. J. 2007. Cellulosomes: microbial nanomachines that display plasticity in quaternary structure. *Mol. Microbiol.* 63:1568–1576.

Grépinet, O., M.-C. Chebrou, and P. Béguin. 1988. Nucleotide sequence and deletion analysis of the xylanase gene (*xynZ*) of *Clostridium thermocellum*. *J. Bacteriol.* 170:4582–4588.

Guedon, E., M. Desvaux, and H. Petitdemange. 2002. Kinetic analysis of *Clostridium cellulolyticum* carbohydrate metabolism: importance of glucose 1-phosphate and glucose 6-phosphate branch points for distribution of carbon fluxes inside and outside cells as revealed by steady-state continuous culture. *Appl. Environ. Microbiol.* 68:53–58.

Guglielmi, G., and P. Béguin. 1998. Cellulase and hemicellulase genes of *Clostridium thermocellum* from five independent collections contain few overlaps and are widely scattered across the chromosome. *FEMS Microbiol. Lett.* 161:209–215.

Halstead, J. R., P. E. Vercoe, H. J. Gilbert, K. Davidson, and G. P. Hazlewood. 1999. A family 26 mannanase produced by *Clostridium thermocellum* as a component of the cellulosome contains a domain which is conserved in mannanases from anaerobic fungi. *Microbiology* 45:3101–3108.

Han, S. O., H. Y. Cho, H. Yukawa, M. Inui, and R. H. Doi. 2004. Regulation of expression of cellulosomes and noncellulosomal (hemi)cellulolytic enzymes in *Clostridium cellulovorans* during growth on different carbon sources. *J. Bacteriol.* 186:4218–4227.

Han, S. O., H. Yukawa, M. Inui, and R. H. Doi. 2005a. Effect of carbon source on the cellulosomal subpopulations of *Clostridium cellulovorans*. *Microbiology* 151:1491–1497.

Han, S. O., H. Yukawa, M. Inui, and R. H. Doi. 2005b. Molecular cloning and transcriptional and expression analysis of *engO*, encoding a new noncellulosomal family 9 enzyme, from *Clostridium cellulovorans*. *J. Bacteriol.* 187:4884–4889.

Han, S. O., H. Yukawa, M. Inui, and R. H. Doi. 2003a. Regulation of expression of cellulosomal cellulase and hemicellulase genes in *Clostridium cellulovorans*. *J. Bacteriol.* 185:6067–6075.

Han, S. O., H. Yukawa, M. Inui, and R. H. Doi. 2003b. Transcription of *Clostridium cellulovorans* cellulosomal cellulase and hemicellulase genes. *J. Bacteriol.* 185:2520–2527.

Hazlewood, G. P., M. P. M. Romaniec, K. Davidson, O. Grépinet, P. Béguin, J. Millet, O. Raynaud, and J.-P. Aubert. 1988. A catalogue of *Clostridium thermocellum* endoglucanase, β-glucosidase and xylanase genes cloned in *Escherichia coli*. *FEMS Microbiol. Lett.* 51:231–236.

Henrissat, B. 1991. A classification of glycosyl hydrolases based on amino acid sequence similarities. *Biochem. J.* 280:309–316.

Henrissat, B., and A. Bairoch. 1993. New families in the classification of glycosyl hydrolases based on amino acid sequence similarities. *Biochem. J.* 293:781–788.

Henrissat, B., P. M. Coutinho, E. Deleury, and G. J. Davies. 2003. Sequence families and modular organisation of carbohydrate-active enzymes, p. 15–34. *In* A. Svendsen (ed.), *Enzyme Functionality: Design, Engineering and Screening*. Marcel Dekker, New York, NY.

Henrissat, B., and G. J. Davies. 2000. Glycoside hydrolases and glycosyltransferases. Families, modules, and implications for genomics. *Plant Physiol.* 124:1515–1519.

Henrissat, B., and A. Romeu. 1995. Families, superfamilies and subfamilies of glycosyl hydrolases. *Biochem. J.* 311:350–351.

Herrero, A. A., and R. F. Gomez. 1980. Development of ethanol tolerance in *Clostridium thermocellum*: effect of growth temperature. *Appl. Environ. Microbiol.* 40:571–577.

Herrero, A. A., R. F. Gomez, and M. F. Roberts. 1982. Ethanol-induced changes in the membrane lipid composition of *Clostridium thermocellum*. *Biochim. Biophys. Acta* 693:195–204.

Himmel, M. E., S.-Y. Ding, D. K. Johnson, W. S. Adney, M. R. Nimlos, J. W. Brady, and T. D. Foust. 2007. Biomass recalcitrance: engineering plants and enzymes for biofuels production. *Science* 315:804–807. (Erratum, 316:982.)

Hughes, S. R., S. B. Riedmuller, J. A. Mertens, X. L. Li, K. M. Bischoff, N. Qureshi, M. A. Cotta, and P. J. Farrelly. 2006. High-throughput screening of cellulase F mutants from multiplexed plasmid sets using an automated plate assay on a functional proteomic robotic workcell. *Proteome Sci.* 4:10.

Ito, J., Y. Fujita, M. Ueda, H. Fukuda, and A. Kondo. 2004. Improvement of cellulose-degrading ability of a yeast strain displaying *Trichoderma reesei* endoglucanase II by recombination of cellulose-binding domains. *Biotechnol. Prog.* 20:688–691.

Jindou, S., Q. Xu, R. Kenig, Y. Shoham, E. A. Bayer, and R. Lamed. 2006. Novel architectural theme of family-9 glycoside hydrolases identified in cellulosomal enzymes of *Acetivibrio cellulolyticus* and *Clostridium thermocellum*. *FEMS Microbiol. Lett.* 254:308–316.

Johnson, E. A., and A. L. Demain. 1984. Probable involvement of sulfhydryl groups and a metal as essential components of the cellulase of *Clostridium thermocellum*. *Arch. Microbiol.* 137:135–138.

Johnson, E. A., E. T. Reese, and A. L. Demain. 1982a. Inhibition of *Clostridium thermocellum* cellulase by end products of cellulolysis. *J. Appl. Biochem.* 4:64–71.

Johnson, E. A., M. Sakojoh, G. Halliwell, A. Madia, and A. L. Demain. 1982b. Saccharification of complex cellulosic substrates by the cellulase system from *Clostridium thermocellum*. *Appl. Environ. Microbiol.* 43:1125–1132.

Jones, R. P. 1989. Biological principles for the effects of ethanol. *Enzyme Microb. Technol.* 11:130–153.

Kakiuchi, M., A. Isui, K. Suzuki, T. Fujino, E. Fujino, T. Kimura, S. Karita, K. Sakka, and K. Ohmiya. 1998. Cloning and DNA sequencing of the genes encoding *Clostridium josui* scaffolding protein CipA and cellulase CelD and identification of their gene products as major components of the cellulosome. *J. Bacteriol.* **180:** 4303–4308.

Karita, S., K. Sakka, and K. Ohmiya. 1997. Cellulosomes, cellulase complexes, of anaerobic microbes: their structure models and functions, p. 47–57. *In* R. Onodera, H. Itabashi, K. Ushida, H. Yano, and Y. Sasaki (ed.), *Rumen Microbes and Digestive Physiology in Ruminants*, vol. 14. Japan Scientific Society Press, Tokyo, Japan.

Katahira, S., Y. Fujita, A. Mizuike, H. Fukuda, and A. Kondo. 2004. Construction of a xylan-fermenting yeast strain through codisplay of xylanolytic enzymes on the surface of xylose-utilizing *Saccharomyces cerevisiae* cells. *Appl. Environ. Microbiol.* 70:5407–5414.

Katahira, S., A. Mizuike, H. Fukuda, and A. Kondo. 2006. Ethanol fermentation from lignocellulosic hydrolysate by a recombinant xylose- and cellooligosaccharide-assimilating yeast strain. *Appl. Microbiol. Biotechnol.* 72:1136–1143.

Kim, A. Y., G. T. Attwood, S. M. Holt, B. A. White, and H. P. Blaschek. 1994. Heterologous expression of endo-beta-1,4-D-glucanase from *Clostridium cellulovorans* in *Clostridium acetobutylicum* ATCC 824 following transformation of the *engB* gene. *Appl. Environ. Microbiol.* 60:337–340.

Kirby, J., J. C. Martin, A. S. Daniel, and H. J. Flint. 1997. Dockerin-like sequences in cellulases and xylanases from the rumen cellulolytic bacterium *Ruminococcus flavefaciens*. *FEMS Microbiol. Lett.* 149:213–219.

Kosugi, A., Y. Amano, K. Murashima, and R. H. Doi. 2004. Hydrophilic domains of scaffolding protein CbpA promote glycosyl hydrolase activity and localization of cellulosomes to the cell surface of *Clostridium cellulovorans*. *J. Bacteriol.* 186:6351–6359.

Kosugi, A., K. Murashima, Y. Tamaru, and R. H. Doi. 2002. Cell-surface-anchoring role of N-terminal surface layer homology domains of *Clostridium cellulovorans* EngE. *J. Bacteriol.* 184:884–888.

Koukiekolo, R., H. Y. Cho, A. Kosugi, M. Inui, H. Yukawa, and R. H. Doi. 2005. Degradation of corn fiber by *Clostridium cellulovorans* cellulases and hemicellulases and contribution of scaffolding protein CbpA. *Appl. Environ. Microbiol.* 71:3504–3511.

Lamed, R., and E. A. Bayer. 1991. Cellulose degradation by thermophilic anaerobic bacteria, p. 377-410. *In* C. H. Haigler and P. J. Weimer (ed.), *Biosynthesis and Biodegradation of Cellulose and Cellulose Materials*. Marcel Dekker, New York, NY.

Lamed, R., and E. A. Bayer. 1993. The cellulosome concept—a decade later!, p. 1-12. *In* K. Shimada, S. Hoshino, K. Ohmiya, K. Sakka, Y. Kobayashi, and S. Karita (ed.), *Genetics, Biochemistry and Ecology of Lignocellulose Degradation*. Uni Publishers Co., Ltd., Tokyo, Japan.

Lamed, R., and E. A. Bayer. 1988a. The cellulosome concept: exocellular/extracellular enzyme reactor centers for efficient binding and cellulolysis, p. 101–116. *In* J.-P. Aubert, P. Beguin, and J. Millet (ed.), *Biochemistry and Genetics of Cellulose Degradation*. Academic Press, London, United Kingdom.

Lamed, R., and E. A. Bayer. 1988b. The cellulosome of *Clostridium thermocellum*. *Adv. Appl. Microbiol.* 33:1–46.

Lamed, R., E. A. Bayer, B. C. Saha, and J. G. Zeikus. 1988. Biotechnological potential of enzymes from unique thermophiles, p. 371–383. *In* G. Durand, L. Bobichon, and J. Florent (ed.), *Proceedings of the 8th International Biotechnology Symposium*, vol. 1. Société Française de Microbiologie, Paris, France.

Lamed, R., R. Kenig, E. Setter, and E. A. Bayer. 1985. Major characteristics of the cellulolytic system of *Clostridium thermocellum* coincide with those of the purified cellulosome. *Enzyme Microb. Technol.* 7:37–41.

Lamed, R., E. Morag (Morgenstern), O. Mor-Yosef, and E. A. Bayer. 1991. Cellulosome-like entities in *Bacteroides cellulosolvens*. *Curr. Microbiol.* 22:27–33.

Lamed, R., J. Naimark, E. Morgenstern, and E. A. Bayer. 1987. Specialized cell surface structures in cellulolytic bacteria. *J. Bacteriol.* 169:3792–3800.

Lamed, R., E. Setter, and E. A. Bayer. 1983a. Characterization of a cellulose-binding, cellulase-containing complex in *Clostridium thermocellum*. *J. Bacteriol.* 156:828–836.

Lamed, R., E. Setter, R. Kenig, and E. A. Bayer. 1983b. The cellulosome—a discrete cell surface organelle of *Clostridium thermocellum* which exhibits separate antigenic, cellulose-binding and various cellulolytic activities. *Biotechnol. Bioeng. Symp.* 13:163–181.

Lamed, R., and J. G. Zeikus. 1980. Ethanol production by thermophilic bacteria: relationship between fermentation product yields of and catabolic enzyme activities in *Clostridium thermocellum* and *Thermoanaerobium brockii*. *J. Bacteriol.* 144:569–578.

Leatherwood, J. M. 1969. Cellulase complex of *Ruminococcus* and a new mechanism for cellulose degradation. *Adv. Chem. Ser.* 95:53–59.

Leibovitz, E., and P. Béguin. 1996. A new type of cohesin domain that specifically binds the dockerin domain of the *Clostridium thermocellum* cellulosome-integrating protein CipA. *J. Bacteriol.* **178:** 3077–3084.

Leibovitz, E., H. Ohayon, P. Gounon, and P. Béguin. 1997. Characterization and subcellular localization of the *Clostridium thermocellum* scaffoldin dockerin binding protein SdbA. *J. Bacteriol.* 179:2519–2523.

Lemaire, M., I. Miras, P. Gounon, and P. Béguin. 1998. Identification of a region responsible for binding to the cell wall within the S-layer protein of *Clostridium thermocellum*. *Microbiology* **144:** 211–217.

Lemaire, M., H. Ohayon, P. Gounon, T. Fujino, and P. Béguin. 1995. OlpB, a new outer layer protein of *Clostridium thermocellum*, and binding of its S-layer-like domains to components of the cell envelope. *J. Bacteriol.* 177:2451–2459.

Levasseur, A., D. Navarro, P. J. Punt, J. P. Belaich, M. Asther, and E. Record. 2005. Construction of engineered bifunctional enzymes and their overproduction in *Aspergillus niger* for improved enzymatic tools to degrade agricultural by-products. *Appl. Environ. Microbiol.* 71:8132–8140.

Levasseur, A., S. Pages, H. P. Fierobe, D. Navarro, P. Punt, J. P. Belaich, M. Asther, and E. Record. 2004. Design and production in *Aspergillus niger* of a chimeric protein associating a fungal feruloyl esterase and a clostridial dockerin domain. *Appl. Environ. Microbiol.* 70:6984–6991.

Linder, M., and T. T. Teeri. 1997. The roles and function of cellulose-binding domains. *J. Biotechnol.* 57:15–28.

Lopez-Contreras, A. M., K. Gabor, A. A. Martens, B. A. Renckens, P. A. Claassen, J. Van Der Oost, and W. M. De Vos. 2004. Substrate-induced production and secretion of cellulases by *Clostridium acetobutylicum*. *Appl. Environ. Microbiol.* 70:5238–5243.

Lopez-Contreras, A. M., A. A. Martens, N. Szijarto, H. Mooibroek, P. Claassen, J. van der Oost, and W. M. de Vos. 2003. Production by *Clostridium acetobutylicum* ATCC 824 of CelG, a cellulosomal glycoside hydrolase belonging to family 9. *Appl. Environ. Microbiol.* 69:869–877.

Lupas, A., H. Engelhardt, J. Peters, U. Santarius, S. Volker, and W. Baumeister. 1994. Domain structure of the *Acetogenium kivui* surface layer revealed by electron crystallography and sequence analysis. *J. Bacteriol.* **176:**1224–1233.

Lynd, L. R., W. H. van Zyl, J. E. McBride, and M. Laser. 2005. Consolidated bioprocessing of cellulosic biomass: an update. *Curr. Opin. Biotechnol.* **16:**577–583.

Lytle, B., C. Myers, K. Kruus, and J. H. D. Wu. 1996. Interactions of the CelS binding ligand with various receptor domains of the *Clostridium thermocellum* cellulosomal scaffolding protein, CipA. *J. Bacteriol.* **178:**1200–1203.

Maamar, H., L. Abdou, C. Boileau, O. Valette, and C. Tardif. 2006. Transcriptional analysis of the *cip-cel* gene cluster from *Clostridium cellulolyticum. J. Bacteriol.* **188:**2614–2624.

Maamar, H., P. de Philip, J. P. Belaich, and C. Tardif. 2003. ISCce1 and ISCce2, two novel insertion sequences in *Clostridium cellulolyticum. J. Bacteriol.* **185:**714–725.

Maamar, H., O. Valette, H. P. Fierobe, A. Belaich, J. P. Belaich, and C. Tardif. 2004. Cellulolysis is severely affected in *Clostridium cellulolyticum* strain cipCMut1. *Mol. Microbiol.* **51:**589–598.

Mandels, M., and E. T. Reese. 1957. Induction of cellulase in *Trichoderma viride* as influenced by carbon sources and metals. *J. Bacteriol.* **73:**269–278.

Mesnage, S., T. Fontaine, T. Mignot, M. Delepierre, M. Mock, and A. Fouet. 2000. Bacterial SLH domain proteins are non-covalently anchored to the cell surface via a conserved mechanism involving wall polysaccharide pyruvylation. *EMBO J.* **19:**4473–4484.

Mesnage, S., E. Tosi-Couture, M. Mock, and A. Fouet. 1999. The S-layer homology domain as a means for anchoring heterologous proteins on the cell surface of *Bacillus anthracis. J. Appl. Microbiol.* **87:**256–260.

Mingardon, F., S. Perret, A. Belaich, C. Tardif, J. P. Belaich, and H. P. Fierobe. 2005. Heterologous production, assembly, and secretion of a minicellulosome by *Clostridium acetobutylicum* ATCC 824. *Appl. Biochem. Biotechnol.* **71:**1215–1222.

Morag, E., E. A. Bayer, and R. Lamed. 1990. Relationship of cellulosomal and noncellulosomal xylanases of *Clostridium thermocellum* to cellulose-degrading enzymes. *J. Bacteriol.* **172:**6098–6105.

Morag, E., I. Halevy, E. A. Bayer, and R. Lamed. 1991. Isolation and properties of a major cellobiohydrolase from the cellulosome of *Clostridium thermocellum. J. Bacteriol.* **173:**4155–4162.

Morrison, M., and J. Miron. 2000. Adhesion to cellulose by *Ruminococcus albus:* a combination of cellulosomes and Pil-proteins? *FEMS Microbiol. Lett.* **185:**109–115.

Murashima, K., C. L. Chen, A. Kosugi, Y. Tamaru, R. H. Doi, and S. L. Wong. 2002a. Heterologous production of *Clostridium cellulovorans engB*, using protease-deficient *Bacillus subtilis*, and preparation of active recombinant cellulosomes. *J. Bacteriol.* **184:**76–81.

Murashima, K., A. Kosugi, and R. H. Doi. 2002b. Thermostabilization of cellulosomal endoglucanase EngB from *Clostridium cellulovorans* by in vitro DNA recombination with non-cellulosomal endoglucanase EngD. *Mol. Microbiol.* **45:**617–626.

Murray, W. D., L. C. Sowden, and J. R. Colvin. 1986. Symbiotic relationship of *Bacteroides cellulosolvens* and *Clostridium saccharolyticum* in cellulose fermentation. *Appl. Environ. Microbiol.* **51:**710–715.

Navarre, W. W., and O. Schneewind. 1994. Proteolytic cleavage and cell wall anchoring at the LPXTG motif of surface proteins in gram-positive bacteria. *Mol. Microbiol.* **14:**115–121.

Navarre, W. W., and O. Schneewind. 1999. Surface proteins of gram-positive bacteria and mechanisms of their targeting to the cell wall envelope. *Microbiol. Mol. Biol. Rev.* **63:**174–229.

Nechooshtan, R. 2007. *Characterization of Regulatory Elements of the Cellulase System in* Clostridium thermocellum. Technion, Haifa, Israel.

Newcomb, M., C. Y. Chen, and J. H. Wu. 2007. Induction of the *celC* operon of *Clostridium thermocellum* by laminaribiose. *Proc. Natl. Acad. Sci. USA* **104:**3747–3752.

Ng, T. K., T. K. Weimer, and J. G. Zeikus. 1977. Cellulolytic and physiological properties of *Clostridium thermocellum. Arch. Microbiol.* **114:**1–7.

Ng, T. K., and J. G. Zeikus. 1982. Differential metabolism of cellobiose and glucose by *Clostridium thermocellum* and *Clostridium thermohydrosulfuricum. J. Bacteriol.* **150:**1391–1399.

Ng, T. K., and J. G. Zeikus. 1981. Purification and characterization of an endoglucanase (1,4-β-D-glucan glucanohydrolase) from *Clostridium thermocellum. Biochem. J.* **199:**341–350.

Nolling, J., G. Breton, M. V. Omelchenko, K. S. Makarova, Q. Zeng, R. Gibson, H. M. Lee, J. Dubois, D. Qiu, J. Hitti, Y. I. Wolf, R. L. Tatusov, F. Sabathe, L. Doucette-Stamm, P. Soucaille, M. J. Daly, G. N. Bennett, E. V. Koonin, and D. R. Smith. 2001. Genome sequence and comparative analysis of the solvent-producing bacterium *Clostridium acetobutylicum. J. Bacteriol.* **183:**4823–4838.

Ohmiya, K., K. Sakka, T. Kimura, and K. Morimoto. 2003. Application of microbial genes to recalcitrant biomass utilization and environmental conservation. *J. Biosci. Bioeng.* **95:**549–561.

Pagès, S., A. Belaich, J.-P. Belaich, E. Morag, R. Lamed, Y. Shoham, and E. A. Bayer. 1997a. Species-specificity of the cohesin-dockerin interaction between *Clostridium thermocellum* and *Clostridium cellulolyticum:* prediction of specificity determinants of the dockerin domain. *Proteins* **29:**517–527.

Pagès, S., A. Belaich, H.-P. Fierobe, C. Tardif, C. Gaudin, and J.-P. Belaich. 1999. Sequence analysis of scaffolding protein CipC and ORFXp, a new cohesin-containing protein in *Clostridium cellulolyticum*: comparison of various cohesin domains and subcellular localization of ORFXp. *J. Bacteriol.* **181:**1801–1810.

Pagès, S., A. Belaich, C. Tardif, C. Reverbel-Leroy, C. Gaudin, and J.-P. Belaich. 1996. Interaction between the endoglucanase CelA and the scaffolding protein CipC of the *Clostridium cellulolyticum* cellulosome. *J. Bacteriol.* **178:**2279–2286.

Pagès, S., L. Gal, A. Belaich, C. Gaudin, C. Tardif, and J.-P. Belaich. 1997b. Role of scaffolding protein CipC of *Clostridium cellulolyticum* in cellulose degradation. *J. Bacteriol.* **179:**2810–2816.

Pagès, S., O. Valette, L. Abdou, A. Belaich, and J.-P. Belaich. 2003. A rhamnogalacturonan lyase in the *Clostridium cellulolyticum* cellulosome. *J. Bacteriol.* **185:**4727–4733.

Perret, S., A. Belaich, H. P. Fierobe, J. P. Belaich, and C. Tardif. 2004a. Towards designer cellulosomes in clostridia: mannanase enrichment of the cellulosomes produced by *Clostridium cellulolyticum. J. Bacteriol.* **186:**6544–6552.

Perret, S., L. Casalot, H.-P. Fierobe, C. Tardif, F. Sabathe, J.-P. Belaich, and A. Belaich. 2004b. Production of heterologous and chimeric scaffoldins by *Clostridium acetobutylicum* ATCC 824. *J. Bacteriol.* **186:**253–257.

Perret, S., H. Maamar, J.-P. Belaich, and C. Tardif. 2004c. Use of antisense RNA to modify the composition of cellulosomes produced by *Clostridium cellulolyticum. Mol. Microbiol.* **51:**599–607.

Prates, J. A., N. Tarbouriech, S. J. Charnock, C. M. Fontes, L. M. Ferreira, and G. J. Davies. 2001. The structure of the feruloyl esterase module of xylanase 10B from *Clostridium thermocellum* provides insights into substrate recognition. *Structure* **9:**1183–1190.

Ragauskas, A. J., C. K. Williams, B. H. Davison, G. Britovsek, J. Cairney, C. A. Eckert, W. J. Frederick, Jr., J. P. Hallett, D. J. Leak, C. L. Liotta, J. R. Mielenz, R. Murphy, R. Templer, and T. Tschaplinski. 2006. The path forward for biofuels and biomaterials. *Science* **311:**484–489.

Rani, K. S., and G. Seenayya. 1999. High ethanol tolerance of new isolates of *Clostridium thermocellum* strains SS21 and SS22. *World J. Microbiol. Biotechnol. Adv.* **15:**173–178.

Reese, E. T. 1975. Enzyme systems for cellulose. *Biotechnol. Bioeng. Symp.* **5:**77–80.

Reese, E. T. 1976a. History of the cellulase program at the U.S. Army Natick Development Center. *Biotechnol. Bioeng. Symp.* **6:** 9–20.

Reese, E. T. 1976b. Cellulase production. *Biotechnol. Bioeng. Symp.* 6:91–93.

Reese, E. T., and M. Mandels. 1980. Stability of the cellulase of *Trichoderma reesei* under use conditions. *Biotechnol. Bioeng.* 22: 323–335.

Reese, E. T., R. G. H. Siu, and H. S. Levinson. 1950. The biological degradation of soluble cellulose derivatives and its relationship to the mechanism of cellulose hydrolysis. *J. Bacteriol.* 59:485–497.

Rincon, M. T., T. Cepeljnik, J. C. Martin, Y. Barak, R. Lamed, E. A. Bayer, and H. J. Flint. 2007. A novel cell surface-anchored cellulose-binding protein encoded by the *sca* gene cluster of *Ruminococcus flavefaciens*. *J. Bacteriol.* 189:4774–4783.

Rincon, M. T., T. Cepeljnik, J. C. Martin, R. Lamed, Y. Barak, E. A. Bayer, and H. J. Flint. 2005. Unconventional mode of attachment of the *Ruminococcus flavefaciens* cellulosome to the cell surface. *J. Bacteriol.* 187:7569–7578.

Rincon, M. T., S.-Y. Ding, S. I. McCrae, J. C. Martin, V. Aurilia, R. Lamed, Y. Shoham, E. A. Bayer, and H. J. Flint. 2003. Novel organization and divergent dockerin specificities in the cellulosome system of *Ruminococcus flavefaciens*. *J. Bacteriol.* 185:703–713.

Rincon, M. T., J. C. Martin, V. Aurilia, S. I. McCrae, G. Rucklidge, M. Reid, E. A. Bayer, R. Lamed, and H. J. Flint. 2004. ScaC, an adaptor protein carrying a novel cohesin that expands the dockerin-binding repertoire of the *Ruminococcus flavefaciens* 17 cellulosome. *J. Bacteriol.* 186:2576–2585.

Rincon, M. T., S. I. McCrae, J. Kirby, K. P. Scott, and H. J. Flint. 2001. EndB, a multidomain family 44 cellulase from *Ruminococcus flavefaciens* 17, binds to cellulose via a novel cellulose-binding module and to another *R. flavefaciens* protein via a dockerin domain. *Appl. Environ. Microbiol.* 67:4426–4431.

Sabathe, F., A. Belaich, and P. Soucaille. 2002. Characterization of the cellulolytic complex (cellulosome) of *Clostridium acetobutylicum*. *FEMS Microbiol. Lett.* 217:15–22.

Sabathe, F., and P. Soucaille. 2003. Characterization of the CipA scaffolding protein and in vivo production of a minicellulosome in *Clostridium acetobutylicum*. *J. Bacteriol.* 185:1092–1096.

Saddler, J. N., and M. K.-H. Chan. 1984. Conversion of pretreated lignocellulosic substrates to ethanol by *Clostridium thermosaccharolyticum* and *Clostridium thermohydrosulphuricum*. *Can. J. Microbiol.* 30:212–220.

Salamitou, S., O. Raynaud, M. Lemaire, M. Coughlan, P. Béguin, and J.-P. Aubert. 1994. Recognition specificity of the duplicated segments present in *Clostridium thermocellum* endoglucanase CelD and in the cellulosome-integrating protein CipA. *J. Bacteriol.* 176:2822–2827.

Salamitou, S., K. Tokatlidis, P. Béguin, and J.-P. Aubert. 1992. Involvement of separate domains of the cellulosomal protein S1 of *Clostridium thermocellum* in binding to cellulose and in anchoring of catalytic subunits to the cellulosome. *FEBS Lett.* 304:89–92.

Schneewind, O., A. Fowler, and K. F. Faull. 1995. Structure of the cell wall anchor of surface proteins in *Staphylococcus aureus*. *Science* 268:103–106.

Schubert, C. 2006. Can biofuels finally take center stage? *Nat. Biotechnol.* 24:777–784.

Schumacher, M. A., G. S. Allen, M. Diel, G. Seidel, W. Hillen, and R. G. Brennan. 2004. Structural basis for allosteric control of the transcription regulator CcpA by the phosphoprotein HPr-Ser46-P. *Cell* 118:731–741.

Schumacher, M. A., G. Seidel, W. Hillen, and R. G. Brennan. 2007. Structural mechanism for the fine-tuning of Ccpa function by the small molecule effectors glucose 6-phosphate and fructose 1,6-bisphosphate. *J. Mol. Biol.* 368:1042–1050.

Schwarz, W. H. 2001. The cellulosome and cellulose degradation by anaerobic bacteria. *Appl. Microbiol. Biotechnol.* 56:634–649.

Schwarz, W. H., V. V. Zverlov, and H. Bahl. 2004. Extracellular glycosyl hydrolases from clostridia. *Adv. Appl. Microbiol.* 56:215–261.

Shimon, L. J. W., S. Pagès, A. Belaich, J. P. Belaich, E. A. Bayer, R. Lamed, Y. Shoham, and F. Frolow. 2000. Structure of a family IIIa scaffoldin CBD from the cellulosome of *Clostridium cellulolyticum* at 2.2 Å resolution. *Acta Crystallogr. D Biol. Crystallogr.* 56:1560–1568.

Shinmyo, A., D. V. Garcia-Martinez, and A. L. Demain. 1979. Studies on the extracellular cellulolytic enzyme complex produced by *Clostridium thermocellum*. *J. Appl. Biochem.* 1:202–209.

Shoham, Y., R. Lamed, and E. A. Bayer. 1999. The cellulosome concept as an efficient microbial strategy for the degradation of insoluble polysaccharides. *Trends Microbiol.* 7:275–281.

Shoseyov, O., M. Takagi, M. A. Goldstein, and R. H. Doi. 1992. Primary sequence analysis of *Clostridium cellulovorans* cellulose binding protein A. *Proc. Natl. Acad. Sci. USA* 89:3483–3487.

Slapack, G. E., I. Russell, and C. G. Stewart. 1987. *Thermophilic Microbes in Ethanol Production*. CRC Press, Boca Raton, FL.

Stevenson, D. M., and P. J. Weimer. 2005. Expression of 17 genes in *Clostridium thermocellum* ATCC 27405 during fermentation of cellulose or cellobiose in continuous culture. *Appl. Environ. Microbiol.* 71:4672–4678.

Tamaru, Y., and R. H. Doi. 2001. Pectate lyase A, an enzymatic subunit of the *Clostridium cellulovorans* cellulosome. *Proc. Natl. Acad. Sci. USA* 20:4125–4129.

Tarbouriech, N., J. A. Prates, C. M. Fontes, and G. J. Davies. 2005. Molecular determinants of substrate specificity in the feruloyl esterase module of xylanase 10B from *Clostridium thermocellum*. *Acta Crystallogr. D* 61:194–197.

Tardif, C., H. Maamar, M. Balfin, and J. P. Belaich. 2001. Electrotransformation studies in *Clostridium cellulolyticum*. *J. Ind. Microbiol. Biotechnol.* 27:271–274.

Tokatlidis, K., P. Dhurjati, and P. Béguin. 1993. Properties conferred on *Clostridium thermocellum* endoglucanase CelC by grafting the duplicated segment of endoglucanase CelD. *Protein Eng.* 6:947–952.

Tokatlidis, K., S. Salamitou, P. Béguin, P. Dhurjati, and J.-P. Aubert. 1991. Interaction of the duplicated segment carried by *Clostridium thermocellum* cellulases with cellulosome components. *FEBS Lett.* 291:185–188.

Tomme, P., R. A. J. Warren, R. C. Miller, D. G. Kilburn, and N. R. Gilkes. 1995. Cellulose-binding domains—classification and properties, p. 142–161. *In* J. M. Saddler and M. H. Penner (ed.), *Enzymatic Degradation of Insoluble Polysaccharides*. American Chemical Society, Washington, DC.

Ton-That, H., L. A. Marraffini, and O. Schneewind. 2004. Protein sorting to the cell wall envelope of Gram-positive bacteria. *Biochim. Biophys. Acta* 1694:269–278.

Tormo, J., R. Lamed, A. J. Chirino, E. Morag, E. A. Bayer, Y. Shoham, and T. A. Steitz. 1996. Crystal structure of a bacterial family-III cellulose-binding domain: a general mechanism for attachment to cellulose. *EMBO J.* 15:5739–5751.

U.S. Department of Energy. 2006. *Breaking the Biological Barriers to Cellulosic Ethanol: A Joint Research Agenda*. DOE/SC-0095. U.S. Department of Energy Office of Science and Office of Energy Efficiency and Renewable Energy, Rockville, MD.

Venkateswaren, S., and A. L. Demain. 1986. The *Clostridium thermocellum-Clostridium thermosaccharolyticum* ethanol production process: nutritional studies and scale-down. *Chem. Eng. Commun.* 45:53–60.

Wang, D. I. C., G. C. Avgerinos, I. Biocic, S.-D. Wang, and H.-Y. Fang. 1983. Ethanol from cellulosic biomass. *Philos. Trans. R. Soc. Lond. B* 300:323–333.

Wang, W. K., K. Kruus, and J. H. D. Wu. 1993. Cloning and DNA sequence of the gene coding for *Clostridium thermocellum* cellulase S_S (CelS), a major cellulosome component. *J. Bacteriol.* 175:1293–1302.

Warner, J. B., and J. S. Lolkema. 2003. CcpA-dependent carbon catabolite repression in bacteria. *Microbiol. Mol. Biol. Rev.* 67: 475–490.

Weimer, P. J., and J. G. Zeikus. 1977. Fermentation of cellulose and cellobiose by *Clostridium thermocellum* in the absence of *Methanobacterium thermoautotrophicum*. *Appl. Environ. Microbiol.* 33:289–297.

Wiegel, J., C. P. Mothershed, and J. Puls. 1985. Differences in xylan degradation by various noncellulolytic thermophilic anaerobes and *Clostridium thermocellum*. *Appl. Environ. Microbiol.* 49:656–659.

Williams, T. I., J. C. Combs, B. C. Lynn, and H. J. Strobel. 2007. Proteomic profile changes in membranes of ethanol-tolerant *Clostridium thermocellum*. *Appl. Microbiol. Biotechnol.* 74:422–423.

Wu, J. H. D. 1993. *Clostridium thermocellum* cellulosome—new mechanistic concept for cellulose degradation, p. 251–264. *In* M. E. Himmel and G. Georgiou (ed.), *Biocatalyst Design for Stability and Specificity*. ACS Symposium Series No. 516. American Chemical Society, Washington, DC.

Wu, J. H. D., W. H. Orme-Johnson, and A. L. Demain. 1988. Two components of an extracellular protein aggregate of *Clostridium thermocellum* together degrade crystalline cellulose. *Biochemistry* 27:1703–1709.

Xu, Q., Y. Barak, R. Kenig, Y. Shoham, E. A. Bayer, and R. Lamed. 2004a. A novel *Acetivibrio cellulolyticus* anchoring scaffoldin that bears divergent cohesins. *J. Bacteriol.* 186:5782–5789.

Xu, Q., E. A. Bayer, M. Goldman, R. Kenig, Y. Shoham, and R. Lamed. 2004b. Architecture of the *Bacteroides cellulosolvens* cellulosome: description of a cell-surface anchoring scaffoldin and a family 48 cellulase. *J. Bacteriol.* 186:968–977.

Xu, Q., W. Gao, S.-Y. Ding, R. Kenig, Y. Shoham, E. A. Bayer, and R. Lamed. 2003. The cellulosome system of *Acetivibrio cellulolyticus* includes a novel type of adaptor protein and a cell-surface anchoring protein. *J. Bacteriol.* 185:4548–4557.

Yaron, S., E. Morag, E. A. Bayer, R. Lamed, and Y. Shoham. 1995. Expression, purification and subunit-binding properties of cohesins 2 and 3 of the *Clostridium thermocellum* cellulosome. *FEBS Lett.* 360:121–124.

Zeikus, J. G., A. Ben-Bassat, T. K. Ng, and R. J. Lamed. 1981. Thermophilic ethanol fermentations. *Basic Life Sci.* 18:441–461.

Zhang, Y. H., M. E. Himmel, and J. R. Mielenz. 2006. Outlook for cellulase improvement: screening and selection strategies. *Biotechnol. Adv.* 24:452–481.

Zhang, Y. H., and L. R. Lynd. 2005a. Cellulose utilization by *Clostridium thermocellum*: bioenergetics and hydrolysis product assimilation. *Proc. Natl. Acad. Sci. USA* 102:7321–7325.

Zhang, Y. H., and L. R. Lynd. 2005b. Regulation of cellulase synthesis in batch and continuous cultures of *Clostridium thermocellum*. *J. Bacteriol.* 187:99–106.

Zhao, G., E. Ali, M. Sakka, T. Kimura, and K. Sakka. 2006. Binding of S-layer homology modules from *Clostridium thermocellum* SdbA to peptidoglycans. *Appl. Microbiol. Biotechnol.* 70:464–469.

Zverlov, V. V., K.-P. Fuchs, and W. H. Schwarz. 2002. Chi18A,the endochitinase in the cellulosome of the thermophilic, cellulolytic bacterium *Clostridium thermocellum*. *Appl. Environ. Microbiol.* 68:3176–3179.

Zverlov, V. V., K. P. Fuchs, W. H. Schwarz, and G. Velikodvorskaya. 1994. Purification and cellulosomal localization of *Clostridium thermocellum* mixed linkage b-glucanase LicB (1,3-1,4-β-D-glucanase). *Biotechnol. Lett.* 16:29–34.

Zverlov, V. V., J. Kellermann, and W. H. Schwarz. 2005. Functional subgenomics of *Clostridium thermocellum* cellulosomal genes: identification of the major catalytic components in the extracellular complex and detection of three new enzymes. *Proteomics* 5:3646–3653.

Zverlov, V. V., G. V. Velikodvorskaya, W. H. Schwarz, K. Bronnenmeier, J. Kellermann, and W. L. Staudenbauer. 1998. Multidomain structure and cellulosomal localization of the *Clostridium thermocellum* cellobiohydrolase CbhA. *J. Bacteriol.* 180:3091–3099.

Bioenergy
Edited by J. Wall et al.

Chapter 7

Cellulosomes from Mesophilic Bacteria

Roy H. Doi

Mesophilic bacteria that produce extracellular enzyme complexes called cellulosomes (Bayer et al., 2004; Doi and Kosugi, 2004) have been found in a wide variety of habitats including soil, wood chip piles, bovine and sheep rumen, termite gut, compost piles, decaying trees, sewage sludge, bovine feces, and swamp mud (Doi et al., 2003). Cellulosomes have been found to be present only with anaerobic bacteria and fungi; however, a few aerobic bacteria have been reported to produce enzyme complexes similar to cellulosomes. These so-called cellulosomes have not been fully characterized and may in fact be aggregates of extracellular enzymes with hydrolytic activity. A considerable amount of information concerning the structure and function of cellulosomes has been obtained during the past 25 years, and there have been several recent reviews on cellulosomes from thermophilic and mesophilic microorganisms (Demain et al., 2005; Desvaux, 2005; Doi and Kosugi, 2004; Bayer et al., 2004; Doi et al., 2003; Lynd et al., 2002; Schwarz, 2001).

In this review, only fully characterized cellulosomes from mesophilic bacteria consisting of two major components are discussed. These components include (i) one or more scaffolding proteins called scaffoldins (Shoseyov et al., 1992; Pages et al., 1996; Gal et al., 1997; Ding et al., 1999) that contain enzyme binding sites called cohesins and (ii) cellulosomal enzymes containing dockerin domains (Bayer et al., 1994). The cohesin-dockerin interaction between the scaffolding protein and cellulosomal enzymes allows the assembly of the extracellular multisubunit enzyme complex called the cellulosome (Fig. 1). This discussion is focused primarily on the enzymes and catalytic properties of the cellulosome.

The scaffoldins minimally contain a carbohydrate binding module (CBM) and enzyme binding domains called cohesins (Bayer et al., 1994). In addition, depending on the scaffoldin, there may be hydrophilic domains, dockerins, even an enzyme function, and other domains of unknown function (Doi and Kosugi, 2004). In the simplest format a single scaffoldin containing a CBM and multiple cohesins binds a number of cellulosomal enzymes through a cohesin-dockerin interaction resulting in a relatively large extracellular enzyme complex that is capable of simultaneously binding to the cell surface and to the lignocellulosic substrate (Fig. 2). In a much more complex cellulosome organization including multiple scaffolding proteins, it is possible that up to 96 enzyme molecules are present in a single large enzyme complex of *Acetivibrio cellulolyticus* (Xu et al., 2003).

The cellulosome is capable of hydrolyzing the cellulose and hemicellulose components of plant cell walls quite efficiently to hexoses and pentoses. This occurs through the tethering of various cellulases and hemicellulases to the scaffoldin, which binds the cellulosome to the substrate, usually through a CBM, and which concentrates the enzymes at the substrate surface and facilitates the synergistic action of the enzymes. The cellulosome also is bound to the cell surface, which allows the products of the cellulosome to be readily consumed by the cell.

The cellulosomes of mesophilic bacteria must play a major role in the turnover of carbon in nature, since most of their anaerobic habitats are in the temperate range of 15 to 45°C in contrast to thermophilic organisms that require temperatures above 50°C. Bacteria in rumens, termite guts, animal intestinal tracts, sewage, manure, compost piles, swamps, and soil, on fallen and decaying trees, and on plants probably never or rarely encounter thermophilic conditions. In certain natural situations it is likely that mesophilic activity occurs initially and then is followed by thermophilic activity such as in a compost pile.

The question then arises, "How will basic research on cellulosomes from mesophilic organisms contribute to the energy needs of our society?" Studies on cellulosomes should provide information on the following: (i) degradation of biomass by synergistic action of cellulosomal and noncellulosomal enzymes, (ii) construction of designer cellulosomes with specified functions,

Roy H. Doi • Section of Molecular & Cellular Biology, University of California, Davis, CA 95616.

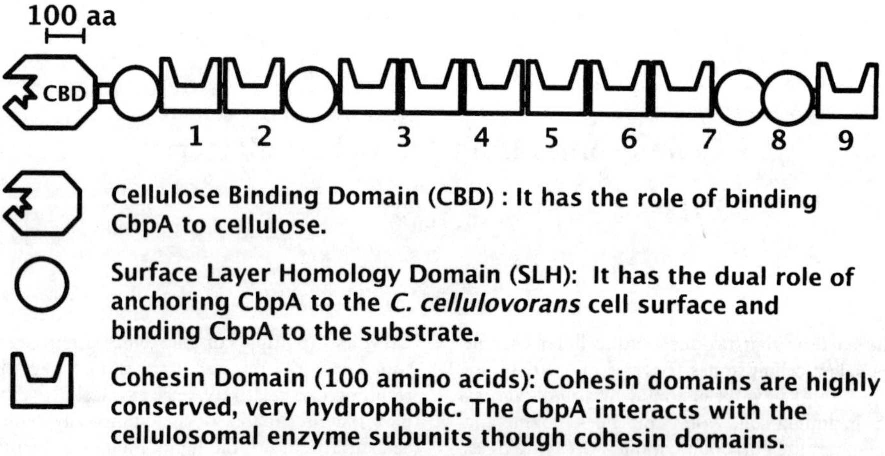

Figure 1. Model of the scaffolding protein CbpA of *Clostridium cellulovorans*.

(iii) utilization of cellulosomal enzymes for the autolysis of plant cell walls, and (iv) heterologous expression of cellulosomal enzymes for productive purposes.

MESOPHILIC CELLULOSOME PRODUCERS AND THEIR CELLULOSOMAL ENZYMES

A list of several mesophilic bacterial cellulosome producers is presented in Table 1. The bacteria that produce cellulosomes are anaerobes. The best-studied group includes *Clostridium* species and other bacterial species found in sewage, rumen, and soil. Not all *Clostridium* species produce an active cellulosome. However, it is likely that most *Clostridium* species contain genes for some, if not all, of the cellulosomal components. The list does not include *Clostridium thermocellum*, a thermophile, which is probably the best studied of all cellulosome producers. Furthermore, the list does not include a number of mesophilic bacterial species that are aerobic cellulose degraders but do not produce cellulosomes.

Another group that synthesizes cellulosomes includes *Ruminococcus albus*, which has been shown to contain a cellulolytic complex containing at least 15 proteins with cellulase and xylanase activities (Ohara et al., 2000; Morrison and Miron, 2000). Since immunological studies with antidockerin showed reactions with several of the *R. albus* proteins, it is likely that the cel-

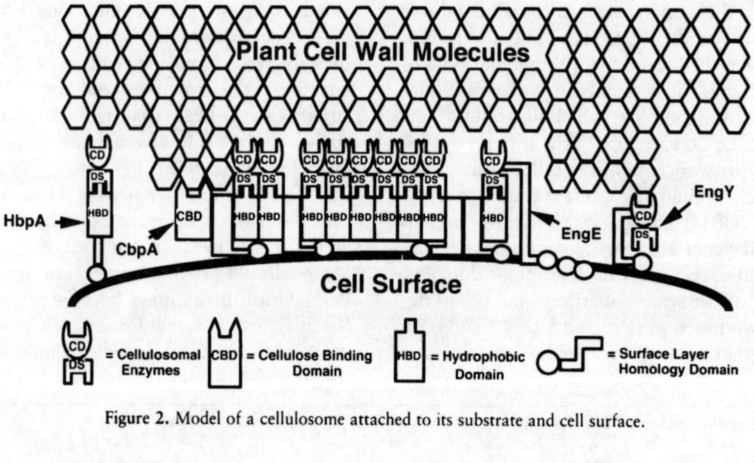

Figure 2. Model of a cellulosome attached to its substrate and cell surface.

Table 1. Cellulosome-producing mesophilic microorganisms

Acetivibrio cellulolyticus
Bacteroides cellulosolvens
Butyrivibrio fibrisolvens
Clostridium cellulolyticum
Clostridium cellulovorans
Clostridium josui
Clostridium papyrosolvens
Ruminococcus albus
Ruminococcus flavefaciens

lulosome is assembled by cohesin-dockerin interactions similar to that observed with clostridia.

Ruminococcus flavefaciens (Aurilia et al., 2000) also has multidomain dockerin-containing enzymes with xylanase and esterase activities, indicating that they are part of cellulosomes.

A. cellulolyticus is another cellulosome producer that forms a very complex cellulosome complex containing a scaffolding protein, CipV (or ScaA); an adaptor protein, ScaB, and an anchoring protein, ScaC (Xu et al., 2003). The intercalation of the three multiple cohesin-containing scaffoldins results in an amplification of the number of enzymes that can be bound in the complex. At least 96 enzymes can be bound in this large cellulosome.

The types of cellulolytic and hemicellulolytic enzymes synthesized by mesophilic cellulosome producers include endoglucanase, exoglucanase, xylanase, pectate lyase, mannanase, rhamnogalacturonan lyase, acetyl xylan esterase, α-galactosidase, and acetyl esterase. There are a large number of endoglucanases (families 5 and 9) and exoglucanases (family 48) that attack cellulose, the major component of plant cell walls. The major core enzymes for cellulosomes include a family 48 glycoside hydrolase (Liu and Doi, 1998; Reverbel-Leroy et al., 1997; Devillard et al., 2004) and endoglucanases of family 5 (Tamaru and Doi, 1999) or of family 9 (Gaudin et al., 2000).

The cellulosomal enzymes also include a variety of enzyme types that are capable of degrading hemicelluloses such as xylan, pectin, and mannan, and these include xylanases (Kosugi et al., 2002a; Han et al., 2004a, 2004b), mannanase (Tamaru and Doi, 2000; Perret et al., 2004a, 2004b), pectate lyase (Tamaru and Doi, 2001), rhamnogalacturonan lyase (Pages et al., 2003), acetyl xylan esterase (Kosugi et al., 2002a), α-galactosidase (Jindou et al., 2002), and acetyl esterase (Aurilia et al., 2000). Feruloyl esterase has been reported to be present in *C. thermocellum* cellulosomes (Blum et al., 2000), but so far not in mesophile cellulosomes. Feruloyl esterase may aid in the release of lignin from hemicellulose and may be involved in lignin solubilization. In the analysis of *C. thermocellum* and *Clostridium acetobutylicum*, it has been estimated that 50 to 60 cellulosomal enzyme genes are present in their genomes. It is interesting that the efficient degradation of a variety of plant cell walls requires this array of enzymes, suggesting that structural differences of the substrates necessitate this variety of enzymes. A more detailed account of mesophilic cellulosomal enzymes is presented in recent reviews (Doi et al., 2003; Doi and Kosugi, 2004).

The presence of a large number of cellulosomal cellulases seems redundant. However, when a group of family 9 enzymes from *Clostridium cellulovorans* were compared for their hydrolytic properties, it was found that each of the enzymes attacked substrates such as acid-swollen cellulose and microcrystalline cellulose (Avicel) in a slightly different manner. This was evident from the pattern of products that were derived (Arai et al., 2006). Thus, the battery of cellulases apparently is necessary for attacking various natural substrates and the presence of so many cellulases becomes more understandable. The presence of five family 9 enzymes in the *Clostridium cellulolyticum* gene cluster also suggests that the multiplicity of family 9 enzymes plays a role in attacking natural substrates in multiple ways (Gaudin et al., 2000; Belaich et al., 2002).

Mesophilic cellulosome producers also synthesize noncellulosomal enzymes that are involved in cellulose and hemicellulose degradation, such as α-arabinofuranosidase, β-galactosidase, xylanase, endoglucanase, exoglucanase, and β-glucan glucohydrolase. α-Arabinofuranosidase and β-galactosidase (Kosugi et al., 2002b) and xylanase (Foong et al., 1992) degrade xylan, as do family 5 and 9 cellulases (Sheweita et al., 1996; Ichiishi et al., 1998; Han et al., 2005a). One of the family 5 enzymes, EngD, has both cellulase and xylanase activities (Hamamoto et al., 1990; Foong et al., 1992). What is interesting is that although cellulosomes play a major role in degrading plant cell walls, the noncellulosomal enzymes also play an important cooperative role. This has been shown in studies on synergy between cellulosomes and noncellulosomal enzymes that are described below. The noncellulosomal enzymes may, by their size, free-acting ability, and preliminary action on natural substrates, play an important role in facilitating the access of the larger cellulosomes to the substrate.

CLUSTERING OF CELLULOSOMAL GENES

In mesophilic cellulosome producers, there are several cases in which several genes for cellulosome components have been found in clusters. This organization of genes seems reasonable since it may facilitate the coordinate expression and synthesis of the major subunits of the cellulosome. This cluster usually contains the gene for the scaffolding protein, the major endoglucanase (family 5 or 9), and the exoglucanase of family 48. In the case of *C. cellulovorans* the gene cluster includes the gene for the scaffolding protein CbpA, for a family 5 endoglucanase (EngE), for an exoglucanase (ExgS), several other family 5 and family 9 endoglucanases, and a mannanase (Tamaru et al., 2000). This cluster in *C. cellulovorans* has the composition *cbpA-exgS-engH-engK-hbpA-engL-manA-engM-engN* and is about 22 kb in length. In *C. cellulolyticum* the gene cluster consists of *cipC-cel48F-cel8C-cel9G-cel9E-orfX-cel9H-cel9J-man5K-cel9M-rglllY-cel9N* (Belaich et al., 2002; Pages et al., 2003; Perret et al., 2004a, 2004b). Similar but slightly varying situations occur with *Clostridium josui* (Kakiuchi et al., 1998) and *C. acetobutylicum* (Sabathe et al., 2002), in which gene clusters of varying sizes and compositions occur.

On the other hand, many of the genes for cellulosomal enzymes are scattered throughout the genome. In the case of *C. cellulovorans*, the gene for one of the major enzymatic subunits (EngE) found in virtually all the cellulosomes is not present in the gene cluster. Thus, the transcriptional regulation of these "cellulosomal" genes must be complex, since some type of coordinate expression followed by translation and secretion would be necessary for the orderly and efficient assembly of the various subunits into the cellulosome.

An analysis of the expression of cellulosomal genes of *C. cellulovorans* revealed that many of the cellulase and hemicellulase genes were expressed coordinately when cells were grown on cellulose or cellobiose (Han et al., 2003). Growth on cellulose, xylan, and pectin resulted in abundant expression of the genes present in the gene cluster containing *cbpA*, *exgS*, *engH*, *manA*, *engM*, and *xynA* (Tamaru et al., 2000) and other noncluster genes such as *engE* and *pelA*. The expressions of *xynA* and *pelA* were specifically induced at a higher level when cells were grown on xylan and pectin, respectively. Very little expression of the genes in the gene cluster was observed when cells were grown on glucose, suggesting that catabolite repression controlled the expression of cellulase and hemicellulase genes.

In the case of *C. cellulolyticum*, transcriptional analyses indicated that a 14-kbp transcript carried the coding sequence for *cipC*, *cel48F*, *cel8C*, *cel9G*, and *cel9E* (Maamar et al., 2006). Other smaller transcripts of 12 kbp for the genes in the 3′ part of the cluster and monocistronic transcripts for *cipC*, *cel48F*, and *cel9E* were also found. It was suggested that the presence of smaller transcripts allowed a modulated expression of the key subunits of *C. cellulolyticum* cellulosomes.

The gene cluster of cellulosomal subunits only partially explains how expression of cellulosomal genes may be coordinately regulated, since so many cellulosomal genes including genes for major components of the cellulosome are dispersed around the genome. Not only are coordinate expression and catabolite repression poorly understood at this time, but also the secretion and assembly of the cellulosomal subunits extracellularly need much further study. In order for proper assembly to occur, several proteins are secreted, properly folded into their mature form, and then bound to the scaffolding protein to form the mature cellulosome. The cellulosome is then attached to the cell surface and to the cellulosic substrate (Kosugi et al., 2004).

MULTICELLULOSOME POPULATIONS

One of the basic properties of cellulosome is their subunit compositional heterogeneity. In the case of *C. cellulovorans*, assuming that there are about 50 cellulosomal enzymes (based on studies with the *C. thermocellum* genome) and that they are all being expressed, there is the possibility of the presence of 50^9 different cellulosomes based on random binding of the subunits to the nine-cohesin-scaffolding protein. The actual number of subpopulations of cellulosomes at any one time of growth is probably less, since some subunits appear to be present in all cellulosomes, thus making them less diverse than any theoretical number. However, the diverse cellulosome population would allow the cell to degrade a variety of plant cell walls of various structure and biochemical compositions.

The heterogeneity of cellulosomes was illustrated most graphically by the results of Pohlschröder et al. (1994, 1995) with *Clostridium papyrosolvens*. They demonstrated that its cellulosome population could be fractionated into several parts depending on their charge and that the morphology of the cellulosomal subpopulations was significantly different when examined by electron microscopy. The one common subunit among the different cellulosome populations was the scaffolding protein. These key observations have been affirmed subsequently by other investigations.

A similar type of analysis with *C. cellulovorans* demonstrated that its cellulosome population could be fractionated according to differences in charge and that the fractions themselves included a mixture of cellulosome subpopulations, since each fraction contained a large number of different proteins. If a fraction based on charge was a pure population, one would expect from 2 to 10 different subunits in a homogeneous population. However, two-dimensional sodium dodecyl sulfate polyacrylamide gel electrophoresis of cellulosomes indicated the presence of 30 or more subunits in a particular charge fraction (Han et al., 2005b). Thus, each fraction was in itself a mixture of cellulosome subpopulations with differing subunit compositions. In the C.

cellulovorans cellulosomal subpopulations, three subunits were always present and included the scaffolding protein (CbpA), endoglucanase EngE (family 5), and exoglucanase ExgS (family 48). These three proteins therefore appeared to represent a "core" group of proteins for all cellulosomes and would allow the binding of seven other protein subunits.

In addition to the heterogeneity of the subpopulations of *C. cellulovorans* cellulosomes, the enzyme composition of the subpopulations was shown to change depending on the carbon substrate for growth of the bacterium (Han et al., 2004a, 2005b). For instance, the pattern of subunits in the cellulosomes changed when either cellulose, xylan, or pectin was the sole carbon source. This indicated that different carbon substrates could induce the expression of different cellulosomal genes and alter the subunit composition of cellulosomes. In a natural environment the subpopulations of cellulosomes would change according to the substrate that was present.

Another factor that may determine the enzyme composition of cellulosomes has been demonstrated by overexpressing the gene encoding Man5K of *C. cellulolyticum* (Perret et al., 2004a, 2004b). In this case it was found that the cellulosomes produced had greater mannanase activity and less cellulase activity than the native cellulosome. It was also shown that the cellulosome had a relatively larger amount of mannanase protein present in the cellulosome than that found in native cellulosomes, where mannanase protein was a very minor component.

In an analogous experiment, antisense RNA was used to limit the production of Cel48F, a major processive cellulase of *C. cellulolyticum* cellulosomes (Perret et al., 2004a, 2004b). The antisense RNA was made to block translation of the *cel48F* mRNA, and this strategy resulted in the reduced production of cellulosomes and a lower activity of the cellulosome on Avicel, although growth on cellulose was not affected in this strain. The results indicate that the amount of Cel48F in the cellulosomes was probably reduced but that the cell may have compensated for this loss by up-production of two other cellulolytic proteins.

The relative amount of a cellulosomal enzyme subunit appears to be controlled in part by the expression level of a particular cellulosomal enzyme gene. This may lead in part to an approach for designing cellulosomes with a particular set of functions.

Thus, the large number and variety of cellulosomes available to the cell make the cellulosomes an efficient plant cell wall-degrading enzyme complex. This mixture of cellulosomes makes the purification of a natural pure cellulosome preparation virtually impossible. All studies of natural cellulosomes therefore include a mixture of cellulosomes with varying functions. The study of pure cellulosome functions will require the assembly of cellulosomes with a known mixture of scaffolding protein and purified cellulosomal enzymes, i.e., the assembly of a designer cellulosome. This is discussed in further detail below.

SYNERGY OF CELLULOSOMAL ENZYMES

Early observations indicated that synergy existed between the cellulosome enzymes/subunits (Wu et al., 1988). A systematic analysis of synergy between various components of the cellulosome has revealed that synergy exists among the cellulosomal cellulases (Murashima et al., 2002; Han et al., 2004a) and between the cellulosomal cellulases and hemicellulases (Murashima et al., 2003; Han et al., 2004b). The complexity of studying synergism was illustrated during the analysis of synergy between cellulosomal cellulases. It was observed that attainment of maximum synergy between the cellulases depended on the proper ratio of cellulases in the enzyme mixture (Murashima et al., 2002). Furthermore, the sequential addition of the enzyme to the reaction mixture also had an effect on synergy, which suggested that some enzymes might facilitate the action of other enzymes by "preparing or exposing" the substrate. For instance endoglucanases could nick the cellulose substrate and allow exoglucanases to liberate cellobiose from these nicks. In the case of *C. cellulovorans*, where the scaffolding protein contains nine cohesins, one wonders whether maximum cellulase synergy requires an optimum number and type of cellulases to make a highly efficient cellulosome, or whether random occupation of a large number of scaffolding proteins by the cellulases, creating a wide diversity of cellulosome types, somehow is sufficient to create maximum synergy among the cellulases and among cellulosomes.

Since cellulosomes contain both cellulases and hemicellulases, it was of interest to determine whether synergism existed in the degradation of substrates such as corn cell wall by these enzymes. By constructing minicellulosomes containing either xylanase or cellulase, mixtures of these minicellulosomes were used to test their degradation of corn cell walls. The mixtures were found to have synergy at a molar ratio of 1:2 of cellulosomal xylanase to cellulase (Murashima et al., 2003). Since the two enzymes attacked different components of the cell wall, i.e., xylan or cellulose, it suggested that the synergism in this case was the result of each type of enzyme degrading its own substrate and liberating or exposing the substrate for the other enzyme. No synergistic effect with sequential reactions was observed in these studies.

SYNERGY OF CELLULOSOMES, CELLULOSOMAL ENZYMES, AND NONCELLULOSOMAL ENZYMES

Although much emphasis has been placed on the structure and function of cellulosomes, more recent studies suggest that bacteria may use not only cellulosomes, but also extracellular noncellulosomal degradative enzymes in a cooperative fashion (Han et al., 2004a). When cellulosomes and noncellulosomal fractions were incubated with carboxymethyl cellulose (CMC), Avicel, pectin, xylan, cellulose arabinoxylan (CAX), and corn fiber, the highest degree of synergy was observed against natural substrates such as corn fiber.

Synergy was also observed when cellulosomal xylanase, XynA, was used in a mixture with two noncellulosomal enzymes, arabinofuranosidase (ArfA) and β-galactosidase (BgaA). This mixture showed synergistic activity against corn fiber gum and corn stem powder (Kosugi et al., 2002b). This indicated that a cellulosomal hemicellulase could act synergistically with extracellular noncellulosomal enzymes.

In further investigations, cellulosomal cellulases and hemicellulases and noncellulosomal hemicellulases were tested for their cooperative activity on treated corn stover preparations such as CAX, corn fiber, and corn stem fiber (Doner and Johnston, 2001). The degradation of these substrates proceeded much more efficiently when XynA and ArfA were present in both simultaneous and sequential reactions (Koukiekolo et al., 2005). In sequential reactions, pretreatment of CAX with XynA and ArfA followed by treatment with EngL resulted in significant synergy. Thus, these studies indicate that cellulosomal and noncellulosomal hemicellulases and cellulosomal cellulases may play a sequential and cooperative role in the degradation of complex substrates.

In a further study, the action of cellulosomes with an extracellular noncellulosomal enzyme revealed that the products of cellulosomal activity could be further degraded prior to consumption by the cell. An extracellular enzyme, β-glucan hydrolase (BglA), that had higher activity on long gluco-oligomers than on cellobiose was found (Kosugi et al., 2006). Cello-oligosaccharides produced by the action of cellulosomes were further degraded to glucose and cellobiose by BglA. This cooperative action of cellulosomes with BglA illustrates again that cellulosomes do not act alone in the degradation of cellulose.

These examples of noncellulosomal enzymes cooperating with the cellulosome demonstrate a more prominent role for noncellulosomal enzymes in the degradation of plant cell wall materials than previously emphasized. It is likely that a sequential series of degradative reactions occur, initiated by noncellulosomal enzymes, and this is followed by the action of a battery of enzymes in the cellulosome.

DESIGNER CELLULOSOMES FOR MAXIMAL DEGRADATIVE ACTIVITY

It had been envisioned that highly active designer cellulosomes that could degrade lignocellulosic materials very efficiently could be developed (Bayer et al., 1994). This is based on the observation that cohesin-dockerin interactions are very firm and on the idea that specific scaffolding proteins could be constructed that would have cohesins that interacted specifically with cognate dockerins of active cellulases and hemicellulases, since it was found that there was a high degree of species specificity between cohesins and dockerins (Pages et al., 1997). Therefore, a construction of a scaffoldin with one cohesin from *C. thermocellum* and one cohesin from *C. cellulolyticum* would allow the formation of a minicellulosome that would bind one enzyme with a *C. thermocellum* dockerin and another enzyme with a *C. cellulolyticum* dockerin. Thus, there would be control of the enzyme composition and the sequence of the enzymes in the minicellulosome.

The formation of designer cellulosomes has been carried out successfully and applied to the study of cellulosome structure and function. In initial experiments, the miniscaffolding protein contained a cellulose binding domain and two cohesins from different species. When an endoglucanase and an exoglucanase were mixed with the miniscaffolding protein, chimeric minicellulosomes were obtained and the activity of the minicellulosomes exceeded the activity of the free enzymes, indicating a synergistic effect (Fierobe et al., 2001).

In subsequent experiments three different dockerin-engineered hybrid cellulases were constructed along with five chimeric scaffoldins and mixed with five native enzymes, resulting in the assembly of 75 different chimeric cellulosomes (Fierobe et al., 2002). It was again demonstrated that these minicellulosomes had much higher activity on a recalcitrant substrate such as Avicel than on a tractable substrate such as CMC. Also, the minicellulosomes had higher activity than the free enzymes, indicating that colocalization of the enzymes and targeting of the enzymes to the substrate by the cellulose binding domain played a role in the efficiency of the cellulosome complex. Although the minicellulosomes had more activity than free enzymes, when the activity of these minicellulosomes was compared to that of native cellulosomes on Avicel, they had only 1/10 of the activity of native cellulosomes. These studies indicated that enzyme proximity and substrate targeting were important in maximizing the activity of the minicellulosomes.

With a trifunctional designer cellulosome containing a miniscaffoldin with three cohesins from *C. thermocellum*, *C. cellulolyticum*, and *Ruminococcus flavefaciens* and *C. cellulolyticum* enzymes Cel48F, Cel9E, and Cel9G with cognate dockerins for the cohesins (Fierobe et al., 2005), it was found that the trifunctional cellulosome had an enhanced activity over the corresponding free enzymes on microcrystalline cellulose; that two Cel48F plus Cel9G did not have more activity than one Cel48F and two Cel9G; that enzyme location in the miniscaffoldin did not have an effect on maximum activity; that activity on hatched straw was less than that on microcrystalline cellulose, but greater than the activity of the free enzymes; and that XynZt when used as the third enzyme did not interfere with the cellulolytic activity of Cel48F and Cel9G, but did enhance the activity on hatched straw.

To illustrate the efficiency of the native cellulosome of *C. cellulolyticum*, the native cellulosome was found to be 5 times more active than the trifunctional cellulosome on microcrystalline cellulose and 6.5 times more active on straw (Fierobe et al., 2005). One has to remember that the native cellulosome contains a scaffolding protein with eight cohesins and is a mixture of cellulosomes with varying composition. This apparently gives a significant advantage to the larger native cellulosomes over the smaller designer minicellulosomes.

Designer minicellulosomes were used extensively to study synergistic actions between various cellulosomal cellulases (Murashima et al., 2002), between cellulosomal cellulases and xylanase (Murashima et al., 2003), and between cellulosomal enzymes and noncellulosomal enzymes (Koukiekolo, 2005). In these experiments, miniscaffolding proteins and defined enzymes were mixed to form minicellulosomes of known enzymatic composition, and the minicellulosomes were mixed to test for synergistic activity. In most cases, under the proper conditions synergy was obtained (see "Synergy of Cellulosomal Enzymes" and "Synergy of Cellulosomes, Cellulosomal Enzymes, and Noncellulosomal Enzymes" above).

The experiments to date with designer cellulosomes demonstrate their relative weakness compared to native cellulosome preparations. This could be attributed to the few cohesins present on the mini-CbpA and therefore the limited number of attached enzymes, the absence of correct ratios and kinds of enzymes in the minicellulosomes, and the lack of widespread subpopulations of cellulosomes in the native cellulosome population. In order to make an efficient cellulosome, one must consider the makeup of cellulosome in terms of scaffoldin size, correct mix and ratio of enzymes that constitute an active cellulosome, and the possible requirement of a very heterogeneous cellulosome population, especially when native substrates are to be degraded, since a variety of lignocellulosic structures have to be considered as the substrate. Another factor to consider in terms of efficient designer cellulosomes is their interaction with noncellulosomal enzymes that may facilitate the interaction of the cellulosome with the complex native substrate.

EXPRESSION OF CELLULOSOMAL GENES IN HETEROLOGOUS ORGANISMS

One of the problems in the utilization of anaerobic mesophilic cellulosome producers has been their slow growth and production of cellulosomes. One way to improve on this situation is to convert a fast-growing noncellulosome-producing aerobic bacterium into an efficient cellulosome producer. In this regard attempts are being made to convert *Bacillus subtilis*, an aerobic microorganism, into a cellulosome producer. Genes for a miniscaffolding protein, mini-CbpA, containing one cohesin, and an endoglucanase, EngB, from *C. cellulovorans* have been introduced into *B. subtilis* and successfully expressed to produce an extracellular minicellulosome consisting of mini-CbpA and EngB (Cho et al., 2004). This minicellulosome was able to degrade CMC but not Avicel.

In a further study of the *B. subtilis* system for expressing cellulase and hemicellulase genes, an intercellular complementation strategy has been used to express mini-CbpA by one strain and either EngB or XynB by another strain (Arai et al., 2007). When the mini-CbpA and EngB strains were cocultured, a minicellulosome was produced in the growth medium containing mini-CbpA and EngB. Coculturing of the mini-CbpA and XynB resulted in the formation of a minicellulosome containing mini-CbpA and XynB. The two minicellulosomes exhibited their respective enzyme activities. This system when expanded can be used to form designer cellulosomes with specific functions and with greater activity.

The introduction of larger miniscaffolding genes into *B. subtilis* containing four or more cohesins should allow the assembly of a larger cellulosome with additional activities such as an exoglucanase, xylanase, mannanase, and endoglucanases. In addition, the genes for noncellulosomal hemicellulolytic enzymes may have to be introduced to facilitate the synergistic action between noncellulosomal hemicellulolytic enzymes and the cellulosome.

The expression of mesophilic cellulosomal enzymes in other organisms has been proposed for a number of different reasons (Stephanopoulos, 2007). One potentially practical use would be to transform yeasts to be cellulose users in order to create an organism that could utilize cellulosic biomass and convert the resultant cellobiose and xylose to ethanol. In this regard, the mesophilic cellulases and hemicellulases would be very useful.

Another approach, which would lead to useful products, would be to convert organisms such as *Corynebacterium* and *Streptomyces* to cellulose utilizers. This would allow these organisms to use relatively inexpensive feedstock to produce amino acids, organic acids, and antibiotics.

As the rate-limiting step in the conversion of cellulosic biomass to ethanol is the degradation of plant cell walls to cellobiose, if plant cell walls can be made weaker or if plants can be engineered to express cellulases and hemicellulases under controlled conditions, it may be possible to convert plant cell walls more readily to cellobiose and xylose by a self-destructive mechanism (deconstruction) (Himmel et al., 2007). For instance if cellulases and hemicellulases in rice straw can be activated after rice has been harvested, the rice straw may be weakened to be readily attacked by further treatment with suitable degradative enzymes. The plant molecular biologist could use genes for mesophilic cellulases and hemicellulases for this purpose. The genes would have to be targeted to cells that would allow the enzymes to attack the plant cell walls and have the genes under the control of promoters that could be activated at the right moment, e.g., after the harvest. This same type of strategy could be applied to corn, switchgrass, and other biomass sources.

OTHER SOURCES OF CARBON SUBSTRATE—MARINE PLANTS—A COROLLARY

Although corn is the current favorite source of carbon for conversion to ethanolic biofuel, land availability, water, and soil conditions limit the amount of corn that can be grown for this purpose. Also, corn is an important commodity that might better be used as food. One source of carbon material that has not been emphasized is marine plants. The ocean covers about two-thirds of the earth's surface, and there are marine plants that consume much CO_2 and grow very rapidly. If a nutrient such as Fe is provided, can the ocean provide the fixed carbon that could be converted to ethanol? Can mesophilic cellulosomes degrade marine plant cell walls efficiently? Is there less or no lignin in marine plants? Would it be possible to harness marine plants (algae) as the source of carbon for conversion to ethanol? It would be an interesting feat for molecular biologists, marine plant biologists, and engineers to produce enough marine plants for production of feedstocks and biofuels.

ROLE OF MESOPHILIC CELLULOSOMES FOR BIOFUELS

What roles can mesophilic cellulosomes play in the production of biofuels? I assume that the cellulosomes will be used under mesophilic conditions to make them more competitive than thermophilic cellulosomes.

(i) If cellulosomes and cellulosome-producing organisms can be improved by genetic engineering, it is possible that a large amount of very efficient cellulosomes with highly synergistic enzymes could be produced for the large-scale degradation of lignocellulose to sugars. The current studies on construction of designer cellulosomes and on synergy of cellulolytic and hemicellulolytic enzymes are providing clues on how to make more-efficient and functionally specific designer cellulosomes. One question remains: will designer cellulosomes be better than the normally occurring mixture of cellulosomes produced by these mesophilic microorganisms?

(ii) If aerobic microorganisms can be transformed with specific genes to be designer-cellulosome producers, they may be useful for the production of copious amounts of specific functional cellulosomes that could be used for the conversion of biomass to sugars. By inserting specific cellulolytic and hemicellulolytic genes into an aerobic microorganism, such as *B. subtilis*, it should be possible to produce not only larger quantities of cellulosomes, but cellulosomes with known enzymatic compositions. These cellulosomes of known composition could be mixed to obtain a highly synergistic mixture of cellulosomes. This may allow aerobic microorganisms to play a major role in the conversion of cellulose to sugars.

(iii) If genes for mesophilic cellulases and hemicellulases are introduced into suitable plants for deconstruction purposes, plant cell walls could be made much more amenable for rapid hydrolysis and may require less pretreatment by chemicals or high pressure prior to enzymatic digestion. Since most plants grow under mild temperature conditions, mesophilic enzymes will be suitably active in plants. For instance, if cellulolytic and hemicellulolytic enzymes could be activated in the stalks of corn or rice straw after harvesting of the ears and grains, this could lead to partial hydrolysis of stalks in the field under mesophilic conditions. The use and activation of suitable postharvest promoters and the location of the degradative gene products near the plant cell walls should be feasible for plant molecular biologists.

CONCLUSION

Mesophilic cellulosomes and their enzymes can play a role in the future for biofuel production as a better understanding of the structure and function of the cellulosome is attained. The construction and utilization of more-efficient cellulosomes based on their enzymatic composition and synergy properties, the use of cellulosomal cellulases and hemicellulases for deconstruction of plant cell walls, and the conversion of aerobic microorganisms into cellulosome producers all have potential for improving the conversion of plant cell wall materials into sugars.

Acknowledgments. Research reported from my laboratory has been supported in part by the U.S. Department of Energy and by the Research Institute of Innovative Technology for the Earth (RITE Institute).

REFERENCES

Arai, T., A. Kosugi, H. Chan, R. Koukiekolo, H. Yukawa, M. Inui, and R. H. Doi. 2006. Properties of cellulosomal family 9 cellulases from *Clostridium cellulovorans*. *Appl. Microbiol. Biotechnol.* **71:** 654–660.

Arai, T., S. Matsuoka, H.-Y. Cho, H. Yukawa, M. Inui, S.-L. Wong, and R. H. Doi. 2007. Synthesis of *Clostridium cellulovorans* minicellulosomes by intercellular complementation. *Proc. Natl. Acad. Sci. USA* **104:**1456–1460.

Aurilia, V., J. C. Martin, S. I. McCrae, K. P. Scott, M. T. Rincon, and H. J. Flint. 2000. Three multidomain esterases from the cellulolytic rumen anaerobe *Ruminococcus flavefaciens* 17 that carry divergent dockerin sequences. *Microbiology* **146:**1391–1397.

Bayer, E. A., J. P. Belaich, Y. Shoham, and R. Lamed. 2004. The cellulosomes: multienzyme machines for degradation of plant cell wall polysaccharides. *Annu. Rev. Microbiol.* **58:**521–554.

Bayer, E. A., E. Morag, and R. Lamed. 1994. The cellulosome—a treasure trove for biotechnology. *Trends Biotechnol.* **12:**378–386.

Belaich, A., G. Parsiegla, L. Gal, C. Villard, R. Haser, and J.-P. Belaich. 2002. Cel9M, a new family 9 cellulase of the *Clostridium cellulolyticum* cellulosome. *J. Bacteriol.* **184:**1378–1384.

Blum, D. L., I. A. Kataeva, X.-L. Li, and L. G. Ljungdahl. 2000. Feruloyl esterase activity of the *Clostridium thermocellum* cellulosome can be attributed to previously unknown domains of XynY and XynZ. *J. Bacteriol.* **182:**1346–1351.

Cho, H.-Y., H. Yukawa, M. Inui, R. H. Doi, and S.-L. Wong. 2004. Production of minicellulosomes from *Clostridium cellulovorans* in *Bacillus subtilis* WB800. *Appl. Environ. Microbiol.* **70:**5704–5707.

Demain, A. L., M. Newcomb, and J. H. Wu. 2005. Cellulase, clostridia, and ethanol. *Microbiol. Mol. Biol. Rev.* **69:**124–154.

Desvaux, M. 2005. *Clostridium cellulolyticum*: model organism of mesophilic cellulolytic clostridia. *FEMS Microbiol. Rev.* **29:**741–764.

Devillard, E., D. B. Goodheart, S. K. R. Karnati, E. A. Bayer, R. Lamed, J. Miron, K. E. Nelson, and M. Morrison. 2004. *Ruminococcus albus* 8 mutants defective in cellulose degradation are deficient in two processive endocellulases, Cel48A and Cel9B, both of which possess a novel modular architecture. *J. Bacteriol.* **186:** 136–145.

Ding, S.-Y., E. A. Bayer, D. Steiner, Y. Shoham, and R. Lamed. 1999. A novel cellulosomal scaffoldin from *Acetivibrio cellulolyticus* that contains a family 9 glycosyl hydrolase. *J. Bacteriol.* **181:**6720–6729.

Doi, R. H., A. Kosugi, K. Murashima, Y. Tamura, and S.-O. Han. 2003. Cellulosomes from mesophilic bacteria. *J. Bacteriol.* **185:** 5907–5914.

Doi, R. H., and A. Kosugi. 2004. Cellulosomes: plant cell wall degrading enzyme complexes. *Nat. Rev. Microbiol.* **2:**541–551.

Doner, L. W., and D. B. Johnston. 2001. Isolation and characterization of cellulose/arabinoxylan residual mixtures from corn fiber gum processes. *Cereal Chem.* **78:**200–204.

Fierobe, H.-P., E. A. Bayer, C. Tardif, M. Czjzek, A. Mechaly, A. Belaich, R. Lamed, Y. Shoham, and J.-P. Belaich. 2002. Degradation of cellulose substrates by cellulosome chimerase: substrate targeting versus proximity of enzyme components. *J. Biol. Chem.* **277:**49621–49630.

Fierobe, H.-P., A. Mechaly, C. Tardif, A. Belaich, R. Lamed, Y. Shoham, J.-P. Belaich, and E. A. Bayer. 2001. Design and production of active cellulosome chimeras: selective incorporation of dockerin-containing enzymes into defined functional complexes. *J. Biol. Chem.* **276:**21257–21261.

Fierobe, H.-P., F. Mingardon, A. Mechaly, A. Belaich, M. Rincon, S. Pages, R. Lamed, C. Tardif, J.-P. Belaich, and E. A. Bayer. 2005. Action of designer cellulosomes on homogeneous versus complex substrates: controlled incorporation of three distinct enzymes into a defined tri-functional scaffoldin. *J. Biol. Chem.* **280:**16325–16334.

Foong, F. C.-F., and R. H. Doi. 1992. Characterization and comparison of *Clostridium cellulovorans* endoglucanases-xylanases EngB and EngD expressed in *Escherichia coli*. *J. Bacteriol.* **174:**1403–1409.

Gal, L., S. Pages, C. Gaudin, A. Belaich, C. Reverbel-Leroy, C. Tardif, and J. P. Belaich. 1997. Characterization of the cellulolytic complex (cellulosome) produced by *Clostridium cellulolyticum*. *Appl. Environ. Microbiol.* **63:**903–909.

Gaudin, C., A. Belaich, S. Champ, and J. P. Belaich. 2000. CelE, a multidomain cellulase from *Clostridium cellulolyticum*: a key enzyme in the cellulosome? *J. Bacteriol.* **182:**1910–1915.

Hamamoto, T., O. Shoseyov, F. Foong, and R. H. Doi. 1990. A *Clostridium cellulovorans* gene, *engD*, codes for both endo-β-1,4-glucanase and cellobiosidase activities. *FEMS Microbiol. Lett.* **72:**285–288.

Han, S.-O., H.-Y. Cho, H. Yukawa, M. Inui, and R. H. Doi. 2004a. Regulation of expression of cellulosomes and noncellulosomal (hemi)cellulolytic enzymes in *Clostridium cellulovorans* during growth on different carbon sources. *J. Bacteriol.* **186:**4218–4227.

Han, S.-O., H. Yukawa, M. Inui, and R. H. Doi. 2003. Regulation of expression of cellulosomal cellulase and hemicellulase genes in *Clostridium cellulovorans*. *J. Bacteriol.* **1185:**6067–6075.

Han, S.-O., H. Yukawa, M. Inui, and R. H. Doi. 2004b. Isolation and expression of the *xynB* gene and its product, XynB, a consistent component of the *Clostridium cellulovorans* cellulosome. *J. Bacteriol.* **186:**8347–8355.

Han, S.-O., H. Yukawa, M. Inui, and R. H. Doi. 2005a. Molecular cloning, transcriptional and expression analysis of *engO*, encoding a new noncellulosomal family 9 enzyme from *Clostridium cellulovorans*. *J. Bacteriol.* **187:**4884–4889.

Han, S.-O., H. Yukawa, M. Inui, and R. H. Doi. 2005b. Effect of carbon source on the cellulosomal subpopulations of *Clostridium cellulovorans*. *Microbiology* **151:**1491–1497.

Himmel, M. E., S.-Y. Ding, D. K. Johnson, W. S. Adney, M. R. Nimlos, J. W. Brady, and T. D. Foust. 2007. Biomass recalcitrance: engineering plants and enzymes for biofuels production. *Science* **315:**804–807.

Ichi-ishi, A., S. Sheweita, and R. H. Doi. 1998. Characterization of EngF from *Clostridium cellulovorans* and identification of a novel cellulose binding domain. *Appl. Environ. Microbiol.* **64:**1086–1090.

Jindou, S., S. Karita, E. Fujino, T. Fujino, H. Hayashi, T. Kimura, K. Sakka, and K. Ohmiya. 2002. α-Galactosidase Aga27A, an enzymatic component of the *Clostridium josui* cellulosome. *J. Bacteriol.* **184:**600–604.

Kakiuchi, M., A. Isui, K. Suzuki, T. Fujino, E. Fujino, T. Kimura, S. Karita, K. Sakka, and K. Ohmiya. 1998. Cloning and DNA sequencing of the genes encoding *Clostridium josui* scaffolding protein CipA and cellulase CelD and identification of their gene products as major components of the cellulosome. *J. Bacteriol.* 180:4303–4308.

Kosugi, A., Y. Amano, K. Murashima, and R. H. Doi. 2004. Hydrophilic domains of scaffolding protein CbpA promote glycosyl hydrolase activity and localization of cellulosomes to the cell surface of *Clostridium cellulovorans*. *J. Bacteriol.* 186:6351–6359.

Kosugi, A., T. Arai, and R. H. Doi. 2006. Degradation of cellulosome-produced cello-oligosaccharides by an extracellular non-cellulosomal β-glucan glucohydrolase, BglA, from *Clostridium cellulovorans*. *Biochem. Biophys. Res. Commun.* 349:20–23.

Kosugi, A., K. Murashima, and R. H. Doi. 2002a. Xylanase and acetyl xylan esterase activities of XynA, a key subunit of the *Clostridium cellulovorans* cellulosome for xylan degradation. *Appl. Environ. Microbiol.* 68:6399–6402.

Kosugi, A., K. Murashima, and R. H. Doi. 2002b. Characterization of two noncellulosomal subunits, ArfA and BgaA, from *Clostridium cellulovorans* that cooperate with the cellulosome in plant cell wall degradation. *J. Bacteriol.* 184:6859–6865.

Koukiekolo, R., H.-Y. Cho, A. Kosugi, M. Inui, H. Yukawa, and R. H. Doi. 2005. Degradation of corn fiber by *Clostridium cellulovorans* cellulases and hemicellulases and contribution of scaffolding protein CbpA. *Appl. Environ. Microbiol.* 71:3504–3511.

Liu, C.-C., and R. H. Doi. 1998. Properties of *exgS*, a gene for a major subunit of the *Clostridium cellulovorans* cellulosome. *Gene* 211:39–47.

Lynd, L. R., P. J. Weimer, W. H. van Zyl, and I. S. Pretorius. 2002. Microbial cellulose utilization: fundamentals and biotechnology. *Microbiol. Mol. Biol. Rev.* 66:506–577.

Maamar, H., L. Abdou, C. Boileau, O. Valetter, and C. Tardif. 2006. Transcriptional analysis of the *cip-cel* gene cluster from *Clostridium cellulolyticum*. *J. Bacteriol.* 188:2614–2624.

Morrison, M., and J. Miron. 2000. Adhesion to cellulose by *Ruminococcus albus*: a combination of cellulosomes and Pil-proteins? *FEMS Microbiol. Lett.* 185:109–115.

Murashima, K., A. Kosugi, and R. H. Doi. 2002. Synergistic effects on crystalline degradation between cellulosomal cellulases from *Clostridium cellulovorans*. *J. Bacteriol.* 184:5088–5095.

Murashima, K., A. Kosugi, and R. H. Doi. 2003. Synergistic effects of cellulosomal xylanase and cellulases from *Clostridium cellulovorans* on plant cell wall degradation. *J. Bacteriol.* 185:1518–1524.

Ohara, H., S. Karita, T. Kimura, K. Sakka, and K. Ohmiya. 2000. Characterization of the cellulolytic complex (cellulosome) from *Ruminococcus albus*. *Biosci. Biotechnol. Biochem.* 64:254–260.

Pages, S., A. Belaich, J.-P. Belaich, E. Morag, R. Lamed, Y. Shoham, and E. A. Bayer. 1997. Species-specificity of the cohesin-dockerin interaction between *Clostridium thermocellum* and *Clostridium cellulolyticum*: prediction of specificity determinants of the dockerin domain. *Proteins* 29:517–527.

Pages, S., A. Belaich, C. Tardif, C. Reverbel-Leroy, C. Gaudin, and J.-P. Belaich. 1996. Interaction between the endoglucanase CelA and the scaffolding protein CipC of the *Clostridium cellulolyticum* cellulosome. *J. Bacteriol.* 178:2279–2286.

Pages, S., O. Valette, L. Abdou, A. Belaich, and J.-P. Belaich. 2003. A rhamnogalacturonan lyase in the *Clostridium cellulolyticum* cellulosome. *J. Bacteriol.* 185:4727–4733.

Perret, S., A. Belaich, H.-P. Fierobe, J.-P. Belaich, and C. Tardif. 2004a. Towards designer cellulosomes in clostridia: mannanase enrichment of the cellulosomes produced by *Clostridium cellulolyticum*. *J. Bacteriol.* 186:6544–6552.

Perret, S., H. Maamar, J.-P. Belaich, and C. Tardif. 2004b. Use of antisense RNA to modify the composition of cellulosomes produced by *Clostridium cellulolyticum*. *Mol. Microbiol.* 51:599–607.

Pohlschröder, M., E. Canale-Parola, and S. B. Leschine. 1995. Ultrastructural diversity of the cellulase complexes of *Clostridium papyrosolvens* C7. *J. Bacteriol.* 177:6625–6629.

Pohlschröder, M., S. B. Leschine, and E. Canale-Parola. 1994. Multicomplex cellulase-xylanase system of *Clostridium papyrosolvens* C7. *J. Bacteriol.* 176:70–76.

Reverbel-Leroy, C., S. Pages, A. Belaich, J.-P. Belaich, and C. Tardif. 1997. The processive endocellulase CelF, a major component of the *Clostridium cellulolyticum* cellulosome: purification and characterization of the recombinant form. *J. Bacteriol.* 179:46–52.

Sabathe, F., A. Belaich, and P. Soucaille. 2002. Characterization of the cellulolytic complex (cellulosome) of *Clostridium acetobutylicum*. *FEMS Microbiol. Lett.* 217:15–22.

Schwarz, W. H. 2001. The cellulosome and cellulose degradation by anaerobic bacteria. *Appl. Microbiol. Biotechnol.* 56:634–649.

Sheweita, S.A., A. Ichi-ishi, J.-S. Park, C.-C. Liu, L. M. Malburg, Jr., and R. H. Doi. 1996. Characterization of *engF*, a gene for a non-cellulosomal *Clostridium cellulovorans* endoglucanase. *Gene* 182:163–167.

Shoseyov, O., M. Takagi, M. Goldstein, and R. H. Doi. 1992. Primary sequence analysis of *Clostridium cellulovorans* cellulose binding protein A (CbpA). *Proc. Natl. Acad. Sci. USA* 89:3483–3487.

Stephanopoulos, G. 2007. Challenges in engineering microbes for biofuels production. *Science* 315:801–804.

Tamaru, Y., and R. H. Doi. 1999. Three surface layer homology domains at the N terminus of the *Clostridium cellulovorans* major cellulosomal subunit EngE. *J. Bacteriol.* 181:3270–3276.

Tamaru, Y., and R. H. Doi. 2000. The *engL* gene cluster of *Clostridium cellulovorans* contains a gene for cellulosomal ManA. *J. Bacteriol.* 182:244–247.

Tamaru, Y., and R. H. Doi. 2001. Pectate lyase A, an enzymatic subunit of the *Clostridium cellulovorans* cellulosome. *Proc. Natl. Acad. Sci. USA* 98:4125–4129.

Tamaru, Y., S. Karita, A. Ibrahim, H. Chan, and R. H. Doi. 2000. A large gene cluster for the *Clostridium cellulovorans* cellulosome. *J. Bacteriol.* 182:5906–5910.

Wu, J. H. D., W. H. Orme-Johnson, and A. L. Demain. 1988. Two components of an extracellular protein aggregate of *Clostridium thermocellum* together degrade crystalline cellulose. *Biochemistry* 27:1703–1709.

Xu, Q., W. Gao, S.-Y. Ding, R. Kenig, Y. Shoham, E. A. Bayer, and R. Lamed. 2003. The cellulosome system of *Acetivibrio cellulolyticus* includes a novel type of adaptor protein and a cell surface anchoring protein. *J. Bacteriol.* 185:4548–4557.

Bioenergy
Edited by J. Wall et al.

Chapter 8

Cohesin-Dockerin Interactions and Folding

J. H. DAVID WU, MICHAEL NEWCOMB, AND KAZUO SAKKA

STUDIES ON COHESIN-DOCKERIN INTERACTIONS

As described in preceding chapters, the cellulosome consists of a noncatalytic scaffolding protein and many catalytic subunits, held together through the interaction between a cohesin of the scaffolding protein and the dockerin borne on each catalytic subunit (Color Plate 2) (Demain et al., 2005; Gilbert, 2007; Kataeva, 2006). The cellulosome assembly thus depends on the cohesin-dockerin interaction. A scaffolding protein typically contains multiple cohesins, and a cellulosomal catalytic subunit typically contains only one dockerin. In earlier studies, the importance of a large scaffolding protein in cell adhesion to cellulose and in maintaining the cellulosome integrity was recognized by the observation that a mutant strain of *Clostridium thermocellum* deficient in this protein showed weaker cell adhesion to cellulose and a decreased cell-associated cellulolytic activity (Bayer et al., 1983). An anchor-enzyme model was subsequently proposed to explain the synergism between the scaffolding protein and a major catalytic subunit (CelS or Cel48A) in degrading crystalline cellulose, providing the first insight into the cellulosome assembly (Wu and Demain, 1988; Wu et al., 1988). However, the mechanism of the assembly was not elucidated until the structures of the genes encoding the scaffolding proteins of *Clostridium cellulovorans* (Shoseyov et al., 1992) and *C. thermocellum* (Gerngross et al., 1993) were determined. The existence of the dockerin was first reported when a duplicated sequence was found at the C terminus of the *C. thermocellum celA* gene encoding a catalytic subunit (Grepinet et al., 1988). The duplicated segment (now known as the dockerin) was first thought to be involved in substrate binding. However, a far-Western blotting analysis showed that the dockerin of *C. thermocellum* CelD or XynZ binds to the scaffolding protein (Tokatlidis, 1991). It was then obvious that the dockerin binds to the cohesin, and such an interaction is the key to the cellulosome assembly. Many investigations were subsequently conducted to characterize the cohesin-dockerin interaction. These studies include measuring cohesin-dockerin affinity, identifying critical amino acid residues by site-directed mutagenesis, and determining the molecular structures of the dockerin, cohesin, and its complex.

PRIMARY STRUCTURE OF DOCKERIN AND COHESIN

As described above, the scaffolding protein and catalytic subunits are assembled into the cellulosome by the cohesin-dockerin interaction and therefore each cellulosomal catalytic subunit is presumed to contain a dockerin. The dockerin consists of a pair of well-conserved, 22-residue repeats spaced by a linker of 8 to 18 residues. Some of the dockerin sequences from different microorganisms are aligned in Fig. 1. Based on sequence homology, dockerins have been classified into three types (Xu et al., 2004): types I, II, and III. Type I is further divided into subtypes Ia, Ib, and Ic. The type Ia dockerins include those from *C. thermocellum*. The type Ib dockerins include those from mesophilic clostridia such as *C. cellulolyticum* and *C. josui*. The type Ic dockerins include those from *Acetivibrio cellulolyticus* and *Bacteroides cellulosolvens*. The type II dockerins are exemplified by those from the *C. thermocellum* and *A. cellulolyticus* scaffolding proteins as well as a *B. cellulosolvens* cellulase. Examples of the type III dockerins include those from the *Ruminococcus flavefaciens* enzymes and scaffolding protein. A characteristic of the dockerin is that it contains a segment resembling the EF-hand Ca^{2+}-binding motif in each of its repeated sequences (Fig. 1) (Chauvaux et al., 1990;

J. H. David Wu and Michael Newcomb • Department of Chemical Engineering, University of Rochester, Rochester, NY 14627.
Kazuo Sakka • Graduate School of Bioresources, Mie University, Tsu 514-8507, Japan.

```
                                 α1           α2
     Ct Xyn10B    733 GDVNGDGTINSTDLTMLKRSVLRAITLTDDAKAR
Ia   Ct Cel48A    675 GDVNDDGKVNSTDAVALKRYVLRSGISINTDN
     Ct Xyn10C    559 GDVNGDGRVNSSDLTLMKRYLLKSISDFPTPEGKIA
     Ct Xyn11A    393 GDLNGDKQVNSTDYTALKRHLLNITRLSGTALAN
     Cj Aga27A    405 GDVDGNGMIDALDYSLVKRYLLGQISDCPDSKGKLA
Ib   Cj Cel8A     402 GDVNNDGAIDALDIAALKKAILTQSTSNINLTN
     Cc Cel5A     414 GDYNNDGNVDALDFAGLKKYIMAADHAYVKN
     Cc Cel5N     479 GDVNDDGNVDALDLAVMKQYILGANPKISLTN
Ic   Ac Cel9B     889 GDVNGDGSLNSIDFGVMRKYLLGMIKEFSYENGLKA
     Ac ScaB      872 GDVDGNGSVRINDAVLIRDYVLGKINEFPYEYGMLA
II   Ct CipA     1791 GDIVKDNSINLLDVAEVIRCFNATKGSANYVEE
     Ac ScaA     1859 QDGAINLEDILEICKAFNTSSTDAKYQVG
III  Rf EndA      692 GDGVVDLADAIFIMQCLANPNKYQLSDAGRVA
     Rf XynB      713 DDGNLDLADAILIMQALANPNKYGLEGTAEKHLTEQGKVN

                                 α3
     Ct Xyn10B    767 ADVDKNGSINSTDVLLLSRYLL
Ia   Ct Cel48A    712 ADLNEDGRVNSTDLGILKRYILKEIDTLPYKN*
     Ct Xyn10C    595 ADLNEDGKVNSTDLLALKKLVLREL*
     Ct Xyn11A    427 ADLNGDGKVDSTDLMILHRYLLGIISSFPRSNPQ
     Cj Aga27A    441 ADVDGDQQITALDFSLIKQYLLGTINKFPAQTASKIKP*
Ib   Cj Cel8A     438 ADMNNDGNIDAIDFAQLKVKLLN*
     Cc Cel5A     445 LDVNLDNEVNAFDLAILKKYLLGMVSKLPSN*
     Cc Cel5N     511 SDMNSDGDINALDFAMLKAKVLGK*
Ic   Ac Cel9B     925 GDVDGNGMFNSLDFAYMRQYMLGIISKFPVQK*
     Ac ScaB      908 ADVDGNGSIKINDAVLVRDYVLGKIFLFPVEE*
II   Ct CipA     1824 LDINRNGAINMQDIMIVHKHFGATSSDYDAQ*
     Ac ScaA     1888 LDLNRDGAISLEDVMIVAKHFNKVSSDY*
III  Rf EndA      724 ADVYGDDGVTGDDAMAIQLLLIKKISSLPVPSGTVL*
     Rf XynB      753 ADVDKSTKGLTADDALMIQEFLIGKVAS*
```

Figure 1. Alignment of amino acid sequences of some dockerin modules classified as types Ia, Ib, Ic, II, and III. Three α helices are indicated with arrows, and the Ca^{2+}-binding loops homologous to the EF-hand motif are indicated with horizontal bars. Amino acid positions 11 and 12, including the highly conserved ST motif in type Ia and AL motif in type II, are boxed. Ct, *C. thermocellum*; Cj, *C. josui*; Cc, *C. cellulolyticum*; Ac, *A. cellulolyticus*; Rf, *R. flavefaciens*.

Lytle et al., 1996). In fact, the cohesin-dockerin interaction was shown to be Ca^{2+}-dependent (Choi and Ljungdahl, 1996; Yaron et al., 1995). Another characteristic is that the amino acid residues at positions 11 and 12 of the types Ia and Ib dockerins are conserved: "ST" or "SS" (type Ia) or "AL" (type Ib). In contrast, the amino acid residues in these two positions are not conserved in the other types of dockerin (Fig. 1). The importance of this conserved motif of two amino acids in cohesin-dockerin recognition is discussed below.

In general, a scaffolding protein that tethers catalytic components is a multidomain protein consisting of multiple cohesins and a cellulose-binding module (CBM) (Fig. 2). A scaffolding protein may also contain a type II or III dockerin. Cohesins are highly conserved within the same scaffolding protein, with sequence identities higher than 50%. Parallel to the classification of dockerins, cohesins are also classified into three types and the type I cohesins are further divided into subtypes Ia, Ib, and Ic. Sequence identities among cohesins of different subtypes (Ia, Ib, and Ic) are about 30 to 40%.

COHESIN-DOCKERIN BINDING SPECIFICITY

Various methods have been used for detecting cohesin-dockerin interactions. The methods for qualitative detection include far-Western blotting analysis using a membrane and gel retardation assays with nondenaturing polyacrylamide gel electrophoresis (PAGE). The methods for quantitative detection include isothermal titration, surface plasmon resonance (SPR), and enzyme-linked interaction assays.

In an earlier study employing the far-Western blotting technique (Yaron et al., 1995), a cloned fragment of *C. thermocellum* CipA, cohesin 2-cohesin 3, was found to bind to all but one cellulosomal catalytic component, suggesting a lack of selectivity in cohesin-dockerin interaction. However, the C-terminal dockerin of CipA

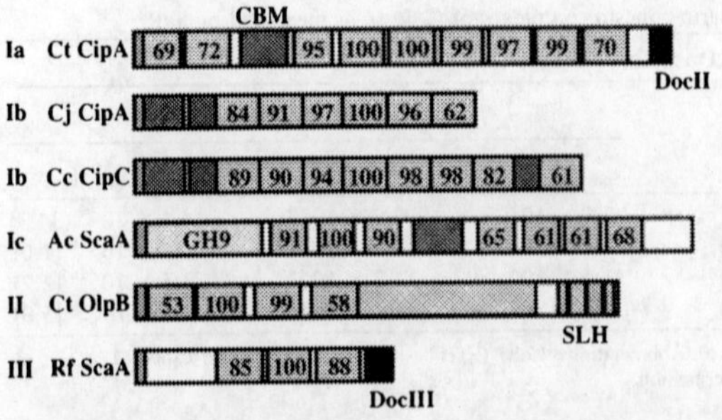

Figure 2. Schematic drawings of some scaffolding proteins from *C. thermocellum*, *C. josui*, *C. cellulolyticum*, *A. cellulolyticus*, and *R. flavefaciens*. Numerical values are sequence identities (in percentages) within each scaffolding protein. CBM, cellulose-binding module; GH9, glycosyl hydrolase family 9; SLH, surface layer homology; Doc, dockerin.

(type II) failed to recognize its own cohesin (type I). It was later found to bind to a different type (i.e., type II) of cohesin present in a few surface layer proteins such as SdbA, OlpB, and Orf2P (Leibovitz and Beguin, 1996). The interactions between the cohesin of the *C. thermocellum* CipA scaffolding protein and the dockerin of cellulosomal catalytic components are categorized as type I, and the interactions between the dockerin of CipA and its counterpart are categorized as type II (Color Plate 2). The interactions are specific since no type I-type II interactions have been reported. This type of specificity is hereafter referred to as the type I-type II specificity or intraspecies specificity.

The type I cohesins and dockerins include those modules from different microorganisms. This classification, however, is based on sequence homology and does not necessarily imply recognition among the same type of modules. Indeed, interspecies specificity was demonstrated using cohesins and dockerins from *C. thermocellum* and *C. cellulolyticum*, respectively (Pages et al., 1997). In this work, Pages et al. found that *C. thermocellum* Cel48A (i.e., CelS), with its dockerin, did not recognize cohesin 1 of the *C. cellulolyticum* scaffolding protein CipC. Furthermore, the *C. cellulolyticum* Cel5A, also with its dockerin, did not recognize cohesin 2 of the *C. thermocellum* scaffolding protein CipA. Interspecies specificity among the type I modules was also observed between *C. thermocellum* and *C. josui* (Jindou et al., 2004). These observations suggest that subtle differences in the structures of cohesins and dockerins may affect recognition despite highly conserved sequences.

As shown in Fig. 2, among the cohesins of one scaffolding protein, the sequence identity can be as low as 50 to 70%, suggesting that binding preferences may exist. This issue has been studied in the *C. thermocellum* CipA by using a "gel-shift" assay. In such an assay, the recombinant dockerin and cohesin are subject to nondenaturing PAGE. Complex formation between the two modules results in a shift of the mobility in the gel. Such analysis revealed that the dockerin of *C. thermocellum* Cel48A (i.e., CelS) forms a stable complex with at least cohesins 1, 2, 3, 4, and 9 of *C. thermocellum* CipA without discrimination (Lytle and Wu, 1998). Conversely, the dockerin of CipA (type II) does not bind to any of the cohesins tested. The lack of binding selectivity among the CipA cohesins was confirmed by quantitative analysis using SPR, which revealed that all the cohesins tested (cohesins 1, 2, 3, 4, and 7) of *C. thermocellum* CipA form a complex with the dockerin of Xyn10C or Xyn11A. The dissociation constants (K_D) of these cohesin-dockerin pairs were all so low that they were below the detection limit of SPR (K_D $<10^{-10}$) (Jindou et al., 2004).

Nondenaturing PAGE analysis using recombinant CelS dockerin and two split halves of the dockerin revealed that both duplicated sequences within the dockerin are required for docking to the cohesin (Lytle and Wu, 1998). A stable complex could be formed only when both of the duplicated sequences were present with the cohesin. The dockerin-cohesin complex migrated with a higher mobility on the gel than the dockerin alone, suggesting that the dockerin undergoes a conformation change to a more compact shape upon complex formation with the cohesin. Despite the fact that both halves of the dockerin are essential for binding, the two halves are homologous enough to replace each other for complex formation with the cohesin, as demonstrated in *C. cellulolyticum* (Fierobe et al., 1999).

COHESIN-DOCKERIN BINDING AFFINITY

The affinity between the cohesin and dockerin has been determined using SPR. For *C. josui* (Jindou et al., 2004), the dockerin from Aga27A, which recognizes the *C. josui* cohesin but not the *C. thermocellum* cohesin, binds to cohesins 1, 2, 5, and 6 of the *C. josui* scaffolding protein with k_{on} values of 1.8×10^5 to 6.9×10^5 and k_{off} values of 1.3×10^{-4} to 3.4×10^{-4}, leading to the K_D values ranging from 1.9×10^{-9} to 2.4×10^{-10}. It showed the highest affinity toward cohesin 2 ($K_D = 2.4 \times 10^{-10}$) and the lowest affinity toward cohesin 1 ($K_D = 1.9 \times 10^{-9}$), resulting in an eightfold difference in the affinity (Table 1). The dockerin of Cel8A similarly showed binding preference for cohesin 2 ($K_D = 1.3 \times 10^{-10}$) over cohesin 1 ($K_D = 4.4 \times 10^{-9}$) with a 34-fold difference in the affinity. The dockerin of *C. cellulolyticum* CelA bound to cohesin 1 of CipC with a K_D value of 2.5×10^{-10} (Fierobe et al., 1999). Thus, the affinity between cohesin and dockerin seems to be universally high, explaining the stability of the cellulosome in the extracellular environment.

Table 1. Association and dissociation constants of various dockerin-cohesin complexes of *C. josui* as measured by SPR[a]

Cohesin	Dockerin					
	Cj-rAga27ADoc			Cj-rCel8ADoc		
	k_{on} s^{-1} M^{-1}	k_{off} s^{-1}	K_D M	k_{on} s^{-1} M^{-1}	k_{off} s^{-1}	K_D M
Coh1-Cj	1.8×10^5	3.4×10^{-4}	1.9×10^{-9} (7.9)	2.1×10^5	9.0×10^{-4}	4.4×10^{-9} (34.3)
Coh2-Cj	6.9×10^5	1.7×10^{-4}	2.4×10^{-10} (1.0)	1.1×10^6	1.4×10^{-4}	1.3×10^{-10} (1.0)
Coh5-Cj	3.9×10^5	2.5×10^{-4}	6.3×10^{-10} (2.6)	1.4×10^6	4.7×10^{-4}	3.5×10^{-10} (2.7)
Coh6-Cj	2.1×10^5	1.3×10^{-4}	6.5×10^{-10} (2.7)	1.2×10^6	7.4×10^{-4}	6.3×10^{-10} (5.0)

[a]Numbers in parentheses indicate values relative to the smallest K_D value, defined as 1.0. Abbreviations: Coh1-Cj, cohesin 1 of *C. josui* CipA; Cj-rAga27ADoc, recombinant dockerin of *C. josui* Aga27A. Reproduced from Jindou et al., 2004, with permission.

ST MOTIF OF THE DOCKERIN AND INTERSPECIES SPECIFICITY

As mentioned above, interspecies specificity was observed for the cohesin-dockerin interaction between *C. thermocellum* (type Ia) and mesophilic clostridia such as *C. cellulolyticum* and *C. josui* (both type Ib). Since ST or SS at positions 11 and 12 in type Ia dockerins of *C. thermocellum* and AL at the same positions of type Ib dockerins of *C. cellulolyticum* and *C. josui* are highly conserved, these residues were presumed to play an important role in determining the interspecies specificity. In fact, when an ST motif in the first segment of the *C. thermocellum* CelS dockerin was changed to AL by site-directed mutagenesis, the mutant protein acquired new affinity ($K_D = 4.5 \times 10^{-9}$) toward cohesin 1 (type 1b) of *C. cellulolyticum* CipC scaffolding protein while maintaining the affinity toward cohesin 2 (type 1a) of *C. thermocellum* CipA (Mechaly et al., 2001). Mutations of ST motifs in both reiterated segments into AL reduced the affinity of the dockerin toward the *C. thermocellum* cohesin by about 2 orders of magnitude but did not completely abolish it, suggesting that additional amino acid residues are important for determining the interspecies binding specificity. The involvement of additional amino acid residues in determining the interspecies specificity may be further illustrated by a surprising finding of an exception to the rule of interspecies specificity. The *C. thermocellum* Xyn11A dockerin having two conserved ST motifs showed a high affinity toward both the *C. josui* and *C. thermocellum* cohesins (Table 2) (Jindou et al., 2004).

Table 2. Association and dissociation constants of the complexes between the recombinant dockerin of *C. thermocellum* Xyn11A and various cohesins of *C. josui* CipA, measured by SPR[a]

Cohesin	Dockerin, Cj-rXyn11ADoc		
	k_{on} s^{-1} M^{-1}	k_{off} s^{-1}	K_D M
Coh1-Cj	8.0×10^5	5.5×10^{-3}	6.9×10^{-9} (1.1)
Coh2-Cj	2.5×10^5	2.3×10^{-3}	9.2×10^{-9} (1.4)
Coh5-Cj	9.3×10^5	5.9×10^{-3}	6.4×10^{-9} (1.0)
Coh6-Cj	2.6×10^5	4.4×10^{-3}	1.7×10^{-8} (2.7)

[a]Numbers in parentheses indicate values relative to the smallest K_D value, defined as 1.0. Reproduced from Jindou et al., 2004, with permission.

It thus appears that the amino acid residues at positions 11 and 12 are not the sole determinants of interspecies specificity between type 1a and type 1b modules. However, these two amino acid residues seem to play a more important role in determining the specificity between types I and II modules within one species. When amino acids in positions 11 and 12 of *C. thermocellum* CelD dockerin (ST or SS; type I) were swapped with their equivalents in the CipA dockerin (LL or MQ; type II), the mutant dockerin still bound to cohesin 7 of CipA, if the swapping occurred in only one of the two duplicated segments (Schaeffer et al., 2002). On the other hand, swapping in both segments completely abolished the affinity toward cohesin 7 (type I) as expected but did not acquire a new affinity toward the type II cohesin of SdbA. Conversely, the mutant type II dockerin of CipA bound to the type II cohesin of SdbA on the condition that the second segment was intact. None of the mutated dockerins displayed detectable binding to the noncognate cohesin. These findings suggest that type I-type II intraspecies specificity is more stringently controlled than the type Ia-type Ib interspecies specificity.

MOLECULAR STRUCTURES OF COHESINS

The extremely high affinity between cohesin and dockerin and their important roles in cellulosome assembly have prompted interest in determining their molecular structures to elucidate the molecular mechanism of the cohesin-dockerin recognition. The first reported molecular structure revealed by X-ray diffraction was that of cohesin 2 of *C. thermocellum* CipA (Shimon et al., 1997). The cohesin was found to be an elongated, conical molecule formed by a nine-stranded β sandwich with a jelly roll topology without an apparent binding pocket or cleft. This structure is similar to the fold displayed by its neighboring CBM of family 3, which has a flat surface for binding to crystalline cellulose.

The lack of binding selectivity among the nine cohesins of *C. thermocellum* CipA suggests that the amino acid residues responsible for molecular recognition are well conserved. A survey of amino acid sequences of these cohesins indicates that the loop regions, in particular loops 4-5 and 6-7, are totally conserved in all of the nine cohesins of CipA. In contrast, the cohesins from *C. cellulolyticum* have different amino acid sequences in these loop regions, suggesting that the loop regions are responsible for interaction with the dockerin and determining the interspecies specificity.

The crystal structure of cohesin 7 of *C. thermocellum* CipA was reported (Tavares et al., 1997). The overall structure of cohesin 7 was highly similar to that of cohesin 2. β-Strands 5, 6, 3, and 8 form one of the two β-sheets on one face, and β-strands 4, 7, 2, and 9 form another on the other face. Site-directed mutagenesis in cohesin 7 showed that four individual mutations drastically affected its affinity toward the CelD dockerin and that these were located in the β-sheet consisting of the strands 5, 6, 3, and 8 and loops 4-5, 6-7, and 8-9 (Miras et al., 2002). These observations strongly suggest that the four-stranded 5, 6, 3, and 8 β-sheet is the dockerin-binding interface.

The X-ray structure of cohesin 1 of *C. cellulolyticum* CipC was also determined to be a β-barrel with a jelly roll topology (Spinelli et al., 2000). The cohesin formed a dimer in the crystal form as well as in solution. The dimer was dissociated by the addition of dockerin-containing proteins, suggesting that the dimerization interface and the cohesin-dockerin interface overlap with each other. This interface corresponds to the four-stranded 5, 6, 3, and 8 β-sheet of the *C. thermocellum* cohesin.

The crystal structure of type II cohesin of *C. thermocellum* SdbA has also been reported. Its overall structure is similar to that of type I cohesin. The dockerin binding site, also located on the β-sheet consisting of strands 5, 6, 3, and 8, is likely conserved as expected from the conserved amino acid sequence (Carvalho et al., 2005). The overall jelly roll topology of the type II cohesin from *Bacillus cellulosolvens* is quite similar to that of the type I cohesins, although the *Bacillus* cohesin possesses three additional secondary structures: an α-helix and two "β-flaps" that disrupt the normal course of a β-strand (Noach et al., 2005).

MOLECULAR STRUCTURE OF THE DOCKERIN

Crystallization of the dockerin appears to be difficult to achieve. The solution structure of the dockerin was thus determined using two-dimensional nuclear magnetic resonance spectroscopy on an isolated type I dockerin from *C. thermocellum* CelS (Lytle et al., 2001; Volkman et al., 2004). The structure consisted of two Ca^{2+}-binding loop-helix motifs connected by a linker; the E helices entering each loop of the classical EF-hand motif were absent from the dockerin (Color Plate 3). Each dockerin's Ca^{2+}-binding subdomain was stabilized by a cluster of buried hydrophobic side chains. Structural comparisons revealed that, in its noncomplexed state, the dockerin fold displayed a dramatic departure from that of Ca^{2+}-bound EF-hand domains (Color Plate 4). Normally in Ca^{2+}-bound EF-hand domains, the loops extend above the helices. In the *C. thermocellum* dockerin, the loops are buried within the structure. Another difference is the lack of a short, antiparallel β-sheet interaction between the position 7 and 9 residues of the Ca^{2+}-binding loop. All paired EF-hand domains studied have this feature. Yet another difference is the orientation of the helices in the dockerin. The helices are almost antiparallel at an angle of 157°, very different from a typical EF-hand domain. Therefore, the dockerin module is a unique Ca^{2+}-binding module despite some similarities to EF-hand domains. The dockerin is one of the very few proteins whose folding is dependent on Ca^{2+}(Lytle et al., 2000).

MOLECULAR STRUCTURES OF THE COHESIN-DOCKERIN COMPLEXES

The first success in X-ray crystallography of a cohesin-dockerin complex was brought about by using a crystal of the dockerin of *C. thermocellum* Xyn10B and cohesin 2 of CipA (Carvalho et al., 2003), coexpressed in *Escherichia coli* and purified as a complex, yielding a good crystal suitable for X-ray analysis. The structure shows that the cohesin-dockerin recognition is mediated mainly by hydrophobic interactions between one face (β-sheet) of the cohesin and α-helices 1 and 3 of the dockerin (Color Plate 5a). Several hydrophobic residues participating in complex formation are found on β-strands 3, 5, and 6 of the cohesin and on α-helices 1 and 3 of the dockerin domain, as predicted by site-specific mutagenesis experiments. Although the structure of the cohesin in the complex remains essentially unchanged, the loop-helix-helix-loop-helix motif of the dockerin undergoes conformation change and ordering when compared with its solution structure. There are relatively few direct hydrogen bonds between the cohesin and dockerin. The conserved amino acids

ST in the second segment of the dockerin dominate the hydrogen-bonding network between the dockerin and cohesin. On the other hand, the ST motif in the first segment does not play a direct role in cohesin-dockerin recognition in this complex although this motif is highly conserved in type I dockerins. This observation does not necessarily mean that the ST motif in the first segment is not important for the cohesin-dockerin interaction. Since the tandem repeats of the dockerin form near-perfect internal twofold symmetry (Color Plate 5b), the "symmetric" binding mode featuring the ST motif in the first segment is expected.

The symmetric binding mode was confirmed by the experiment using a mutant dockerin lacking the ST in the second segment (Carvalho et al., 2007). This mutant still formed a complex with the cohesin; the roles of segments 1 and 2 (i.e., α-helices 1 and 3) in the wild-type complex were reversed in the mutant complex, and the ST motif in the first segment of the mutant dockerin came to dominate the hydrogen-bonding network between the dockerin and cohesin (Color Plate 5c). The association constant ($K_a = 7 \times 10^7 M^{-1}$) for the mutant complex is comparable to that for the wild-type complex ($K_a = 8 \times 10^7 M^{-1}$), suggesting that the symmetric binding mode is possible in the native cellulosome assembly process and is predicted to impart significant plasticity to the orientation of the cellulosomal catalytic subunits.

For the type II cohesin-dockerin complex, the cohesin of SdbA undergoes very little conformation change upon binding to the dockerin (Adams et al., 2006). The type II cohesin in the complex forms an elongated nine-stranded β-sandwich in a classical jelly roll topology with an extensive hydrophobic core. The type II dockerin on the surface of type II cohesin has a distinct orientation compared to the type I dockerin in the complex. The type II dockerin is in a parallel/antiparallel arrangement, whereas the orientation of the type I dockerin is rotated clockwise 20°. In addition, both helices of the type II dockerin contact the cohesin surface over their entire length, whereas the type I dockerin contacts the cohesin with the entire length of helix 3 and only the C terminus of helix 1.

CONCLUDING REMARKS

After about 2 decades of investigation on the cellulosome by many research groups, the mechanism of the cellulosome assembly has surfaced. In *C. thermocellum*, subtle differences between the type I and type II cohesin-dockerin complex confer intraspecies binding specificity, leading to an orderly arrangement of the cellulosome. The cellulosomal catalytic subunits are assembled into the cellulosome by the interaction between the type I dockerin borne on each catalytic subunit and the type I cohesion of CipA. On the other hand, the cellulosome is anchored onto the cell surface by the interaction between the type II dockerin of CipA and the type II cohesin of a surface layer protein such as SdbA. It is remarkable that slight variations of the cohesin and dockerin sequences lead to both intraspecies and interspecies specificities that form the foundation of the cellulosome assembly. Much remains to be learned about possible interactions between the cellulosomal proteins, which are independent from the cohesin and dockerin, and possible higher order structures of the cellulosome. The cellulosome as a molecular machine for cellulose degradation will continue to be an intriguing subject of study from both the fundamental and applied science perspectives.

Acknowledgment. Research reported from our laboratories has been supported in part by grants from the U.S. Department of Energy (DE-FG02-94ER20155) to J.H.D.W. and from the Research Institute of Innovative Technology for the Earth (RITE) to K.S.

REFERENCES

Adams, J. J., G. Pal, Z. Jia, and S. P. Smith. 2006. Mechanism of bacterial cell-surface attachment revealed by the structure of cellulosomal type II cohesin-dockerin complex. *Proc. Natl. Acad. Sci. USA* 103:305–310.

Bayer, E. A., R. Kenig, and R. Lamed. 1983. Adherence of *Clostridium thermocellum* to cellulose. *J. Bacteriol.* 163:552–559.

Carvalho, A. L., F. M. Dias, T. Nagy, J. A. Prates, M. R. Proctor, N. Smith, E. A. Bayer, G. J. Davies, L. M. Ferreira, M. J. Romao, C. M. Fontes, and H. J. Gilbert. 2007. Evidence for a dual binding mode of dockerin modules to cohesins. *Proc. Natl. Acad. Sci. USA* 104:3089–3094.

Carvalho, A. L., F. M. V. Dias, J. A. M. Prates, T. Nagy, H. J. Gilbert, G. J. Davies, L. M. A. Ferreira, M. J. Romao, and C. M. G. A. Fontes. 2003. Cellulosome assembly revealed by the crystal structure of the cohesin-dockerin complex. *Proc. Natl. Acad. Sci. USA* 100:13809–13814.

Carvalho, A. L., V. M. Pires, T. M. Gloster, J. P. Turkenburg, J. A. Prates, L. M. Ferreira, M. J. Romao, G. J. Davies, C. M. Fontes, and H. J. Gilbert. 2005. Insights into the structural determinants of cohesin-dockerin specificity revealed by the crystal structure of the type II cohesin from *Clostridium thermocellum* SdbA. *J. Mol. Biol.* 349:909–915.

Chauvaux, S., P. Beguin, J.-P. Aubert, K. M. Bhat, L. A. Gow, T. M. Wood, and A. Bairoch. 1990. Calcium-binding affinity and calcium-enhanced activity of *Clostridium thermocellum* endoglucanase D. *Biochem. J.* 265:261–265.

Choi, S. K., and L. G. Ljungdahl. 1996. Structural role of calcium for the organization of the cellulosome of *Clostridium thermocellum*. *Biochemistry* 35:4906–4910.

Demain, A. L., M. Newcomb, and J. H. Wu. 2005. Cellulase, clostridia, and ethanol. *Microbiol. Mol. Biol. Rev.* 69:124–154.

Fierobe, H. P., S. Pages, A. Belaich, S. Champ, D. Lexa, and J. P. Belaich. 1999. Cellulosome from *Clostridium cellulolyticum*: molecular study of the dockerin/cohesin interaction. *Biochemistry* 38:12822–12832.

Gerngross, U. T., M. P. M. Romainiec, N. S. Huskisson, and A. L. Demain. 1993. Sequencing of a *Clostridium thermocellum* gene (*cipA*) encoding the cellulosomal SL-protein reveals an unusual degree of internal homology. *Mol. Microbiol.* 8:325–334.

Gilbert, H. J. 2007. Cellulosomes: microbial nanomachines that display plasticity in quaternary structure. *Mol. Microbiol.* 63:1568–1576.

Grepinet, O., M.-C. Chebrou, and P. Beguin. 1988. Nucleotide sequence and deletion analysis of the xylanase gene (*xynZ*) of *Clostridium thermocellum*. *J. Bacteriol.* 170:4582–4588.

Jindou, S., A. Soda, S. Karita, T. Kajino, P. Beguin, J. H. D. Wu, M. Inagaki, T. Kimura, K. Sakka, and K. Ohmiya. 2004. Cohesin-dockerin interactions within and between *Clostridium josui* and *Clostridium thermocellum*: binding selectivity between cognate dockerin and cohesin domains and species specificity. *J. Biol. Chem.* 279:9867–9874.

Kataeva, I. A. (ed.). 2006. *Cellulosome*. Nova Science Publishers, Hauppauge, NY.

Leibovitz, E., and P. Beguin. 1996. A new type of cohesin domain that specifically binds the dockerin domain of the *Clostridium thermocellum* cellulosome-integrating protein CipA. *J. Bacteriol.* 178:3077–3084.

Lytle, B., C. Myers, K. Kruus, and J. H. D. Wu. 1996. Interactions of the CelS binding ligand with various receptor domains of the *Clostridium thermocellum* cellulosomal scaffolding protein, CipA. *J. Bacteriol.* 178:1200–1203.

Lytle, B., and J. H. D. Wu. 1998. Involvement of both dockerin subdomains in assembly of the *Clostridium thermocellum* cellulosome. *J. Bacteriol.* 180:6581–6585.

Lytle, B., B. F. Volkman, W. M. Westler, and J. H. D. Wu. 2000. Secondary structure and calcium-induced folding of the *Clostridium thermocellum* dockerin domain determined by NMR spectroscopy. *Arch. Biochem. Biophys.* 379:237–244.

Lytle, B. L., B. F. Volkman, W. M. Westler, M. P. Heckman, and J. H. D. Wu. 2001. Solution structure of a type I dockerin domain, a novel prokaryotic, extracellular calcium-binding domain. *J. Mol. Biol.* 307:745–753.

Mechaly, A., H.-P. Fierobe, A. Belaich, J.-P. Belaich, R. Lamed, Y. Shoham, and E. A. Bayer. 2001. Cohesin-dockerin interaction in cellulosome assembly. A single hydroxyl group of a dockerin domain distinguishes between nonrecognition and high affinity recognition. *J. Biol. Chem.* 276:9883–9888.

Miras, I., F. Schaeffer, P. Beguin, and P. M. Alzari. 2002. Mapping by site-directed mutagenesis of the region responsible for cohesin-dockerin interaction on the surface of the seventh cohesin domain of *Clostridium thermocellum* CipA. *Biochemistry* 41:2115–2119.

Noach, I., F. Frolow, H. Jakoby, S. Rosenheck, L. W. Shimon, R. Lamed, and E. A. Bayer. 2005. Crystal structure of a type-II cohesin module from the *Bacteroides cellulosolvens* cellulosome reveals novel and distinctive secondary structural elements. *J. Mol. Biol.* 348:1–12.

Pages, S., A. Belaich, J.-P. Belaich, E. Morag, R. Lamed, Y. Shoham, and E. A. Bayer. 1997. Species-specificity of the cohesin-dockerin interaction between *Clostridium thermocellum* and *Clostridium cellulolyticum*: prediction of specificity determinants of the dockerin domain. *Proteins* 29:517–527.

Schaeffer, F., M. Matuschek, G. Guglielmi, I. Miras, P. M. Alzari, and P. Beguin. 2002. Duplicated dockerin subdomains of *Clostridium thermocellum* endoglucanase CelD bind to a cohesin domain of the scaffolding protein CipA with distinct thermodynamic parameters and a negative cooperativity. *Biochemistry* 41:2106–2114.

Shimon, L. J. W., E. A. Bayer, E. Morag, R. Lamed, S. Yaron, Y. Shoham, and F. Frolow. 1997. A cohesin domain from *Clostridium thermocellum*: the crystal structure provides new insights into cellulosome assembly. *Structure* 5:381–390.

Shoseyov, O., M. Takagi, M. A. Goldstein, and R. H. Doi. 1992. Primary sequence analysis of *Clostridium cellulovorans* cellulose binding protein A. *Proc. Natl. Acad. Sci. USA* 89:3483–3487.

Spinelli, S., H.-P. Fierobe, A. Belaich, J.-P. Belaich, B. Henrissat, and C. Cambillau. 2000. Crystal structure of a cohesin module from *Clostridium cellulolyticum*: implications for dockerin recognition. *J. Mol. Biol.* 304:189–200.

Tavares, G. A., P. Beguin, and P. M. Alzari. 1997. The crystal structure of a type I cohesin domain at 1.7 A resolution. *J. Mol. Biol.* 273:701–713.

Tokatlidis, K., S. Salamitou, P. Beguin, P. Dhurjati, and J. P. Aubert. 1991. Interaction of the duplicated segment carried by *Clostridium thermocellum* cellulases with cellulosome components. *FEBS Lett.* 291:185–188.

Volkman, B. F., B. L. Lytle, and J. H. D. Wu. 2001. Dockerin domains, p. 617–628. *In* A. Messerschmidt, W. Bode, and M. Cygler (ed.), *Metalloproteins*, vol. 3. John Wiley & Sons, Ltd., Chichester, United Kingdom.

Wu, J. H. D., and A. L. Demain. 1988. Proteins of the *Clostridium thermocellum* complex responsible for degradation of crystalline cellulose, p. 117–131. *In* J. P. Aubert, P. Beguin, and J. Millet (ed.), *Biochemistry and Genetics of Cellulose Degradation*. Academic Press, New York, NY.

Wu, J. H. D., W. H. Orme-Johnson, and A. L. Demain. 1988. Two components of an extracellular protein aggregate of *Clostridium thermocellum* together degrade crystalline cellulose. *Biochemistry* 27:1703–1709.

Xu, Q., E. A. Bayer, M. Goldman, R. Kenig, Y. Shoham, and R. Lamed. 2004. Architecture of the *Bacteroides cellulosolvens* cellulosome: description of a cell surface-anchoring scaffoldin and a family 48 cellulase. *J. Bacteriol.* 186:968–977.

Yaron, S., E. Morag, E. A. Bayer, R. Lamed, and Y. Shoham. 1995. Expression, purification and subunit-binding properties of cohesins 2 and 3 of the *Clostridium thermocellum* cellulosome. *FEBS Lett.* 360:121–124.

Bioenergy
Edited by J. Wall et al.

Chapter 9

Contribution of Domain Interactions and Calcium Binding to the Stability of Carbohydrate-Active Enzymes

LARS G. LJUNGDAHL, IRINA A. KATAEVA, AND VLADIMIR N. UVERSKY

Production of ethanol from renewable plant biomass and its use as an alternative fuel or as additive to traditional fuels will partially contribute to the solution of the "global warming" problem. Biomass consists mainly of cellulose, hemicelluloses, and lignin. The conventional method to release the sugars in cellulose and hemicelluloses involves acidic hydrolysis. Recent research is focused on more "eco-friendly" enzymatic processes involving carbohydrate active enzymes that attack plant biomass efficiently, leading to high yields of sugars released not only from cellulose and hemicellulose, but also from pectin and other carbohydrate polymers present in plant cell walls. The sugars are then fermented to ethanol. In the industrial environment, strains of the yeast *Saccharomyces cerevisiae* have been used. Unfortunately, *S. cerevisiae* ferments only glucose, the product of cellulose hydrolysis. Thus, xylose and other sugars, derived from hemicelluloses and constituting a significant part of the biomass, are not fermented. Breakthroughs in fermentation technology in the past decade, i.e., the discovery and constructions of new yeast and bacterial strains with the ability to ferment the full spectrum of available sugars into ethanol, have led to the possible commercialization of biomass conversion technology (Kerr, 2004).

Industrial enzymatic hydrolysis of plant biomass is still more expensive than chemical hydrolysis. A search for cellulases and other carbohydrate-active enzymes with high specific activities and stabilities with regard to pH and temperature is under way. In addition, methods to recycle the enzymes are being investigated and if successful, will make industrial processes more productive and cost-effective. In this regard, the stability of the carbohydrate-active enzymes for use in bioreactors is one of the most important properties to be considered for scaling up the process.

Most carbohydrate-active enzymes are modular proteins, usually with calcium presented in some of the domains. The final fold, enzymatic properties, and stability of these enzymes are significantly affected by specific interactions between domains. Calcium plays a structural role and mediates proper intradomain interactions. In this chapter, we focus on the contributions that domain interactions and calcium have on the properties of carbohydrate-active enzymes.

PROTEIN DOMAINS: AN OVERVIEW

Many proteins in nature are constructed from a relatively small number of domains or modules. One domain definition reads: "Domains are topological entities which, at the atomic level, exhibit more pronounced interactions within the structural unit than with other parts of the polypeptide chain" (Janin and Wodak, 1983). Domains have an intrinsic capacity to form their native fold spontaneously and autonomously, often mediate specific biological functions, and combine to form larger multidomain proteins with segregated functions. A specific domain can be found in different proteins, and several different domains can be found within a given protein. Proteins thus can be viewed as being built of a finite set of domains, which are joined together in diverse combinations. Thus, domains may be considered basal units of the structure, function, and evolution of proteins.

Domains range in size from 25 to 500 amino acid residues. The robustness of the characteristic size of a domain suggests a simple underlying physical principle

Lars G. Ljungdahl • Department of Biochemistry and Molecular Biology, A216 Fred Davison Life Sciences Complex, University of Georgia, Athens, GA 30602. Irina A. Kataeva • Department of Biochemistry and Molecular Biology, B222A Fred Davison Life Sciences Complex, University of Georgia, Athens, GA 30602. Vladimir N. Uversky • Department of Biochemistry and Molecular Biology, Center for Computational Biology and Bioinformatics, Indiana University School of Medicine, 635 Narnhill Dr. MS 4021, Indianapolis, IN 46202-3763.

that is determined by only a few parameters (Shen et al., 2005). The optimal size of a globular protein domain, estimated by using a sphere-packing model, ranges from 117 to 213 residues, with an average of 165 residues. The model takes into account four parameters: (i) hydrophilic amino acid content, (ii) protein size, (iii) surface polarity, and (iv) eccentricity of a protein. These parameters are in good agreement with numerous experimental data (Shen et al., 2005). There are two points of view on folding of multidomain proteins: (i) simultaneous domain folding when a long polypeptide chain starts to fold in several places independently (Jaenicke, 1999), and/or (ii) sequential or cotranslational domain folding when domains fold one by one starting with an N-terminal domain (Frydman et al., 1999; Maity et al., 2005; Rumbley et al., 2001). At any rate, the modularity of proteins is a great achievement of evolution as it minimizes protein misfolding due to wrong intradomain interactions and provides biological systems with a convenient way of presenting binding sites on a stable protein scaffold, in the correct position of function; it also allows regulation by modular rearrangement (Campbell, 2003; Jaenicke, 1999).

Knowledge about modular proteins is increasing rapidly due to good databases and more-systematic approaches to protein expression and structure resolution. Several classification systems have attempted to combine domains sharing a similar sequence/fold. Some of the commonly used databases are InterPro, InterDom, ProDom, BLOCKS, Pfam, PRINTS, SBASE, SMART, TIGRFAM, and EVEREST. Studies show that the main task of structural biology is to move from determination of structures of individual modules to the task of assessing the ways by which these modules interact and bind their various ligands. Since modules occur in many proteins from various genomes, information about their structure and function can have an impact in a wide range of fields.

MODULARITY OF CARBOHYDRATE-ACTIVE ENZYMES AND DOMAIN CLASSIFICATION

Plant biomass is composed mainly of insoluble carbohydrates. Enzymes involved in the degradation of these compounds display modular architecture. In the simplest case, an enzyme is composed of only the catalytic module. However, in most cases, catalytic domains are surrounded by different noncatalytic ancillary modules, the total number of which within one protein can be eight and even more (Ahsan et al., 1996; Kataeva et al., 1999; Zverlov et al., 1998). The most common modules flanking catalytic domains from N or C termini are carbohydrate-binding modules (CBMs) to bring catalytic centers in close contact with the substrate. Other known domains are immunoglobulin (Ig)-like modules, fibronectin type 3 (Fn3)-like modules now designated X-modules, a second catalytic domain of similar or different activity, more than one CBM of similar or different binding affinity, and dockerin domains. The last named are found in catalytically active subunits of exocellular protein complexes called cellulosomes, which are produced by many anaerobic bacteria and fungi (Bayer et al., 2004; Doi and Kosugi, 2004; Ljungdahl et al., 2007). The function of the dockerins is to bind the catalytic subunits to special domains called cohesins of the special scaffolding proteins of the cellulosomes. Dockerin and cohesin domains are discussed in chapter 8 and also in reviews by Bayer et al. (2000, 2004). The biological function of some domains remains unknown (Bayer et al., 2000, 2004; Doi and Kosugi, 2004; Kataeva and Ljungdahl, 2003; Zverlov and Schwarz, 2007).

Several thousand sequences of carbohydrate-active enzymes have been described. For convenience, a classification system of catalytic domains of these enzymes based on amino acid sequence similarity was constructed in the early nineties (Henrissat, 1991). The classification system of catalytic and CBMs is available at the Carbohydrate-Active Enzymes (CAZy) server (http://afmb.cnrs-mrs.fr/~cazy/CAZY). The database includes glycoside hydrolases (GH) (108 families), glycosyltransferases (87 families), polysaccharide lyases (18 families), carbohydrate esterases (14 families), and CBMs (48 families). The family classification of catalytic modules and CBMs is expected to (i) aid in their identification, (ii) predict catalytic activity or binding specificity, (iii) aid in identifying functional residues, (iv) reveal evolutionary relationships, and (v) be predictive of polypeptide folds. The database is updated periodically and includes the GenBank/GenPept, Swiss-Prot, and PDB/3D entries. Information regarding catalytic domains, when they are known, includes the nature of residues participating in catalysis, molecular mechanism of the reaction, and availability of three-dimensional structure. Representatives of a particular family have highly conserved residues involved in catalysis, hydrolyze substrates by the same mechanism, and possess a similar fold allowing a homology modeling of known sequences (Henrissat and Davies, 2000). With a growing number of new sequences and three-dimensional protein structures, it has been noted that sometimes a new sequence shares similarity to sequences of different families, and vice versa, i.e., sequence-unrelated proteins display similar fold and mechanistic properties (Dominguez et al., 1995; Henrissat and Romeu, 1995). Such observations have led to the enzymes of some families being grouped into "clans."

DOMAIN INTERACTIONS IN CARBOHYDRATE-ACTIVE ENZYMES

Little is known about domain communications in carbohydrate-active enzymes, although the importance of such interactions has been demonstrated in various mammalian, fungal, bacterial, and viral proteins (Berr et al., 2000; Clout et al., 2000; Jaenicke, 1999; Smith et al., 2007; Wassenberg et al., 1997, 1999; Wenk and Jaenicke, 1998, 1999). It has been noted that modules in carbohydrate-active enzymes are not randomly combined. They often are located in a particular place within a protein and are associated with a particular type of neighboring module(s). Thus, CBMs of a particular family have a tendency to be associated with catalytic domains of a particular family and to be located in a specific place on the polypeptide chain (Tomme et al., 1995, 1998). Ig-like domains often, but not always, precede catalytic domains belonging to family GH9 (Bayer et al., 2000; Zverlov and Schwarz, 2007). Fn3-like domains, also designated X1, are common in microbial chitinases, often are duplicated or triplicated, and are located between catalytic modules and CBMs (Little et al., 1994; Shen et al., 1995). Surface layer homologous domains are often in duplicate and triplicate and located at C termini (Ohmiya et al., 1997). Dockerin domains are usually located at the C termini of polypeptide chains (Bayer et al., 2000, 2004). It has been demonstrated that deletion of a CBM from a catalytic domain drastically decreases catalytic activity against insoluble substrates (Bolam et al., 1998; Kataeva et al., 1999, 2001a, 2001b; Srisodsuk et al., 1993) and, in some cases, the thermostability of the residual protein (Araki et al., 2006; Fontes et al., 1995; Hayashi et al., 1997; Kataeva et al., 1999, 2005). Fusion of catalytic and CBM domains of different origins decreased the thermostability of the chimeric polypeptide. Attachment of an N-terminal CBM to the C terminus of the catalytic domain led to a decrease in binding affinity, enzymatic capacity, and thermostability (Kataeva et al., 2001a, 2001b). Transposition of the domains within the polypeptide chain resulted in a drop of thermostability (Shin et al., 2002). These observations imply the existence of specific interactions between domains in the carbohydrate-active enzymes similar to those found in some other modular proteins.

ROLE OF LINKERS LOCATED BETWEEN FUNCTIONAL DOMAINS

Many carbohydrate-active enzymes contain linkers of different length, which are rich in hydroxylic amino acid residues (threonine, serine, and proline), two of which, Ser and Pro, are disorder-promoting residues (Dunker et al., 2001). Linker sequences are often highly O glycosylated (Gilkes et al., 1991). It has been shown that a deletion of a 23-residue linker separating the catalytic module from the CBM in CenC from *Cellulomonas fimi* changed the relative orientation of these domains and affected both catalytic activity and desorption from cellulose, although it did not affect the overall fold of each domain (Shen et al., 1991). Similarly, the deletion of part of the linker located between the catalytic module and CBM in CBHI from *Trichoderma reesei* had a serious effect on binding and activity against crystalline cellulose (Srisodsuk et al., 1993). These findings imply that linker sequences serve to set proper interactions between domains. As the presence of linkers negatively affects crystallization of modular proteins, only a limited number of structures of domain combinations have been resolved so far. As a result, although structural and functional properties of many individual domains are well studied, very little is known about the role and structural aspects of the linker peptide.

More information about conformational versatility of linkers in carbohydrate-active proteins was obtained by exploring a small-angle X-ray scattering technique (Receveur et al., 2002; von Ossowski et al., 2005). Cel45 from *Humicola insolens* is composed of catalytic domain GH45 and a CBM1 separated by a 36-residue glycosylated linker. Combination of light-scattering data with a known crystal structure of the GH45 catalytic domain and a modeled structure of CBM1 (based on a 45% identity to Cel7A CBM1) helped to evaluate hydrodynamic dimensions of the enzyme. The average size of a protein is estimated by measuring its radius of gyration (R_g). The R_g of Cel45 was much higher than what would be expected for a spherical protein with a similar number of residues. Cel45 had an elongated shape in solution despite the fact that its large catalytic domain is a globular polypeptide and that a small-size CBM1 did not greatly affect R_g. The data showed that the dimension of the linker is very large with respect to its mass. As R_g reflects a mixture of all possible conformations the protein can adopt in solution, one might conclude that the linker has some internal flexibility.

Further progress in understanding the role of linkers in carbohydrate-active enzymes was made by studying hydrodynamic properties of the chimeric fungal cellulase Cel6AB (von Ossowski et al., 2005). Wild-type Cel6A and Cel6B are both domain proteins. Cel6A contains an N-terminal CBM followed by a 52-residue linker and a C-terminal catalytic module (CBM-L_{52}-$GH6_A$). Cel6B is composed of an N-terminal catalytic module and a C-terminal CBM separated by a 36-residue linker ($GH6_B$-L_{36}-CBM). In wild-type proteins linkers and CBDs are small and they are difficult to distinguish clearly. To further evaluate the linkers and CBMs, the CBMs were removed and the catalytic modules with natural linkers were fused together so that the two linkers formed a combined linker of 88 residues in total length located between the catalytic modules ($GH6_B$-L_{36}-L_{52}-$GH6_A$). The extended conformation of the linker within the chimeric construct was confirmed by measuring the R_g value and the distance distribution function $P(r)$. The shape of the experimental distance distribution profile $P(r)$ indicated that there is a distribution of conformations with various distances between the $GH6_A$ and $GH6_B$ domains. Interestingly, molecular modeling of the linker length in combination with the small-angle X-ray scattering data implied that the linker is not just an extended polypeptide chain; it adopts a much more compact conformation, which is the most stable. It is, however, with a relatively low energy cost able to unwind, forming a stretched-out longer peptide. This suggests that the linker may function as a "molecular spring" between the two functional domains. With regard to cellulose hydrolysis, the linker would thus give the catalytic site the ability to reach out and hydrolyze new glycosidic bonds while the CBM is still bound to the substrate surface (Receveur et al., 2002; von Ossowski et al., 2005).

DOMAINS IN THE *CLOSTRIDIUM THERMOCELLUM* CELLULOSOME

Clostridium thermocellum, a thermophilic, anaerobic bacterium, produces an exocellular multiprotein complex cellulosome highly active against plant cell wall carbohydrates (Bayer et al., 2000). Besides the cellulosome, the bacterium synthesizes several noncellulosomal enzymes with different hydrolytic activities (Zverlov and Schwarz, 2007). The assembly of the cellulosome occurs by specific interactions between cohesin domains of scaffoldin protein CipA and dockerin domains (DD) of the enzymes (Bayer et al., 2000; Béguin and Lemaire, 1996). Analysis of the whole genome of *C. thermocellum* revealed over 70 putative open reading frames bearing DD and several other components without dockerins (Zverlov and Schwarz, 2007). Below we focus on the interactions between domains in one of the largest cellulosomal catalytic components, cellobiohydrolase A (CbhA) (Schubot et al., 2004; Zverlov et al., 1998). It has recently been shown to also possess endoglucanase activity (McGrath, 2007). This enzyme is a thermostable, multimodular, calcium-containing protein. It has a molecular mass of 135 kDa and is composed of seven modules: an N-terminal CBM4, an Ig-like domain, a GH9 domain, two Fn3-like domains ($X1_1$ and $X1_2$), a CBM3, and a DD (Fig. 1). All modules, except the Ig-like domain, bind calcium. One major role of calcium is to stabilize native folds of polypeptides. For this reason, calcium is a constituent of many thermostable proteins (Medved et al., 1995; Notenboom et al., 2001; Wenk and Jaenicke, 1998, 1999). CbhA is the only *C. thermocellum* cellulosomal enzyme whose domain interactions and role of calcium have been studied in detail by different techniques including genetic manipulations, crystallography, circular dichroism (CD) spectroscopy, and differential scanning calorimetry (DSC).

INTERACTIONS BETWEEN THREE INTERNAL CbhA MODULES PROBED BY CD SPECTROSCOPY

Three domains of CbhA, $X1_1$, $X1_2$, and CBM3, each bind 1 mol of calcium (Fig. 1). Expressed either individually or in combinations $X1_1X1_2$ and $X1_1X1_2$-CBM3, all variants, when saturated with calcium, are designated holoproteins. When calcium is reversibly removed by incubating the proteins with Chelex-100, they are designated apoproteins. Thermal unfolding of all variants was totally reversible (Kataeva et al., 2003). The recorded near-UV CD spectra of domain combinations were compared with the spectra calculated as simple weighted sums of $X1_1$ + $X1_2$ ($X1_1X1_2$ construct) and $X1_1$ + $X1_2$ + CBM3 or $X1_1X1_2$ + CBM3 ($X1_1X1_2$-CBD3 construct) (Fig. 2). In the case of total domain independency, the recorded spectra of $X1_1X1_2$ and $X1_1X1_2$-CBD3 were expected to be equal to the spectra calculated as weighted sums of the recorded spectra of individual domains. When domain interactions induced changes in the tertiary structure, differences between the observed and calculated spectra were expected. Figure 2 demonstrates that domain interactions play an important role in the conformation of $X1_1X1_2$-CBM3 either in the presence (Fig. 2B) or in the absence (Fig. 2D) of Ca^{2+}. As for the $X1_1X1_2$ construct, a considerable difference between recorded and calculated spectra was observed in the absence of Ca^{2+} (Fig. 2C), whereas upon calcium binding, this difference was considerably reduced or eliminated (Fig. 2A). Thus, (i) the three domains interact, and (ii) Ca^{2+} is involved as mediator of these interdomain interactions (Kataeva et al., 2003).

To gain insight into the mechanisms of thermal denaturation of individual and linked domains, phase-diagram (PD) analyses were performed with spectroscopic data. Such an analysis is extremely sensitive for the detection of intermediate states (Kuznetsova et al., 2002). The essence of this method is to build up the diagram of $I_{\lambda 1}$ versus $I_{\lambda 2}$, where $I_{\lambda 1}$ and $I_{\lambda 2}$ are the spectral intensity values measured at wavelength λ1 and λ2, under different experimental conditions for a protein undergoing structural transformations. The relationship $I_{\lambda 1} = f\,[I_{\lambda 2}]$ is linear if changes in protein environment lead to an all-or-nothing transition between two different conformations. Alternatively, nonlinearity of this function reflects the sequential character of structural transformations; and each linear portion of the $I_{\lambda 1} = f\,[I_{\lambda 2}]$ dependence describes an individual all-or-nothing transition. Phase diagrams of individual domains and domain combinations are given in Fig. 3. Figure 3A shows that the PD of holo-$X1_1$ consists of two linear parts. Removal of Ca^{2+} dramatically changes the shape of the PD, although the plot still has two linear parts indicating the existence of two independent transitions. Figure 3B shows that the denaturation of holo- and apo-$X1_2$ most likely represents an all-or-nothing transition. The denaturation of CBM3 is Ca^{2+} sensitive (Fig. 4C). Thus, apo-CBM3 denatures according to the three-state model (the PD has two linear parts), whereas holo-CBM3 denatures as an all-or-nothing transition. Finally, panels D and E of Fig. 3 show that removal of calcium has no effect on the mechanisms of $X1_1X1_2$ and $X1_1X1_2$-CBM3 melting, respectively, with the apo- and holoforms being denatured by the two-state scheme. Thermodynamic analysis proved the assumption that the stability of individual $X1_1$ is Ca^{2+} dependent. Removal of calcium resulted in a 4.2 kcal/mol decrease in ΔG value and a 20°C decrease of T_m of the domain (82°C versus 61.8°C, respectively). In contrast, the individual $X1_2$ domain and the $X1_1X1_2$ domain combination are relatively calcium independent with T_ms of 75.8 and 78.3°C ($X1_2$ apo and holo) and 78.7 and 83.0°C ($X1_1X1_2$ apo and holo), respectively. This leads to the conclusion that the individual $X1_1$ is stabilized by calcium whereas in the $X1_1X1_2$ and $X1_1X1_2$-CBM3 constructs, the stabilization is by interaction of $X1_1$ with $X1_2$. Finally, in the presence of calcium, the stabilities of the domains are relatively independent, while in the absence of Ca^{2+}, domain interactions play a stabilizing role (Kataeva et al., 2003).

DOMAIN COUPLING IN CbhA PROBED BY DSC

Comparison of thermal denaturation of truncated variants of CbhA of increasing length as detected by DSC yields a good illustration of domain coupling (Fig.

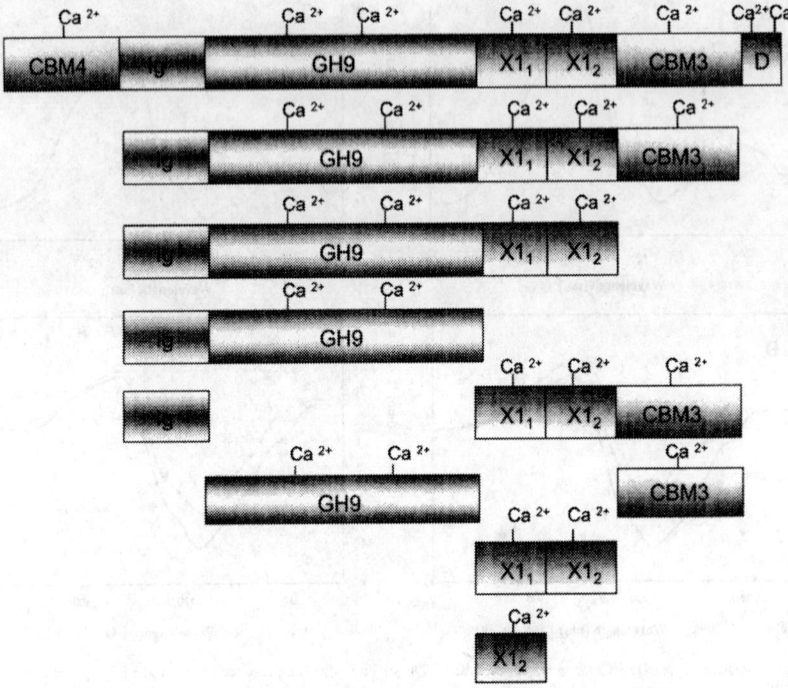

Figure 1. Domain structure of CbhA and its truncated variants. Abbreviations: CBD4 and CBD3, carbohydrate-binding domains of family 4 and 3, respectively; Ig, immunoglobulin-like domain; GH9, catalytic domain of family 9 glycoside hydrolases; $X1_1$ and $X1_2$, X domains of family 1; DD, duplicated dockerin domain. The content of calcium is also shown. (From Kataeva et al., 2005.)

4) (Kataeva et al., 2005). As all domains except the Ig-like domain bind calcium (Fig. 1) (Béguin and Alzari, 1998; Henrissat and Romeu, 1995), the thermal denaturation was studied (i) with 20 mM sodium phosphate buffer, pH 6.0; (ii) with 2 mM calcium; and also (iii) by heating with 2 mM EDTA. Individual domains possessed relatively independent folding since they unfolded as cooperative units (Fig. 4A, C, E, F). As expected, the Ig-like domain was insensitive to Ca^{2+} and EDTA (Fig. 4E). The effects of calcium and EDTA on thermal denaturation of the other proteins can be summarized as follows: (i) in buffer (i.e., only endogenous calcium present), all proteins except CBM3 and $X1_1X1_2$-CBM3 denatured less cooperatively with lower mid-point temperature (T_d) and higher number of cooperative units (r) than in the presence of Ca^{2+}; (ii) calcium increased T_d and the cooperativeness of unfolding; (iii) EDTA decreased T_d and, in most cases, r values. In other words, Ca^{2+} significantly stabilizes the proteins. As in buffer, all constructs contain Ca^{2+}, and the stabilizing effect of external calcium occurs due to the shift of the dissociation equilibrium towards association of Ca^{2+} and proteins. The higher cooperativeness of thermal unfolding in the presence of EDTA in comparison to that in buffer is a result of lower heterogeneity of the protein population. Thermal unfolding in buffer assumes coexistence of protein molecules (i) with calcium bound to all binding sites, (ii) with partially lost calcium, and (iii) without calcium. In the presence of EDTA, only apoprotein molecules are present.

Measured calorimetric denaturation enthalpy ($\Delta_d H_{cal}$)of domain combinations was higher than the sum of $\Delta_d H_{cal}$s of individual domains (Table 1), indicating the importance of domain communications. The number of cooperative units in the system r, calculated as $\Delta_d H_{cal}/\Delta_d H_{v.H}$, is often used as a measure of cooperativeness of thermal transitions (Kozhevnikov et al., 2001). In an ideal case, the polypeptide is one cooperative unit, with an r value of 1. In Table 1 the r values deviate from 1. For individual domains ($X1_2$, Ig-like, and GH9), this means that although they preserved a relatively independent fold, the cooperativeness of unfolding is low. Upon gradual association of domains ($X1_1$, $X1_1$-$X1_2$, and $X1_1X1_2$-CBM3), the r value de-

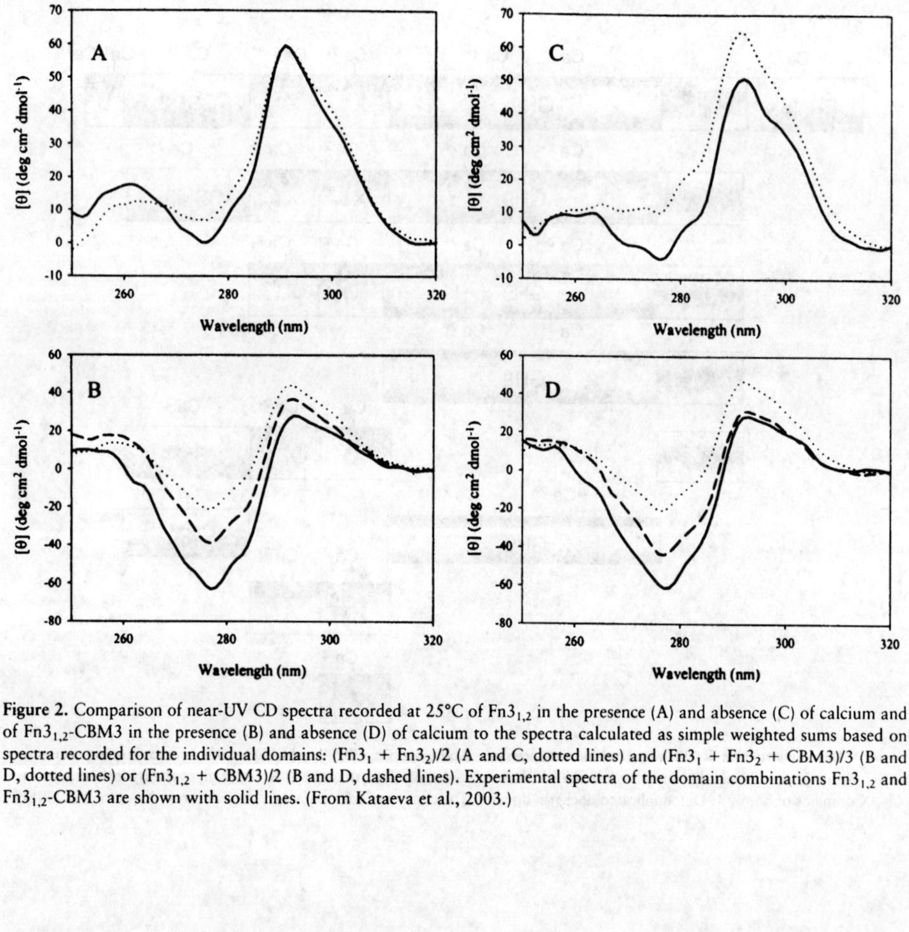

Figure 2. Comparison of near-UV CD spectra recorded at 25°C of $Fn3_{1,2}$ in the presence (A) and absence (C) of calcium and of $Fn3_{1,2}$-CBM3 in the presence (B) and absence (D) of calcium to the spectra calculated as simple weighted sums based on spectra recorded for the individual domains: ($Fn3_1$ + $Fn3_2$)/2 (A and C, dotted lines) and ($Fn3_1$ + $Fn3_2$ + CBM3)/3 (B and D, dotted lines) or ($Fn3_{1,2}$ + CBM3)/2 (B and D, dashed lines). Experimental spectra of the domain combinations $Fn3_{1,2}$ and $Fn3_{1,2}$-CBM3 are shown with solid lines. (From Kataeva et al., 2003.)

creases and reaches 1 in the three-domain construct which unfolds as one cooperative unit. The large GH9 domain is the least-stable construct, with a highly asymmetric thermogram (Fig. 4; Table 1), r value of 2.8, and the lowest T_d (68.3°C). Addition of the Ig-like domain decreased the r value in buffer to 1.83. Addition of other domains to GH9 led to an increase of cooperativeness of thermal unfolding. Binding calcium further increased the cooperativity of thermal denaturation. Thus, in a four- and a five-domain construct, the r values in the presence of calcium are 1.12 and 1.86, respectively. In other words, these multidomain polypeptides thermally unfold by a more simple mechanism than expected from their domain architecture. This phenomenon, first observed in some proteins from thermophiles and called "domain coupling" (Wassenberg et al., 1999), is clear evidence of domain interactions.

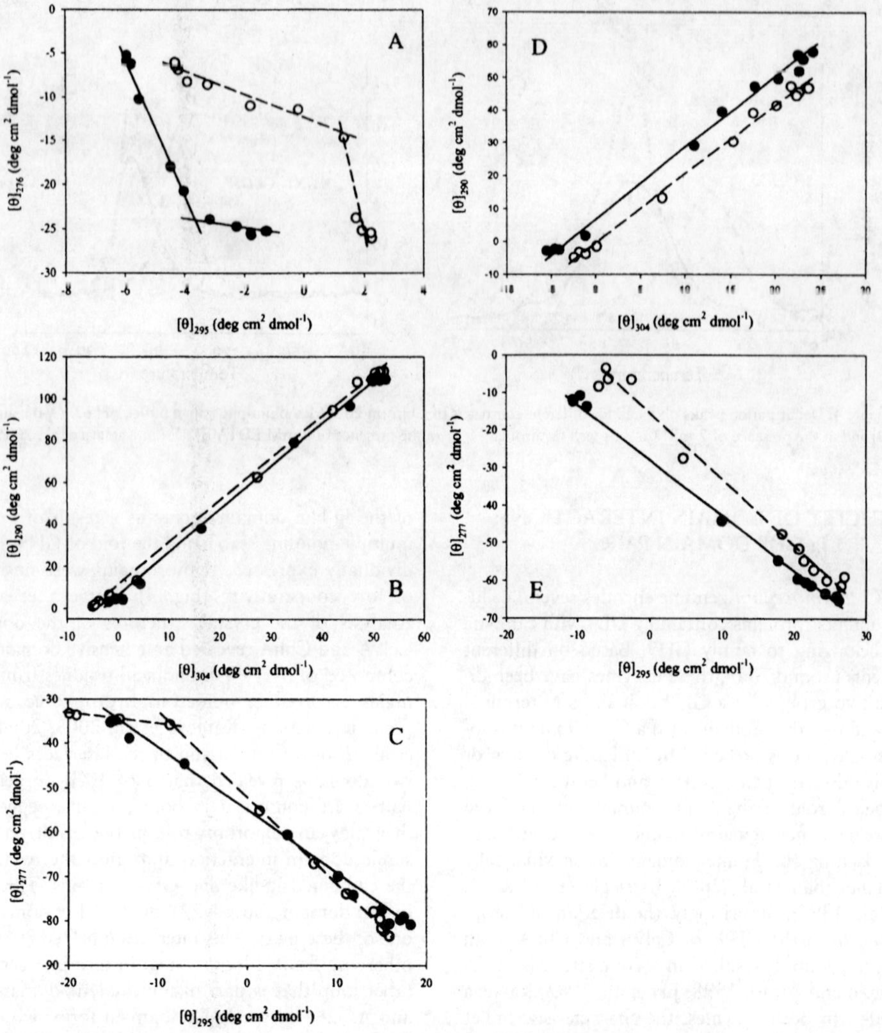

Figure 3. Phase diagrams based on $[\theta]_{\lambda 1}$ versus $[\theta]_{\lambda 2}$ (see the text for details) characterize heat-induced denaturation of different domains of CbhA, based on the temperature-induced changes in the near-UV CD spectra of individual domains $Fn3_1$ (A), $Fn3_2$ (B), and CBM3 (C) and of domain combinations $Fn3_{1,2}$ (D) and $Fn3_{1,2}$-CBM3 (E). Data for holo- and apoproteins are given with closed symbols and solid lines, and open symbols and dashed lines, respectively. (From Kataeva et al., 2003.)

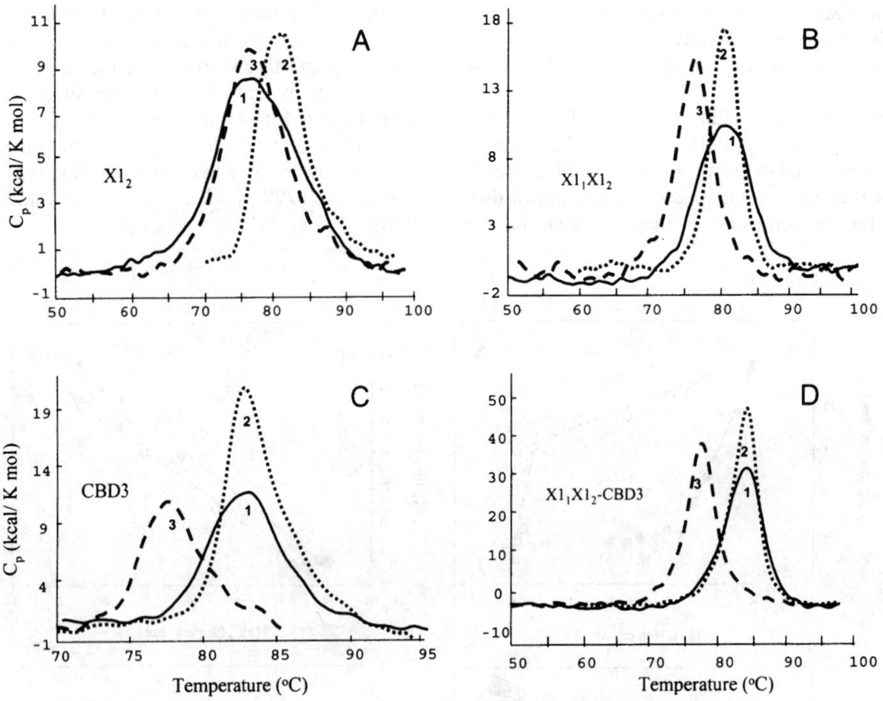

Figure 4. Denaturation peaks obtained for different constructs of CbhA in 20 mM sodium-phosphate buffer, pH 6.0 (A through D) and in the presence of 2 mM Ca^{2+} as well (E through H) or in the presence of 2 mM EDTA (I). (From Kataeva et al., 2005.)

SPECIFICITY OF DOMAIN INTERACTIONS: Ig-GH9 DOMAIN PAIR

The *C. thermocellum* genome encodes several cellulosomal subunits (proteins containing DD) with catalytic domains belonging to family GH9. Based on different arrangements around GH9, these enzymes have been divided into two groups: (i) a GH9 domain is N terminal and followed by other domains; (ii) a GH9 domain is located internally and is preceded by an Ig-like domain or CBM-Ig-like domain pair (Zverlov and Schwarz, 2007). The biological role of the Ig-like domain in plant cell wall active enzymes remains unknown. The enzymes originally lacking the Ig-like domain are enzymatically active (Mandelman et al., 2003; Parsiegla et al., 2002; Sakon et al., 1997). Surprisingly, the deletion of the Ig-like domain from the GH9s of Cel9A and CbhA, both belonging to group 2, resulted in a complete loss of activity (Béguin and Alzari, 1998; Juy et al., 1992; Kataeva et al., 2004). In both enzymes, the characteristic linker sequence between the Ig-like domain and the GH9 is missing. This leads to the possibility of a tight interaction between the two domains, and it has been assumed that the inactivation of the GH9 domain upon deletion of the Ig-like domain occurs as a result of breakage of multiple bonding stabilizing the fold of GH9. Indeed, individually expressed, both domains were unstable units of low cooperativity (Fig. 3) (Kataeva et al., 2004). Analysis of the crystal structures of the domain pair Cel9A and CbhA revealed an extensive domain interface composed of over 40 amino acid residues from both domains involved in numerous hydrophobic and hydrophilic interactions (Kataeva et al., 2003, 2004; Schubot et al., 2004). Comparison of the interfaces between the two domains revealed that 3 of 10 hydrogen bonding pairs were conserved in both proteins, suggesting that they play an important role in the maintenance of the stable domain interaction and affect the overall fold of the combined Ig-like and GH9 domains. Thr230 of the Ig-like domain and Gly221 of the GH9 domain formed one of these pairs. This interaction helped to stabilize an otherwise flexible loop, which in turn interacts with another loop that is part of the catalytic domain. Asp262 and Asp264 of the Ig-like domain form two other conserved hydrogen bonds with Gly221 and Tyr676, respectively, of the GH9 domain. To alter the H-bonding network between the domains and evaluate the importance of domain interactions in the domain pair, these residues

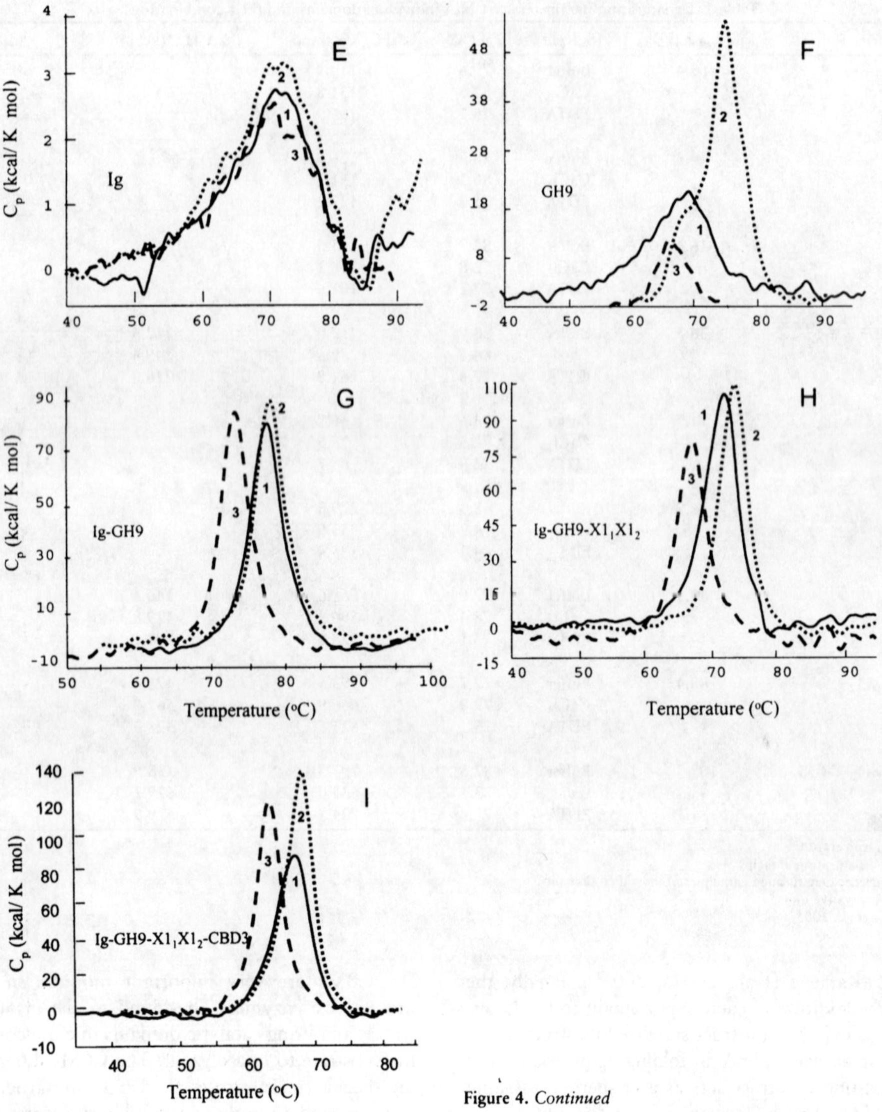

Figure 4. *Continued*

were replaced with alanine residues, giving a double mutant, T230A/D262A, and a single mutant, D264A. Figure 5 shows thermograms of the Ig-GH9 domain pair and its T230A/D262A and D264A mutants. Both mutants significantly differed from the original Ig-GH9 by their denaturation temperatures (T_d) and calorimetric enthalpies ($\Delta_d H_{cal}$). The T_d values of Ig-GH9, T264A, and T230A/D262A were 77.9, 71.9, and 69.2°C, respectively; and the $\Delta_d H_{cal}$ values were 530.5, 469.8, and 296.4 kcal/mol^{-1}, respectively. Enzymatic properties of the mutants were comparable with those of the Ig-GH9. The values of K_M and k_{cat} for Ig-GH9, D264A, and T230A/D262A were 2.0 ± 0.15, 2.3 ± 2.25, and 2.0 ± 1.73 mM and 18.4 ± 2.11, 16.7 ± 2.42, and 19.0 ± 2.35 s^{-1}, respectively. Correspondingly, the efficiencies of catalysis were similar. In summary, mutation of two and even one residue from the domain interface significantly destabilized the combined Ig-GH9 protein structure. The data imply that the domain interactions are very specific and cannot be transferred in artificial fusion

Table 1. Denaturation parameters of CbhA individual domains and their combinations[e]

Protein	Mol mass (kDa)	Solvent	$T_d{}^a$ (°C)	$\Delta_d H_{cal}{}^b$(kcal mol^{-1})	$\Sigma\Delta_d H_{cal}{}^c$(kcal mol^{-1})	$r^d(\Delta_d H_{cal}/\Delta_d H^{v.H.})$
$X1_2$	9.4	Buffer	77.6	110.9		1.72
		$CaCl_2$	81.3	131.8		1.26
		EDTA	76.1	106.9		1.09
$X1_1X1_2$	19.6	Buffer	78.7	126.9		1.31
		$CaCl_2$	80.5	131.2		1.03
		EDTA	75.2	127.2		0.93
CBD3	16.4	Buffer	82.7	85.5		0.59
		$CaCl_2$	82.8	116.2		0.71
		EDTA	77.2	101.6		0.74
$X1_1X1_2$-CBD3	36.5	Buffer	83.7	252.0	212.4	0.99
		$CaCl_2$	84.5	259.5	249.4	1.04
		EDTA	77.6	261.8	228.8	0.84
Ig	11.7	Buffer	73.6	66.7		1.41
		$CaCl_2$	72.5	75.9		1.37
		EDTA	73.8	65.7		1.29
GH9	56.9	Buffer	68.3	279.8		2.80
		$CaCl_2$	74.2	353.9		1.71
		EDTA	64.2	135.4		0.91
Ig-GH9	68.5	Buffer	77.9	565.6	346.5	1.83
		$CaCl_2$	78.4	596.2	429.8	1.76
		EDTA	73.1	526.1	201.1	2.11
Ig-GH9-$X1_1X1_2$	88.4	Buffer	77.7	593.5	473.4	2.28
		$CaCl_2$	79.0	666.9	561.0	1.12
		EDTA	75.7	625.5	328.3	2.81
Ig-GH9-$X1_1X1_2$-CBD3	105.3	Buffer	82.3	797.3	558.9	2.14
		$CaCl_2$	83.1	835.0	679.2	1.86
		EDTA	77.4	734.5	429.9	2.90

[a]Denaturation temperature.
[b]Calorimetric denaturation enthalpy.
[c]Sum of calorimetric denaturation enthalpies of individual domains.
[d]Number of cooperative units.
[e]From Kataeva et al., 2005.

constructs (Kataeva et al., 2001a, 2001b). Finally, the fact that the addition of each next domain to the C terminus of the existing construct stabilized its structure indicates that at least CbhA is folding sequentially and each preexisting construct acts as a chaperone assisting correct folding of each C-terminal fused domain.

BIOLOGICAL FUNCTIONS OF INTERNAL MODULES

Although it is clear that the Ig domain stabilizes the GH9 structure of CbhA, where they are tightly bound to each other, and appears to retain the configuration of the active site of the GH9, allowing it to be enzymatically active (Kataeva et al., 2004; Schubot et al., 2004), not much is known about the biological activity of Ig-like modules in other enzymes.

CBMs are very important modules in carbohydrate-active enzymes. They bind to the insoluble substrates and bring catalytic domains in close contact with the substrate to hydrolyze it. The CBMs differ by binding specificity (Tomme et al., 1995). In particular, some of them bind amorphous insoluble substrates like cellulose or xylan (Brun et al., 2000; Johnson et al., 1996). Some of the CBMs binding amorphous insoluble carbohydrates also bind some soluble substrates such as soluble xylans, lichenan, laminarin, beta-glucan, mannans, or carboxymethyl cellulose (Arai et al., 2003; Kataeva et al., 2001a, 2001b). Other CBMs have the ability to bind highly crystalline recalcitrant substrates (Tomme et al., 1996). It has been shown that the binding specificity of CBMs depends on the topology of the binding surface. In those CBMs binding amorphous soluble and insoluble substrates, the binding surface is a cleft accommodating a single carbohydrate chain (Brun et al.,

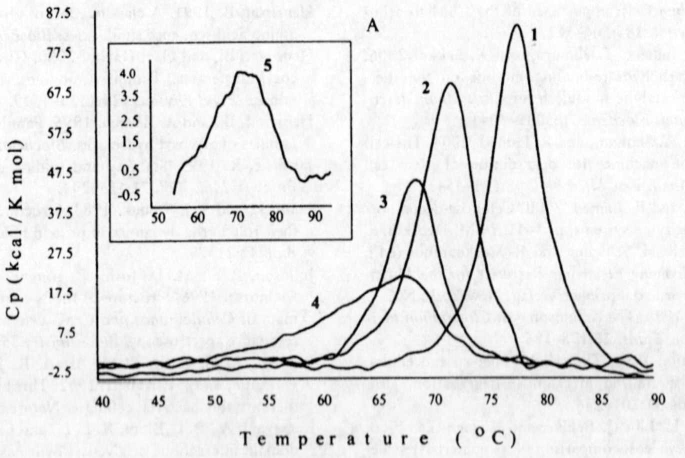

Figure 5. DSC thermograms of Ig-GH9 module pair (1), its D264A (2) and T230/D262A (3) mutants, and individual GH9 module (4) and Ig-like module (5). The protein concentrations and scan rate were 6 mg/ml and 60°C/h, respectively. All thermal transitions were completely irreversible, so that second scans of the proteins were used as baselines. (From Kataeva et al., 2004.)

2000, Johnson et al., 1996). The CBMs binding to the crystalline cellulose with carbohydrate chains connected by numerous hydrogen bonds have flat binding surfaces fitting very well the surface of the crystalline carbohydrate (Tomme et al., 1996). In some cases, the carbohydrate-active enzyme might contain more than one CBM with different binding specificities, the presence of which increases the hydrolytic potential of the enzyme.

Another interesting group of domains classified based on sequence similarity into family "X1" (B. Henrissat, personal communication) are less common. They are present in some bacterial chitinases and a few cellulases (Kataeva et al., 2001a, 2001b). These domains have a beta-sandwich fold typical of representatives of the Ig superfamily. This superfamily incorporates proteins involved in different binding functions. The duplicated version of the X1 domains, $X1_1$ and $X1_2$, is present in CbhA (Fig. 1). It has been demonstrated that the $X1_1X1_2$ domain pair of CbhA is involved in a nonhydrolytic loosening of the cellulose surface, thus enhancing catalytic activity of CbhA against insoluble cellulose. This effect is even more pronounced when the $X1_1X1_2$ domain pair is attached to the CBM3 in the three-domain construct $X1_1X1_2$-CBM3 (Kataeva et al., 2002). It is still unclear if this function is general for all X1 domains.

CONCLUSIONS

This chapter describes modular architectures of carbohydrate-active enzymes and analyzes the role of interdomain interactions in the structure, stability, and functionality of these interesting and important proteins. Although the majority of these proteins are complex machines, in which catalytic domains are surrounded by different noncatalytic ancillary modules, these modules are not randomly combined. They are located in a particular place within a protein often being associated with a particular type of neighboring modules. Modules vary in length and possess diverse functions. They are involved in complex communications and interactions. Entire proteins are greatly stabilized via a series of specific interdomain interactions. Structures and communications between modules in large cellulosomal catalytic enzymes like CbhA from *C. thermocellum* are modulated by calcium binding. The complex nature of interdomain interactions and cross talking has been confirmed by structural and conformational analyses. Even linkers between domains are crucial for the functionality of carbohydrate-active enzyme in that they serve as "molecular springs" allowing catalytic sites to reach and hydrolyze new glycosidic bonds, while the CBM is still bound to the substrate surface. As a general conclusion one may say that domain interactions in modular carbohydrate hydrolytic enzymes enhance the activity of the catalytic domains of these enzymes.

REFERENCES

Ahsan, M. M., T. Kimura, S. Karita, K. Sakka, and K. Ohmiya. 1996. Cloning, DNA sequencing, and expression of the gene encoding *Clostridium thermocellum* cellulase CelJ, the largest catalytic component of the cellulosome. *J. Bacteriol.* 178:5732–5740.

Arai, T., R. Araki, A. Tanaka, S. Karita, T. Kimura, K. Sakka, and K. Ohmiya. 2003. Characterization of a cellulose containing a family 30 carbohydrate-binding module (CBM) derived from

Clostridium thermocellum CelJ: importance of the CBM to cellulose hydrolysis. *J. Bacteriol.* **185**:504–512.

Araki, R., S. Karita, A. Tanaka, T. Kimura, and K. Sakka. 2006. Effect of family 22 carbohydrate-binding module on the thermostability of Xyn10B catalytic module from *Clostridium stercorarium. Biosci. Biotechnol. Biochem.* **70**:3039–3041.

Bayer, E. A., J.-P. Belaich, Y. Shoham, and R. Lamed. 2004. The cellulosomes: multienzyme machines for degradation of plant cell wall polysaccharides. *Annu. Rev. Microbiol.* **58**:521–554.

Bayer, E. A., Y. Shoham, and R. Lamed. 2000. Cellulose-decomposing bacteria and their enzyme systems, p. 1–41. *In* M. Dvorkin, S. Falkow, E. Rosenberg, K.-H. Schleifer, and E. Stackebrandt (ed.), *The Prokaryotes, an Evolving Electronic Resource for the Microbiological Community,* 3rd ed. Springer Verlag, New York, NY.

Béguin, P., and P. Alzari. 1998. The cellulosome of *Clostridium thermocellum. Biochem. Soc. Trans.* **26**:178–185.

Béguin, P., and M. Lemaire. 1996. The cellulosome: an exocellular, multiprotein complex specialized in cellulose degradation. *Crit. Rev. Biochem. Mol. Biol.* **31**:201–236.

Berr, K., D. Wassenberg, H. Lilie, J. Behlke, and R. Jaenicke. 2000. ϵ-Crystallin from duck eye lens: comparison of its quaternary structure and stability with other lactate dehydrogenases and complex formation with α-crystallin. *Eur. J. Biochem.* **267**:5413–5420.

Bolam, D. N., A. Ciruela, S. McQueen-Mason, P. Simpson, M. P. Williamson, J. E. Rixon, A. Boraston, J. P. Hazlewood, and J. Harry. 1998. *Pseudomonas* cellulose-binding domains mediate their effects by increasing enzyme substrate proximity. *Biochem. J.* **331**:775–781.

Brun, E., P. E. Johnson, A. L. Creagh, P. Tomme, P. Webster, C. A. Haynes, and I. P. McIntosh. 2000. Structure and binding specificity of the second N-terminal cellulose-binding domain from *Cellulomonas fimi* endoglucanase C. *Biochemistry* **39**:2445–2450.

Campbell, I. D. 2003. Modular proteins at the cell surface. *Biochem. Soc. Trans.* **31**:1107–1114.

Clout, N. J., A. Basak, K. Wieligmann, O. A. Bateman, R. Jaenicke, and C. Slingsby. 2000. The N-terminal domain of betaB2-crystallin resembles the putative ancestral homodimer. *J. Mol. Biol.* **304**: 253–257.

Doi, R. H., and A. Kosugi. 2004. Cellulosomes: plant-cell-wall-degrading enzyme complexes. *Nat. Rev.* **2**:541–551.

Dominguez, R., H. Souchon, S. Spinelli, Z. Dauter, K. S. Wilson, S. Chauvaux, and P. Béguin. 1995. A common protein fold and similar active site in two distinct families of beta-glycanases. *Nat. Struct. Biol.* **2**:569–576.

Dunker, A. K., J. D. Lawson, C. J. Brown, R. M. Williams, P. Romero, J. S. Oh, C. J. Oldfield, A. M. Campen, C. M. Ratliff, K. W. Hipps, J. Ausio, M. S. Nissen, R. Reeves, C. Kang, C. R. Kissinger, R. W. Bailey, M. D. Griswold, W. Chiu, E. C. Garner, and Z. Obradovic. 2001. Intrinsically disordered protein. *J. Mol. Graph. Model.* **19**:26–59.

Fontes, C. M., G. P. Hazlewood, E. Morag, J. Hall, B. H. Hirst, and H. J. Gilbert. 1995. Evidence for a general role for non-catalytic thermostabilizing domains in xylanases from thermophilic bacteria. *Biochem. J.* **307**:151–158.

Frydman, J., H. Erdjument-Bromage, P. Tempst, and F. U. Hartl. 1999. Co-translational domain folding as the structural basis for the rapid *de novo* folding of firefly luciferase. *Nat. Struct. Biol.* **6**:697–705.

Gilkes, N. R., B. Henrissat, D. G. Kilburn, R. C. Miller, and R. A. J. Warren. 1991. Domains in microbial beta-1,4-glycanases: sequence conservation, function, and enzyme families. *Microbiol. Rev.* **55**: 303–315.

Hayashi, H., K. I. Takagi, M. Fukumura, T. Kimura, S. Karita, K. Sakka, and K. Ohmiya. 1997. Sequence of xynC and properties of XynC, a major component of the *Clostridium thermocellum* cellulosome. *J. Bacteriol.* **179**:4246–4253.

Henrissat, B. 1991. A classification of glycosyl hydrolases based on amino acid sequence similarities. *Biochem. J.* **280**:309–316.

Henrissat, B., and G. J. Davies. 2000. Glycoside hydrolases and glycosyltransferases. Families, modules, and implications for genomics. *Plant Physiol.* **124**:1515–1519.

Henrissat, B., and A. Romeu. 1995. Families, superfamilies and subfamilies of glycosyl hydrolases. *Biochem. J.* **311**:350–351.

Jaenicke, R. 1999. Stability and folding of domain proteins. *Prog. Biophys. Mol. Biol.* **71**:155–241.

Janin, J., and S. J. Wodak. 1983. Structural domains in protein and their role in the dynamics of protein function. *Prog. Biophys. Mol. Biol.* **42**:21–78.

Johnson, P. E., M. D. Joshi, P. Tomme, D. G. Kilburn, and I. P. McIntosh. 1996. Structure of the N-terminal cellulose-binding domain of *Cellulomonas fimi* CenC determined by nuclear magnetic resonance spectroscopy. *Biochemistry* **35**:14383–14394.

Juy, M., A. G. Amit, P. M. Alzari, R. J. Poljak, M. Claeyssens, P. Béguin, and J.-P. Aubert. 1992. Three-dimensional structure of a thermostable bacterial cellulase. *Nature* **357**:89–91.

Kataeva, I. A., D. L. Blum, X.-L. Li, and L. G. Ljungdahl. 2001a. Do domain interactions of glycosyl hydrolases from *Clostridium thermocellum* cellulosome contribute to protein thermostability? *Protein Eng.* **14**:167–172.

Kataeva, I. A., J. M. Brewer, V. N. Uversky, and L. G. Ljungdahl. 2005. Domain coupling in a multimodular cellobiohydrolase CbhA from *Clostridium thermocellum. FEBS Lett.* **579**:4367–4373.

Kataeva, I. A., X.-L. Li, H. Chen, and L. G. Ljungdahl. 1999. CelK—a new cellobiohydrolase from *Clostridium thermocellum* cellulosome: role of N-terminal cellulose-binding domain, p. 454–460. *In* K. Ohmiya, K. Sakka, S. Karita, M. Hayashi, Y. Kobayashi, and T. Kimura (ed.), *Genetics, Biochemistry and Ecology of Cellulose Degradation.* UniPublishers Co., Tokyo, Japan.

Kataeva, I. A., and L. G. Ljungdahl. 2003. The *Clostridium thermocellum* cellulosome: a multi-protein complex of domain-composed components, p. 651–666. *In* V. Uversky (ed.), *Protein Structures. Kaleidoscope of Structural Properties and Functions.* Research Signpost, Kerala, India.

Kataeva, I. A., R. D. Siedel III, X.-L. Li, and L. G. Ljungdahl. 2001b. Properties and mutation analysis of the CelK cellulose-binding domain from the *Clostridium thermocellum* cellulosome. *J. Bacteriol.* **183**:1552–1559.

Kataeva, I. A., R. D. Seidel III, A. Shah, L. T. West, X.-L. Li, and L. G. Ljungdahl. 2002. The fibronectin type 3-like repeat from the *Clostridium thermocellum* cellobiohydrolase CbhA promotes hydrolysis of cellulose by modifying its substrate. *Appl. Environ. Microbiol.* **68**:4292–4300.

Kataeva, I. A., V. N. Uversky, J. M. Brewer, F. Schubot, J. Rose, B.-C. Wang, and L. G. Ljungdahl. 2004. Interactions between immunoglobulin-like and catalytic modules in *Clostridium thermocellum* cellulosomal cellobiohydrolase CbhA. *Protein Eng. Des. Sel.* **17**:759–769.

Kataeva, I. A., V. N. Uversky, and L. G. Ljungdahl. 2003. Calcium and domain interactions contribute to the thermostability of domains of the multimodular cellobiohydrolase, CbhA, a subunit of the *Clostridium thermocellum* cellulosome. *Biochem. J.* **372**:151–161.

Kerr, E. 2004. Broadened applicability of use for industrial enzymes. *Genet. Eng. News* **24**:1–5.

Kozhevnikov, G. O., A. N. Danilenko, E. E. Braudo, and K. D. Schwenke. 2001. Comparative studies of thermodynamic characteristics of pea ligumin and ligumin-T thermal transition. *Int. J. Biol. Macromol.* **29**:225–236.

Kuznetsova, I. M., O. V. Stepanenko, K. K. Zhu, J. M. Zhou, A. L. Fink, and V. N. Uversky. 2002. Unraveling multistate unfolding of rabbit muscle creatine kinase. *Biochim. Biophys. Acta* **1596**:138–155.

Little, E., P. Bork, and R. F. Doolittle. 1994. Tracing and spread of fibronectin type III domain in bacterial glycohydrolases. *J. Mol. Evol.* 39:631–643.

Ljungdahl, L. G., H. J. M. Op den Camp, H. J. Gilbert, H. R. Harangi, P. J. M. Steenbakkers, and X.-L. Li. 2007. Cellulosomes of anaerobic fungi, p. 271–303. *In* V. Uversky and I. A. Kataeva (ed.), *Cellulosome.* Nova Science Publishers, Inc, New York, NY.

Maity, H., M. Maity, M. M. G. Krishna, L. Mayne, and S. W. Englander. 2005. Protein folding; the step-wise assembly of foldon units. *Proc. Natl. Acad. Sci. USA* 102:4741–4746.

Mandelman, D., A. Belaich, J.-P. Belaich, N. Aghajari, H. Driguez, and R. Haser. 2003. X-Ray crystal structure of the multidomain endoglucanase Cel9G from *Clostridium cellulolyticum* complexed with natural and synthetic cello-oligosaccharides. *J. Bacteriol.* **185:**4127–4135.

McGrath, C. 2007. Mechanistic and functional characterization of glycosyl hydrolases involved in biomass degradation: *Thermobifida fusca* LAM81A and CHI18A and *Clostridium thermocellum* CbhA. Thesis. Cornell University, Ithaca, NY.

Medved, L. V., C. L. Orthner, H. Lubon, T. K. Lee, W. N. Drohan, and K. C. Ingham. 1995. Thermal stability and domain-domain interactions in natural and recombinant protein C. *J. Biol. Chem.* **270:**13652–13659.

Notenboom, V., A. B. Boraston, D. G. Kilburn, and D. R. Rose. 2001. Crystal structures of the family 9 carbohydrate-binding module from *Thermotoga maritima* xylanase 10A in native and ligand bound forms. *Biochemistry* 40:6248–6256

Ohmiya, K., K. Sakka, S. Karita, and T. Kimura. 1997. Structure of cellulases and their applications. *Biotechnol. Genet. Eng. Rev.* **14:** 365–414.

Parsiegla, G., A. Belaich, J.-P. Belaich, and R. Haser. 2002. Crystal structure of the cellulase Cel9M enlightens structure/function relationships of the variable catalytic modules in glycoside hydrolases. *Biochemistry* **41:**11134–11142.

Receveur, V., M. Czjzek, M. Schulein, P. Panine, and B. Henrissat. 2002. Dimension, shape, and conformational flexibility of a two domain fungal cellulase in solution probed by small angle X-ray scattering. *J. Biol. Chem.* **273:**40888–40892.

Rumbley, J., L. Hoang, L. Mayne, and S. W. Englander. 2001. An amino acid code for protein folding. *Proc. Natl. Acad. Sci. USA* 98:105–112.

Sakon, J., D. Irwin, D. B. Wilson, and P. A. Karplus. 1997. Structure and mechanism of endo/exocellulase E4 from *Thermomonospora fusca. Nat. Struct. Biol.* 4:810–818.

Schubot, F. D., I. A. Kataeva, J. Chang, A. K. Shah, L. G. Ljungdahl, J. P. Rose, and B.-C. Wang. 2004. Structural basis for the exocellulase activity of the cellobiohydrolase CbhA from *Clostridium thermocellum. Biochemistry* 41:1163–1170.

Shen, H., N. R. Gilkes, D. G. Kilburn, R. C. Miller, Jr., and R. A. J. Warren. 1995. Cellobiohydrolase B, a second exo-cellobiohydrolase from cellulolytic bacterium *Cellulomonas fimi. Biochem. J.* 311:67–74.

Shen, H., M. Schmuck, I. Pilz, N. R. Gilkes, D. G. Kilburn, R. C. Miller, and R. A. J. Warren. 1991. Deletion of the linker connecting the catalytic and cellulose-binding domains of endoglucanase A (CenA) of *Cellulomonas fimi* alters its conformation and catalytic activity. *J. Biol. Chem.* **266:**11335–11340.

Shen, M.-Y., F. P. Davis, and A. Sali. 2005. The optimal size of a globular protein domain: a simple sphere-packing model. *Chem. Phys. Lett.* 405:224–228.

Shin, E.-S., M.-J. Yang, K.-H. Jung, E.-J. Kwon, J. S. Jung, S. Park, J. Kim, H. D. Yun, and H. Kim. 2002. Influence of the transposition of the thermostabilizing domain of *Clostridium thermocellum* xylanase (XynX) on xylan binding and thermostabilization. *Appl. Environ. Microbiol.* **68:**3496–3501.

Smith, M. A., O. A. Bateman, R. Jaenicke, and C. Slingsby. 2007. Mutation of interfaces in domain-swapped human β-B2-crystallin. *Protein Sci.* **16:**615–625.

Srisodsuk, M., T. Reinikainen, M. Penttila, and T. T. Teeri. 1993. Role of the interdomain linker peptide of *Trichoderma reesei* cellobiohydrolase I in its interaction with crystalline cellulose. *J. Biol. Chem.* **268:**20756–20761.

Tomme, P., A. Boraston, B. McLean, J. Kormos, A. L. Creagh, K. Sturch, N. R. Gilkes, C. Haynes, R. A. J. Warren, and D. G. Kilburn. 1998. Characterization and affinity applications of cellulose-binding domains. *J. Chromatogr.* 715:283–296.

Tomme, J., R. Lamed, A. J. Chirino, E. Morag, E. A. Bayer, Y. Shoham, and T. A. Steitz. 1996. Crystal structure of a bacterial family III cellulose-binding domain: a general mechanism for attachment to cellulose. *EMBO J.* 15:5739–5751.

Tomme, P., R. A. J. Warren, R. C. Miller, D. G. Kilburn, and N. R. Gilkes. 1995. Cellulose-binding domains: classification and properties, p. 142–163. *In* J. M. Saddler and M. H. Penner (ed.), *Enzymatic Degradation of Insoluble Polysaccharides.* American Chemical Society, Washington, DC.

von Ossowski, I., J. T. Eaton, M. Czjzek, S. J. Perkins, T. P. Frandsen, P. M. Schulein, P. Panine, B. Henrissat, and V. Receveur-Brechot. 2005. Protein disorder: conformational distribution of the flexible linker in a chimeric double cellulose. *Biochem. J.* **88:** 2823–2832.

Wassenberg, D., H. Schurig, W. Liebl, and R. Jaenicke. 1997. Xylanase XynA from the hyperthermophilic bacterium *Thermotoga maritima:* structure and stability of the recombinant enzyme and its isolated cellulose-binding domain. *Protein Sci.* **6:**1718–1726.

Wassenberg, D., C. Welker, and R. Jaenicke. 1999. Thermodynamics of the unfolding of the cold-shock protein from *Thermotoga maritima. J. Mol. Biol.* 289:187–193

Wenk, M., and R. Jaenicke. 1998. Kinetic stabilization of a modular protein by domain interactions. *FEBS Lett.* **438:**127–130.

Wenk, M., and R. Jaenicke. 1999. Calorimetric analysis of the Ca2+-binding βγ -crystallin homolog protein S from *Myxococcus xanthus:* intrinsic stability and mutual stabilization of domains. *J. Mol. Biol.* **293:**117–124.

Zverlov, V. V., and W. H. Schwarz. 2007. The *C. thermocellum* cellulosome: novel components and insights from the genomic sequence, p. 119–151. *In* I. A. Kataeva (ed.), *Cellulosome.* Nova Science Publishers, Inc., Hauppage, NY.

Zverlov, V. V., G. V. Velikodvorskaya, W. H. Schwarz, K. Bronnenmeier, J. Kellermann, and W. L. Staudenbauer. 1998. Multidomain structure and cellulosomal localization of the *Clostridium thermocellum* cellobiohydrolase CbhA. *J. Bacteriol.* 180:3091–3099.

Bioenergy
Edited by J. Wall et al.

Chapter 10

The Development of Ethanologenic Bacteria for Fuel Ethanol Production

GREGORY W. LULI, LAURA JARBOE, AND LONNIE O. INGRAM

Recent events have stimulated the development of technologies for the utilization of plant-derived materials as a raw material for the production of liquid fuels. Foremost of these events is the rapid increase in the cost of petroleum due not only to the instability and uncertainty of supply but also to the rapidly increasing demand for automotive fuel worldwide.

Although the current interest in the development of a renewable domestic fuel supply is now influencing governmental and financial communities to take action, the development of the underlying technologies required to achieve this goal has been under way for many years.

It is not within the scope of this review to cover all of the developments for the wide variety of microbiological activities in this area. Nor is there room enough to discuss the development of the physical and chemical processes needed to convert plant material to liquid fuel. We have focused here on the development of two related bacteria that have proven to be effective in a variety of these processes: *Escherichia coli* and *Klebsiella oxytoca*.

BACKGROUND

In 1976, one of us published a paper (Ingram, 1976) on the toxicity of alcohols and the adaptive response of *E. coli* to the changes in membrane fluidity caused by increasing concentrations of alcohol. Even though this paper dealt with the response of bacteria to their environments, it started Ingram down the path of ethanol research.

After a series of papers on the effects of ethanol on membrane fluidity, Ingram and coworkers began utilizing *Zymomonas mobilis* as a model for study because of its natural resistance to high concentrations of ethanol (Carey and Ingram, 1983). *Z. mobilis* was seen as an "evolutionary adaptation for survival in the presence of ethanol." From that observation, because this bacterium was more tolerant to temperature and ethanol than traditional yeast, it seemed to be useful for the commercial production of ethanol. They sought genetic tools to modify this bacterium and to expand its range of sugars fermented (Carey and Ingram, 1983; Walia et al., 1984a, 1984b; Carey et al., 1983, 1984; Ingram et al., 1984; Byun, et al., 1986; Conway et al., 1987).

Although they were successful in genetically manipulating *Z. mobilis*, its increased tolerance of ethanol (relative to *E. coli*) was temperature dependent. Tolerance to ethanol and temperature is related to membrane fluidity. So even though *Z. mobilis* was more heat tolerant than yeast, a quality thought necessary for large-scale ethanol production, the tolerance to ethanol concentration was inversely related (Fig. 1).

These studies led to the realization that ethanol production is affected not only by ethanol tolerance but also by growth temperature and osmotic stresses present in the growth medium. In 1984, it was recognized that the ability to produce ethanol from inexpensive agricultural material would be advantageous. This led to the understanding that an improved microbe for the production of ethanol from these sources should have several key traits as shown in Table 1 (Ingram et al., 1984).

In the development of genetic tools to add desirable traits to *Z. mobilis*, it was discovered that the gene encoding the enzyme necessary for the initial step in the ethanol pathway in *Z. mobilis* (pyruvate decarboxylase [PDC]) was initiated by promoters not recognized by *E. coli*. Despite obtaining expression of *Z. mobilis* PDC from other *E. coli* promoters, little expression was obtained using the native *Z. mobilis* promoter region (Conway et al., 1987).

Although improvements were made in glycolytic flux and ethanol production in *Z. mobilis* in 1987 (Osman et al., 1987; Osman and Ingram, 1987), it was

Gregory W. Luli • Verenium Corp., 4955 Directors Place, San Diego, CA 92121. **Laura Jarboe and Lonnie O. Ingram** • Department of Microbiology and Cell Science, University of Florida, Gainesville, FL 32611.

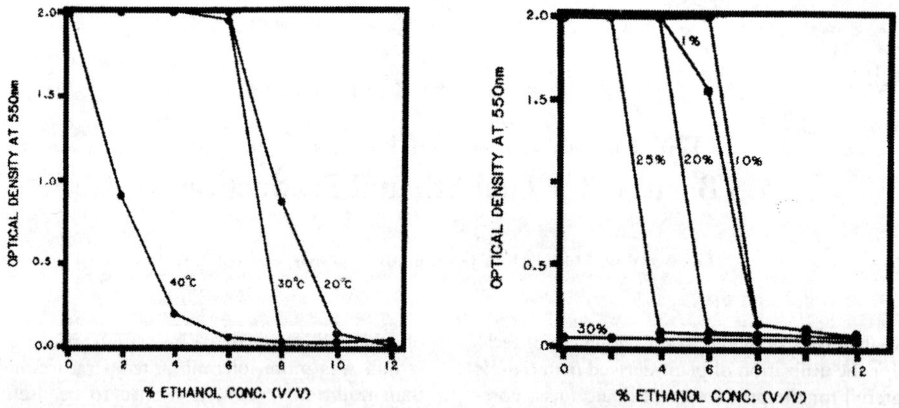

Figure 1. (Left) Effects of growth temperature on the alcohol tolerance of *Z. mobilis* strain CP4. Cells were grown as described previously (Carey et al., 1984). An overnight culture was diluted 1:200 into fresh medium containing a series of concentrations of ethanol and incubated for 48 h. Growth was measured as optical density at 550 nm. (Right) Effects of initial glucose concentration on the alcohol tolerance of *Z. mobilis* strain CP4. An overnight culture was diluted 1:200 into fresh medium containing 1, 10, 20, or 25% glucose and a series of concentrations of ethanol. Cultures were incubated at 30°C for 48 h, and growth was measured as optical density at 550 nm (Ingram et al., 1984).

recognized that many of the desired traits sought for this bacterium were already present in another common laboratory bacterium, *E. coli*, which already had well-established genetics, grew more robustly, and was capable of utilizing a wide variety of sugar substrates, including pentose sugars such as xylose and arabinose.

Because the genes for ethanol production from pyruvic acid, coding for PDC and alcohol dehydrogenase (ADH), were already isolated from *Z. mobilis* and expressed in *E. coli* separately, it was a relatively straightforward experiment to express both of these genes from a *lac* promoter on plasmid pUC18 in *E. coli* strain TC4 (Ingram et al., 1987). Surprisingly, not only did the recombinant strain express the two genes, but also its cells grew more densely than the parent plasmid-free host cells. This was attributed to the reduction in acidic by-products made by TC4pLOI295 because the need to regenerate glycolytic intermediates (NAD^+) was met by ethanol production.

Table 1. Desirable microbial traits for conversion of biomass to ethanol[a]

Identification of useful traits	
	Alcohol tolerance genes
	Lactose operon (galactose operon)
	Xylose operon
	Secreted enzymes: cellulose, etc.
Development of vector systems	
	Insertion (conjugate, transform)
	Stability
Improvement of gene expression	
	Operator/promoter mutations
	Cloned promoters
	Excretion sequences

[a]Data from Ingram et al., 1984.

The successful expression of *Z. mobilis pdc* and *adh* in *E. coli* represented a breakthrough in the development of a microbe that could utilize all the major sugars in low-cost, abundant agricultural material and produce ethanol at high rates and yields. This technology was recognized by the issuance of landmark U.S. patent 5,000,000 in 1991. The recombinant strain produced up to 6.1% vol/vol ethanol, with yields of >90% of the theoretical yield on glucose and xylose (Ingram et al., 1991).

By 1989, it was recognized that recombinant *E. coli* containing the *Z. mobilis pdc* and *adh* genes represented an opportunity to utilize low-cost waste materials as a source of carbohydrate for ethanol production in the United States (Alterthum and Ingram, 1989). Glucose, lactose, and xylose were shown to be effectively converted to ethanol by the recombinant *E. coli*. Using the approach used with Ingram's constructs, Tolan and Finn demonstrated xylose conversion to ethanol in 1987 in similar gram-negative bacteria, *Erwinia chrysanthemi* and *Klebsiella planticola* (Tolan and Finn, 1987a, 1987b).

A variety of *E. coli* host strains were evaluated, and ATCC strains 11303 (Luria strain B) and 15224 (Kepes strain ML308) were found have the highest PDC activity and ethanol productivity while ATCC 9637 (Waksman strain W) had superior environmental hardiness (Alterthum and Ingram, 1989). Although it was originally reported that strain B was used for strain development, it has recently been discovered that the ethanologenic strains developed by the Ingram lab are derivatives of strain W (L. Jarboe and L. O. Ingram, unpublished data).

In order to produce ethanol on a commercial scale, plasmid-borne traits are often unstable, especially in extended fermentations such as fed-batch or continuous feed processes. Although integration of genes into the chromosome was common in yeast and bacteria at the time (1991), it was thought that the integration of single copies of the PET operon (*pdc, adh,* and *cat*) would not be sufficient for ethanol production given the high level of expression in the original *Z. mobilis* host. Surprisingly, Ingram and coworkers achieved sufficient expression to produce high yields of ethanol from glucose and xylose from a single integration (followed by mutation) of the PET operon into the pyruvate-formate lyase gene in *E. coli* W (Ohta et al., 1991b). The *pfl* gene was chosen because it is highly expressed constitutively in *E. coli*. The method of integration was also novel in that the *pdc* promoter was removed due to poor expression in *E. coli* and the native host *pfl* promoter was used by virtue of integration into the gene downstream. The integration event alone was not sufficient for high levels of ethanol production. Random mutations were induced, and mutants that exhibited high levels of ethanol production were selected. Among the resulting constructs was strain KO11, and this strain was embodied in U.S. patent 5,821,093 in 1998 (Ingram et al., 1998).

A similar approach was taken for the development of another related bacterium, *Klebsiella oxytoca* M5a1 (Ohta et al., 1991a). Strains harboring the PET operon on a plasmid were stable for 60 generations without antibiotic selective pressure. In 1991, the PET operon was integrated into the chromosome of M5a1 by using the same approach used to construct *E. coli* KO11. The resulting construct, strain P2, was shown to achieve high ethanol yields and concentrations from glucose and xylose.

With these two modified bacteria, Ingram and coworkers believed they had developed ethanol-producing microbes that met many of the desirable traits they sought in 1984 (Table 1). A notable exception was tolerance to high ethanol concentrations.

DEMONSTRATION OF BIOMASS CONVERSION

Hemicellulose Sugar Fermentation

Development of recombinant microbes that utilize a variety of sugars for ethanol production in laboratory media under optimum growth conditions has been repeated by several laboratories around the world. However, economic production of ethanol from lignocellulosic waste materials will require efficient ethanol production under conditions that are not optimal and often very toxic to growth of the ethanologenic microbe.

During the mid-to-late 1990s, strains KO11 and P2 were shown to ferment the mixture of sugars from a variety of biomass sources that had to be treated with various chemical processes to produce the hemicellulose sugars. Because of its more efficient ability to co-utilize pentose and hexose sugars, *E. coli* strain KO11 was primarily used to ferment the mixture of sugars found in hemicellulose hydrolysates. Alternatively, *K. oxytoca* strain P2 and derivatives were used in combination with enzymes to convert the cellulosic portions of biomass because of their preference for lower pH values (pH 5.5 versus 6.5) and their ability to metabolize cellobiose and cellotriose (see below).

A summary of the early work is shown in Table 2. Fermentation of pure sugars in laboratory media has not been included in Table 2 with the exception of the work by Yomano (Yomano et al., 1998), which demonstrated the isolation of a strain derived from KO11 that is more tolerant to ethanol concentrations (up to 6.3% [wt/vol]), and that of Wood (Wood et al., 2005) demonstrating increased ethanol yield from an improved strain of *K. oxytoca*.

Verenium Corp. has exclusive rights to the commercialization of these two bacteria and has focused on the utilization of sugarcane bagasse due to a commercial opportunity in southern Louisiana. The use of dilute acid at temperatures above 140°C is effective for the hydrolysis of hemicellulose in bagasse without significant loss of sugars or the production of degraded by-products. Verenium has used dilute acid hydrolysis of hemicellulose in order to produce high concentrations of hemicellulose sugars for fermentation by *E. coli* strain KO11-RD1. This variant of KO11 has been adapted to toxicity of dilute acid hydrolysates of bagasse. An example of a fed-batch fermentation process for the fermentation of bagasse hemicellulose hydrolysate to ethanol is shown in Fig. 2. The fed-batch process was used to reduce the negative effects of toxic by-products found in dilute acid hydrolyzed biomass.

In the example in Fig. 2, the acidic hydrolysate contained 99.7 g of total sugars per liter after neutralization. Therefore, glucose was added to the hydrolysate in order to provide sufficient sugars to reach 50 g of ethanol per liter after dilution due to nutrient and inoculum additions. This process achieves >90% of the theoretical ethanol yield from total sugar.

More recently, this strain has also been demonstrated to achieve high ethanol yields at the 1,000-liter scale by using dilute acid hydrolysate from mixed waste wood (Okuda et al., 2007).

E. coli KO11-RD1 represents a commercially ready strain for first-of-its-kind biomass-to-ethanol facilities.

Table 2. Fermentation of hemicellulose sugars from various lignocellulosic biomasss sources by strains of ethanologenic *E. coli* and *K. oxytoca*[a]

Strain	Biomass	Treatment	Fermentation vol. (liter)	Ethanol concn. (% [wt/vol])	Ethanol yield (% of theoretical)	Time (h)	Nutrient	Reference
KO11	Mixed pine chips	Dilute acid	0.35	3.6	91	48	YE, Tryp	Barbosa et al., 1992
KO11	Corn cobs, hulls	Dilute acid	0.35	3.8	>100	48	CSL	Beall et al., 1992
KO11	Orange peels	Enzymatic	0.35	2.7	80	72	YE, Tryp	Grohmann et al., 1994
P2	Sugarcane bagasse	Dilute acid	0.35	3.9	70	168	YE, Tryp	Doran et al., 1994
P2	Sugarcane bagasse	AFEX	0.35	2.5	70	168	YE, Tryp	Doran et al., 1994
P2	Mixed waste paper	Dilute acid	0.35	4.4	90	104	YE, Tryp	Brooks and Ingram, 1995
KO11	Corn hulls/fiber	Dilute acid	0.35	4.4	94	72	CSL, YA	Asghari et al., 1996
KO11	Corn stover	Dilute acid	25.0	3.5	100	48	CSL, YA	Asghari et al., 1996
KO11	Sugarcane bagasse	Dilute acid	25.0	3.6	100	48	CSL, YA	Asghari et al., 1996
KO11	Sugarcane bagasse	Dilute acid	150.0	4.1	88	48	YE, Tryp	Asghari et al., 1996
KO11	Corn fiber	AFEX	0.20	2.7	92	120	YE, Tryp	Moniruzzaman et al., 1996
P2	Corn fiber	AFEX	0.20	2.0	69	120	YE, Tryp	Moniruzzaman et al., 1996
KO11	Corn fiber	Dilute acid	0.2	3.47	89	72	YE, Tryp	Dien et al., 1997
KO11	Corn germ meal	Dilute acid	0.2	3.42	89	72	YE, Tryp	Dien et al., 1997
KO11	Corn Diet Fiber	Dilute acid	0.2	3.76	89	72	YE, Tryp	Dien et al., 1997
LY01	Pure sugar	NA	0.35	6.3	89	96	YE, Tryp	Yomano et al., 1998
KO11	Rice hulls	Dilute acid	0.30	4.0	87	72	CSL	Moniruzzaman and Ingram, 1998
BW21	Pure sugar	NA	0.35	4.0	92	48	OUM1[a]	Wood et al., 2005
KO11-RD1	Waste wood	Dilute acid	1,000.0	3.0	88	50	CSL	Okuda et al., 2007

[a]Abbreviations: AFEX, ammonia fiber expansion; YE, yeast extract; CSL, corn steep liquor; Tryp, tryptone; YA, yeast autolysate; OUM1, minimal salts medium plus 0.5% CSL.

It has been shown to be robust with respect to toxic byproducts found in dilute acid hydrolysates and to process upsets of temperature and pH (Asghari et al., 1996).

Cellulose Fiber Conversion to Ethanol

The conversion of cellulose to ethanol can be traced back to the turn of the 20th century. In fact, many of the techniques employed today can be found

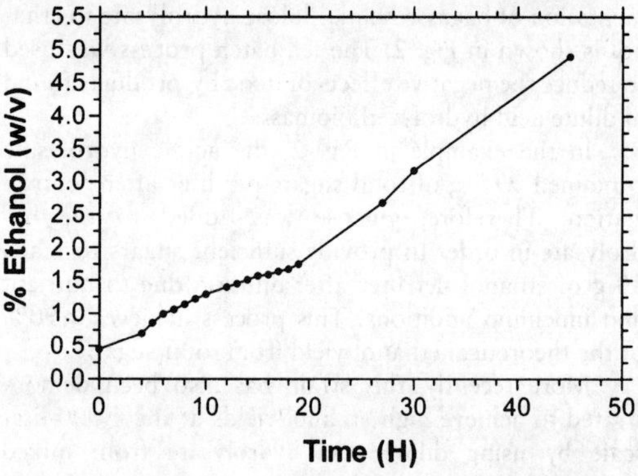

Figure 2. A 200-liter fed-batch fermentation of dilute acid hydrolysate of sugarcane bagasse by *E. coli* strain KO11-RD1. Final ethanol yield was 93% of theoretical.

in the work of Heinrich Scholler in Germany from 1932 through 1940 (Scholler, 1932, 1935, 1938, 1940). He used dilute acid to break down hemicellulose and cellulose in wood to sugars and ferment the resulting sugars to ethanol using traditional yeast, *Saccharomyces cerevisiae*. Because traditional yeasts do not ferment the pentose sugars, Scholler referred to the sugars produced by the acid hydrolysis as "fermentable" and "non-fermentable." A two-stage process was used to maximize hemicellulose sugar yields at about 150°C and cellulose sugar yields at about 200°C. He also demonstrated the use of lime to neutralize the acidic hydrolysates as well as other techniques employed today such as flowthrough percolation and shrinking bed reactor designs. He achieved approximately 25 liters of ethanol from 100 kg of wood (Scholler, 1935) in 1935, or about 60 gal per U.S. ton. Current ethanol yield targets set by the U.S. Department of Energy are about 60 gal per U.S. ton (U.S. Department of Energy, 2006).

The yield of ethanol from acidic hydrolysis of cellulose is limited due to the poor recovery of glucose during the acid hydrolysis process. The degradation of glucose occurs very rapidly under conditions necessary for cellulose hydrolysis. This limits the recovery of glucose to about 60% of the theoretical amount present in the cellulosic fibers (see review by Lee et al., 1999).

Therefore, the use of cellulolytic enzymes has been pursued for several decades as a means of increasing the ethanol yield from cellulose. Much of the work has been done using a yeast-based process that combines the enzymatic hydrolysis of cellulose to glucose and the subsequent fermentation of glucose to ethanol in a single reactor. This process has been called simultaneous saccharification and fermentation (SSF). The early work is best exemplified in the 1976 patent by Gauss et al. (1976). In this work, they used an enzyme produced by *Trichoderma viride* QM-9414 in combination with yeast to ferment sugars produced from wood pulp to ethanol at about 88% of the theoretical yield. They also demonstrated that the addition of the whole fungal broth to the SSF reaction produced higher yields than filtrates of the fungal broth.

This approach was demonstrated at pilot plant scale by Katzen and coworkers during the 1970s and 1980s (Katzen and Monceaux, 1995). Even though fungal cellulases are produced commercially for a variety of applications including textiles and beverages, they are too costly for economic viability for these biomass conversion processes. Methods to reduce the cost have been pursued, including fungal fermentation medium reformulation, use of whole fungal broth rather than filtered enzyme product, genetic improvement of the enzyme or producing fungi, coproduction of enzymes by the ethanologen, and on-site production of the enzymes.

The authors have pursued the SSF process model for conversion of the cellulose-rich solids produced by dilute acid hydrolysis of the hemicellulose. The SSF model has the following advantages over the sequential hydrolysis and fermentation process model: (i) lower enzyme dosages required for efficient conversion, (ii) compatiblility with coproduction of enzymes during ethanol fermentation, and (iii) lower free-sugar concentrations during the SSF process.

The coproduction of enzymes by the ethanologenic microbe represents a key breakthrough in cost reduction. The more enzymes produced by the ethanologenic microbe during ethanol fermentation, the less externally added fungal enzyme mixture is required. This approach was taken in the development of ethanologenic microbes for biomass conversion to ethanol with the engineering of *K. oxytoca* strain P2 to produce and secrete endoglucanases from *Erwinia chrysanthemi* (Zhou and Ingram, 1999, 2001; Zhou et al., 2001). The advantage of *K. oxytoca* over yeast (as well as the *E. coli* KO11 used for hemicellulose fermentation) is that it can naturally metabolize cellobiose and cellotriose (Wood and Ingram, 1992). This reduces the dependence on the enzyme mixture for beta-glucosidase (Wood and Ingram, 1992).

A summary of the early work by Ingram and coworkers during the 1990s using *K. oxytoca* strain P2 in combination with added fungal enzymes is given in Table 3. The results obtained from this work indicated that the ethanologenic *K. oxytoca* strain is suitable for SSF processes in conversion of cellulose to ethanol using added fungal enzymes.

The results of work on the development of a strain of *K. oxytoca* that could produce and secrete sufficient amounts of cellulolytic enzymes to impact the yield of the SSF process to convert a variety of polymers representing biomass to ethanol are shown in Table 4.

K. oxytoca strain SZ21 was used in SSF processes for conversion of dilute acid hydrolyzed sugarcane bagasse. Using this strain, the coproduction of CelY and CelZ from *E. chrysanthemi* has been shown to reduce the amount of added fungal cellulase enzymes by up to 50% (G. W. Luli, B. E. Wood, and Y. Y. Lee, unpublished data).

Table 3. Summary of cellulose conversion to ethanol using *K. oxytoca* strain P2 in SSF processes with added enzymes

Substrate	Treatment	Vol. (liter)	Enzyme loading (FPU/g)	Substrate loading (%)	Time (h)	Ethanol concn. (% [wt/vol])	Fiber conversion (%)	Reference
SigmaCell	Autoclaved	0.35	25	10	87	4.5	81	Doran and Ingram, 1993
Sugarcane bagasse	Dilute acid	0.35	10	10	144	2.4	63	Doran et al., 1994
Sugarcane bagasse	AFEX[a]	0.35	10	10	144	2.1	53	Doran et al., 1994
Office paper	Water pulped	0.50	8.3	12	96	4.1	67	Brooks and Ingram, 1995
Office paper	Acid pulped	0.50	8.3	12	96	4.1	67	Brooks and Ingram, 1995
Office paper	Acid pulped	0.50	8.3	12	96 + 12[c]	4.4	71	Brooks and Ingram,1995
Office paper	Autoclaved	10.0	10	10	96	3.5	69	Wood et al., 1997
Office paper	Autoclaved + ultrasound[b]	10	5	10	96	3.6	70	Wood et al., 1997

[a]AFEX, ammonia fiber expansion.
[b]Ultrasonic energy was applied to disrupt fiber structure.
[c]Includes 12 h of preincubation with enzyme.

Table 4. Coproduction of ethanol plus enzymes by bacteria for the conversion of various polymers

Strain	Enzyme	Enzyme source	Substrate	Enzyme loading	Substrate concn. (%)	Ethanol concn. (% [wt/vol])	Fiber conversion (%)	Reference
M5a1	XynZ + XylB	*C. thermocellum*	Xylan	None	4	0.8	35	Burchhardt and Ingram, 1992
pLOI555		*B. fibrisolvens*						
P2	CelD	*C. thermocellum*	Solka Floc	1%	5	1.7	65	Wood and Ingram, 1992
KO11	Amy	*B. stearothermophilus*	Starch	None	4.5	1.6	78	Guimaraes et al., 1992
	Pul	*T. brockii*						
P2	CelZ	*E. chrysanthemi*	Amorphous cellulose	None	5	0.39	14	Zhou and Ingram, 1999
SZ21	CelY + CelZ	*E. chrysanthemi*	SigmaCell	10 ml	10	2.7		Zhou et al., 2001
SZ21	CelY + CelZ	*E. chrysanthemi*	Amorphous cellulose	None	1.5	0.78	70	Zhou and Ingram, 2001

In Fig. 3, the effects of hemicellulose hydrolysis conditions on the effectiveness of a later SSF process for conversion of the fibers, as studied by Verenium Corp., are shown. Hydrolysis of hemicellulose at relatively low temperatures (circa 140 to 160°C) resulted in lignocellulosic fibers that could be converted to ethanol with about 75% efficiency in 120 h in SSF processes using enzyme dosages of 12 to 13 filter paper units (FPU)/g of glucan (Fig. 3, squares). However, when the hemicellulose hydrolysis conditions were modified to higher temperatures (circa 170 to 190°C), the resulting lignocellulosic fibers could be more rapidly converted to ethanol using lower enzyme dosages (Fig. 3, circles). In fact, the same conversion yield (circa 75%) could be achieved in 48 h using the higher hydrolysis temperature versus 120 h using standard conditions.

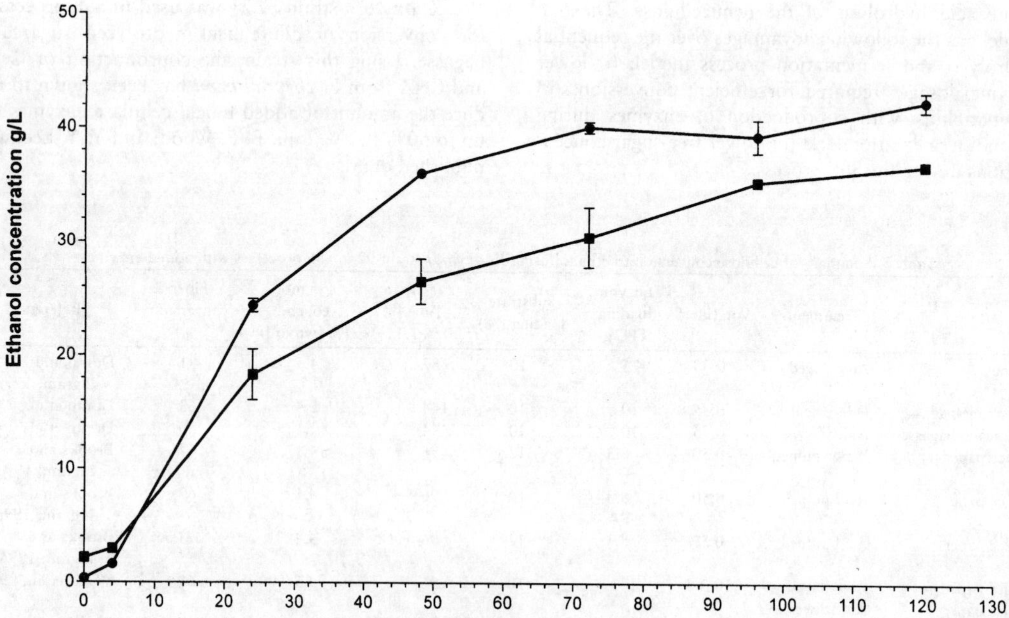

Figure 3. SSF of sugarcane bagasse fiber after dilute acid hydrolysis of hemicellulose under low- and high-temperature conditions, using *K. oxytoca* strain SZ21.

Because of its short duration and low enzyme dosage, this SSF-type of process shows promise for first-of-its-kind commercial facilities for ethanol production from lignocellulosic biomass.

Improved Strains

The development of *E. coli* and *K. oxytoca* strains for ethanol production continues to reduce both capital and variable cost of the biomass-to-ethanol process.

While KO11's ethanol production rate is as high as that of yeast, its ethanol tolerance is not (Yomano et al., 1998). Spontaneous mutants of KO11 with increased ethanol tolerance in liquid medium and increased ethanol production on solid medium were selected during 3 months of serial transfers in rich medium. One of the resulting strains, LY01, was able to grow in the presence of more than 50-g liter^{-1} ethanol and had greater than 80% survival from 30 s of exposure to 100-g liter^{-1} ethanol (Yomano et al., 1998). Fermentation data for LY01 are listed in Table 1.

Expression changes that accompanied increased ethanol tolerance were identified by comparing the transcriptomes of KO11 and LY01 during exposure to 0, 10, or 20 g of ethanol per liter in rich media with 100 g of glucose or xylose per liter (Gonzales et al., 2003). In this study, 205 genes with significantly altered expression were identified by the Student *t* test; expression of 49 of these genes differed more than twofold. The differentially expressed genes have a wide variety of functions, including amino acid biosynthesis, cell processes, cell structure, central intermediary metabolism, and energy metabolism as well as acid and osmotic stress. Further physiological analysis supported three major physiological differences in LY01 relative to KO11: increased glycine degradation; increased expression of genes related to uptake and synthesis of protective osmolytes, such as betaine; and lack of FNR (fumarate nitrate reduction) regulatory function. FNR normally regulates the expression of fermentation and anaerobic respiration genes (reviewed in Lynch and Lin, 1996). Pyruvate availability and distribution are affected by both glycine metabolism and expression of the FNR regulon. Therefore, it appears that pyruvate partitioning and osmotic protection are two of the multiple physiological factors that contribute to the increased ethanol tolerance of LY01.

Ideally, ethanologenic biocatalysts rapidly and efficiently produce high ethanol titers from lignocellulosic materials with minimum nutritional supplementation. While KO11 and LY01 have successfully utilized a variety of biomass sources, both also displayed an undesirable dependence on complex nutritional supplementation for rapid and efficient attainment of high ethanol titers (Asghari et al., 1996; Martinez et al., 1999; Underwood et al., 2004; York and Ingram, 1996). In contrast, the KO11 derivative SZ110, which was engineered for lactic acid production (Zhou et al., 2005), rapidly produced lactic acid with high yield and titer without dependence on complex nutrient supplementation.

Because SZ110 showed the properties desired in an ethanologenic biocatalyst, SZ110 was reengineered for ethanol production (L. P. Yomano, S. P. York, S. Zhou, K. T. Shanmugam, and L. O. Ingram, submitted for publication). This conversion consisted of both directed engineering and metabolic evolution. Directed engineering included removal of the lactic acid production pathway, insertion of the *Pseudomonas putida* short-chain esterase *estZ* to lower ethyl acetate levels, deletion of methylglyoxyal synthase *mgsA* to eliminate lactate production, and restoration of the native *pfl* gene. The *Z. mobilis* PET operon, which was inserted at the *pfl* locus in KO11, was eliminated during construction of SZ110. To allow for optimal reintegration, the *Z. mobilis* homoethanol pathway was randomly inserted by transposon. Because the donor and recipient strains both had specific growth requirements, direct functional selection in minimal medium without antibiotics was possible. Serial transfers in mineral salts medium enabled selection and further evolution of strains that express the homoethanol pathway and grow rapidly without complex nutrient supplementation. One strain was selected for further analysis and engineering; it was determined that in this strain, the *Z. mobilis* genes integrated within 23S ribosomal RNA subunit *rrlE*. Because ribosomal RNA transcription regulation ensures high expression at high growth rates and basal expression at low growth rates and during stationary phase (reviewed by Paul et al., 2004, and by Dennis et al., 2004), this is an excellent site for PET integration. The final strain, LY168, performs equally well in rich or mineral salts medium supplemented with 1 mM betaine and converted 90 g of xylose per liter to 45 g of ethanol per liter in 48 h, >99% of the theoretical yield (Yomano et al., submitted). LY168 also produced ethanol from 90 g of xylose per liter at >95% of the theoretical yield in a mineral salts medium containing only 4.2 g of total salts per liter and 1 mM betaine (Martinez et al., 2007).

The *Z. mobilis* PET operon has also been used to engineer a series of ethanologenic K-12 derivatives by the Fermentation Biochemistry Research Unit, termed the FBR strains. The design of these strains has focused on maximization of the strain stability (Hespell et al., 1996; Dien et al., 2000). FBR5, the most recent strain in this series, produced ethanol at 86 to 92% of the theoretical yield from a variety of substrates and maintained these yields during 26 days of continuous culture (Dien et al., 2000; Martin et al., 2006). However, the final titers and yields produced by FBR5 in rich media are lower than those from LY168 in minimal medium and the strains have an undesirable dependence on complex nutritional supplementation and plasmids.

As described above, *K. oxytoca* has been engineered to produce ethanol from lignocellulosic materials by use of the *Z. mobilis* PET operon and also to produce and secrete endoglucanases from *E. chrysanthemi*. The desire to further reduce the process cost led to the design of Optimized Urea Medium (OUM1) specifically for *K. oxytoca* (Wood et al., 2005). *K. oxytoca* is able to utilize urea as the sole nitrogen source, which adds to the cost reduction by reducing media acidification (Teixeira de Mattos and Neijssel, 1997). In addition to urea, OUM1 medium contains corn steep liquor and mineral salts. Though *K. oxytoca* P2 grew well in OUM1 medium, a substantial amount of carbon was lost in the form of 2,3-butanediol and acetoin. Following elimination of the butanediol pathway, *K. oxytoca* P2 derivative strain BW21 produced >40 g of ethanol per liter from 90 g of glucose per liter (0.47 g of ethanol per g of glucose) within 48 h in OUM1 medium (Wood et al., 2005).

SUMMARY AND CONCLUSIONS

The development of biocatalysts that can efficiently convert all of the predominant sugars found in biomass sources is a key step in commercial viability of ethanol production from renewable resources. However, the overall process must be taken into account when evaluating the economics of such future facilities. Some have taken the path of choosing a process, and then attempting to engineer biocatalysts that optimally perform in the preconceived process. This approach is well suited for the design of fully optimized "*n*th" plants. However, we have chosen to start with an efficient microbe and build a process that takes advantage of the microbe's characteristics. We believe this path is the shortest to economically viable "first-of-its-kind" lignocellulosic ethanol factories.

REFERENCES

Alterthum, F., and L. O. Ingram. 1989. Efficient ethanol production from glucose, lactose, and xylose by recombinant *Escherichia coli*. *Appl. Environ. Microbiol.* **55:**1943–1948.

Asghari, A., R. J. Bothast, J. B. Doran, and L. O. Ingram. 1996. Ethanol production from hemicellulose hydrolysates of agricultural residues using genetically engineered *Escherichia coli* strain KO11. *J. Ind. Microbiol.* **16:**42–47.

Barbosa, M. F., M. J. Beck, J. E. Fein, D. Potts, and L. O. Ingram. 1992. Efficient fermentation of *Pinus sp.* acid hydrolysates by an ethanologenic strain of *Escherichia coli*. *Appl. Environ. Microbiol.* **58:**1382–1384.

Beall, D. S., L. O. Ingram, A. Ben-Bassat, J. B. Doran, D. E. Fowler, R. G. Hall, and B. E. Wood. 1992. Conversion of hydrolysates of corn cobs and hulls into ethanol by recombinant *Escherichia coli* B containing integrated genes for ethanol production. *Biotechnol. Lett.* **14:**857–862.

Brooks, T. A., and L. O. Ingram. 1995. Conversion of mixed waste office paper to ethanol by genetically engineered *Klebsiella oxytoca* strain P2. *Biotechnol. Prog.* **11:**619–625.

Burchhardt, G., and L. O. Ingram. 1992. Conversion of xylan to ethanol by ethanologenic strains of *Escherichia coli* and *Klebsiella oxytoca*. *Appl. Environ. Microbiol.* **58:**1128–1133.

Byun, M. O.-K., J. B. Kaper, and L. O. Ingram. 1986. Construction of a new vector for the expression of foreign genes in *Zymomonas mobilis*. *J. Ind. Microbiol.* **1:**9–15.

Carey, V. C., S. K. Walia, and L. O. Ingram. 1983. Expression of a lactose transposon (TN951) in *Zymomonas mobilis*. *Appl. Environ. Microbiol.* **46:**1163–1168.

Carey, V. C., and L. O. Ingram. 1983. Lipid composition of *Zymomonas mobilis*: effects of ethanol and glucose. *J. Bacteriol.* **154:**1291–1300.

Carey, V. C., S. K. Walia, and L. O. Ingram. 1984. Expression of the lactose operon in *Zymomonas mobilis*. *Dev. Ind. Microbiol.* **25:** 589–596.

Conway, T., Y. A. Osman, J. I. Konnan, E. M. Hoffmann, and L. O. Ingram. 1987. Promoter and nucleotide sequence of the *Zymomonas mobilis* pyruvate decarboxylase. *J. Bacteriol.* **169:**949–954.

Dennis, P. P., M. Ehrenberg, and H. Bremer. 2004. Control of rRNA synthesis in *Escherichia coli*: a systems biology approach. *Microbiol. Mol. Biol. Rev.* **68:**639–668.

Dien, B. S., R. B. Hespell, L. O. Ingram, and R. J. Bothast. 1997. Conversion of corn milling fibrous co-products into ethanol by recombinant *Escherichia coli* strains KO11 and SL40. *World J. Microbiol. Biotechnol.* **13:**619–625.

Dien, B. S., N. N. Nichols, P. J. O'Bryan, and R. J. Bothast. 2000. Development of new ethanologenic *Escherichia coli* strains for fermentation of lignocellulosic biomass. *Appl. Biochem. Biotechnol.* **84:**181–196.

Doran, J. B., and L. O. Ingram. 1993. Fermentation of crystalline cellulose to ethanol by *Klebsiella oxytoca* containing chromosomally integrated *Zymomonas mobilis* genes. *Biotechnol. Prog.* **9:** 533–538.

Doran, J. B., H. C. Aldrich, and L. O. Ingram. 1994. Saccharification and fermentation of sugar cane bagasse by *Klebsiella oxytoca* P2 containing chromosomally integrated genes encoding the *Zymomonas mobilis* ethanol pathway. *Biotechnol. Bioeng.* **44:**240–247.

Gauss, W. F., S. Suzuki, and M. Takagi. November 1976. Manufacture of alcohol from cellulosic materials using plural ferments. U.S. patent 3,990,994.

Gonzalez, R., H. Tao, J. E. Purvis, S. W. York, K. T. Shanmugam, and L. O. Ingram. 2003. Gene array-based identification of beneficial changes that contribute to increased ethanol tolerance in ethanologenic *Escherichia coli* LY01. *Biotechnol. Prog.* **19:**612–623.

Grohmann, K., E. A. Baldwin, B. S. Buslig, and L. O. Ingram. 1994. Fermentation of galacturonic acid and other sugars in orange peel hydrolysates by the ethanologenic strain of *Escherichia coli*. *Biotechnol. Lett.* **16:**281–286.

Guimaraes, W. V., K. Ohta, G. Burchhardt, and L. O. Ingram. 1992. Ethanol production from starch by recombinant *Escherichia coli* containing integrated genes for ethanol production and plasmid genes for saccharification. *Biotechnol. Lett.* **14:**415–420.

Hespell, R. B., H. Wyckoff, B. S. Dien, and R. J. Bothast. 1996. Stabilization of pet operon plasmids and ethanol production in *Escherichia coli* strains lacking lactate dehydrogenase and pyruvate-formate lyase activities. *Appl. Environ. Microbiol.* **62:**4594–4597.

Ingram, L. O. 1976. Adaptation of membrane lipids to alcohols. *J. Bacteriol.* **125:**670–678.

Ingram, L. O., A. S. Holt, V. C. Carey, K. M. Dombek, W. A. Holt, Y. A. Osman, and S. K. Walia. 1984. Biochemical and genetic improvement of *Zymomonas mobilis*. *Biomass* **6:**131–143.

Ingram, L. O., T. Conway, D. P. Clark, G. W. Sewell, and J. F. Preston. 1987. Genetic engineering of ethanol production in *Escherichia coli*. *Appl. Environ. Microbiol.* 53:2420–2425.

Ingram, L. O., T. Conway, and F. Alterthum. March 1991. Ethanol production by *Escherichia coli* strains co-expressing *Zymomonas* PDC and ADH genes. U.S. patent 5,000,000.

Ingram, L. O., K. Ohta, and B. Wood. October 1998. Recombinant cells that highly express chromosomally integrated heterologous genes. U.S. patent 5,821,093.

Katzen, R., and D. A. Monceaux. 1995. Development of bioconversion of cellulosic wastes. *Appl. Biochem. Biotechnol.* 51/52:585–592.

Lee, Y. Y., P. Iyer, and R. W. Torget. 1999. Dilute-acid hydrolysis of lignocellulosic biomass. *Adv. Biochem. Eng. Biotechnol.* 65:93–115.

Lynch, A. S., and E. C. C. Lin. 1996. Responses to molecular oxygen, p. 1526–1538. *In* F. C. Neidhart, R. Curtiss III, J. L. Ingraham, E. C. C. Lin, K. B. Low, B. Magasanik, W. S. Reznikoff, M. Riley, M. Schaechter, and H. E. Umbarger (ed.), Escherichia coli *and* Salmonella: *Cellular and Molecular Biology*, 2nd ed. ASM Press, Washington, DC.

Martin, G. J. O., A. K. Bin Zhou, and N. B. Pamment. 2006. Performance and stability of ethanologenic *Escherichia coli* strain FBR5 during continuous culture on xylose and glucose. *J. Ind. Microbiol. Biotechnol.* 33:834–844.

Martinez, A., S. W. York, L. P. Yomano, V. L. Pineda, F. C. Davis, J. C. Shelton, and L. O. Ingram. 1999. Biosynthetic burden and plasmid burden limit expression of chromosomally integrated heterologous genes (*pdc, adhB*) in *Escherichia coli*. *Biotechnol. Prog.* 15: 891–897.

Martinez, A., T. B. Grabar, K. T. Shanmugam, L. Yomano, S. W. York, and L. O. Ingram. 2007. Low salt medium for lactate and ethanol production by recombinant *Escherichia coli* B. *Biotechnol. Lett.* 29:394–404.

Moniruzzaman, M., B. S. Dien, B. Ferrer, R. B. Hespell, B. E. Dale, L. O. Ingram, and R. J. Bothast. 1996. Ethanol production from AFEX pretreated corn fiber by recombinant bacteria. *Biotechnol. Lett.* 18:955–990.

Moniruzzaman, M., and L. O. Ingram. 1998. Ethanol production from dilute acid hydrolysate of rice hulls using genetically engineered *Escherichia coli*. *Biotechnol. Lett.* 20:943–947.

Ohta, K., D. S. Beall, J. P. Mejia, K. T. Shanmugam, and L. O. Ingram. 1991a. Metabolic engineering of *Klebsiella oxytoca* M5A1 for ethanol production from xylose and glucose. *Appl. Environ. Microbiol.* 57:2810–2815.

Ohta, K., D. S. Beall, K. T. Shanmugam, and L. O. Ingram. 1991b. Genetic improvement of *Escherichia coli* for ethanol production: chromosomal integration of *Zymomonas mobilis* genes encoding pyruvate decarboxylase and alcohol dehydrogenase II. *Appl. Environ. Microbiol.* 57:893–900.

Okuda, N., K. Ninomiya, M. Takao, Y. Katakura, and S. Shioya. 2007. Microaeration enhances productivity of bioethanol from hydrolysate of waste house wood using ethanologenic *Escherichia coli* KO11. *J. Biosci. Bioeng.* 103:350–357.

Osman, Y. A., T. Conway, S. J. Bonetti, and L. O. Ingram. 1987. Glycolytic flux in *Zymomonas mobilis*: enzyme and metabolite levels during batch fermentation. *J. Bacteriol.* **169**:3726–3736.

Osman, Y. A., and L. O. Ingram. 1987. *Zymomonas mobilis* mutants with an increased rate of alcohol production. *Appl. Environ. Microbiol.* 53:1425–1432.

Paul, B. J., W. Ross, T. Gaal, and R. L. Gourse. 2004. rRNA transcription in *Escherichia coli*. *Annu. Rev. Genet.* 38:749–770.

Scholler, H. December 1932. Process and device for the saccharification of cellulose and the like. U.S. patent 1,890,304.

Scholler, H. February 1935. Process of converting cellulose and the like into sugar with dilute acids under pressure. U.S. patent 1,990,097.

Scholler, H. July 1938. Process of saccharification of cellulose. U.S. patent 2,123,211.

Scholler, H. January 1940. Fermentation process of treating sugar worts. U.S. patent 2,188,193.

Teixeira de Mattos, M. J., and O. M. Neijssel. 1997. Bioenergetic consequences of microbial adaptation to low-nutrient environments. *J. Biotechnol.* 59:117–126.

Tolan, J. S., and R. K. Finn. 1987a. Fermentation of D-xylose and L-arabinose to ethanol by *Erwinia chrysanthemi*. *Appl. Environ. Microbiol.* 53:2033–2038.

Tolan, J. S., and R. K. Finn. 1987b. Fermentation of D-xylose to ethanol by genetically modified *Klebsiella planticola*. *Appl. Environ. Microbiol.* 53:2039–2044.

Underwood, S. A., M. L. Buszko, K. T. Shanmugam, and L. O. Ingram. 2004. Lack of protective osmolytes limits final cell density and volumetric productivity of ethanologenic *Escherichia coli* KO11 during xylose fermentation. *Appl. Environ. Microbiol.* **70**: 2734–2740.

United States Department of Energy. 2006. *Financial Assistance Funding Opportunity Announcement: DE-PS36-07GO97003.*

Walia, S. K., B. P. All III, V. C. Carey, and L. O. Ingram. 1984a. Self-transmissible plasmid in *Zymomonas mobilis* carrying antibiotic resistance. *Appl. Environ. Microbiol.* 47:198–200.

Walia, S. K., B. P. All III, V. C. Carey, and L. O. Ingram. 1984b. Occurrence of R plasmid in *Zymomonas mobilis*. *Dev. Ind. Microbiol.* 25:485–490.

Wood, B. E., and L. O. Ingram. 1992. Ethanol production from cellobiose, amorphous cellulose, and crystalline cellulose by recombinant *Klebsiella oxytoca* containing chromosomally integrated *Zymomonas mobilis* genes for ethanol production and plasmids expressing thermostable cellulase genes from *Clostridium thermocellum*. *Appl. Environ. Microbiol.* 58:2103–2110.

Wood, B. E., H. C. Aldrich, and L. O. Ingram. 1997. Ultrasound stimulates ethanol production during the simultaneous saccharification and fermentation of mixed waste office paper. *Biotechnol. Prog.* 13:232–237.

Wood, B. E., L. P. Yomano, S. W. York, and L. O. Ingram. 2005. Development of industrial-medium-required elimination of the 2,3 butanediol fermentation pathway to maintain ethanol yield in an ethanologenic strain of *Klebsiella oxytoca*. *Biotechnol. Prog.* **21**: 1366–1372.

Yomano, L. P., S. W. York, and L. O. Ingram. 1998. Isolation and characterization of ethanol tolerant mutants of *Escherichia coli* KO11 for fuel ethanol production. *J. Ind. Microbiol.* 20:132–138.

York, S. W., and L. O. Ingram. 1996. Ethanol production by recombinant *Escherichia coli* KO11 using crude yeast autolysate as a nutrient supplement. *Biotechnol. Lett.* 18:683–688.

Zhou, S., and L. O. Ingram. 1999. Engineering endoglucanase-secreting strains of ethanologenic *Klebsiella oxytoca* P2. *J. Ind. Microbiol. Biotechnol.* 22:600–607.

Zhou, S., and L. O. Ingram. 2001. Simultaneous saccharification and fermentation of amorphous cellulose to ethanol by recombinant *Klebsiella oxytoca* SZ21 without supplemental cellulase. *Biotechnol. Lett.* 23:1455–1462.

Zhou, S., F. C. Davis, and L. O. Ingram. 2001. Gene integration and expression and extracellular secretion of *Erwinia chrysanthemi* endoglucanase CelY (*celY*) and CelZ (*celZ*) in ethanologenic *Klebsiella oxytoca* P2. *Appl. Environ. Microbiol.* 67:6–14.

Zhou, S., L. P. Yomano, K. T. Shanmugam, and L. O. Ingram. 2005. Fermentation of 10% (w/v) sugar to D(-)-lactate by engineered *Escherichia coli* B. *Biotechnol. Lett.* 27:1891–1896.

Bioenergy
Edited by J. Wall et al.

Chapter 11

Chrysosporium lucknowense Cellulases and Xylanases in Cellulosic Biofuels Production

MARCO A. BÁEZ-VÁSQUEZ AND ARKADY P. SINITSYN

Cellulose and hemicellulose are the major organic components of plant cell wall polysaccharides in lignocellulosic biomass, accounting for approximately 70% of plant biomass. They represent a major renewable energy source (Bhat et al., 1998). Agricultural biomass is an inexpensive, abundant, and renewable natural feedstock supply that can be converted into liquid biofuels and chemicals (Demain et al., 2005). The conversion of wood, municipal cellulosic wastes, and agricultural residues to fermentable sugars continues to be studied intensively because cellulosic ethanol can be obtained as liquid motor fuel. However, the rate-limiting step to convert lignocellulose into fuel grade ethanol is overcoming the recalcitrant nature of cell wall structures and intrinsic resistance of the lignin complex to enzymatic attack. It is well known that a good pretreatment selection might help to amplify reactivity of the cellulose and hemicellulose fractions, enabling better accessibility and susceptibility to enzymatic attack (Sinitsyn et al., 1981). However, it is thought that the level of cellulose crystallinity is increased after pretreatment (Wyman et al., 2005). Cellulases and hemicellulases are very active depolymerizing enzyme systems that work through concerted action and dynamic synergism and are responsible for conversion of biomass polysaccharides into fermentable monosaccharides, such as glucose, xylose, and other related hexose and pentose oligomers (Sinitsyn et al., 1983). Cellulases and hemicellulases have applications in many industries: food, beverages, textiles, animal feed, pulp, and paper (Himmel et al., 1999; Suurnäkki et al., 2004).

Filamentous fungi are the major sources of glycosyl hydrolases. Different strains of *Trichoderma* sp., *Penicillium* sp., and *Humicola* sp. have been traditionally considered overproducers of a complex array of glycosyl hydrolases such as cellulases, xylanases, and accessory enzymes which actively depolymerize crystalline cellulose, amorphous cellulose, and insoluble xylan (Rabinovich et al., 2002; Kabel et al., 2005).

In spite of the success achieved in recent decades in elucidating the mechanisms of enzymatic cellulose and xylan degradation and expanding cellulase and xylanase production, the task of finding new improved strains remains active and important.

Chrysosporium lucknowense is a fungal overproducer of cellulases and xylanases. It is classified as an ascomycete and was isolated from Far East alkaline soil at the Institute of Biochemistry and Physiology of Microorganisms of the Russian Academy of Sciences (Bukhtojarov et al., 2004). Mutant strains, characterized by increased secretion level of cellulases and xylanases, were further obtained using mutagenesis (Emalfarb et al., 2000, 2001, 2003) and are being currently used for industrial production of cellulases and xylanases. *C. lucknowense* produces more than 100 g of protein per liter in fermentation broth (accompanied by low viscosity of the fermentation medium). The major portion of total secreted protein is represented by cellulases and xylanases. For comparison, *Trichoderma* strains provide 45 to 40 g of protein per liter (Durand et al., 1988), and *Penicillium* sp. produces up to 47 g/liter (Solovieva et al, 2005). Enzyme preparations from *C. lucknowense* have exhibited high efficiency in cellulose saccharification (Gusakov et al., 2007) as well as in treatment of cotton textiles (Gusakov et al., 1998; Sinitsyn et al., 2001; Miettinen-Oinonen and Suominen, 2002; Miettinen-Oinonen et al., 2004). In general, cellulases and xylanases of *C. lucknowense* have a great potential for degradation of lignocellulosic materials.

CELLULASES OF *C. LUCKNOWENSE*

Endoglucanases

The strategy for isolation of individual cellulases has been described (Bukhtojarov et al., 2004; Gusakov et al., 2005) and is based on the sequential use of different purification techniques, such as anion-exchange

Marco A. Báez-Vásquez • 396 Mallard Point, Jupiter, FL 33458-8353. Arkady P. Sinitsyn • Department of Chemical Enzymology, Moscow State University, Moscow, Russia 119899.

Table 1. Biochemical characteristics of cellulases from *C. lucknowense*

Enzyme	Glycoside hydrolase family	Systematic name	Presence of CBM[a]
EG 1	7	Cel7C	–
EG 2	5	Cel5A	+
EG 3	12	Cel12A	–
EG 5	45	Cel45A	–
EG 6	6	Cel6C	–
CBH 1a	7	Cel7A	+
CBH 1b	7	Cel7B	–
CBH 2a	6	Cel6A	–
CBH 2b	6	Cel6B	+

[a]+, present; –, absent.

chromatography, gel filtration, and chromatofocusing. The cellulase system of *C. lucknowense* consists of several endoglucanases, cellobiohydrolases, and β-glucosidase. Major properties and classification of these cellulases are summarized in Tables 1 and 2. *C. lucknowense* produces five enzymes which have a high level of CMCase activity and have been characterized as endo-1,4-β-glucanases (EC 3.2.1.4): EG 1 (Cel7C), EG 2 (Cel5A), EG 3 (Cel12A), EG 5 (Cel45A), and EG 6 (Cel6C).

Typical values of CMCase activity for purified fungal endoglucanases fall within the range of 10 to 50 U of protein/mg. *C. lucknowense* endoglucanase EG 2 has a specific activity of 52 U/mg, which indicates a high level of activity. Some *C. lucknowense* endoglucanases have additional activities toward other polysaccharide substrates. For example, EG 3 and EG 5 can hydrolyze xyloglucan, i.e., β-1,4-glucan-containing xylose residues as constituents in the side chain. The specific xyloglucanase activities of these enzymes are 16 and 2 U/mg, respectively (Bukhtojarov et al., 2004). *C. lucknowense* also produces true hemicellulase exhibiting an exo-type of attack on xyloglucan (Grishutin, 2004). This enzyme has fairly high specific xyloglucanase activity (75 U/mg) and very low CMCase activity (4 U/mg). EG 3 exhibits high specific activity on β-glucan (125 U/mg), but this enzyme does not have laminarinase activity. These endoglucanases exhibit maximal activity in the pH range of 4.7 to 6 and retain high activity at increased pH (50% of maximal activity at pH 7 to 8). The temperature optima for these endoglucanases are within the range of 60 to 70°C, and they retain high activity at elevated temperature (50% at 70 to 80°C). Enzyme EG 2 exhibits the highest specific activity among endoglucanases toward crystalline cellulose such as Avicel (Table 2). Thus, EG 2 can be most useful for cellulose degradation and saccharification (Bukhtojarov et al., 2004).

Other endoglucanases of *C. lucknowense* show highest washing performance (ability to remove indigo from cellulose fibers of denim). At the same time, they do not bind strongly to cellulose, which prevents damage of the cellulosic fibers, and do not decrease the mechanical firmness of fabrics. Such endoglucanases exhibit hydrophobic clusters on the surface of the molecule capable of binding to indigo, which, along with low adsorbability of these enzymes, prevents redeposition of indigo on denim and leads to low backstaining (Gusakov et al., 2000). Thus, from the point of view of biotechnological importance, these endoglucanases are useful enzymes for the biostoning processes, i.e., they provide high washing performance and low backstaining.

Other fungal strains such as *Trichoderma reesei* secrete a complex array of glucosyl hydrolases. So far, the identified enzymes are eight endoglucanases (EG1 [Cel7B], EGII [Cel5A], EGIII [Cel12A], EGIV [Cel61A],

Table 2. Properties of cellulases from *C. lucknowense*[a]

Substrate	Enzyme								
	EG 1	EG 2	EG 3	EG 5	EG 6	CBH 1a	CBH 1b	CBH 2a	CBH 2b
Sp act (U/mg)									
CMC	12	52	11	17	14	0.2	0.2	0.1	0.1
β-Glucan	20	75	125	5	18	0.1	0.1	2	0.2
Xylan	0	0.1	0.2	0	0	0	0	0	0
Xyloglucan	0	0	16	2	0	0	0	0	0
Avicel	0.1	0.2	0.01	0.01	0.01	0.25	0.1	0.1	0.25
p-Nitrophenyl-cellobioside	0.2	0	0	0	0	0.025	0.025	0	0
p-Nitrophenyl-lactoside	0.3	0	0	0	0	0.12	0.09	0	0
Optima for activity									
pH	4.8	4.7	5.3	5.5	6.0	5.0	4.7	5.0	5.0
Temp (°C)	60	70	60	65	70	60	65	60	60

[a]Activity against polysaccharide substrates was determined at 50°C and pH 5.0 according to the initial rates of reducing sugar formation using the Somogyi-Nelson assay. The activity against *p*-nitrophenyl-β-D-cellobioside and *p*-nitrophenyl-β-D-lactoside was determined by initial rates on *p*-nitrophenyl formation at 40°C and pH 5.0.

EGV [Cel45A], EG [Cel74A], EG [Cel61B], and EG [Cel5B]) (Pentillä et al., 1986; Saloheimo et al., 1988, 1997; Okada et al., 1998; Foreman et al., 2003), two cellobiohydrolases, and seven β-glucosidases (Barnett et al., 1991; Foreman et al., 2003). *Humicola insolens* makes six different endoglucanases (EG, EGI [Cel7], EGII [Cel5], EGIII [Cel12], EGV [Cel45], EGVI [Cel6]) (Dalboge and Heldt-Hansen, 1994; Schülein, 1997).

Cellobiohydrolases

C. lucknowense produces four cellobiohydrolases (EC 3.2.1.91: CBH 1a [Cel7A], CBH 1b [Cel7B], CBH 2a [Cel6A], and CBH 2b [Cel6B]). CBH 2b and CBH 1a have the highest specific activity toward crystalline cellulose (Avicel) (Table 2). Cellobiohydrolases show a low ability to hydrolyze polysaccharides other than insoluble cellulose, such as soluble cellulose (CMC) and β-glucan. CBH 1a and CBH 1b possess activity toward *p*-nitrophenyl-β-D-cellobioside and *p*-nitrophenyl-β-D-lactobioside (Gusakov et al., 2005, 2007).

Cellobiohydrolases exhibit maximal activity at pH 5 and retain high activity at increased pH (50% of maximal activity at pH 6.5). The temperature optima for the activities of cellobiohydrolases are within the range of 60 to 65°C and retain high activity at elevated temperature (50% at 75 to 80°C). A study of enzyme adsorption on Avicel, carried out at pH 5.0 and 60°C (Gusakov et al., 2005), revealed that of all the cellobiohydrolases, only CBH 1a and CBH 2b possess a cellulose-binding module (CBM).

It should be noted that the most studied fungus, *T. reesei*, has only two cellobiohydrolases: CBH 1 (Cel7A) and CBH 2 (Cel6A) (Teeri, 1997, 1983, 1987). *H. insolens* also secretes only two cellobiohydrolases (Cel7A and Cel6A) (Dalboge and Heldt-Hansen, 1994; Schälein, 1997). *Phanerochaete chrysosporium*, on the other hand, produces at least seven different cellobiohydrolases, of which six belong to the GH7 family (Covert et al., 1992; Muñoz et al., 2001). All the enzymes mentioned above, except for the *P. chrysosporium* CBH 1-1 (Cel7A), possess a CBM.

Cellobiohydrolases are the key components of multienzymatic cellulose complexes which are responsible for conversion of cellulose to soluble sugars (Teeri, 1997). Table 3 shows the results of Avicel hydrolysis by the purified *C. lucknowense* cellobiohydrolases, at equal protein concentration (0.1 mg/ml). For a comparison, hydrolysis of Avicel by purified *T. reesei* CBH 1 and CBH 2 at the same protein concentration was examined. The enzymes from *T. reesei* were isolated as described by Markov et al. (2005). Hydrolysis of Avicel was carried out in the presence of purified β-glucosidase (BGL) from *Aspergillus japonicus* (0.5 U/ml) to eliminate the effect of cellobiose inhibition. The highest conversion was observed in the case of *C. lucknowense* CBH 2b: 3.2 mg of glucose per ml, i.e., 58% conversion after 5 days of hydrolysis. *T. reesei* CBH 1 and CBH 2 and *C. lucknowense* CBH 1a (all these enzymes have a CBM) were less effective, showing cellulose conversions of 50, 36, and 46%, respectively. The *C. lucknowense* cellobiohydrolases without a CBM (CBH 1b and CBH 2a) showed the lowest ability to hydrolyze Avicel. Conversions were 23 and 21%, respectively, after the same reaction time.

Table 3. Results of Avicel hydrolysis by individual cellobiohydrolases from two fungal sources[a]

Enzyme	Source organism	Glucose concn (mg/ml)		
		24 h	48 h	120 h
CBH 1a	*C. lucknowense*	1.4	2.0	2.5
CBH 1b	*C. lucknowense*	0.6	0.8	1.3
CBH 2a	*C. lucknowense*	0.6	0.8	1.2
CBH 2b	*C. lucknowense*	1.5	2.1	3.2
CBH 1	*T. reesei*	1.5	2.3	2.8
CBH 2	*T. reesei*	1.2	1.5	2.0

[a]The enzymes are equalized by protein concentration (0.1 mg/ml), 40°C, pH 5.0. Hydrolysis of Avicel (5 mg/ml) was carried out in the presence of purified BGL from *A. japonicus* (0.5 U/ml).

Synergism between *C. lucknowense* Cellulases

Synergistic interaction between different cellulases in the course of saccharification of crystalline cellulose is a well-known phenomenon for multienzymatic cellulase systems, especially between cellobiohydrolases and endoglucanases (Nidetzky et al., 1994; Henrissat et al., 1985; Kleman-Leyer et al., 1996). Data describing results of hydrolysis of dewaxed cotton by different combinations of *C. lucknowense* CBH 2b with other *C. lucknowense* cellulases are given in Table 4. The hydrolysis was carried out in the presence of purified BGL from *A. japonicus* to minimize the inhibitory effect of cellobiose. The coefficient of synergism (K_{syn}) was calculated as the ratio of experimental glucose concentration to the theoretical sum of glucose concentrations.

Individual cellulases do not hydrolyze native cotton effectively. The most active cellulase, CBH 2b, yielded 1.1 mg of glucose per ml after 120 h of hydrolysis, which corresponds to a substrate conversion of 20%. The highest glucose yield of 4.6 mg/ml, representing 83.6% conversion, was achieved with combinations of CBH 2b with EG 2. The coefficient of synergism varied in the range of 2.14 to 3.40. A strong synergism (K_{syn}, 2.86) was also observed between CBH 2a and CBH 2b 9 (data not shown). This combination of two cellobiohydrolases (with BGL) provided 5.3 mg of glucose per ml, which corresponds to almost complete conversion (96%) of cotton cellulose to glucose after 120 h of hydrolysis.

Table 4. Synergism between *C. lucknowense* cellulases in hydrolysis of dewaxed cotton cellulose[a]

Enzyme	Glucose concn (mg/ml) after 120 h		K_{syn}
	Experimental	Theoretical[b]	
CBH 1a	0.75		
CBH 2b	1.10		
EG2	0.60		
EG 5	0.65		
EG 6	0.38		
CBH 1a + EG 2	4.00	1.35	2.96
CBH 1a + EG 5	3.50	1.40	2.50
CBH 1a + EG 6	3.85	1.13	3.40
CBH 2b + EG 2	4.60	1.70	2.70
CBH 2b + EG 5	3.75	1.75	2.14
CBH 2b + EG 6	4.00	1.48	2.70
CBH 1a + CBH 2b	5.30	1.85	2.86

[a]The hydrolytic depolymerization of dewaxed cotton cellulose (5 mg/ml) at 40°C and pH 5.0 was carried out in the presence of 0.5 U of purified BGL from *A. japonicus* per ml. In all cases, the CBH concentration was 0.15 mg/ml and the EG concentration was 0.05 mg/ml.
[b]Calculated as the sum of glucose concentrations obtained under the action of individual enzymes.

It should be emphasized that although the synergistic interaction between endoglucanases and cellobiohydrolases has been known for quite some time, the synergy between cellobiohydrolases has not yet been explained. Besides cellobiohydrolases of *C. lucknowense*, the synergistic effect has been clearly demonstrated for CBH1 and CBH2 of *T. reesei* (Nidetzky et al., 1994; Medve et al., 1994) as well as for CBH 2 and CBH2 of *H. insolens* (Boisset et al., 2000).

Cellulolytic Potential of *C. lucknowense* Cellulases

Based on the results of the synergistic interaction between cellulases of *C. lucknowense*, it was possible to design the composition of a *C. lucknowense* cellulolytic complex which could be successfully used for efficient hydrolysis of cellulosic substrates to glucose (Gusakov et al., 2007). The cellulolytic potential of *C. lucknowense* cellulases was studied with dewaxed cotton as the substrate representing pure crystalline cellulose. The total protein concentration in the reaction system was 0.5 mg/ml. The multienzyme composition from *C. lucknowense* contained CBH 1a, CBH 2b, EG 2, and BGL. A crude *Trichoderma longibrachiatum* cellulase preparation (BioACE from Dyadic International, diluted so that the protein concentration in the reaction system would also be 0.5 mg/ml) was used for comparison. Since the *T. longibrachiatum* preparation is deficient in BGL, the hydrolysis experiments were carried out in the absence and also in the presence of added BGL from *A. japonicus* (0.5 U/ml).

The above multienzyme complex of purified *C. lucknowense* cellulases was found to provide a much higher glucose yield in the hydrolysis of nonpretreated cotton after 72 h (23.4 mg/ml, i.e., 84% substrate conversion) than a crude multienzyme *T. longibrachiatum* preparation. In the latter case, the glucose yield was 12.0 mg/ml (43% substrate conversion) when hydrolysis was carried out with extra BGL. Thus, an artificially designed *C. lucknowense* cellulase complex prepared with CBH 1a (40% of the total protein), CBH 2b (40%), EG 2 (16%), and BGL (4%) demonstrated a high ability to convert crystalline cellulose to glucose. As mentioned above, the best *C. lucknowense* mutant strains secrete more than 100 g of extracellular protein per liter. The fungal genome of the *C. lucknowense* was sequenced in 2005 (http://www.dyadic-group.com/wt/dyad/pr_1115654417) and then annotated in collaboration with The Scripps Research Institute in Florida, Bioinformatics Department. These developments make possible the creation and design of *C. lucknowense* strains expressing a desired optimal composition of *C. lucknowense* cellulases (and related accessory enzymes) that could amount to >90% of the secreted protein. Thus, this fungus has great potential as a producer of enzymes for saccharification of cellulosic feedstocks for production of renewable bioethanol from lignocellulosic biomass.

XYLANASES AND HEMICELLULASES OF *C. LUCKNOWENSE*

Xylanases are known for the production of fermentable monomeric xylose for production of biofuel and xylooligosaccharides for nutraceutical applications. An important breakthrough has been the use of xylanolytic enzyme systems for biobleaching of wood pulp. This has allowed the worldwide pulp and paper industry to experience a substantial reduction in the consumption of chlorine bleach (Polizeli, 2005). The Kraft pulping process accounts for 80% of the total pulp production in the United States and developed countries. It creates an alkaline background that requires xylanases which are active and stable at alkaline pH values. By cleaving the secondary xylan that cross-links lignin chromophores to cellulose in the crude paper pulps, fungal xylanases help to release the xylan-lignin complex at the bleaching stage. This improves the brightness of pulps and reduces the amount of chlorine compounds (elemental chlorine, Cl_2, and chlorine dioxide, ClO_2) used for bleaching by up to 25 to 40% (Daneault et al., 1994). The cereal-based raw materials used in animal feed contain not only carbohydrates and proteins that are nutritionally complete but also fiber-rich nonstarch polysaccharides (NSP) consisting of cellulose and hemicelluloses. Because of limited digestion of NSP by monogastric animals, the feed is not com-

Table 5. Biochemical characteristics of xylanases from *C. lucknowense*

Enzyme	Glycoside hydrolase family	Systematic name
Xyl 1	10	Xyn10A
Xyl 3	10	Xyn10B
Xyl 4	10	Xyn10C
Xyl 5	11	Xyn11C
Xyl 6	11	Xyn11B

pletely utilized. Xylanases have been proposed to improve the digestibility of animal feeds by breaking down arabinoxylans, which represent one of the major classes of NSPs in cereals, particularly in wheat and rye (Bok et al., 1994).

Xylanases increase the digestibility of feed by lowering viscosity in the intestinal tract, decreasing absorption and water-holding capacity of feeds with significant content of NSP by destruction of arabinoxylan. They also reduce the ingredient variations between feed samples containing different concentrations of NSP (Twomey et al., 2003). Xylanases can be usefully employed (especially when used in conjunction with other hydrolytic enzymes) for bioconversion of forestry and agricultural wastes to fermentable sugars for production of bioethanol (Bok et al., 1994). Feed and bioconversion applications require acid-neutral active and stable xylanases. Multiple xylanases tend to cleave glycosidic bonds between unsubstituted xylose units. Multiple accessory enzymes are needed to enhance the hydrolytic depolymerization of highly decorated NSPs such as β-xylosidases, ferulolyl and acetylxylan esterases, α-L-arabinofuranosidases, ferrulic acid esterase, *p*-coumaric acid esterase, α-glucoronidases, β-mannanase, and so forth (Ustinov et al., 2005).

The isolation strategy for individual xylanases of *C. lucknowense* is based on the sequential use of different purification techniques, such as anion-exchange chromatography, hydrophobic chromatography, and gel filtration. *C. lucknowense* produces five enzymes (Table 5) possessing high activity toward xylans which belong to endo-1,4-β-xylanases (EC 3.2.1.8): Xyl 1 (Xyn10A), Xyl 3 (Xyn10B), Xyl 4 (Xyn10C), Xyl 5 (Xyn11C), and Xyl 6 (Xyn11B). Xylanases with higher molecular weights belong to family 10 (Family F) of glycosyl hydrolases (Xyl 1, Xyl 3, and Xyl 4), whereas those enzymes with lower molecular weights belong to family 11 (Family G) (Xyl 5 and Xyl 6).

All *C. lucknowense* xylanases are characterized by high specific activity toward birchwood xylan and wheat arabinoxylan (Table 6). The highest activity on birchwood xylan is shown by Xyl 5 (300 U/mg) and on wheat arabinoxylan by Xyl 6 (270 U/mg). All the xylanases tested show little to no activity on soluble (CMC) and insoluble (Avicel) cellulose and hence are very specific in cleaving the xylosyl β-1,4-linkage. This makes *C. lucknowense* xylanases potential candidates for an application such as pulp biobleaching (Bucheret, 1994). The action of all the xylanases led to a significant decrease in viscosity of birchwood and arabinoxylan, which reflects an endwise type of action on polymeric substrates. Xyl 3 and Xyl 4 showed some exoxylanase activity, being active toward *p*-nitrophenyl-β-D-xyloside (Table 6). The ability of xylanases to efficiently decrease the viscosity of arabinoxylan makes them potential candidates to degrade NSPs and to be used as feed additives, and also to depolymerize pretreated biomass in the production of cellulosic ethanol. All xylanases produce xylose, xylobiose (major product), and xylotriose as final products of arabinoxylan degradation. In addition, Xyl 1, Xyl 3, and Xyl 4 generate xylotriose with an attached arabinose residue, and Xyl 4 and Xyl 6 generate xylotetraose with an attached arabinose residue.

Table 6. Properties of xylanases from *C. lucknowense*[a]

Substrate	Enzyme				
	Xyl 1	Xyl 3	Xyl 4	Xyl 5	Xyl 6
Sp act (U/mg)					
Birchwood xylan	83	85	32	300	170
Wheat arabinoxylan	100	98	40	140	270
CMC	0.1	0.2	0	0	0
Avicel	0	0	0	0	0
p-Nitrophenyl-β-D-xylopyranoside	0	0.6	0.3	0	0
p-Nitrophenyl-β-D-cellobioside	0.4	1.6	4.0	0	0
Optima for activity					
pH	5.5–7.0	5.5–6.5	5.0	4.5	6.0
Temp (°C)	65–70	80–85	80	65	65–70

[a]Activity against polysaccharide substrates was determined at 50°C and pH 5.0 according to the initial rates of reducing sugar formation using the Somogyi-Nelson assay. The activity against *p*-nitrophenyl-β-D-xylopyranoside and *p*-nitrophenyl-β-D-cellobioside was determined by initial rates of *p*-nitrophenyl formation at 40°C and pH 5.0.

Almost all *C. lucknowense* xylanases exhibit maximal activity in the pH range of 5 to 7; only Xyl 5 has a pH optimum at pH 4.5. The temperature optima for the activities of Xyl 1, Xyl 5, and Xyl 6 are within the range of 65 to 70°C, whereas those for the activities of Xyl 3 and Xyl 4 are within the range of 80 to 85°C (Table 6). Xyl 1, Xyl 3, and Xyl 4 are fairly stable and retain 100% activity during 1 h at 50 and 60°C (pH 5) and 70 to 92% of activity at these same temperatures at pH 7; Xyl 5 and Xyl 6 are less stable.

CONCLUDING REMARKS

As we gradually move into a plant biomass-based economy, multiple strategic approaches have been taken to increase enzyme catalytic efficiency and increase fermentation yield in order to reduce manufacturing costs in the production of renewable fuels and valuable chemicals. Filamentous fungi are the most common natural degraders of lignocellulosic biomass. These fungal cell factories are the best hyperproducers of a complex array of glycosyl hydrolases such as cellulases, hemicellulases, and accessory enzymes. We have been able to show the hydrolytic efficiency of fungal multienzyme complexes in the saccharification of lignocellulosic biomass. This synergistic mechanism depends highly on the properties of each individual enzyme component and their optimal ratios in the multienzyme preparation. The cost-effective integration of pretreatment, enzymatic saccharification, and fermentation will make possible the path to commercialization of cellulosic ethanol and alternatively other renewable motor biofuels and chemicals. These collective efforts will successfully catalyze the transition from a petroleum-based to a biomass-based economy.

REFERENCES

Barnett, C., R. Berka, and T. Fowler. 1991. Cloning and amplification of the gene encoding an extracellular β-glucosidase from *Trichoderma reesei*: evidence for improved rates of saccharification of cellulosic substrates. *Bio/Technology* 9:562–567.

Bhat, K., N. J. Parry, S. Kalogiannis, D. E. Beever, E. Owen, M. Nerinckx, and M. Claeyssens. 1998. Biochemical characterization of cellulases and xylanases from *Thermoascus aurantiacus*, and carbohydrases from *Trichoderma reesei* and other microorganisms, p. 102–112. *In* M. Claeyssens, W. Nerinckx, and K. Piens (ed.), *Structures, Biochemistry, Genetics and Applications*. Royal Society of Chemistry, Cambridge, United Kingdom.

Boisset, C., M. Fraschini, M. Schülein, B. Henrissat, and H. Chanzy. 2000. Imaging the enzymatic digestion of bacterial cellulose ribbons reveals the endo character of the cellobiohydrolase Cel6A from *Humicola insolens* and its mode of synergy with cellobiohydrolase Cel7A. *Appl. Environ. Microbiol.* 66:1444–1452.

Bok, J. D., S. K. Goer, and D. E. Eveleigh. 1994. Cellulase and xylanase systems of *Thermatoga neapolitana*, p. 54–65. *In* M. Himmel, J. Baker, and R. Overend (ed.), *Enzymatic Conversion of Biomass for Fuels Production*, ACS Symposium 566. American Chemical Society, Washington, DC.

Bucheret, J., M. Tenkanen, A. Kanteleinen, and L. Viikari. 1994. Application of xylanases in pulp and paper industry. *Bioresour. Technol.* 50:65–72.

Bukhtojarov, F. E., B. B. Ustinov, T. N. Salanowich, A. I. Antonov, A. V. Gusakov, O. N. Okunev, and A. P. Sinitsyn. 2004. Cellulase complex of the fungus *Chrysosporum lucknowense*: isolation and characterization of endoglucanases and cellobiohydrolases. *Biochemistry* (Moscow) 69:542–551.

Covert, S. F., A. van den Wymelenberg, and D. Cullen. 1992. Structure, organization, and transcription of a cellobiohydrolase gene cluster from *Phanerochaete chrysosporium*. *Appl. Environ. Microbiol.* 58:2168–2175.

Dalboge, H., and H. P. Heldt-Hansen. 1994. A novel method for efficient expression cloning of fungal enzyme genes. *Mol. Gen. Genet.* 243:253–260.

Daneault, C., C. Leduc, and J. L. Valade. 1994. The use of xylanases in kraft pulp bleaching: a review. *Tappi J.* 77:125–131.

Demain, A. L., M. Newcomb, and D. J. H. Wu. 2005. Cellulase, clostridia, and ethanol. *Microbiol. Mol. Biol. Rev.* 69:124–154.

Durand, H., M. Clanet, and G. Tiraby. 1988. Genetic improvement of *Trichoderma reesei* for large scale cellulase production. *Enzyme Microb. Technol.* 101:40–60.

Emalfarb, M. A., A. Ben-Bassat, and A. P. Sinitsyn. January 2000. Treating cellulosic materials with cellulases from chrysosporium. U.S. patent 6015707.

Emalfarb, M. A., et al. June 2003. Transformation system in the field of filamentous fungal hosts. U.S. patent 6573086.

Emalfarb, M. A., P. J. Punt, C. van Zeijl, and C. van den Hondel. 2001. High-throughput screening of expressed DNA libraries in filamentous fungi. Foreign patent WO 01/79558.

Foreman, P., D. Brown, L. Dankmyer, R. Dean, S. Diener, N. Dunn-Coleman, F. Goedegebuur, T. Houfek, G. England, A. Kellwy, H. Meerman, T. Mitchell, C. Mitchinson, H. Olivares, P. Teunissen, J. Yao, and M. Ward. 2003. Transcriptional regulation of biomass-degrading enzymes in the filamentous fungus *Trichoderma reesei*. *J. Biol. Chem.* 278:31988–31997.

Grishutin, S. G., A. V. Gusakov, A. V. Markov, B. B. Ustinov, M. V. Semenova, and A. P. Sinitsyn. 2004. Specific xyloglucanases as a new class of polysaccharide-degrading enzymes. *Biochim. Biophys. Acta* 1674:268–281.

Gusakov, A. V., A. P. Sinitsyn, A. G. Berlin, N. N. Popova, A. V. Markov, O. N. Okunev, D. F. Tikhomirov, and M. A. Emalfarb. 1998. Interaction between indigo and adsorbed protein as a major factor causing backstaining during cellulose treatment of cotton fabrics. *Appl. Biochem. Biotechnol.* 75:279–293.

Gusakov, A. V., A. P. Sinitsyn, A. G. Berlin, A. V. Markov, and N. V. Ankudimova. 2000. Discrimination of surface hydrophobic amino acid residues in cellulase molecules as a structural factor responsible for their high abrasive activity in the denim washing process. *Enzyme Microb. Technol.* 27:664–671.

Gusakov, A. V., T. N. Salanovich, A. I. Antonov, B. B. Ustinov, O. N. Okunev, R. Burlingame, M. Emalfarb, M. Baez, and A. P. Sinitsyn. 2007. Design of highly efficient cellulase mixtures for enzymatic hydrolysis of celluloses. *Biotechnol. Bioeng.* 97:1028–1038.

Gusakov, A. V., A. P. Sinitsyn, T. N. Salanovich, F. E. Bukhtojarov, A. V. Markov, B. B. Ustinov, C. van Zeijl, P. Punt, and R. Burlingame. 2005. Purification, cloning and characterization of two forms of thermostable and highly active cellobiohydrolase I (Cel7A) produced by industrial strain of *Chrysosporium lucknowense*. *Enzyme Microb. Technol.* 36:57–69.

Henrissat, B., H. Driguez, C. Viet, and M. Schülein. 1985. Synergism of cellulases from *Trichoderma reesei* in degradation of cellulose. *Bio/Technology* 3:722–726.

Himmel, M., M. Ruth, and C. Wayman. 1999. Cellulase for commodity products from cellulosic biomass. *Curr. Opin. Biotechnol.* 10:358–364.

Kabel, M. A., M. J. E. C. van der Maarel, G. Klip, A. G. J. Voragen, and H. A. Schols. 2005. Standard assays do not predict the efficiency of commercial cellulase preparations towards plant materials. *Biotechnol. Bioeng.* 93:56–63.

Kleman-Leyer, K. M., M. Siika-Aho, T. T. Teeri, and T. K. Kirk. 1996. The cellulases endoglucanase I and cellobiohydrolase II of *Trichoderma reesei* act synergistically to solubilize native cotton cellulose but not decrease its molecular size. *Appl. Environ. Microbiol.* 62:2883–2887.

Markov, A. V., A. V. Gusakov, E. G. Kondratieva, J. N. Okunev, A. O. Bekkarevich, and A. P. Sinitsyn. 2005. New effective method for analysis of the component composition of enzyme complexes from *Trichoderma reesei*. *Biochemistry* (Moscow) 70:657–663.

Mathlouthi, N., L. Saulnier, B. Quemener, and M. Larbier. 2002. Xylanase, β-glucanase and other side enzymatic activities have greater effect on the viscosity of several foodstuffs than xylanase and β-glucanase used alone or in combination. *J. Agric. Food Chem.* 50:5121–5127.

Medve, J., J. Ståhlberg, and F. Tjerneld. 1994. Adsorption and synergism of cellobiohydrolase I and II of *Trichoderma reesei* during hydrolysis of microcrystalline cellulose. *Biotechnol. Bioeng.* 44: 1064–1073.

Miettinen-Oinonen, A., and P. Suominen. 2002. Enhanced production of *Trichoderma reesei* endoglucanases and use of the new cellulase preparations in producing the stonewashed effect on denim fabric. *Appl. Environ. Microbiol.* 68:3956–3964.

Miettinen-Oinonen, A., J. Londesborough, V. Joutsjoki, R. Lantto, and J. Vehmaanpera. 2004. Three cellulases from *Melanocarpus albomyces* for textile treatment at neutral pH. *Enzyme Microb. Technol.* 34:332–341.

Muñoz, I. G., W. Ubhayasekera, H. Henriksson, I. Szabo, G. Pettersson, G. Johansson, S. L. Mowbray, and J. Ståhlberg. 2001. Family 7 cellobiohydrolases from *Phanerochaete chrysosporium*: crystal structure of the catalytic module of Cel7D (CBH58) at 1.32 angstrom resolution and homology models of the isozymes. *J. Mol. Biol.* 314:1097–1111.

Nidetzky, B., W. Steiner, M. Hayn, and M. Claeyssen. 1994. Cellulose hydrolysis by the cellulases from *Trichoderma reesei*: a new model for synergistic interaction. *Biochem. J.* 298:705–710.

Okada, H., K. Tada, T. Sekiya, K. Yokoyama, A. Takahashi, H. Tohda, H. Kumagai, and Y. Morikawa. 1998. Molecular characterization and heterologous expression of the gene encoding a low-molecular-mass endoglucanase from *Trichoderma reesei* QM9414. *Appl. Environ. Microbiol.* 64:555–563.

Pentillä, M., P. Lehtovaara, H. Nevalainen, R. Bhikhabhai, and J. Knowles. 1986. Homology between cellulose genes of *Trichoderma reesei*: complete nucleotide sequence of the endoglucanase I gene. *Gene* 45:253–263.

Polizeli, M. L., T. M. Rizzatti, A. C. S. Monti, R. Terenzi, J. A. Jorge, and D. S. Amorim. 2005. Xylanases from fungi: properties and industrial applications. *Appl. Microbiol. Biotechnol.* 67:577–591.

Rabinovich, M. L., M. S. Meinik, and A. V. Bolobova. 2002. Microbial cellulases. *Appl. Biochem. Microbiol.* 38:305–321.

Saloheimo, M., P. Lehtovaara, M. Pentillä, T. T. Teeri, J. Ståhlberg, G. Johansson, G. Pettersson, M. Claeyssens, P. Tomme, and J. Knowles. 1988. EGIII, a new endoglucnase from *Trichoderma reesei*: the characterization of both gene and enzyme. *Gene* 63: 11–12.

Saloheimo, M., T. Nakari-Setälä, M. Tenkanen, and M. Penttilä. 1997. cDNA cloning of a *Trichoderma reesei* cellulase and demonstration of endoglucanase activity by expression in yeast. *Eur. J. Biochem.* 249:584–591.

Schülein, M. 1997. Enzymatic properties of cellulases from *Humicola insolens*. *J. Biotechnol.* 57:71–81.

Sinitsyn, A. P., B. Jadjemy, and A. A. Klyosov. 1981. Enzymatic conversion of cellulose to glucose: effect of product inhibition and decrease of substrate reactivity on the rate of enzymatic hydrolysis. *Microbiology* 17:315–321.

Sinitsyn, A. P., H. R. Bungay, and L. S. Clesceri. 1983. Enzyme management in the Iotech process. *Biotechnol. Bioeng.* 25:1393–1399.

Sinitsyn, A. P., A. V. Gusakov, S. G. Grishutin, O. A. Sinitsyna, and N. V. Ankudimova. 2001. Application of microassay for investigation of cellulose abrasive activity and backstaining. *J. Biotechnol.* 89:233–238.

Solovieva, I. V., O. N. Okunev, V. V. Velkov, A. V. Koshelev, T. V. Bubnova, E. G. Kondratieva, A. A. Skomarovsky, and A. P. Sinitsyn. 2005. The selection and properties of *Penicillium verruculosum* mutants with enhanced production of cellulases and xylanases. *Microbiology* (Moscow) 74:141–146.

Suurnäkki, A., M.-L. Niiu-Paavola, J. Buchert, and L. Viikari. 2004. Enzymes in pulp and paper processing, p. 232–244, 437–439. *In* W. Achel (ed.), *Enzymes in Industry.* Wiley-VCH, Weinheim, Germany.

Teeri, T. 1997. Crystalline cellulose degradation: new insight into the function of cellobiohydrolase. *Trends Biotechnol.* 15:160–167.

Teeri, T., I. Salovouri, and J. Knowles. 1983. The molecular cloning of the major cellulose gene from *Trichoderma reesei*. *Bio/Technology* 1:696–699.

Teeri, T., P. Lehtovaara, S. Kauppinen, I. Salovuori, and J. Knowles. 1987. Homologous domains in *Trichoderma reesei* cellulolytic enzymes: gene sequence and expression of cellobiohydrolase II. *Gene* 51:43–52.

Twomey, L. N., J. R. Pluske, J. B. Rowe, M. Chost, W. Brown, M. F. McConnell, and D. W. Pethick. 2003. The effect of increasing levels of soluble non-starch polysaccharides and inclusion of feed enzymes in dog diets on faecal quality and digestibility. *Anim. Feed Sci. Technol.* 108:71–82.

Ustinov, B. B., A. I. Antonov, S. G. Grishutin, E. I. Dzedsiulia, A. V. Gusakov, and A. P. Sinitsyn. 2005. Possibility of application of *Chrysosporium lucknowense* xylanases in pulp and paper industry and as feed additive, p. 218. *In Abstracts of Second International Congress "Biotechnology: State and Perspectives,"* March 14–18, 2005, Moscow, Russia.

Wyman, C. E. 1994. Ethanol from lignocellulosic biomass: technology, economics, and opportunities. *Bioresour. Technol.* 50:3–16.

Wyman, C. E., B. D. Dale, R. T. Elander, M. Hotzapple, M. R. Ladish, and Y. Y. Lee. 2005. Coordinated development of leading biomass pretreatment technologies. *Bioresour. Technol.* 96:1959–1966.

Bioenergy
Edited by J. Wall et al.

Chapter 12

Production of Ethanol from Synthesis Gas

RALPH S. TANNER

After the oil supply shocks in the 1970s, research on development of alternative sources of liquid fuels was intensified. The initial emphasis was on processes that used coal as a feedstock, but investigations into the direct fermentation of lignocellulosic biomass were also undertaken. Most of this research was in direct fermentation, whereby biomass is first treated to convert it to fermentable sugars, followed by the familiar alcoholic fermentation, generally by yeast. Great progress has been made in this area (Mielenz, 2001), and a number of ethanol plants based on direct fermentation of biomass are being actively considered (www.energy.gov/news/4827.htm). An alternative method to convert lignocellulosic biomass to ethanol is indirect fermentation (Klasson et al., 1992). In indirect fermentation, biomass is first pyrolyzed in a gasifier to convert it to synthesis gas (McKendry, 2002), which consists primarily of carbon monoxide, carbon dioxide, and hydrogen (also nitrogen if air rather than oxygen is used in the gasifier). Synthesis gas can be fermented or otherwise converted to a number of desired products, including ethanol (Table 1). Another interesting potential use of carbon monoxide is in a biofuel cell for the production of electricity (Colby et al., 1985).

There are several advantages to indirect fermentation compared to direct fermentation. The pyrolysis of biomass to synthesis gas (Van Der Drift et al., 2001) is considered by many to be a mature technology, which is certainly the case for coal gasification. Pyrolysis is a rapid, exothermic process, and a relatively small unit can process a large amount of biomass. For example, a pilot plant gasifier with an inner diameter of 25 cm can pyrolyze almost 1,000 kg of biomass per day (Datar et al., 2004). This is in contrast to the large tanks and relatively long treatment times required to saccharify biomass for direct fermentation. Even after saccharification, the lignin component of biomass remains, but this lignin is converted to synthesis gas along with the cellulose and hemicellulose in a gasifier. This illustrates the fact that gasifiers can produce synthesis gas from just about any combustible material and even from sources not generally considered combustible or otherwise useful for an ethanol-producing fermentation, such as sewage sludge (Midilli et al., 2002). In theory a metric ton of dry biomass could be converted to 580 liters of ethanol by indirect fermentation. Realistic yields are closer to 290 to 440 liters of ethanol per ton of biomass (Spath and Dayton, 2003), but this is still enough to pursue indirect fermentation as a viable process for ethanol production.

Synthesis gas can be chemically converted to liquid fuels (e.g., Fischer-Tropsch synthesis). However, these chemical processes require elevated temperatures and pressures, complex catalysts, removal of inactivators, such as hydrogen sulfide, from the synthesis gas, and treatment of the synthesis gas to alter it to specific carbon monoxide:hydrogen ratios (Bredwell et al., 1999). Underlying assumptions for the biological fermentation of synthesis gas to ethanol include relative insensitivity to minor components of the produced gas, including hydrogen sulfide, and insensitivity to the carbon monoxide:hydrogen ratios for producing ethanol (Bredwell et al., 1999; Grethlein and Jain, 1992; Levy et al., 1981; Spath and Dayton, 2003). The limited information reported to date supports these assumptions (below).

The current disadvantages for the fermentation of synthesis gas to ethanol include relatively low production rates and low product concentration in an aqueous phase compared to a yeast fermentation. The latter can be addressed by new product recovery technologies, such as membrane-based separations (Dürre, 1998). These disadvantages are not surprising given the very modest level of support that indirect fermentation has garnered (Spath and Dayton, 2003). In addition, most of this research was conducted out of engineering programs rather than microbiology programs. However, enough progress has been made that funds have been committed to build a production plant using, at least in part, indirect fermentation of lignocellulosic biomass after gasification (www.energy.gov/news/4827.htm).

Ralph S. Tanner • Department of Botany & Microbiology, University of Oklahoma, Norman, OK 73019.

Table 1. Microbial products from CO fermentation[a]

CO → X	Microorganism (example[s])
$CO + H_2O \rightarrow H_2 + CO_2$	*Rubrivivax gelatinosus, Citrobacter* sp. Y19, *Carboxydothermus hydrogenoformans*
$CO + H_2 \rightarrow HCOOH$	*Methanosarcina acetivorans*
$4\ CO + 2\ H_2O \rightarrow CH_4 + 3\ CO_2$	*Methanosarcina acetivorans, Methanothermobacter thermoautotrophicus*
$4\ CO + 2\ H_2O \rightarrow CH_3COOH + 2\ CO_2$	*Eubacterium limosum, "Butyribacterium methylotrophicum," Peptostreptococcus productus, Clostridium ljungdahlii, Moorella thermoacetica, Sporomusa termitida*[b]
$6\ CO + 3\ H_2O \rightarrow CH_3CH_2OH + 4\ CO_2$	*Clostridium ljungdahlii, "Butyribacterium methylotrophicum," Clostridium carboxidivorans*[c]
$10\ CO + 4\ H_2O \rightarrow CH_3(CH_2)_2COOH + 6\ CO_2$	*"Butyribacterium methylotrophicum," Clostridium carboxidivorans*[c]
$12\ CO + 5\ H_2O \rightarrow CH_3(CH_2)_2CH_2OH + 8\ CO_2$	*"Butyribacterium methylotrophicum," Clostridium carboxidivorans*[c]

[a]Data from Sipma et al., 2006.
[b]Data from Breznak et al., 1988.
[c]Data from Liou et al., 2005.

MICROBIOLOGY OF SYNTHESIS GAS FERMENTATION

The fact that anaerobic bacteria can ferment carbon monoxide to ethanol has only been recognized for a relatively short time. In a chapter published in 1990 on the metabolism of C_1 substrates by nonmethanogenic anaerobes (Heijthuijsen and Hansen, 1990), only acetate and butyrate are noted as primary end products from hydrogen:carbon dioxide or carbon monoxide: carbon dioxide. No mention is made of ethanol. While there are a handful of bacteria known which can carry out this reaction (below), not much is known about the precise biochemistry of the process. These bacteria are all acetogens, capable of growth on hydrogen:carbon dioxide with the production of acetate and presumably using the Wood/Ljungdahl pathway (Drake, 1994; Grethlein and Jain, 1992). It is also presumed that the pathway of ethanol production is analogous to that found in *Clostridium acetobutylicum* (Jones and Woods, 1986). A reduction in pH is important for switching from producing acids to alcohols as end products (Bredwell et al., 1999; Grethlein et al., 1990, 1991; Jones and Woods, 1986; Phillips et al., 1994; Sakai et al., 2004; Worden et al., 1991). *C. acetobutylicum* can take up and directly reduce acetate and butyrate to the corresponding alcohols (Hartmanis et al., 1984). There is indirect evidence that *Clostridium carboxidivorans* also does this (Datar et al., 2004). Interestingly, the acetogens *Clostridium formicaceticum* and *Moorella thermoacetica* have been shown to reduce acids to the corresponding alcohols with carbon monoxide as the reductant (Fraisse and Simon, 1988; Simon et al., 1987). The biological significance of that reduction is uncertain. However, there is more to the fermentation of carbon monoxide to ethanol than what is presumed above. The presence of carbon monoxide itself is important in enabling acetogens to produce ethanol rather than acetate. For example, *C. carboxidivorans* fermented 100 mmol of fructose to 23 mmol of ethanol, 81 mmol of acetate, and 4 mmol of butanol (Liou et al., 2005). Upon fermentation of an equivalent amount of carbon monoxide (600 mmol), the end products produced shifted to 96 mmol of ethanol, 12 mmol of acetate, and 24 mmol of butanol. The presence of reducing agents alone can shift products of fermentation from acetate to ethanol (Klasson et al., 1992). It may be that carbon monoxide acts just as an effective reductant. Research into the biochemistry of carbon monoxide fermentation by acetogens is needed.

"Butyribacterium methylotrophicum"

"Butyribacterium methylotrophicum" was isolated from a sewage digester using methanol as the substrate (Zeikus et al., 1980). Soon after, it was shown that this microorganism could also ferment carbon monoxide (Lynd et al., 1982). The products of fermentation were acetate and butyrate. Further work showed that *"B. methylotrophicum"* produced ethanol and butanol, in addition to acetate and butyrate, from carbon monoxide (Grethlein et al., 1990), although the concentration of these alcohols produced was initially quite low, about 1 mM each. The fermentation of carbon monoxide to higher concentrations of ethanol (7 mM) and butanol (36 mM) was reported later (Grethlein et al., 1991). A possible pathway for this alcohol production was proposed in the 1991 paper, as well, based on pathways known for carbon monoxide disproportionation and the ethanol-butanol pathway in *C. acetobutylicum* (above). The ability to produce butanol from synthesis gas appears to be the most attractive feature of this microorganism.

It has always been recognized that *"B. methylotrophicum"* is physiologically very similar to *Eubacterium limosum* (Heijthuijsen and Hansen, 1989, 1990; Worden et al., 1991), and these are phylogenetically indistinguishable (GenBank accession numbers M59120 and AF064241). Currently, neither the genus name "*Butyribacterium*" nor the species name "*B. methylotrophicum*" (never validly published) have standing in the nomenclature.

Clostridium ljungdahlii

Clostridium ljungdahlii was isolated from chicken yard waste based on its ability to ferment synthesis gas to ethanol (Barik et al., 1988; Tanner et al., 1993). This was the first acetogen described that was truly a species of *Clostridium* as it is currently understood (Collins et al., 1994). When this species is grown as a normal acetogen on fructose or hydrogen:carbon dioxide, acetate is the only end product (Tanner et al., 1993), but when it was cultured on synthesis gas, ethanol was an additional product (Barik et al., 1988). *C. ljungdahlii* is also unusual in that it can utilize ethanol as a substrate (Tanner et al., 1993) in addition to producing it as an end product. This phenomenon was also observed in *C. carboxidivorans* (Liou et al., 2005), and an analogous observation was made for *Acetobacterium woodii* (Buschhorn et al., 1989).

C. ljungdahlii has been the microbial catalyst used in the best collection of studies on the fermentation of carbon monoxide to ethanol. This research is illustrative of both the problems and promise of this process.

C. ljungdahlii was placed in the patent depository at ATCC under accession number 49587^T, and this was the reference number used in the initial patent (Gaddy and Clausen, 1992). This culture was lost, however, probably due to problems with culture pH, and *C. ljungdahlii* was redeposited under accession number 55383^T, where it is found now. Reports on ethanol production vary widely, particularly when examined as the final concentration of ethanol in culture achieved. For example, in one paper fermentation of carbon monoxide resulted in a final ethanol concentration in culture of <1 mM (Najafpour and Younesi, 2006). However, in the same paper, it was reported that a final concentration of 330 mM ethanol was achieved in another culture, but the data were not shown. The same group had reported earlier that *C. ljungdahlii* fermented carbon monoxide to a final concentration of 13 mM ethanol, along with acetate as a product (Younesi et al., 2005). These results were not very promising. However, much higher final concentrations of ethanol have been reported by a research group in Arkansas, as follows. An ethanol concentration of 460 mM was reported in a U.S. patent (Gaddy, 2000). More impressively, after a lengthy incubation of *C. ljungdahlii* with synthesis gas a final ethanol concentration of 1,040 mM was attained (Phillips et al., 1993). This is a high enough concentration for product recovery.

The differences in these reported fermentations reflect the differences among different laboratories and the extensive culture adaptation, medium development, etc., conducted by the investigators in Arkansas. The first commercial fermentation of synthesis gas to ethanol will probably use *C. ljungdahlii* as the microbial catalyst.

Clostridium carboxidivorans

Clostridium carboxidivorans was isolated from a settling lagoon near the dairy barns at Oklahoma State University based on its ability to ferment carbon monoxide to ethanol and acetate (Liou et al., 2005). In addition to its ability to indirectly ferment biomass via gasification, *C. carboxidivorans* can directly ferment biomass, at least the cellulose fraction (Liou et al., 2005). *C. carboxidivorans* is a good ethanol producer, achieving ethanol concentrations of 175 mM from carbon monoxide in early studies (Liou and Tanner, 2001).

Fermentation results from another laboratory illustrate the differences in performance between groups. *C. carboxidivorans* was grown in a bubble column bioreactor with synthesis gas as the substrate (Rajagopalan et al., 2002). A steady-state concentration of 3 mM ethanol was achieved in this system. A few years later in a similar bioreactor, a concentration of 90 mM ethanol was reached (Datar et al., 2004). More importantly, however, actual synthesis gas from the pyrolysis of switchgrass was the substrate for *C. carboxidivorans*. This was the first publication showing ethanol production from producer gas (Datar et al., 2004).

Several other acetogens capable of producing ethanol from carbon monoxide have appeared in the literature. "*Clostridium autoethanogenum*" was isolated from rabbit feces based on its ability to produce ethanol from carbon monoxide (Abrini et al., 1994). The species name "*C. autoethanogenum*" does not have standing in the nomenclature. A *Moorella* sp. that could produce ethanol from hydrogen:carbon dioxide was isolated from sediment in Japan (Sakai et al., 2004). Since the great majority of *Moorella* strains can also ferment carbon monoxide (M. W. Maune and R. S. Tanner, unpublished results), this strain should be able to produce ethanol from carbon monoxide. *Clostridium* strain P11 was reported to produce up to 200 mM ethanol from carbon monoxide, dependent upon elevated levels of zinc in the medium (Saxena and Tanner, 2006).

While there has been a continuous, albeit low-level, effort to develop indirect fermentation as a viable process for the last 2 decades, there is still a long way to go. Regardless, this process should go to commercial demonstration in the very near future. There will be much more known about this microbiology and process in the near future, especially if research support is expanded.

Acknowledgments. I thank Matt Maune, Erin Arms, Paul Lawson, Ray Huhnke, and the others who helped me assemble the disparate collection of literature used to prepare this chapter.

REFERENCES

Abrini, J., H. Naveau, and E.-J. Nyns. 1994. *Clostridium autoethanogenum*, sp. nov., an anaerobic bacterium that produces ethanol from carbon monoxide. *Arch. Microbiol.* **161:**345–351.

Barik, S., S. Prieto, S. B. Harrison, E. C. Clausen, and J. L. Gaddy. 1988. Biological production of alcohols from coal through indirect liquefaction. *Appl. Biochem. Biotechnol.* **18:**363–378.

Bredwell, M. D., P. Srivastava, and R. M. Worden. 1999. Reactor design issues for synthesis-gas fermentations. *Biotechnol. Prog.* **15:** 834–844.

Breznak, J. A., J. M. Switzer, and H.-J. Seitz. 1988. *Sporomusa termitida* sp. nov., an H_2/CO_2-utilizing acetogen isolated from termites. *Arch. Microbiol.* **150:**282–288.

Buschhorn, H., P. Dürre, and G. Gottschalk. 1989. Production and utilization of ethanol by *Acetobacterium woodii*. *Appl. Environ. Microbiol.* **55:**1835–1840.

Colby, J., E. Williams, and A. P. F. Turner. 1985. Applications of CO-utilizing microorganisms. *Trends Biotechnol.* **3:**12–17.

Collins, M. D., P. A. Lawson, A. Willems, J. J. Cordoba, J. Fernandez-Garayzabal, P. Garcia, H. Hippe, and J. A. Farrow. 1994. The phylogeny of the genus *Clostridium*: proposal of five new genera and eleven new species combinations. *Int. J. Syst. Bacteriol.* **44:**812–826.

Datar, R. P., R. M. Shenkman, B. G. Cateni, R. L. Huhnke, and R. S. Lewis. 2004. Fermentation of biomass-generated producer gas to ethanol. *Biotechnol. Bioeng.* **86:**587–594.

Drake, H. L. 1994. *Acetogenesis*. Chapman & Hall, New York, NY.

Dürre, P. 1998. New insights and novel developments in clostridial acetone/butanol/isopropanol fermentation. *Appl. Microbiol. Biotechnol.* **49:**639–648.

Fraisse, L., and H. Simon. 1988. Observations on the reduction of non-activated carboxylates by *Clostridium formicoaceticum* with carbon monoxide or formate and the influence of various viologens. *Arch. Microbiol.* **150:**381–386.

Gaddy, J. L. October, 2000. Biological production of ethanol from waste gases with *Clostridium ljungdahlii*. U.S. patent 6,136,577.

Gaddy, J. L., and E. C. Clausen. December, 1992. *Clostridium ljungdahlii*, an anaerobic ethanol and acetate producing microorganism. U.S. patent 5,173,429.

Grethlein, A. J., and M. K. Jain. 1992. Bioprocessing of coal-derived synthesis gas by anaerobic bacteria. *Trends Biotechnol.* **10:**418–423.

Grethlein, A. J., R. M. Worden, M. K. Jain, and R. Datta. 1990. Continuous production of mixed alcohols and acids from carbon monoxide. *Appl. Biochem. Biotechnol.* **24/25:**875–884.

Grethlein, A. J., R. M. Worden, M. K. Jain, and R. Datta. 1991. Evidence for production of *n*-butanol from carbon monoxide by *Butyribacterium methylotrophicum*. *J. Ferment. Bioeng.* **72:** 58–60.

Hartmanis, M. G. N., T. Klason, and S. Gatenbeck. 1984. Uptake and activation of acetate and butyrate in *Clostridium acetobutylicum*. *Appl. Microbiol. Biotechnol.* **20:**66–71.

Heijthuijsen, J. H. F. G., and T. A. Hansen. 1989. Selection of sulphur sources for the growth of *Butyribacterium methylotrophicum* and *Acetobacterium woodii*. *Appl. Microbiol. Biotechnol.* **32:**186–192.

Heijthuijsen, J. H. F. G., and T. A. Hansen. 1990. C_1-metabolism in anaerobic non-methanogenic bacteria, p. 163–191. *In* G. A. Codd, L. Dijkhuizen, and F. R. Tabita (ed.), *Autotrophic Microbiology and One-Carbon Metabolism*. Kluwer Academic Publishers, Dordrecht, The Netherlands.

Jones, D. T., and D. R. Woods. 1986. Acetone-butanol fermentation revisited. *Microbiol. Rev.* **50:**484–524.

Klasson, K. T., M. D. Ackerson, E. C. Clausen, and J. L. Gaddy. 1992. Bioconversion of synthesis gas into liquid or gaseous fuels. *Enzyme Microb. Technol.* **14:**602–608.

Levy, P. F., G. W. Barnard, D. V. Garcia-Martinez, J. E. Sanderson, and D. L. Wise. 1981. Organic acid production from CO_2/H_2 and CO/H_2 by mixed-culture anaerobes. *Biotechnol. Bioeng.* **23:**2293–2306.

Liou, J. S.-C., D. L. Balkwill, G. R. Drake, and R. S. Tanner. 2005. *Clostridium carboxidivorans* sp. nov., a solvent-producing clostridium isolated from an agricultural lagoon, and reclassification of *Clostridium scatologenes* strain SL1 as *Clostridium drakei* sp. nov. *Int. J. Syst. Evol. Microbiol.* **55:**2085–2091.

Liou, S., and R. S. Tanner. 2001. Production of acids and alcohols from CO by clostridial strain P7, abstr. O-11, p. 533. *Abstr. 101st Annu. Meet. Am. Soc. Microbiol. 2001*. American Society for Microbiology, Washington, DC.

Lynd, L., R. Kerby, and J. G. Zeikus. 1982. Carbon monoxide metabolism of the methylotrophic acidogen *Butyribacterium methylotrophicum*. *J. Bacteriol.* **149:**255–263.

McKendry, P. 2002. Energy production from biomass (part 3): gasification technologies. *Bioresour. Technol.* **83:**55–63.

Midilli, A., M. Dogru, G. Akay, and C. R. Howarth. 2002. Hydrogen production from sewage sludge via a fixed bed gasifier product gas. *Int. J. Hydrogen Energy* **27:**1035–1041.

Mielenz, J. R. 2001. Ethanol production from biomass: technology and commercialization status. *Curr. Opin. Microbiol.* **4:**324–329.

Najafpour, G., and H. Younesi. 2006. Ethanol and acetate synthesis from waste gas using batch culture of *Clostridium ljungdahlii*. *Enzyme Microb. Technol.* **38:**223–228.

Phillips, J. R., E. C. Clausen, and J. L. Gaddy. 1994. Synthesis gas as substrate for the biological production of fuels and chemicals. *Appl. Biochem. Biotechnol.* **45/46:**145–157.

Phillips, J. R., K. T. Klasson, E. C. Clausen, and J. L. Gaddy. 1993. Biological production of ethanol from coal synthesis gas: medium development studies. *Appl. Biochem. Biotechnol.* **39/40:**559–571.

Rajagopalan, S., R. P. Datar, and R. S. Lewis. 2002. Formation of ethanol from carbon monoxide via a new microbial catalyst. *Biomass Bioenerg.* **23:**487–493.

Sakai, S., Y. Nakashimida, H. Yoshimoto, S. Watanabe, H. Okada, and N. Nishio. 2004. Ethanol production from H_2 and CO_2 by a newly isolated thermophilic bacterium, *Moorella* sp. HUC22-1. *Biotechnol. Lett.* **26:**1607–1612.

Saxena, J., and R. S. Tanner. 2006. Effect of trace metals on ethanol production by *Clostridium* strain P11, abstr. O-006, p. 422. *Abstr. 106th Annu. Meet. Am. Soc. Microbiol. 2006*. American Society for Microbiology, Washington, DC.

Simon, H., H. White, H. Lebertz, and I. Thanos. 1987. Reduktion von 2-Enoaten und Alkanoaten mit Kohlenmonoxid oder Formiat, Viologenen, und *Clostridium thermoaceticum* zu gesättigten Säuren und ungesättigten bzw. gessätigten Alkoholen. *Angew. Chem.* **99:** 785–787.

Sipma, J., A. M. Henstra, S. N. Parshina, P. N. L. Lens, G. Lettinga, and A. J. M. Stams. 2006. Microbial CO conversions with applications in synthesis gas purification and bio-desulfurization. *Crit. Rev. Biotechnol.* **26:**41–65.

Spath, P. L., and D. C. Dayton. 2003. *Preliminary Screening—Technical and Economic Assessment of Synthesis Gas to Fuels and Chemical with Emphasis on the Potential for Biomass-Derived Syngas*. NREL/TP-510-34929. NTIS, Springfield, VA.

Tanner, R. S., L. M. Miller, and D. Yang. 1993. *Clostridium ljungdahlii* sp. nov., an acetogenic species in clostridial rRNA homology group I. *Int. J. Syst. Bacteriol.* 43:232–236.

Van Der Drift, A., J. Van Doorn, and J. W. Vermeulen. 2001. Ten residual biomass fuels for circulating fluidized-bed gasification. *Biomass Bioenerg.* 20:45–56.

Worden, R. M., A. J. Grethlein, M. K. Jain, and R. Datta. 1991. Production of butanol and ethanol from synthesis gas via fermentation. *Fuel* 70:615–619.

Younesi, H., G. Najafpour, and A. R. Mohamed. 2005. Ethanol and acetate production from synthesis gas via fermentation processes using anaerobic bacterium, *Clostridium ljungdahlii*. *Biochem. Eng. J.* 27:110–119.

Zeikus, J. G., L. H. Lynd, T. E. Thompson, J. A. Krzycki, P. J. Weimer, and P. W. Hegge. 1980. Isolation and characterization of a new, methylotrophic, acidogenic anaerobe, the Marburg strain. *Curr. Microbiol.* 3:381–386.

2. METHANE PRODUCTION

Bioenergy
Edited by J. Wall et al.

Chapter 13

Acetate-Based Methane Production

JAMES G. FERRY

Approximately 1 billion metric tons of methane are produced each year in the Earth's biosphere (Thauer, 1998), the end product of the decomposition of complex organic matter in oxygen-free (anaerobic) environments. The process is an essential link in the global carbon cycle (Fig. 1). Carbon dioxide, fixed into complex organic matter by photosynthesis (reaction A), is regenerated upon decomposition of the organic matter by O_2-requiring (aerobic) microorganisms in oxygenated habitats (reaction B). However, a portion of the organic matter is also deposited in diverse anaerobic habitats, where microbes (anaerobes) decompose it to methane and CO_2 (reactions C and D) in a process called biomethanation. The methane escapes into aerobic habitats, where it is oxidized to CO_2 by O_2-requiring methylotrophic microbes (reaction E), thereby closing the global carbon cycle. The discovery of biological methane is often credited to Alessandro Volta, the Italian physicist who also invented the battery. A modern-day reenactment of the 1776 discovery is illustrated in Color Plate 6. Methane trapped in the sediment of a freshwater pond is released by disturbing the sediment below the water column with a pole. The rising gas bubbles are collected in a submerged bottomless vessel filled with water. The collected methane is ignited upon release by tipping the funnel, which creates a dramatic flare.

The biomethanation process is harnessed for disposal of organic waste and production of methane as a biofuel. Municipal sewage treatment plants exploit the process for the large-scale disposal of domestic and industrial waste, reducing the volume of volatile solids by up to 75% and utilizing the methane to power electrical generators that serve the facility. The methane produced in municipal landfills also makes a significant contribution to local energy needs. In developing countries, the process is harnessed for the small-scale disposal of domestic waste and use of the methane for home heating and cooking. Thus, biomethanation of sustainable plant biomass has the potential for an alternative to fossil fuels. Methane has a weight for weight energy content three times that of hydrogen, and methane is easily stored and transported. It is also possible, by using a catalytic reformation process, to convert the methane into methanol as a feedstock, which further reduces the dependence on petroleum. This chapter explores the microbiology and biochemistry of acetate conversion to methane, a key component of biomethanation. The chapter provides a fundamental background appropriate for stimulating advances to improve the process that will ensure biomethanation among the competitive alternatives to fossil fuels.

ECOLOGY AND MICROBIOLOGY

Biomethanation of organic matter in nature occurs in diverse habitats such as freshwater sediments, rice paddies, sewage digesters, the rumen, the lower intestinal tract of monogastric animals, landfills, hydrothermal vents, coastal marine sediments, and the subsurface (Whitman et al., 1998). At least three interacting metabolic groups of anaerobes constitute a consortium necessary to convert complex organic matter to the most oxidized (CO_2) and reduced (CH_4) forms of carbon (Fig. 2). Groups I and II are primarily from the *Bacteria* domain. The fermentative group decomposes complex organic matter to acetate, higher volatile fatty acids (mainly propionate and butyrate), H_2, and CO_2. The H_2-producing acetogenic group decomposes the higher volatile fatty acids to acetate, H_2, and CO_2. The methanogenic groups III and IV (methanogens) convert the metabolic products of the first two groups to CH_4 by two major pathways. Conversion of the methyl group of acetate by Group III accounts for at least two-thirds of the CH_4 produced in nature. The methanogenic Group IV produces approximately one-third by reducing CO_2 with electrons supplied from the oxidation of H_2. Thus, methanogens are dependent on the first two groups to supply substrates for their growth.

James G. Ferry • Department of Biochemistry and Molecular Biology, The Pennsylvania State University, University Park, PA 16802.

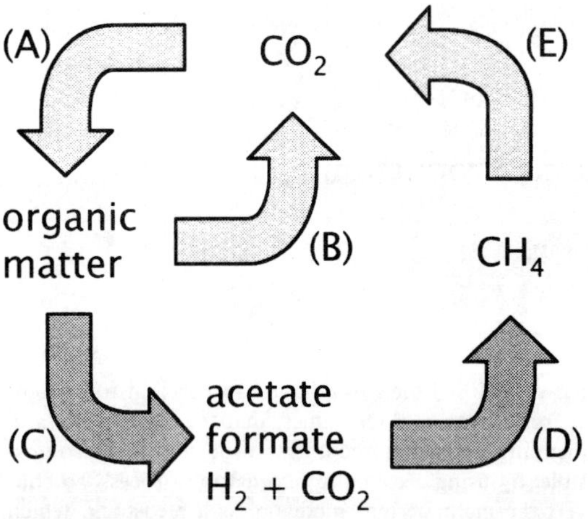

Figure 1. The global carbon cycle in nature. (A) Fixation of CO_2 into organic matter; (B) aerobic decomposition of organic matter to CO_2; (C) anaerobic decomposition of organic matter to fermentative end products; (D) anaerobic conversion of fermentative end products to methane and escape into aerobic environments; (E) aerobic oxidation of methane to CO_2 by O_2-requiring methylotrophs.

The production of H_2 by Groups I and II is thermodynamically unfavorable, and growth of these groups depends on the CO_2-reducing methanogenic Group IV to maintain low concentrations of H_2. Further, the acetate-utilizing methanogens of Group III maintain a neutral pH that is necessary for optimal metabolism and survival of the other groups. For example, inhibition of Group III results in acidification of municipal sewage digesters and complete cessation of biomethanation termed a "stuck digester" that requires considerable downtime to restart and stabilize. Thus, acetate conversion to methane is the major route for methanogenesis and key to maintaining the stability of large-scale digesters.

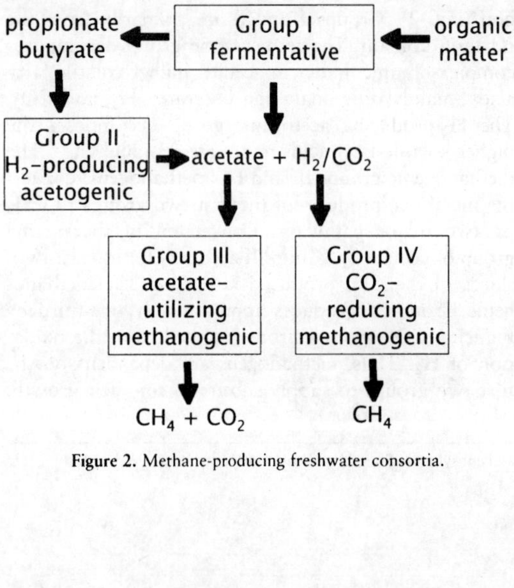

Figure 2. Methane-producing freshwater consortia.

Methanogens are found exclusively in the *Archaea* domain and are the main constituency of the *Euryarchaeota* kingdom, which contains five orders. Acetate-utilizing methanogens are found only in the order *Methanosarcinales*, which contains two families, *Methanosarcinaceae*, and *Methanosaetaceae*.

Methanosarcinaceae is the most metabolically versatile family, containing species with the ability to grow by producing methane from the methyl groups of acetate and several methylated compounds such as methanol, and reduction of CO_2 to methane with H_2 or CO as the electron donor. The *Methanosarcinaceae* family is comprised of eight genera, although *Methanosarcina* is the only genus containing acetate-utilizing species. All nine species of *Methanosarcina* grow by producing methane from methylated compounds. However, only *Methanosarcina acetivorans* (Sowers et al., 1984), *Methanosarcina baltica* (von Klein et al., 2002; Singh et al., 2005), *Methanosarcina barkeri* (Mah et al., 1978; Bryant and Boone, 1987; Maestrojuan and Boone, 1991), *Methanosarcina mazei* (Mah and Kuhn, 1984; Maestrojuan and Boone, 1991), *Methanosarcina siciliae* C2J (Elberson and Sowers, 1997), *Methanosarcina thermophila (Msr. thermophila)* (Zinder and Mah, 1979; Murray and Zinder, 1985; Zinder et al., 1985), and *Methanosarcina vacuolata* (Zhilina and Zavarzin, 1979, 1987; Maestrojuan and Boone, 1991) are capable of growth and methane formation with acetate. Cell morphologies are coccoidal or pseudosarcinal, as shown for *M. mazei* (Fig. 3). The proteinaceous cell wall is devoid of peptidoglycan or pseudomurein, and some species form large aggregates surrounded by a sheath as shown in Fig. 4 for *Msr. thermophila*. Acetate-utilizing species have been isolated from a variety of habitats. Most are mesophilic, although *Msr. thermophila* was isolated from a 55°C anaerobic digester and has an optimum temperature for growth at 50°C (Zinder et al., 1985). Interestingly, no extremely thermophilic acetate-utilizing species has been reported. *M. baltica* was isolated from the Gotland Deep of the Baltic Sea and grows at temperatures between 4 and 27°C with an optimum at 25°C. All species are neutralophilic with pH optima for growth between 6.5 and 7.5. Most are autotrophic, although cofactor supplements such as biotin are required by some. The major nitrogen and sulfur sources are ammonia and cysteine.

The family *Methanosaetaceae* is comprised of three species within the single genus *Methanosaeta*. The genus was previously named *Methanothrix*; however, the type species *Methanothrix soehngenii* was believed to be contaminated, and thus, the genus was re-

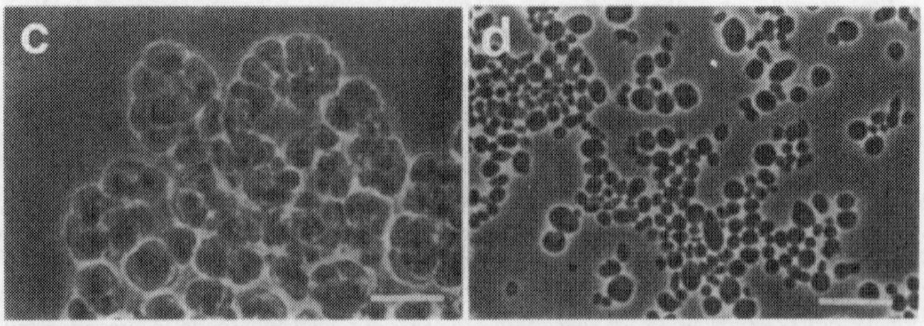

Figure 3. Phase-contrast micrographs of *Methanosarcina mazei* showing cells in pseudosarcinal aggregates (left) and single cells (right). Bars, 10 μm. Reprinted from *Applied and Environmental Microbiology* (Sowers et al., 1993) with permission of the publisher.

named *Methanosaeta* (Boone and Kamagata, 1998). Several species have been obtained in pure culture. *Methanosaeta concilii*, isolated from sewage sludge, is mesophilic and has an optimum pH for growth at 7.0 (Patel, 1984; Patel and Sprott, 1990). *Methanosaeta thermophila (Mse. thermophila)*, isolated from a thermophilic digester, has an optimum temperature and pH for growth near 55°C and 7.0, respectively (Kamagata and Mikami, 1991). *Methanosaeta harundinacea* was isolated from a reactor treating beer-manufacturing wastewater (Ma et al., 2006). Acetate is the only substrate for growth and methanogenesis for all species. Minimal media support growth of all species. *Methanosaeta* spp. have a higher affinity for acetate than *Methanosarcina* and likely dominate in environments where acetate is in low concentrations (Min and Zinder, 1989; Westermann et al., 1989). Cells are nonmotile, straight rods of unusual width (~0.5 to 1.0 μm) that often grow in long filaments (Fig. 5) composed of cells encased in a tubular proteinaceous sheath constructed with stacked annular hoops (Beveridge et al., 1986). Structures called spacer plugs separate individual cells (Fig. 6) and also cap the ends of the sheath (Beveridge et al., 1986). Lying between the sheath and the cell is an amorphous granular matrix encapsulating each cell (Fig. 6).

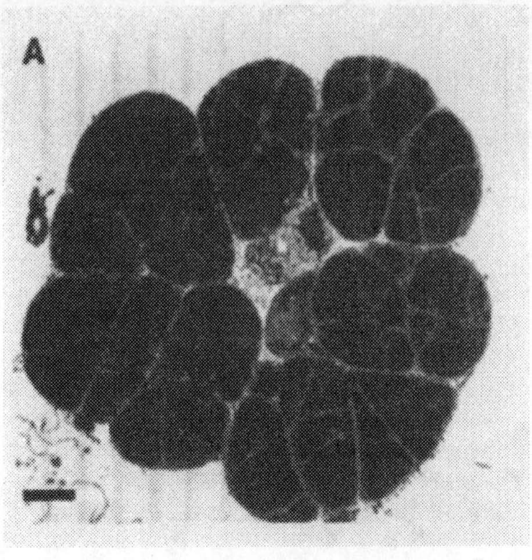

Figure 4. Thin-section electron micrograph of *Methanosarcina thermophila* showing a cell aggregate enclosed by an outer membrane. Bar, 1 μm. Reprinted from *International Journal of Systematic Bacteriology* (Zinder et al., 1985) with permission of the publisher.

BIOCHEMISTRY

The production of methane is the energy-yielding metabolism of all methanogens. Two major pathways account for most of the methane produced in nature. Approximately two-thirds derive from the methyl group of acetate (reaction 1) by the "aceticlastic" pathway, and the remaining third is produced by the reduction of CO_2 with electrons derived from oxidation of

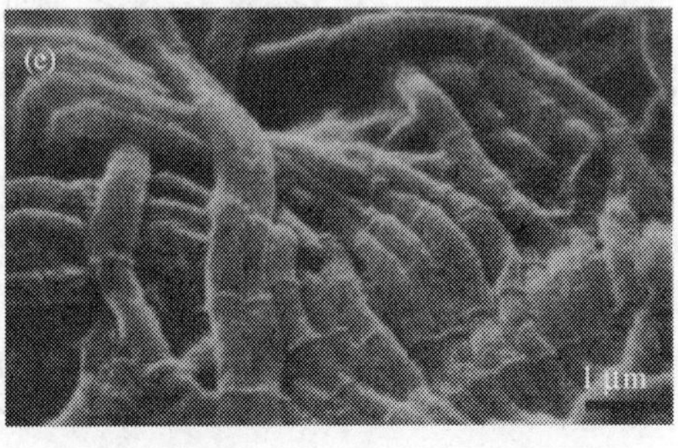

Figure 5. Scanning electron micrograph of *Methanosaeta concilii* strain T-3 filaments. Reprinted from *Bioscience, Biotechnology, and Biochemistry* (Mizukami et al., 2006) with permission of the publisher.

Figure 6. Thin section of "*Methanothrix*" (*Methanosaeta*) *concilii*. Spacer plugs are indicated by arrows, and "M" indicates the amorphous granular layer. Bar, 1 μm. Reprinted from *Canadian Journal of Microbiology* (Beveridge et al., 1986) with permission of the publisher.

either H_2, formate, or CO (reactions 2a, 2b, and 2c) in the "CO_2 reduction" pathway.

$$CH_3COO^- + H^+ \rightarrow CH_4 + CO_2 \qquad (1)$$

$$CO_2 + 4H_2 \rightarrow CH_4 + 2H_2O \qquad (2a)$$

$$4HCO_2H \rightarrow 3CO_2 + CH_4 + 2H_2O \qquad (2b)$$

$$4CO + 2H_2O \rightarrow 3CO_2 + CH_4 \qquad (2c)$$

The reader is referred to several recent reviews describing the CO_2 reduction pathway and other pathways for methanogenesis in greater detail (Grahame and Gencic, 2000; Deppenmeier, 2002; Shima et al., 2002; Ferry, 2003; Deppenmeier, 2004; Hedderich, 2004).

Reactions Leading to Methane That Are Common to the CO_2 Reduction and Aceticlastic Pathways

Figure 7 shows the structures of the cofactors and coenzymes involved in the aceticlastic and CO_2 reduction pathways. Factor III is the cofactor of proteins

2-Mercaptoethanesulfonic acid
(Coenzyme M, HS-CoM)

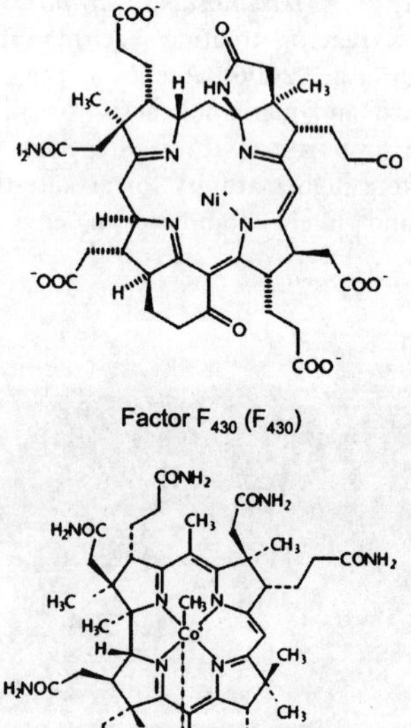

Factor F_{430} (F_{430})

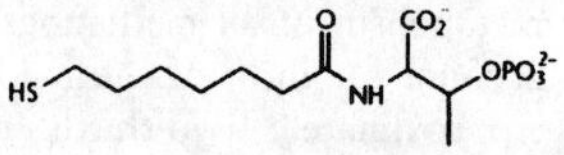

7-Mercaptoheptanoylthreonine phosphate
(coenzyme B, HS-CoB)

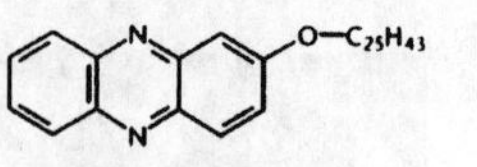

Methanophenazine

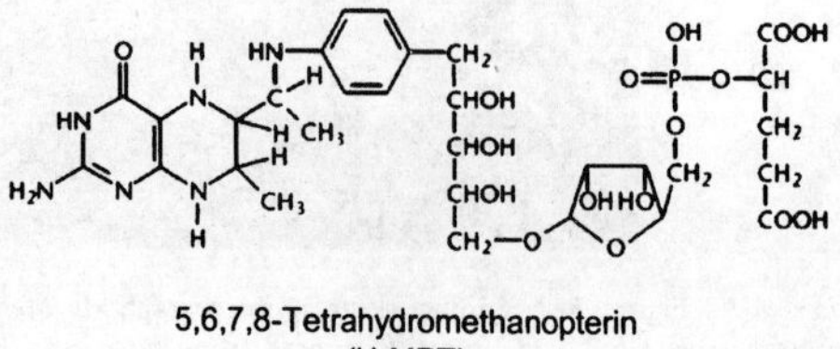

5,6,7,8-Tetrahydromethanopterin
(H_4MPT)

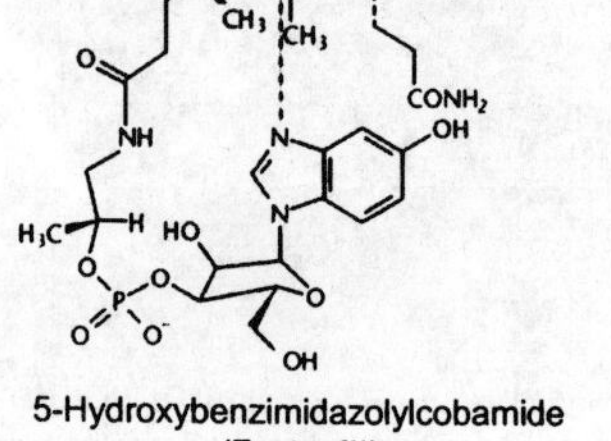

5-Hydroxybenzimidazolylcobamide
(Factor III)

Figure 7. Cofactors and coenzymes utilized in the pathway for aceticlastic methanogenesis.

that carry a methyl group attached to the upper axial ligand of the cobalt atom (Fig. 7). The structure of Factor III is similar to that of vitamin B_{12}, except that Factor III contains a 5-hydroxybenzimidazolyl base (Scherer et al., 1984). Tetrahydromethanopterin (H_4MPT) is the coenzyme of a membrane-bound methyltransferase that carries a methyl group bound to the ^5N position (Fig. 7). *Methanosarcina* species synthesize tetrahydrosarcinapterin (H_4SPT) which serves the same function as H_4MPT. H_4SPT has the same structure as H_4MPT except for an added terminal α-linked glutamate (van Beelen et al., 1984). Coenzyme M (HS-CoM) also carries a methyl group and is a coenzyme of methylreductase, which catalyzes the reductive demethylation of CH_3-S-CoM to methane. Cofactor F_{430} (F_{430}) is the methyl-accepting cofactor of the methylreductase, and coenzyme B (CoB-SH) is the electron donor. Each of these coenzymes and cofactors functions in both pathways. Methanophenazine (Abken et al., 1998) is a quinone-like 2-hydroxyphenazine derivative which is hypothesized to function in the membrane of acetate-utilizing methanogens as an electron carrier. A role for methanophenazine in the CO_2 reduction pathway has not been reported.

Each of the coenzymes and cofactors shown in Fig. 7 are reported to function exclusively in methanogens with the exception of H_4MPT and coenzyme M. H_4MPT-dependent enzymes function in sulfate-reducing *Archaeoglobus* species in the *Archaea* domain (Moller-Zinkhan et al., 1989). H_4MPT-dependent enzymes also function in aerobic proteobacterial methylotrophs from the *Bacteria* domain (Vorholt et al., 1999). Coenzyme M is the coenzyme of an enzyme catalyzing aliphatic epoxide carboxylation in *Xanthobacter* strain Py2, an aerobe from the *Bacteria* domain that metabolizes short-chain aliphatic alkenes to beta-ketoacids (Allen et al., 1999). Interestingly, an F_{430}-like cofactor has been reported in microbial mats oxidizing methane anaerobically, which suggests a possible role for the novel cofactor in activation of methane for oxidation (Kruger et al., 2003).

Figure 8 compares and contrasts the aceticlastic and CO_2 reduction pathways. Reactions 3 to 5 are common to the two pathways, which differ primarily in the steps by which the methyl group is generated and passed to $H_4M(S)PT$ to form ^5N-methyl-$H_4M(S)PT$ and the source of electrons for generating HS-CoB.

$$CH_3\text{-}H_4M(S)PT + HS\text{-}CoM \rightarrow CH_3\text{-}S\text{-}CoM + H_4M(S)PT \quad (3)$$

$$CH_3\text{-}S\text{-}CoM + HS\text{-}CoB \rightarrow CoMS\text{-}SCoB + CH_4 \quad (4)$$

$$CoMS\text{-}SCoB + 2e^- + 2H^+ \rightarrow HS\text{-}CoB + HS\text{-}CoM \quad (5)$$

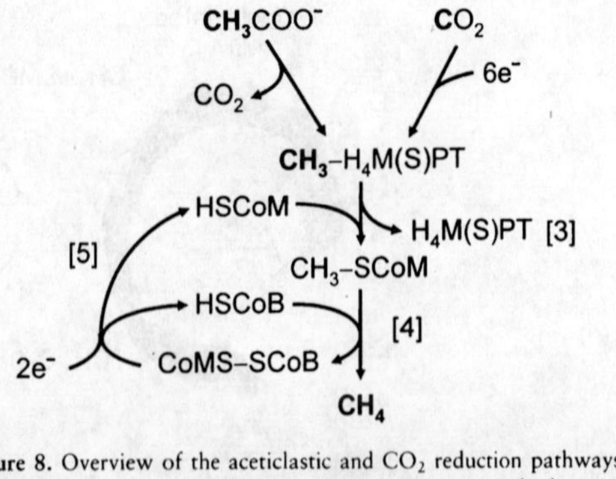

Figure 8. Overview of the aceticlastic and CO_2 reduction pathways. $H_4M(S)PT$, tetrahydromethanopterin (H_4MPT) or tetrahydrosarcinapterin (H_4SPT); CoM, coenzyme M; CoB, coenzyme B.

The membrane-bound methyl-$H_4M(S)PT$:coM methyltransferase catalyzes reaction 3. Comprised of eight nonidentical subunits (MtrA through H) (Harms et al., 1995), the enzyme couples the exergonic methyl transfer to generation of a sodium ion gradient (high outside the cytoplasmic membrane) that is postulated to drive various energy-requiring reactions. The MtrA subunit contains Factor III, which is proposed to extend into the cytoplasm where the methyl group of CH_3-$H_4M(S)PT$ is transferred to Factor III catalyzed by MtrH (Fig. 9) (Sachs et al., 2006). The cobalt atom of methylated Factor III is ligated in the lower axial position to a histidine residue of MtrA (Sauer and Thauer, 1998). MtrE is postulated to demethylate Factor III, resulting in loss of the lower axial ligand, which induces a conformational change in MtrA (Gottschalk and Thauer, 2001). It is hypothesized that this conformational change in MtrA is transmitted to MtrE, which drives the translocation of sodium. The demethylation of Factor III is shown coupled with vectorial sodium ion translocation, supported by the observation that demethylation is dependent on sodium ions (Gartner et al., 1994; Weiss et al., 1994).

Reaction 4 is catalyzed by the methyl-CoM reductase (Mcr) with HS-CoB serving as the electron donor. In addition to methane, the heterodisulfide CoMS-SCoB is a product of the reaction. The enzyme from the CO_2-reducing species *Methanothermobacter marburgensis* (*Methanobacterium thermoautotrophicum* strain Marburg) contains three nonidentical subunits with an $\alpha_2\beta_2\gamma_2$ arrangement revealed by the crystal structure (Ermler et al., 1997). A concerted mechanism is proposed (Grabarse et al., 2001; Finazzo et al., 2003) wherein specific conformational changes ensure entry of CH_3-S-CoM adjacent to F_{430} and before entry of HS-CoB in the narrow active-site channel (Fig. 10, upper left). Nucleophilic attack of $[F_{430}]Ni(I)$ on CH_3-S-CoM

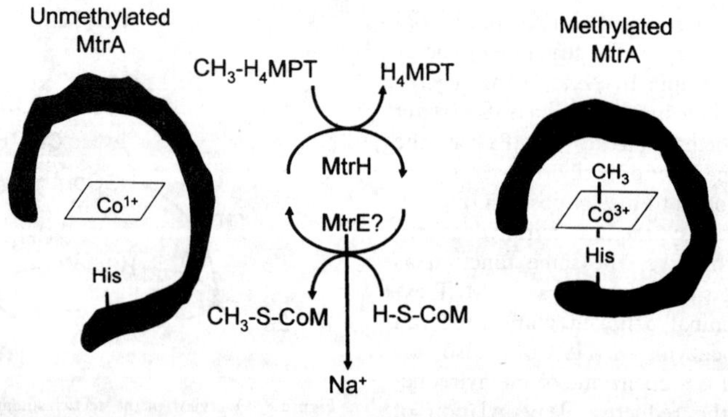

Figure 9. Proposed mechanism of sodium translocation by the methyl-$H_4M(S)PT$:coenzyme M methyltransferase. Reprinted from *Biochimica et Biophysica Acta* (Gottschalk and Thauer, 2001) with permission of the publisher.

forms an $[F_{430}]Ni(III)$-CH_3 intermediate and HSCoM (Fig. 10, upper right). Transfer of electrons from HS-CoM to Ni(III) produces the thiyl radical ·S-CoM and $[F_{430}]Ni(II)$-CH_3 (Fig. 10, lower right). In the final step, protonolysis of $[F_{430}]Ni(II)$-CH_3 produces methane. An alternative to this mechanism has been proposed (Goenrich et al., 2004) in which Ni(I) attacks the sulfur of CH_3-S-CoM, producing a free methyl radical reacting with HS-CoB to produce methane and the thiyl radical ·S-CoB. In both proposals, ·S-CoM is coupled with $^-$S-CoB to form CoB-S-S-CoM followed by one-electron reduction of Ni(II) to Ni(I) regenerating the active form of F_{430}. The crystal structure reveals two F_{430} cofactors embedded between subunits α, α', β, and γ and α', α, β', and γ' with each cofactor separated by 50 Å, indicating two independent active sites (Ermler et al., 1997). The dependence of temperature on activity indicates an alternating-sites mechanism for release of CoMS-SCoB that is driven by conformational changes transmitted from the adjacent monomer (Goenrich et al., 2005).

Heterodisulfide reductase (Hdr) is essential to both pathways, catalyzing reduction of the disulfide bond of CoM-S-S-CoB and regenerating the active sulfhydryl forms of the coenzymes (reaction 5). In both pathways, CoM-S-S-CoB is the terminal electron acceptor of a membrane-bound electron transport chain coupled to formation of an electrochemical proton and/or sodium gradient which drives ATP synthesis (Peinemann et al., 1988; Deppenmeier, 2002; Li et al., 2006). A two-subunit HdrDE and a three-subunit HdrABC have been described, the latter of which functions in the CO_2-reduction pathway (Hedderich et al., 1990, 1994). The HdrDE type is found in acetate-grown cells of *Msr. thermophila* (Simianu et al., 1998) and *Methanosarcina barkeri* (Heiden et al., 1993; Kunkel et al., 1997), consistent with a role in acetate metabolism. A water-soluble analog of methanophenazine, 2-hydroxyphenazine (Murakami et al., 2001), is an electron donor to the *Msr. thermophila* enzyme consistent with a role for methanophenazine in electron transport. Studies with the *M. barkeri* enzyme show that the HdrE subunit contains cytochrome *b*, which is proposed to transfer electrons to the catalytic subunit HdrD (Hedderich et al., 1994; Kunkel et al., 1997). EPR studies identify a 4Fe-4S center in HdrD, where it is proposed that reduction of CoM-S-S-CoB occurs in two one-electron steps involving a thiyl radical intermediate (Hedderich et al., 2005). Although involvement of an Fe_4S_4 cluster is also proposed for the HdrDE from acetate-grown *Msr. thermophila*, kinetic studies and inhibition experiments indicate that the order of electron flow is 2-hydroxyphenazine $\rightarrow$ high-potential Fe_4S_4 cluster $\rightarrow$ low-potential heme $\rightarrow$ CoM-S-S-CoB (Murakami et al., 2001).

As described above, the aceticlastic and CO_2 reduction pathways generate primary sodium and proton gradients that are the only possible driving forces for ATP synthesis. No reactions in either pathway support a substrate-level phosphorylation of ADP. The "archaeal" (Muller, 2004; Lewalter and Muller, 2006) A_1A_0-type ATP synthase is abundant in acetate-grown *M. acetivorans* (Li et al., 2007) and *M. mazei* (Hovey et al., 2005) consistent with a role for the enzyme in ATP synthesis during growth with acetate. The A_1A_0-type ATP synthase is also found in CO_2-reducing methanogens (Muller, 2004; Lewalter and Muller, 2006). The A_1A_0-type is a new class of ATP synthase that is structurally similar to the eukaryotic V_1V_0 ATPases which hydrolyze ATP to generate ion gradients (Muller, 2004; Lewalter and Muller, 2006). Although sequence analysis identifies a putative Na^+ binding site in the A_1A_0-type of methanogens, including *M. acetivorans* and *M. mazei*, the coupling ion is unknown (Muller, 2004). Clearly, further research is necessary to elucidate the mechanism by which sodium and proton gradients drive ATP synthesis.

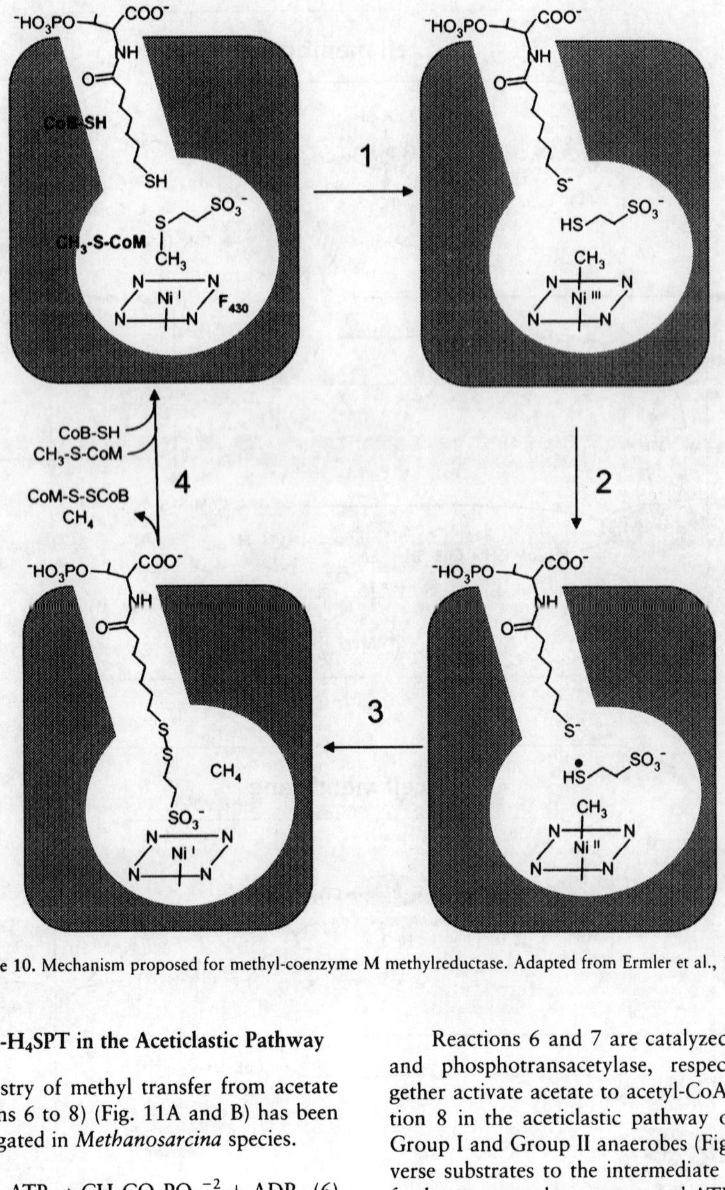

Figure 10. Mechanism proposed for methyl-coenzyme M methylreductase. Adapted from Ermler et al., 1997.

Synthesis of CH_3-H_4SPT in the Aceticlastic Pathway

The biochemistry of methyl transfer from acetate to H_4SPT (reactions 6 to 8) (Fig. 11A and B) has been extensively investigated in *Methanosarcina* species.

$$CH_3COO^- + ATP \rightarrow CH_3CO_2PO_3^{-2} + ADP \quad (6)$$

$$CH_3CO_2PO_3^{-2} + HS\text{-}CoA \rightarrow CH_3COSCoA + Pi \quad (7)$$

$$CH_3COSCoA + H_4SPT + H_2O + Fd^{ox} \rightarrow CH_3\text{-}H_4SPT + Fd^{red} + CO_2 + HS\text{-}CoA \quad (8)$$

Homologs of the enzymes catalyzing these reactions are widespread in diverse anaerobes, which has significantly impacted an understanding of the broader field of procaryotic biology.

Reactions 6 and 7 are catalyzed by acetate kinase and phosphotransacetylase, respectively, which together activate acetate to acetyl-CoA required for reaction 8 in the aceticlastic pathway of methanogenesis. Group I and Group II anaerobes (Fig. 2) metabolize diverse substrates to the intermediate acetyl-CoA that is further converted to acetate and ATP by reversal of reactions 6 and 7. The ATP produced by Group I and Group II is a major source of energy for growth. Overproduction in *Escherichia coli* of acetate kinase and phosphotransacetylase from the aceticlastic methanogen *Msr. thermophila* has established these enzymes as models for investigation (Latimer and Ferry, 1993). The crystal structure (Buss et al., 2001) of the enzyme from *Msr. thermophila* suggests that acetate kinase is the founding member of the ASKHA (acetate

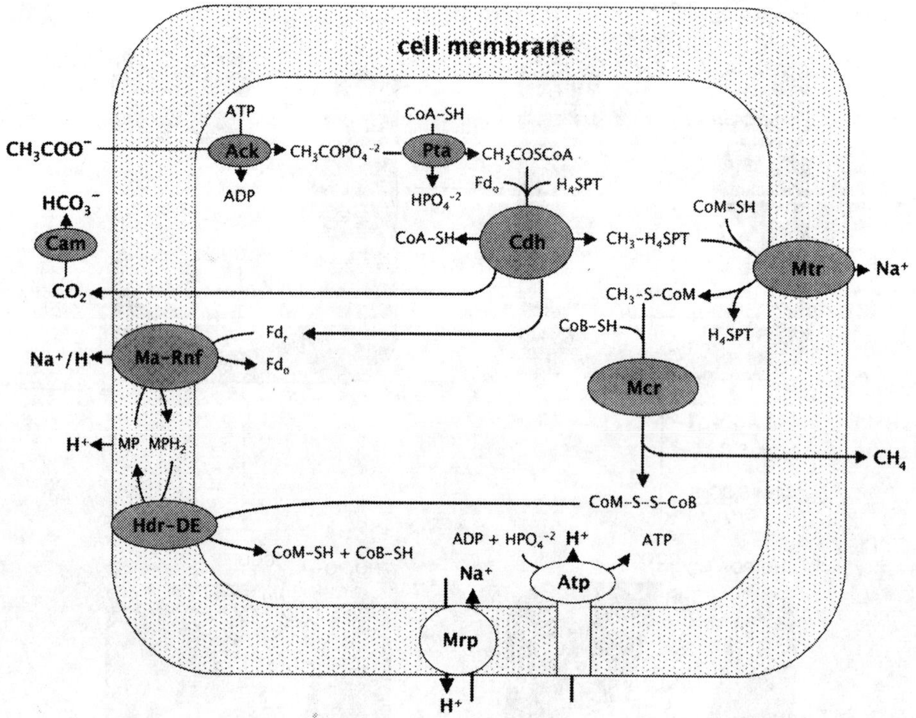
cell membrane
ATP
Ack
ADP
CoA-SH
Pta
HPO4-2
CH3COO−
CH3COPO4-2
CH3COSCoA
Fdo
H4SPT
HCO3−
Cam
CO2
CoA-SH
Cdh
CH3-H4SPT
CoM-SH
Mtr
Na+
CH3-S-CoM
H4SPT
Fdr
Fdo
Na+/H+
Ma-Rnf
CoB-SH
Mcr
H+
MP MPH2
CH4
Hdr-DE
CoM-S-S-CoB
CoM-SH + CoB-SH
ADP + HPO4-2
H+
ATP
Atp
Na+
Mrp
H+

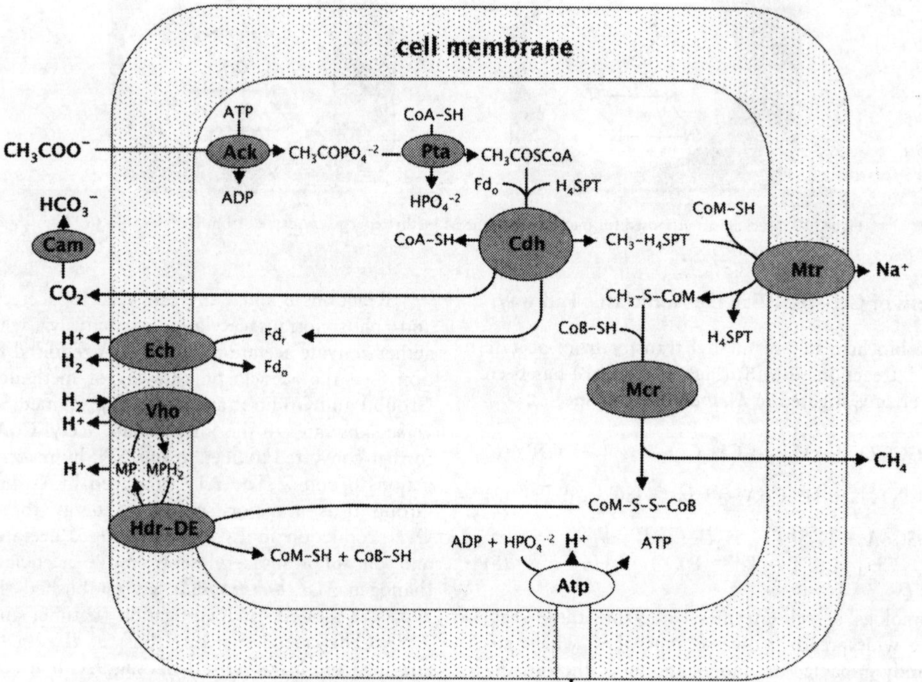
cell membrane
ATP
Ack
ADP
CoA-SH
Pta
HPO4-2
CH3COO−
CH3COPO4-2
CH3COSCoA
Fdo
H4SPT
HCO3−
Cam
CO2
CoA-SH
Cdh
CH3-H4SPT
CoM-SH
Mtr
Na+
CH3-S-CoM
H4SPT
H+
H2
Ech
Fdr
Fdo
H2
H+
Vho
CoB-SH
Mcr
H+
MP MPH2
CH4
Hdr-DE
CoM-S-S-CoB
CoM-SH + CoB-SH
ADP + HPO4-2
H+
ATP
Atp

and sugar kinase/Hsc70/actin) superfamily of phosphotransferases. Kinetic and biochemical studies of the wild-type and site-specific amino acid variants of the enzyme support a direct in-line mechanism for transfer of the phosphate from ATP to acetate (Singh-Wissmann et al., 1998, 2000; Ingram-Smith et al., 2000, 2005; Miles et al., 2001, 2002; Gorrell et al., 2005). The crystal structure of the *Msr. thermophila* phosphotransacetylase has been obtained with CoA-SH bound (Iyer et al., 2004; Lawrence et al., 2006), revealing the active site. A catalytic mechanism is proposed based on features of the active site and kinetic analyses of the wild-type (Lawrence and Ferry, 2006) and site-specific amino acid variants (Rasche et al., 1997; Iyer and Ferry, 2001; Lawrence and Ferry, 2006; Lawrence et al., 2006). In the mechanism, base catalysis generates $^{-}$S-CoA followed by nucleophilic attack of the thiolate anion on the carbonyl carbon of acetyl phosphate yielding acetyl-CoA and inorganic phosphate (Lawrence et al., 2006).

In *Methanosaeta* species, acetyl-CoA synthetase functions as the acetate activating enzyme (reaction 9) converting acetate to acetyl-CoA in one step (Jetten et al., 1990; Eggen et al., 1991a).

$$CH_3COO^- + ATP + CoASH \rightarrow CH_3COSCoA + AMP + P_iP_i \quad (9)$$

The enzyme has a lower K_m for acetate (40 to 90 μM) (Jetten et al., 1990) than for acetate kinase (2.7 mM) (Gorrell et al., 2005), a basis for the existence of *Methanosaeta* rather than *Methanosarcina* species in environments where the acetate concentrations are low.

Reaction 8 is catalyzed by the CO dehydrogenase/acetyl-CoA synthase (Cdh) that cleaves the C-C and C-S bonds of acetyl-CoA, yielding methyl and carbonyl groups (Abbanat and Ferry, 1990; Raybuck et al., 1991). The enzyme has been investigated from both *Methanosarcina* and *Methanosaeta* species (Terlesky et al., 1986; Jetten et al., 1989, 1991a, 1991b; Abbanat and Ferry, 1990; Eggen et al., 1991b, 1996; Raybuck et al., 1991; Grahame, 1993; Grahame and Demoll, 1995). The methyl group is transferred to H_4SPT (Fischer and Thauer, 1989; Grahame, 1991) for eventual conversion to methane (reactions 3 to 5), and the carbonyl group is oxidized to CO_2 with reduction of ferredoxin (Fig. 11A and B). In addition to aceticlastic methanogens, a diversity of anaerobes employ Cdh homologs in energy-yielding pathways and pathways for generating acetyl-CoA from CO_2 for cell biosynthesis (Grahame, 2003). The enzyme is thought to be of ancient origin and to have played a central role in the early evolution of life (Lindahl and Chang, 2001; Ferry and House, 2006; Martin and Russell, 2007). The bulk of mechanistic investigations have focused on the Cdh of *Msr. thermophila* and *M. barkeri*, a five-subunit (α, β, γ, δ, and ϵ) complex (Terlesky et al., 1986) resolvable into a Ni/Fe-S component (α and ϵ subunits), a corrinoid/Fe-S component (γ and δ subunits), and the β subunit (Abbanat and Ferry, 1991; Kocsis et al., 1999). The Ni/Fe-S component catalyzes oxidation of exogenous CO and presumably the carbonyl group of acetyl-CoA (Abbanat and Ferry, 1991). The electron acceptor is ferredoxin (Terlesky and Ferry, 1988a, 1988b; Fischer and Thauer, 1990). The β subunit contains an NiFeS cluster ("A" cluster) that is the proposed site of acetyl-CoA cleavage or synthesis (Grahame and Demoll, 1996; Murakami and Ragsdale, 2000). A crystal structure for Cdh from a methanogen has not been reported; however, the crystal structure of the homolog called acetyl-CoA synthase from the acetogen *Moorella thermoacetica* (Ragsdale, 2004) reveals the A cluster with a 4Fe-4S center bridged to a Ni via sulfur atoms and a metal (Mp) proximal to the 4Fe-4S center (Fig. 12). The identity of the proximal metal is controversial, although it appears that Ni is the most likely candidate. Assembly or cleavage of acetyl-CoA is proposed to occur on the Ni distal to the 4Fe-4S center. In addition to CO oxidation, the Ni/Fe-S component of Cdh from methanogens is required for acetyl-CoA cleavage or synthesis (Murakami and Ragsdale, 2000); however, published reports on the type of metal clusters present and their function are contradictory (Lu et al., 1994; Murakami and Ragsdale, 2000). The corrinoid/Fe-S component contains Factor III and is involved in transfer of the methyl group of acetyl-CoA to H_4SPT (Grahame, 1991, 1993; Jablonski et al., 1993).

Energy Conservation in the Aceticlastic Pathway

The conversion of acetate to CH_4 and CO_2 provides only a marginal amount of energy available for ATP synthesis ($\Delta G^{\circ\prime} = -36$ kJ/CH_4) under standard conditions of equimolar reactants and products. Thus, it is proposed that efficient removal of CH_4 and CO_2 from the cytoplasm is essential to create thermodynamic conditions necessary for growth with acetate (Alber and

Figure 11. Pathways for conversion of acetate to methane by *M. mazei* (A) and *M. acetivorans* (B). Ack, acetate kinase; Pta, phosphotransacetylase; CoA-SH, coenzyme A; H_4SPT, tetrahydrosarcinapterin; Fd_r, reduced ferredoxin; Fd_o, oxidized ferredoxin; Cdh, CO dehydrogenase/acetyl-CoA synthase; CoM-SH, coenzyme M; Mtr, methyl-H_4SPT:CoM-SH methyltransferase; CoB-SH, coenzyme B; Cam, carbonic anhydrase; Ech, H_2-evolving hydrogenase; Vho, H_2-consuming hydrogenase; Ma-Rnf, *M. acetivorans* Rnf; MP, methanophenazine; Hdr-DE, heterodisulfide reductase; Mrp, multiple resistance/pH regulation Na^+/H^+ antiporter; Atp, H^+-translocating A_1A_0 ATP synthase. Adapted from Li et al., 2006.

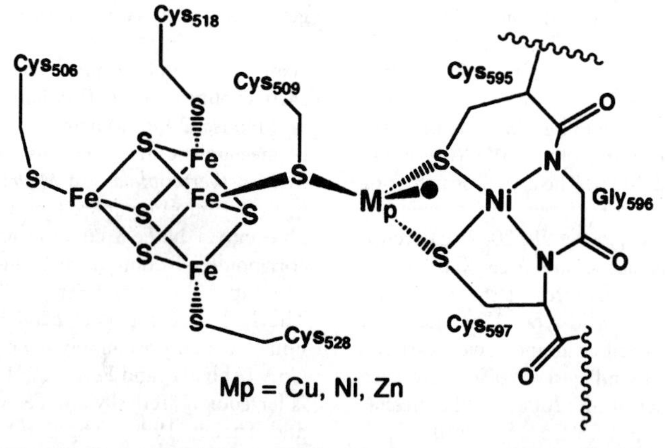

Figure 12. The A cluster of Acs from *Moorella thermoacetica*. Reprinted from *Critical Reviews in Biochemistry and Molecular Biology* (Ragsdale, 2004) with permission of the publisher.

Ferry, 1994; Liu et al., 2001). It is postulated that a carbonic anhydrase (Cam) isolated from *Msr. thermophila* is located outside the cell membrane (Fig. 11A and B), where it converts CO_2 to membrane-impermeable HCO_3^- (reaction 10), thereby facilitating removal of CO_2 from the cytoplasm (Alber and Ferry, 1994).

$$CO_2 + H_2O \rightarrow HCO_3^- + H^+ \quad (10)$$

Indeed, *Msr. thermophila* (Alber and Ferry, 1994), *M. acetivorans* (Li et al., 2006), and *M. mazei* (Hovey et al., 2005) up-regulate the synthesis of Cam when switched from growth on methanol to growth on acetate, a result consistent with a role during growth on acetate. Cam from *Msr. thermophila* is the prototype of an independently evolved class of carbonic anhydrases (γ class) with a novel left-handed β-helical fold and is the first carbonic anhydrase shown to function with iron in the active site (Kisker et al., 1996; Tripp et al., 2004). Although independently evolved, the catalytic mechanism of Cam is fundamentally similar to the well-studied α class from mammals that contains zinc at the active site. In the mechanism, a water molecule bound to the metal is deprotonated, yielding a metal-bound hydroxyl that attacks CO_2, producing bicarbonate (Zimmerman and Ferry, 2006). Cam homologs are widely distributed in the *Bacteria* and *Archaea* domains (Smith et al., 1999) suggesting functions in procaryotes other than that proposed for acetotrophic methanogens.

In the aceticlastic pathway of *Methanosarcina*, a membrane-bound Fd^{red}:CoMS-SCoB oxidoreductase system generates a electrochemical gradient which drives ATP synthesis (Fig. 11). The Fd^{red} is produced by oxidation of the carbonyl group of acetyl-CoA catalyzed by Cdh. In the freshwater species *M. barkeri* (Fig. 11A), it is proposed that Fd^{red} donates electrons to Ech hydrogenase which produces H_2 and pumps protons (Meuer et al., 2002; Hedderich, 2004). A requirement for Ech is supported by gene knockout experiments showing that Ech is essential for growth with acetate (Meuer et al., 2002). Further, a cytochrome *b*-containing H_2:heterodisulfide oxidoreductase complex was purified from acetate-grown *M. barkeri*, consistent with a role for HdrDE and H_2 as an intermediate in electron transport (Heiden et al., 1993). It is hypothesized that a hydrogenase (Vho/Vht) reoxidizes the H_2, and methanophenazine mediates electron transfer to the heterodisulfide reductase (Meuer et al., 2002) (Fig. 11A). The HdrDE purified from acetate-grown cells of the freshwater isolate *Msr. thermophila* is reported to have hydrogenase activity (Simianu et al., 1998), consistent with an H_2:CoMS-SCoB oxidoreductase system for this species. A role for Isf (iron-sulfur flavoprotein) has been proposed for *Msr. thermophila* wherein Isf transfers electrons from Fd^{red} to the membrane-bound electron transport chain (Latimer et al., 1996). The crystal structure of Isf shows an unusual compact cysteine motif ligating the 4Fe4S cluster that is proposed to accept electrons from Fd^{red} and donate to the flavin (flavin mononucleotide) cofactor of Isf (Andrade et al., 2005). The electron acceptor for Isf is not known. In yet another freshwater species, *M. mazei*, expression of genes encoding Ech are up-regulated in acetate- versus methanol-grown cells (Hovey et al., 2005), prompting the suggestion that Ech functions as proposed for *M. barkeri*. The mechanism for transfer of electrons from Ech to CoMS-SCoB in *M. mazei* is unknown; however, a role for an Isf homolog is postulated based on up-regulation of the encoding gene in response to growth on acetate (Hovey et al., 2005). Thus, although a role for Ech is well supported, the mechanism by which electrons are transported from Ech to CoMS-SCoB and the generation of ion gradients are largely unknown for freshwater *Methanosarcina* species.

The marine species *M. acetivorans* evolved a different mechanism for oxidizing Fd^{red} and reducing CoMS-SCoB (Li et al., 2006) that does not involve H_2 (Fig. 11B). The genome does not encode a functional Ech (Galagan et al., 2002), and acetate-grown cells have very low H_2-dependent methyl-CoM methylreductase activity (Nelson and Ferry, 1984). It is proposed that a homolog of Rnf oxidizes Fd^{red} and that methanophenazine mediates electron transfer between reduced Rnf and HdrDE. Rnf was first discovered in *Rhodobacter capsulatus* where the six-subunit membrane-bound electron transfer complex supplies Fd^{red} to nitrogenase (Schmehl et al., 1993; Saeki and Kumagai, 1998). It has been suggested that Rnf homologs also function to couple electron transport to generation of an Na^+ gradient that drives energy-requiring processes (Bruggemann et al., 2003; Boiangiu et al., 2005). Proteomic analyses indicate that subunits of an Rnf homolog are at least 10-fold greater in acetate-grown than in methanol-grown *M. acetivorans*, consistent with the proposed function during growth on acetate (Li et al., 2006). Indeed, deletion of the genes encoding Rnf confirm that it is essential for growth on acetate (W. Metcalf, personal communication). The subunits are encoded in a transcriptional unit containing the *rnfCDGEAB* genes and two additional flanking open reading frames, one of which encodes a cytochrome *c*, which spectroscopic analysis of membrane fractions show is in greater abundance in acetate-grown than in methanol-grown cells (Li et al., 2006). Thus, it is postulated that the cytochrome and Rnf function to transport electrons from Fd^{red} to methanophenazine coupled to generation of an Na^+ gradient (high outside the membrane). Proteomic analyses show that subunits of an Na^+/H^+ antiporter (Mrp) are at least 30-fold more abundant in acetate-grown than in methanol-grown *M. acetivorans*, consistent with a role during growth on acetate (Li et al., 2006). It is postulated that Mrp functions to exchange the Na^+ gradient for a proton gradient (high outside the membrane) that drives ATP synthesis by the A_1A_0 ATP synthase (Fig. 11B) (Li et al., 2006).

Biochemical investigations of *Methanothrix soehngenii* and genomic analysis of *Mse. thermophila* indicate that, except for the activation of acetate to acetyl-CoA, enzymes and core reactions in the pathway for conversion of acetate to methane are similar to those of *Methanosarcina* species (Jetten et al., 1992; Smith and Ingram-Smith, 2007). However, it is reported that genes encoding Ech and Rnf are absent in the genome of *Mse. thermophila*, suggesting an alternate mechanism for energy conservation. The presence of genes in the genome that encode a proton-translocating ATP synthetase is consistent with generation of a proton gradient that drives ATP synthesis (Smith and Ingram-Smith, 2007). A candidate for generating the proton gradient is a putative $F_{420}H_2$ dehydrogenase. A gene cluster in the genome of *Mse. thermophila* (loci 1050 to 1060) encodes 10 putative proteins with 42 to 67% deduced identity to the deduced sequence of proteins comprising the $F_{420}H_2$ dehydrogenase complex of *M. acetivorans* (Lessner et al., 2006).

REGULATION

Growth and methanogenesis with methanol or acetate each require a unique set of genes subject to regulation of expression dependent on the growth substrate. Thus, regulations of genes specific to acetate-dependent growth and methanogenesis in *Methanosarcina* species grown on methanol and on acetate have been investigated and compared. A plasmid-mediated *lacZ* fusion reporter system, based on the promoter sequence upstream of the *Msr. thermophila cdh* operon, shows a 54-fold down-regulation in *M. acetivorans* grown on methanol compared to that grown on acetate, consistent with a role exclusive to the aceticlastic pathway (Apolinario et al., 2005). However, *M. acetivorans* (Galagan et al., 2002) and *M. mazei* (Deppenmeier et al., 2002) each contain duplicate *cdh* operons with >95% identity. A proteomic analysis of *M. acetivorans* indicates that the two Cdh complexes are approximately 200-fold and 10-fold less abundant in cells grown on methanol than on acetate (Li et al., 2006). The results suggest that both Cdh complexes function during growth on acetate although one is dominant. Conversely, global transcriptional profiling of *M. mazei* Gö1 suggests that duplicate *cdh* operons are transcribed at approximately equal levels (Hovey et al., 2005). Genes encoding ferredoxin, acetate kinase, and phosphotransacetylase in *Msr. thermophila* are also down-regulated in methanol- versus acetate-grown cells (Clements and Ferry, 1992; Singh-Wissmann and Ferry, 1995). Global proteomic and DNA microarray analyses of *M. mazei* and *M. acetivorans* have confirmed these results and significantly extended an understanding of gene expression in acetate-grown *Methanosarcina* (Hovey et al., 2005; Li et al., 2005; Li et al., 2006, 2007). Proteomic analyses identified 71 proteins that were more abundant in acetate-grown than in methanol-grown *M. acetivorans* (Li et al., 2007). Microarray analyses identified 200 genes up-regulated in acetate- versus methanol-grown *M. acetivorans* (Li et al., 2007), a result similar to that reported for *M. mazei* (Hovey et al., 2005). The results indicate that many more proteins, in addition to those shown in Fig. 11, are either essential or important for growth and methanogenesis with acetate. For example, global analyses indicate an enhanced stress response in acetate-grown *M. acetivorans* compared to methanol-grown *M. acetivorans*, which included enzymes specific for polyphosphate accumulation and oxidative stress (Li et al., 2007). These studies also suggested roles for two-component regulatory systems specific for growth with acetate, although their function is unknown. Indeed, global analyses have identified many proteins and genes implicated in aceticlastic methanogenesis that have no known function. Finally, although global analyses show extensive gene regulation in response to growth with acetate, the mechanism of regulation has not been investigated.

CONCLUSIONS

The production of acetate from complex biomass by fermentative and acetogenic anaerobes and the subsequent conversion of acetate to methane by aceticlastic methanogens are of primary importance in the biomethanation process. Aceticlastic methanogenesis is the major factor controlling the rate and reliability of the process; thus, a comprehensive understanding of these methanogens is paramount for developing an efficient process for biomethanation of renewable and waste biomass for use as a biofuel. Although the enzymology of reactions leading from acetate to methane by *Methanosarcina* species is fairly well understood, there have been only a few investigations reported on the mechanism of energy conservation and regulation of gene expression. Further, global proteomic and microarray analyses have identified a host of proteins and genes in *Methanosarcina* species, many with unknown functions, that may be important or essential. Fortunately, robust genetic systems for *Methanosarcina* species are in hand that offer considerable promise for rectifying these deficiencies in understanding and also allow genetic manipulation for process development (Metcalf et al., 1997). On the other hand, very little is known concerning the microbiology, biochemistry, molecular biology, and genetics of *Methanosaeta* species. Thus, the research opportunities are clearly defined that will lead to the level of understanding of aceticlastic methanogenesis necessary to maximize and control processes for conversion of biomass to methane as a biofuel.

Acknowledgments. Research in the laboratory of J.G.F. has been supported by the NIH, DOE, NSF, and NASA.

REFERENCES

Abbanat, D. R., and J. G. Ferry. 1991. Resolution of component proteins in an enzyme complex from *Methanosarcina thermophila* catalyzing the synthesis or cleavage of acetyl-CoA. *Proc. Natl. Acad. Sci. USA* 88:3272–3276.

Abbanat, D. R., and J. G. Ferry. 1990. Synthesis of acetylcoenzyme A by carbon monoxide dehydrogenase complex from acetate-grown *Methanosarcina thermophila. J. Bacteriol.* 172:7145–7150.

Abken, H.-J., M. Tietze, J. Brodersen, S. Baumer, U. Beifuss, and U. Deppenmeier. 1998. Isolation and characterization of methanophenazine and the function of phenazines in membrane-bound electron transport of *Methanosarcina mazei* Go1. *J. Bacteriol.* 180:2027–2032.

Alber, B. E., and J. G. Ferry. 1994. A carbonic anhydrase from the archaeon *Methanosarcina thermophila. Proc. Natl. Acad. Sci. USA* 91:6909–6913.

Allen, J. R., D. D. Clark, J. G. Krum, and S. A. Ensign. 1999. A role for coenzyme M (2-mercaptoethanesulfonic acid) in a bacterial pathway of aliphatic epoxide carboxylation. *Proc. Natl. Acad. Sci. USA* 96:8432–8437.

Andrade, S. L. A., F. Cruz, C. L. Drennan, V. Ramakrishnan, D. C. Rees, J. G. Ferry, and O. Einsle. 2005. Structures of the iron-sulfur flavoproteins from *Methanosarcina thermophila* and *Archaeoglobus fulgidus. J. Bacteriol.* 187:3848–3854.

Apolinario, E. E., K. M. Jackson, and K. R. Sowers. 2005. Development of a plasmid-mediated reporter system for in vivo monitoring of gene expression in the archaeon *Methanosarcina acetivorans. Appl. Environ. Microbiol.* 71:4914–4918.

Beveridge, T. J., G. B. Patel, B. J. Harris, and G. D. Sprott. 1986. The ultrastructure of *Methanothrix concilii*, a mesophilic aceticlastic methanogen. *Can. J. Microbiol.* 32:703.

Boiangiu, C. D., E. Jayamani, D. Brugel, G. Herrmann, J. Kim, L. Forzi, R. Hedderich, I. Vgenopoulou, A. J. Pierik, J. Steuber, and W. Buckel. 2005. Sodium ion pumps and hydrogen production in glutamate fermenting anaerobic bacteria. *J. Mol. Microbiol. Biotechnol.* 10:105–119.

Boone, D. R., and Y. Kamagata. 1998. Rejection of the species *Methanothrix soehngenii*^{VP} and the genus *Methanothrix*^{VP} as *nomina confusa*, and transfer of *Methanothrix thermophila*^{VP} to the genus *Methanosaeta*^{VP} as *Methanosaeta thermophila* comb. nov. Request for an Opinion. *Int. J. Syst. Bacteriol.* 48:1079–1080.

Bruggemann, H., S. Baumer, W. F. Fricke, A. Wiezer, H. Liesegang, I. Decker, C. Herzberg, R. Martinez-Arias, R. Merkl, A. Henne, and G. Gottschalk. 2003. The genome sequence of *Clostridium tetani*, the causative agent of tetanus disease. *Proc. Natl. Acad. Sci. USA* 100:1316–1321.

Bryant, M. P., and D. R. Boone. 1987. Emended description of strain MS^T (DSM 800^T), the type strain of *Methanosarcina barkeri. Int. J. Syst. Bacteriol.* 37:169–170.

Buss, K. A., D. R. Cooper, C. Ingram-Smith, J. G. Ferry, D. A. Sanders, and M. S. Hasson. 2001. Urkinase: structure of acetate kinase, a member of the ASKHA superfamily of phosphotransferases. *J. Bacteriol.* 183:680–686.

Clements, A. P., and J. G. Ferry. 1992. Cloning, nucleotide sequence, and transcriptional analyses of the gene encoding a ferredoxin from *Methanosarcina thermophila. J. Bacteriol.* 174:5244–5250.

Deppenmeier, U. 2004. The membrane-bound electron transport system of *Methanosarcina* species. *J. Bioenerg. Biomembr.* 36:55–64.

Deppenmeier, U. 2002. The unique biochemistry of methanogenesis. *Prog. Nucleic Acid Res. Mol. Biol.* 71:223–283.

Deppenmeier, U., A. Johann, T. Hartsch, R. Merkl, R. A. Schmitz, R. Martinez-Arias, A. Henne, A. Wiezer, S. Baumer, C. Jacobi, H. Bruggemann, T. Lienard, A. Christmann, M. Bomeke, S. Steckel, A. Bhattacharyya, A. Lykidis, R. Overbeek, H. P. Klenk, R. P. Gunsalus, H. J. Fritz, and G. Gottschalk. 2002. The genome of *Methanosarcina mazei*: evidence for lateral gene transfer between *Bacteria* and *Archaea. J. Mol. Microbiol. Biotechnol.* 4:453–461.

Eggen, R. I. L., A. C. M. Geerling, A. B. P. Boshoven, and W. M. Devos. 1991a. Cloning, sequence analysis, and functional expression of the acetyl coenzyme A synthetase gene from *Methanothrix soehngenii* in *Escherichia coli. J. Bacteriol.* 173:6383–6389.

Eggen, R. I. L., A. C. M. Geerling, M. S. M. Jetten, and W. M. Devos. 1991b. Cloning, expression, and sequence analysis of the genes for carbon monoxide dehydrogenase of *Methanothrix soehngenii*. *J. Biol. Chem.* **266:**6883–6887.

Eggen, R. I. L., R. Vankranenburg, A. J. M. Vriesema, A. C. M. Geerling, M. F. J. M. Verhagen, W. R. Hagen, and W. M. Devos. 1996. Carbon monoxide dehydrogenase from *Methanosarcina frisia* Go1. Characterization of the enzyme and the regulated expression of two operon-like *cdh* gene clusters. *J. Biol. Chem.* **271:** 14256–14263.

Elberson, M. A., and K. R. Sowers. 1997. Isolation of an aceticlastic strain of *Methanosarcina siciliae* from marine canyon sediments and emendation of the species description for *Methanosarcina siciliae*. *Int. J. Syst. Bacteriol.* **47:**1258–1261.

Ermler, U., W. Grabarse, S. Shima, M. Goubeaud, and R. K. Thauer. 1997. Crystal structure of methyl-coenzyme M reductase: the key enzyme of biological methane formation. *Science* **278:**1457–1462.

Ferry, J. G. 2003. One-carbon metabolism in methanogenic anaerobes, p. 143–156. *In* L. G. Ljungdahl, M. W. Adams, L. L. Barton, J. G. Ferry, and M. K. Johnson (ed.), *Biochemistry and Physiology of Anaerobic Bacteria*. Springer-Verlag, New York, NY.

Ferry, J. G., and C. H. House. 2006. The stepwise evolution of early life driven by energy conservation. *Mol. Biol. Evol.* **23:**1286–1292.

Ferry, J. G., and K. Kastead. 2007. Methanogenesis, p. 288–314. *In* R. Cavicchioli (ed.), *Archaea: Molecular Cell Biology*. ASM Press, Washington, DC.

Finazzo, C., J. Harmer, C. Bauer, B. Jaun, E. C. Duin, F. Mahlert, M. Goenrich, R. K. Thauer, S. Van Doorslaer, and A. Schweiger. 2003. Coenzyme B induced coordination of coenzyme M via its thiol group to Ni(I) of F_{430} in active methyl-coenzyme M reductase. *J. Am. Chem. Soc.* **125:**4988–4989.

Fischer, R., and R. K. Thauer. 1989. Methyltetrahydromethanopterin as an intermediate in methanogenesis from acetate in *Methanosarcina barkeri*. *Arch. Microbiol.* **151:**459–465.

Fischer, R., and R. K. Thauer. 1990. Ferredoxin-dependent methane formation from acetate in cell extracts of *Methanosarcina barkeri* (strain MS). *FEBS Lett.* **269:**368–372.

Galagan, J. E., C. Nusbaum, A. Roy, M. G. Endrizzi, P. Macdonald, W. FitzHugh, S. Calvo, R. Engels, S. Smirnov, D. Atnoor, A. Brown, N. Allen, J. Naylor, N. Stange-Thomann, K. DeArellano, R. Johnson, L. Linton, P. McEwan, K. McKernan, J. Talamas, A. Tirrell, W. Ye, A. Zimmer, R. D. Barber, I. Cann, D. E. Graham, D. A. Grahame, A. M. Guss, R. Hedderich, C. Ingram-Smith, H. C. Kuettner, J. A. Krzycki, J. A. Leigh, W. Li, J. Liu, B. Mukhopadhyay, J. N. Reeve, K. Smith, T. A. Springer, L. A. Umayam, O. White, R. H. White, E. C. de Macario, J. G. Ferry, K. F. Jarrell, H. Jing, A. J. Macario, I. Paulsen, M. Pritchett, K. R. Sowers, R. V. Swanson, S. H. Zinder, E. Lander, W. W. Metcalf, and B. Birren. 2002. The genome of *M. acetivorans* reveals extensive metabolic and physiological diversity. *Genome Res.* **12:**532–542.

Gartner, P., D. S. Weiss, U. Harms, and R. K. Thauer. 1994. N^5-methyltetrahydromethanopterin:coenzyme M methyltransferase from *Methanobacterium thermoautotrophicum*. Catalytic mechanism and sodium ion dependence. *Eur. J. Biochem.* **226:**465–472.

Goenrich, M., E. C. Duin, F. Mahlert, and R. K. Thauer. 2005. Temperature dependence of methyl-coenzyme M reductase activity and of the formation of the methyl-coenzyme M reductase red2 state induced by coenzyme B. *J. Biol. Inorg. Chem.* **10:**333–342.

Goenrich, M., F. Mahlert, E. C. Duin, C. Bauer, B. Jaun, and R. K. Thauer. 2004. Probing the reactivity of Ni in the active site of methyl-coenzyme M reductase with substrate analogues. *J. Biol. Inorg. Chem.* **9:**691–705.

Gorrell, A., S. H. Lawrence, and J. G. Ferry. 2005. Structural and kinetic analyses of arginine residues in the active-site of the acetate kinase from *Methanosarcina thermophila*. *J. Biol. Chem.* **280:** 10731–10742.

Gottschalk, G., and R. K. Thauer. 2001. The Na^+ translocating methyltransferase complex from methanogenic archaea. *Biochim. Biophys. Acta.* **1505:**28–36.

Grabarse, W., F. Mahlert, E. C. Duin, M. Goubeaud, S. Shima, R. K. Thauer, V. Lamzin, and U. Ermler. 2001. On the mechanism of biological methane formation: structural evidence for conformational changes in methyl-coenzyme M reductase upon substrate binding. *J. Mol. Biol.* **309:**315–330.

Grahame, D. A. 2003. Acetate C-C bond formation and decomposition in the anaerobic world: the structure of a central enzyme and its key active-site metal cluster. *Trends Biochem. Sci.* **28:**221–224.

Grahame, D. A. 1991. Catalysis of acetyl-CoA cleavage and tetrahydrosarcinapterin methylation by a carbon monoxide dehydrogenase-corrinoid enzyme complex. *J. Biol. Chem.* **266:**22227–22233.

Grahame, D. A. 1993. Substrate and cofactor reactivity of a carbon monoxide dehydrogenase corrinoid enzyme complex. Stepwise reduction of iron sulfur and corrinoid centers, the corrinoid $Co^{2+/1+}$ redox midpoint potential, and overall synthesis of acetyl-CoA. *Biochemistry* **32:**10786–10793.

Grahame, D. A., and E. Demoll. 1996. Partial reactions catalyzed by protein components of the acetyl-CoA decarbonylase synthase enzyme complex from *Methanosarcina barkeri*. *J. Biol. Chem.* **271:** 8352–8358.

Grahame, D. A., and E. Demoll. 1995. Substrate and accessory protein requirements and thermodynamics of acetyl-CoA synthesis and cleavage in *Methanosarcina barkeri*. *Biochemistry* **34:**4617–4624.

Grahame, D. A., and S. Gencic. 2000. Methane biochemistry, p. 188–198. *In Encyclopedia of Microbiology*, 2nd ed, vol. 3. Academic Press, New York, NY.

Harms, U., D. S. Weiss, P. Gartner, D. Linder, and R. K. Thauer. 1995. The energy conserving N^5-methyltetrahydromethanopterin: coenzyme M methyltransferase complex from *Methanobacterium thermoautotrophicum* is composed of eight different subunits. *Eur. J. Biochem.* **228:**640–648.

Hedderich, R. 2004. Energy-converting [NiFe] hydrogenases from *Archaea* and extremophiles: ancestors of complex I. *J. Bioenerg. Biomembr.* **36:**65–75.

Hedderich, R., A. Berkessel, and R. K. Thauer. 1990. Purification and properties of heterodisulfide reductase from *Methanobacterium thermoautotrophicum* (strain Marburg). *Eur. J. Biochem.* **193:**255–261.

Hedderich, R., N. Hamann, and M. Bennati. 2005. Heterodisulfide reductase from methanogenic archaea: a new catalytic role for an iron-sulfur cluster. *Biol. Chem.* **386:**961–970.

Hedderich, R., J. Koch, D. Linder, and R. K. Thauer. 1994. The heterodisulfide reductase from *Methanobacterium thermoautotrophicum* contains sequence motifs characteristic of pyridine-nucleotide-dependent thioredoxin reductases. *Eur. J. Biochem.* **225:**253–261.

Heiden, S., R. Hedderich, E. Setzke, and R. K. Thauer. 1993. Purification of a cytochrome-*b* containing H_2-heterodisulfide oxidoreductase complex from membranes of *Methanosarcina barkeri*. *Eur. J. Biochem.* **213:**529–535.

Hovey, R., S. Lentes, A. Ehrenreich, K. Salmon, K. Saba, G. Gottschalk, R. P. Gunsalus, and U. Deppenmeier. 2005. DNA microarray analysis of *Methanosarcina mazei* Go1 reveals adaptation to different methanogenic substrates. *Mol. Genet. Genomics* **273:** 225–239.

Ingram-Smith, C., R. D. Barber, and J. G. Ferry. 2000. The role of histidines in the acetate kinase from *Methanosarcina thermophila*. *J. Biol. Chem.* **275:**33765–33770.

Ingram-Smith, C., A. Gorrell, S. H. Lawrence, P. Iyer, K. Smith, and J. G. Ferry. 2005. Identification of the acetate binding site in the *Methanosarcina thermophila* acetate kinase. *J. Bacteriol.* **187:** 2386–2394.

Iyer, P. P., and J. G. Ferry. 2001. Role of arginines in coenzyme A binding and catalysis by the phosphotransacetylase from *Methanosarcina thermophila*. *J. Bacteriol.* 183:4244–4250.

Iyer, P. P., S. H. Lawrence, K. B. Luther, K. R. Rajashankar, H. P. Yennawar, J. G. Ferry, and H. Schindelin. 2004. Crystal structure of phosphotransacetylase from the methanogenic archaeon *Methanosarcina thermophila*. *Structure* 12:559–567.

Jablonski, P. E., W. P. Lu, S. W. Ragsdale, and J. G. Ferry. 1993. Characterization of the metal centers of the corrinoid/iron-sulfur component of the CO dehydrogenase enzyme complex from *Methanosarcina thermophila* by EPR spectroscopy and spectroelectrochemistry. *J. Biol. Chem.* 268:325–329.

Jetten, M. S. M., A. J. M. Stams, and A. J. B. Zehnder. 1989. Purification and characterization of an oxygen-stable carbon monoxide dehydrogenase of *Methanothrix soehngenii*. *FEBS Lett.* 181:437–441.

Jetten, M. S. M., W. R. Hagen, A. J. Pierik, A. J. M. Stams, and A. J. B. Zehnder. 1991a. Paramagnetic centers and acetyl-coenzyme A/CO exchange activity of carbon monoxide dehydrogenase from *Methanothrix soehngenii*. *Eur. J. Biochem.* 195:385–391.

Jetten, M. S. M., A. J. Pierik, and W. R. Hagen. 1991b. EPR characterization of a high-spin system in carbon monoxide dehydrogenase from *Methanothrix soehngenii*. *Eur. J. Biochem.* 202:1291–1297.

Jetten, M. S. M., A. J. M. Stams, and A. J. B. Zehnder. 1990. Acetate threshold values and acetate activating enzymes in methanogenic bacteria. *FEMS Microbiol. Ecol.* 73:339–344.

Jetten, M. S. M., A. J. M. Stams, and A. J. B. Zehnder. 1992. Methanogenesis from acetate. A comparison of the acetate metabolism in *Methanothrix soehngenii* and *Methanosarcina* spp. *FEMS Microbiol. Rev.* 88:181–198.

Kamagata, Y., and E. Mikami. 1991. Isolation and characterization of a novel thermophilic *Methanosaeta* strain. *Int. J. Syst. Bacteriol.* 41:191–196.

Kisker, C., H. Schindelin, B. E. Alber, J. G. Ferry, and D. C. Rees. 1996. A left-handed beta-helix revealed by the crystal structure of a carbonic anhydrase from the archaeon *Methanosarcina thermophila*. *EMBO J.* 15:2323–2330.

Kocsis, E., M. Kessel, E. DeMoll, and D. A. Grahame. 1999. Structure of the Ni/Fe-S protein subcomponent of the acetyl-CoA decarbonylase/synthase complex from *Methanosarcina thermophila* at 26-A resolution. *J. Struct. Biol.* 128:165–174.

Kruger, M., A. Meyerdierks, F. O. Glockner, R. Amann, F. Widdel, M. Kube, R. Reinhardt, J. Kahnt, R. Bocher, R. K. Thauer, and S. Shima. 2003. A conspicuous nickel protein in microbial mats that oxidize methane anaerobically. *Nature* 426:878–881.

Kunkel, A., M. Vaupel, S. Heim, R. K. Thauer, and R. Hedderich. 1997. Heterodisulfide reductase from methanol-grown cells of *Methanosarcina barkeri* is not a flavoenzyme. *Eur. J. Biochem.* 244: 226–234.

Latimer, M. T., and J. G. Ferry. 1993. Cloning, sequence analysis, and hyperexpression of the genes encoding phosphotransacetylase and acetate kinase from *Methanosarcina thermophila*. *J. Bacteriol.* 175:6822–6829.

Latimer, M. T., M. H. Painter, and J. G. Ferry. 1996. Characterization of an iron-sulfur flavoprotein from *Methanosarcina thermophila*. *J. Biol. Chem.* 271:24023–24028.

Lawrence, S. H., and J. G. Ferry. 2006. Steady-state kinetic analysis of phosphotransacetylase from *Methanosarcina thermophila*. *J. Bacteriol.* 188:1155–1158.

Lawrence, S. H., K. B. Luther, H. Schindelin, and J. G. Ferry. 2006. Structural and functional studies suggest a catalytic mechanism for the phosphotransacetylase from *Methanosarcina thermophila*. *J. Bacteriol.* 188:1143–1154.

Lessner, D. J., L. Li, Q. Li, T. Rejtar, V. P. Andreev, M. Reichlen, K. Hill, J. J. Moran, B. L. Karger, and J. G. Ferry. 2006. An unconventional pathway for reduction of CO_2 to methane in CO-grown *Methanosarcina acetivorans* revealed by proteomics. *Proc. Natl. Acad. Sci. USA* 103:17921–17926.

Lewalter, K., and V. Muller. 2006. Bioenergetics of archaea: ancient energy conserving mechanisms developed in the early history of life. *Biochim. Biophys. Acta* 1757:437–445.

Li, L., Q. Li, L. Rohlin, U. Kim, K. Salmon, T. Rejtar, R. P. Gunsalus, B. L. Karger, and J. G. Ferry. 2007. Quantitative proteomic and microarray analysis of the archaeon *Methanosarcina acetivorans* grown with acetate versus methanol. *J. Proteome Res.* 6:759–771.

Li, Q., L. Li, T. Rejtar, B. L. Karger, and J. G. Ferry. 2005. The proteome of *Methanosarcina acetivorans*. Part II. Comparison of protein levels in acetate- and methanol-grown cells. *J. Proteome Res.* 4:129–136.

Li, Q., L. Li, T. Rejtar, D. J. Lessner, B. L. Karger, and J. G. Ferry. 2006. Electron transport in the pathway of acetate conversion to methane in the marine archaeon *Methanosarcina acetivorans*. *J. Bacteriol.* 188:702–710.

Lindahl, P. A., and B. Chang. 2001. The evolution of acetyl-CoA synthase. *Orig. Life Evol. Biosph.* 31:403–434.

Liu, J. S., I. W. Marison, and U. von Stockar. 2001. Microbial growth by a net heat up-take: a calorimetric and thermodynamic study on acetotrophic methanogenesis by *Methanosarcina barkeri*. *Biotechnol. Bioeng.* 75:170–180.

Lu, W. P., P. E. Jablonski, M. Rasche, J. G. Ferry, and S. W. Ragsdale. 1994. Characterization of the metal centers of the Ni-Fe-S component of the carbon-monoxide dehydrogenase enzyme complex from *Methanosarcina thermophila*. *J. Biol. Chem.* 269:9736–9742.

Ma, K., X. Liu, and X. Dong. 2006. *Methanosaeta harundinacea* sp. nov., a novel acetate-scavenging methanogen isolated from a UASB reactor. *Int. J. Syst. Evol. Microbiol.* 56:127–131.

Maestrojuan, G. M., and D. R. Boone. 1991. Characterization of *Methanosarcina barkeri* MST and 227, *Methanosarcina mazei* S-6T, and *Methanosarcina vacuolata* Z-761T. *Int. J. Syst. Bacteriol.* 41:267–274.

Mah, R. A., and D. A. Kuhn. 1984. Transfer of the type species of the genus *Methanococcus* to the genus *Methanosarcina*, naming it *Methanosarcina mazei* (Barker, 1936) comb. nov. et emend. and conservation of the genus *Methanococcus* (Approved Lists 1980) with *Methanococcus vannielii* (Approved Lists 1980) as the type species. *Int. J. Syst. Bacteriol.* 34:263–265.

Mah, R. A., M. R. Smith, and L. Baresi. 1978. Studies on an acetate-fermenting strain of *Methanosarcina*. *Appl. Environ. Microbiol.* 35:1174–1184.

Martin, W., and M. J. Russell. 2007. On the origin of biochemistry at an alkaline hydrothermal vent. *Philos. Trans. R. Soc. B* doi: 10.1098/rstb.2002.1183.

Metcalf, W. W., J. K. Zhang, E. Apolinario, K. R. Sowers, and R. S. Wolfe. 1997. A genetic system for *Archaea* of the genus *Methanosarcina*. Liposome-mediated transformation and construction of shuttle vectors. *Proc. Natl. Acad. Sci. USA* 94:2626–2631.

Meuer, J., H. C. Kuettner, J. K. Zhang, R. Hedderich, and W. W. Metcalf. 2002. Genetic analysis of the archaeon *Methanosarcina barkeri* Fusaro reveals a central role for Ech hydrogenase and ferredoxin in methanogenesis and carbon fixation. *Proc. Natl. Acad. Sci. USA* 99:5632–5637.

Miles, R. D., A. Gorrell, and J. G. Ferry. 2002. Evidence for a transition state analog, MgADP-aluminum fluoride-acetate, in acetate kinase from *Methanosarcina thermophila*. *J. Biol. Chem.* 277: 22547–22552.

Miles, R. D., P. P. Iyer, and J. G. Ferry. 2001. Site-directed mutational analysis of active site residues in the acetate kinase from *Methanosarcina thermophila*. *J. Biol. Chem.* 276:45059–45064.

Min, H., and S. H. Zinder. 1989. Kinetics of acetate utilization by two thermophilic acetotrophic methanogens: *Methanosarcina* sp.

strain CALS-1 and *Methanothrix* sp. strain CALS-1. *Appl. Environ. Microbiol.* 55:488–491.

Mizukami, S., K. Takeda, S. Akada, and T. Fujita. 2006. Isolation and characteristics of *Methanosaeta* in paddy field soils. *Biosci. Biotechnol. Biochem.* 70:828–835.

Moller-Zinkhan, D., G. Borner, and R. K. Thauer. 1989. Function of methanofuran, tetrahydromethanopterin, and coenzyme-F_{420} in *Archaeoglobus fulgidus*. *Arch. Microbiol.* 152:362–368.

Muller, V. 2004. An exceptional variability in the motor of Archaeal A_1A_0 ATPases: from multimeric to monomeric rotors comprising 6–13 ion binding sites. *J. Bioenerg. Biomembr.* 36:115–125.

Murakami, E., U. Deppenmeier, and S. W. Ragsdale. 2001. Characterization of the intramolecular electron transfer pathway from 2-hydroxyphenazine to the heterodisulfide reductase from *Methanosarcina thermophila*. *J. Biol. Chem.* 276:2432–2439.

Murakami, E., and S. W. Ragsdale. 2000. Evidence for intersubunit communication during acetyl-CoA cleavage by the multienzyme CO dehydrogenase/acetyl-CoA synthase complex from *Methanosarcina thermophila*. Evidence that the beta subunit catalyzes C-C and C-S bond cleavage. *J. Biol. Chem.* 275:4699–4707.

Murray, P. A., and S. H. Zinder. 1985. Nutritional requirements of *Methanosarcina* sp. strain TM-1. *Appl. Environ. Microbiol.* 50: 49–55.

Nelson, M. J. K., and J. G. Ferry. 1984. Carbon monoxide-dependent methyl coenzyme M methylreductase in acetotrophic *Methanosarcina* spp. *J. Bacteriol.* 160:526–532.

Patel, G. B. 1984. Characterization and nutritional properties of *Methanothrix concilii* sp. nov., a mesophilic, aceticlastic methanogen. *Can. J. Microbiol.* 30:1383–1396.

Patel, G. B., and G. D. Sprott. 1990. *Methanosaeta concilii* gen. nov., sp. nov. ("*Methanothrix concilii*") and *Methanosaeta thermoacetophila* nom. rev., comb. nov. *Int. J. Syst. Bacteriol.* 40:79–82.

Peinemann, S., V. Muller, M. Blaut, and G. Gottschalk. 1988. Bioenergetics of methanogenesis from acetate by *Methanosarcina barkeri*. *J. Bacteriol.* 170:1369–1372.

Ragsdale, S. W. 2004. Life with carbon monoxide. *Crit. Rev. Biochem. Mol. Biol.* 39:165–195.

Rasche, M. E., K. S. Smith, and J. G. Ferry. 1997. Identification of cysteine and arginine residues essential for the phosphotransacetylase from *Methanosarcina thermophila*. *J. Bacteriol.* 179:7712–7717.

Raybuck, S. A., S. E. Ramer, D. R. Abbanat, J. W. Peters, W. H. Orme-Johnson, J. G. Ferry, and C. T. Walsh. 1991. Demonstration of carbon-carbon bond cleavage of acetyl coenzyme A by using isotopic exchange catalyzed by the CO dehydrogenase complex from acetate-grown *Methanosarcina thermophila*. *J. Bacteriol.* 173:929–932.

Sachs, G., J. A. Kraut, Y. Wen, J. Feng, and D. R. Scott. 2006. Urea transport in bacteria: acid acclimation by gastric *Helicobacter* spp. *J. Membr. Biol.* 212:71–82.

Saeki, K., and H. Kumagai. 1998. The *rnf* gene products in *Rhodobacter capsulatus* play an essential role in nitrogen fixation during anaerobic DMSO-dependent growth in the dark. *Arch. Microbiol.* 169:464–467.

Sauer, K., and R. K. Thauer. 1998. His(84) rather than His(35) is the active site histidine in the corrinoid protein MtrA of the energy conserving methyltransferase complex from *Methanobacterium thermoautotrophicum*. *FEBS Lett.* 436:401–402.

Scherer, P., V. Hollriegel, C. Krug, M. Bokel, and P. Renz. 1984. On the biosynthesis of 5-hydroxybenzimidazolylcobamide (vitamin B_{12}-factor III) in *Methanosarcina barkeri*. *Arch. Microbiol.* 138: 354–359.

Schmehl, M., A. Jahn, A. Meyer zu Vilsendorf, S. Hennecke, B. Masepohl, M. Schuppler, M. Marxer, J. Oelze, and W. Klipp. 1993. Identification of a new class of nitrogen fixation genes in *Rhodobacter capsulatus*: a putative membrane complex involved in electron transport to nitrogenase. *Mol. Gen. Genet.* 241:602–615.

Shima, S., E. Warkentin, R. K. Thauer, and U. Ermler. 2002. Structure and function of enzymes involved in the methanogenic pathway utilizing carbon dioxide and molecular hydrogen. *J. Biosci. Bioeng.* 93:519–530.

Simianu, M., E. Murakami, J. M. Brewer, and S. W. Ragsdale. 1998. Purification and properties of the heme- and iron-sulfur-containing heterodisulfide reductase from *Methanosarcina thermophila*. *Biochemistry* 37:10027–10039.

Singh, N., M. M. Kendall, Y. Liu, and D. R. Boone. 2005. Isolation and characterization of methylotrophic methanogens from anoxic marine sediments in Skan Bay, Alaska: description of *Methanococcoides alaskense* sp. nov., and emended description of *Methanosarcina baltica*. *Int. J. Syst. Evol. Microbiol.* 55:2531–2538.

Singh-Wissmann, K., and J. G. Ferry. 1995. Transcriptional regulation of the phosphotransacetylase-encoding and acetate kinase-encoding genes (*pta* and *ack*) from *Methanosarcina thermophila*. *J. Bacteriol.* 177:1699–1702.

Singh-Wissmann, K., C. Ingram-Smith, R. D. Miles, and J. G. Ferry. 1998. Identification of essential glutamates in the acetate kinase from *Methanosarcina thermophila*. *J. Bacteriol.* 180:1129–1134.

Singh-Wissmann, K., R. D. Miles, C. Ingram-Smith, and J. G. Ferry. 2000. Identification of essential arginines in the acetate kinase from *Methanosarcina thermophila*. *Biochemistry* 39:3671–3677.

Smith, K. S., and C. Ingram-Smith. 2007. *Methanosaeta*, the forgotten methanogen? *Trends Microbiol.* 7:150–155.

Smith, K. S., C. Jakubzick, T. C. Whittam, and J. G. Ferry. 1999. Carbonic anhydrase is an ancient enzyme widespread in prokaryotes. *Proc. Natl. Acad. Sci. USA* 96:15184–15189.

Sowers, K. R., J. E. Boone, and R. P. Gunsalus. 1993. Disaggregation of *Methanosarcina* spp and growth as single cells at elevated osmolarity. *Appl. Environ. Microbiol.* 59:3832–3839.

Sowers, K. R., S. F. Baron, and J. G. Ferry. 1984. *Methanosarcina acetivorans* sp. nov., an acetotrophic methane-producing bacterium isolated from marine sediments. *Appl. Environ. Microbiol.* 47:971–978.

Terlesky, K. C., and J. G. Ferry. 1988a. Ferredoxin requirement for electron transport from the carbon monoxide dehydrogenase complex to a membrane-bound hydrogenase in acetate-grown *Methanosarcina thermophila*. *J. Biol. Chem.* 263:4075–4079.

Terlesky, K. C., and J. G. Ferry. 1988b. Purification and characterization of a ferredoxin from acetate-grown *Methanosarcina thermophila*. *J. Biol. Chem.* 263:4080–4082.

Terlesky, K. C., M. J. K. Nelson, and J. G. Ferry. 1986. Isolation of an enzyme complex with carbon monoxide dehydrogenase activity containing a corrinoid and nickel from acetate-grown *Methanosarcina thermophila*. *J. Bacteriol.* 168:1053–1058.

Thauer, R. K. 1998. Biochemistry of methanogenesis: a tribute to Marjory Stephenson. *Microbiology* 144:2377–2406.

Tripp, B. C., C. B. Bell, F. Cruz, C. Krebs, and J. G. Ferry. 2004. A role for iron in an ancient carbonic anhydrase. *J. Biol. Chem.* 279:6683–6687.

van Beelen, P., J. F. Labro, J. T. Keltjens, W. J. Geerts, G. D. Vogels, W. H. Laarhoven, W. Guijt, and C. A. Haasnoot. 1984. Derivatives of methanopterin, a coenzyme involved in methanogenesis. *Eur. J. Biochem.* 139:359–365.

von Klein, D., H. Arab, H. Volker, and M. Thomm. 2002. *Methanosarcina baltica*, sp. nov., a novel methanogen isolated from the Gotland Deep of the Baltic Sea. *Extremophiles* 6:103–110.

Vorholt, J. A., L. Chistoserdova, S. M. Stolyar, R. K. Thauer, and M. E. Lidstrom. 1999. Distribution of tetrahydromethanopterin-dependent enzymes in methylotrophic bacteria and phylogeny of methenyl tetrahydromethanopterin cyclohydrolases. *J. Bacteriol.* 181:5750–5757.

Weiss, D. S., P. Gartner, and R. K. Thauer. 1994. The energetics and sodium-ion dependence of N^5-methyltetrahydromethanopterin: coenzyme M methyltransferase studied with cob(I)alamin as methyl acceptor and methylcob(III) alamin as methyl donor. *Eur. J. Biochem.* 226:799–809.

Westermann, P., B. K. Ahring, and R. A. Mah. 1989. Threshold acetate concentrations for acetate catabolism by aceticlastic methanogenic bacteria. *Appl. Environ. Microbiol.* 55:514–515.

Whitman, W. B., D. C. Coleman, and W. J. Wiebe. 1998. Prokaryotes: the unseen majority. *Proc. Natl. Acad. Sci. USA* 95:6578–6583.

Zhilina, T. N., and G. A. Zavarzin. 1979. Comparative cytology of *Methanosarcinae* and description of *Methanosarcina vacuolata* sp. nova. *Microbiology* 48:223–228.

Zhilina, T. N., and G. A. Zavarzin. 1987. *Methanosarcina vacuolata* sp. nov., a vacuolated *Methanosarcina*. *Int. J. Syst. Bacteriol.* 37: 281–283.

Zimmerman, S. A., and J. G. Ferry. 2006. Proposal for a hydrogen bond network in the active site of the prototypic gamma-class carbonic anhydrase. *Biochemistry* 45:5149–5157.

Zinder, S. H., and R. A. Mah. 1979. Isolation and characterization of a thermophilic strain of *Methanosarcina* unable to use H_2-CO_2 for methanogenesis. *Appl. Environ. Microbiol.* 38:996–1008.

Zinder, S. H., K. R. Sowers, and J. G. Ferry. 1985. *Methanosarcina thermophila* sp. nov., a thermophilic, acetotrophic, methane-producing bacterium. *Int. J. Syst. Bacteriol.* 35:522–523.

Bioenergy
Edited by J. Wall et al.

Chapter 14

Energetic Aspects of Methanogenic Feeding Webs

BERNHARD SCHINK

The degradation of organic matter by methanogenesis is the most complex and the most efficient way of transforming organic matter into an energetically useful product. It is the dominant process in every anoxic environment where nitrate, sulfate, oxidized iron, or manganese species have been reduced, and where only CO_2 is available as a terminal electron acceptor. Moreover, methanogenesis is the process of organic matter degradation that most likely evolved first in the Earth's biosphere and therefore had the longest time for evolutionary optimization. The complex interplay of different metabolic types (guilds) of microorganisms in the overall process represents an evolutionary strategy quite different from that of aerobic degradation. Whereas aerobic organisms typically oxidize complex organic substrates including polymers in one single cell all the way down to CO_2 and inorganic residues with oxygen as electron acceptor, anaerobic degradative processes depend on cooperation between different guilds of microbes, and the complexity increases from nitrate-dependent over iron- or sulfate-dependent oxidation (typically a two-step process) to methanogenesis, in which at least three types of organisms have to cooperate for efficient substrate conversion.

The technological advantage of methanogenic degradation of biomass is obvious: a complex methanogenic community can degrade basically all types of major constituents of biomass, including polysaccharides, proteins, lipids, and nucleic acids finally to methane plus CO_2, with few exceptions. The formed product, a mixture of methane, CO_2, and traces of hydrogen sulfide, is gaseous and separates easily from the fermentation broth, without specific efforts involving distillation or transfer across membranes, etc. The advantage of methanogenesis over other fuel-producing processes becomes obvious from a comparison of energetics (all calculations are from tables published in Thauer et al., 1977):

Aerobic oxidation:

$$C_6H_{12}O_6 + 6\ O_2 \rightarrow 6\ CO_2 + 6\ H_2O \quad \Delta G^{0'} = -2{,}870 \text{ kJ per mol of glucose} \quad (1)$$

Methanogenic fermentation:

$$C_6H_{12}O_6 + 14\ H_2O \rightarrow 3\ CH_4 + 3\ CO_2 \quad \Delta G^{0'} = -390 \text{ kJ per mol glucose} \quad (2)$$

As exemplified by these reactions, methanogenic conversion of biomass (using hexose as a representative) consumes only about 14% of the energy that is released in aerobic oxidation of sugar. The difference between the two is that 86% of the energy originally available in glucose oxidation is stored in the methane molecule and can be released later by burning this gas:

$$3\ CH_4 + 6\ O_2 \rightarrow 3\ CO_2 + 6\ H_2O \quad \Delta G^{0'} = -2{,}480 \text{ kJ per mol of glucose equivalent} \quad (3)$$

Of course, biomass and other waste materials do not consist only of sugars. With proteins and, even more so, with lipids the overall redox balance shifts towards more methane than CO_2 formation (reaction 4):

$$2\ C_{16}H_{32}O_2 + 14\ H_2O \rightarrow 23\ CH_4 + 9\ CO_2 \quad \Delta G^{0'} = -385 \text{ kJ per mol of palmitate} \quad (4)$$

Thus, highly reduced compounds such as lipids, oils, fats, etc., are even better energy carriers in methane production than sugars and other biomass constituents, and the aerobic oxidation requires even higher amounts of oxygen, i.e., aeration in a conventional sewage treatment plant and, with this, higher waste treatment costs:

$$C_{16}H_{32}O_2 + 23\ O_2 \rightarrow 16\ CO_2 + 16\ H_2O \quad \Delta G^{0'} = -9{,}800 \text{ kJ per mol of palmitate} \quad (5)$$

Bernhard Schink • Department of Biology, University of Konstanz, 78457 Konstanz, Germany.

It is for this reason that especially high-load organic wastewaters and semisolid waste materials have found their way recently preferentially into anaerobic, methanogenic treatment systems (Lettinga, 1995), and new treatment systems are being developed at present for treatment of average- and low-load municipal wastewaters (Aiyuk et al., 2006).

This chapter deals with the complex cooperations in methanogenic microbial communities under different treatment regimes and the technological perspectives in the optimization of energy recovery from biomass treatment in the present and in the future.

THE BASICS

Conversion of biomass to methane and carbon dioxide depends on a complex cooperation of numerous partner organisms of different metabolic capacities (Fig. 1). In this network, primary fermenting bacteria, often including strictly anaerobic protozoa, convert complex polymeric biomass (polysaccharides, proteins, nucleic acids, lipids, etc.) via their respective monomeric constituents (sugars and disaccharides, amino acids and oligopeptides, nucleosides, fatty acids, and glycerol) by fermentation to classical fermentation products such as H_2, CO_2, formate and other C_1 compounds, acetate, and a series of reduced fermentation products including fatty acids, alcohols, succinate or lactate, aromatic fatty acids, and branched-chain fatty acids. Since the last partners in the overall process, the methanogens, use only very few substrates, namely C_1 compounds, hydrogen, and acetate for methanogenesis, the majority of the classical fermentation products have to be further modified in so-called secondary fermentations to produce compounds that can be used by the classical methanogens.

The free energy available in the total methanogenic conversion of sugars (reaction 2) allows the synthesis of six ATP units at maximum, and the lion's share of this (up to four ATP units) is used up by the primary fermenting organisms, leaving only little energy for the partner organisms further down in the feeding chain.

The physiology and biochemistry of the secondary fermenting bacteria are of special interest because they have to live on an extremely small energy budget, catalyzing reactions that are endergonic under standard conditions and that are facilitated by close cooperation with the methanogenic partners which keep the con-

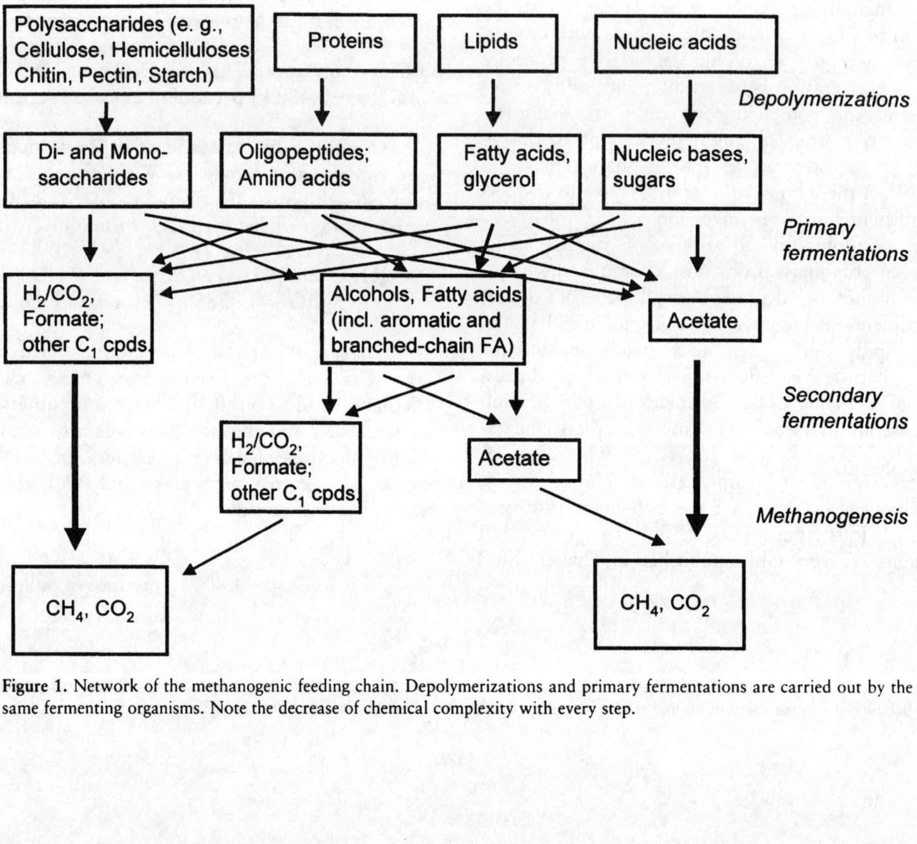

Figure 1. Network of the methanogenic feeding chain. Depolymerizations and primary fermentations are carried out by the same fermenting organisms. Note the decrease of chemical complexity with every step.

Table 1. Changes of Gibbs free energy for secondary fermentations and methane-forming reactions

Reaction	$\Delta G^{0'}$ (kJ/mol)	$\Delta G'$ (kJ/mol)[a]
$CH_3CH_2OH + H_2O \rightarrow CH_3COO^- + H^+ + 2\,H_2$	+9.6	−41
$CH_3CH_2CH_2COO^- + 2\,H_2O \rightarrow 2\,CH_3COO^- + H^+ + 2\,H_2$	+48	−26
$CH_3CH_2COO^- + 2\,H_2O \rightarrow CH_3COO^- + CO_2 + 3\,H_2$	+76	−11
$4\,H_2 + CO_2 \rightarrow CH_4 + 2\,H_2O$	−131	−21
$CH_3COO^- + H^+ \rightarrow CH_4 + CO_2$	−35	−13

[a]Under the following conditions: [acetate], 100 μM; [ethanol], [propionate], and [butyrate], 10 μM; CO_2, 1 atm; CH_4, 1 atm; H_2, 10^{-5} atm.

centrations of their reaction products (hydrogen, acetate, and formate) low. Some of these reactions are listed in Table 1, comparing energetics under standard conditions and under in situ conditions. It is obvious that these reactions can run and even provide energy for the catalyzing bacterium only under conditions of low product concentrations that are maintained by the methanogenic partners, and this specific type of symbiotic interrelationship has been termed "syntrophy." On average, these secondary fermentation reactions produce energy amounts in the range of −15 to −25 kJ per mol, which is at the lowermost limit of energy increments that are needed to synthesize fractions of an ATP unit at all. If we assume that −60 to −70 kJ is needed for synthesis of 1 mol of ATP under physiological conditions (Thauer et al., 1977), and if we assume, according to our present-day understanding of the membrane-bound F_1F_0-ATPase apparatus, that 3 to 4 protons cross the cytoplasmic membrane per ATP unit formed or released, the minimum amount of energy required for ATP synthesis is equivalent to one-third to one-fourth of one ATP unit, i.e., −15 to −20 kJ per mol (Schink, 1997; Schink and Stams, 2001). This is probably the minimum energy quantum needed for ATP synthesis in general, and there is no reliable report in the literature which documents that ATP is synthesized in growing cultures with smaller energy increments than this.

The mechanism by which syntrophically oxidizing bacteria organize their energy metabolism is only generally understood. The basic concept in those cases studied appears to be that one ATP unit per reaction run is synthesized by substrate-level phosphorylation, e.g., via phosphotransacetylase and acetate kinase in the release of an acetate residue. However, the bacteria cannot keep this full ATP unit but have to invest part of it into a reversed electron transport process in order to release electrons from unfavorable substrate oxidation steps at a comparably positive redox potential as molecular hydrogen (Fig. 2). Thus, part of this ATP unit is hydrolyzed again by a membrane-bound ATPase enzyme, thus providing a proton potential sufficient to drive a reversed electron flow towards hydrogen release, and only a fraction of an ATP unit remains for biosynthetic purposes. This basic concept (Thauer and Morris, 1984) was demonstrated first with suspensions of intact cells of *Syntrophomonas wolfei* in butyrate oxidation (Wallrabenstein and Schink, 1994) and has been confirmed in more detail with the glycolate-oxidizing bacterium *Syntrophobotulus glycolicus*. In the latter case, a tight coupling between glycolate oxidation, hydrogen release, and ATP hydrolysis could be shown convincingly with membrane vesicles (Friedrich and Schink, 1993). The reaction could even be reverted, as demonstrated through hydrogen- and glyoxylate-dependent ATP synthesis with the same membrane vesicle preparations. In both cases, the coupling between the two steps was alleviated by protonophores (Friedrich and Schink, 1995), indicating that they are coupled via a proton gradient system. At least in propionate oxidation, also formate appears to play an important role as an electron carrier (de Bok et al., 2002).

The narrow range of energy available in the respective catalytic steps in the methanogenic feeding chain (not "food chain," because the bacteria do not eat each other) also explains why the concentrations of the

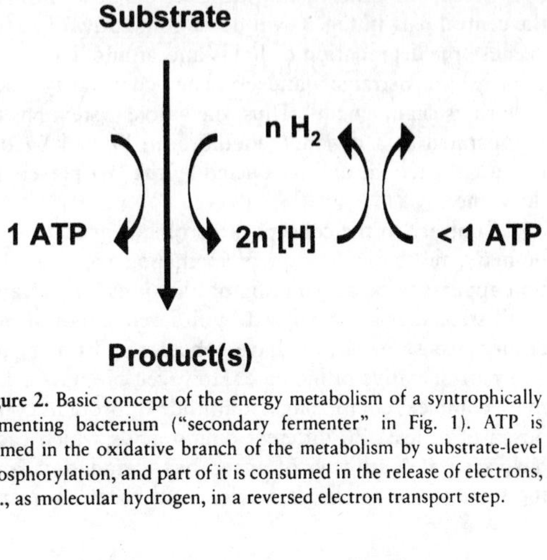

Figure 2. Basic concept of the energy metabolism of a syntrophically fermenting bacterium ("secondary fermenter" in Fig. 1). ATP is formed in the oxidative branch of the metabolism by substrate-level phosphorylation, and part of it is consumed in the release of electrons, e.g., as molecular hydrogen, in a reversed electron transport step.

fermentation intermediates can change only within narrow ranges, i.e., 1 order of magnitude at maximum, because otherwise the energy budget for one of the individual partners in this feeding chain will become too low for ATP synthesis and, with that, for growth. This energetic limitation is also the main reason for the limited flexibility of methanogenic biomass degradation. Low energy yields in the secondary fermentations produce low amounts of biomass, thus causing slow growth and, with this, slow adaptation to environmental changes. These factors dictate that the elasticity of a methanogenic bioreactor is limited and that major fluctuations in feed composition are not always easy to balance. Although the rate-limiting steps in the overall processes are the depolymerization reactions at the beginning of the reaction chain, the terminal methanogenic reactions are the ones that determine the overall efficiency of electron flow through the inner versus the outer frame of the scheme in Fig. 1, and the methanogens are the "conductors" in the process (Zeikus, 1977). As long as hydrogen- and acetate-utilizing methanogens operate at high efficiency, the carbon and electron flow proceeds mainly directly towards acetate, CO_2, and hydrogen, channeling only a small fraction of carbon and electrons to run through the central part (long-chain fatty acids etc.), which requires the energetically difficult secondary fermentations for complete degradation. If the hydrogen-utilizing methanogens are inhibited, for example, by increased acidity (pH <6.5), the product pattern of the primary fermenting organisms shifts to enhanced formation of long-chain fatty acids, and the further syntrophic oxidation of fatty acids is inhibited by enhanced hydrogen accumulation. This in turn leads to further accumulation of fatty acids, further acidification, and, finally, to entire reactor failure with little methane formation and accumulation of ill-smelling long-chain fatty acids ("reactor turnover"). On the other hand, the electron flow through the central part in Fig. 1 will never come down to zero because the degradation of lipids and amino acids necessarily forms straight- and branched-chain fatty acids of various chain lengths. Thus, the whole system has to be operated at a dynamic equilibrium, limited by the initiating reactions and conducted by the last players in the game.

Contrary to the concept of aerobic degradation of biomass, the basic concept of methanogenic degradation appears to be a resolution of functions into several small steps that are catalyzed by different types of microorganisms. So far, we have concentrated on sugars as a representative of biomass. However, biomass is far more complex, comprising a multitude of proteins composed of at least 20 different amino acids which each have separate pathways of anaerobic degradation, and the same is true for lipids, nucleic acid bases, aromatic residues, and many more (Fig. 1). It appears that the methanogenic degradation of all these different types of compounds is carried out by different microorganisms which cooperate in a complex network of separate modules that are performed by many different players. Obviously, the evolution of anaerobic degradative communities followed a strategy different from that of aerobic ones. The multitude of functions (catalytic steps and enzymes) needed for the degradation of complex biomass are not combined in one single organism, but in separate organisms which have to cooperate via metabolite transfer. The question arises as to why nature did not develop a methanogen that could grow on sugars or amino acids directly and, with this, combine a major part of the metabolic functions in one organism. Considerations on the efficiency of metabolic activities in long reaction chains suggest that separation into different organisms appears to be advantageous once too many reaction steps have to be lined up in a reaction sequence. Similar considerations were applied to the phenomenon that aerobic oxidation of ammonia to nitrate is always carried out by two separate bacteria acting cooperatively, never by one alone (Costa et al., 2006).

This modular arrangement of a complex cooperation network by partitioning functions into different organisms also requires that these organisms cooperate in a spatially optimized manner so that the diffusion distances for metabolites remain as short as possible. The implications of this strategy for the spatial arrangement of microbial communities have been discussed repeatedly (Schink, 1997), but no decisive answer has been found so far to the question of how optimally short diffusion distances are maintained in growing communities in which every single organism forms nests because it produces only offspring of its own.

ALTERNATIVES

Hydrogen

Energy-rich products other than methane can be produced fermentatively from biomass. Like methane, hydrogen is a gaseous product that easily separates from the fermenting broth. Nonetheless, fermentation of biomass to hydrogen has to face severe energetic limitations as exemplified here, again with glucose as a representative substrate:

$$C_6H_{12}O_6 + 6\ H_2O \rightarrow 12\ H_2 + 6\ CO_2$$
$$\Delta G^{0\prime} = -26\text{ kJ per mol of glucose} \quad (6)$$

Up to now there is no organism known that would live with this small energy yield, transforming 12 electron pairs to 12 H_2 molecules and still growing under these conditions. Instead, the sugar fermentation process with the highest hydrogen yield observed proceeds according to the following reaction:

$$C_6H_{12}O_6 + 2\,H_2O \rightarrow 2\,CH_3COO^- + 2\,H^+ + 2\,CO_2 + 4\,H_2 \quad \Delta G^{0\prime} = -216 \text{ kJ per mol of glucose} \quad (7)$$

Even this overall balance is reached only if the hydrogen partial pressure is kept low, e.g., by methanogenic partner organisms (see above). Under standard conditions, the free-energy change is not sufficient for the formation of four ATP units, which is intrinsically linked to this fermentation pattern through substrate-level phosphorylation in glycolysis (two ATP units) and acetate release in the acetate kinase reaction (a further two ATP units). The energetic situation improves at higher temperature, and thermophilic fermenting bacteria (at >70°C) may be better suited for catalyzing this process, especially since several thermophiles and hyperthermophiles ferment sugars through pathways different from glycolysis, with smaller ATP yields (Siebers and Schönheit, 2005). Indeed, enhanced hydrogen formation close to that described in equation 7 has so far been observed only with thermophiles (Claassen et al., 1999; Hallenbeck, 2005; Kanai et al., 2005).

Nonetheless, during this type of fermentation two-thirds of the overall electron content of the sugar molecule remains in the two acetate molecules, from which hydrogen can be released only at very low hydrogen pressure:

$$CH_3COO^- + H^+ + 2\,H_2O \rightarrow 2\,CO_2 + 4\,H_2 \quad \Delta G^{0\prime} = +96 \text{ kJ per mol} \quad (8)$$

This fermentation (inversion of homoacetogenesis!) becomes exergonic only at hydrogen pressures lower than 10^{-5} atm (Zinder and Koch, 1984) and cannot be applied efficiently for technical hydrogen production, therefore. It is the specific advantage of methanogenesis that it can transfer the electrons bound in acetate quantitatively to formation of methane, yielding sufficient energy for the respective bacteria involved for growth:

$$CH_3COO^- + H^+ \rightarrow CH_4 + CO_2 \quad \Delta G^{0\prime} = -35 \text{ kJ per mol} \quad (9)$$

Therefore, only one-third of the overall electron freight of sugar (as representative of biomass in general) is released as hydrogen under optimal conditions, and the rest stays with the acetate which cannot be used as a fuel itself.

Ethanol, Butanol, and Acetone

Contrary to methane and hydrogen, the energy-rich fermentation products ethanol, butanol, and acetone dissolve easily in an aqueous medium, and their recuperation from the medium always requires energy-intensive separation efforts such as distillation, stripping, or vacuum treatment. Fermentation of a sugar solution with yeasts or with the bacterium *Zymomonas mobilis* produces fermentation broths with alcohol contents in the range of 10 to 12% (vol/vol). For distillation of such a broth to reach an alcohol content of about 90%, a substantial amount of energy is needed, which is in the same range as that finally available in the produced ethanol. Butanol and acetone fermentations reach even lower concentrations in the fermentation broth (4 to 5%). Moreover, only a small part of biomass, especially monomeric sugars and di- and oligosaccharides, can be transformed to alcohols and acetone; starch fermentation already requires enzymatic pretreatment for this process. Other polysaccharides such as cellulose, hemicelluloses, proteins, and lipids are hardly ever accessible to alcoholic fermentation.

Obviously, the ethanol yield from specifically raised sugar-rich crop plants such as sugarcane, corn, or sugar beets is comparably high, but only minor parts of these plants are converted, and the remnants (leaves, stems, etc.) remain as waste products. Moreover, the efforts invested into the agricultural production of such energy plants have to be considered, including the unavoidable consequences of landscape devastation and soil degradation and erosion. For these reasons, production of energy-rich fermentation products should be confined primarily to the utilization of waste materials, including general household and industrial wastes and wastewaters, and methanogenesis appears to be the most efficient way of converting the majority of this waste biomass efficiently into an energy-rich product that separates easily from its broth origin.

LIMITATIONS AND SOLUTIONS

Reactor Performance

The enormous efficiency of the methanogenic feeding chain described above, which employs a complex network of many different metabolic types of microorganisms that perform their activities at a minimum gain of energy of about −20 kJ per mol of reaction—all this implicates certain limitations to methanogenic transformations in general. Compared to aerobic degradation processes, anaerobic transformations are usually considered to be slow. This is not necessarily true. The metabolic activity of single cells of strict anaerobes are not necessarily lower than those of aerobes if calculated per unit of cell protein; usually, we observe even for slow-growing anaerobes substrate transformation rates in the range of several hundred nanomoles per milligram of protein, which are similar to transformation rates for aerobes. However, the small energy gains, especially of some key groups in the feeding chain, imply that the cell yields for these organisms are low, i.e., that at similar substrate transformation rates only small cell yields can be obtained and that therefore the increase in biomass is slow. Thus, a methanogenic reactor system requires extensive start-up times for the establishment or reestablishment of a certain transformation process, and it will react only slowly to changes in the feed quality that require growth of new metabolic representatives (Schink, 1988). It is common practice in aerobic wastewater treatment systems employing the (aerobic!) "activated-sludge" process that the growing biomass is retained in the system through sludge recycling, thus maintaining a microbial community at about 10 times lower growth rates than the flow rate of the treated water (Schink, 1999). This sludge recycling uncouples growth of the active biomass in the activated sludge process from the wastewater flow through the system and allows also the maintenance of slow-growing microorganisms in the system, or of those which are needed only occasionally as a consequence of changes in the feed composition.

Of course, this necessity is even greater for an anaerobic waste treatment system due to the slow growth and low growth yields, especially of the secondary fermenting bacteria. Sludge recycling, analogous to the aerobic activated-sludge process, is not easy to apply in the strictly anaerobic world because the selection for easily settling sludge flocs in activated sludge depends on the presence of protozoa, especially ciliates, which graze preferentially on free-floating single cells and select for easily settling bacterial aggregates. Although strictly anaerobic flagellates and ciliates feeding on bacterial cells have been described, especially in nutrient-rich, permanently anoxic environments (Embley et al., 1995), there is no indication so far that such organisms can be enriched and maintained in a productive manner in technical waste treatment systems.

Strategies to increase the retention time of microbes in a strictly anaerobic fermentor system have started with providing surfaces on which the microbial communities can establish themselves and thus can easily be withheld within the reactor. Depending on the waste load of the water to be treated, either fixed-bed reactors with low strength or fluidized bed reactor systems with high strength have been designed for this purpose, with varying success. A real breakthrough in the treatment of high-strength wastewaters was the development of the Upflow Anaerobic Sludge Blanket (UASB) technology developed by G. Lettinga and his coworkers in Wageningen, The Netherlands, in the 1980s (Lettinga, 1995; Stams and Oude Elferink, 1997). This reactor type was developed primarily for the anaerobic treatment of high-load wastewaters, especially from the paper and fiber industry or from sugar manufacturing. Through a specific arrangement of separation blades in the upper part of the reactor, an efficient separation of gas bubbles from bacterial aggregates was obtained, and the aggregates grew to about 2 mm in diameter. These aggregates sedimented easily enough to be withheld and were retained in the reactor for sufficient times to maintain transformation efficiency and reactor performance stability. Although the reason for the development of these microbial aggregates remains obscure, the UASB technology has proven to be applicable to many different types of high-load wastewaters and has conquered the market in this field nearly worldwide. With this technology the major obstacle of anaerobic treatment processes, i.e., the slow growth of anaerobic microbial communities, could be overcome by maintenance of high densities of complex microbial communities for long periods. In a similar manner, methanogenic technologies can be applied for the treatment also of semisolid waste materials (Korner et al., 2003).

Besides methanogens, homoacetogenic bacteria and sulfate reducers can also cooperate in anaerobic fermentation, keeping the hydrogen partial pressure and the concentration of formate low and contributing to acetate formation and degradation. Obviously, sulfate-reducing bacteria depend on sulfate as an external electron acceptor which is usually available only at limiting amounts, in the range of a few hundred micromolars at maximum. In certain cases, especially in the degradation of marine waste material (algae, chitin shells of shrimp, etc.) the high sulfate content of seawater leads to enhanced sulfate reduction which outcompetes methanogens for their key substrates. Enhanced proton activity, enhanced sulfate content, and enhanced ammonium concentrations, e.g., in treatment of poultry waste, all can impede the activity of hydrogen-oxidizing methanogens and, with this, the overall operation of a methanogenic bioreactor.

One-Step versus Two-Step Technology

From the discussion above, it should appear obvious that the efficiency of methanogenic transformations of biomass depends intrinsically on a close cooperation between all partners involved and that the maintenance of low hydrogen and acetate concentrations by the methanogens also improves the overall efficiency of all members in the game, including the primary fermenting bacteria, which under these conditions produce only small amounts of reduced intermediates. Nonetheless, some strategies in anaerobic biomass treatment prefer to run the primary fermentation first and quickly into an acidic stage and feed the reaction products of the first step slowly into a second, methanogenic step in which these classical fermentation intermediates are then degraded by syntrophic cooperations as described above. At first sight, such a strategy does not look promising because the efficiency of the primary fermenting organisms is soon inhibited by the accumulation of acids and other fermentation products. Nonetheless, operation of the process in two steps appears to be advisable in certain specific cases, such as in treatment of the easily fermented sugar-rich wastes of the sugar production industry. This offal becomes available only a few weeks per year at high amounts, and its methanogenic treatment would require for short periods high capacities of biogas reactors which are not needed for the rest of the year. In this case, it appears to be advisable to stabilize the primary waste materials by an acidogenic step at the time when they become available and to further treat the prefermented material in a medium-scale methanogenic biomass system over a longer period of time subsequent to the main sugar production campaign.

Depolymerization of Biomass

Polysaccharides including starch, celluloses, hemicelluloses, pectin, chitin, and many others as well as proteins, lipids, and nucleic acids are depolymerized extracellularly by the action of hydrolytic enzymes. With few minor exceptions, such hydrolytic depolymerases operate in the presence or absence of oxygen, basically through identical chemical strategies, no matter whether aerobic or anaerobic bacteria are involved. The situation is basically different with lignin, a polymer of aromatic residues which is produced by higher plants specifically as a solid structure able to resist microbial degradation for many years. The specialists for lignin degradation are fungi, especially white-rot fungi, which employ molecular oxygen in an extracellular lignin depolymerization reaction in which hydroxyl radicals attack the phenolic polymer in a random manner, releasing mono- and oligoaromatic derivatives (Leonowicz et al., 2001; Kirby, 2006; Baldrian, 2006). So far, there is no reliable indication that lignin can be degraded in the absence of oxygen at any substantial rate (Hackett et al., 1977). Unfortunately, lignin not only is a major constituent of wood but also is found in grasses and other green plant parts, of which it constitutes 10 to 20% (dry weight) of mass. Since lignin is typically tightly connected with the cellulosic and hemicellulosic parts of the plant cell wall, there is no easy way to separate lignin efficiently from the nonligninaceous plant material. Wood-feeding termites found an efficient way to separate them and utilize the cellulosic part of wood while excreting the lignin residues, but this process also requires oxygen as a cosubstrate (Brune and Friedrich, 2000). Thus, lignin is the key impediment in the degradation of organic matter, especially of residues of higher plants. This serious limitation of anaerobic degradation capacities is exemplified by the preservation of wooden ships and other wooden constructions in anoxic sediments, dating back to the Middle Ages, Roman times, and even to the Neolithic, i.e., more than 4,000 years ago. Technical strategies to destabilize the intrinsic linkages between lignin and cellulose in wood such as steam explosion have been applied successfully in paper manufacturing, but it is probably not feasible to apply this technology to lignocellulosic waste materials. So far, there is no convincing solution to overcome this obvious limitation of methanogenic degradation of natural biomass.

Biochemical Challenges

Our knowledge of the biochemistry of the anaerobic breakdown of low-molecular-mass derivatives of biomass has grown dramatically in the last 20 years. While textbooks in the 1970s still maintained the view that aromatic compounds cannot be degraded in the absence of oxygen, this view has changed dramatically, and there is hardly any type of compound that cannot be degraded in the absence of oxygen, including single-ring aromatics, condensed aromatics (napththalene), aromatic and aliphatic hydrocarbons, straight- and branched-chain fatty acids, nitrogen- and oxygen-containing heterocycles, and many others. Halogen and nitro substituents can be removed reductively or hydrolytically in the absence of oxygen, and in some cases anaerobic, reductive approaches have proven to be more successful in the removal of such substituents than the strategies used by aerobic bacteria (Holliger and Schink, 1997). A whole new biochemistry has been discovered to be involved in the transformation of such comparably stable compounds, and contrary to earlier assumptions many of these reactions employ radical mechanisms (Spormann and Widdel, 2000; Boll et al., 2005; Buckel and Golding, 2006).

CONCLUSIONS

Methanogenic degradation of organic matter appears to be the superior process to convert the majority of organic waste materials to an energy fuel at high efficiency with respect to the overall fuel yields. In this respect, methanogenic degradation is far more efficient than other fermentative processes leading to hydrogen, ethanol, butanol, or acetone. All major constituents of living biomass or organic waste materials can be converted to methane plus CO_2, with the only exceptions being lignin and lignocellulose. This high conversion efficiency is secured through the cooperation of many different types of fermentative bacteria and methanogenic archaeobacteria that accomplish the multitude of specific reactions converting the enormous diversity of major biomass constituents to less and less complex reaction intermediates by a modular arrangement of metabolic functions. Since the overall energy yield of methanogenic fermentation is small, only a very small amount of energy is available to the single organisms catalyzing its specific reaction step within this cross-feeding network. Several steps in this process yield just enough energy to supply the catalyzing organism with the absolute minimum of energy needed for ATP synthesis and growth, i.e., an energy amount in the range of -15 to -20 kJ per mol reaction. It is this energetic efficiency that makes the overall process possible at all, but it also limits its kinetic and dynamic versatility.

REFERENCES

Aiyuk, S., I. Forrez, K. de Lieven, A. van Haandel, and W. Verstraete. 2006. Anaerobic and complementary treatment of domestic sewage in regions with hot climates—a review. *Bioresour. Technol.* 97:2225–2241.

Baldrian, P. 2006. Fungal laccases—occurrence and properties. *FEMS Microbiol. Rev.* 30:215–242.

Boll, M., B. Schink, A. Messerschmidt, and P. M. Kroneck. 2005. Novel bacterial molybdenum and tungsten enzymes: three-dimensional structure, spectroscopy, and reaction mechanism. *Biol. Chem.* 386:999–1006.

Brune, A., and M. Friedrich. 2000. Microecology of the termite gut: structure and function on a microscale. *Curr. Opin. Microbiol.* 3:263–269.

Buckel, W., and B. T. Golding. 2006. Radical enzymes in anaerobes. *Annu. Rev. Microbiol.* 60:27–49.

Claassen, P. A. M., J. B. van Lier, A. M. L. Contreras, E. W. J. van Niel, L. Sijtsma, A. J. M. Stams, S. S. de Vries, and R. A. Weusthuis. 1999. Utilisation of biomass for the supply of energy carriers. *Appl. Microbiol. Biotechnol.* 52:741–755.

Costa, E., J. Perez, and J. U. Kreft. 2006. Why is metabolic labour divided in nitrification? *Trends Microbiol.* 14:213–219.

de Bok, F. A., M. L. Luijten, and A. J. Stams. 2002. Biochemical evidence for formate transfer in syntrophic propionate-oxidizing cocultures of *Syntrophobacter fumaroxidans* and *Methanospirillum hungatei*. *Appl. Environ. Microbiol.* 68:4247–4252.

Embley, T. M., B. J. Finlay, P. L. Dyal, R. P. Hirt, M. Wilkinson, and A. G. Williams. 1995. Multiple origins of anaerobic ciliates with hydrogenosomes within the radiation of aerobic ciliates. *Proc. Biol. Sci.* 262:87–93.

Friedrich, M., and B. Schink. 1993. Hydrogen formation from glycolate driven by reversed electron transport in membrane vesicles of a syntrophic glycolate-oxidizing bacterium. *Eur. J. Biochem.* 217: 233–240.

Friedrich, M., and B. Schink. 1995. Electron transport phosphorylation driven by glyoxylate respiration with hydrogen as electron donor in membrane vesicles of a glyoxylate-fermenting bacterium. *Arch. Microbiol.* 163:268–275.

Hackett, W. F., W. J. Connors, T. K. Kirk, and J. G. Zeikus. 1977. Microbial decomposition of synthetic ^{14}C-labeled lignins in nature: lignin biodegradation in a variety of natural materials. *Appl. Environ. Microbiol.* 33:43–51.

Hallenbeck, P. C. 2005. Fundamentals of the fermentative production of hydrogen. *Water Sci. Technol.* 52:21–29.

Holliger, C., and B. Schink. 1997. Anaerober Abbau organischer Schadstoffe, p. 83–90. *In* C. Knoll and T. von Schell (ed.), *Mikrobieller Schadstoffabbau*. Vieweg, Braunschweig, Germany.

Kanai, T., H. Imanaka, A. Nakajima, K. Uwamori, Y. Omori, T. Fukui, H. Atomi, and T. Imanaka. 2005. Continuous hydrogen production by the hyperthermophilic archaeon, *Thermococcus kodakarensis* KOD1. *J. Biotechnol.* 116:271–282.

Kirby, R. 2006 Actinomycetes and lignin degradation. *Adv. Appl. Microbiol.* 58:125–168.

Korner, I., J. Braukmeier, J. Herrenklage, K. Leikam, M. Ritzkowski, M. Schlegelmilch, and R. Stegmann. 2003. Investigation and optimization of composting processes—test systems and practical examples. *Waste Manag.* 23:17–26.

Leonowicz, A., N. S. Cho, J. Luterek, A. Wilkolazka, M. Wojtas-Wasilewska, A. Matuszewska, M. Hofrichter, D. Wesenberg, and J. Rogalski. 2001. Fungal laccase: properties and activity on lignin. *J. Basic Microbiol.* 41:185–227.

Lettinga, G. 1995. Anaerobic digestion and wastewater treatment systems. *Antonie Leeuwenhoek* 67:3–28.

Schink, B. 1988. Principles and limits of anaerobic degradation—environmental and technological aspects, p. 771–846. *In* A. J. B. Zehnder (ed.), *Biology of Anaerobic Microorganisms*. John Wiley and Sons, New York, NY.

Schink, B. 1997. Energetics of syntrophic cooperations in methanogenic degradation. *Microbiol. Mol. Biol. Rev.* 61:262–280.

Schink, B. 1999. Prokaryotes in environmental processes, p. 900–912. *In* J. W. Lengeler, G. Drews, and H. G. Schlegel (ed.), *Biology of the Prokaryotes*. Thieme, Stuttgart, Germany.

Schink, B., and A. J. M. Stams. 2001. Syntrophism among prokaryotes. *In* M. Dworkin, S. Falkow, E. Rosenberg, K.-H. Schleifer, and E. Stackebrandt (ed.), *The Prokaryotes: an Evolving Electronic Resource for the Microbiological Community*, 3rd ed., Springer-Verlag, New York, NY.

Siebers, B., and P. Schonheit. 2005. Unusual pathways and enzymes of central carbohydrate metabolism in Archaea. *Curr. Opin. Microbiol.* 8:695–705.

Spormann, A. M., and F. Widdel. 2000. Metabolism of alkylbenzenes, alkanes, and other hydrocarbons in anaerobic bacteria. *Biodegradation* 11:85–105.

Stams, A. J., and S. J. Oude Elferink. 1997. Understanding and advancing wastewater treatment. *Curr. Opin. Biotechnol.* 8:328–334.

Thauer, R. K., and J. G. Morris. 1984. Metabolism of chemotrophic anaerobes: old views and new aspects, p. 123–168. *In* D. P. Kelly and N. G. Carr (ed.), *The Microbe, Part II. Prokaryotes and Eukaryotes*. Cambridge University Press, Cambridge, United Kingdom.

Thauer, R. K., K. Jungermann, and K. Decker. 1977. Energy conservation in chemotrophic anaerobic bacteria. *Bacteriol. Rev.* 41:100–180.

Wallrabenstein, C., and B. Schink. 1994. Evidence of reversed electron transport involved in syntrophic butyrate and benzoate oxidation by *Syntrophomonas wolfei* and *Syntrophus buswellii*. *Arch. Microbiol.* 162:136–142.

Zeikus, J. G. 1977. The biology of methanogenic bacteria. *Bacteriol. Rev.* 41:514–541.

Zinder, S. H., and M. Koch. 1984. Non-aceticlastic methanogenesis from acetate: acetate oxidation by a thermophilic syntrophic coculture. *Arch. Microbiol.* 138:263–272.

Bioenergy
Edited by J. Wall et al.

Chapter 15

Optimizing Mixed-Culture Bioprocessing To Convert Wastes into Bioenergy

LARGUS T. ANGENENT AND BRIAN A. WRENN

MIXED-CULTURE PROCESSES THAT EXTRACT ENERGY FROM COMPLEX WASTES

Symbiotic Associations between Microbial Communities and Animals

The most efficient systems for biodegradation of polymeric organic compounds, such as the cellulose and hemicellulose components of lignocellulose, are mixed cultures that have evolved in some insect and mammalian guts. These diverse microbial communities have evolved in symbiotic relationship with the host organism and have developed the ability to extract energy from hard-to-degrade lignocellulosic feedstocks. The host provides a suitable environment for the microbial community, and the microbial community provides fermentation end products that can be assimilated by the host. Two model communities with distinct characteristics have been extensively studied: the termite hindgut and the rumen. In both communities, the recalcitrant polymers that make up the bulk of solid plant material are hydrolyzed to water-soluble products by a mixture of periplasmic and extracellular enzymes (Leschine, 1995). The water-soluble products are then metabolized by the processes described below.

Termite hindgut

The termite hindgut is a potent lignocellulosic bioconversion system, which involves symbiotic relationships among a variety of microorganisms, including facultatively anaerobic bacteria, strictly anaerobic bacteria, and anaerobic protozoa (Breznak and Brune, 1994). The main end product of this microbial community is acetate, which constitutes 94 to 98 mol% of the volatile fatty acids that are formed (Odelson and Breznak, 1983). Some of the hydrogen that is produced as a by-product by cellulose-degrading protozoa, which contain endosymbiotic bacteria, is oxidized by spirochetes through homoacetogenesis (i.e., formation of acetate from hydrogen and carbon dioxide) (Leadbetter et al., 1999). Removal of hydrogen by these spirochetes reduces product inhibition and improves the overall efficiency of the cellulose fermentation. The protozoan *Mixotrichia paradoxa* derives an additional benefit from its interaction with these spirochetes, because the bacteria attach to its cell surface and create a propelling force (Cleveland and Grimstone, 1964; Wenzel et al., 2003). Any hydrogen that is not oxidized by these hydrogenotrophic spirochetes diffuses outward toward the epithelium, where a dense and diverse population of hydrogenotrophic methanogens consumes it (Leadbetter and Breznak, 1996; Leadbetter et al., 1998). It is somewhat surprising that the strictly anaerobic methanogens grow in the microoxic zone near or attached to the gut epithelium, but the oxygen levels in this microoxic zone are kept low enough for methanogenesis to occur by facultative bacteria, which grow by degrading the toxic aromatic by-products of lignin degradation (Brune et al., 1995a, 1995b).

Mammalian rumen

The rumen (i.e., foregut organ) contains a diverse community of anaerobic bacteria, archaea, fungi, and protozoa that efficiently degrades cellulose and hemicellulose and produces fermentation products that support the nutritional requirements of the host (Leschine, 1995). Anaerobic bacteria, which hydrolyze plant material to oligo- and monosaccharides, are the most important hydrolytic organisms in the rumen (Akin and Benner, 1988). Other fermenting and anaerobically respiring bacteria then convert these sugars into volatile fatty acids (e.g., acetate, butyrate, and propionate), which are readily assimilated by the host. As in the termite hindgut, spirochetes are present at high numbers in the rumen. These bacteria are noncellulolytic, but they

Largus T. Angenent • Department of Energy, Environmental & Chemical Engineering, Washington University in St. Louis, St. Louis, MO 63130. **Brian A. Wrenn** • National Corn-to-Ethanol Research Center, Southern Illinois University Edwardsville, Edwardsville, IL 62025.

consume the products of cellulolytic bacteria (e.g., sugars) and, therefore, reduce product inhibition and improve the overall conversion of cellulose (Stanton and Canale-Parola, 1980). In anaerobic environments, such as the rumen, cellulolytic enzymes exist as membrane-bound multiprotein enzyme complexes rather than as simple extracellular enzymes (Raphael et al., 1991; Leschine, 1995; Malherbe and Cloete, 2002). Therefore, cellulolytic bacteria attach to the insoluble cellulose particles as biofilms (O'Sullivan et al., 2005). The motile spirochete *Treponema bryantii* is chemotactically attracted to *Fibrobacter succinogenes* and attaches to the surface of this nonmotile cellulose-degrading bacterium to facilitate the transport toward their nondiffusible substrate (Stanton and Canale-Parola, 1980). Like the termite hindgut, the metabolism of cellulose is enhanced by the presence of hydrogenotrophic microbes. In the rumen, hydrogenotrophic bacteria (sulfate reducers and homoacetogens) and archaea (methanogens) play a role similar to that of hindgut spirochetes by scavenging hydrogen and shifting the fermentation from lactate and succinate toward acetate, which is readily assimilated by the host.

Engineered Bioprocesses

Because wastewater consists of a complex mixture of compounds, such as carbohydrates, proteins, and lipids, environmental engineers use mixed cultures that can cope with the complexity. The activated sludge process, which was developed in the 1910s, is the most widely used bioprocess for domestic and industrial wastewater treatment because aerobic bacteria can remove soluble organic pollutants to very low concentrations even at ambient temperatures (15°C) (Rittmann and McCarty, 2001). A major disadvantage of activated sludge processes is the amount of electricity needed to provide oxygen, which is equivalent to approximately 2% of the total U.S. electricity consumption (Electric Power Research Institute, 2002). The energy cost for aeration is proportional to the concentration of organic material in wastewater, and therefore, anaerobic digestion is attractive for high-strength wastewater (>2,000 mg of chemical oxygen demand [COD]/liter) because enough methane can be generated to offset the energy required to heat the system to mesophilic temperatures (35°C) (Speece, 1996; Tchobanoglous et al., 2003). Thousands of microbial species create a food web in which complex molecules, such as cellulose, proteins, and lipids, are converted to methane in these anaerobic digesters (Gujer and Zehnder, 1983; McCarty and Smith, 1986). Therefore, a century of work with anaerobic digesters has provided environmental engineers with abundant experience in the use of mixed cultures for production of bioenergy from complex organic wastes. Anaerobic bioprocesses that can convert organic wastes to energy-rich products that have advantages over methane, such as hydrogen, electricity, solvents, and liquid fuels, are currently being investigated. These alternative processes, which are collectively called dark-fermentation processes because they do not depend on the input of light energy, are the focus of this chapter.

Characteristics of Potential Waste-Derived Feedstocks for Biofuels Production

The efficiency and economic viability of converting organic wastes to biofuels depends on the characteristics of the waste material, especially the chemical composition and the concentrations of the components that can be converted into products that can be used as fuels. Wastes can be categorized in several ways, with source, strength, and composition being the most common factors. Table 1 lists the characteristics of some important wastes, using these factors as the basis for classification.

With respect to source, wastes are most commonly divided into domestic/municipal, agricultural, and industrial categories. The source may be correlated with composition and strength, which are described in more detail below, but it also has important implications for the costs of collection and transportation to the waste-to-energy facility. If wastes from a specific source (e.g., municipal sewage) are already collected and transported to a central location for treatment or disposal, the only additional costs for converting them to energy will be those associated with constructing and operating the new facility. For wastes that are not currently collected (e.g., agricultural wastes), the costs for waste-to-energy conversion will include development of new collection, transportation, and (possibly) storage infrastructure in addition to the costs of construction and operation of the facility.

The strength of organic waste streams can be characterized based on the biochemical oxygen demand (BOD) or COD. Waste strength is important because the substrate concentration will determine the concentration of products that can be formed, which will affect the costs of product separation. In general, low-strength wastes lead to low product concentrations and high separation costs, but some products are more easily recovered from aqueous process media than others. For example, gases (e.g., hydrogen and methane) can be recovered from aqueous media relatively easily, whereas water-soluble products (e.g., ethanol) are more difficult to recover. Therefore, the waste strength is a more important consideration for some energy products than for others.

Wastes can also be categorized based on their physical and chemical properties, with the most important physical characteristic being whether the waste

Table 1. Composition and concentration of potential waste-derived feedstocks for biofuels production

Strength	Source	Concn	Composition	Reference(s)
Low	Municipal wastewater	250–1,000 mg COD/liter; 110–400 mg BOD_5/liter	Carbohydrate (25–50%), protein (40–60%), fat (10%)	Tchobanoglous et al., 2003
High	Brewery wastewater	3,000–5,000 mg COD/liter; 1,500–2,400 mg BOD_5/liter	Protein (12–19%)	Diaz et al., 2006; Vijayaraghavan et al., 2006
	Slaughterhouse wastewater	1,400–2,700 mg COD/liter	Protein (30–65%), fat (45–75%)	Batstone et al., 2001
	Fruit/vegetable canning	100–50,000 mg COD/liter	Carbohydrate (50–60%)	Lepisto and Rintala, 1997
	Whey	60,000–80,000 mg COD/liter; 30,000–50,000 mg BOD_5/liter	Carbohydrate (67%), protein (13%), fat (18%)	Siso, 1996
Solid	Potato wastes	~20% solids	Carbohydrate (>95%)	Russ and Meyer-Pittroff, 2004
	Apple pomace	20–35% solids	Carbohydrate (50%), protein (2–7%), fat (2–10%)	Kennedy et al., 1999
	Cattle manure	27–31% solids	Carbohydrate (47–48%), protein (10–12%), lignin (20%)	Hao et al., 2005; Chen et al., 1989; Miller and Varel, 2003; Moral et al., 2005
	Swine manure	27% solids	Carbohydrate (42%), protein (20%)	Miller and Varel, 2003

must be handled as a solid or as a liquid. This physical characteristic will determine the methods that must be used to transport the waste material from its site of generation to the waste-to-energy processing facility, but it may also have important consequences with respect to process kinetics and storability. When the waste material consists of particles, the reaction kinetics may be limited by interfacial area, and therefore, a size reduction step may be required near the front end of the process. On the other hand, waste materials that contain a relatively large amount of water, even when the material would otherwise be characterized as a solid (e.g., potato waste and apple pomace), have poor storage characteristics and are likely to be unstable during collection and transportation. In this case, the waste-to-energy facility will have to be located close to the source of the waste. A clear advantage of solid wastes relative to liquid wastes is the strength: even very-high-strength liquid wastes, like whey, have concentrations of <100,000 mg of COD/liter, whereas most organic solid wastes have COD values of >100,000 mg/kg.

The chemical composition of waste materials is extremely important from the perspective of biofuels production because it determines which products can be made and which processes must be used to make them. With respect to chemical composition, the relative proportions of carbohydrate, protein, and oil or fat are most important, but the presence of other components, such as polyphenols and furfurals, must also be considered because these compounds can have profound effects on microbial activity. Due to the thermodynamic and physiological limitations of some of the intermediate metabolic reactions, carbohydrates tend to be more easily converted into high-value biofuel products than proteins and fats, but a significant fraction of the energy available in some wastes can be present in the latter compounds. The carbohydrate fraction of the waste must be further subdivided into easily degradable and recalcitrant fractions. In general, carbohydrates whose biological function is to provide energy, such as simple sugars and starch, fall into the easily degradable category, whereas compounds that perform a structural function, such as cellulose and hemicellulose, tend to be more recalcitrant.

Low-strength wastes

Municipal sewage is an example of an abundant low-strength wastewater. Municipal wastewater is a mixture of domestic (i.e., residential), commercial (e.g., schools and businesses), and compatible industrial wastewater with a 5-day BOD (BOD_5) ranging between about 110 and 400 mg/liter (Tchobanoglous et al., 2003). Municipal sewage contains a wide variety of organic compounds and is dominated by compounds that occur in food, including carbohydrates (25 to 50%), protein (40 to 60%), and fats and oils (10%). The high concentrations of protein and fat relative to carbohydrate and the overall low substrate concentrations make municipal wastewater a poor candidate for production of liquid biofuels, which are produced by direct fermentation of carbohydrates and are relatively difficult to separate from the water.

High-strength wastes

Wastewater produced by the food-processing industry has significantly higher concentrations of degradable substrates than municipal wastewater, and many of these wastewaters are rich in carbohydrates and organic acids with relatively low concentrations of organic nitrogen (e.g., proteins) (Nemerow and Desgupta, 1991; Lepisto and Rintala, 1997; Fuchs et al., 2003; Mulkerrins et al., 2004; Diaz et al., 2006; Gohil and Nakhla, 2006). Important examples include brewery, canning, and dairy wastewaters. Although the concentration of brewery wastewater varies, it is typically in the range of 3,000 to 5,000 mg of COD/liter (i.e., about 10 times more concentrated than domestic wastewater) (Rodriguez-Martinez et al., 2005; Diaz et al., 2006; Vijayaraghavan et al., 2006). Wastewater produced during canning of fruits and vegetables can vary over a wider range (e.g., <1,000 to >10,000 mg of COD/liter) depending on the nature of the product and the process (Nemerow and Desgupta, 1991; Lepisto and Rintala, 1997; Contreras et al., 2000). Whey, which is the dominant liquid waste product of the dairy industry, has even higher substrate concentration (approximately 60,000 to 80,000 mg of COD/liter) (Siso, 1996). Lactose constitutes about two-thirds of the COD of whole whey, whereas fat and protein represent about 18 and 13% of the COD, respectively. Due to its high nutritional value, whey protein is often recovered, leaving a product that is about 80% lactose and 20% fat. The carbohydrates in these wastewaters can be fermented to a variety of products, including several potential biofuels (e.g., ethanol, butanol, electric current, hydrogen, and methane) (Jones and Woods, 1986; Siso, 1996; Fang and Yu, 2001; Rao et al., 2007), but recovery of water-soluble products, such as ethanol, may be uneconomical due to the low product concentrations that can be achieved. Although the concentrations of fermentable substrates in these wastewaters may be high relative to wastewater, they are low relative to the substrate concentrations used to produce many biofuels. Thus, anaerobic systems that produce easily separable gases (hydrogen and methane) or electric current may be more appropriate for this type of wastewater.

Livestock-related industries (e.g., feedlots, slaughterhouses, and packinghouses) produce wastewater of similar strength to those produced by brewery and fruit and vegetable canning (Nemerow and Desgupta, 1991), but the relative abundance of protein and fat or oil is much greater (Batstone and Keller, 2001; Batstone et al., 2001). Proteins can be degraded by hydrolysis followed by fermentation of the amino acids to volatile fatty acids (McInerney, 1988; Yu and Fang, 2003), but fatty acids, which are produced by hydrolysis of lipids (i.e., fats and oils), cannot be degraded by fermentation (McInerney, 1988). The nonfermentability of fatty acids restricts the processes through which they can be converted to biofuels to those involving interspecies hydrogen transfer or transfer of the reducing equivalents to an external electron acceptor (e.g., respiration). Thus, these wastewaters (e.g., animal wastes) can be converted to methane in farm-based anaerobic digesters (Fig. 1) (Angenent et al., 2002), but conversion to liq-

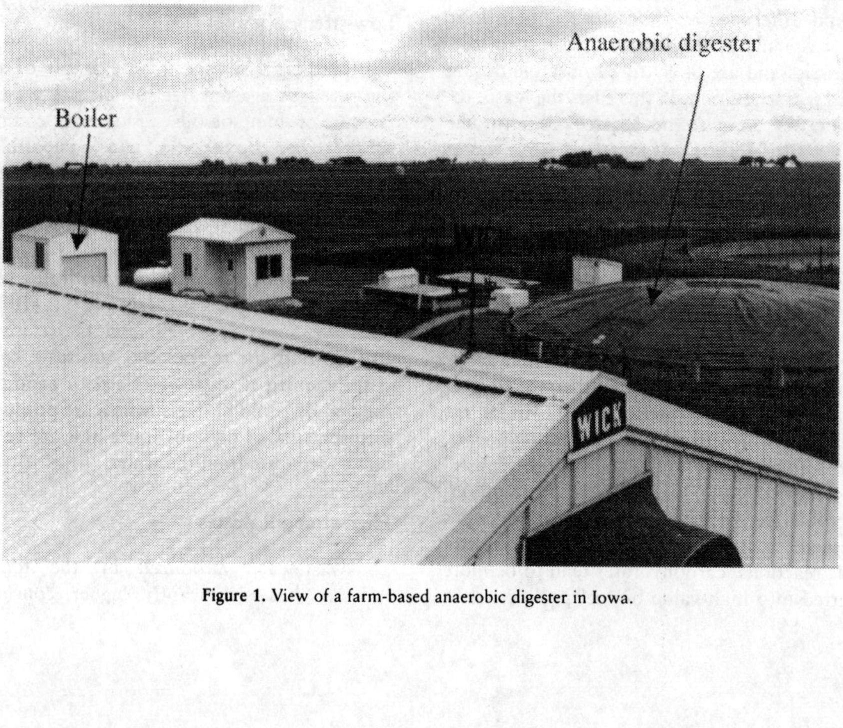

Figure 1. View of a farm-based anaerobic digester in Iowa.

uid biofuels may require subsequent conversion through a nonbiological process (Holtzapple et al., 1999).

Solid wastes

The solids concentrations of solid wastes are generally high enough to be considered for production of liquid biofuels that are difficult to separate from the fermentation broth (if the substrates in the waste are suitable for such biofuel production). Like wastewaters, solid wastes can be categorized based on chemical composition. Solid wastes produced by processing of fruits and vegetables are the most suitable for biofuels production because they have high concentrations of readily fermentable substrates, such as simple sugars and starch (Russ and Meyer-Pittroff, 2004). Examples include pomace, which can be obtained from production of fruit juices. Despite the possibility to produce liquid biofuels, methane generation from solid wastes, such as maize silage (a mix of stover and kernel), is currently being implemented in some European countries (e.g., Germany and Austria) at a large scale because of subsidies, the maturity of the technology, and a relative high biogas yield (e.g., 230 m^3 of biogas per ton of wet maize silage) (Weiland, 2006).

LIKELY IMPORTANT BIOFUELS FROM MIXED CULTURES

Because mixed-culture fermentation involves large microbial communities, only certain compounds can be produced. Some products cannot be generated because they are converted to other compounds by the mixed culture more quickly than they are formed. Production of specific desired products can be optimized when clear criteria for selection of appropriate microbial communities can be identified and exploited in bioreactor design (Kleerebezem and van Loosdrecht, 2007). Ideally, a direct selective pressure favoring community members that catalyze the desired reactions should be applied, but such criteria have not yet been identified for some desired products (e.g., alcohols that can be used as biofuels). As described above, the characteristics of the mixed-culture system depend on the characteristics of organic wastes as well as the desired end product. Processes that can be used to produce several useful biologically derived energy carriers, including methane, alcohol fuels, and electricity, are described below.

Methane

Anaerobic digestion converts organic substrates to methane, carbon dioxide, and other organic and inorganic compounds. The fundamentals of anaerobic digestion were reviewed by McCarty in 1964 (McCarty, 1964a, 1964b, 1964c, 1964d). The anaerobic food web through which biological polymers are converted to methane requires four major steps as in Fig. 2 (Gujer and Zehnder, 1983; McCarty and Smith, 1986): (1) hydrolysis of polymers, (2) fermentation of the monomers to volatile fatty acids and alcohols, (3) conversion of the intermediates to hydrogen and acetic acid (i.e., acetogenesis), and (4) formation of methane (i.e., methanogenesis) from hydrogen and carbon dioxide or acetate. This food web involves five functionally distinct groups of microorganisms: (a) fermentative bacteria, (b) obligatory H_2-producing acetogenic bacteria, (c) H_2-utilizing acetogenic (also known as homoacetogenic) bacteria, (d) hydrogenotrophic methanogenic archaea, and (e) aceticlastic methanogenic archaea. In conventional anaerobic digesters and typical methanogenic natural environments, most of the methane (~70%) is formed by aceticlastic methanogens (group e).

Due to the complex interrelationships that exist in this anaerobic food web, digester failure can occur when one group of microorganisms is inhibited or overloaded (Duran and Speece, 1998; Azbar et al., 2001). In fact, a rather small inhibition of one group of organisms can result in an inhibition cascade that can cause the entire system to crash. For example, accumulation of intermediate products (e.g., propionate, butyrate, or acetate) can inhibit acetogenesis and methanogenesis by reducing the pH to less than the optimal range for these microorganisms. Alternatively, accumulation of hydrogen or formate can reduce the free-energy change of

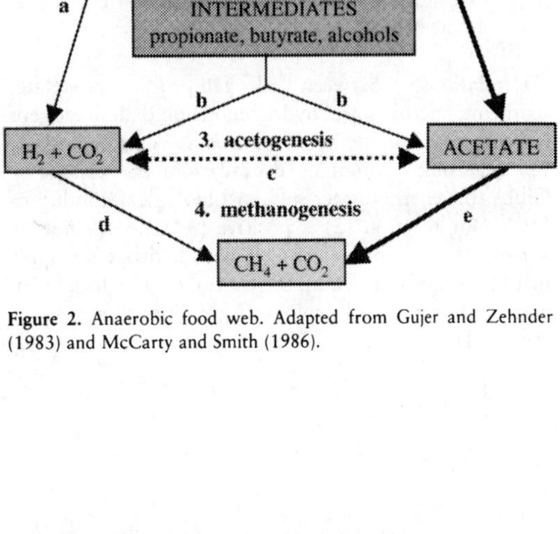

Figure 2. Anaerobic food web. Adapted from Gujer and Zehnder (1983) and McCarty and Smith (1986).

important metabolic reactions and make them thermodynamically infeasible. An example of the latter case is provided by propionate oxidation, which is oxidized through the following reaction in anaerobic systems:

$$\underset{\text{propionate}}{CH_3CH_2COOH} + 2\,H_2O \rightarrow \underset{\text{acetate}}{CH_3COOH} + CO_2 + 3\,H_2 \qquad (1)$$

Under standard biological conditions (i.e., all reactants and products at unit activity except pH of 7.0), the Gibbs free-energy change for reaction 1 is +76.1 kJ mol^{-1} (Thauer et al., 1977; Stams, 1994). For the free-energy change to be negative under ambient conditions, the hydrogen partial pressure must be $<10^{-4}$ atm, but a minimum free-energy change of about 17 to 20 kJ mol^{-1} is required to sustain growth of propionate-oxidizing bacteria (Schink, 1992). Thus, the sustained conversion of propionate can occur only when the hydrogen partial pressure is $<10^{-5}$ atm. In stable anaerobic systems, hydrogen-consuming microorganisms, such as hydrogenotrophic methanogens, keep the hydrogen partial pressure low enough to allow propionate degradation to proceed. This obligatory interaction between propionate oxidation and hydrogenotrophic methanogenesis is known as interspecies hydrogen transfer, and the anaerobic oxidation of propionate is a syntrophic process (Thauer et al., 1977; Gujer and Zehnder, 1983; Harper and Pohland, 1986; Schmidt and Ahring, 1993). In unstable reactors, acetate and hydrogen uptake may cease and propionate will accumulate. Propionate conversion is, therefore, a keystone process.

In some stressed anaerobic digestion systems, acetate is converted to methane through a two-step process involving oxidation of acetate to hydrogen and carbon dioxide by homoacetogens (group c) and conversion of these products to methane by hydrogenotrophic methanogens (group e). Note that oxidation of acetate by homoacetogens is—at least conceptually—a reversal of the process for which these organisms are best known and is made possible by their syntrophic relationship with the hydrogenotrophic methanogens:

$$\underset{\text{acetate}}{CH_3COOH} + 2\,H_2O \rightarrow 2\,CO_2 + 4\,H_2 \qquad (2)$$

The relationship between hydrogen-producing acetate-oxidizing microbes and hydrogenotrophic methanogens is syntrophic for the same reasons as described above for anaerobic propionate degradation: the change in Gibbs free energy under standard biological conditions ($\Delta G^{0'}$) for reaction (2) is positive (+104.6 kJ mol^{-1}). When this reaction is coupled with hydrogenotrophic methanogenesis, however, the overall change in free energy under standard biological conditions is −31 kJ mol^{-1} (Thauer et al., 1977).

The production of methane from acetate through this syntrophic consortium has been observed in thermophilic (Zinder and Koch, 1984; Angelidaki and Ahring, 1994; Zinder, 1994) and stressed mesophilic systems (Angenent et al., 2002). One stressor that has been shown to shift mesophilic methanogenic food webs from one involving aceticlastic methanogens to one involving hydrogen-producing acetate oxidizers is the accumulation of high concentrations of ammonium species (i.e., the sum of ammonia and ammonium) (Angenent et al., 2002; Petersen and Ahring, 1992; Schnürer et al., 1994, 1999). Another stressor is high concentrations of sulfate (Petersen and Ahring, 1992; Schnürer et al., 1994, 1999). Under these conditions, aceticlastic methanogens are inhibited. For example, in an anaerobic swine waste digester, aceticlastic methanogens (*Methanosarcina* spp.) were inhibited severely by a total ammonium concentration of approximately 3,500 mg of NH_4^+-N/liter, whereas hydrogenotrophic methanogens (*Methanomicrobiales*) were less sensitive (Angenent et al., 2002). This alternative pathway for conversion of acetate to methane is not unique to engineered systems. It has also been observed in the sediments of a subtropical lake (Nüsslein et al., 2001). Acetate-oxidizing bacteria that grow in syntrophic association with hydrogenotrophic methanogens and that have been isolated or enriched from anaerobic digesters include the thermophilic bacteria "*Reversibacter*" (Zinder and Koch, 1984; Lee and Zinder, 1988) and *Thermoacetogenium phaeum* (Hattori et al., 2000), and the mesophilic bacteria *Clostridium ultunense* (Schnürer et al., 1996, 1997) and *Geobacter sulfurreducens* (Cord-Ruwisch et al., 1998).

Because of improved methane yields and production rates due to favorable kinetics at higher temperatures, thermophilic digestion (55°C) of wastes and wastewaters with a very high organic content may be advantageous over mesophilic digestion (Vandevoorde and Verstraete, 1987; Mackie and Bryant, 1995). The higher rates can directly result in smaller reactor volumes, especially important when using recalcitrant substrates, such as cellulose and lignocellulose-based materials. Carbonaceous wastes are treated successfully under thermophilic conditions; however, the formation of higher levels of the inhibiting nondissociated ammonia at thermophilic than at mesophilic temperatures may prohibit thermophilic digestion of protein-rich wastes.

Carboxylic Acids, Hydrogen, and Alcohols

Carboxylic acid mixtures as fuel precursors

When methanogenesis is inhibited in anaerobic digesters (e.g., by addition of selective inhibitors of methanogenesis, such as bromoethanesulfonic acid, bromoform [$CHBr_3$], or iodoform [CHI_3], at near neutral pH), mixed-culture fermentation can be used to produce mixtures of carboxylic acids, including acetic, propionic, butyric, valeric, caproic, and heptanoic acids, and hydrogen instead of methane (Thanakoses et al., 2003a, 2003b; Agbogbo and Holtzapple, 2007). Because the hydrogen-consuming syntrophic partner is not present in these communities, all reactions that require low hydrogen partial pressure are eliminated from the anaerobic food web. The mixture of soluble organic acids that is produced must be converted into useful bioenergy carriers through other mechanisms (e.g., abiotic conversion to alcohol fuels [Holtzapple et al., 1999]). Although hydrogen is a coproduct of this fermentation process, its energy content is less than that of the organic acids.

This process has been evaluated for several organic waste streams, including municipal solid waste, sugarcane bagasse, a mixture of paper fiber and industrial biosludge, a mixture of corn stover and swine waste, and a mixture of rice straw and chicken manure (Chan and Holtzapple, 2003; Thanakoses et al., 2003a, 2003b; Domke et al., 2004; Aiello-Mazzarri et al., 2005; Agbogbo and Holtzapple, 2006). This shows that carboxylic acids can be produced from protein-rich wastes. Depending on the substrate, pretreatment may be required to improve the fermentation kinetics and product yield (Mosier et al., 2005).

Mixtures of carboxylic acids can also be produced in the first phase of a two-phase bioreactor that is designed to produce hydrogen and methane as the main energy carriers (Zhang and Zhong, 2002). Thus, the mixture of acids is fed to the second phase and a precursor to methane production. In these systems, acidogenesis and methanogenesis can be optimized independently. For example, hydrolysis rates in the acidogenic first phase are maximized by conducting the reactions at a slightly acidic pH (e.g., about 5.6), especially when the process is operated under thermophilic conditions (55°) (Chyi and Dague, 1994), whereas the methanogenic second phase would be operated at near-neutral pH. Because the low pH and high concentration of organic acids partially inhibit methanogenesis, hydrogen is the main gaseous product of the first-phase reactor. This configuration, therefore, allows production of separate hydrogen and methane product streams. Hydrogen yields in the first-phase reactor are limited by product inhibition, however, resulting in a maximum hydrogen yield of about 2 mol of hydrogen per mol of glucose (Angenent et al., 2004). Therefore, this process is less efficient than hydrogen production in microbial fuel cells (MFCs), which is described below. From an energy perspective, it seems more logical to focus this type of fermentation process on maximizing the production of organic acids to produce methane with hydrogen being considered a valuable coproduct, rather than to focus on improving the hydrogen generation.

Production of butyrate and butanol

When glucose-containing waste streams, such as those that are high in starch or cellulose, are used to produce bioenergy, butyrate may be one of the most important organic acid products (Fang and Liu, 2002; Zhang et al., 2003; Yu et al., 2004; Kraemer and Bagley, 2005; Whang et al., 2006). Butyrate production is linked to hydrogen production, because hydrogen partial pressures above 60 Pa inhibit proton reduction by NADH during acidogenesis, shifting the fermentation toward butyrate or other reduced products instead of toward acetate (Angenent et al., 2004). Kinetic and thermodynamic modeling has also shown that production of butyrate by mixed-culture fermentation is feasible (Kleerebezem and Stams, 2000; Rodriguez et al., 2006). Fermentations optimized for the production of butyrate may be important for biofuels production because butyrate can be converted to butanol in a two-stage bioprocess (Fig. 3). A system involving two different clostridia, which produce butyrate and butanol in separate pure-culture bioprocesses placed in series, has been patented (Ramey, 1998). In the first bioprocess, *Clostridium thermobutyricum* converts carbohydrates to (mainly) butyrate, and the butyrate is transported to the second bioreactor, where it is converted to butanol by, for example, *Clostridium acetobutylicum*. A mixed-culture bioprocess for conversion of complex mixtures of carbohydrates (e.g., lignocellulosic hydrolysates and brewery wastes) to butyrate in the first-stage bioprocess might be a superior alternative to the pure-culture process.

Although its production through biological processes decreased dramatically in the latter half of the twentieth century, butanol was widely produced on an industrial scale using batch cultures of *C. acetobutylicum*. In this process, starch or simple sugars are converted to organic acids, particularly butyrate, during the first stage of the fermentation. During the second stage, solventogenesis (i.e., the formation of butanol, acetone, and ethanol) is initiated after the pH of the culture medium decreases due to acid accumulation (Zverlov et al., 2006). Hexose sugars, such as glucose, sucrose, and starch, are fermented via the Embden-Meyerhof pathway (i.e., glycolysis), and fermentation of pentoses occurs via the pentose phosphate and glycolytic pathways. The distinctive metabolic activity of butanol-producing bacteria is the rapid conversion of butyrate to butanol through solventogenesis using electrons derived from oxidation of carbohydrates through glycolysis (Gottschalk, 1986). Therefore, butanol can be efficiently produced by feeding butyrate along with carbohydrates to certain clostridia (e.g., *C. aceto-*

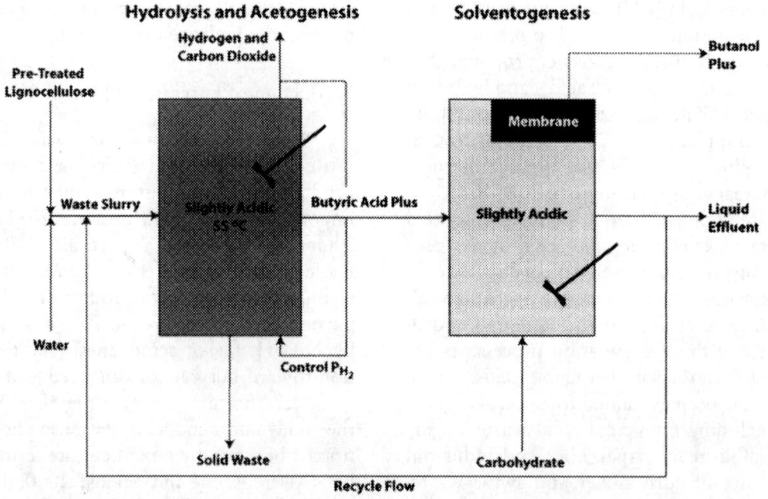

Figure 3. Schematic of biological butanol production from waste slurries.

butylicum, *Clostridium beijerinckii*, and *Clostridium tetanomorphum*) instead of by feeding carbohydrates only (Gottschalk, 1986). In addition, Tashiro et al. (2004) showed that butanol production increased and ethanol production decreased when butyrate and some glucose were fed to a pH-stat, fed-batch culture of *Clostridium saccharoperbutylacetonicum* N1-4 in the solventogenic phase. Furthermore, completely continuous cultures of *C. acetobutylicum* showed excellent butanol production in the laboratory by feeding a mixture of butyrate and glucose (Bahl et al., 1982a, 1982b; Grupe and Gottschalk, 1992). Similar results were achieved by limiting nutritional factors, such as phosphorus and iron (Bahl et al., 1982a, 1982b; Bahl et al., 1986). The yield of butanol from continuous processes could be maximized by feeding butyrate instead of sugars only because the butyrate was almost completely removed, whereas a maximum of 80% removal was observed for glucose (Bahl et al., 1982a, 1982b).

Microbial Fuel Cells: Hydrogen or Electric Power

The hydrogen yield in mixed-culture bioprocessing can be increased by physically separating the anaerobic oxidation of sugars from hydrogen production by conducting the reactions in the anode and cathode, respectively, of an MFC. This process is, in essence, biocatalyzed electrolysis. Physical separation of these processes prevents product inhibition of the fermentation reactions by hydrogen (Fig. 4). Hydrogen yields of 7 to 8 mol per mol of glucose have been observed when the potential was artificially increased using a potentiostat (Liu et al., 2005; Rozendal et al., 2006). This represents about two-thirds of the theoretical hydrogen yield for complete oxidation of glucose to carbon dioxide and a fourfold increase over the maximum hydrogen yield observed for standard dark-fermentation reactions.

These results prove that MFCs can be used to produce hydrogen, but they are better known for their ability to produce electric power. This power is generated when anaerobic bacteria growing in an anodic biofilm oxidize organic material in the anode chamber and transfer the reducing equivalents (i.e., the electrons) to the electrode instead of to an electron-acceptor molecule. An electric current results when the electrons flow through an external electrical circuit between the anode and cathode electrodes. The driving force for movement of electrons through this circuit is the redox potential difference between the anode and cathode. To maintain electroneutrality, cations or anions move through a semipermeable membrane that separates the two chambers of the fuel cell. Hydrogen can be produced at the cathode by catalytic reduction of protons (e.g., with platinum [Pt]) when the potential is externally supplemented to about 0.6 V, or electric power can be generated when an electron acceptor (e.g., oxygen or the electron mediator ferricyanide) is reduced at the cathode (Fig. 4). Although reduction of oxygen would be ideal for many current-generating applications, this reaction occurs efficiently only when the cathode is doped with an expensive noble metal (e.g., Pt) or the surface area of the carbon electrode is enormous. Generation of electric power by reduction of oxygen with sustainable biocathodes is an attempt to improve upon the current state of cathode technology for MFCs (He and Angenent, 2006).

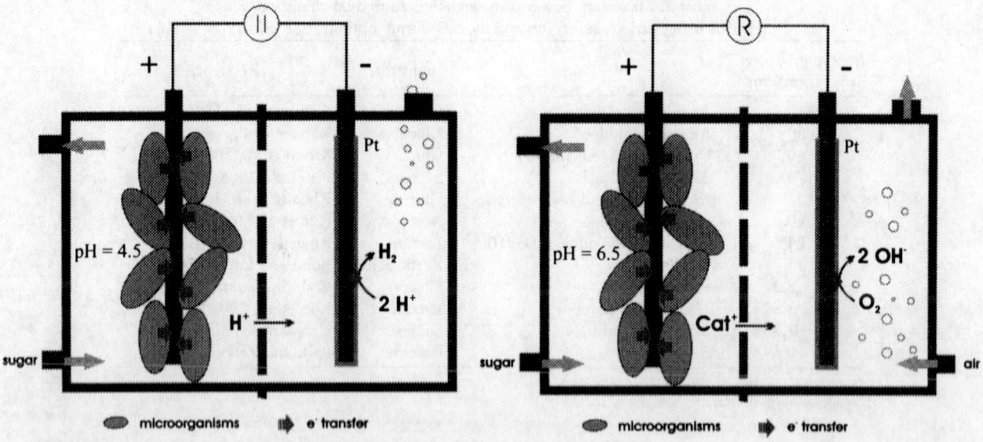

Figure 4. Schematic of a microbial fuel cell for hydrogen generation in the cathode (left panel) and electrical power generation (right panel). The surface of the anodic electrode is positive, and the surface of the cathodic electrode is negative. Electron current is from the anode to the cathode.

Microbiologists, biochemists, and electrochemists have used MFCs to generate electricity since the 1910s (Ieropoulos et al., 2005). Environmental engineers recently adapted MFC technology for laboratory-scale treatment of wastewater with promising results (Logan, 2004; Rabaey et al., 2004; He et al., 2005), but MFCs have not been optimized for wastewater treatment, and it is not clear how these small-scale systems can be scaled up to the level that will be required for practical application of this technology. Because MFCs are based on the smaller-scale technology of hydrogen fuel cells, fundamental issues limiting scale-up are becoming apparent.

Effective scale-up of MFCs requires a better understanding of the factors that limit performance. Electrochemical techniques, such as impedance spectroscopy, can be used to identify these limitations (Katz and Willner, 2003). Impedance spectroscopy can provide detailed information on the contribution of different sources of internal resistance in electrochemical cells (Barsoukov and Macdonald, 2005). The slow movement of ions between the electrode chambers (electrolyte resistance of 8.6 Ω at a volumetric loading rate of 3,400 mg COD/liter/day) limited the performance of a mixed-culture upflow MFC more than the cathode reaction rates (cathodic charge transfer resistance of 4.8 Ω) and the biofilm anode oxidation rates (anodic charge transfer resistance of 2.4 Ω). The slow diffusion of reactive species to the electrode surface (i.e., diffusion resistance) was the least important factor (1.5 Ω) (He et al., 2006). Thus, reactor configuration and the cathode reaction rate limit the power output of MFCs more than microbial capabilities for a mixed-culture MFC.

MFCs containing mixed-culture anodic biofilms produce greater power densities than MFCs with pure-culture anodes (Table 2). Particularly effective mixed cultures were selected in a batch-fed MFC by operating under an optimum anode potential (Rabaey et al., 2003, 2004). This effort led to the highest reported short-term power density (4,310 mW/m^2 of anode surface area; Table 2) and volumetric power (479 W/m^3 of anode chamber volume). The mechanisms by which mixed-culture MFCs generate more power than single-culture MFCs are not clear, but cyclic voltammetry showed that the anodic biofilm that developed in a mixed-culture MFC after a 3-month period of operation was more electrochemically active than its inoculum (anaerobic granular biomass), which indicates that electricigens were enriched by operation of the MFC (He et al., 2005). The food web through which anodic microbial communities convert organic compounds into harvestable electrons is not understood. By analogy to other anaerobic ecosystems, we anticipate that electrochemical active bacteria oxidize fermentation products that are produced by organisms that cannot transfer electrons to the anode effectively. However, some fermenters may also be able to transfer electrons directly to the anode (Kim et al., 2006).

Several mechanisms can be used to transfer electrons from microbial cells to solid electron acceptors, such as metal oxides or electrodes. One mechanism, which we call direct electron transfer, involves direct physical contact between the bacteria and the solid electron acceptor. Some dissimilatory metal-reducing bacteria use this mechanism, and outer membrane *c*-type cytochromes appear to function as the reductase

Table 2. Maximum power densities achieved in dual-chamber MFCs using various substrates and inocula during optimization efforts

Maximum power density[a] (mW/m^2)	Inoculum[b]	Substrate	Reference
4,310	Anaerobic granules	Glucose	Rabaey et al., 2004
3,600	Anaerobic sludge	Glucose	Rabaey et al., 2003
560	Unknown, mixed community	Glucose	Moon et al., 2006
170	Anaerobic sludge	Sucrose	He et al., 2005
127	*Klebsiella* sp. and *Leptothrix* sp.	Glucose	Rhoads et al., 2005
40	Anaerobic sludge	Acetic acid	Kim et al., 2005
24	***Shewanella oneidensis* DSP10**	Lactate	Ringeisen et al., 2006
15	***Geobacter* sp.**	Acetic acid	Bond and Lovley, 2003
8.2	***Rhodoferax* sp.**	Glucose	Chaudhuri and Lovley, 2003
3.6	***Shewanella* sp.**	Lactate	Kim et al., 2002
2.8	Mix of >2 cultures	Sucrose	Ieropoulos et al., 2005
0.5	***E. coli***	Glucose	Park and Zeikus, 2000

[a]Power per unit area of anode electrode surface.
[b]Single-culture MFCs are in boldface.

(Magnuson et al., 2000; Nevin and Lovley, 2000; Magnuson et al., 2001). *Geobacter* spp., for example, must grow on the electrode surface to transfer electrons (Bond and Lovley, 2003). A related mechanism, nanopili electron transfer, involves transfer of electrons through conductive nanopili, which make contact with each other or with the electrode (Reguera et al., 2005; Gorby et al., 2006). Although this mechanism involves direct contact of cell structures (i.e., the nanopili) with the electrode surface, the cells themselves can be present in the outer layers of the biofilm. The third mechanism does not involve direct contact of the cell or any cell surface structures with the electrode because the electrons are transferred to the solid acceptor through diffusible intermediate electron carriers, such as quinones or other redox-active compounds (e.g., pyocyanin) (Newman and Kolter, 2000; Nevin and Lovley, 2002; Rabaey et al., 2004; Bond and Lovley, 2005). This electron transfer mechanism, which we call mediated electron transfer, can be used by dissimilatory metal-reducing bacteria (e.g., *Shewanella* spp.) or other *Gammaproteobacteria* (e.g., *Pseudomonas* spp.). Rabaey et al. (2004) showed that bacterial mediators are released into solution, allowing bacteria that do not produce the mediators to benefit from them. For the latter two electron transfer mechanisms, bacteria in biofilms can transfer electrons without contacting the electrode directly. The relative importance of these mechanisms in mixed cultures is not understood.

THE ADVANTAGES OF A DIVERSE MICROBIAL COMMUNITY

The food web in anaerobic digesters, which is well understood, depends on the beneficial interactions between different trophic groups. Although the food web in MFCs has not been studied in much detail, it is clear that MFCs containing mixed cultures are capable of producing more electric power than single-species systems. Therefore, some important advantages of mixed cultures are described in this section.

Food Webs

Bioconversion yields of pure-culture batch processes decrease with time due to either inhibition by pretreatment by-products (Zverlov et al., 2006) or accumulation of intermediate products of the desired bioconversion reaction (Ezeji et al., 2004). These problems can be avoided when food webs incorporating trophic groups that can extract energy from the pretreatment by-products and metabolic intermediates are an integral part of the bioprocess. Members of these trophic groups remove the inhibitors, which should increase the yield of the desired product. As described earlier, this is a critical function of the anaerobic food webs in conventional anaerobic digestion processes, where specific groups of organisms maintain low concentrations of critical intermediate products and promote flux of carbon and electrons from the feedstock material to the desired end product by reducing direct inhibition of microbial activity by metabolic intermediates or by increasing the Gibbs free energy released by intermediated metabolic reactions (Gujer and Zehnder, 1983; McCarty and Smith, 1986).

Metabolic Flexibility

Mixed cultures have the ability to handle variations in substrate composition because alternative metabolic pathways are available among different members of the microbial community. Such metabolic resilience requires large metagenomes, such as those available in the human gut to extract energy from polysaccharides (Ley et al., 2006; Turnbaugh et al., 2006). The stability—possibly due to metabolic flexibility—of *aerobic* biological waste treatment systems has been correlated with increased biodiversity of the microbial community (Yu and Mohn, 2001; Egli et al., 2003; Pynaert et al., 2003). In *anaerobic* digestion systems, however, a direct link between functional stability, community stability, and higher biodiversity has not been demonstrated, but functional stability (i.e., performance) appears to be correlated with community flexibility (i.e., the ability of a system to shift the flow of electrons and carbon to the same products through alternative pathways) (Fernandez et al., 2000; Hashsham et al., 2000). Therefore, in anaerobic systems, flexibility may be more important than diversity.

Open Cultures

Because foreign bacteria cannot become established due to effective competition by members of the microbial community, mixed-culture fermentations do not require expensive procedures for preventing contamination, such as addition of antibiotics or sterilization of the fermenter and feedstock. This is especially important for agricultural waste streams, which contain high concentrations of bacteria (Angenent et al., 2004). An additional advantage of the open culture is that mixed-culture fermentations can be operated as continuous rather than batch processes. Because contamination by foreign microbes is unlikely, mixed-culture processes may be operated indefinitely. For pure-culture bioprocessing, the chances of contamination increase as the period of operation increases, and therefore, almost all industrial bioprocessing is performed in a batch mode (Kleerebezem and van Loosdrecht, 2007).

Reduction of Nonspecific Binding

The efficiency of enzymatic hydrolysis of cellulose is reduced by the strong intermolecular interactions of cellulose with other common fiber polysaccharides, such as pectin and hemicellulose. These interactions prevent binding of cellulases to their substrate. This problem can be circumvented by appropriate pretreatment and use of enzyme cocktails that contain "accessory" enzymes, such as xylanase and pectinase which degrade these noncellulosic polysaccharides (Berlin et al., 2007). Similar enzyme cocktails (e.g., mixtures of cellulases and xylanase) are naturally present in the microbial communities that are present in the gut of the termite *Odontotermes obesus* and in the mammalian rumen (Paul et al., 1993; Ekinci et al., 2001). In addition, noncarbohydrate polymers, such as lignin, bind cellulases nonproductively (Sutcliffe and Saddler, 1986). In the laboratory, this problem has been circumvented by adding bovine serum albumin, which coats lignin and reduces its interaction with cellulases (Yang and Wyman, 2006). It seems likely that extracellular proteins produced by mixed-culture lignocellulose-degrading biofilms may also reduce the nonspecific binding of lignin with cellulases.

Oxygen Removal and Growth-Promoting Factors

Strictly anaerobic cellulolytic bacteria benefit from the presence of facultatively anaerobic, noncellulolytic bacteria that remove oxygen and produce growth factors. Kato et al. (2005) showed that the cellulose-degrading capabilities of a defined, five-species mixed culture were reduced by eliminating a noncellulolytic bacterium from the community.

Improved Resistance to Bacteriophages

In the early and mid 1900s, industrial production of butanol with pure cultures of *C. acetobutylicum* was plagued by bacteriophage infections. To save the industry, microbiologists searched for new bacterial isolates that were more resistant to bacteriophage attack (Zverlov et al., 2006). Diverse microbial communities with metabolic flexibility should be more resistant to bacteriophage attack because different species or strains with similar metabolic functions can take over.

WAYS TO IMPROVE BIOFUELS GENERATION WITH MIXED CULTURES

Historically, bioprocess engineers treated mixed-culture systems as black boxes and were unaware of the effects of process modifications on the structure of the microbial communities. In the absence of a fundamental understanding of the microbiology of these systems, these engineers have attempted to maximize conversion rates by optimizing reactor configuration and/or reactor operation. Among other alternatives, the latter approach includes appropriate selection of electron acceptor and addition of inhibitors that target a specific trophic group (e.g., methanogens). Recent advances in molecular biology, however, have provided new tools that can be used to manipulate the composition of the microbial community, monitor the effects of process modifications on the community composition, select for specific fermentation products, and optimize the conversion rates and yields. This section describes two methods that can be used to manipulate the composition of the microbial community: (i) bioaugmentation, which may include horizontal gene transfer; and (ii) phage biocontrol.

Bioaugmentation

Bioaugmentation can be used when modeling or systems biology analysis shows that a metabolic pathway that is needed to produce a useful energy carrier or its precursor is missing from the community metabolome (Dejonghe et al., 2001). Bioaugmentation is the artificial introduction of specific organisms to a microbial community. For example, Boon et al. (2002) successfully introduced a 3-chloroanaline-degrading microbe to activated sludge. An alternative outcome from bioaugmentation was demonstrated after introduction of a bacterium to an endophytic (i.e., living in a plant) microbial community in poplar (Taghavi et al., 2005). These investigators showed that genes for toluene degradation remained in the community even though the introduced species disappeared. This phenomenon was explained as movement of the plasmid between species through horizontal gene transfer, through the same mechanism by which antibiotic resistance genes move between species in mixed cultures (Shoemaker et al., 2001). Workers have already used horizontal gene transfer to add function to a bioprocess (Bathe et al., 2004), and this approach is called gene bioaugmentation (Aspray et al., 2005). Horizontal gene transfer may thus circumvent problems of survivability of genetically modified microorganisms in a complex milieu (Sayler and Ripp, 2000) and may provide a mechanism for the genetic modification of mixed cultures with the goal of producing alternative energy carriers or their precursors.

Phage Biocontrol

When modeling or systems biology analysis shows that a specific metabolic pathway must be inhibited to obtain a certain product from a mixed culture, bacteriophages may be useful for control of community function. Bacteriophages are viruses that infect bacteria. These viruses can control the levels of specific bacterial species by causing infected cells to lyse. This approach is sometimes called "biocontrol." Due to the increasing prevalence of antibiotic resistance among human pathogens, bacteriophages are being investigated as an alternative to antibiotics for the treatment of bacterial infections in humans (Goodridge and Adebon, 2003). This approach is called "phage therapy." Biocontrol has also been used to control the composition of microbial communities in the gut (Verthe et al., 2004), for food safety applications (O'Flynn et al., 2004), and in bioprocess engineering (Lu et al., 2003). For example, O'Flynn et al. (2004) used a cocktail of three bacteriophages to specifically lyse and kill *Escherichia coli* O157:H7 to prevent illness resulting from consumption of contaminated meat. They also found that bacteriophage-insensitive mutants (i.e., *E. coli* strains that are insensitive to bacteriophages) developed following the challenge, but these insensitive *E. coli* strains reverted to bacteriophage sensitivity within 50 generations. Based on further studies, the authors concluded that insensitive mutants should not hinder the use of bacteriophages as biocontrol agents (O'Flynn et al., 2004).

Biocontrol has also been studied for control of unwanted species in mixed-culture bioprocesses. The importance of phage ecology in controlling conventional bioprocesses, such as sauerkraut fermentation and activated sludge systems, has been understood for many years (Hantula et al., 1991; Lu et al., 2003), and exogenous biocontrol of these systems was proposed. For example, Thomas et al. (2002) propagated bacteriophages with the ultimate goal to control filamentous bacteria, including specific mycobacteria, in activated-sludge systems to prevent excessive biological foaming. These successful applications of biocontrol suggest that it could also be used to manipulate the flux of carbon and electrons through metabolic pathways in mixed-culture fermentations to improve the yield of useful energy carriers.

OUTLOOK

Mixed-culture processing of waste to generate bioenergy has numerous advantages over single- or co-culture processing. However, to choose the proper biofuel product or its precursor from wastes, the process engineer must understand the limitations of the food web in accumulating fermentation end products and the influence of waste strength and composition on the economic viability of the process. For example, for low-strength wastes it would not make sense to produce soluble biofuels that are expensive to separate. For such wastes, the engineer should choose to produce gaseous products, such as methane, which freely bubbles out of the bioreactor liquor. Another option would be to produce electric power with mixed-culture MFCs; however, the food web in this process is not well understood, and the effect of waste composition on the coulombic efficiency is not completely known. If scale-up of the MFC is possible, it seems likely that engineers will first focus on easily degradable soluble components of wastewaters, such as carbohydrates or carboxylic acids. New research techniques that are being developed by system biologists in conjunction with powerful modeling efforts and lab-scale experiments are necessary to completely understand the anaerobic food web and to be truly in control of the process. Then, tools such as bioaugmentation and phage control could aid to produce additional fermentation end products besides methane, hydrogen, carboxylic acids, and electric current.

Acknowledgments. L.T.A. was supported by the National Research Initiative of the USDA Cooperative State Research, Education and Extension Service, grant number 2004-35504-14896.

We thank Miriam Rosenbaum (Washington University) for providing Fig. 4.

REFERENCES

Agbogbo, F. K., and M. T. Holtzapple. 2006. Fermentation of rice straw/chicken manure to carboxylic acids using a mixed culture of marine mesophilic microorganisms. *Appl. Biochem. Biotechnol.* **129-132:**997–1014.

Agbogbo, F. K., and M. T. Holtzapple. 2007. Fixed-bed fermentation of rice straw and chicken manure using a mixed culture of marine mesophilic microorganisms. *Bioresour. Technol.* **98:**1586.

Aiello-Mazzarri, C., G. Coward-Kelly, F. K. Agbogbo, and M. T. Holtzapple. 2005. Conversion of municipal solid waste into carboxylic acids by anaerobic countercurrent fermentation: effect of using intermediate lime treatment. *Appl. Biochem. Biotechnol.* **127:**79–94.

Akin, D. E., and R. Benner. 1988. Degradation of polysaccharides and lignin by ruminal bacteria and fungi. *Appl. Environ. Microbiol.* **54:**1117–1125.

Angelidaki, I., and B. K. Ahring. 1994. Anaerobic thermophilic digestion of manure at different ammonia loads: effect of temperature. *Water Res.* **28:**727–731.

Angenent, L. T., K. Karim, M. H. Al-Dahhan, B. A. Wrenn, and R. Domínguez-Espinosa. 2004. Production of bioenergy and biochemicals from industrial and agricultural wastewater. *Trends Biotechnol.* **22:**477–485.

Angenent, L. T., S. Sung, and L. Raskin. 2002. Methanogenic population dynamics during startup of a full-scale anaerobic sequencing batch reactor treating swine waste. *Water Res.* **36:**4648–4654.

Aspray, T. J., S. K. Hansen, and R. G. Burns. 2005. A soil-based microbial biofilm exposed to 2,4-D: bacterial community development and establishment of conjugative plasmid pJP4. *FEMS Microbiol. Ecol.* **54:**317–327.

Azbar, N., P. Ursillo, and R. E. Speece. 2001. Effect of process configuration and substrate complexity on the performance of anaerobic processes. *Water Res.* **35:**817–829.

Bahl, H., W. Andersch, K. Braun, and G. Gottschalk. 1982a. Effect of pH and butyrate concentration on the production of acetone and butanol by *Clostridium acetobutylicum* grown in continuous culture. *Appl. Microbiol. Biotechnol.* **V14:**17.

Bahl, H., W. Andersch, and G. Gottschalk. 1982b. Continuous production of acetone and butanol by *Clostridium acetobutylicum* in a two-stage phosphate limited chemostat. *Appl. Microbiol. Biotechnol.* **V15:**201.

Bahl, H., M. Gottwald, A. Kuhn, V. Rale, W. Andersch, and G. Gottschalk. 1986. Nutritional factors affecting the ratio of solvents produced by *Clostridium acetobutylicum. Appl. Environ. Microbiol.* **52:**169–172.

Barsoukov, E., and J. R. Macdonald. 2005. Impedance spectroscopy: theory, experiment, and applications. John Wiley & Sons, Hoboken, NJ.

Bathe, S., T. V. Mohan, S. Wuertz, and M. Hausner. 2004. Bioaugmentation of a sequencing batch biofilm reactor by horizontal gene transfer. *Water Sci. Technol.* **49:**337–344.

Batstone, D. J., and J. Keller. 2001. Variation of bulk properties of anaerobic granules with wastewater type. *Water Res.* **35:**1723–1729.

Batstone, D. J., J. Keller, R. B. Newell, and M. Newland. 2001. Modelling anaerobic degradation of complex wastewater. II. Parameter estimation and validation using slaughterhouse effluent. *Bioresour. Technol.* **75:**75–85.

Berlin, A., V. Maximenko, N. Gilkes, and J. Saddler. 2007. Optimization of enzyme complexes for lignocellulose hydrolysis. *Biotechnol. Bioeng.* **97:**287–296.

Bond, D. R., and D. R. Lovley. 2003. Electricity production by *Geobacter sulfurreducens* attached to electrodes. *Appl. Environ. Microbiol.* **69:**1548–1555.

Bond, D. R., and D. R. Lovley. 2005. Evidence for involvement of an electron shuttle in electricity generation by *Geothrix fermentans. Appl. Environ. Microbiol.* **71:**2186–2189.

Boon, N., L. De Gelder, H. Lievens, and W. Verstraete. 2002. Bioaugmenting bioreactors for the continuous removal of 3-chloroaniline by slow release approach. *Environ. Sci. Technol.* **36:**4698.

Breznak, J. A., and A. Brune. 1994. Role of microorganisms in the digestion of lignocellulose by termites. *Annu. Rev. Entomol.* **39:**453–487.

Brune, A., D. Emerson, and J. A. Breznak. 1995a. The termite gut microflora as an oxygen sink: microelectrode determination of oxygen and pH gradients in guts of lower and higher termites. *Appl. Environ. Microbiol.* **61:**2681–2687.

Brune, A., E. Miambi, and J. A. Breznak. 1995b. Roles of oxygen and the intestinal microflora in the metabolism of lignin-derived phenylpropanoids and other monoaromatic compounds by termites. *Appl. Environ. Microbiol.* **61:**2688–2695.

Chan, W. N., and M. T. Holtzapple. 2003. Conversion of municipal solid wastes to carboxylic acids by thermophilic fermentation. *Appl. Biochem. Biotechnol.* **111:**93–112.

Chaudhuri, S. K., and D. R. Lovley. 2003. Electricity generation by direct oxidation of glucose in mediatorless microbial fuel cells. *Nat. Biotechnol.* **21:**1229–1232.

Chen, Y., Y. Inbar, Y. Hadar, and R. L. Malcolm. 1989. Chemical properties and solid-state CPMAS ^{13}C-NMR of composted organic matter. *Sci. Total Environ.* **81/82:**201–208.

Chyi, Y. T., and R. R. Dague. 1994. Effects of particulate size in anaerobic acidogenesis using cellulose as a sole carbon source. *Water Environ. Res.* **66:**670–678.

Cleveland, L., and A. Grimstone. 1964. The fine structure of the flagellate *Mixotricha paradoxa* and its associated micro-organisms. *Proc. R. Soc. London B* **159:**668.

Contreras, E. M., L. Giannuzzi, and N. E. Zaritzky. 2000. Growth kinetics of the filamentous microorganism *Spaerotilus natans* in a model system of a food industry wastewater. *Water Res.* **34:**4455–4463.

Cord-Ruwisch, R., D. R. Lovley, and B. Schink. 1998. Growth of *Geobacter sulfurreducens* with acetate in syntrophic cooperation with hydrogen-oxidizing anaerobic partners. *Appl. Environ. Microbiol.* **64:**2232–2236.

Dejonghe, W., N. Boon, D. Seghers, E. M. Top, and W. Verstraete. 2001. Bioaugmentation of soils by increasing microbial richness: missing links. *Environ. Microbiol.* **3:**649.

Diaz, E. E., A. J. M. Stams, R. Amils, and J. L. Sanz. 2006. Phenotypic properties and microbial diversity of methanogenic granules from a full-scale upflow anaerobic sludge bed reactor treating brewery wastewater. *Appl. Environ. Microbiol.* **72:**4942–4949.

Domke, S. B., C. Aiello-Mazzarri, and M. T. Holtzapple. 2004. Mixed acid fermentation of paper fines and industrial biosludge. *Bioresour. Technol.* **91:**41.

Duran, M., and R. E. Speece. 1998. Staging of anaerobic processes for reduction of chronically high concentrations of propionic acid. *Water Environ. Res.* **70:**241–248.

Egli, K., C. Langer, H. R. Siegrist, A. J. Zehnder, M. Wagner, and J. R. Van Der Meer. 2003. Community analysis of ammonia and nitrite oxidizers during start-up of nitritation reactors. *Appl. Environ. Microbiol.* **69:**3213–3222.

Ekinci, M. S., N. Ozcan, E. Ozkose, and H. J. Flint. 2001. A study on cellulolytic and hemicellulolytic enzymes of anaerobic rumen bacterium *Ruminococcus flavefaciens* strain 17. *Turk. J. Vet. Anim. Sci.* 25:703–709.

Electric Power Research Institute. 2002. *Water and Sustainability*, vol. 1. *Research Plan.* Electric Power Research Institute, Palo Alto, CA.

Ezeji, T. C., N. Qureshi, and H. P. Blaschek. 2004. Butanol fermentation research: upstream and downstream manipulations. *Chem. Rec.* 4:305–314.

Fang, H. H., and H. Liu. 2002. Effect of pH on hydrogen production from glucose by a mixed culture. *Bioresour. Technol.* 82:87–93.

Fang, H. H. P., and H. Q. Yu. 2001. Acidification of lactose in wastewater. *J. Environ. Eng.* 127:825–831.

Fernandez, A. S., S. A. Hashsham, S. L. Dollhopf, L. Raskin, O. Glagoleva, F. B. Dazzo, R. F. Hickey, C. S. Criddle, and J. M. Tiedje. 2000. Flexible community structure correlates with stable community function in methanogenic bioreactor communities perturbed by glucose. *Appl. Environ. Microbiol.* 66:4058–4067.

Fuchs, W., H. Binder, G. Mavrias, and R. Braun. 2003. Anaerobic treatment of wastewater with high organic content using a stirred tank reactor coupled with a membrane filtration unit. *Water Res.* 37:902–908.

Gohil, A., and G. Nakhla. 2006. Treatment of food industry waste by bench-scale upflow anaerobic sludge blanket-anoxic-aerobic system. *Water Environ. Res.* 78:974–985.

Goodridge, L., and S. T. Adebon. 2003. Bacteriophage biocontrol and bioprocessing: application of phage therapy to industry. *SIM News* 53:254–262.

Gorby, Y. A., S. Yanina, J. S. McLean, K. M. Rosso, D. Moyles, A. Dohnalkova, T. J. Beveridge, I. S. Chang, B. H. Kim, K. S. Kim, D. E. Culley, S. B. Reed, M. F. Romine, D. A. Saffarini, E. A. Hill, L. Shi, D. A. Elias, D. W. Kennedy, G. Pinchuk, K. Watanabe, S. Ishii, B. Logan, K. H. Nealson, and J. K. Fredrickson. 2006. Electrically conductive bacterial nanowires produced by *Shewanella oneidensis* strain MR-1 and other microorganisms. *Proc. Natl. Acad. Sci. USA* **103**:11358–11363.

Gottschalk, G. 1986. *Bacterial Metabolism.* Springer-Verlag, New York, NY.

Grupe, H., and G. Gottschalk. 1992. Physiological events in *Clostridium acetobutylicum* during the shift from acidogenesis to solventogenesis in continuous culture and presentation of a model for shift induction. *Appl. Environ. Microbiol.* 58:3896–3902.

Gujer, W., and A. J. B. Zehnder. 1983. Conversion processes in anaerobic digestion. *Water Sci. Technol.* 15:127–167.

Hantula, J., A. Kurki, P. Vuoriranta, and D. H. Bamford. 1991. Ecology of bacteriophages infecting activated sludge bacteria. *Appl. Environ. Microbiol.* 57:2147–2151.

Hao, X., P. S. Mir, M. A. Shah, and G. R. Travis. 2005. Influence of canola and sunflower diet amendments on cattle feed lot manure. *J. Environ. Qual.* 34:1439–1445.

Harper, S. R., and F. G. Pohland. 1986. Recent developments in hydrogen management during anaerobic biological wastewater treatment. *Biotechnol. Bioeng.* 28:585–602.

Hashsham, S. A., A. S. Fernandez, S. L. Dollhopf, F. B. Dazzo, R. F. Hickey, J. M. Tiedje, and C. S. Criddle. 2000. Parallel processing of substrate correlates with greater functional stability in methanogenic bioreactor communities perturbed by glucose. *Appl. Environ. Microbiol.* 66:4050–4057.

Hattori, S., Y. Kamagata, S. Hanada, and H. Shoun. 2000. *Thermacetogenium phaeum* gen. nov., sp. nov., a strictly anaerobic, thermophilic, syntrophic acetate-oxidizing bacterium. *Int. J. Syst. Evol. Microbiol.* 50:1601–1609.

He, Z., and L. T. Angenent. 2006. Application of bacterial biocathodes in microbial fuel cells. *Electroanalysis* 18:2009–2015.

He, Z., S. D. Minteer, and L. T. Angenent. 2005. Electricity generation from artificial wastewater with an upflow microbial fuel cell. *Environ. Sci. Technol.* 39:5262–5267.

He, Z., N. Wagner, S. D. Minteer, and L. T. Angenent. 2006. The upflow microbial fuel cell with an interior cathode: assessment of the internal resistance by impedance spectroscopy. *Environ. Sci. Technol.* 40:5212–5217.

Holtzapple, M. T., R. R. Davison, M. K. Ross, S. Albrett-Lee, M. Nagwani, C. M. Lee, C. Lee, S. Adelson, W. Kaar, D. Gaskin, H. Shirage, N. S. Chang, V. S. Chang, and M. E. Loescher. 1999. Biomass conversion to mixed alcohol fuels using the MixAlco process. *Appl. Biochem. Biotechnol.* 77–79:609–631.

Ieropoulos, I. A., J. Greenman, C. Melhuish, and J. Hart. 2005. Comparative study of three types of microbial fuel cell. *Enzyme Microb. Technol.* 37:238–245.

Jones, D. T., and D. R. Woods. 1986. Acetone-butanol fermentation revisited. *Microbiol. Rev.* 50:484–524.

Kato, S., S. Haruta, Z. J. Cui, M. Ishii, and Y. Igarashi. 2005. Stable coexistence of five bacterial strains as a cellulose-degrading community. *Appl. Environ. Microbiol.* 71:7099–7106.

Katz, E., and I. Willner. 2003. Probing biomolecular interactions at conductive and semiconductive surfaces by impedance spectroscopy: routes to impedimetric immunosensors, DNA-sensors, and enzyme biosensors. *Electroanalysis* 15:913–947.

Kennedy, M., D. List, and Y. Lu. 1999. Apple pomace and products derived from apple pomace: uses, composition, and analysis, p. 75–120. *In* H.-F. Linskens and J. E. Jackson (ed.), *Modern Methods of Plant Analysis.* Springer-Verlag, Berlin, Germany.

Kim, G. T., G. Webster, J. W. T. Wimpenny, B. H. Kim, H. J. Kim, and A. J. Weightman. 2006. Bacterial community structure, compartmentalization and activity in a microbial fuel cell. *J. Appl. Microbiol.* 101:698–710.

Kim, H. J., H. S. Park, M. S. Hyun, I. S. Chang, M. Kim, and B. H. Kim. 2002. A mediator-less microbial fuel cell using a metal reducing bacterium, *Shewanella putrefaciens. Enzyme Microb. Technol.* 30:145–152.

Kim, J. R., B. Min, and B. E. Logan. 2005. Evaluation of procedures to acclimate a microbial fuel cell for electricity production. *Appl. Microbiol. Biotechnol.* 68:23–30.

Kleerebezem, R., and A. J. M. Stams. 2000. Kinetics of syntrophic cultures: a theoretical treatise on butyrate fermentation. *Biotechnol. Bioeng.* 67:529–543.

Kleerebezem, R., and M. C. M. van Loosdrecht. 2007. Mixed culture biotechnology for bioenergy production. *Curr. Opin. Biotechnol.* 18:207–212.

Kraemer, J. T., and D. M. Bagley. 2005. Continuous fermentative hydrogen production using a two-phase reactor system with recycle. *Environ. Sci. Technol.* 39:3819–3825.

Leadbetter, J. R., and J. A. Breznak. 1996. Physiological ecology of *Methanobrevibacter cuticularis* sp. nov. and *Methanobrevibacter curvatus* sp. nov., isolated from the hindgut of the termite *Reticulitermes flavipes. Appl. Environ. Microbiol.* 62:3620–3631.

Leadbetter, J. R., L. D. Crosby, and J. A. Breznak. 1998. *Methanobrevibacter filiformis* sp. nov., a filamentous methanogen from termite hindguts. *Arch. Microbiol.* 169:287–292.

Leadbetter, J. R., T. M. Schmidt, J. R. Graber, and J. A. Breznak. 1999. Acetogenesis from H_2 plus CO_2 by spirochetes from termite guts. *Science* 283:686–689.

Lee, M. J., and S. H. Zinder. 1988. Isolation and characterization of a thermophilic bacterium which oxidizes acetate in syntrophic association with a methanogen and which grows acetogenically on H_2-CO_2. *Appl. Environ. Microbiol.* 54:124–129.

Lepisto, S. S., and J. A. Rintala. 1997. Start-up and operation of laboratory-scale thermophilic upflow anaerobic sludge blanket reactors treating vegetable processing wastewaters. *J. Chem. Technol. Biotechnol.* 68:331–339.

Leschine, S. B. 1995. Cellulose degradation in anaerobic environments. *Annu. Rev. Microbiol.* 49:399–426.

Ley, R. E., P. J. Turnbaugh, S. Klein, and J. I. Gordon. 2006. Microbial ecology: human gut microbes associated with obesity. *Nature* 444:1022–1023.

Liu, H., S. Grot, and B. E. Logan. 2005. Electrochemically assisted microbial production of hydrogen from acetate. *Environ. Sci. Technol.* 39:4317–4320.

Logan, B. E. 2004. Biologically extracting energy from wastewater; biohydrogen production and microbial fuel cells. *Environ. Sci. Technol.* 38:160A–167A.

Lu, Z., F. Breidt, V. Plengvidhya, and H. P. Fleming. 2003. Bacteriophage ecology in commercial sauerkraut fermentations. *Appl. Environ. Microbiol.* 69:3192–3202.

Mackie, R. I., and M. P. Bryant. 1995. Anaerobic digestion of cattle waste at mesophilic and thermophilic conditions. *Appl. Microbiol. Biotechnol.* 43:346–350.

Magnuson, T. S., A. L. Hodges-Myerson, and D. R. Lovley. 2000. Characterization of a membrane-bound NADH-dependent Fe(3+) reductase from the dissimilatory Fe(3+)-reducing bacterium *Geobacter sulfurreducens*. *FEMS Microbiol. Lett.* 185:205–211.

Magnuson, T. S., N. Isoyama, A. L. Hodges-Myerson, G. Davidson, M. J. Maroney, G. G. Geesey, and D. R. Lovley. 2001. Isolation, characterization and gene sequence analysis of a membrane-associated 89 kDa Fe(III) reducing cytochrome c from *Geobacter sulfurreducens*. *Biochem. J.* 359:147–152.

Malherbe, S., and T. E. Cloete. 2002. Lignocellulose biodegradation: fundamental and applications. *Rev. Environ. Sci. Technol.* 1:105 114.

McCarty, P. L. 1964a. Anaerobic waste treatment fundamentals, part four: process design. *Public Works* **1964**:94–99.

McCarty, P. L. 1964b. Anaerobic waste treatment fundamentals, part one: chemistry and microbiology. *Public Works* **1964**:107–112.

McCarty, P. L. 1964c. Anaerobic waste treatment fundamentals, part three: toxic materials and their control. *Public Works* **1964**:91–94.

McCarty, P. L. 1964d. Anaerobic waste treatment fundamentals, part two: environmental requirements and control. *Public Works* **1964**: 123–126.

McCarty, P. L., and D. P. Smith. 1986. Anaerobic wastewater treatment. *Environ. Sci. Technol.* 20:1200–1206.

McInerney, M. J. 1988. Anaerobic hydrolysis and fermentation of fats and proteins. *In* A. J. B. Zehnder (ed.), *Biology of Anaerobic Microorganisms*. John Wiley and Sons, Inc., New York, NY.

Miller, D. N., and V. H. Varel. 2003. Swine manure composition affects the biochemical origins, composition, and accumulation of odorous compounds. *J. Anim. Sci.* 81:2131–2138.

Moon, H., I. S. Chang, and B. H. Kim. 2006. Continuous electricity production from artificial wastewater using a mediator-less microbial fuel cell. *Bioresour. Technol.* 97:621–627.

Moral, R., J. Moreno-Caselles, M. D. Perez-Murcia, A. Perez-Espinosa, B. Rufete, and C. Paredes. 2005. Characterization of the organic matter pool in manures. *Bioresour. Technol.* 96:153–158.

Mosier, N. S., R. Hendrickson, M. Brewer, N. Ho, M. Sedlak, R. Dreshel, G. Welch, B. S. Dien, A. Aden, and M. R. Ladisch. 2005. Industrial scale-up of pH-controlled liquid hot water pretreatment of corn fiber for fuel ethanol production. *Appl. Biochem. Biotechnol.* 125:77–97.

Mulkerrins, D., E. O'Connor, B. Lawlee, P. Barton, and A. Dobson. 2004. Assessing the feasibility of achieving biological nutrient removal from wastewater at an Irish food processing factory. *Bioresour. Technol.* 91:207–214.

Nemerow, N. L., and A. Desgupta. 1991. *Industrial and Hazardous Waste Treatment*. Van Nostrand Reinhold, New York, NY.

Nevin, K. P., and D. R. Lovley. 2000. Lack of production of electron-shuttling compounds or solubilization of Fe(III) during reduction of insoluble Fe(III) oxide by *Geobacter metallireducens*. *Appl. Environ. Microbiol.* 66:2248–2251.

Nevin, K. P., and D. R. Lovley. 2002. Mechanisms for accessing insoluble Fe(III) oxide during dissimilatory Fe(III) reduction by *Geothrix fermentans*. *Appl. Environ. Microbiol.* 68:2294–2299.

Newman, D. K., and R. Kolter. 2000. A role for excreted quinones in extracellular electron transfer. *Nature* 405:94–97.

Nüsslein, B., K. J. Chin, W. Eckert, and R. Conrad. 2001. Evidence for anaerobic syntrophic acetate oxidation during methane production in the profundal sediment of subtropical Lake Kinneret (Israel). *Environ. Microbiol.* 3:460–470.

Odelson, D. A., and J. A. Breznak. 1983. Volatile fatty acid production by the hindgut microbiota of xylophagous termites. *Appl. Environ. Microbiol.* 45:1602–1613.

O'Flynn, G., R. P. Ross, G. F. Fitzgerald, and A. Coffey. 2004. Evaluation of a cocktail of three bacteriophages for biocontrol of *Escherichia coli* O157:H7. *Appl. Environ. Microbiol.* 70:3417–3424.

O'Sullivan, C. A., P. C. Burrell, W. P. Clarke, and L. L. Blackall. 2005. Structure of a cellulose degrading bacterial community during anaerobic digestion. *Biotechnol. Bioeng.* 92:871–878.

Park, D. H., and J. G. Zeikus. 2000. Electricity generation in microbial fuel cells using neutral red as an electronophore. *Appl. Environ. Microbiol.* 66:1292–1297.

Paul, J., S. Saxena, and A. Varma. 1993. Ultrastructural studies of the termite (*Odontotermes obesus*) gut microflora and its cellulolytic properties. *World J. Microbiol. Biotechnol.* 9:108–112.

Petersen, S. P., and B. K. Ahring. 1992. The influence of sulfate on substrate utilization in a thermophilic sewage sludge digester. *Appl. Microbiol. Biotechnol.* 36:805–809.

Pynaert, K., B. F. Smets, S. Wyffels, D. Beheydt, S. D. Siciliano, and W. Verstraete. 2003. Characterization of an autotrophic nitrogen-removing biofilm from a highly loaded lab-scale rotating biological contactor. *Appl. Environ. Microbiol.* 69:3626–3635.

Rabaey, K., N. Boon, S. D. Siciliano, M. Verhaege, and W. Verstraete. 2004. Biofuel cells select for microbial consortia that self-mediate electron transfer. *Appl. Environ. Microbiol.* 70:5373–5382.

Rabaey, K., G. Lissens, S. D. Siciliano, and W. Verstraete. 2003. A microbial fuel cell capable of converting glucose to electricity at high rate and efficiency. *Biotechnol. Lett.* 25:1531–1535.

Ramey, D. E., and Environmental Energy, Inc. May 1998. Continuous two stage, dual path anaerobic fermentation of butanol and other organic solvents using two different strains of bacteria. U.S. patent 5,753,474.

Rao, K., V. Chaudhari, S. Varanasi, and D.-S. Kim. 2007. Enhanced ethanol fermentation of brewery wastewater using the genetically modified strain *E. coli* KO11. *Appl. Microbiol. Biotechnol.* 74:50–60.

Raphael, L., M. Ely, M.-Y. Orly, and A. B. Edward. 1991. Cellulosome-like entities in *Bacteroides cellulosolvens*. *Curr. Microbiol.* 22:27.

Reguera, G., K. D. McCarthy, T. Mehta, J. S. Nicoll, M. T. Tuominen, and D. R. Lovley. 2005. Extracellular electron transfer via microbial nanowires. *Nature* 435:1098–1101.

Rhoads, A., H. Beyenal, and Z. Lewandowski. 2005. Microbial fuel cell using anaerobic respiration and biomineralization manganese as a cathodic reactant. *Environ. Sci. Technol.* 39:4666–4671.

Ringeisen, B. R., E. Henderson, P. K. Wu, J. Peietron, R. Ray, B. Little, J. C. Biffinger, and J. M. Jones-Meehan. 2006. High power density from a miniature microbial fuel cell using *Shewanella oneidensis* DSP10. *Environ. Sci. Technol.* 40:2629–2634.

Rittmann, B. E., and P. L. McCarty. 2001. *Environmental Biotechnology*. McGraw Hill, Boston, MA.

Rodriguez, J., R. Kleerebezem, J. M. Lema, and M. C. M. van Loosdrecht. 2006. Modeling product formation in anaerobic mixed culture fermentations. *Biotechnol. Bioeng.* 93:592–606.

Rodriguez-Martinez, J., S. Y. Martinez-Amador, and Y. Garza-Garcia. 2005. Comparative anaerobic treatment of wastewater from pharmaceutical, brewery, paper and amino acid industries. *J. Ind. Microbiol. Biotechnol.* 32:691–696.

Rozendal, R. A., H. V. M. Hamelers, G. J. W. Euverink, S. J. Metz, and C. J. N. Buisman. 2006. Principle and perspectives of hydrogen production through biocatalyzed electrolysis. *Int. J. Hydrogen Energy* 31:1632–1640.

Russ, W., and R. Meyer-Pittroff. 2004. Utilizing waste products from the food production and processing industries. *Crit. Rev. Food Sci. Nutr.* 44:57–62.

Sayler, G. S., and S. Ripp. 2000. Field applications of genetically engineered microorganisms for bioremediation processes. *Curr. Opin. Biotechnol.* 11:286.

Schink, B. 1992. Syntrophism among prokaryotes, p. 276–299. *In* A. Balows, H. G. Trüper, M. Dworkin, W. Harder, and K. H. Schleifer (ed.), *The Prokaryotes*. Springer Verlag, New York, NY.

Schmidt, J. E., and B. K. Ahring. 1993. Effects of hydrogen and formate on the degradation of propionate and butyrate in thermophilic granules from an upflow anaerobic sludge blanket reactor. *Appl. Environ. Microbiol.* 59:2546–2551.

Schnürer, A., F. P. Houwen, and B. H. Svensson. 1994. Mesophilic syntrophic acetate oxidation during methane formation by a triculture at high ammonia concentration. *Arch. Microbiol.* **162:**70–74.

Schnürer, A., B. Schink, and B. H. Svensson. 1996. *Clostridium ultunense* sp nov, a mesophilic bacterium oxidizing acetate in syntrophic association with a hydrogenotrophic methanogenic bacterium. *Int. J. Syst. Bacteriol.* 46:1145–1152.

Schnürer, A., B. H. Svensson, and B. Schink. 1997. Enzyme activities in and energetics of acetate metabolism by the mesophilic syntrophically acetate-oxidizing anaerobe *Clostridium ultunense*. *FEMS Microbiol. Lett.* 154:331–336.

Schnürer, A., G. Zellner, and B. H. Svensson. 1999. Mesophilic syntrophic acetate oxidation during methane formation in biogas reactors. *FEMS Microbiol. Ecol.* 29:249–261.

Shoemaker, N. B., H. Vlamakis, K. Hayes, and A. A. Salyers. 2001. Evidence for extensive resistance gene transfer among *Bacteroides* spp. and among *Bacteroides* and other genera in the human colon. *Appl. Environ. Microbiol.* 67:561–568.

Siso, M. I. G. 1996. The biotechnological utilization of cheese whey: a review. *Bioresour. Technol.* 57:1–11.

Speece, R. E. 1996. *Anaerobic Biotechnology for Industrial Wastewaters*. Archaea Press, Nashville, TN.

Stams, A. J. M. 1994. Metabolic interactions between anaerobic bacteria in methanogenic environments. *Antonie Leeuwenhoek* **66:** 271–294.

Stanton, T. B., and E. Canale-Parola. 1980. *Treponema bryantii* sp. nov., a rumen spirochete that interacts with cellulolytic bacteria. *Arch. Microbiol.* 127:145.

Sutcliffe, R., and J. N. Saddler. 1986. The role of lignin in the adsorption of cellulases during enzymatic treatment of lignocellulosic material. *Biotechnol. Bioeng. Symp.* 17:749–762.

Taghavi, S., T. Barac, B. Greenberg, B. Borremans, J. Vangronsveld, and D. van der Lelie. 2005. Horizontal gene transfer to endogenous endophytic bacteria from poplar improves phytoremediation of toluene. *Appl. Environ. Microbiol.* 71:8500–8505.

Tashiro, Y., K. Takeda, G. Kobayashi, K. Sonomoto, A. Ishizaki, and S. Yoshino. 2004. High butanol production by *Clostridium saccharoperbutylacetonicum* N1-4 in fed-batch culture with pH-stat continuous butyric acid and glucose feeding method. *J. Biosci. Bioeng.* 98:263–268.

Tchobanoglous, G., F. L. Burton, and H. D. Stensel. 2003. *Wastewater Engineering, Treatment and Reuse*, 4th ed. Metcalf & Eddy, McGraw Hill, New York, NY.

Thanakoses, P., A. S. Black, and M. T. Holtzapple. 2003a. Fermentation of corn stover to carboxylic acids. *Biotechnol. Bioeng.* 83:191–200.

Thanakoses, P., N. A. Mostafa, and M. T. Holtzapple. 2003b. Conversion of sugarcane bagasse to carboxylic acids using a mixed culture of mesophilic microorganisms. *Appl. Biochem. Biotechnol.* 105–108:523–546.

Thauer, R. K., K. Jungermann, and K. Decker. 1977. Energy conservation in chemotrophic anaerobic bacteria. *Bacteriol. Rev.* **41:**100–180.

Thomas, J. A., J. A. Soddell, and D. I. Kurtböke. 2002. Fighting foam with phages? *Water Sci. Technol.* 46:511–518.

Turnbaugh, P. J., R. E. Ley, M. A. Mahowald, V. Magrini, E. R. Mardis, and J. I. Gordon. 2006. An obesity-associated gut microbiome with increased capacity for energy harvest. *Nature* **444:** 1027–1031.

Vandevoorde, L., and W. Verstraete. 1987. Anaerobic solid state fermentation of cellulosic substrates with possible application to cellulase production. *Appl. Microbiol. Biotechnol.* **V26:**479.

Verthe, K., S. Possemiers, N. Boon, M. Vaneechoutte, and W. Verstraete. 2004. Stability and activity of an *Enterobacter aerogenes*-specific bacteriophage under simulated gastro-intestinal conditions. *Appl. Microbiol. Biotechnol.* 65:465–472.

Vijayaraghavan, K., D. Ahmad, and R. Lesa. 2006. Electrolytic treatment of beer brewery wastewater. *Ind. Eng. Chem. Res.* **45:**6854–6859.

Weiland, P. 2006. Biomass digestion in agriculture: a successful pathway for the energy production and waste treatment in Germany. *Eng. Life Sci.* 6:302–309.

Wenzel, M., R. Radek, G. Brugerolle, and H. Konig. 2003. Identification of the ectosymbiotic bacteria of *Mixotricha paradoxa* involved in movement symbiosis. *Eur. J. Protistol.* 39:11–23.

Whang, L.-M., C.-J. Hsiao, and S.-S. Cheng. 2006. A dual-substrate steady-state model for biological hydrogen production in an anaerobic hydrogen fermentation process. *Biotechnol. Bioeng.* **95:**492–500.

Yang, B., and C. E. Wyman. 2006. BSA treatment to enhance enzymatic hydrolysis of cellulose in lignin containing substrates. *Biotechnol. Bioeng.* **94:**611–617.

Yu, H. Q., and H. H. P. Fang. 2003. Acidogenesis of gelatin-rich wastewater in an upflow anaerobic reactor: influence of pH and temperature. *Water Res.* 37:55–66.

Yu, H. Q., Y. Mu, and H. H. P. Fang. 2004. Thermodynamic analysis of product formation in mesophilic acidogenesis of lactose. *Biotechnol. Bioeng.* 87:813–822.

Yu, Z., and W. W. Mohn. 2001. Bacterial diversity and community structure in an aerated lagoon revealed by ribosomal intergenic spacer analyses and 16S ribosomal DNA sequencing. *Appl. Environ. Microbiol.* 67:1565–1574.

Zhang, R., Z. Zhang, and The Regents of the University of California. January 2002. Biogasification of solid waste with an anaerobic-phased solids digester system. U.S. patent 6,342,378.

Zhang, T., H. Liu, and H. H. Fang. 2003. Biohydrogen production from starch in wastewater under thermophilic condition. *J. Environ. Manage.* 69:149–156.

Zinder, S. H. 1994. Syntrophic acetate oxidation and "reversible acetogenesis," p. 386–415. *In* H. L. Drake (ed.), *Acetogenesis*. Chapman and Hall, New York, NY.

Zinder, S. H., and M. Koch. 1984. Non-aceticlastic methanogenesis from acetate: acetate oxidation by a thermophilic syntrophic coculture. *Arch. Microbiol.* 138:263–272.

Zverlov, V. V., O. Berezina, G. A. Velikodvorskaya, and W. H. Schwarz. 2006. Bacterial acetone and butanol production by industrial fermentation in the Soviet Union: use of hydrolyzed agricultural waste for biorefinery. *Appl. Microbiol. Biotechnol.* **71:** 587–597.

Bioenergy
Edited by J. Wall et al.

Chapter 16

Biomethane from Biomass, Biowaste, and Biofuels

Ann C. Wilkie

Biofuel production from agricultural, municipal, and industrial wastes is efficiently accomplished through conversion to biogas, a mixture of mostly methane (CH_4) and carbon dioxide (CO_2), via anaerobic digestion. Anaerobic digestion is a process by which a complex mixture of symbiotic microorganisms transforms organic materials under oxygen-free conditions into biogas, nutrients, and additional cell matter, leaving salts and refractory organic matter. In practice, microbial anaerobic conversion to methane is a process for both effective waste treatment and sustainable energy production. In waste treatment, this process can provide a source of energy while reducing the pollution and odor potential of the substrate. Unlike fossil fuels, use of renewable methane represents a closed carbon cycle and thus does not contribute to increases in the atmospheric concentration of carbon dioxide (Wilkie, 2005).

Microbial methane production has the potential for reducing the demand for fossil fuels like coal, oil, and natural gas that have provided the power for developing and maintaining the technologically advanced modern world. However, fossil resources are finite, and their continued recovery and use significantly impact our environment and affect the global climate. Shortages of oil and gas are predicted to occur within our lifetimes or those of our children. To prepare for a transition to more sustainable sources of energy, viable alternatives for conservation, supplementation, and replacement must be explored, posthaste.

Biogas production from agricultural, municipal, and industrial wastes can contribute to sustainable energy production, especially when nutrients conserved in the process are returned to agricultural production (Fig. 1). Little energy is consumed in the process, and consequently the net energy from biogas production is high compared to other conversion technologies. The technology for methane production is scalable and has been applied globally to a broad range of organic waste feedstocks, most commonly animal manures (Wilkie et al., 2004). However, methane production is not limited to conversion of animal manures. Biogas can be made from most biomass and waste materials regardless of the composition and over a large range of moisture contents, with limited feedstock preparation. The feedstocks for this omnivorous process can be composed of carbohydrates, lignocellulosics, proteins, fats, or mixtures of these components. The process is suitable for conversion of liquid, slurry, and solid wastes; it can even be employed for the conversion of gaseous combustion products (synthesis gas) from thermochemical gasification systems. In addition, methane production can be effectively applied to improve energy yields from other biofuel production processes including bioethanol, biodiesel, and biohydrogen production. Implementation of digestion technology at agricultural, municipal, and industrial facilities allows efficient decentralized energy generation and distribution to local markets. While traditionally applied to wastes and wastewaters, the anaerobic digestion of energy crops can also be employed in a sustainable bioenergy system.

ANAEROBIC MICROBIOLOGY

Methane is the end product of anaerobic metabolism—a metabolic sequence carried out by communities of hydrolytic bacteria and fungi, acid-producing intermediary organisms, and finally, methanogenic *Archaebacteria*. Methane-producing communities are very stable and resilient, but they are also complex and largely undefined.

Buswell and Sollo (1948) demonstrated the treatability of a range of wastes and emphasized the concept of an acid phase versus a methane phase, showing the importance of volatile organic acids as intermediates in the process. They also demonstrated the applicability of a stoichiometric equation that balanced carbon, hydrogen, and oxygen (equation 1) to predict the amount

Ann C. Wilkie • Soil and Water Science Department, University of Florida—Institute of Food and Agricultural Sciences, Gainesville, FL 32611.

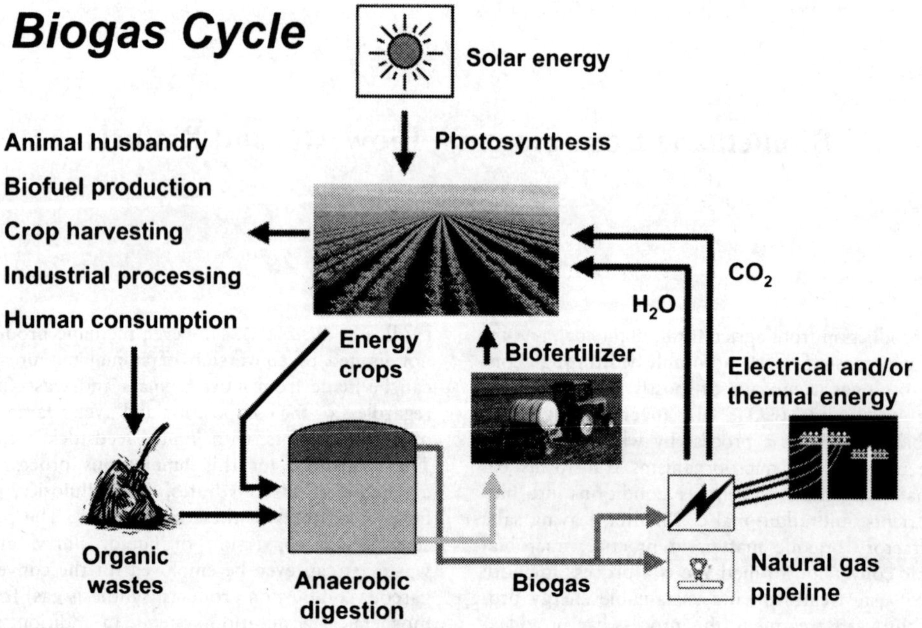

Figure 1. Biogas cycle

of methane and carbon dioxide evolved from conversion of organic compounds with a known empirical formula. Later, ^{14}C tracers were used to show that acetate was indeed cleaved to form methane and carbon dioxide, suggesting that acids were important intermediates in the conversion process.

$$C_nH_aO_b + (n - a/4 - b/2)H_2O \rightarrow (n/2 - a/8 + b/4)CO_2 + (n/2 + a/8 - b/4)CH_4 \quad (1)$$

Of great importance to the understanding of anaerobic microbiology was the discovery of Bryant et al. (1967), through isolating the elusive "S-organism" from *Methanobacterium omelianskii*, that the conversion of ethanol to methane was accomplished with a mixed culture. The discovery of other cocultures quickly followed, and the number of species isolated in pure culture increased. With the identification of closely coupled syntrophic cocultures of methanogens and other species, the earlier hypothesis of an acid phase followed by a methanogenic phase developed into a more descriptive scheme that embraces the importance of hydrogen as an intermediate in the process.

First the fermentative, or hydrolytic, bacteria and fungi hydrolyze complex organic polymers and ferment them to organic acids, hydrogen (or formate), and carbon dioxide. The hydrogen-producing acetogenic bacteria ferment the larger acids to a combination of acetic acid, one-carbon compounds, hydrogen, and carbon dioxide. The homoacetogenic bacteria synthesize acetic acid by utilizing hydrogen/carbon dioxide or one-carbon compounds or by hydrolyzing multicarbon compounds.

The methanogenic *Archaebacteria* uniquely catabolize acetic acid and one-carbon compounds to methane. The methanogens are obligate anaerobes that can pick up electrons from dead-end fermentations, through interspecies hydrogen transfer, and shuttle these electrons through a unique form of respiration which results in the reduction of carbon dioxide to methane. The organisms that use hydrogen to reduce CO_2 are commonly regarded as the earliest life forms due to their chemoautotrophic abilities.

All morphological forms are represented among the methanogens including rods, cocci, spirals, sarcinae, and filamentous organisms. Surprisingly, this diverse group of organisms is known to metabolize a very limited number of substrates including acetate, formate, methanol, acetone, methylamines, carbon monoxide, and H_2/CO_2. The substrates for methanogenesis divide these organisms into groups, two of which are notably important in active digesters: the aceticlastic methanogens, which cleave acetic acid, and the hydrogen-utilizing methanogens, which utilize hydrogen and one-carbon compounds. However, this distinction is not always useful since some species may metabolize both substrates.

For aceticlastic methanogens, low levels of acetate (<50 mg/liter) favor the growth of more-filamentous organisms (e.g., *Methanosaeta*) that must rely on a larger surface-to-volume ratio in order to improve substrate diffusion rates. High levels of acetate favor the predominance of clusters of aceticlastic methanogens (e.g., *Methanosarcina*), which have lower surface-to-volume ratios that serve to protect them from the inhibitory nature of high organic acid concentrations. Differences in maximum growth rate and substrate utilization affinities can be exploited to select for predominant methanogens. Organisms such as *Methanosarcina* should be favored for selection if high conversion rates of high-strength wastes are the primary goal, whereas *Methanosaeta* should be favored if low effluent biochemical oxygen demand is more important. In addition, these attributes can be exploited together by staging an anaerobic process with the first stage favoring high conversion rates and the next stage favoring effluent quality.

PROCESS

In practice, anaerobic digestion is the engineered methanogenic decomposition of organic matter, carried out in reactor vessels, called digesters, that may be mixed or unmixed and heated or unheated. The process uses a mixed culture of ubiquitous organisms, and due to its mixed-culture nature, there are no requirements for feedstock sterilization and no contamination concerns. Stable digester operation requires that the bacterial groups be in dynamic equilibrium, as some of the intermediate metabolites (hydrogen, propionate, ammonia, and sulfide) can be inhibitory and the pH of the system must remain near neutral. Also, methane is sparingly soluble, such that end product recovery is efficient and economical as the gas separates itself from the aqueous phase and is easily removed from the digester through piping that conveys it to storage for final use.

Current commercial-scale methods of methane production yield from 50 to 97% conversion of substrate to methane on an energy basis, depending on the feedstock. The mean oxidation state of the feedstock determines the stoichiometry of the end products. Carbohydrate substrates yield 50% methane and 50% carbon dioxide, while more-reduced feedstocks (e.g., lipids) yield higher proportions of methane. The theoretical methane yield of carbohydrates, proteins, and fats is given in Table 1. Also, carbohydrate-rich substrates yield more methane than do feedstocks with high concentrations of lignocellulose.

Table 1. Theoretical methane yield of biomass components

Component	Methane yield (liter/g of VS)
Carbohydrates	0.35
Proteins (leucine)	0.57
Fats (lauric acid)	0.95

The natural assemblages in the mixed culture have evolved to form robust and stable cultures with extremely broad substrate utilization capabilities. There is no requirement for genetically modified organisms to extend catabolic activity, so sterilization of process residuals is not necessary. Although the free energy available from anaerobic conversion of substrates to methane is low, causing low microbial-growth rates, the activity and turnover rates of substrates are higher than for aerobic metabolism. Also, anaerobic respiration of methanogens results in the production of a noninhibitory product, methane, that moves into the gaseous phase, which contrasts with other fermentation processes that produce inhibitory final products (e.g., ethanol) that remain in solution. This gives the process a distinct advantage for either continuous or batch conversion of substrates to energy products.

Of significance for the application of anaerobic digestion is the high level of energy recovery in the biogas compared to the energy content of the substrate utilized. The efficiency of this conversion is directly related to the low level of free energy of reaction available for microbial synthesis. Rather than transforming the energy in waste into sludge as in aerobic processes, a minimal amount of this energy is consumed by anaerobic cell synthesis and the rest is retained in the methane end product. This also explains the low rates of microbial growth in anaerobic systems compared to aerobic processes.

Chemical oxygen demand (COD) is a convenient measurement to estimate the organic content in a wastewater or biomass sample and, theoretically, 0.35 liter of methane is formed from 1 g of COD digested. In aerobic processes such as the activated-sludge process, the sludge yield can be as high as 0.5 kg of dry solids per kg of COD utilized, whereas the sludge yields for anaerobic processes range from 0.03 to 0.15 kg of dry solids per kg of COD consumed depending on the substrate. Sludge by-product from aerobic processes requires further treatment for stabilization in order to reduce odor and pollution potential. Furthermore, after stabilization, the sludge still requires final disposal. Anaerobic treatment processes, in contrast, produce a relatively small amount of sludge by-product which is more stable, less capable of causing odor or pollution problems, and ready for final disposition in sustainable crop production. Also, anaerobic treatment results in pathogen decimation through microbial competition and starvation.

Nutrients contained in the organic matter are conserved and mineralized to more soluble and biologically available forms, providing a more predictable biofertilizer. Since sludge production in anaerobic digestion is minimal, virtually all of the nitrogen and phosphorus contained in the original waste is retained in the treated effluent. By recycling the treated effluents back to productive agricultural lands at appropriate rates, the crops benefit from the presence of these important plant nutrients. Where insufficient cropland is available, other nutrient recovery technologies may be employed to reduce the nutrient content of the digested wastewater.

DESIGNS

The construction of anaerobic digesters for biogas production has little in common with that of industrial fermentors. The low value of energy products compared with fermentation products necessitates low-cost construction and materials. While industrial fermentation vessels are often jacketed stainless steel tanks, with baffles, agitators, and clean-in-place systems, and constructed on elevated stands, anaerobic digesters are often simple insulated concrete or carbon steel tanks constructed with low-cost materials either on or below the surface. Without the need for efficient aeration, the requirement for mixing must only meet the needs for microbial contact with substrate, uniform temperature, and prevention of solids accumulation. Since sterility is not a concern, no clean-in-place systems or provisions to prevent microbial contamination are required. Unlike other fermentations, no specific process or equipment is required for product recovery, since methane is relatively insoluble and therefore separates spontaneously.

Anaerobic digesters must be essentially gas-tight vessels with a provision for introducing feedstock and removing effluent and biogas. The classical anaerobic digester is essentially a chemostat. Tanks with rigid tops must have provisions for pressure and vacuum relief, and biogas piping must meet safety standards. Tank tops may also be floating rigid tops or flexible membrane materials. Simple heat exchangers may be placed internally or external to the tank, and mixing can employ agitators, simple recirculation of the mixed liquor, or injection of compressed biogas.

There are two broad classifications of digesters, those that rely on suspended growth of microorganisms and those that employ a mechanism for immobilization to retain active microbial biomass within the vessel. With feedstocks containing high levels of suspended solids, nonimmobilized designs are generally used including covered anaerobic lagoons, complete-mix reactors, plug-flow reactors, and anaerobic contact reactors. These digesters require relatively long hydraulic retention times (HRT) of 15 to 60 days and moderate organic loading rates (OLR), typically expressed as weight of organic matter (volatile solids [VS] or COD) per culture volume of reactor per day. The maximum OLR and minimum HRT that can be applied are dependent on operating temperature, waste characteristics, and reactor type.

Feeds with low concentrations of suspended solids (<2%) can be digested in high-rate immobilized reactors such as the upflow anaerobic sludge-bed digester (UASB), anaerobic filter (AF), and fixed-film systems. These reactors retain high concentrations of immobilized microorganisms, permit low HRT without organism washout, and are particularly suited for treatment of soluble wastewaters. The tendency of microbial consortia to adhere to surfaces and grow as a biofilm spurred the development of both the aerobic trickling filter and the AF reactor (also called a packed bed). While the principle of filling a reactor with a packing media is straightforward, the selection of packing material and operational strategies may have significant effects on performance and costs. Media used for packing have included natural materials such as stones, clay, wood, bamboo, and reeds, as well as polymers made of polyvinyl chloride, polyethylene, and polypropylene. Polymers shaped as rings, bio-balls, and oriented modular media have been used in various applications. In some cases, AFs rely on trapping microbial solids within the media rather than using a true biofilm for microbial activity. Thus, the term "fixed-film digesters" should be used to designate true biofilm reactor designs.

In immobilized reactors where a highly degradable, high-COD wastewater (>20 g/liter) is fed, effluent recycling can be employed to overcome localized acidification of the microbial biomass. In addition, highly acidic wastewater can benefit from effluent recycle to minimize the requirement for added alkali by using the alkalinity of the effluent. Phased digestion is often employed for highly degradable waste, where a primary acidogenic reactor is operated at short HRT to form intermediate acids, which are then fed into a methanogenic reactor. This approach can control sharp pH swings, enhance biofilm and granular sludge activity, and lower overall process HRT. A further refinement involves staging, where reactors are employed in series to achieve higher treatment efficiencies. The first reactor is optimized to maximize biogas production (higher OLR), whereas the second reactor is optimized for treatment efficiency (lower OLR).

INOCULATION

The use of a source high in anaerobic microbes (e.g., digester effluent) to start up an anaerobic system is called inoculation. The quality and quantity of inoculum are critical to the performance, time required, and stability of biomethanogenesis during commissioning (start-up) or restart of an anaerobic digester. Much agricultural processing occurs on a seasonal basis, and at the start of a new campaign, the anaerobic treatment operation must be restarted after a period of being idled. In addition, a digester may need inoculation after maintenance operations. In manures and some wastes, the microbes needed for digestion may be already present in the waste in small numbers, albeit sufficient to act as an inoculum, and will develop into a fully functional bacterial population if the right conditions are provided, including a suitable temperature and retention time. Other wastes, especially from industry, may be relatively sterile and require the addition of inoculum. With batch and plug flow designs, inoculum must be added with the feed and low inoculum levels may lead to imbalance due to the more rapid growth rate of acid-forming bacteria (compared to methanogens) and depression of pH. Depending on the buffering capacity (alkalinity), a digester may be able to overcome low inoculum rates.

Granular sludge, the microbial by-product from UASB reactors, has been shown to be a practical source for inoculum due to its stability in storage, microbial density, and availability. Granular sludge may be used to enhance methanogen populations for start-up of complete-mix reactors and anaerobic filters as well as UASBs. Start-up of immobilized systems requires that biofilm or granule growth be optimized to achieve design performance quickly. During start-up, performance parameters (methane gas content, ratio of acids-to-alkalinity, and pH) should be carefully monitored to ensure that performance is not deteriorating. In many applications, a high inoculation rate is not feasible or digester effluent is not available. Under such circumstances, one must obtain inoculum from an anaerobic environment (anaerobic sediments or animal manure) and gradually develop and acclimate the inoculum to the level needed. The major obstacle to overcome is the fact that, during growth toward a mature population, acid formers may grow faster than methanogens, leading to an increase in volatile organic acids, reduced pH, and loss of methane production. This can be prevented by buffering the system and/or reducing the feed loading rate.

NUTRITION

Nitrogen and phosphorus are the major nutrients required for anaerobic digestion. These elements are building blocks for cell synthesis, and their requirements are directly related to the microbial growth in anaerobic digesters. An empirical formula for a typical anaerobic bacterium is $C_5H_7O_2NP_{0.06}$ (Speece, 1996). Thus, the nitrogen and phosphorus requirements for cell growth are 12 and 2%, respectively, of the volatile solids converted to cell biomass. If 10% of the degradable solids are converted into microbial biomass, this would be equivalent to a requirement of 1.2 and 0.2% of the biodegradable volatile solids, respectively, for nitrogen and phosphorus. Ammonia is also an important contributor to the buffering capacity in digesters but can be toxic to the process at high levels.

Methane production and volatile acid utilization may be enhanced when micronutrients are added to nutrient-deficient substrates. Requirements for several micronutrients have been identified, including iron, copper, manganese, zinc, molybdenum, nickel, and vanadium (Wilkie et al., 1986; Speece, 1996). Available forms of these nutrients may be limiting because of their ease of precipitation and removal by reactions with phosphate and sulfide. Limitations of these micronutrients have been demonstrated in reactors in which the analytical procedures failed to distinguish between available and sequestered forms. Other nutrients needed in intermediate concentrations include sodium, potassium, calcium, magnesium, and sulfur. Combining wastes is an effective means of overcoming nutrient limitations. Codigestion with manure often enhances the conversion of other biomass and waste feedstocks through balancing micronutrients.

CONTROL/TOXICITY

Biological methanogenesis has been reported at temperatures ranging from 2°C (in marine sediments) to over 100°C (in geothermal areas). Most applications of this fermentation have been performed under ambient (15 to 25°C), mesophilic (30 to 40°C), or thermophilic (50 to 60°C) temperatures. In general, the overall process kinetics doubles for every 10-degree increase in operating temperature, up to some critical temperature (about 60°C) above which a rapid drop-off in microbial activity occurs. Most commercial anaerobic digesters are operated at mesophilic or ambient temperatures. A higher operating temperature permits reduced reactor size.

Thermophilic digesters exhibit some differences compared to mesophilic digesters. The microbial populations operating in the thermophilic range are genetically unique, do not survive well at lower temperatures, and can be more sensitive to temperature fluctuations outside their optimum range. Also, ammonia is more toxic in thermophilic digesters due to a higher proportion of free ammonia. Although thermophilic digesters have higher energy requirements, heat losses can be minimized through effective insulation and use of heat exchangers to reduce effluent heat losses. Thermophilic operation is practiced when the reduced reactor size justifies the higher energy requirements and added effort to ensure stable performance, when process wastewater is already hot, or when pathogen removal is of greater concern.

Biomethanogenesis is sensitive to several groups of inhibitors including alternate electron acceptors (oxygen, nitrate, and sulfate), sulfides, heavy metals, halogenated hydrocarbons, volatile organic acids, ammonia, and cations. The intermittent presence of microbial inhibitors in the wastewater stream can lead to serious process upsets and failure. The toxic effect of an inhibitory compound depends on its concentration and the ability of the bacteria to acclimate to its effects. The inhibitory concentration depends on different variables, including pH, HRT, temperature, and the ratio of the toxic substance concentration to the bacterial mass concentration. Antagonistic and synergistic effects are also common. Methanogenic populations are usually influenced by dramatic changes in their environment but can be acclimated to otherwise toxic concentrations of many compounds.

Organic acids, pH, and alkalinity are related parameters that influence digester performance. Under conditions of overloading and the presence of inhibitors, methanogenic activity cannot remove hydrogen and organic acids as fast as they are produced. The result is accumulation of acids, depletion of buffer, and depression of pH. If uncorrected via pH control and reduction in feeding, pH will drop to levels that stop the fermentation. Independent of pH, extremely high volatile acid levels (>10,000 mg/liter) also inhibit performance. The major alkalis contributing to alkalinity are ammonia and bicarbonate. The most common chemicals for pH control are sodium hydroxide, lime, magnesium hydroxide, and sodium bicarbonate. Lime produces calcium bicarbonate up to the point of solubility of 1,000 mg/liter. Sodium bicarbonate adds directly to the bicarbonate alkalinity without reacting with carbon dioxide. However, precautions must be taken not to add this chemical to a level of sodium toxicity (>3,500 mg/liter). Currently, the control of feed rate to an anaerobic digester most often relies on off-line measurements of volatile organic acids to prevent process upset through manual intervention. Several investigators have advocated control schemes based on biogas production rate, alkalinity, liquid-phase hydrogen, pH, and digester substrate concentration.

BIOGAS USE

Biogas is a flexible form of renewable energy that may be used directly for process heat and steam or converted to electricity in reciprocating engines, gas turbines, or fuel cells. Biogas is composed mostly of methane, as is natural gas, but may contain some impurities such as hydrogen sulfide. Biogas can be used readily in all applications designed for natural gas such as direct combustion for absorption heating and cooling, cooking, space and water heating, and drying. Biogas can also be upgraded to natural gas specifications and injected into the existing network of natural gas pipelines. Biogas may also be catalytically transformed into hydrogen, ethanol, or methanol.

If cogeneration is employed in the biogas conversion system, heat normally wasted may be recovered and used for hot water production. In gas turbines, the waste heat may be used to make steam and drive an additional steam turbine, with the final waste heat going to hot water production. This is termed a combined cycle cogeneration system. Combining hot water recovery with electricity generation, biogas can provide an overall conversion efficiency of 65 to 85%.

For smaller biogas installations, shaft horsepower and electrical generation are most effectively achieved by the use of a stationary internal combustion engine. Adequate removal of hydrogen sulfide is important to reduce engine maintenance requirements. If compressed for use as an alternative transportation fuel in light and heavy-duty vehicles, biogas can use the same existing technique for fueling as currently used for compressed-natural-gas vehicles. In many countries, biogas is viewed as an environmentally attractive alternative to diesel and gasoline for operating buses and other local transit vehicles. The exhaust fume emissions from methane-powered engines are lower than the emissions from diesel and gasoline engines. Also, the sound level generated by methane-powered engines is generally lower than that generated by diesel engines.

WASTE RESOURCES

Recently, anaerobic wastewater pretreatment has attained extensive acceptance for a variety of industrial wastewaters associated with food processing, beverages, breweries, distilleries, and most recently pulp and paper production. Batch operation of the production sequence is common in these industries, producing a wastewater of variable strength and quantity, complicating the operation of a continuous biological treatment system. A few examples of agricultural and industrial waste streams are identified in Table 2. Traditionally, treatment of manures and municipal sludge have been the most prominent applications of anaerobic digestion, and there is currently a resurgence in the promotion of on-farm biogas production from animal manure (see the Agstar Program, U.S. Environmental Protection Agency; http://www.epa.gov/agstar/). Anaerobic digestion of municipal sludge is applied at many municipal wastewater treatment plants. However, the pretreat-

Table 2. Examples of agricultural and industrial wastewater strength

Feed source	Wastewater COD (mg/liter)
Beef processing	7,500
Beverage	1,600
Brewery	1,900–2,400
Clam	3,500
Confectionery	9,500
Dairy	1,900–5,260
Distillery	95,000
Ice cream	29,063
Municipal	200
Pharmaceutical	9,985
Pork processing	1,572
Potato	2,000–10,500
Pulp and paper	1,600–16,400
Rendering	8,800
Sauerkraut	10,000
Starch	8,800–11,400
Sugar beet refining	5,000–20,000
Vegetable	2,300–10,000
Whey	8,900
Yeast	30,000

ment of municipal wastewater by high-rate anaerobic treatment offers a new application of biogas production in municipal wastewater treatment works (van Haandel and Lettinga, 1994). Conversion of soluble biochemical oxygen demand in municipal wastewater to biogas avoids much of the costs of aeration and the production of residual sludge requiring disposal.

The organic fraction of municipal solid wastes (MSW) also has a high potential for biogas production. A majority of MSW is disposed of in landfills, many of which are implementing biogas recovery systems. However, nutrients contained in MSW are sequestered in landfills and the land area for these operations is often not suitable for economic development. Separation of the organic fraction of MSW and conversion to biogas can produce compost residuals that are suitable for crop production, which results in a sustainable solid-waste recycling system.

CODIGESTION

Digestion of a given waste can often benefit from codigestion with other waste streams that are locally available. There are many reasons for considering codigestion, including the potential to reach a more favorable economy of scale due to materials handling or optimal production and utilization of biogas. Codigestion may provide increased revenues from tipping fees as well as from enhanced biogas production. Very dry feedstocks may be blended with wastewaters to facilitate handling and digestion. Waste high in protein, which could suffer from ammonia toxicity, can be blended with lignocellulosic materials, which are low in nitrogen, to improve digestion rates. Household or other waste streams can be blended with manure to improve the microbial diversity and contribute essential micronutrients. The organic fraction of MSW is suitable for codigestion with farm and industrial wastes, and many successful examples can be found in Europe (see the European Anaerobic Digestion Network; http://www.adnett.org/). Conversion of agricultural, municipal, and industrial wastes to biogas offers a sustainable means for biofuels production, yet the role of biogas production in the production of other biofuels (e.g., alcohols, biodiesel, hydrogen, and syngas) is also an application worthy of exploitation.

METHANE IN BIOETHANOL PRODUCTION

The production of biomethane at bioethanol production facilities can contribute to the energy requirements of ethanol production or to increasing the energy yields from substrates for sale to local markets as fuel or electricity. Depending on the feedstock and process design, ethanol production results in several by-products (Fig. 2) which may include crop residues, stillage, evaporator condensate, condensed solubles, spent cake and/or distillers' grains, all of which have a high potential for methane production (Table 3). Stillage, a residual of the distillation of ethanol from fermentation liquor, contains a high level of biodegradable COD as well as nutrients and has a high pollution potential (Wilkie et al., 2000). Up to 20 liters of stillage may be generated for each liter of ethanol produced. Conversion of stillage to biogas and application of effluent to croplands results in a more sustainable ethanol production system.

Many ethanol plants minimize effluent discharges by evaporation of the stillage to produce evaporator condensate (used partially for makeup water) and condensed solubles. The evaporator condensate contains volatile fermentation products that can inhibit ethanol fermentation. Anaerobic digestion can remove these fermentation products and provide a liquid more suitable for process recycling. The distillers' grains and condensed solubles are normally blended for use in animal feed as dried distillers' grains and solubles. However, the current rapid expansion of ethanol production could lead to saturation of the feed market with dried distillers' grains and solubles, affecting the sale value of this by-product. Thus, there is an opportunity for biogas production from these by-products to offset facility energy requirements. In cellulosic ethanol production, nonfermentable hydrolysis products can also be converted to methane. Finally, crop residues may also be harnessed for biogas production, which can greatly improve the energy yield from ethanol production.

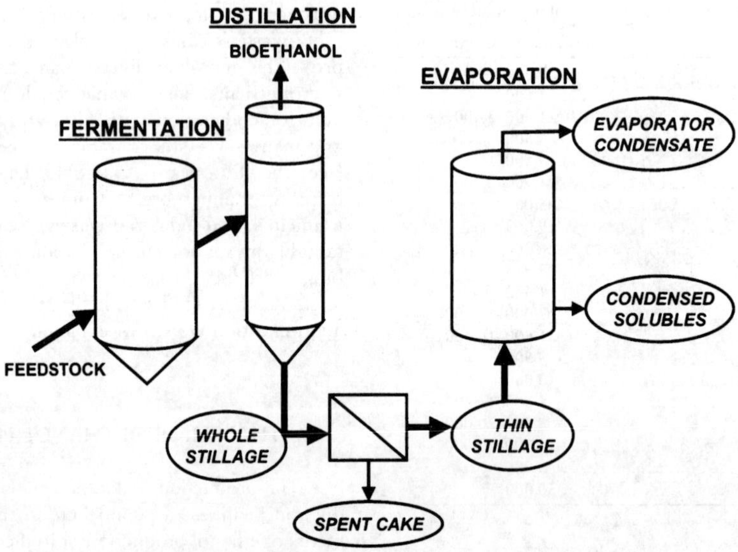

Figure 2. Potential biogas feedstocks from bioethanol production.

METHANE IN BIODIESEL PRODUCTION

Biodiesel is normally produced from either virgin plant oils or waste vegetable oils through a catalytic transesterification process. The typical biodiesel production process uses an alkaline hydrolysis reaction to convert vegetable oil into biodiesel by using methanol, potassium hydroxide, and heat. A transesterification reaction splits the glycerol group from the triglyceride oils, producing methyl esters (biodiesel) and glycerol by-product (Fig. 3). To purify the biodiesel, a washing process is employed to remove soaps, free fatty acids, and excess methanol, producing a washwater by-product. While process yields and inputs depend largely on oil type and quality, for every 100 liters of oil, approximately 25 liters of methanol and 0.8 kg of KOH/NaOH are consumed, yielding around 75 liters of biodiesel and 25 liters of crude glycerol. The washing process produces another 30 liters of biodiesel washwater. Both the crude glycerol and the biodiesel washwater have significant methane production potential. When vegetable oil is pressed from seeds (or algae), there is also a press cake by-product along with crop residues from harvesting that are both amenable to biogas production (Table 3). Conversion of biodiesel by-products to methane offers a sustainable treatment solution, while also providing additional energy. Methane can also be converted to methanol, an ingredient used in biodiesel production. Also, digester effluent could be used to grow oleaginous algae for biodiesel production.

Table 3. COD of some bioenergy by-products

Feedstock	COD (g/kg)
Ethanol thin stillage from corn	56.0–64.5
Ethanol stillage from beet molasses	55.5–116.0
Ethanol stillage from cane juice	22.0–45.0
Ethanol stillage from cane molasses	22.5–118.0
Ethanol stillage from cellulosics	19.1–140.0
Evaporator condensate	2.6–5.7
Condensed solubles	724
Dried distillers' grains	368
Crude glycerol from biodiesel production	1,800–2,600
Washwater from biodiesel	40.1–161.0
Press cake from oil crops	1,570

METHANE IN THE HYDROGEN ECONOMY

Hydrogen is often considered as a long-term solution to dwindling petroleum supplies and the environmental consequences of petroleum use in the transportation sector. However, hydrogen production and storage are still very expensive. Since water is the primary product of H_2 combustion, the fuel is viewed as a means to eliminate CO_2 emissions. Yet, if H_2 production is from fossil sources, it will still result in significant CO_2 emissions. Only the production of H_2 from renewable energy sources can result in reduced greenhouse gas emissions. One means of renewable H_2 production is through fermentation of organic matter.

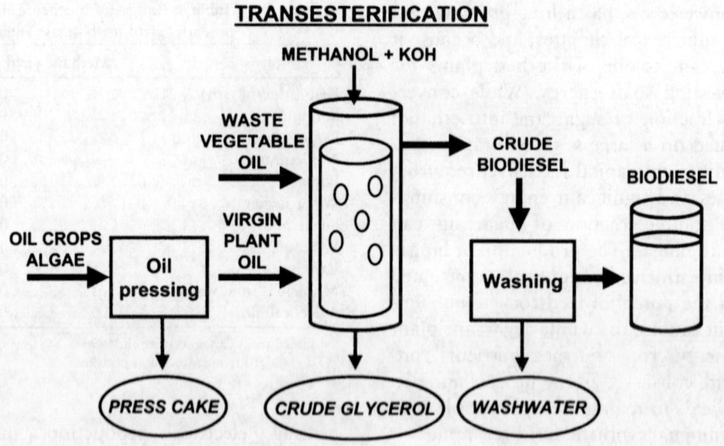

Figure 3. Potential biogas feedstocks from biodiesel production.

However, theoretically, only 33% of the energy in carbohydrates is available for microbial H_2 production due to the requirement to regenerate metabolic reducing potential (Angenent et al., 2004; Hungate, 1974). This means that 66% of the carbohydrate feedstock remains in the fermentation effluent and requires further processing. Anaerobic digestion can easily convert this residual carbon to biomethane, and the methane could then be converted to hydrogen catalytically. Still, the efficiency of conversion for methane production suggests that it is easier to convert all of the carbohydrate directly to methane rather than suffer the low yields of microbial hydrogen production. This methane could be upgraded to natural gas or converted into electricity, both of which are easier to transport than H_2.

There are other means by which methane factors into the hydrogen economy. First, the energy density of H_2 is four times less than that of CH_4 on a molar or volume basis, suggesting that methane could serve as a more efficient storage vector for hydrogen. Secondly, there is an existing infrastructure of pipelines for transporting CH_4 that are not suitable for moving H_2. Capitalizing on this network of pipelines, methane could be transported to regions of demand and converted to H_2 locally as required. Renewable methane, therefore, is an appropriate energy vector even if hydrogen is a desirable replacement fuel.

SYNTHESIS GAS

Another renewable fuel source that could integrate with methane production is the production of synthesis gas (syngas) through thermochemical gasification of biomass. Wastes and biomass crops can be gasified in a reduced atmosphere combustion process to convert the biomass into a mixture of CH_4, CO_2, CO, and H_2. While catalytic conversion of syngas to methanol has historical application for producing "wood alcohol," the H_2, CO_2, and CO in syngas can be used as a feedstock in methane production. Currently, catalysts for conversion of syngas to mixed higher alcohols (ethanol, propanol, and butanol) are in development, but in any of the catalytic processes, H_2S is problematic for catalyst longevity. Anaerobic digestion could serve as a process for syngas cleanup to convert the mixture to CH_4 (Sipma et al., 2006) and allow more-efficient catalytic conversion to further products (H_2, ethanol, or methanol). Pure-culture fermentation of syngas to ethanol is also in development (Younesi et al., 2005), a process which also generates acetate that may in turn be converted to CH_4 via anaerobic digestion.

ENERGY CROPS

Meeting the demand for alternative fuels from seasonal crops grown for bioenergy is potentially tenuous. Storage of crops can result in losses of carbohydrates available for fermentation to ethanol. Direct methane production from energy crops can overcome these losses because of the ability of the anaerobic digestion process to use fermentation intermediates as substrates. Harvested crops can be ensiled to preserve overall energy content, using technology with which farmers are already familiar. Further, any improvement in conversion efficiency that enhances cellulosic ethanol yields is equally applicable for biomass conversion to methane.

Sugarcane, a power crop, has a long growing season in tropical and subtropical climates, and because it is a C4 plant, sugarcane is one of the best plants for collecting and harvesting solar energy. While conversion of the soluble fraction of sugarcane into ethanol has been implemented on a large scale in Brazil, ethanol production facilities are capital intensive, requiring several unit processes and significant energy consumption. However, the soluble fraction of sugarcane can also be converted into biogas. The production of biogas requires much less investment, little energy is consumed in the process, and the potential feedstock is not limited to the sugars but can use the whole sugarcane plant as well as other energy crops. Further, nutrients contained in the cane are conserved in the process and can be returned to the fields to maintain a sustainable production cycle with minimal synthetic fertilizer inputs.

Cane juice can be digested directly to produce methane, without the need for alcohol fermentation, centrifugation and distillation, and the consumption of high-grade energy associated with these processes. Some 47% of the total energy in cane would be present in the biogas produced. The remaining bagasse could still be used for energy production through combustion, as currently implemented in the sugar industry. A further reduction of investment and operational costs and an increase in energy output could be obtained by subjecting not only the juice but the whole cane to anaerobic digestion. Assuming that 70% of the bagasse can be converted into methane, which is a realistic figure for a low-lignin (only 6.3%) plant such as sugarcane, then the energy conversion efficiency would increase to 80% of the energy content of cane (Chynoweth et al., 1993; Pate et al., 1984; van Haandel, 2005). By comparison, only 40% of the energy content of cane is actually converted into alcohol, consuming 24% of the cane energy in the process, while 12% is discharged as wastewater (stillage) and 24% remains in the excess bagasse (van Haandel, 2005).

Corn has also undergone whole-plant conversion to methane. Methane yields for corn at varying harvest times have ranged from 268 to 366 liters/kg of VS (Amon et al., 2007). Table 4 gives ranges of methane yield for various terrestrial and marine energy crops. Methane yields from seaweeds, grasses, and crops all approach theoretical yields, such that as much as 80% of biomass energy content could be recovered in methane.

Table 4. Ranges of biochemical methane potential yield for biomass energy crops[a]

Sample	Methane yield (at STP[b]) (liter/g of VS)
Kelp (*Macrocystis*)	0.39–0.41
Sorghum	0.26–0.39
Sargassum	0.26–0.38
Napiergrass	0.19–0.34
Poplar	0.23–0.32
Water hyacinth	0.19–0.32
Sugarcane	0.23–0.30
Willow	0.13–0.30
Laminaria	0.26–0.28
Municipal solid waste	0.20–0.22
Avicel cellulose	0.37

[a]Modified from Chynoweth et al., 1993.
[b]STP, standard temperature and pressure.

SUMMARY

Biogas production from agricultural, municipal, and industrial wastes is a sustainable means for producing a useful biofuel that can be used for process heating, electrical production, and vehicular fuel. Biogas can be upgraded and injected into natural gas pipelines, leveraging the existing distribution infrastructure. Liquids, slurries, solid wastes, and gaseous waste can all be processed by anaerobic digestion to form biogas. Several digester designs have been developed to optimize processing of different feedstocks. Digester size can be scaled to match the application, and centralized plants, codigesting a mixture of wastes, can be utilized to achieve economies of scale and improved performance. Methane production can be integrated into biorefineries since by-products from production of bioethanol, biodiesel, biohydrogen, and syngas are also suitable for anaerobic digestion, thus increasing net energy yields and recycling valuable nutrients for crop production. Finally, processing of terrestrial and marine energy crops to biomethane can result in higher energy yields than that of other biofuels. Given the diversity of feedstocks and ease of product recovery, methane from organic wastes and energy crops offers a major sustainable energy solution that is renewable, carbon dioxide neutral, and locally based, thereby protecting the environment, creating jobs, and strengthening local economies.

REFERENCES

Amon, T., B. Amon, V. Kryvoruchko, W. Zollitsch, K. Mayer, and L. Gruber. 2007. Biogas production from maize and dairy cattle manure—influence of biomass composition on the methane yield. *Agric. Ecosystems Environ.* **118:**173–182.

Angenent, L. T., K. Karim, M. H. Al-Dahhan, B. A. Wrenn, and R. Domíguez-Espinosa. 2004. Production of bioenergy and biochemicals from industrial and agricultural wastewater. *Trends Biotechnol.* **22:**477–485.

Bryant, M. P., E. A. Wolin, M. J. Wolin, and R. S. Wolfe. 1967. *Methanobacillus omelianskii*, a symbiotic association of two species of bacteria. *Arch. Mikrobiol.* **59:**20.

Buswell, A. M., and F. W. Sollo. 1948. The mechanism of the methane fermentation. *J. Am. Chem. Soc.* **70:**1778–1780.

Chynoweth, D. P., C. E. Turick, J. M. Owens, D. E. Jerger, and M. W. Peck. 1993. Biochemical methane potential of biomass and waste feedstocks. *Biomass Bioenergy* **5:**95–111.

Hungate, R. E. 1974. Potentials and limitations of microbial methanogenesis. *ASM News* 40:833–838.

Pate, F. M., J. Alvarez, J. D. Phillips, and B. R. Eiland. 1984. Sugarcane as a cattle feed: production and utilization. *In Florida Sugarcane Handbook. Bulletin 884.* Institute of Food and Agricultural Sciences, University of Florida, Gainesville.

Sipma, J., A. M. Henstra, S. N. Parshina, P. N. L. Lens, G. Lettinga, and A. J. M. Stams. 2006. Microbial CO conversions with applications in synthesis gas purification and bio-desulfurization. *Crit. Rev. Biotechnol.* 26:41–65.

Speece, R. E. 1996. *Anaerobic Biotechnology for Industrial Wastewaters.* Archae Press, Nashville, TN.

van Haandel, A. C. 2005. Integrated energy production and reduction of the environmental impact at alcohol distillery plants. *Water Sci. Technol.* 52:49–57.

van Haandel, A. C., and G. Lettinga. 1994. *Anaerobic Sewage Treatment.* John Wiley & Sons, Ltd., Chichester, United Kingdom.

Wilkie, A. C. 2005. Anaerobic digestion: biology and benefits, p. 63–72. *In Dairy Manure Management: Treatment, Handling, and Community Relations.* NRAES-176. Natural Resource, Agriculture, and Engineering Service, Cornell University, Ithaca, NY.

Wilkie, A., M. Goto, F. M. Bordeaux, and P. H. Smith. 1986. Enhancement of anaerobic methanogenesis from Napiergrass by addition of micronutrients. *Biomass* 11:135–146.

Wilkie, A. C., K. J. Riedesel, and J. M. Owens. 2000. Stillage characterization and anaerobic treatment of ethanol stillage from conventional and cellulosic feedstocks. *Biomass Bioenergy* 19:63–102.

Wilkie, A. C., H. F. Castro, K. R. Cubinski, J. M. Owens, and S. C. Yan. 2004. Fixed-film anaerobic digestion of flushed dairy manure after primary treatment: wastewater production and characterisation. *Biosystems Eng.* 89:457–471.

Younesi, H., G. Najafpour, and A. R. Mohamed. 2005. Ethanol and acetate production from synthesis gas via fermentation processes using anaerobic bacterium, *Clostridium ljungdahlii. Biochem. Eng. J.* 27:110–119.

Bioenergy
Edited by J. Wall et al.

Chapter 17

Methane from Gas Hydrates

BAHMAN TOHIDI AND ROSS ANDERSON

Included within this volume on bioenergy, methane hydrates might be considered something of an enigma. In terms of production, hydrates are more analogous to the fossil fuels; they occur as natural, essentially finite, subsurface deposits, and must first be prospected for and then (potentially) drilled and produced from wells in a way similar to oil or natural gas. However, they can be arguably classified as a bioenergy resource, their formation being primarily the result of recent (on geological timescales) and currently active biological processes.

Like methane production from biomass, but on a much grander scale, methane trapped in gas hydrates is primarily the product of the natural bacterial breakdown of organic matter. Washed from land and deposited in modern and ancient ocean basins, detrital organic material in sediments forms the basic feedstock for a variety of active subsurface bacterial communities. Under anaerobic conditions, through various intermediate metabolic pathways, different bacterial species cooperate to progressively reduce large, complex organic molecules into methane gas. In the low-temperature, high-pressure environment of the deep ocean and arctic permafrost, this methane combines with sediment pore waters to form clathrate hydrates, i.e., ice-like crystalline compounds composed of water molecule "cages" accommodating large volumes of "guest" gases. Trapped as hydrates growing in sediment pores, considerable quantities of methane can accumulate into potentially exploitable deposits.

Methane is a low-carbon fuel, producing much less CO_2 for the same energy yield than coal, oil, or the higher-hydrocarbon gases (e.g., propane and butane); thus, utilizing gas hydrates as an energy resource could help significantly in reducing global atmospheric carbon emissions. Furthermore, if the CO_2 produced (from combustion or reformation to produce hydrogen) were captured and stored, such as in subsurface geological formations (e.g., aging oil reservoirs, saline aquifers, or trapped as CO_2 hydrates beneath the seafloor), then methane hydrates would present an alternative, clean, potentially enormous, untapped resource of low-carbon biogas for energy generation which could last well beyond conventional reserves.

The aim of this chapter is to provide an introduction to naturally occurring methane hydrates: their physical properties, formation, occurrence, and distribution in the natural environment, current estimates of resources, and potential for exploitation.

PHYSICAL PROPERTIES OF GAS HYDRATES

The term "gas hydrates" is that commonly used in reference to the clathrate hydrates; more specifically, the clathrate hydrates of natural gases. Clathrate hydrates are a group of ice-like crystalline compounds formed through the combination of water and suitably sized guest molecules, typically under low-temperature and elevated-pressure conditions (Sloan, 1998). Within the hydrate lattice, water molecules form a network of hydrogen-bonded polyhedral cavity structures that enclose the guests; the latter are generally comprised of low-molecular-diameter gases (e.g., methane and CO_2) and/or organic liquids (e.g., tetrahydrofuran). The clathrate hydrates can be distinguished from other hydrated compounds (e.g., hydrated salts) in that guests are not physically bonded with the water lattice but rather are held within—and lend stability to—clathrate cavities primarily through Van der Waals interactions ("clathrate" coming from the Latin *clathratus*, to be encaged).

There are three principal clathrate hydrate structures; I, II, and H, as illustrated in Fig. 1. The clathrate hydrate structure (or structures) formed in a given system is primarily dependent on the composition of the

Bahman Tohidi and Ross Anderson • Centre for Gas Hydrate Research, Institute of Petroleum Engineering, Heriot-Watt University, Edinburgh, EH 14 4AS, United Kingdom.

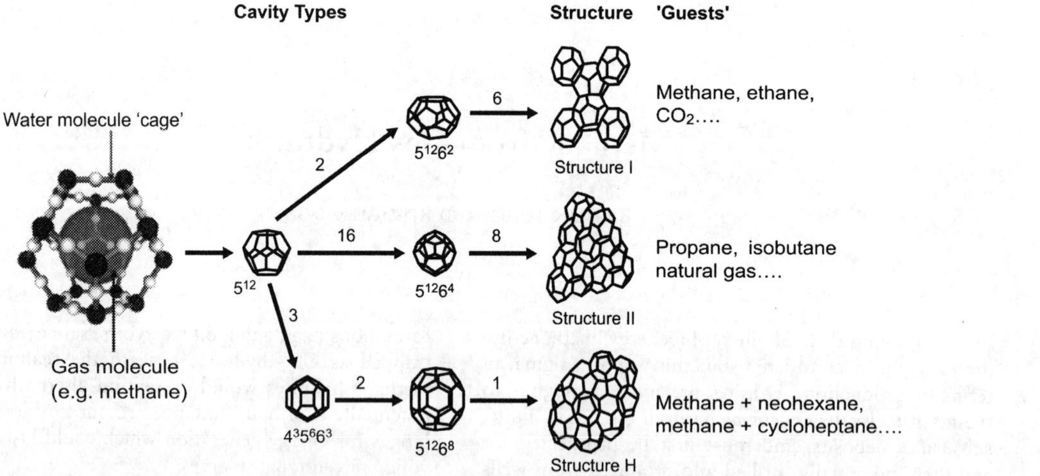

Figure 1. Common clathrate hydrate structures, cavity types, and guest examples.

gas and the respective molecular diameters of its constituent components. There are a large number of gases and organic polar and nonpolar liquids which can form clathrate hydrates, these varying in molecular diameter from as small as that of hydrogen (Mao et al., 2002) to as large as that of *cis*-1,2-dimethylcyclohexane (Ripmeester and Ratcliffe, 1990; Sloan, 1998). Typically, smaller gas molecules, such as methane and CO_2 (and their mixtures), form structure I clathrates in both simple (single guest) and binary (or greater, with two or more guests) gas systems, occupying both the large ($5^{12}6^2$) and small (5^{12}) cavities (Fig. 1). The presence of larger molecules, such as propane and butane, allows stabilization of $5^{12}6^4$ cavities and the formation of structure II hydrates. The formation of structure H hydrates specifically requires the presence of small helper gases (e.g., methane) in addition to larger heavy hydrate formers (e.g., neohexane and cycloheptane); filling of small 5^{12} and $4^35^66^3$ cavities by these helper gases provides the structural stability needed to allow large $5^{12}6^8$ cage occupancy (Ripmeester et al., 1987; Ripmeester and Ratcliffe, 1990; Sloan, 1998). Gas hydrates are generally considered to be nonstoichiometric, with cavity occupancy varying to a degree as a function of pressure, higher pressures normally giving greater cavity filling (Sloan, 1998). In gas mixtures, guest ratios and respective cage occupancies are a function of gas composition, pressure, and temperature, as generally described by the classical solid solution theory of Van der Waals and Platteeuw (1959).

Clathrate hydrates can hold very large volumes of gas within their structure. For methane hydrate (structure I), complete filling of both small and large cavities by single CH_4 molecules would give a hydrate of formula $CH_4 \cdot 5.75H_2O$. However, as noted, occupancy varies to a degree with pressure and complete filling is unusual; analyses of methane hydrate typically yield a hydration ratio of around 1:6 (i.e., $CH_4 \cdot 6H_2O$), which is equivalent to 96% cage occupancy (Handa, 1986; Reuff and Sloan, 1988). Based on this figure, 1 m^3 of clathrate hydrate can hold around 167 m^3 (compared to 175 m^3 if all cavities are filled) of methane (standard temperature and pressure [STP])—a considerable volume.

NATURE AND ORIGINS OF GAS HYDRATES IN THE SUBSURFACE

Although gas hydrates, due to their tendency to block hydrocarbon pipelines, have been the subject of research since the early part of the last century, it was not until the 1970s that they were first recognized to occur naturally (Vasil'ev et al., 1970; Makogon et al., 1971). The formation of gas hydrates requires low temperatures, elevated (hydrostatic) pressures, water, and a source of gas in sufficient concentrations. In the natural environment, these conditions are found in sediments of the deep sea, deep-lake bottoms, and the subsurface of arctic permafrost regions. As shown in Fig. 2, hydrate deposits have now been identified in sediments of the ocean floor along most of the world's continental margins, in deeper inland seas (e.g., Caspian Sea and

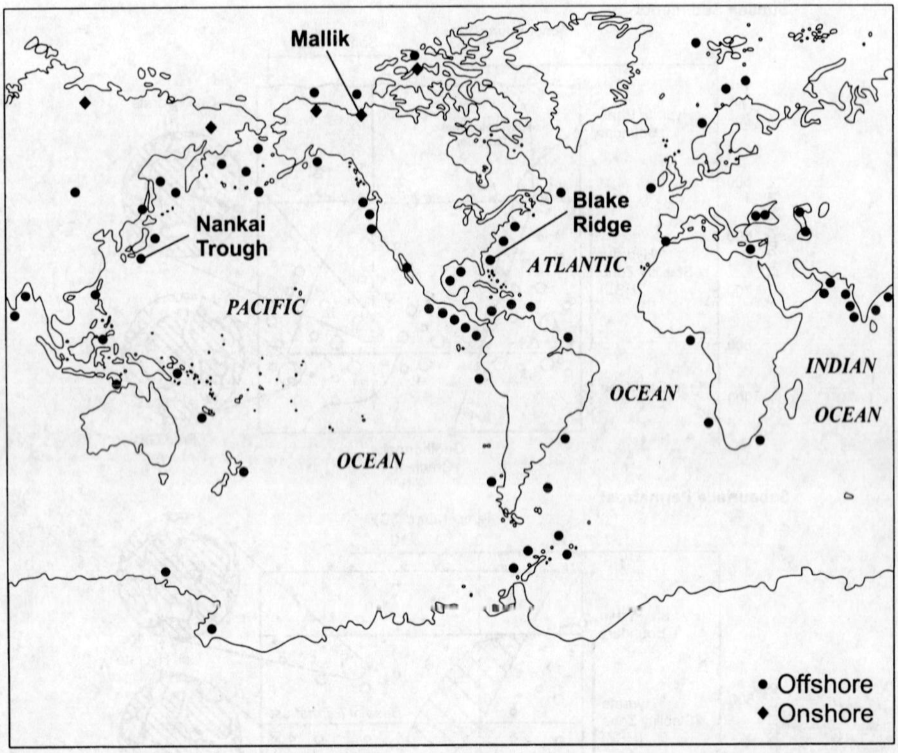

Figure 2. World map of known and inferred gas hydrate occurrences in aquatic (sea and lake floor) and continental (permafrost) environments. Locations updated from Kvenvolden (1998).

Black Sea), and beneath the permafrost of Siberia and North America.

HSZ in Marine and Permafrost Environments

The hydrate stability zone (HSZ) in marine and subsurface permafrost sediments can be delineated as the overlap of the pressure-temperature region of hydrate thermodynamic stability—the limits which are given by the hydrate phase boundary—and the hydrothermal (in the case of marine gas hydrates)/geothermal gradients, as illustrated in Fig. 3. The position of the hydrate phase boundary is primarily a function of the gas composition and pore water salinity (dissolved salt inhibiting stability) but may also be influenced by host sediment properties (Clennell et al., 1999; Anderson et al., 2003).

In both marine and permafrost sediments, within the region of pressure-temperature stability, gas hydrates may form wherever the concentration of methane exceeds solubility in pore waters (Tohidi et al., 1997; Zatsepina and Buffet, 1997; Clennell et al., 1999; Tohidi et al., 2001). In the marine environment, the upper limit of hydrate thermodynamic stability may lie above the seafloor (Fig. 3). However, as seawater is undersaturated with respect to methane, hydrate does not generally form in this region, the exception being where there is a continuous gas supply, such as near seafloor gas seeps, where transient (i.e., remaining as long as the seep is active) features such as hydrate mounds may form (for an example, see Vardaro et al., 2006). In the case of permafrost hydrates, gas concentrations may be such that the zone of hydrate occurrence overlaps the permafrost, meaning hydrate will be present in pores in equilibrium with ice. While increasing (hydrostatic) pressures with depth may be sufficient to stabilize hydrates, the base of the HSZ is ultimately limited by the rising temperatures associated with the local geothermal gradient.

If methane is present in sufficient concentrations below the base of the HSZ, then free gas may exist in

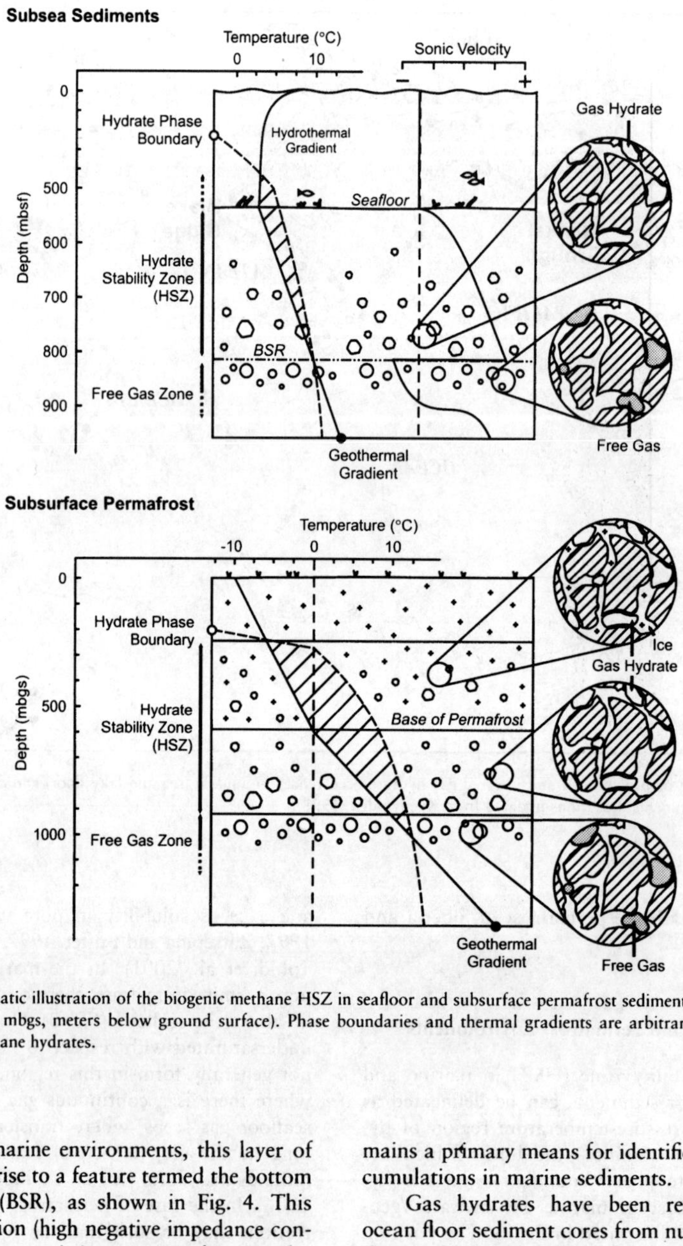

Figure 3. Schematic illustration of the biogenic methane HSZ in seafloor and subsurface permafrost sediments (mbsf, meters below seafloor; mbgs, meters below ground surface). Phase boundaries and thermal gradients are arbitrary examples for structure I methane hydrates.

sediment pores. In marine environments, this layer of gas commonly gives rise to a feature termed the bottom simulating reflector (BSR), as shown in Fig. 4. This strong seismic reflection (high negative impedance contrast) typically cross-cuts existing stratigraphy at a relatively constant subbottom depth (hence the term "bottom simulating") and is understood to arise from the rapid drop in sonic velocities across the transition from sediments hosting solid hydrate to those containing appreciable (~5% of pore volume or greater) free gas (Shipley et al., 1979; Kvenvolden and Lorenson, 2001). The appearance of a strong BSR on seismic profiles remains a primary means for identification of hydrate accumulations in marine sediments.

Gas hydrates have been recovered in shallow ocean floor sediment cores from numerous sites around the world. Field studies have included three Ocean Drilling Program (ODP) Legs: 146, Cascadia Margin (Westbrook et al., 1994); 164, Blake Ridge, offshore South Carolina (Paull et al., 1996); and 204, Hydrate Ridge, offshore Oregon (Tréhu et al., 2003). Sediments hosting gas hydrates are generally characterized by organic-matter-rich fine-grained silts, muds, and clays, with lesser, coarser sandy layers present at some sites.

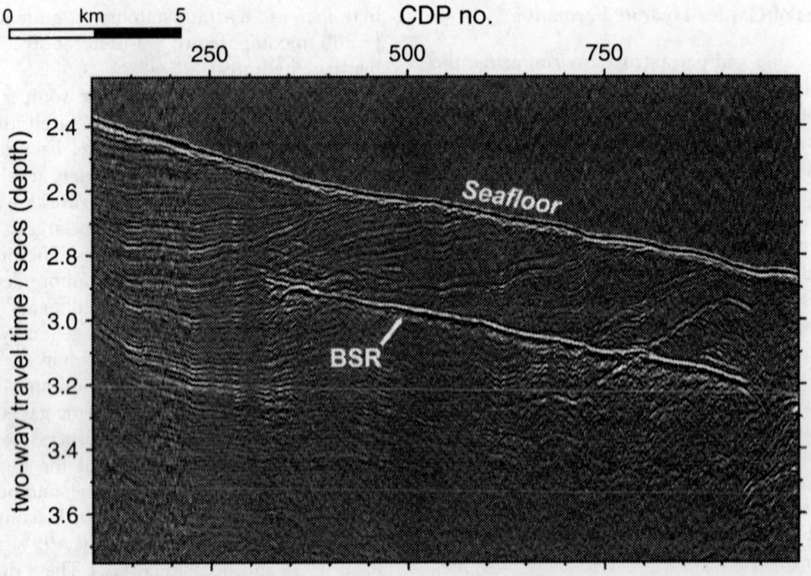

Figure 4. Seismic section showing a prominent BSR at the base of the gas HSZ, Lima Basin, offshore Peru. Note how the strong BSR reflection cross-cuts weaker reflections from sedimentary layers, mimicking the seafloor topography at a constant depth. Courtesy of Ingo Pecher, Heriot-Watt University, United Kingdom.

Hydrates commonly display a wide range of growth habits and are often patchily distributed within the host sediment according to texture (Booth et al., 1996). In fine-grained sediments, hydrates are generally found in the form of segregated nodules, lenses, pellets, or sheets (Fig. 5). Where coarser layers are present, these appear to present a preferential site for hydrate accumulation, with clathrates forming an interstitial pore fill between sediment grains. This variation in growth patterns according to sediment type suggests that sediment properties likely play an important role in controlling hydrate morphology and distribution within the subsurface (Clennell et al., 1999; Anderson et al., 2003).

Depending on fluid flux, sediment properties, and other factors, the timescale for the accumulation of biogenic gas hydrate deposits likely ranges from thousands to potentially millions of years (Nimblett and Ruppel, 2003). While considerably less than that required for the formation of coal, oil, and natural gas deposits (hundreds of millions of years), the time needed for regenerating gas hydrate to concentrations of 10 to 15% of pore space in marine sediments with fluid fluxes similar to those of localities such as the Blake Ridge has been estimated as 10^3 to 10^5 years (Nimblett and Ruppel, 2003), meaning that hydrate deposits cannot be viewed as a renewable resource.

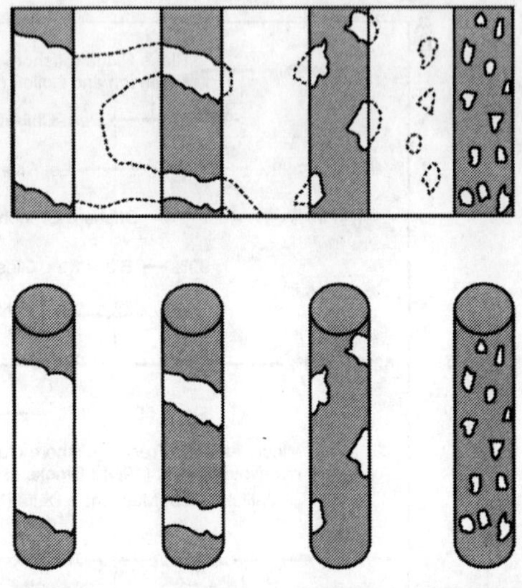

Figure 5. Typical morphologies of gas hydrate within sediments. In coarse-grained sediments (e.g., sands), hydrates tend to grow as a disseminated interstitial pore fill, while in finer-grained sediments (silts, muds, and clays) segregation (nodules and layers) is common.

Origins of Gas for Hydrate Formation

In both oceanic and permafrost environments, the gas for hydrate formation originates from one or a combination of two principal sources: (i) thermogenic and (ii) biogenic. Thermogenic gases are those derived from the inorganic high-temperature breakdown (thermocatalytic decarboxylation) of sediment organic material buried to great depths. As sediments undergo diagenesis, large complex organic molecules are cracked into increasingly lighter hydrocarbon fractions (Barnes et al., 1990; Hesse, 1990). Through bouncy driven flow, these migrate upward, where they may pool in natural geological traps, giving rise to conventional oil and/or natural gas reservoirs. However, it is common that reservoir cap rocks do not form perfect seals and considerable volumes of gas may escape through time along structural conduits such as faults and fractures towards the seafloor (e.g., Bernard et al., 1976). As this gas rises into increasingly cooler sediments close to the seabed, it can become trapped as gas hydrate. The presence of higher hydrocarbons (e.g., propane and butane), as is typical of thermogenic gases, generally results in a change to the more stable structure II and may increase hydrate stability to quite shallow depths (~300 m compared to ~500 m for structure I biogenic methane hydrates).

In contrast to thermogenic sources, biogenic hydrate gases are those generated purely through natural biological processes, namely, the bacterial breakdown of sedimentary organic material at shallow depths (hundreds of meters). Biogenic gases are understood to form the source for by far the largest proportion of global methane hydrate deposits. The processes which produce this gas are explored in more detail below (see "Bacterial Methanogenesis in Sediments").

Biogenic hydrate gases are generally distinguished from those of a thermogenic origin by two principal characteristics: (i) gas composition and (ii) carbon isotopic composition. As thermogenic gases are generated by inorganic thermal cracking processes, they generally contain appreciable fractions of the higher hydrocarbon gases such as ethane, propane, and butane (Bernard et al., 1976; Kvenvolden, 1998). In contrast, biogenic hydrates are almost ubiquitously >99% methane, with lesser CO_2 and occasional H_2S. These differing origins are also reflected in carbon isotopic compositions; the lighter (most common) ^{12}C isotope of carbon is more

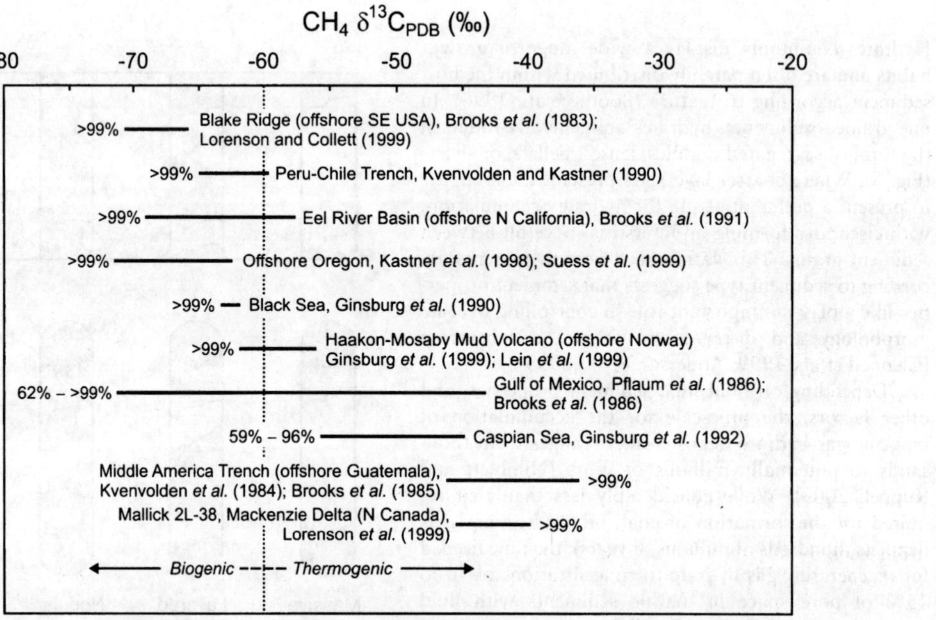

Figure 6. Carbon isotopic compositions (per mill relative to the Pee Dee Belemnite standard) and concentrations (percent of hydrocarbon gases) of methane in natural gas hydrates from various global localities. Values lighter than −60‰ indicate a biogenic origin, while heavier values suggest a thermogenic source. Overlap is common on both local and regional scales due to mixing of different gas sources.

favorably metabolized by bacteria, meaning that methane produced by biological processes is generally depleted in the heavier ^{13}C isotope, with biogenic methane—and thus, hydrates—generally having $\delta^{13}C$ values less than −60‰ relative to the Pee Dee Belemnite standard (Bernard et al., 1976; Kvenvolden, 1998).

Figure 6 shows a comparison of percent methane and methane carbon isotopic composition ($\delta^{13}C$ ‰) for natural gas hydrates from various global localities. For regions lacking significant hydrocarbon deposits, such as the Blake Ridge (offshore southeastern United States [Fig. 2]), Peru-Chile Trench, and Offshore Oregon, hydrate gas is typically >99% methane with $\delta^{13}C$‰ values generally <−60‰, indicating a dominantly biogenic origin. In contrast, for hydrocarbon-rich basins such as the Gulf of Mexico and Caspian Sea, hydrate methane contents as low as ~60%, coupled with $\delta^{13}C$ ‰ values as heavy as ~−45‰, reflect a large thermogenic input. Contributions from both sources are common on both local and regional scales, resulting in overlapping or median values (e.g., for the Gulf of Mexico).

Bacterial Methanogenesis in Sediments

As discussed, biogenic methane forms by far the largest gas source for gas hydrate formation, particularly in the marine environment. Methane is produced through the metabolic activity of methanogens, a morphologically diverse group of anaerobic *Archaea* distinct from other bacteria. Many methanogens are extremophiles, capable of living at high temperature, pressure, salt concentrations, and pH. However, they are obligate anaerobes, being killed by the presence of oxygen. Methanogens can utilize only a very narrow range of small molecules for metabolism. These are generally supplied as the end product of the metabolic activity of other bacteria, meaning that methanogens form only a subset of the microbial population of their environment (Hesse, 1990; Parkes et al., 2000; Wellsbury and Parkes, 2000).

Figure 7 illustrates the primary zones of organic matter oxidation in sediments. Organic matter represents highly reduced carbon compounds, which are among the most powerful reductants in newly deposited sediments. Depending on the source of the oxidant (electron acceptor), six zones or stages can be distinguished: zone 1, oxidation; zone 2, nitrate reduction; zone 3, sulfate reduction; zone 4, carbonate reduction; zone 5, fermentation; and zone 6, thermocatalytic decarboxylation (Claypool and Kaplan, 1974; Hesse, 1990).

Organic matter degradation in the first five of these zones requires the presence of bacteria. In zone 6, bacterial activity ceases and thermocatalytic reactions dominate, the latter being ultimately responsible for the generation of natural gas, i.e., the thermogenic input for hydrate formation (Barnes et al., 1990; Hesse, 1990).

In the oxidation zone (zone 1), free dissolved oxygen is present, originating either from seawater trapped in depositing sediments or through diffusion from proximal bottom waters. Microbial oxidation of organic matter typically leads to loss of organic functional groups, in particular, hydrocarbon chains and carboxyl groups, with the resulting smaller molecules being partly converted to CO_2 according to an overall reaction which may be written as

$$CH_2O + O_2 \rightarrow CO_2 + H_2O$$

where the bulk composition of organic matter has been simplified to CH_2O (Hesse, 1990). The efficient use of oxygen by aerobic respirators rapidly leads—particularly where sedimentation rates are high—to O_2 depletion, and the oxidation zone may only be a few centimeters or millimeters in thickness. In stagnant or euxinic basins, the oxidation zone is absent.

When the concentration of dissolved oxygen drops below ~0.5 ml of $O_2\cdot(\text{liter of } H_2O)^{-1}$ (Devol, 1978), nitrate reduction commences (zone 2). This environment may be termed "suboxic" and is populated by microaerobic bacteria that live where oxygen concentrations are between ~1.0 and 0.1 ml of $O_2\cdot(\text{liter of } H_2O)^{-1}$. Nitrate concentrations, which tend to rise from ambient seawater levels to a peak in the oxidation zone (where ammonia released from organic matter decomposition is oxidized [Froelich et al., 1979]), decrease toward zero at the lower boundary of the nitrate reduction zone. The base of the nitrate reduction zone is characterized by a change from positive to negative electrochemical potentials and the appearance of obligately anaerobic bacteria; this boundary marks the true transition between oxic and anoxic conditions.

In the underlying sulfate reduction zone (zone 3), anaerobic sulfate-reducing species such as *Desulfovibrio desulfuricans* dominate. These oxidize relatively small organic molecules such as dicarboxylic (fatty) acids and alcohols, which are produced by fermenting bacteria whose symbiosis with sulfate reducers is required. A simplified equation for the process is

$$2CH_2O + SO_4^{2-} \rightarrow S^{2-} + 2H^+ + 2HCO_3^-$$

This reaction produces dissolved sulfide; however, as soluble ferrous iron is generally available (from reduction of ferric oxyhydroxides and oxides in the sulfate and lower nitrate reduction zones), concentrations remain relatively low due to the precipitation of metastable iron monosulfides, which later convert to pyrite.

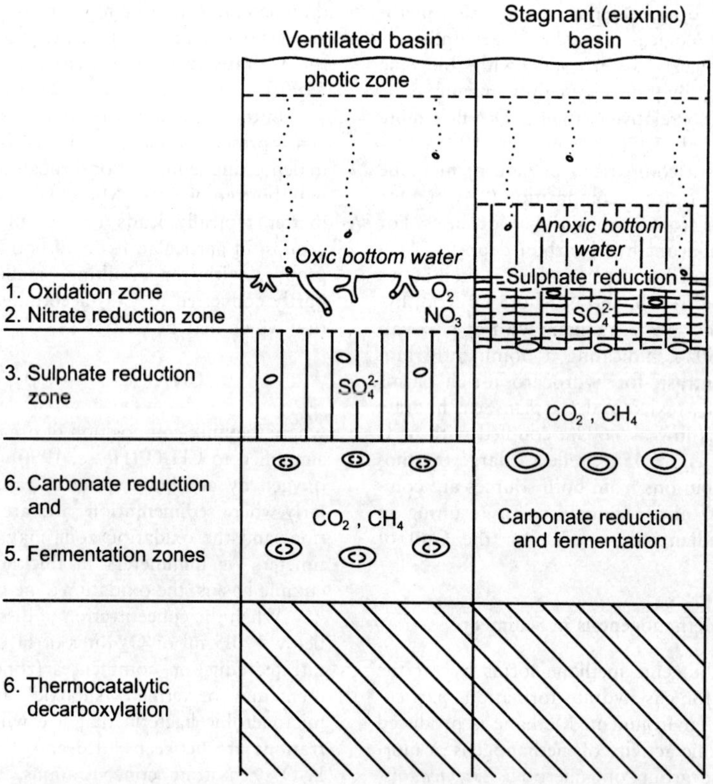

Figure 7. Principal zones of organic matter oxidation in marine sediments. Based on Claypool and Kaplan (1974).

In zones 1 to 3, one of the main products of bacterial organic matter degradation is carbonate (CO_3^{2-}) or carbonic acid (HCO_3^-). Below the sulfate reduction zone, this carbonate in turn becomes one of the principal oxidants for further bacterial degradation of organic matter. In the carbonate reduction zone (zone 4), bacterial carbonate reduction leads, for the first time, to the production of methane.

Figure 8 summarizes the principal metabolic pathways involved in methanogenesis. The most important substrates for bacterial methanogenesis are acetate and H_2/CO_2 (Parkes et al., 2000; Wellsbury and Parkes, 2000). Methane generation from H_2 and CO_2 (carbon reduction) is a nearly universal process in isolated cultures of methanogenic bacteria, and while formation from acetate or acetoclastic methanogenesis is restricted to fewer species, it is still an important process (Wellsbury and Parkes, 2000). As shown in Fig. 8, metabolic intermediates (e.g., fatty acids, sugars, amino acids, hydroxyl fatty acids, and alcohols) produced by hydrolyzing/fermenting bacteria, in turn form the feedstock for further bacterial species, notably the acetogens (Berner, 1980; Parkes et al., 2000; Wellsbury and Parkes, 2000), which produce acetate through the process of acetogenesis. For example, oxidation of ethanol and oxidation of butanoic acid (simple fatty acid) yield acetic acid and hydrogen according to the following respective reactions:

$$CH_3CH_2OH + H_2O \rightarrow CH_3COOH + 2H_2$$
$$CH_3CH_2CH_2COO^- + 2H_2O \rightarrow 2CH_3COO^- + 2H_2 + H^+$$

A number of these metabolic intermediate compounds may be utilized by other species of bacteria. H_2 and acetate in particular are examples of such "competitive substrates." For example, sulfate-reducing bacteria have a higher affinity for H_2/acetate than do the methanogens (Schonheit et al., 1982; Lovley et al., 1994) and will outcompete the methanogens as long as sulfate concentrations remain high. In some rare cases,

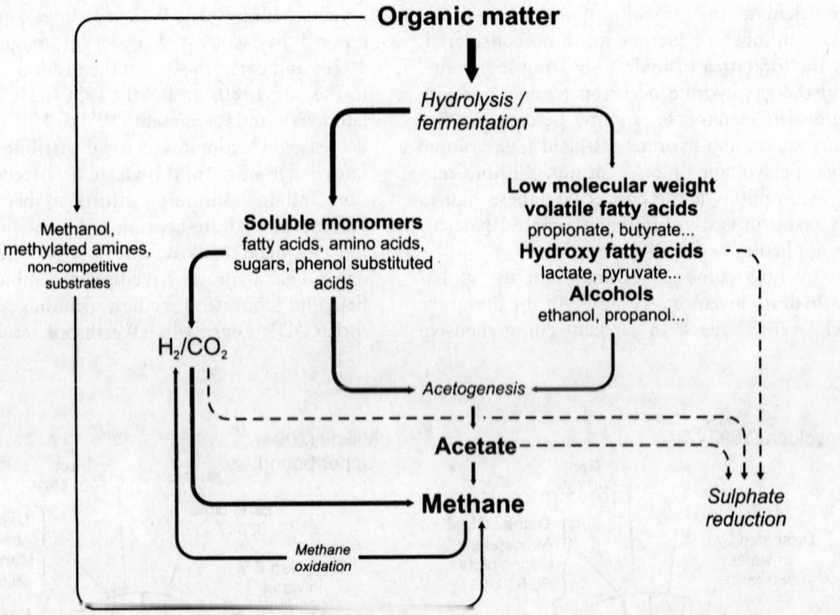

Figure 8. Illustration of main pathways in the production of methane by bacterial degradation of marine sedimentary organic matter. The thickness of arrows relates to relative importance of processes. Solid arrows indicate pathways to (and from) methane, while dashed arrows show substrate pathways to sulfate reduction. Based on Wellsbury and Parkes (2000).

where organic matter concentrations are very high, competitive substrates for methanogenesis may be at concentrations sufficient to support both methanogenesis and sulfate reduction (Wellsbury and Parkes, 2000).

The hydrogen produced in acetogenesis can be utilized by true methanogenic bacteria such as *Methanobacterium thermoautotrophicum* to reduce carbonate generated in previous zones (through the reaction of the general form $CH_2O + H_2O \rightarrow CO_2 + 4H$, which is enzyme catalyzed and yields atomic H) by the redox reaction

$$8H + CO_2 \rightarrow CH_4 + 2H_2O$$

The overall acetoclastic methanogenic reaction can be expressed as

$$CH_3COOH \rightarrow CH_4 + CO_2$$

Ultimately, it is this methane, generated in situ or through upward migration from greater depths (where temperatures are too high for clathrate stability), which reacts with cool pore waters to become concentrated as clathrate hydrates in sediments.

RESOURCE ESTIMATION

If natural gas hydrates are to be developed as a viable future energy resource, then clearly a sound estimate of reserves is essential. Since oceanic and permafrost hydrates were first recognized in the 1970s, a strong driving force for research has been the assumption that they (in particular marine gas hydrates) contain huge volumes of natural gas, predominantly in the form of methane (Kvenvolden, 1988, 1999). Until recently, probably the most widely cited global estimate of hydrate-bound gas was 21×10^{15} m^3 of methane (STP) or ~10,000 gigatons of methane carbon. Based on an analysis of a subset of the global estimates, Kvenvolden (1999) proposed this as a "consensus value" because some independent estimates converged at around that value (Kvenvolden, 1999; Kvenvolden and Lorenson, 2001). However, overall estimates of hydrate methane in oceanic sediments vary by several orders of magnitude and are believed to be highly uncertain (Kvenvolden, 1999; Lerche, 2000; Milkov, 2004). This can be largely attributed—particularly for estimates made in the past—to a lack of fundamental information concerning the concentration and distribution of hydrate within sediments on both local and regional scales. To calculate the global volume of hydrate-bound gas, a number of factors must be considered, principally the total area of hydrate-bearing sediments, thickness of the gas hydrate occurrence zone, porosity of hydrate-bearing sediments, volume percent gas hydrate in pore space, and hydrate gas yield (i.e., volume at STP of gas per volume of gas hydrate). Without reliable data, assumptions made concerning these figures can readily result in vastly different estimates, particularly when applied on a global scale.

There are now some 20 estimates of the global marine gas hydrate inventory available in the literature dating back over 30 years. In a recent comprehensive review, Milkov (2004) details how estimates have decreased by at least 1 order of magnitude from the 1970s and early 1980s (on the order of 10^{17} to 10^{18} m^3) to the late 1980s and early 1990s (10^{16} m^3) and to the late 1990s and the present (10^{14} to 10^{15} m^3) (Fig. 9). This decrease in estimates can be attributed to improved knowledge of natural hydrate occurrence and distribution and to continuing efforts to better evaluate the volume of hydrate-bearing sediments and their potential gas yield (Milkov, 2004). Over the past 20 years, natural gas hydrates have been the subject of numerous field and laboratory studies, including, as earlier noted, three ODP Legs: 146 (Westbrook et al., 1994), 164

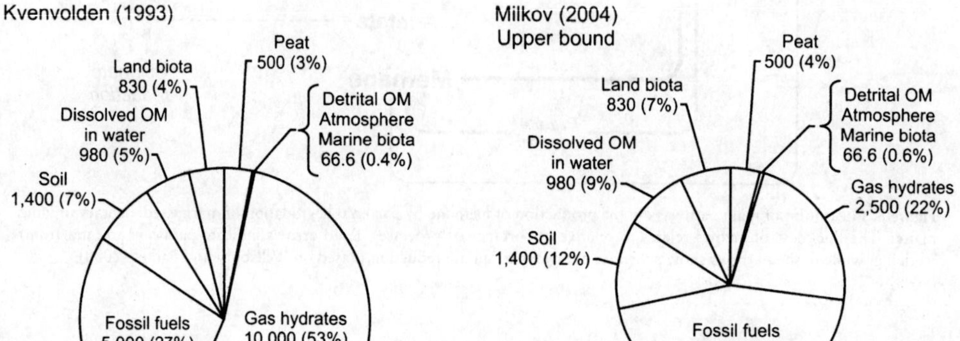

Figure 9. Pie charts show the distribution of organic carbon on Earth (excluding dispersed organic carbon such as kerogen and bitumen) with past (Kvenvolden, 1993, 1999) and more recent (Milkov, 2004) estimates of the global hydrate inventory. Values are given in gigatons (Gt) of carbon. Inset shows global estimates of hydrate-bound gas versus year of estimate. From data compiled by Milkov (2004).

(Paull et al., 1996), and 204 (Tréhu et al., 2003). Through detailed laboratory assessment, retrieval and analyses of hydrate-hosting sediment cores, detailed seismic surveys, and geophysical modeling, the parameters needed for resource evaluation are now far more accurately established than ever before.

Figure 9 shows a comparison of the global distribution of organic carbon and estimated hydrate gas inventory for the previous consensus value of Kvenvolden (1993, 1999) compared to the most recent conservative upper and lower bounds proposed by Milkov (2004). It can be seen that while the latter, more recent estimates are greatly reduced relative to that supposed previously, even assuming that the actual figure lies at the median of suggested bounding values, the volume of hydrate-bound methane still represents a considerable resource, comparable in size to conventional natural gas reserves (the "fossil fuels" rubric in Fig. 9 includes coal, oil, and gas resources).

While estimates appear to be now converging to more realistic values, reliable knowledge of the total global volume of gas as hydrates is not the deciding factor in making gas hydrates a viable future energy resource. The global volume merely represents the total actual hydrate-bound methane in the subsurface, and numerous further factors, both geological and otherwise, define what fraction of available resources could be considered reserves, i.e., be economically exploitable (Milkov and Sassen, 2002; Milkov, 2004). As noted by Milkov and Sassen (2002), only a part of global gas hydrate resources may actually be considered viable reserves, as the largest proportion of gas is potentially in the form of subeconomic stratigraphic hydrate accumulations such as those of the Blake Ridge (ODP Leg 164 [Paull et al., 1996]). What proportion of the global hydrate inventory proves to be a profitable source of gas in the future is difficult to predict. Subsurface accumulations that may be considered uneconomic at present could prove cost-effective in the face of technological development, as has commonly been the case for conventional oil and gas exploration. Probably the biggest potential for exploitation of gas hydrates lies in concentrated accumulations (so-called "hot spots"), such as those associated with structural fluid migration pathways and areas of high gas flux (Hovland et al., 1997; Milkov and Sassen, 2002).

METHANE PRODUCTION FROM HYDRATES: HISTORY AND STATUS QUO

Although the existence of very large volumes of methane in the form of natural gas hydrate was first recognized over 30 years ago, concerted efforts to exploit this resource began in earnest only in the mid-1990s. At this time, Japan and (to a lesser extent) India—two countries with large energy needs but limited domestic hydrocarbon reserves—began aggressive, well-funded, national hydrate research and development programs. In 1999, the United States followed suit; Congress offered legislation that culminated in the Methane Hydrate Research and Development Act and associated well-funded research programs in 2000 (U.S. National Research Council, 2004).

In 1998/1999, international efforts resulted in the drilling of the first wells specifically aimed at assessing methane hydrate-bearing strata. The first of these, the Mallik 3L-18C (Dallimore et al., 1999), was drilled into hydrate-bearing horizons underlying the permafrost of the McKenzie River delta in the Canadian Northwest Territories (Fig. 2). Following this, a series of wells were drilled by a consortium of Japanese government agencies and commercial interests in 945 m of water off the southeast coast of Japan close to a deep ocean trench known as the Nankai Trough (Takahashi et al., 2001) (Fig. 2). In both cases, drilling revealed the presence of high concentrations of methane hydrates, although production tests were not carried out. The Nankai Trough discovery was particularly significant, as hydrate concentrations of up to 80% (of pore space) were found in specific confined reservoir sandstones; this was in stark contrast to the dispersed, shale-hosted, low-saturation deposits described at the well-studied Blake Ridge.

The success of the initial field studies at Mallik led to a second research program, the Mallik 2002 consortium, this time with the aim of investigating methane production. A variety of well tests were carried out, and the hydrate reservoir's response to both thermal stimulation and depressurization was assessed. Both these approaches resulted in hydrate dissociation and significant methane release, with production being sustained over a number of days, thereby confirming the technical feasibility of methane production from hydrates (Dallimore and Collett, 2005).

In 2004, a further 32 exploratory wells were drilled offshore from Japan through hydrate-bearing strata and the underlying BSR bordering the Nankai Trough at water depths of 720 to 2,033 m. Sixteen wells were logged while drilling, 2 wells were wireline logged, 12 wells were cored, and 2 wells were cased and tested to verify technologies for future development plans, including one horizontal well. Data collected during this drilling program are currently being used to plan the next phase of operations (Takahashi, 2005).

There are still significant technological challenges to overcome before hydrate production can be realized. Hydrate reservoirs differ markedly from conventional gas reservoirs, principally because methane is present as an (often disseminated) solid hydrate (rather than mobile gas phase) which must first be dissociated (e.g., by depressurization or heating) before gas is released (Milkov and Sassen, 2002; Johnson and Max, 2006). Hydrate exploitation will require the development of new production methods (e.g., hot saline fluid injection), coupled with adaptation of current (e.g., reservoir fracturing to improve gas permeability) and emerging technologies (e.g., horizontal wells for exploitation of hydrate seams).

Ultimately, the future development of gas hydrates as an energy resource will be driven by market forces. The forecast for future tight supplies of natural gas, along with increasingly higher prices, point to growing demand for alternative supplies. The exploitation of gas hydrates is seen by many as a means to meet this demand, with some analysts suggesting that marine gas hydrates may begin contributing to natural gas markets in less than 10 years (Johnson and Max, 2006).

Acknowledgments. We thank Ingo Pecher for his comments on the manuscript and for providing the BSR image shown in Fig. 4.

Support of the UK Engineering and Physical Sciences Research Council for research into gas hydrates in sediments at Heriot-Watt University (EPSRC Grant no. EP/D013844/1 and EP/D052556/1) is gratefully acknowledged.

REFERENCES

Anderson, R., M. Llamedo, B. Tohidi, and R. W. Burgass. 2003. Experimental measurement of methane and carbon dioxide clathrate hydrate equilibria in mesoporous silica. *J. Phys. Chem. B* 107:3507–3514.

Barnes, M. A., W. C. Barnes, and R. M. Bustin. 1990. Chemistry and diagenesis of organic matter in sediments and fossil fuels, p. 189–204. *In* I. A. Mcllreath and D. W. Morrow (ed.), *Diagenesis: Geoscience Canada Reprint Series 4.* Runge Press, Ontario, Canada.

Bernard, B., J. Brooks, and W. Sackett. 1976. Natural gas seepage in the Gulf of Mexico. *Earth Planet. Sci. Lett.* 31:48–54.

Berner, R. A. 1980. *Early Diagenesis: A Theoretical Approach.* Princeton Series in Geochemistry, vol. 241. Princeton University Press, Princeton, NJ.

Booth, J. S., M. M. Rowe, and K. M. Fisher. 1996. Offshore gas hydrate sample database with an overview and preliminary analysis. *U.S. Geol. Surv. Open-File Rep.* 96–272:17.

Brooks, J. M., L. A. Barnard, D. A. Weisenberg, M. C. Kennicutt III, and K. A. Kvenvolden. 1983. Molecular and isotopic composition of hydrocarbons at Site 533, Deep Sea Drilling Project Leg 76. *Init. Rep. Deep Sea Drill. Proj.* 76:377–389.

Brooks, J. M., B. H. Cox, W. R. Bryant, M. C. Kennicutt III, R. G. Mann, and T. J. MacDonald. 1986. Association of gas hydrates and oil seepage in the Gulf of Mexico. *Org. Geochem.* 10:221–234.

Brooks, J. M., M. E. Field, and M. C. Kennicutt. 1991. Observations of gas hydrates in marine sediments, offshore northern California. *Mar. Geol.* 96:103–109.

Brooks, J. M., A. W. A. Jeffrey, T. J. MacDonald, R. C. Pflaum, and K. A. Kvenvolden. 1985. Geochemistry of hydrate gas and water from Site 570, Deep Sea Drilling Project Leg 84. *Init. Rep. Deep Sea Drill. Proj.* 84:699–703.

Claypool, G. E., and I. R. Kaplan. 1974. The origin and distribution of methane in marine sediments, p. 99–139. *In* I. R. Kaplan (ed.), *Natural Gases in Marine Sediments.* Plenum Press, New York, NY.

Clennell, M. B., M. Hovland, J. S. Booth, P. Henry, and W. J. Winters. 1999. Formation of natural gas hydrates in marine sediments 1: conceptual model of gas hydrate growth conditioned by host sediment properties. *J. Geophys. Res. B* 104:22985–23004.

Dallimore, S. R., and T. S. Collett (ed.). 2005. *Scientific Results from the 2002 Gas Hydrate Production Research Well Program, Mackenzie Delta, Northwest Territories, Canada.* Geological Survey of Canada Bulletin 585. Geological Survey of Canada, Ottawa, Ontario, Canada.

Dallimore, S. R., T. Uchida, and T. S. Collett (ed.). 1999. *Scientific Results from JAPEX/JNOC/GSC Mallik 2L-38 Gas Hydrate Research Well, Mackenzie Delta, Northwest Territories, Canada.* Geological Survey of Canada Bulletin 544. Geological Survey of Canada, Ottawa, Ontario, Canada.

Devol, A. H. 1978. Bacterial oxygen uptake kinetics as related to biological processes in oxygen deficient zones of the oceans. *Deep Sea Res.* 25:137–146.

Froelich, P. N., G. P. Linkhammer, M. L. Bender, N. A. Luedtke, G. R. Heath, D. Cullen, P. Dauphin, D. Hammond, B. Hartman, and V. Maynard. 1979. Early oxidation of organic matter in pelagic sediments of the eastern equatorial Atlantic: suboxic diagenesis. *Geochim. Cosmochim. Acta* 43:1075–1090.

Ginsburg, G. D., R. A. Guseynov, A. A. Dadashev, G. A. Ivanova, S. A. Kazantsev, V. A. Solviev, E. V. Telepnev, P. Y. Askeri-Nasirov, A. A. Yesikov, V. I. Mal'tseva, G. Y. Mashirov, and I. Shabayeva. 1992. Gas hydrates of the southern Caspian. *Int. Geol. Rev.* 43: 765–782.

Ginsburg, G. D., A. N. Kremlev, M. N. Grigorev, G. V. Larkin, A. D. Pavlenkin, and N. A. Saltykova. 1990. Filtrogenic gas hydrates in the Black Sea (twenty-first voyage of the research vessel *Evapatoriya*). *Geol. Geofiz.* 31:10–19.

Ginsburg, G. D., A. V. Milnov, V. A. Solviev, A. V. Egorov, G. A. Cherdashev, P. R. Vogt, K. Crane, T. D. Lorenson, and M. D. Khutorsky. 1999. Gas hydrate accumulation at the Haakon-Mosby mud volcano. *Geo-Mar. Lett.* 19:57–67.

Handa, Y. P. 1986. Compositions, enthalpies of dissociation, and heat-capacities in the range 85-k to 270-k for clathrate hydrates of methane, ethane, and propane, and enthalpy of dissociation of isobutane hydrate, as determined by a heat-flow calorimeter. *J. Chem. Thermodyn.* 18:915–921.

Hesse, R. 1990. Early diagenetic pore water/sediment interaction: modern offshore basins, p. 277–316. *In* I. A. Mcllreath and D. W. Morrow (ed.), *Diagenesis: Geoscience Canada Reprint Series 4.* Runge Press, Ottawa, Ontario, Canada.

Hovland, M., J. W. Gallagher, M. B. Clennell, and K. Lekvam. 1997. Gas hydrate and free gas volumes in marine sediments: example from the Niger Delta front. *Mar. Petrol. Geol.* 14:245–255.

Johnson, A. H., and M. D. Max. 2006. The path to commercial hydrate gas production. *Leading Edge* 25:648–651.

Kastner, M., K. A. Kvenvolden, and T. D. Lorenson. 1998. Chemistry, isotopic composition, and origin of methane-hydrogen sulfide hydrates at the Cascadia subduction zone. *Earth Planet. Sci. Lett.* 156:173–183.

Kvenvolden, K. A. 1988. Methane hydrates—a major reservoir of carbon in the shallow geosphere. *Chem. Geol.* 71:41–51.

Kvenvolden, K. A. 1993. A primer on gas hydrates, p. 279–291. *In* D. G. Howel (ed.), *The Future of Energy Gases.* U.S. Geological Survey Professional Paper, vol. 1570. U.S. Geological Survey, Reston, VA.

Kvenvolden, K. A. 1998. A primer on the geological occurrence of gas hydrate, p. 9–30. *In* J.-P. Henriet and J. Meinert (ed.), *Gas Hydrates: Relevance to World Margin Stability and Climatic Change.* Geological Society of London Special Publication, vol. 137. Geological Society of London, London, United Kingdom.

Kvenvolden, K. A. 1999. Potential effects of gas hydrate on human welfare. *Proc. Natl. Acad. Sci. USA* 96:3420–3426.

Kvenvolden, K. A., G. E. Claypool, C. N. Threlkeld, and E. D. Sloan. 1984. Geochemistry of a naturally occurring massive marine gas hydrate. *Org. Geochem.* **6**:703–713.

Kvenvolden, K. A., and M. Kastner. 1990. Gas hydrates of the Peruvian outer continental margin. *Proc. Ocean Drill. Prog. Sci. Results* **112**:517–526.

Kvenvolden, K. A., and T. D. Lorenson. 2001. The global occurrence of natural gas hydrate, p. 3–18. *In* C. K. Paull and W. P. Dillon (ed.), *Natural Gas Hydrates: Occurrence, Distribution, and Detection.* American Geophysical Union, Geophysical Monograph Series, vol. 124. American Geophysical Union, Washington, DC.

Lein, A., P. Vogt, K. Crane, A. Ergorov, and M. Ivanov. 1999. Chemical and isotopic evidence for the nature of the fluid in CH_4-containing sediments of the Haakon-Mosby mud volcano. *Geo-Mar. Lett.* **19**:76–83.

Lerche, I. 2000. Estimates of worldwide gas hydrate resource. *Energ. Explor. Exploit.* **18**:329-337.

Lorenson, T. D., and T. S. Collett. 1999. Gas content and composition of gas hydrate from sediments of the southeastern North American continental margin. *Proc. Ocean Drill. Prog. Sci. Results* **164**:37–46.

Lorenson, T. D., M. J. Whiticar, A. Waseda, S. R. Dallimore, and T. S. Collet. 1999. Gas composition and isotopic geochemistry of cuttings, core, and gas hydrate from the JAPEX/JNOC/GSC Mallick 2L-38 gas hydrate research well. *Geol. Surv. Canada Res. Bull.* **544**:143–163.

Lovley, D. R., F. H. Chapelle, and J. C. Woodward. 1994. Use of dissolved H_2 concentrations to determine the distribution of microbially catalysed redox reactions in anoxic ground water. *Environ. Sci. Technol.* **28**:1005–1210.

Makogon, Y. F., A. A. Trofimuk, V. P. Tarsev, and N. V. Cherskiy. 1971. Detection of a pool of natural gas in a solid (hydrated gas) state. *Dok. Akad. Nauk. SSSR* **196**:203–206.

Mao, W. L., H. Mao, A. F. Goncharov, V. V. Stuzhkin, Q. Guo, J. Hu, J. Shu, R. J. Hemley, M. Somayazulu, and Y. Zhao. 2002. Hydrogen clusters in clathrate hydrate. *Science* **297**:2247–2249.

Milkov, A. V. 2004. Global estimates of hydrate-bound gas in marine sediments: how much is really out there? *Earth-Sci. Rev.* **66**:183–197.

Milkov, A. V., and R. Sassen. 2002. Economic geology of offshore gas hydrate accumulations and provinces. *Mar. Petrol. Geol.* **19**:1–11.

Nimblett, J., and C. Ruppel. 2003. Permeability evolution during the formation of gas hydrates in marine sediments. *J. Geophys. Res. B.* **108**:2420.

Parkes, R. J., B. A. Cragg, and P. Wellsbury. 2000. Recent studies on bacterial populations and processes in marine sediments: a review. *Hydrogeol. J.* **8**:11–28.

Paull, C. K., R. Matsumoto, and P. J. Wallace (ed.). 1996. *Proceedings of ODP, Initial Reports, vol. 164.* Ocean Drilling Program, College Station, TX.

Pflaum, R. C., J. M. Brooks, H. B. Cox, M. C. Kennicutt, and D. D. Sheu. 1986. Molecular and isotopic analysis of core gases and gas hydrates, Deep Sea Drilling Project Leg 96. *Init. Rep. Deep Sea Drill. Proj.* **96**:781–784.

Reuff, R. M., and E. D. Sloan. 1988. Heat capacity and heat of dissociation of methane hydrates. *AIChE J.* **34**:1468–1476.

Ripmeester, J. A., J. S. Tse, C. I. Ratcliffe, and B. M. Powell. 1987. A new clathrate hydrate structure. *Nature* **325**:135–136.

Ripmeester, J. A., and C. I. Ratcliffe. 1990. ^{129}Xe NMR Studies of clathrate hydrates: new guests for structure II and structure H. *J. Phys. Chem.* **94**:8773–8776.

Schonheit, P., J. K. Kristjensson, and R. K. Thauer. 1982. Kinetic mechanism for the ability of sulphate reducers to out-compete methanogens for acetate. *Arch. Microbiol.* **132**:285–288.

Shipley, T. H., M. H. Houston, R. T. Buffler, F. J. Shaub, K. J. McMillen, J. W. Ladd, and J. L. Worzel. 1979. Seismic reflection evidence for the widespread occurrence of possible gas-hydrate horizons on continental slopes and rises. *Am. Assoc. Pet. Geol. Bull.* **62**:2204–2213.

Sloan, E. D. 1998. *Clathrate Hydrates of Natural Gases*, 2nd ed. Marcel Dekker Inc., New York, NY.

Suess, E., M. E. Torres, G. Bohrmann, R. W. Collier, J. Greinert, P. Linke, G. Rehder, A. Tréhu, K. Wallmann, G. Winckler, and E. Zuleger. 1999. Gas hydrate destabilization; enhanced dewatering, benthic material turnover and large methane plumes at the Cascadian convergent margin. *Earth Planet. Sci. Lett.* **170**:1–15.

Takahashi, H. 2005. Multi-well exploration program in 2004 for natural hydrate in the Nankai-Trough offshore Japan, OTC17162. *2005 Offshore Technology Conference*, Houston, Texas, 2 to 5 May 2005.

Takahashi, H., T. Yonezawa, and Y. Takedomi. 2001. Exploration for natural hydrate in Nankai-Trough wells offshore Japan, OTC13040. *2001 Offshore Technology Conference*, Houston, Texas, 30 April to 3 May 2001.

Tohidi, B., R. Anderson, M. B. Clennell, R. W. Burgass, and A. B. Biderkab. 2001. Visual observation of gas hydrate formation and dissociation synthetic porous media by means of glass micromodels. *Geology* **29**:867–870.

Tohidi, B., A. Danesh, and A. C. Todd. 1997. On the mechanism of gas hydrate formation in subsea sediments. *Abstr. Pap. Am. Chem. Soc.* **213**:37-Fuel.

Tréhu, A. M., G. Bohrmann, F. R. Rack, M. E. Torres, and Shipboard Scientific Party. 2003. *Proceedings of ODP, Initial Reports, vol. 204.* Ocean Drilling Program, College Station, TX.

U.S. National Research Council. 2004. *Charting the Future of Methane Hydrate Research in the United States.* The National Academies Press, Washington, DC.

Van der Waals, J. H., and J. C. Platteeuw. 1959. Clathrate solutions. *Adv. Chem. Phys.* **2**:1–57.

Vardaro, M. F., I. R. MacDonald, L. C. Bender, and N. L. Guinasso, Jr. 2006. Dynamic processes observed at a gas hydrate outcropping on the continental slope of the Gulf of Mexico. *Geo-Mar. Lett.* **26**:6–15.

Vasil'ev, V. G., Y. F. Makogon, F. A. Trebin, A. A. Trofimuk, and N. V. Cherskiy. 1970. The property of natural gases to occur in the Earth crust in a solid state and to form gas hydrate deposits. *Otkrytiya v SSSR* **1968–1969**:15–17.

Wellsbury, P., and R. J. Parkes. 2000. Deep biosphere: source of methane for oceanic hydrate, p. 91–104. *In* M. D. Max (ed.), *Natural Gas Hydrates in Oceanic and Permafrost Environments.* Kluwer Academic Publishers, Dordrecht, The Netherlands.

Westbrook, G. K., B. Carson, R. J. Musgrave, and Shipboard Scientific Party. 1994. *Proceedings of ODP, Initial Reports, vol. 146 (Pt.1).* Ocean Drilling Program, College Station, TX.

Zatsepina, O. Y., and B. A. Buffett. 1997. Phase equilibrium of gas hydrate: implications for the formation of hydrate in the deep sea floor. *Geophys. Res. Lett.* **24**:1567–1570.

3. METHANOL

Bioenergy
Edited by J. Wall et al.

Chapter 18

Methanol from Biomass

MANFRED RINGPFEIL, HANS-JOACHIM SANDER, MATTHIAS GERHARDT, AND MONIKA WOLF

METHANOL PRODUCTION AND USE IN THE CONTEXT OF CLIMATE CONTROL

The Situation

There has been a sharp rise in carbon dioxide in the atmosphere as a consequence of today's oil- and-gas-based economy. Immobile fossil carbon is transformed into mobile carbon, which accumulates in the atmosphere and contributes to global warming. To ensure the provision of enough energy for humankind and to avoid the further release of fossil carbon into the atmosphere, new technologies will need to be developed to make proper use of Earth's carbon cycle based on sun, water, and biomass. Current U.S. efforts are aimed at the wide-scale use of ethanol, a product of biomass. We propose introducing methanol as an alternative product to ethanol, using the chain carbon dioxide/water/sun energy → biomass → methane → methanol.

Properties of Methanol

Methanol is a member of the group of organic compounds that possess only one carbon atom in their molecules, also called C_1 compounds. C_1 compounds are found nowadays in many important technical and economic applications. Among them, methane is of utmost importance because it appears as a fossil product in huge underground and undersea deposits and, more important for future use, in recent natural processes as the final product of the anaerobic microbial metabolism of organic matter. All anaerobically degradable organic substances end up as methane and carbon dioxide. Absolutely necessary for the success of this process is the absence of molecular oxygen. Methane is used in many natural and technical processes and can also be produced under industrial conditions. In the atmosphere, it presently accounts for 20% of the greenhouse effect (Metz et al., 2005; World Resource Institute, http://www.wri.org/climate/pubs_content_text.cfm?cid=2162).

Methanol is the primary oxidation product of methane. The added oxygen atom changes the properties of the methane molecule dramatically. Methanol, as the primary alcohol of methane, appears as a liquid under normal temperature, miscible with water and hydrocarbons. These properties enable its easy handling for technical purposes. Methanol also contains an energy density that corresponds approximately to that of a higher hydrocarbon. Methanol is an easily storable source of hydrogen:

$$CH_4O + H_2O \rightarrow CO_2 + 3\ H_2$$

As a motor fuel, methanol has a favorable octane rating of 110, and most of today's production goes directly into motor fuels or becomes a component of methyl tertiary-butyl ether, a fuel additive, or an esterification component of vegetable fatty acids in biodiesel. As a solvent, methanol is used in many industries, and as a reactant, methanol is oxidized to formaldehyde. Finally, methanol is used as a carbon source for some aerobic bacteria. In water remediation it is added as feed for aerobic microorganisms, whose growth helps to decrease the nitrogen load of sewage.

Manufacture of Methanol

In the past, methanol was produced by dry distillation of wood or hydrolysis of methyl ester, methyl ether, and methyl amine groups of natural substances as lignins and pectins. The resulting methanol contained recent carbon atoms (^{r}C) exclusively, which ultimately originated from atmospheric carbon dioxide ($^{atm}CO_2$) converted into organic substances by plant photosynthesis:

$$^{atm}CO_2 + H_2O \rightarrow [^{r}CH_2O]_n \rightarrow$$
$$R\text{–}O\ ^{r}CH_3 + H_2O \rightarrow {}^{r}CH_3OH + ROH$$

Manfred Ringpfeil, Matthias Gerhardt, and Monika Wolf • BIOPRACT GmbH, Magnusstraße 11, 12489 Berlin, Germany. **Hans-Joachim Sander** • Goldlackweg 5, 06118 Halle(Saale), Germany.

Schink and Zeikus (1980) describe the microbial conversion of the methyl groups from pectin into methanol by *Clostridium butyricum*. Because such methyl groups constitute only a small percentage of these materials, technical manufacturing in this way will never be significant, even when materials based on recent carbon are in demand.

Incomplete combustion (gasification) of carbon-rich natural compounds such as coal, mineral oil, and natural gas yields large amounts of synthesis gas (CO/H_2), which is converted to methanol by using a metal oxide catalyst. Since these industrial methods for methanol production were introduced in the 1920s, methanol has become an important commodity chemical, of which 30 million tons per annum are now produced:

$$\text{Coal, mineral oil, natural gas} + \text{water} + \text{heat} \rightarrow {}^{f}CO + 2H_2 \rightarrow {}^{f}CH_3OH$$

The problem of today's methanol use is caused by the origin of its raw materials. Coal, mineral oil, and natural gas are carbonaceous raw materials containing fossil carbon (^{f}C) exclusively. Methanol produced from them consists of fossil carbon ($^{f}CH_3OH$); therefore, when methanol of this type is burnt, the fossil carbon is carried into the atmosphere in the form of carbon dioxide ($^{f}CO_2$), which increases the concentration of the greenhouse gases. This increase in greenhouse gases can be avoided only if the fossil-carbon-containing methanol is not burnt or if the carbon dioxide that results from burning this methanol is captured and stored outside the atmosphere. Another alternative is the production of methanol from carbonaceous raw materials, namely biomass, that are not of fossil origin. Fortunately, light energy flux from the sun to the Earth is great enough to in theory produce sufficient biomass to serve the energy needs of the human population.

The energetic shifting of biomass to hydrocarbon by communities of microbes occurs according to the following equation:

$$2CH_2O \rightarrow CH_4 + CO_2$$

One half of the carbon of carbohydrates is reduced to the hydrocarbon level, and the other half is oxidized to carbon dioxide. This is performed by anaerobic microbiological processes that consume a bit of energy from the reacting molecule for the maintenance and growth of the involved organisms. Almost all the energy of the original molecule ends up as methane. As an intermediate step in this disproportion, microbes convert carbohydrate molecules to acetic acid molecules:

$$[CH_2O]_6 + 2H_2O \rightarrow 2CH_3COOH + 2CO_2 + 4H_2$$

The acetic acid then undergoes a microbially catalyzed conversion into methane and carbon dioxide.

$$2CH_3COOH + 2CO_2 + 4H_2 \rightarrow 3CO_2 + 3CH_4 + 2H_2O$$

In summary:

$$[CH_2O]_6 \rightarrow 3CO_2 + 3CH_4$$

Methane is the organic molecule with the highest energy load per carbon atom. Because many organic compounds can be degraded to acetic acid, methane can be built from many organic molecules, e.g., from higher fatty acids and amino acids. Methane is the universal end product of all anaerobic microbial metabolism in nature. In practice, the resulting mixture of methane and carbon dioxide is called biogas. It burns when ignited. It separates voluntarily from its liquid production phase. Thus, biogas can be used—and is increasingly being used—directly for heat and power generation.

Methane and carbon dioxide have very different properties: the molecular weight, boiling point, solubility, and electric charge are so different that the separation of the two gases can be performed easily. Ultimately, almost pure methane can be produced from the biogas mixture (W. Graf, personal communication), and biomethane obtained in this way is indistinguishable from the methane found in natural gas. The two can be mixed without any practical disadvantage, the only problem being that methane from natural gas consists of fossil carbon and should not be exploited in the future if enough biomethane can be made available. Natural gas should stay where it is, in the Earth or in the sea. Under these conditions, biomethane can be used for methanol synthesis in order to receive climate-neutral biomethanol. The present opportunities for the use of recent carbon for producing energy-rich compounds, besides the direct thermal conversion of biomass and the ethanol route from carbohydrates, can be expressed in the terse phrase: "from biomass to biogas" (Fig. 1).

The direct thermal conversion of biomass can be applied preferentially with dry biomass, such as wood, straw, or microbial sludges. The degradation of humid biomasses, e.g., grass, maize silage, or mechanically dewatered microbial sludges, can be accelerated by adding preproduced enzymes. Enzymatic degradation increases the amount of substances accessible for uptake into microbes, up to 30% of the total degradable organic substance. It is the rate-limiting step in the conversion of the organic substance to methane. The addition of preproduced enzymes to the biomass to be converted improves the velocity and yield of the whole methane-producing process.

Biomethane can be separated from carbon dioxide and, if still necessary, from traces of other gases. The

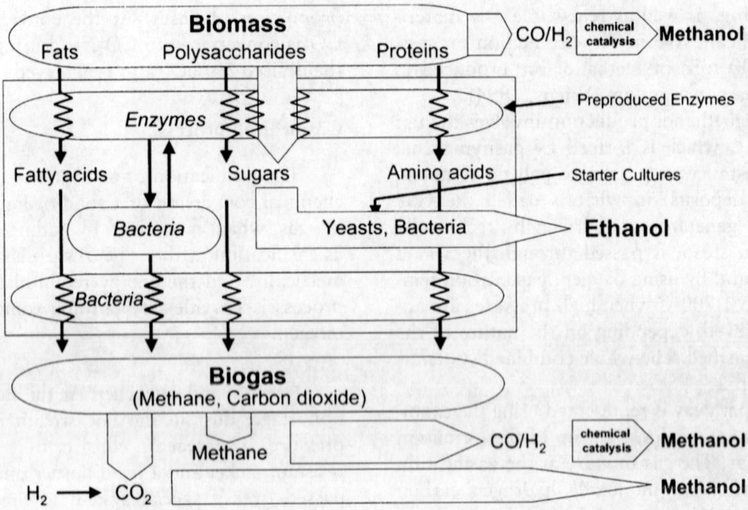

Figure 1. Technology-oriented scheme for microbial biomass conversion under exclusion of oxygen and addition of chemical methanol synthesis.

biomethane is now available for synthesis of climate-neutral biomethanol, independent of the type of processing applied. Similarly, the separated recent carbon-containing CO_2 can be converted into biomethanol if H_2 from nonfossil sources is supplied.

Practical Ways To Convert Carbon-Containing Raw Materials into Methanol

Today, methanol is produced by three main processes that are differentiated by the raw materials used: coal and oil, natural gas, and carbon dioxide.

Coal and oil

There are commercial processes using coal and oil separately and in combination (Supp, 1990; SVZ, 2001). Solid materials containing recent carbon such as wood are in the precommercial phases of industrial testing (Althapp, 2003). Coal is used commercially in mixture with biogenic materials, e.g., microbial sludges, in order to produce methanol that is enriched with methanol-containing recent carbon. It is said that the percentage of recent carbon in these mixtures can be raised to 100%.

Natural gas

There are large commercial processes that use natural gas (Vriens, 1994). Mixing biomethane from biogas into the natural gas is obviously under consideration. Processes using commercially available biogas directly are not yet known. It is unlikely that the sensitive chemical catalysts used for methanol production will function with biogas without prior removal of contaminating substances. Direct oxidation of methane with biochemical oxidases could possibly prevent the need for special gas cleaning units and could produce methanol directly from biogas (see below).

Carbon dioxide

Catalytic reduction of carbon dioxide with hydrogen to methanol is inherent in every commercial methanol production process from coal, oil, or gas. Moreover, 30% of the methane of biogas is formed by microbial carbon dioxide reduction using hydrogen produced as a by-product during the anaerobic formation of organic acids (Heider, 2006a). The development of a separate technology that utilizes large amounts of excess carbon dioxide should be rational, yet this carbon dioxide must originate from renewable sources only. Hydrogen provision could be organized by using nonfossil energy for water hydrolysis, e.g., nuclear power. Hydrogen could also be produced biologically.

POSSIBLE FUTURE ROUTES OF METHANOL PRODUCTION

Methanol Production from Biomass and Coal

The production of methanol using coal decomposition technology has been substantially developed in Germany. The use of waste and residue from petroleum and coal processing, as well as renewable raw materials, plays a significant role in this production process. Around 1,800,000 tons of methanol are produced in this way in Germany per annum (Vriens, 1994).

This kind of methanol production involves the use of a synthesis gas, which is formed by gasifying coal and biogenic substances (e.g., sludge pellets, compost, or urban waste deposits) in what is called the GSP process. The gas generators are driven by technically pure oxygen, and steam is passed through them as a gasification medium. By using oxygen, gasification temperatures of up to 1,700°C when high pressures are applied can be achieved, depending on the nature of the substance being gasified. The whole combined system is represented in Fig. 2.

The process pathway is represented using the example of flue stream gasification right up to the formation of the synthesis gas. The gas mixture at the gasification stage, consisting of carbon monoxide, hydrogen, carbon dioxide, nitrogen, and ammonia, as well as some dust and heavy metal compounds, is fully quenched with excess water, cooling it to the dew point of steam. Once the gas mixture is saturated with steam, it goes through an initial washing stage to remove the dust. Once the desired H_2/CO proportions have been achieved, CO conversion takes place. Using the water vapor absorbed in the quenching room, the conversion reaction can take place with little additional steam. If only a certain proportion of the gas mixture undergoes the conversion stage, carbon oxysulfide must be removed from the remainder by catalytic COS hydrolysis. There are also various processes that can be used to remove the acidic gases, hydrogen sulfide, and carbon dioxide. This also applies to the preparation of quenching water. The hydrogen sulfide is converted (in large units as appropriate) into elementary sulfur in a Claus plant (Peter and Woy, 1969). The gasification pressure can be adjusted to suit the ensuing gas synthesis. The typical chemical reaction from synthesis gas production in the gasifiers to the synthesis of methanol can be represented as follows.

Synthesis gas production in the gasifier:

$$2C + O_2 \rightarrow 2CO$$
$$H_2O + C \rightarrow CO + H_2$$
$$H_2O + CO \rightarrow H_2 + CO_2$$

Methanol production using synthesis gas:

$$2H_2 + CO \rightarrow CH_3OH$$
$$3H_2 + CO_2 \rightarrow CH_3OH + H_2O$$

To synthesize methanol from an H_2/CO-rich gas, a stoichiometric ratio (volume ratio) of around 2 parts H_2 to 1 part CO is required, but the gas mixture after the gasification stage does not have the appropriate proportion of gases. At the conversion stage, excess CO is converted into CO_2, and using a bypass system, the desired H_2/CO ratio is achieved.

Gasification process

The gasification of a substance means the thermochemical conversion of a solid or liquid into a flammable gas, which is attained by adding steam and oxygen as gasification media. The oxygen is added substoichiometrically, and the energy required for the conversion process is provided by burning some of the coal with oxygen.

Reactor and quencher. At the flue stream gasification stage, dust in the flue stream is gasified at high pressure. The reactor consists of a cooled, pressure-resistant jacket and a fixed burner unit. The coal dust is passed over a special system to the burner and converted by the gasification media; this conversion takes place above the melting point of the ashes produced as a result. The latter leave the reactor as liquid slag through the slag removal unit together with the gas mixture, going directly into the quenching system, where the gas is cooled to around 200°C, and the slag is simultaneously cooled and granulated by sprayed water.

Gas purification. After the gas mixture leaves the gasifier, other constituents—dust, ammonia, and hydrogen chloride—are removed. Now that it has been washed, the gas mixture reaches the next preparation stage, CO conversion. The water used to wash it, or quenching water, is circulated in the quenching system. Some of the quenching water is continuously channeled off and prepared for reuse.

CO conversion. To achieve the required H_2/CO ratio for the synthesis of methanol, CO is converted into CO_2 by means of steam decomposition, which also produces hydrogen. The required temperature ranges are 360 to 530°C for high-temperature conversion and 210 to 270°C for low-temperature conversion. Little extra steam is required, and the water in the cycle is warmed by the gas, thus regulating the temperature of the gas.

H_2S/CO_2 removal. To remove the acidic components from the gas mixture (primarily CO_2, H_2S, and COS), an absorption process is normally used. In the contacting section, which is equipped with either packing or trays, a solvent (flowing in the opposite direction) more or less selectively removes one or several components from the gas mixture. The rich solvent is then regenerated and has all the gas components

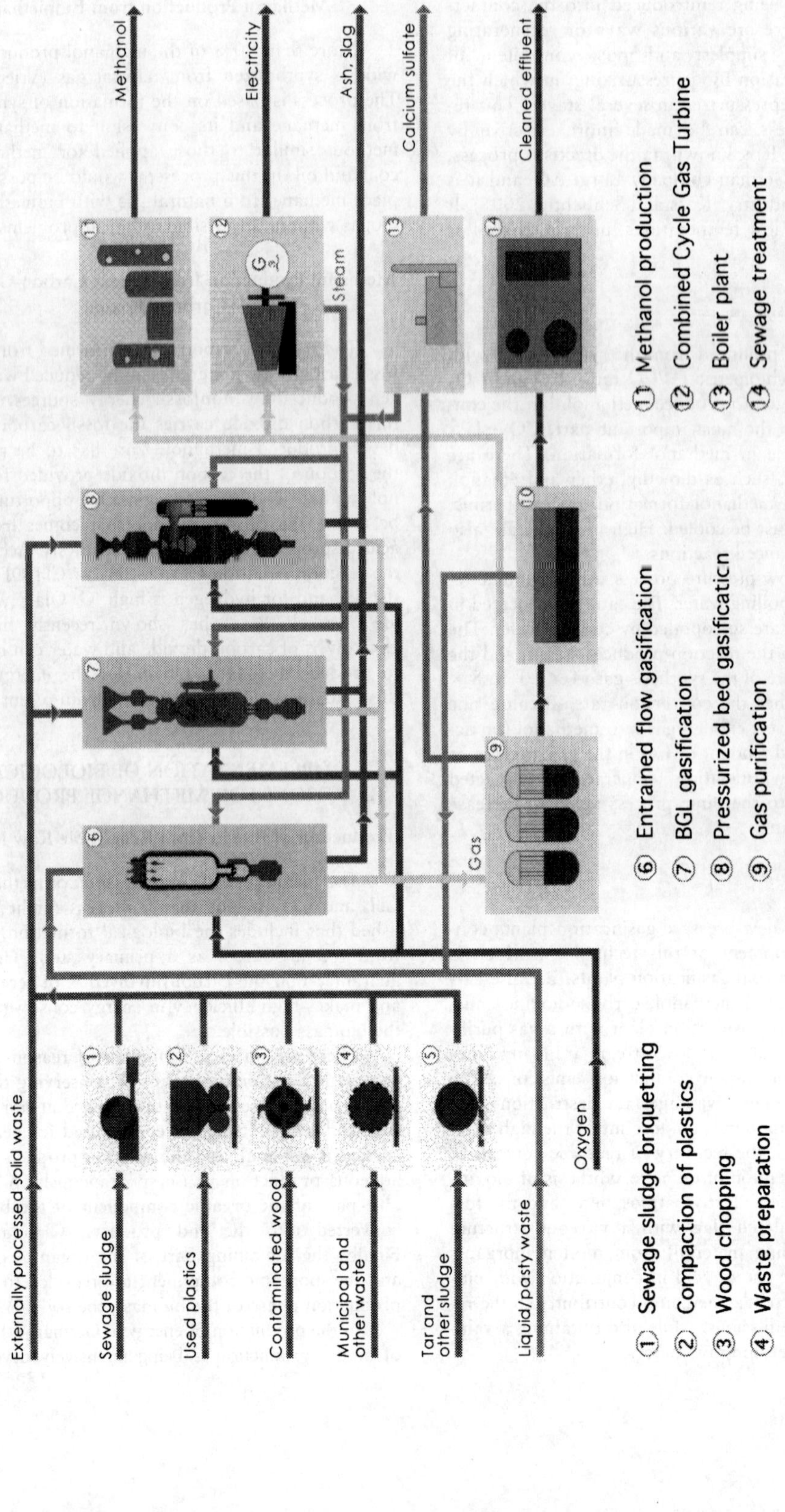

Figure 2. Schematic overview of the treatment of solid and liquid wastes (from SVZ, 2007).

removed before being reintroduced into the contacting section. There are various ways of regenerating the solvent, the simplest and most convenient of which is regeneration by depressurizing, in which the rich solvent is depressurized in several stages. This regeneration process can be made more effective by steam stripping. It is known as the Rectisol process, licensed by the German company Lurgi AG, and it is widely used in industry (Koss and Schlichtig, 2005). It is carried out at low temperatures, using methanol as a solvent.

Methanol synthesis

Methanol is produced from the synthesis gas with the required stoichiometric H_2/CO ratio. CO and CO_2 react with hydrogen to produce methanol, but the conversion of CO is the most important part; CO_2 plays only a minor role in methanol formation. There are also by-products, such as dimethyl ether and ethanol. The reactions in methanol formation are exothermic, so the process must be cooled. High pressures are also favorable for balanced reactions.

The Lurgi low-pressure process uses a tubular reactor cooled by boiling water. The catalyst is located in the tubes, which are surrounded by cooling water. The heat energy from the reaction produces steam, and the higher the pressure of the synthesis gas (4×10^6 to 8×10^6 Pa), the higher the conversion rate. Because one cycle cannot convert enough gas into methanol (on account of a limited amount of CO in the gas mixture or the catalyst's low operating temperature), unreacted gas is subjected to the same process again to increase the conversion rate.

Aims and effects

Establishing new pressure gasification plants continues the development of this technology, which is based on existing coal gasification plants. Because carbon is bound within methanol, carbon dioxide emissions in a pressure gasification plant with a gas purification and methanol plant downstream are many times lower than power station combustion emissions. The process also has other advantages: the destruction of all chloroorganic compounds on account of the high reaction temperatures, the recovery of reusable substances (sulfur), the prevention of de novo synthesis of dioxins and furans by the reductive atmosphere, and the formation of granulated slag with a vitreous structure (used as a building material) from molten inorganic components. The use of coal in conjunction with biogenic waste to produce methanol contributes to the reduction of CO_2 emissions, while also obtaining a valuable energy carrier and raw material.

Methanol Production from Biomethane

More than 90% of the methanol produced worldwide is synthesized from natural gas (Vriens, 1994). The process is based on the formation of synthesis gas from methane and its conversion to methanol, using methods similar to those applied for methanol from coal and oil. In this process, it would be possible to replace methane from natural gas with biomethane from biogas without any visible technical problems.

Methanol Production from Recent Carbon-Containing Carbon Dioxide

The residual carbon dioxide formed from burning fossil carbon raw materials can be reduced with hydrogen produced by nonfossil energy sources. However, this carbon dioxide carries the fossil carbon into the newly produced methanol. Care has to be taken that the carbon of the carbon dioxide provided for methanol synthesis is of recent origin. One opportunity could be to use the carbon dioxide that comes from acetic acid disproportion in methane synthesis. According to the reaction equation $CO_2 + 3H_2 \rightarrow CH_3OH + H_2O$, the demand for hydrogen is high. G. Olah (www.heise.de/tr/artikel/70633) has shown recently that direct electrolysis of carbon dioxide and water can be applied to produce methanol. Obviously, the energy used in this reaction would need to come from recent sources.

IMPLEMENTATION OF BIOLOGICAL REACTIONS FOR METHANOL PRODUCTION

Production of Biogas from Renewable Raw Materials

The chemical-catalytic production of methanol from CH_4 and CO_2 means that a process might be established that includes the biological formation of biogas from recent sources as a primary step. This would guarantee that the carbon involved is of recent nature and makes high efficiency in energy conservation from the biomass possible.

There are, in fact, a number of reasons why this process is a preferable means of conserving the energy of the utilized biomass: it takes place at normal pressure and temperature, the energy used for heating and stirring is minimal, and the reaction products leave the aqueous production phase spontaneously. A considerable part of the organic component of the biomass is converted into the end products, CH_4 and CO_2. Finally, the remaining part of the organic component and the inorganic component are recycled to the solid production phase of the biomass, the soil.

In the production of energy in Germany, the process of biogas production is being intensively developed to produce renewable energy (electric power and heat). With a 30 to 40% annual increase, it is predicted that there will be around 50,000 biogas plants in Germany in 2020 with a capacity of 100,000 MW (Gomez, 2007), equivalent to 450×10^9 m^3 of biogas per year. Alternatively, this biogas might be used to produce biomethane to be fed directly into the natural gas network.

Starting from the various main components of biomasses, it is possible to calculate the stoichiometric formation of methane and carbon dioxide. (Buswell and Mueller, 1952). Here, the stoichiometric figures are compared to the yields actually obtained in practice (Table 1). These data indicate that, depending on the substrates utilized, the biogas process realizes 30 to 95% of the theoretical yield. Reduced conversion rates are mostly a result of incomplete hydrolysis of structural components of the biomass and the trapping of readily fermentable substrates within their matrices, known as the "cage effect" (Aulrich and Flachowsky, 2001).

In the biogas production process, the first step is the hydrolysis of the polymers, of which polysaccharides constitute the greatest part (40 to 85% of the biomass). The hydrolytic enzymes required for this are produced by anaerobic microorganisms in the biogas reaction system. The enzymes are found docked to the cell wall polysaccharides, single or organized as cellulosomes (Bayer et al., 1998; Shoham et al., 1999), loosely adhered to the cell wall (Pell and Schofield, 1993), or (in small quantities) in free, soluble form.

Demonstrable hydrolysis activity in the biogas reactor, for example, is limited (A. Blokesch and J. König, personal communication). It has been shown that the hydrolysis, particularly of structural polysaccharides, can be significantly accelerated by adding externally produced hydrolytic enzymes (V. Pelenc, personal communication). The broad practical application of this process has demonstrated that it is both technically possible and economically advisable to accelerate the hydrolysis by adding preproduced enzymes, which positively influence the velocity of the whole biogas process (Gerhardt, 2006). The yield could be increased by an average of more than 20%. Because of the increase in hydrolysis velocity, delay time can be reduced, and because of a reduction in viscosity resulting from hydrolysis, the concentration of biomass can be increased.

The methane content in the biogas yield is determined largely by the proportions of carbon, hydrogen, and oxygen in the substrates. From polysaccharides, which make up 40 to 85% of vegetable biomasses, 50 to 54 vol% methane is obtainable. Only the substantial fat content in some substrates (e.g., rapeseed) produces higher methane contents in biogas:

$$CH_2O \rightarrow 50\ \text{vol\%}\ CH_4 + 50\ \text{vol\%}\ CO_2$$

$$CH_2O_{0,11} \rightarrow 74\ \text{vol\%}\ CH_4 + 26\ \text{vol\%}\ CO_2$$

Biogas is currently used in special combustion engines with 30 to 40% energy conversion efficiency. Another 30% is converted into heat. There is a far more effective form of energy use that does not convert energy in a power plant. Instead, the methane is separated from the CO_2—by washing, dry adsorption, or membrane filtration—and fed directly into the natural gas network for further use. It can then be used for methane-based chemical methanol synthesis, completing the biomass-to-methanol production process without any restrictions. Germany currently produces about 2 million tons of methanol per annum. To change this production to biomethane by using renewable sources would require 1,500,000 tons of CH_4. Contemporary biogas production could meet this need by using 8,700,000 tons of biomass (dry matter).

Production of H_2 from Biomasses

Hydrogen plays a major role in the production of methanol. It is involved in the formation of both synthesis gas and biogas. Whenever hydrogen is added, more carbon dioxide is involved in the reactions.

Table 1. Conversion of substrates into biogas

Substrate	Stoichiometric values[a] (m^3/ton)		Yields obtained in practice[b] (m^3/ton)		Potential (m^3/ton)	
	Gas	Methane	Gas	Methane	Gas	Methane
Grass (conservation)	513	287	143	82	370	205
Liquid cattle manure	830	476	370	204	460	273
Liquid pig manure	846	490	400	240	446	250
Wheat straw	594	309	370	204	224	105
Soilage	736	387	550	297	186	90
Rapeseed	998	641	770	506	506	228
Maize silage	726	385	600	312	126	73
Grass silage	696	376	627	332	69	44
Wheat	778	411	764	404	14	7

[a]Remaining stoichiometric potential for gas/methane formation.
[b]Data from KTBL, 2005.

Similar to the chemical-catalytic water gas shift reaction, there is an alternative biological route for hydrogen formation:

$$CO + H_2O \rightarrow CO_2 + H_2$$

performed by the thermophilic microorganism *Carboxydothermus hydrogenoformans*, which has been described by Gerhardt and Svetlichny (Gerhardt et al., 1991; Svetlichny et al., 1991a, 1991b). Using only carbon monoxide as a source of carbon and energy, CO and water were converted almost stoichiometrically to hydrogen and CO_2 under strict anaerobic conditions. This fermentation process can be made continuously over a long period of time, promising technological viability: 9.6 to 9.8 mol of hydrogen are produced by converting 10 mol of carbon monoxide. With normal atmospheric pressure and at 75°C, 1.0 liter of H_2 per h has been produced continuously in a 5-liter fermentor (Gerhardt et al., patent application DD 297449, 1990; Gerhardt et al., patent application DD 297450, 1990). This opens a biotechnological way to produce hydrogen from biomass pyrolysis via recent-carbon-containing carbon monoxide.

PROSPECTS AND CONSIDERATIONS ABOUT BIOLOGICAL PATHWAYS FOR CONVERSION OF METHANE TO METHANOL

Oxidation of Methane to Methanol

The fermentation pathway from sugars and other substances (such as fats and amino acids) via acetic acid to methane is very straightforward, although there must be strictly anaerobic conditions in the fermentation system. Methane is the end product of this anaerobic conversion pathway. It can be oxidized to methanol and finally to carbon dioxide (Fig. 3).

Methane oxidation is well understood in oxic sytems (Fig. 3; Hanson and Hanson, 1996).

Biological oxidation is performed by methane monooxygenases. The first evidence for this was published by Leadbetter and Foster (Leadbetter and Foster, 1959), and the enzyme was isolated and first characterized by Tonge et al. (Tonge et al., 1977). One oxygen molecule is attached to methane by the appropriate enzyme: one of the oxygen atoms reacts with methane to form methanol, while the other is captured by the hydrogen of coenzymes to produce water (for a review,

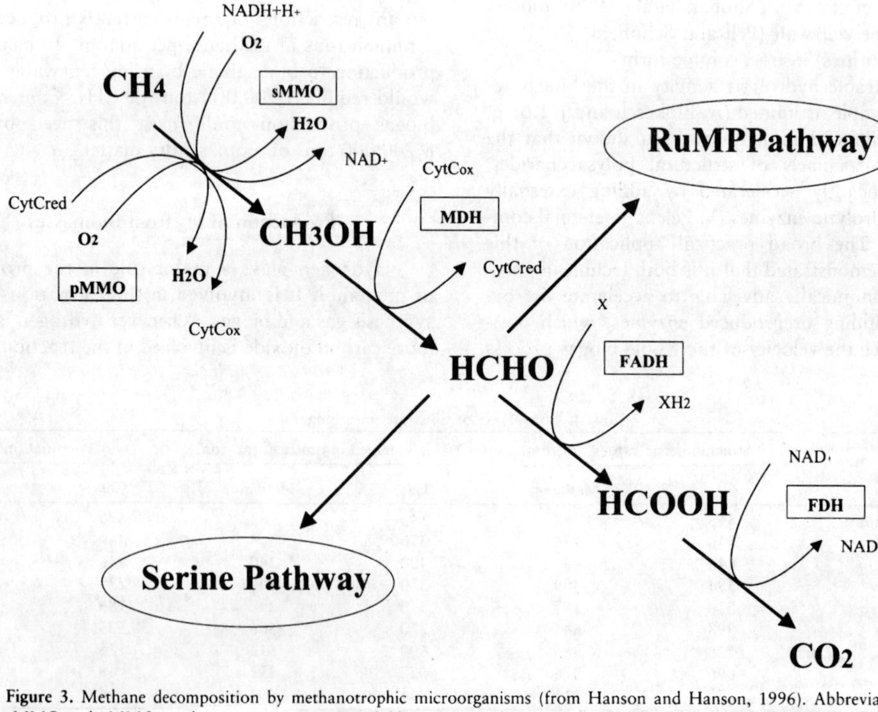

Figure 3. Methane decomposition by methanotrophic microorganisms (from Hanson and Hanson, 1996). Abbreviations: sMMO and pMMO, methane monooxygenase, soluble and particulate, respectively; CycC, cytochrome *c*; FADH, formaldehyde dehydrogenase; GDH, formate dehydrogenase; RuMP, ribulose monophosphate pathway.

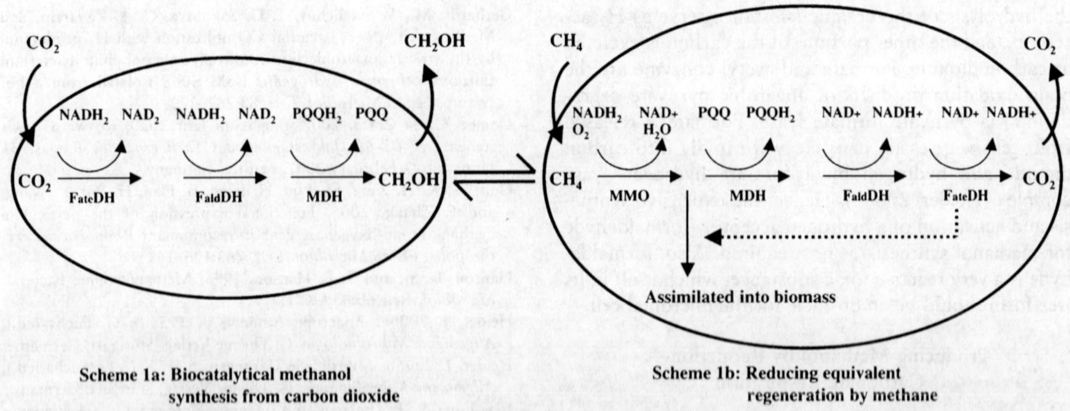

Figure 4. Pathway of methanol synthesis from carbon dioxide (1a) and reducing-equivalent regeneration by methane (1b) (from Xin et al., 2004a).

see Hanson and Hanson, 1996). The latter reaction makes the whole process expensive, because two reduction equivalents are channeled off to remove the remaining oxygen atom.

Thus, methanol can be synthesized from methane by enzymatic oxidation using methane monooxygenases, which are enzymes with the unique property that they can oxidize nonreactive methane in the presence of oxygen and NADH + H^+ or reduced cytochrome *c* (Fig. 3, step 1). The first group of methane monooxygenases is the particulate membrane-bound methane-monooxygenase (pMMO) family, which is Cu^{2+} dependent. Under Cu^{2+} limited conditions a second group, soluble monooxygenases (sMMO), which belongs to the multicomponent di-iron monooxygenase family possessing a hydrolase subunit with a di-iron center, is prominent. Some bacteria can produce both types of monooxygenases, though the pMMO level is dependent on the concentration of Cu^{2+}. Aerobic methylotrophic microorganisms (e.g., *Methylosinus*) might be used to oxidize methane to methanol if the further steps of methanol breakdown by methanol dehydrogenases could be prevented and if a mechanism for reductant generation could be established (Fig. 4). Several studies have demonstrated the feasibility of using resting cells to convert methane to methanol (Xin et al., 2004a, 2004b; Furuto et al., 1999). Also the synthesis of methanol from methane by recombinant particulate methane monooxygenase in *Rhodococcus erythropolis* has been demonstrated (Gou et al., 2006).

Although there has been some effort to develop a biotechnological process for aerobic methanol synthesis, the chemical-catalytic processes to produce methanol from methane via synthesis gas that have been developed are so technologically and commercially successful that there has been no great interest in establishing a biological alternative. This is particularly true since the chemical process can be performed using renewable raw materials, thus supporting efforts to combat climate change.

Recently, the surprising discovery of methane oxidation processes in anoxic systems was made. Methane is recycled by chemolithotrophic microorganisms through coupling with sulfate reduction (Boetius et al., 2000) or nitrification (Islas-Lima et al., 2004) in specialized microbial consortia. Methane oxidation in this case is not performed via oxygenation through a monooxygenase; it was proposed rather that this process could be the reverse of microbial methane synthesis (Chistoserdova et al., 2005).

Producing Methanol from Organic Substances by Reduction—Fermentation

Unlike ethanol, methanol cannot be produced by fermenting sugars. The reductive pathway from sugars to C_1-alcohol apparently does not occur naturally, or at any rate, it has not been discovered yet. Although many pathways in anaerobic microbial energy conversion result in the formation of the C_2 alcohol ethanol, there is no record of the microbial reduction of formaldehyde to form methanol.

Let us consider the nature of fermentation to understand the unusual nature of microbial methanol synthesis. As is generally known, microbial fermentations are substrate-level ATP-generating processes, whereby the hydrogen produced from the oxidation of organic substances must be released from the microbial cell. If the hydrogen gas cannot be released, the products of the hydrolysis of the organic substance serve as H_2 acceptors, and the other portion of the carbon is excreted as carbon dioxide. Formate and acetyl conzyme are the main oxidation products of anaerobic pyruvate degradation via pyruvate formate lyase. Formate is released from cells very easily or is cleaved rapidly into carbon dioxide and hydrogen by a formate hydrogen lyase complex (Heider, 2006b). Hence, the expensive synthesis and activation of a hydrogen acceptor (formaldehyde for methanol synthesis) is not required. Also, formaldehyde is a very reactive toxic substance, which itself in its free form should be an obstacle for the microbial cell.

Producing Methanol by Reduction—Carbonate Respiration

To answer the question of direct microbial synthesis of methanol from C_1 compounds as carbon dioxide, the mechanism by which microbes synthesize methane should be considered. It is estimated that around one-third of the methane on Earth is produced by carbonate respiration. The methane producers capable of carbonate respiration are hydrogen-oxidizing bacteria. They gain their energy from oxidative phosphorylation according to the following overall equation (Daniels et al., 1984):

$$4H_2 + CO_2 \rightarrow CH_4 + 2H_2O \text{ delta } G_0 = -131 \text{ kJ/mol of } CO_2$$

The pathway of methanogenesis, reviewed in chapter 13, does not include methanol as an intermediate under natural conditions.

REFERENCES

Althapp, A. 2003. Kraftstoffe aus Biomasse mit dem Carbo-V®-Vergasungsverfahren. *Fachtagung Regenerative Kraftstoffe,* 13 to 14 November 2003, Stuttgart, Germany.

Aulrich, K., and G. Flachowsky. 2001. Studies on the mode of action of non-starch-polysaccharides (NSP)-degrading enzymes in vitro. 2. Communication: effects on nutrient release and hydration properties. *Arch. Tierernähr.* 54:19–32.

Bayer, E. A., H. Chanzy, R. Lamed, and Y. Shoham. 1998. Cellulose, cellulases and cellulosomes. *Curr. Opin. Struct. Biol.* 8:548–557.

Boetius, A., K. Ravenschlag, C. J. Schubert, D. Rickert, F. Widdel, A. Gieseke, R. Amann, B. B. Jorgensen, U. Witte, and O. Pfannkuche. 2000. A marine microbial consortium apparently mediating anaerobic oxidation of methane. *Nature* 407:623–626.

Buswell, A. M., and H. F. Mueller. 1952. Mechanism of methane fermentation. *Ind. Eng. Chem.* 44:550–552.

Chistoserdova, L., J. A. Vorholt, and M. E. Lidstrom. 2005. A genomic view of methane oxidation by aerobic bacteria and anaerobic archea. *Genome Biol.* 6:208.

Daniels, L., R. Sparling, and G. D. Sprott. 1984. The bioenergetics of methanogenesis. *Biochim. Biophys. Acta* 768:113–163.

Furuto, T., M. Takeguchi, and I. Okura. 1999. Semicontinuous methanol biosynthesis by *Methylosinus trichosporium* OB3b. *J. Mol. Catal. A* **144**:257–261.

Gerhardt, M. 2006. Optimisation of agricultural biogas production plants by application of hydrolytic enzymes, abstr. 61. *Abstr. Third Annual World Congress on Industrial Biotechnology and Bioprocessing,* 11 to 14 July 2006, Toronto, Canada.

Gerhardt, M., V. Svetlichny, T. G. Sokolova, G. A. Zavarzin, and M. Ringpfeil. 1991. Bacterial CO utilization with H_2 production by the strictly anaerobic lithoautotrophic thermophilic bacterium *Carboxydothermus hydrogenus* DSM 6008 isolated from a hot swamp. *FEMS Microbiol. Lett.* **83**:267–272.

Gomez, C. da Costa. 2007. Biogas im Jahr 2020: wo werden wir stehen?, p. 63–68. *Jahrestagung des Fachverbandes Biogas,* 31 January to 2 February 2007, Leipzig, Germany.

Gou, Z., X. H. Xing, M. Luo, H. Jiang, B. Han, H. Wu, L. Wang, and F. Zhang. 2006. Functional expression of the particulate methane mono-oxygenase gene in recombinant *Rhodococcus erythropolis. FEMS Microbiol. Lett.* **263**:136–141.

Hanson, R. S., and T. E. Hanson. 1996. Methanotrophic bacteria. *Microbiol. Rev.* **60**:439–471.

Heider, J. 2006a. Anaerobe Atmung, p. 395. *In* G. Fuchs (ed.), *Allgemeine Mikrobiologie.* G. Thieme Verlag, Stuttgart, Germany.

Heider, J. 2006b. Mikrobielle Gärungen, p. 352. *In* G. Fuchs (ed.), *Allgemeine Mikrobiologie.* G. Thieme Verlag, Stuttgart, Germany.

Islas-Lima, S., F. Thalasso, and J. Gomes-Hernandez. 2004. Evidence of anoxic methane oxidation coupled to denitrification. *Water Res.* 38:13–16.

Koss, U., and H. Schlichtig. 2005. Lurgi's MPG Gasification plus Rectisol gas purification—advanced process combination for reliable syngas production. *Gasification Technologies 2005,* San Francisco, CA, October 2005.

KTBL. 2005. *Gasausbeute in landwirtschaftlichen Biogasanlagen.* Kuratorium für Technik und Bauwesen in der Landwirtschaft e.V., Darmstadt, Germany.

Leadbetter, E. R., and J. W. Foster. 1959. Incorporation of molecular oxygen in bacterial cells utilizing hydrocarbons for growth. *Nature* **184**:1428–1429.

Metz, B., et al. (ed.). 2005. *Special Report on Safeguarding the Ozone Lazer and the Global Climate System: Issues Related Hydrofluorocarbons and Perfluorocarbons.* IPCC/TEAP. Cambridge University Press, Cambridge, United Kingdom.

Pell, A. N., and P. Schofield. 1993. Microbial adhesion and degradation of plant cell walls, p. 397–423. *In* H. G. Jung, D. R. Buxton, R. D. Hartfield, and J. Ralph (ed.), *Forage Cell Walls Structure and Digestibility.* American Society of Agronomy Inc.; Soil Science Society America Inc., Madison, WI.

Peter, S., and H. Woy. 1969. Gewinnung von Schwefel aus Schwefelwasserstoff nach dem Claus-Verfahren. *Chemie Ingenieur Technik—CIT* **41**:1–7.

Schink, B., and G. Zeikus. 1980. Microbial methanol formation: a major end product of pectin metabolism. *Curr. Microbiol.* 4:387–389.

Shoham, Y., R. Lamed, and E. A. Bayer. 1999. The cellulosome concept as an efficient microbial strategy for the degradation of insoluble polysaccharides. *Trends Microbiol.* 7:275–281.

Supp, E. 1990. *How to Produce Methanol from Coal.* Springer-Verlag, Berlin, Germany.

Svetlichny, V. A., T. G. Sokolova, M. Gerhardt, M. Ringpfeil, N. A. Kostrikina, and G. A. Zavarzin. 1991a. *Carboxydothermus hydrogenoformans* gen. nov., sp. nov., a CO-utilizing thermophilic anaerobic bacterium from hydrothermal environments of Kunashir Island. *Syst. Appl. Microbiol.* **14**:254–260.

Svetlichny, V. A., T. G. Sokolova, M. Gerhardt, N. A. Kostrikina, and G. A. Zavarzin. 1991b. Anaerobic extremely thermophilic carboxydotrophic bacteria in hydrotherms of Kuril Islands. *Microb. Ecol.* **21**:1–10.

SVZ. 2001. *News-Letter.* Sonderausgabe 2001. Sekundärrohstoff-Verwertungszentrum Schwarze Pumpe (SVZ), Spreetal, Germany.

Tonge, G. M., E. F. Harrison, and I. J. Higgins. 1977. Purification and properties of the methane mono-oxygenase enzyme system from *Methylosinus trichosporium* OB3b. *Biochem. J.* **161**:333–344.

Vriens, R. 1994. How will Europe survive, methanol supply/demand outlook 1993–1997. *World Methanol Conference*, Geneva, Switzerland.

Xin, J. Y., J. R. Cui, J. Z. Niu, S. F. Hua, C. G. Xia, S. B. Li, and L. M. Zhu. 2004a. Production of methanol from methane by methanotrophic bacteria. *Biocatal. Biotransformation* 22:225–229.

Xin, Y. Y., J. R. Cui, J. Z. Niu, S. F. Hua, C. G. Xia, S. B. Li, and L. M. Zhu. 2004b. Biosynthesis of methanol from CO_2 and CH_4 by methanotrophic bacteria. *Biotechnology* 3:67–71.

Bioenergy
Edited by J. Wall et al.

Chapter 19

Prospects for Methanol Production

BAKUL C. DAVE

Energy is an essential requirement of life. The primary process responsible for generation of energy is through the photosynthetic pathway whereby photoexcitation of electrons is used to oxidize water to dioxygen along with the storage of reductive equivalents in the form of carbohydrates. These carbohydrates are utilized by plants through different biochemical pathways for generation of energy via oxidative processes ultimately to form carbon dioxide as an end product. Essentially, the source of all biomass and bioenergy generated from organic sources can be attributed to this simple process. In general, living systems rely on some fuel source in the form of reduced species which is consumed through oxidative pathways to generate energy with the eventual formation of carbon dioxide as a waste product. An analogous process is also employed on a global scale by societal organizations, in that fuel sources play an important role in their overall development to generate energy, to drive and sustain the socioeconomic infrastructure. At present, the dominant sources of fuel are carbon based, with the unavoidable consequence of CO_2 being a principal component of the waste products. The biomass generated from CO_2 by green plants to a certain extent restores the lost energy resources. The fossil fuel stores also owe their primary origins to biomass generated by photosynthetic processes. However, the growth in human population as well as rapid industrialization and significant deterioration of plant habitat due to deforestation has resulted in a pronounced shift in the overall balance of fuel resources and CO_2 generation pathways such that there appears to be an enhanced rate of CO_2 production that is not sufficiently counterbalanced by natural CO_2 sequestration processes. The levels of CO_2 in the atmosphere have reached record levels in recent times, and CO_2 emissions from human activity have been postulated to be the major contributors to global warming effects (Houghton et al., 2001). The growing use of fossilized carbon content as an energy source by the developing civilization simultaneously depletes the natural energy resources and contributes a major portion of anthropogenic CO_2 in the atmosphere, leading to global warming effects. As such, the energy and global warming crises are intrinsically intertwined.

Most living systems generate energy through oxidation of carbohydrates or hydrocarbons through a series of oxidative metabolic pathways. Combustion or oxidation of reduced hydrocarbon species is also the primary mechanism of energy generation from fossil fuel sources. An important distinction of these processes is that in contrast to industrial energy-generating mechanisms, the biochemical routes employ isothermal conditions for liberation of stored energy. Overall, the liberation of reductive equivalents stored in carbohydrates and hydrocarbons is the primary mode of energy generation, be it through direct oxidative combustion or through chemical, biochemical, or electrochemical means, with the eventual formation of carbon dioxide as a by-product. The liberation of stored energy through oxidative means along with entropic dissipation of CO_2 constitutes intrinsic aspects of energy generation through hydrocarbon sources. On the other hand, photosynthetic pathways are the ones that accompany storage of energy and CO_2 for continued sustainability of life. Therefore, effective energy strategies would need to be able to mimic these biological energy equilibria with mechanisms based on carbon-neutral pathways for long-term sustainability of fuel sources and for limiting global warming.

The utilization of biogenic materials for energy is gaining attention due to increasing concerns driven by anticipated shortages of fossil fuels (Greene et al., 2004; Huber et al., 2006; Petrus and Noordermeer, 2006). The use of biologically derived matter or biomass for energy provides an attractive option because it is a renewable resource made through the photosynthetic route and is believed to be essentially carbon neutral (Okkerse and van Bekkum, 1999; Kheshgi et al.,

Bakul C. Dave • Department of Chemistry and Biochemistry, Southern Illinois University—Carbondale, Carbondale, IL 62901.

2000; Lemus and Lal, 2005). The use of biomass as a fuel source is something that has existed ever since human beings first discovered fire. Direct utilization of biomass for energy generation is carbon intensive and also energy inefficient due to incomplete utilization of the total carbon content stored in the biomass. From a technological standpoint, conversion of biomass to easily storable and transportable fuel provides an effective means of utilization of natural resources. Currently, the generation of fuel through biomass is mainly focused on processing strategies based on enzyme-assisted hydrolysis of carbohydrates to generate sugar followed by fermentation of sugars to produce ethanol. Two main examples of this strategy are the corn-ethanol and the sugarcane-ethanol systems, which essentially utilize energy stored in the biomass through the photosynthetic process. The generation of fuel using this pathway harnesses solar radiation in the form of usable fuel sources. However, long-term sustainability of this technology would depend on the available land resources necessary for cultivating biomass on a global scale for widespread use and consumption as a fuel (Bungay, 2004). Furthermore, the total economics of cultivation, harvesting, processing, and conversion of biomass to fuel must consider other factors. For example, the allocation of land and other resources would compete with the use of these resources for generating food and other essential commodities typically obtained from plant sources. In assessing the realistic potential and associated constraints of this technology, the competing demands of energy versus food would play a critical role, especially in light of the projected population increases. Finally, a consequence of this process is the generation of large amounts of CO_2 during bioethanol production, which reduces the overall energy efficiency of the final fuel product. Overall, the microbial/enzymatic biomass-to-biofuel pathways such as those utilized in the production of ethanol from biomass provide an example of a solution toward bioproduction of renewable fuels.

The use of biological systems—be it either through microbes or through the use of enzymes isolated from living sources—furnishes a particularly attractive strategy for biofuel generation. The metabolic processes of living organisms employ chemical conversions of carbon-based molecules that can not only provide an access to energetic fuels but also effectively combat the growing problems of greenhouse gas accumulations in the environment. While bioproduction of ethanol is currently a mature technology (Moreira and Goldemberg, 1999; Hahn-Hågerdal et al., 2006), biological processes, in principle, may be utilized in formation of other fuel sources such as methanol. Methanol can be made from biomass, but its production requires combustion of biomass and condensation of the resulting gases, which renders the overall process less competitive. On the other hand, methanol—being a part of the single-carbon biotransformations—is particularly well suited for production through biocatalytic pathways. Especially important in this regard are the single-carbon redox transformations between the fully reduced methane and the fully oxidized carbon dioxide with methanol, formaldehyde, and formic acid as intermediates. Methanol is perhaps a unique alcohol which lacks any carbon-carbon bonds and therefore can be produced from available single-carbon precursors without the need for energy-consuming carbon-carbon bond formation steps. Given the predominance of single-carbon biochemical transformations that are mediated by biological entities, bioproduction of methanol will likely play a central role in energy sustainability as well as mitigation of excess of CO_2 in the environment, especially if methanol can be made directly from the reduction of CO_2 in a cost-effective manner. With this in mind, this chapter discusses the potential of methanol as an alternative fuel along with the prospects for its production using biomimetic pathways for efficient conversion of carbon dioxide to methanol based on single-carbon biotransformations. This chapter focuses on methanol production through biocatalysis and is organized in three parts. First, the effects of fuel sources and their influence on the global carbon cycle and atmospheric accumulations of CO_2 are discussed. Second, the potential utility of methanol as an alternative fuel and the scope of different methods for its commercial production are outlined. Finally, the use of biological systems in efficient conversion processes leading to methanol are elucidated with specific emphasis on dehydrogenase-catalyzed synthesis of methanol from carbon dioxide.

CARBON CYCLE AND FUEL ECONOMY

The temporal distribution of carbon species in the global framework along with the different equilibria governing the displacement of carbon between bio- and geosystems as well as their atmospheric and oceanic components are described by what is known as the carbon cycle (Falkowski et al., 2000). Given the closed nature of the system, the total amount of carbon equivalents remains constant but the local spatiotemporal distribution of carbon content is altered between the several subsystems comprised of carbon sources and sinks. The main sources of CO_2 release in the atmosphere include human, societal, and industrial activity, respiratory contributions from living sources, and decomposition of organic matter. Photosynthesis along with incorporation of CO_2 in water, soil, and sediments and the subsequent mineralization of CO_2 into the geological framework are the main pathways for assimilation of environmental CO_2 back into the carbon cycle. Ideally, the two opposing fluxes must balance in order to maintain CO_2 levels in the environment. In addition to the respiratory activity of the ever-increasing human population, the major anthropogenic contributions to CO_2 activity in the environment primarily result from liberation of CO_2 from organic stores of carbon via consumption of organic fossil fuels as well as from inorganic mineralized carbon sources due to calcination and processing of inorganic carbonates in cement and concrete manufacturing for infrastructure development. Therefore, the growing necessities of the human population place critical constraints on both organic and geological reservoirs of carbon with an entropic dissipation of carbon dioxide in the atmosphere. This is further compounded by the increasing proliferation of societies on available land mass and consequent deterioration of vegetative habitat. As such, the cumulative growth and socioeconomic development of human population and its contributions to CO_2 emissions are rather deleterious consequences of normal growth patterns and furthermore appear to be intrinsically correlated. In other words, CO_2 generation caused by humans is an intrinsic index of the essential socioeconomic drivers of a developing civilization.

The socioeconomic indicators of fuel utilization and associated carbon emissions are expressed by the so-called IPAT framework (Ehrlich and Holdren, 1974) which expresses the environmental impact (I) in terms of economic development as $I = P \times A \times T$ (where P is population, A is affluence expressed by per capita consumption, and T is per capita pollution generated from energy consumption due to advancing technologies). This approach is further extended to derive a correlation of CO_2 emissions with fuel sources in terms of gross national product (GNP) as an indicator of economic development as expressed by the Kaya identity (Kaya, 1995):

$$CO_2 \text{ emissions} = (CO_2/\text{energy}) \times (\text{energy/GNP}) \times (\text{GNP/population}) \times \text{population} = \text{carbon intensity} \times \text{energy intensity} \times \text{consumption intensity} \times \text{population}$$

These indices outline an intrinsic correlation of economic development with population increase and fuel utilization. As a result, increasing economic development appears to be strongly intertwined with anthropogenic CO_2 emissions due to the predominant utilization of fossilized carbon sources in economic development. While these trends of economic development will continue in the future, a remedy to CO_2 mitigation involves utilization of fuel sources with reduced carbon intensity. In other words, fuels with reduced carbon content—such that a lesser amount of CO_2 is generated through their combustion—are likely to alleviate the problem of increased CO_2 emissions. Therefore, fuels with a low carbon equivalent (amount of carbon per gram of fuel mass) may constitute preferred energy sources. In this context, methanol can be viewed as an attractive option because of its singular carbon along with a low carbon-to-hydrogen ratio and low carbon content (carbon content of gasoline, 85 to 88%; diesel, 84 to 87%; methane, 75%; propane, 82%; ethanol, 52%; and methanol, 37.5%). Furthermore, if methanol can be made from renewable sources, or better still from CO_2 itself, then it provides a critical technology in reducing the overall carbon footprint in normal fuel usage (Weimer et al., 1996).

POTENTIAL UTILITY OF METHANOL

Methanol—Properties and Characteristics

Methanol is perhaps one of the most industrially important chemicals. Methanol is the simplest of alcohols with a single carbon atom and is characterized by several unique properties and characteristics (Table 1) that make it useful as an alternative fuel. Under room temperature conditions, it is a liquid, which makes it easy to store and transport. It has many uses in industry, primarily as a starting material or feedstock for the synthesis of other industrially relevant compounds,

Table 1. Methanol properties and characteristics

Can be prepared from different carbon sources sources (natural gas, liquid hydrocarbons, coal, sewage sludge, biomass, etc.) via gasification/reforming
Can be made from natural feedstocks such as wood, methane, agricultural waste, etc.
Majority of commercial methanol produced from natural gas
Disperses easily in the environment and is rapidly biodegradable
Vapors in the atmosphere oxidized by oxygen (with the help of sunlight) to CO_2 and water
High solubility in water
Toxic in large doses but noncarcinogenic and nonmutagenic
Easily transportable liquid
Handled as liquid but burns as a gas
High hydrogen-to-carbon ratio with four hydrogens per carbon
High-energy flammable liquid
Contains no carbon-carbon bonds and is a cleaner burning fuel
Partially oxygenated; requires less oxygen for complete combustion
Provides octane enhancement to gasoline fuel; used as octane booster
Used in M85 (85% methanol and 15% unleaded gasoline) blends
Excellent automotive fuel; used in race cars and rocket fuels
Utilized in transesterification of oils for production of biodiesel
Excellent as feedstock in chemical industry for preparation of other commodity chemicals, plastics, etc.
Versatile chemical; used in synthesis of different chemicals and as a solvent

such as formaldehyde, acetic acid, and plastics, among a whole slew of diverse industrially relevant commodity chemicals. It is also widely used as a solvent in different formulations such as paints and varnishes, cleansing agents, and antifreeze. Methanol is also commonly employed in transesterification of oils for production of biodiesel (Meher et al., 2006).

Methanol results naturally in the environment from metabolic processes of certain varieties of bacteria. Furthermore, methanol is biodegradable and is also degraded in the environment by means of light and oxygen. While methanol in large quantities is extremely toxic to humans, it is found naturally in food sources in trace quantities and is metabolized in the body through the action of dehydrogenase enzymes in the liver to formaldehyde and formic acid. Other than its acute toxicity in large doses, methanol has not been found to induce any carcinogenic or mutagenic effects in living systems.

Because of its oxygenated structure, methanol is a clean-burning fuel and has long been used as a component in rocket fuel. Methanol represents the simplest example of an oxygenated hydrocarbon with four hydrogens per carbon. Since methanol lacks carbon-carbon bonds, it does not leave any particulate residues which are formed during combustion. When added to conventional gasoline, it increases the fuel's octane value. Blends of methanol with gasoline such as M85 (85% methanol and 15% gasoline) have been used as automotive fuels, especially in race cars and flexible-fuel vehicles. Since methanol is a liquid, storage, transport, and delivery of methanol is compatible with current means employed with gasoline fuels. In addition to its direct utilization as a combustion engine fuel, an added application of methanol in energy generation involves its use as a material in fuel cells for direct conversion to electrical energy (McGrath et al., 2004). While the use of ethanol as a fuel has become widespread in recent times, it appears that methanol may be better suited as a transportation fuel in the long range (Marsden, 1983). Given its prospects as a source of energy, a methanol economy (Olah, 2005; Olah et al., 2006) has been propounded as an alternative to hydrogen energy (Dunn, 2002). The success of widespread utilization of methanol as a fuel is likely to depend on efficient low-cost "green" manufacturing strategies that maintain the global CO_2 balance.

Methanol Production Pathways

Methanol—also known as wood alcohol—has long been produced by burning of wood chips followed by condensation of the gases to form methanol. However, this process was very inefficient and resulted in only a few gallons of methanol per ton of wood. In general, methanol can be made from any carbon-containing feedstock including biomass, crops, forest residues, and municipal and food-processing waste products. Nowadays, methanol is mainly synthesized from natural gas (mostly methane), which improves the overall yields substantially. Methanol can also be produced from coal. Typically, coal is pulverized and fed to a gasifier where it is treated with oxygen and steam to form syngas. Biogas—which is principally methane and CO_2—made from anaerobic digestion of biomass is also used in manufacture of methanol. The large-scale industrial manufacture of methanol is typically done by making syngas (a mixture of CO and H_2) from gasification and steam reforming of feedstock. The chemical reactions leading to syngas formation are described in Fig. 1. For optimum methanol production the stoichiometry ratio $\{([H_2] - [CO_2])/([CO] + [CO_2])\}$ should be equal to or greater than 2. Since the process involves conversion of a gas to a liquid, the process can be made favorable by carrying out the reaction under high pressure to increase methanol yields. Methanol produced for commercial applications is made predominantly via the syngas pathway, and although the feedstock may vary, natural gas is still the preferred raw material.

BIOLOGICAL METHANOL PRODUCTION

Carbon Biotransformation Pathways as a Means to Methanol Production

Microbial metabolic pathways involve carbon-based molecules for isothermal energy generation. Several classes of microorganisms utilize hydrocarbons

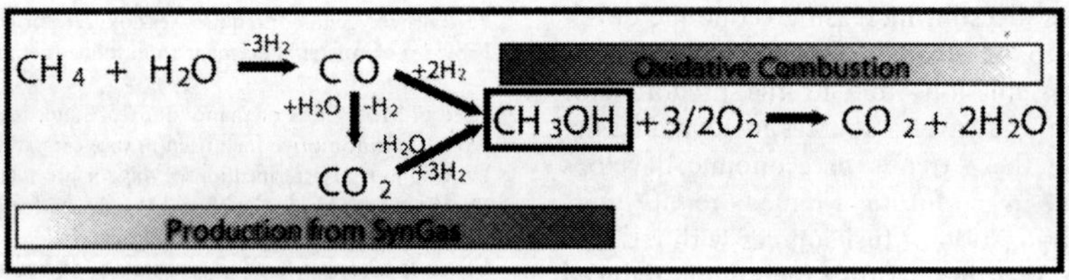

Figure 1. Reactions involved in production and combustion of methanol.

and/or carbohydrates for energy conversion, and therefore, the associated intrinsic metabolic activity can be utilized for conversion of biomass to fuel sources. Microbial digestion of biomass is being increasingly utilized in the manufacture of energy-rich species. Chief among these is the production of bioethanol from cellulosic biomass (Wyman, 1999). Corn and sugarcane have been the principal feedstocks used in bioethanol production; however, any biomass rich in cellulose and/or starch can be used in bioethanol manufacture (Lynd, 1996). The general process is based on chemical and/or biochemical hydrolysis of starches and cellulose to form sugars, followed by fermentation of sugars, usually facilitated by yeast, to ethanol. While the process has been found to be quite efficient for ethanol production, the typically large amounts of CO_2 produced during the fermentation step remain an area of concern. The single-carbon metabolic pathways of methanogenic microbes provide a potentially attractive pathway to production of energy-rich methane via anaerobic fermentation (Zeikus, 1977; Rouviere and Wolfe, 1988; Reeburgh, 2007). The methanogenic archaea can also produce methane as a metabolic product (Schåfer et al., 1999). While methane is a good source of energy, for practical applications, storage and transport of methanol in liquid form offers critical advantages necessary for its practical utility as a fuel.

The use of microbes in methanol production has been a long-sought-after goal. Particularly important in this regard are the redox pathways utilized by methanotrophic bacteria used in metabolism of methane catalyzed by methane monooxygenase (MMO), methanol dehydrogenase (MDH), formaldehyde dehydrogenase ($F_{ald}DH$), and formate dehydrogenase ($F_{ate}DH$) as shown below:

$$CH_4 \xrightarrow{\text{MMO}} CH_3OH \xrightarrow{\text{MDH}} HCHO \xrightarrow{F_{ald}DH} HCOOH \xrightarrow{F_{ate}DH} CO_2$$

The formation of methanol by microbes has been considered through the use of methanotrophic bacteria that use methane as the source of energy to produce CO_2 because methanol is formed as an intermediate in this process (Hanson and Hanson, 1996; Trotsenko and Khymelenina, 2005). A particularly remarkable feature of the metabolism of methanotrophic bacteria is the use of MMO to catalyze oxidation of the very stable C-H bond in methane to C-OH group to form methanol (Lieberman and Rosenzweig, 2004). In principle, this process can be utilized in the manufacture of methanol; however, at present the means to stop the process at the methanol level without the continued formation of oxidative products, i.e., formaldehyde, formic acid, and carbon dioxide, remains an issue. Furthermore, the general pathway used by methanotrophic bacteria is toward oxidation of single-carbon species to CO_2, which, from a fuel perspective, results in loss of overall energy density if this process were to be used for conversion of methane to methanol.

A Strategy for Bioconversion of Carbon Dioxide to Methanol

The single-carbon transformations mediated by microorganisms—spanning the entire range from the hydrogen-rich methane to the oxygen-rich CO_2—offer a potentially powerful approach towards effective bioconversion strategies for production of energetic fuels as well as CO_2 fixation. The general sensitivity of biological cells to methanol has largely precluded use of microbes in methanol production processes, although generation of methanol as a major end product of pectin metabolism by use of several strains of pectinolytic bacteria has been reported (Schink and Zeikus, 1980). While the prospects of using microbes in biomethanol generation appear appealing, there are several additional technological limitations that are likely to prevent their utilization in large-scale manufacturing processes. The metabolic processes of microorganisms usually result in a variety of products, by-products, and intermediates. The lack of effective means for systematic control and regulation of microbial metabolic processes to generate the desired end product while eliminating unwanted side reactions imposes critical restraints on practical applications. Finally, the issues related to microbial growth, limiting batch concentrations, and long-term sustainability of living cells in a dynamic culture environment need to be taken into consideration towards development of economically feasible microbial processes for methanol production.

The stability, reaction control, and sustainability issues related to the use of microorganisms can be effectively overcome by using enzymes to catalyze the desired reactions (Comfort et al., 2004; Held et al., 2000) instead of living cell cultures. While the metabolic activities of living cells in a batch process are characterized by growth, maturity, and eventual decline, enzymatic processes are free from these constraints. The nature, direction, and extent of microbial transformations are dictated by the specific microbial metabolic activities, whereas the molecular nature of enzymes offers considerable flexibility in process design. Microbial processes, in general, are dependent on several factors such as pH, temperature, water activity, and osmotic and hydrostatic pressure along with bioavailability of nutritional components and oxygen to maintain the desired process kinetics. On the other hand, enzyme kinetics can be efficiently modulated and regulated through systematic control of enzyme, substrate, and product concentrations. More importantly, different combinations of enzymes can be used to design novel process sequences, which are typically not possible with living cells due to the intrinsic metabolic restrictions of microorganisms. Enzymes derived from living cells can be used effectively as catalysts for biotransformations if they can be stabilized to withstand the process conditions for prolonged sustained operation.

The enzyme-catalyzed single-carbon redox transformations furnish a facile access to methanol synthesis through use of oxidoreductase enzymes for the synthesis of methanol from carbon dioxide. Oxidases and dehydrogenases are two distinct classes of enzymes that are utilized by living systems to catalyze redox processes (Burton, 2003). Oxidases catalyze these reactions with the help of dioxygen as an electron acceptor, while dehydrogenases typically utilize NAD^+ (or $NADP^+$) as the electron sink. The reactions catalyzed by dehydrogenases are proton-coupled electron transfer processes such that transfer of proton(s) accompanies electron transfer reactions. These enzymes are generally stable under ambient conditions and are utilized in a variety of practical biotechnological applications because of their compatibility with a diverse range of environments. While these enzymes are typically used for oxidative catalysis, according to the principle of microscopic reversibility, these enzymes are capable of catalyzing reactions in both the oxidative and the reductive directions. Thus, the NAD^+/NADH couple can act either as a source or as a sink of electrons depending on the relative concentrations of reactants and products. While the normal biochemical function of dehydrogenases is characterized by oxidative transformation, a reduction of CO_2 to methanol can be accomplished by using an excess of NADH to drive the reaction in the opposite direction. This feature has been effectively used to design a three-enzyme catalytic system for a sequential conversion of carbon dioxide to methanol (Obert and Dave, 1999):

$$CO_2 \xrightarrow{F_{ate}DH} HCOOH \xrightarrow{F_{ald}DH} HCHO \xrightarrow{ADH} CH_3OH$$

The direct reduction of CO_2 to methanol is a six-electron, six-proton reaction, which is a thermodynamically and kinetically unfavorable process. However, the sequential reduction process is made thermodynamically favorable by the excess of NADH present in the system as well as by the use of enzymes to accelerate the reaction kinetics. On the other hand, since the reaction involves sequentially coupled catalytic steps such that the product of one step acts as a substrate for the next step, it makes the probability of individual primary reaction events less probable in the solution media due to Brownian motion and random movement of molecules in the solution phase. As such, the lack of structural organization in the solution phase is something that limits the practical feasibility of utilizing this sequential catalytic system for large-scale biomanufacturing of methanol.

Immobilization of Dehydrogenase Enzymes in Sol-Gel Glasses

Sequential reactions are favored in cells due to the cellular organization provided by the cellular structure. Microorganisms utilize the compartmentalized organization conferred by the cellular structure to carry out complex series of enzymatically catalyzed reactions as part of their metabolic activity (Deamer, 1997). Biological cellular organization is such that the overall state of aggregation of biological cellular components is characterized by molecular assemblies interspersed in aqueous phase to achieve a semiorganized architecture. Such a cellular organization usually confers enhanced stability and selectivity to biomolecules along with control and regulation mechanisms that are usually not accessible to isolated enzymes in solution. Therefore, use of an external organizing matrix is essential. Theoretical analyses show that immobilization of enzymes in a matrix is an effective strategy to enhanced enzyme kinetics (Celayeta et al., 2001). In this context, immobilization of enzymes in sol-gel-derived porous silica glasses (Dave et al., 1994) provides a strategy to overcome limitations of free enzyme mobility and provides unique advantages from the practical standpoints of bioproduction and practical manufacturing of methanol from carbon dioxide.

Unlike isotropic solution media in which the biomolecules have equal mobility and conformational flexibility in all directions, encapsulation constrains them within the porous structure of silica gel. The amorphous glasses incorporate biological molecules dispersed in an aqueous phase in the interconnected porous network. The nature of trapped biomolecules in such an aggregation is characterized by several important factors such as improved stability by virtue of being confined to a pore, preorientation, and limited free volume in the pore, which increases the effective concentration. These specific effects often induce additional selectivity for reaction processes compared to individual enzymes freely moving in the solution. The partial order imposed due to the presence of a rigid cage can result in unique phenomena usually not feasible in solution. Indeed, in spite of general similarities with solution-based reactions, the enzymes in sol-gel matrices are characterized by significant variations in overall stability and reaction kinetics (Dave et al., 1996). The sol-gel system utilizes the properties of spatially isolated molecules in a solvent-rich environment necessary for stability of these biomolecules. The molecular dimensions of the single-carbon reactants and intermediates are considerably smaller than average pore diameters, and these low-molecular-weight species are diffusionally mobile in the solvent-filled pores. The high-molecular-weight biomolecules, on the other hand, are confined within the nanopores of the sol-gel matrix. Overall, the reaction chemistry of biomolecules in gel glasses is analogous to that in solution with the exception that now the system involves a porous silicate matrix. The porous sol-gels reduce the mean free path length of the low-molecular-weight reactants, intermediates, and products while immobilizing the enzymes and making the overall process more favorable.

The sol-gel-derived materials are porous inorganic silica glasses that are prepared by hydrolysis of alkoxysilane precursors (usually tetramethyl orthosilicate). The process of making these begins with aqueous mixtures of precursors along with small amounts of acid or base as a catalyst. Initial hydrolysis of the precursor results in formation a liquid sol which ultimately turns to a solid porous gel as the reaction progresses, with condensation of the silicate species forming a silica polymer. Hydrolysis results in conversion of Si-OR bonds to Si-OH bonds which condense together to form an oxo-bridged polymeric Si-O-Si structure. These reactions occurring in a localized region lead to formation of sol particles. As a result of polycondensation, the degree of cross-linking increases, as does viscosity. This viscous material then solidifies and forms a porous gel. The gel is a two-phase system comprised of a porous solid phase and a trapped aqueous phase. Since the process begins from a liquid solution phase, it is possible to add proteins and enzymes to the liquid sol prior to its gelation. Furthermore, it is possible to buffer the pH of the sol to the optimum necessary for biomolecular stability. The enzymes get encapsulated in the porous structure when the sol turns to a gel (Dave et al., 1998). The as-prepared silica gels are characterized by substantial retention of aqueous phase in the pores, which stabilizes the biomolecules. Additionally, the hydrophilic pore walls present a biocompatible microenvironment that is conducive to biomolecular stability. Finally, the porous structure of the sol-gel enables diffusion of low-molecular-weight substrates and intermediates while preventing movement of the large enzymes.

The stabilization of enzymes in sol-gel materials provides a strategy for efficient utilization of enzymes in conversion of carbon dioxide to methanol. Immobilization of these enzymes confers additional thermal and environmental stability to the enzyme structure due to elimination or minimization of protein unfolding pathways. While there have been several approaches used for immobilization of enzymes in polymeric materials, the use of inorganic sol-gel-derived materials has been found to offer promising technological advantages in practical applications. The water-rich environment available to biomolecules in the pores of silica gels to some extent exemplifies the short-range structural order conferred by an extrinsic organizing medium.

Enzymatic Conversion of Carbon Dioxide to Methanol in Sol-Gel Glasses

The sol-gel materials provide an ideal framework for dehydrogenase enzyme biocatalysts. Enzymatically coupled sequential reduction of carbon dioxide to methanol can be accomplished through a series of reactions catalyzed by three different dehydrogenases such that overall process involves an initial reduction of CO_2 to formate catalyzed by $F_{ate}DH$, followed by reduction of formate to formaldehyde by $F_{ald}DH$, and finally, the reduction of formaldehyde to methanol mediated by ADH (Attwood, 1990; Popov and Lamzin, 1994; Reid and Fewson, 1994). In this process, NADH acts as the terminal electron donor for each dehydrogenase-catalyzed reduction (Fig. 2). The process strategy for CO_2 reduction in sol-gel matrices takes advantage of the fact that dehydrogenases can effectively catalyze the reverse reactions (i.e., reduction) in the presence of an excess of NADH as a terminal electron donor. Additionally, since the process involves a sequential reaction of in situ-generated substrates with three different

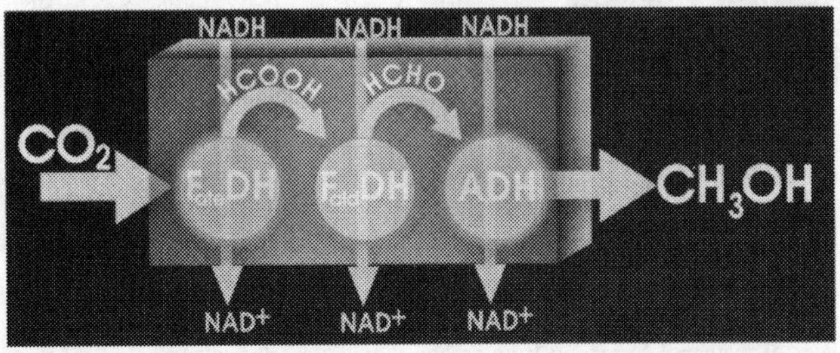

Figure 2. Schematic depiction of enzyme-catalyzed sequential reduction of carbon dioxide to methanol in a sol-gel matrix.

enzymes organized and immobilized in the nanopores of a silica sol-gel, it results in an enhanced probability of primary reaction events due to an overall increase in local concentration of reactants within the nanopores of the sol-gel. As a result, when the enzymes are encapsulated in the porous silica sol-gel matrix, the yield of methanol production is substantially enhanced compared to that in solution media.

According to procedures initially developed in our lab (Obert and Dave, 1999), the three enzymes were encapsulated in silica sol-gel glass and CO_2 was bubbled through the solution containing the glasses in the presence of NADH. Since the pores of the sol-gel are small, the enzymes remained trapped inside while the CO_2 and methanol diffused in and out. The results for methanol production in solution and the sol-gel system are shown in Fig. 3. The moles of methanol produced are plotted as a function of the moles of the terminal electron donor (NADH). Since NADH served as a limiting reagent in the overall reaction, it provides a relative measure of the efficiency of the reaction and the yield of the methanol production. As can be seen from Fig. 2, 3 mol of NADH were consumed per mol of methanol produced. As such, for 100% yield, the moles of methanol produced should be one-third of the NADH added. The overall yield of the reaction in solution was very low. However, in the sol-gels, the production of methanol was substantially enhanced. The results of this study indicated that confinement of the multienzyme system in the nanopores of silica sol-gels alters the reaction thermodynamics and final equilibrium of the reaction. The yield of the reaction was very low in solution phase. However, for the same concentration of NADH, it was seen that in the sol-gel, the production of methanol was significantly enhanced compared to production in solution with yields of up to about 90%, indicating that the overall equilibrium was shifted more towards the products (Fig. 3). The enhancement of methanol production in sol-gel was due to confinement and matrix effects. An immobilized system in the nanopores of a sol-gel matrix is characterized by limited pore volume. As such, increased local concentrations of enzymes and reactants prevail in the porous structure of silica sol-gels. The overall efficiency of enzyme reactions in sol-gels is enhanced such that the final equilibrium is shifted more towards the product. The overall reaction involves conversion of a gas to a liquid, and such a shift in equilibrium towards products is most likely consistent with reduction in available volume as a result of confinement.

The overall yield of the reaction for methanol production through this pathway depended on several factors. Chief among them is the NADH concentration. The percent yield of methanol decreased at higher concentrations of NADH presumably due to an increased tendency of the system to undergo the reverse reaction (i.e., conversion of methanol to carbon dioxide) as NAD^+ accumulated. Therefore, the system is characterized by an optimum concentration of NADH for maximum methanol formation. Another factor that had significant influence on the system was the pH of the medium. A neutral pH was found to be most effective, and the enzymes worked the best when the pH was close to 7. At low pH there was substantial denaturation of the enzymes, while at high pH the reaction yields were diminished, presumably due to lack of free protons that are involved in the proton-coupled electron transfer reactions. Increasing the temperature resulted in reduced yields likely due to reduced solubility of carbon dioxide, whereas the yields were enhanced when the reaction was carried out under ambient conditions or at 4°C. The gels containing the enzymes

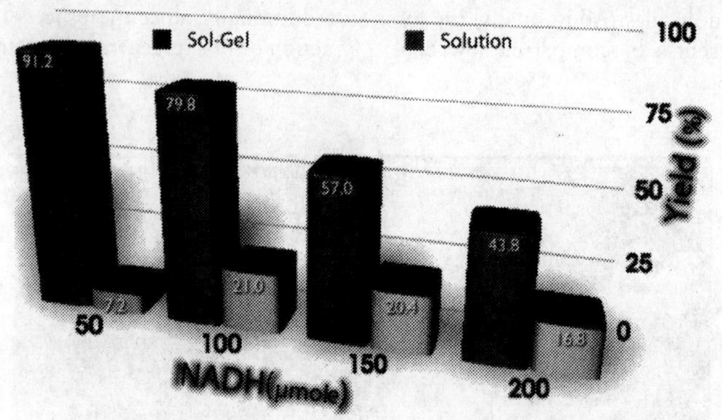

Figure 3. Relative yields of methanol formation from carbon dioxide in solution versus sol-gel matrix (Obert and Dave, 1999).

could be pulverized without loss of enzymatic activity due to the stabilization provided by the sol-gel matrix. The powdered gels increased the surface area and therefore were more suited for practical applications. Taken together, several mechanisms exist for precise tailoring of the reaction system as well as process conditions for efficient production of methanol from carbon dioxide.

The sequential enzymatic conversion pathway to methanol production from CO_2 provides several significant advantages (Table 2). First, it provides a facile biological means of storing reductive equivalents on carbon dioxide to form methanol, which has been found to be an effective fuel in different applications. Second, the reaction sequence mimics the biological mechanism of carbon dioxide fixation, which is analogous to that employed by the photosynthetic apparatus. While photosynthesis stores reductive equivalents on CO_2 in the form of carbohydrates, this process results in storage of reductive equivalents in the form of methanol. Third, the raw materials for this process such as enzymes and NADH are biogenic, and therefore, the process generates practically usable methanol fuel from renewable sources for long-term sustainability. Fourth, the use of enzymes as catalysts immobilized in sol-gel glasses increases their stability for cost-effective production strategies. Immobilization in sol-gel not only provides longevity to enzymes but also significantly enhances the overall thermodynamics of the system for increased methanol production. Fifth, enzymes immobilized in the sol-gel can be used in powdered form and can be easily separated from the reaction mixture such that the enzymes may be reused either as part of a continuous process or as part of different batch processes. Finally, the strategy outlines the tremendous potential and prospects for energy harvesting through biochemical means for the production of methanol. The implications of using biochemical pathways for methanol generation from carbon dioxide can be significant, especially if the process is competitive with existing methods for large-scale production of methanol.

Table 2. Advantages of sol-gel enzymatic methanol production

Carbon-neutral pathway
Recycling of carbon dioxide into a fuel product
Gaseous CO_2 as feedstock for methanol production
Sequential reduction of CO_2 to methanol
Stepwise reduction process optimizes conversion efficiency
Porous sol-gels facilitate diffusion of reactants, intermediates, and products
High efficiency and selectivity due to use of enzymes
Enhanced stability of enzymes due to confinement in sol-gel matrices
Encapsulation in sol-gel prevents enzyme denaturation
Sol-gel catalysts more robust than free enzymes
Enhanced methanol production due to confinement effect
Ready availability of raw materials
Enzymes and coenzymes available from natural sources
Process feasible under ambient conditions
Potential for direct conversion of atmospheric CO_2
Industrially compatible manufacturing process

SUMMARY AND FUTURE PERSPECTIVES

Energy and environment are the most important prerequisites to improved standards of living. The global warming and energy crises have placed a premium on novel innovative technologies that can provide effective carbon-neutral fuel sources essential for maintaining the global balance of equilibria between oxidative and reductive pathways of single-carbon species. Continued population growth and industrialization have resulted in rapid depletion of natural fuel resources due to increasing energy demands. Likewise, the increased utilization of fossil energy has also contributed to global warming scenarios arising from generation of excess CO_2 in the environment. It is expected that the use of fossil fuels will continue for quite some time in the foreseeable future, since currently the alternative fuels are not as economically viable and constitute only a small fraction of the total energy economy. Long-term strategies to reduce global CO_2 emissions must rely on renewable fuel sources in the context of their carbon neutrality. Likewise, the depletion of fossil sources requires alternative energy solutions (Armaroli and Balzani, 2007). In the long range, as the world moves towards displacement of fossil fuels through utilization of alternative fuel sources for abatement of CO_2 levels, innovative sustainable carbon management strategies will be imperative (Aresta, 2003; Mazen, 2006; Song, 2006). However, this would likely require a paradigm shift in future energy technologies. Given the urgency of the dire scenarios of fuel shortages and global warming, radical new innovations are needed as opposed to incremental improvements in bioenergy generation.

The long-term environmental and energy sustainability issues span an intricate balance of interdependent variables, and key future technologies will be based on synergistic approaches that maintain that harmony. Improved renewable fuel sources would be critical to continued economic growth and progress to provide clean, secure, competitive, and reliable means of energy supply without compromising the integrity of the environment. In this context, the biomimetic pathways outlined herein are likely to provide critical solutions necessary to satisfy the energy needs without compromising the integrity of the environment. Conversion of the greenhouse gas CO_2 to a valuable fuel product such as methanol offers a promising method not only for recycling of the gas but also for production of an alternative fuel source (Fig. 4). The enzymatic conversion of

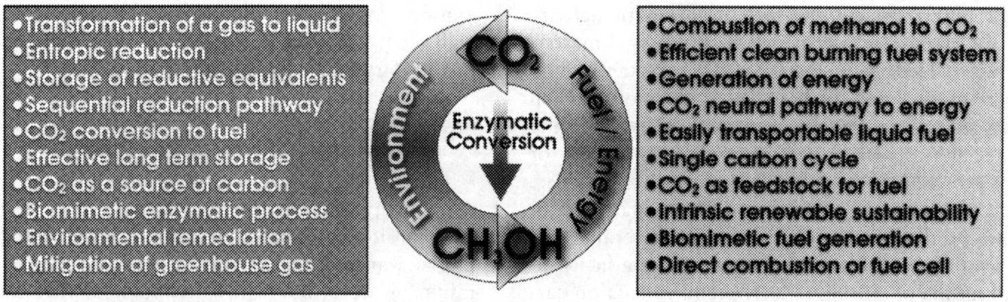

Figure 4. Schematic depiction of implications of enzymatic bioconversion of carbon dioxide to methanol.

CO_2 to methanol to a certain extent is analogous to the photosynthetic process in that the latter leads to conversion of CO_2 to sugar. A sequential reduction of carbon dioxide by three different dehydrogenases encapsulated in a sol-gel matrix results in enhanced yields for generation of methanol. The efficient production of methanol provides a facile pathway not only for on-site generation of methanol from readily available resources but also for potential applications related to energy technology and environmental fixation of carbon dioxide. Enzymatic processes have fundamental advantages over other processes due to their specificity and selectivity and also because of the less energy-intensive ambient temperature-pressure operational conditions. The utilization of microbes and/or enzymes in large-scale generation of fuels is currently in its incipient stages, and significant developments and new breakthroughs await widespread recognition and use of these alternative strategies for energy generation. New green fuel technologies will be successful in the long range only if they are made competitive with existing methods. Long-term economic sustainability will be necessary for the new biomimetic methods to displace the current manufacturing processes. The key drivers of use of enzymes in biomimetic energetic transformations will be the investment of capital and other resources toward a sustained development of essential technological framework for large-scale production in a cost-effective manner. In the long range, with appropriate resource allocations, enzymatic biomethanol production pathways offer appealing prospects for practical development of new self-sustainable technologies.

REFERENCES

Aresta, M. 2003. Carbon dioxide utilization: greening both the energy and chemical industry: an overview. *ACS Symp. Ser.* **852:**2–39.

Armaroli, A., and V. Balzani. 2007. The future of energy supply: challenges and opportunities. *Angew. Chem. Int. Ed.* **46:**52–66.

Attwood, M. M. 1990. Formaldehyde dehydrogenases from methylotrophs. *Methods Enzymol.* **188:**314–327.

Bungay, H. R. 2004. Confessions of a bioenergy advocate. *Trends Biotechnol.* **22:**67–71.

Burton, S. 2003. Oxidizing enzymes as biocatalysts. *Trends Biotechnol.* **21:**543–549.

Celayeta, J. F., A. H. Silva, V. M. Balcao, and F. X. Malcata. 2001. Maximisation of the yield of final product on substrate in the case of sequential reactions catalysed by coimmobilised enzymes: a theoretical analysis. *Bioprocess Biosyst. Eng.* **24:**143–149.

Comfort, D. A., S. R. Chhabra, S. B. Conners, C.-J. Chou, K. L. Epting, M. R. Johnson, K. L. Jones, A. C. Sehgal, and R. M. Kelly. 2004. Strategic biocatalysis with hyperthermophilic enzymes. *Green Chem.* **6:**459–465.

Dave, B. C., B. Dunn, J. S. Valentine, and J. I. Zink. 1994. Sol-gel encapsulation methods for biosensors. *Anal. Chem.* **66:**1120A–1127A.

Dave, B. C., B. Dunn, J. S. Valentine, and J. I. Zink. 1996. Nanoconfined proteins and enzymes: sol-gel based biomolecular materials, p. 351–365. *In* G.-M. Chow and K. E. Gonsalves (ed.), *Nanotechnology*. American Chemical Society, Washington, DC.

Dave, B. C., B. Dunn, J. S. Valentine, and J. I. Zink. 1998. Sol-gel matrices for protein entrapment, p. 113–134. *In* F. S. Ligler and A. E. G. Cass (ed.), *Immobilized Biomolecules in Analysis: A Practical Approach*. Oxford University Press, Oxford, United Kingdom.

Deamer, D. W. 1997. The first living systems: a bioenergetic perspective. *Microbiol. Mol. Biol. Rev.* **61:**239–261.

Dunn, S. 2002. Hydrogen futures: toward a sustainable energy system. *Int. J. Hydrogen Energy* **27:**235–264.

Ehrlich, P. R., and J. P. Holdren. 1974. Human population and the global environment. *Am. Sci.* **May–June:**282–292.

Falkowski, P., R. J. Scholes, E. Boyle, J. Canadell, D. Canfield, J. Elser, N. Gurber, K. Hibbard, P. Hogberg, S. Linder, F. T. Mackenzie, B. Moore, T. Pedersen, Y. Rosenthal, S. Seitzinger, V. Smetacek, and W. Steffen. 2000. The global carbon cycle: a test of our knowledge of earth as a system. *Science* **290:**291–296.

Greene, N., F. E. Celik, B. Dale, M. Jackson, K. Jayawardhana, H. Jin, E. D. Larson, M. Laser, L. Lynd, D. MacKenzie, J. Mark, J. McBride, S. McLaughlin, and D. Saccardi. 2004. *Growing energy: How Biofuels Can Help End America's Oil Dependence.* Natural Resources Defense Council, New York, NY.

Hahn-Hågerdal, B., M. Galbe, M. F. Gorwa-Grauslund, G. Lidén, and G. Zacchi. 2006. Bio-ethanol—the fuel of tomorrow from the residues of today. *Trends Biotechnol.* **24:**549–556.

Hanson, R. S., and T. E. Hanson. 1996. Methanotrophic bacteria. *Microbiol. Rev.* **6:**439–471.

Held, M., A. Schmid, J. B. van Beilan, and B. Witholt. 2000. Biocatalysis: biological systems for the production of chemicals. *Pure Appl. Chem.* **72:**1337–1343.

Houghton, J. T., Y. Ding, D. J. Griggs, M. Noguer, P. J. van der Linden, X. Dai, K. Maskell, and C. A. Johnson (ed.). 2001. *Climate Change 2001: The Scientific Basis. Contribution of Working*

Group I to the Third Assessment Report of the Intergovernmental Panel on Climate Change. IPCC, Cambridge University Press, Cambridge, United Kingdom.

Huber, G. W., S. Iborra, and A. Corma. 2006. Transportation fuels from biomass: chemistry, catalysis, and engineering. *Chem. Rev.* **106:**4044–4098.

Kaya, Y. 1995. The role of CO_2 removal and disposal. *Energy Convers. Mgmt.* **36:**375–380.

Kheshgi, H. S., R. C. Prince, and G. Marlend. 2000. The potential of biomass fuels in the context of global climate change: focus on transportation fuels. *Annu. Rev. Energy. Environ.* **25:**199–244.

Lemus, R., and R. Lal. 2005. Bioenergy crops and carbon sequestration. *Crit. Rev. Plant Sci.* **24:**1–21.

Lieberman, R. L., and A. C. Rosenzweig. 2004. Biological methane oxidation: regulation, biochemistry, and active site structure of particulate methane monooxygense. *Crit. Rev. Biochem. Mol. Biol.* **39:**147–164.

Lynd, L. R. 1996. Overview and evaluation of fuel ethanol from cellulosic biomass: technology, economics, the environment, and policy. *Annu. Rev. Energy Environ.* **21:**403–465.

Marsden, S. S. 1983. Methanol as a viable energy source in today's world. *Annu. Rev. Energy* **8:**333–354.

Mazen, M. A.-K. 2006. Recent progress in CO_2 capture/sequestration: a review. *Energy Sources A* **28:**1261–1279.

McGrath, K. M., G. K. S. Prakash, and G. A. Olah. 2004. Direct methanol fuel cells. *J. Ind. Eng. Chem.* **10:**1063–1080.

Meher, L. C., D. V. Sagar, and S. N. Naik. 2006. Technical aspects of biodiesel production by transesterification—a review. *Renew. Sust. Energ. Rev.* **10:**248–268.

Moreira, J. R., and J. Goldemberg. 1999. The alcohol program. *Energy Policy* **27:**229–245.

Obert, R., and B. C. Dave. 1999. Enzymatic conversion of carbon dioxide to methanol: enhanced methanol production in silica sol-gel matrices. *J. Am. Chem. Soc.* **121:**12192–12193.

Okkerse, C., and H. van Bekkum. 1999. From fossil to green. *Green Chem.* **1:**107–114.

Olah, G. A. 2005. Beyond oil and gas: the methanol economy. *Angew. Chem. Int. Ed.* **44:**2636–2639.

Olah, G. A., A. Goeppert, and G. K. S. Prakash. 2006. *Beyond Oil and Gas: The Methanol Economy*. Wiley-VCH, Weinheim, Germany.

Petrus, L., and M. A. Noordermeer. 2006. Biomass to biofuels: a chemical perspective. *Green Chem.* **8:**861–867.

Popov, V. O., and V. S. Lamzin. 1994. NAD(+) dependent formate dehydrogenase. *Biochem. J.* **301:**625–643.

Reeburgh, W. S. 2007. Oceanic methane biogeochemistry. *Chem. Rev.* **107:**486–513.

Reid, M. F., and C. A. Fewson. 1994. Molecular characterization of microbial alcohol dehydrogenases. *Crit. Rev. Microbiol.* **20:**13–56.

Rouviére, P. E., and R. S. Wolfe. 1988. Novel biochemistry of methanogenesis. *J. Biol. Chem.* **263:**7913–7916.

Schåfer, G., M. Engelhard, and V. Müller. 1999. Bioenergetics of archaea. *Microbiol. Mol. Biol. Rev.* **63:**570–620.

Schink, B., and J. G. Zeikus. 1980. Microbial methanol formation: a major end product of pectin metabolism. *Curr. Microbiol.* **4:**387–389.

Song, C. S. 2006. Global challenges and strategies for control, conversion and utilization of CO2 for sustainable development involving energy, catalysis, adsorption, and chemical processing. *Catal. Today* **115:**2–32.

Trotsenko, Y. A., and V. N. Khymelenina. 2005. Aerobic methanotrophic bacteria of cold ecosystems. *FEMS Microbiol. Ecol.* **53:**15–26.

Weimer, T., K. Schaber, M. Specht, and A. Bandi. 1996. Methanol from atmospheric carbon dioxide: a liquid zero emission fuel for the future. *Energy Convers. Mgmt.* **37:**1351–1356.

Wyman, C. E. 1999. Biomass ethanol: technical progress, opportunities, and commercial challenges. *Annu. Rev. Energy Environ.* **24:**189–226.

Zeikus, J. G. 1977. The biology of methanogenic bacteria. *Bacteriol. Rev.* **41:**514–541.

4. HYDROGEN PRODUCTION

Bioenergy
Edited by J. Wall et al.

Chapter 20

Towards Hydrogenase Engineering for Hydrogen Production

MARC ROUSSET AND LAURENT COURNAC

An option attracting great attention for replacing greenhouse gas-producing fossil fuels in the future is molecular hydrogen, provided that it is produced under a clean and renewable way. Methods for its production, use, and storage are currently the subject of intensive investigations. Development of new biotechnological processes designed to meet future energy demands may take advantage of microbes that have been using H_2 from very early in the evolution of life (Martin and Muller, 1998; Rotte et al., 2000). Many organisms including some bacteria, *Archaea*, and lower eukaryotes have an active hydrogen metabolism, utilizing the cleavage of H_2 into two electrons and two protons to gain energy or producing hydrogen to release reducing power (Casalot and Rousset, 2001). In biological systems, the reaction is catalyzed by metal-containing enzymes called hydrogenases (Vignais and Billoud, 2007).

Three classes of hydrogenases can be distinguished based on their metal content (Vignais et al., 2001; Vignais and Colbeau, 2004). (i) The oxygen-sensitive [FeFe] hydrogenases (EC 1.12.7.2) contain a di-iron active site (Color Plate 7). They have been identified in lower eukaryotes and in gram-positive bacteria, in which they are primarily involved in H_2 production. (ii) The more robust [NiFe] hydrogenases (EC 1.12.2.1) harbor a hetero dinuclear nickel-iron active site (Color Plate 8). These hydrogenases are the most widespread in nature. They are found in many bacteria and *Archaea* genera, in which they are mainly engaged in H_2 uptake. In addition to their bimetallic-active sites, both classes contain iron-sulfur clusters. (iii) Members of the third hydrogenase class function as H_2-forming methylenetetrahydromethanopterine dehydrogenases (EC 1.12.98.2) and are found only in a small group of methanogenic *Archaea* (Shima and Thauer, 2007). They do not have any Fe-S cluster (Fig. 1) but contain an $Fe(CO)_2$ core bound to a novel organic cofactor (Korbas et al., 2006; Lyon et al., 2004; Shima et al., 2004). Based on the metal content of their active site, although in the past they were considered as metal-free hydrogenases, they have been recently designated as [Fe] hydrogenases (Armstrong and Albracht, 2005).

Structural and spectroscopic studies indicate that the three classes of hydrogenases all contain at the active site at least one iron atom coordinated to two diatomic ligands, one CO and one additional CN or CO (Color Plates 7B and 8B). Therefore, this structure, $Fe(CO)_2$ or $Fe(CO\text{-}CN)$, likely represents the minimal cofactor making hydrogenase activity possible. Genetic studies revealed that the three classes are not phylogenetically related, indicating that a convergent evolution phenomenon occurred leading to the selection of three different enzymes catalyzing the same reactions.

Even though [FeFe] hydrogenases may appear to be the best suited for hydrogen production purposes, enzyme engineering studies for these enzymes are still poorly developed because of their great sensitivity to oxidative damage, which makes any biochemical characterization very uncertain. This is why this chapter is mainly focused on [NiFe] hydrogenases, which because of their greater stability is the most studied class. The core of these enzymes consists of two subunits; the large subunit is approximately 60 kDa and houses the Ni-Fe-active site, whereas the small subunit, which can be of variable size, harbors from one to three iron-sulfur clusters (Color Plate 9). In certain enzymes, additional subunits enable the interaction of these clusters with physiological electron carriers such as quinones, pyridine nucleotides, ferredoxins, and cytochromes (Vignais and Colbeau, 2004). Crystal structure analysis of heterodimeric [NiFe] hydrogenases from *Desulfovibrio* species (Higuchi et al., 1997; Matias et al., 2001; Volbeda et al., 1995; Volbeda et al., 2002) (Color Plate 8A), now taken as the prototypical [NiFe] hydrogenases, revealed that the Ni-Fe cofactor is deeply buried in the large subunit. The Ni is coordinated to the protein

Marc Rousset • CNRS, BIP, 31 chemin Joseph Aiguier, 13402 Marseille Cedex 20, France. **Laurent Cournac** • CEA-Cadarache, DSV, IBEB, SBVME, LB3M, UMR 6191 CNRS CEA Université de la Méditerranée, Bt 161, 13108 Saint Paul Lez Durance Cedex, France.

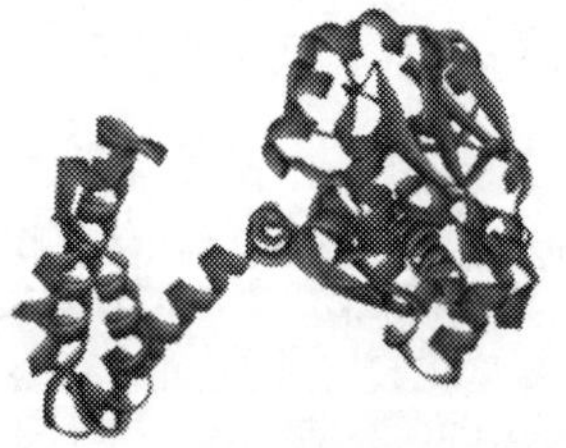

Figure 1. [Fe] hydrogenase from *Methanocaldococcus jannaschii* (Pilak et al., 2006). Protein Data Bank ID 2b0j. Three-dimensional structure of the apoenzyme is shown.

via four thiol groups from conserved cysteine residues; two of these are bridging ligands that coordinate both Fe and Ni (Color Plate 8B) (Higuchi et al., 1997; Matias et al., 2001; Volbeda et al., 1995, 2002). Infrared spectroscopy revealed that the Fe coordination sphere also possesses three diatomic ligands: one CO and two CN molecules (Pierik et al., 1999; Volbeda et al., 1996). The sixth iron coordination position is assumed to be occupied by a bridging hydride between iron and nickel (De Lacey et al., 2007; Pardo et al., 2006). Hydrophobic cavities which channel the gas substrate between the protein surface and the active site (Color Plate 10) (Montet et al., 1997; Teixeira et al., 2006; Volbeda et al., 2002) as well as a proton-conducting channel (Dementin et al., 2004) were identified inside the hydrogenase.

At present, structure-function relationship studies of hydrogenases have mainly remained in the basic research realm, aimed at understanding the enzyme catalytic mechanism (Fernandez et al., 2005). These studies were limited to very few enzyme models because of the great difficulty to produce recombinant hydrogenases in commonly used heterologous hosts such as *Escherichia coli.* Even though [NiFe] hydrogenase operons are highly conserved and exhibit a high degree of similarity, each maturation system is specific to the corresponding enzyme, probably because of tight protein-protein interactions occurring during processing (Leach and Zamble, 2007). This specificity barrier is responsible for the failure of hydrogenases to be matured when produced in heterologous hosts (Casalot and Rousset, 2001). The exploitation of this basic research knowledge is nevertheless a major challenge for a broad range of biotechnological applications. Microorganisms harboring optimized hydrogenases may play a major role in hydrogen generation for fuels. Biofuel cells and biosensors also represent an important potential application of hydrogenases as immobilized enzymes.

In order to obtain a sufficient level of enzyme efficiency and robustness for technological purposes, hydrogenases must be functionally optimized by improving electron transfer from electron donors (i.e., improving kinetic parameters of the enzyme) or improving O_2 tolerance. These challenges can be approached through genetic engineering by two different strategies: at the cellular level, by metabolic engineering, it is possible to create favorable conditions for substrate reduction or to avoid O_2 exposure (Henstra et al., 2007; Kruse et al., 2005; Rupprecht et al., 2006). At the enzyme level, by protein engineering, the goal will be to improve hydrogenases so that they can outcompete other enzymes for substrate utilization, use a thermodynamically more favorable substrate, become more O_2 tolerant, or become more efficient in H_2 production than H_2 uptake. In this chapter, we review hydrogenase structure-function relationship studies in which new properties of modified enzymes might serve as an inspiration source for rational optimization of hydrogenases for biotechnological processes.

MODIFICATIONS OF THE INTRAMOLECULAR ELECTRON TRANSFER PATHWAY INFLUENCE THE ENZYME BIAS

The electrons produced upon H_2 oxidation at the active site are transferred to the redox partner via a chain of closely spaced iron-sulfur (FeS) clusters, at least one of which is exposed at the surface of the enzyme. In group 1 [NiFe] hydrogenases, defined as respiratory enzymes (Vignais et al., 2001; Vignais and Billoud, 2007), the redox chain that mediates the electron transfer between the active site and the redox partner consists of a proximal (to the active site) [4Fe4S] cluster, a median [3Fe4S], and a distal [4Fe4S] that exchanges electrons with the redox partner (Color Plate 10). The redox potentials of these three clusters are quite different. In *Desulfovibrio* [NiFe] hydrogenases, the [3Fe-4S] cluster (Color Plate 9B) exhibits a much higher midpoint potential (from −70 to +65 mV) than the two [4Fe-4S] clusters (Color Plate 9A) (from −290 to −340 mV) (Rousset et al., 1998; Teixeira et al., 1989), the latter being closer to the redox potential of the couple H_2/H^+ at pH 7 ($E°' = -420$ mV) present at the active site during the catalytic cycle. Although the middle position of the [3Fe4S] center suggests an active redox role for this center (Volbeda et al., 1995), it surely represents a potential barrier in the electron pathway.

The [3Fe4S] center has been converted into a [4Fe4S] center in *Desulfovibrio fructosovorans* [NiFe] hydrogenase with the aim of reducing the potential difference (Rousset et al., 1998). This modification induced a decrease from +65 to −250 mV of the midpoint potential. This important redox potential variation of 315 mV resulted in a moderate but significant modification of the enzyme bias, which means a preferential catallysis in one direction. The H_2 oxidation activity of the mutant enzyme, with methyl viologen (MV) or cytochrome c_3 as electron acceptor, exhibited a decrease of 38% compared to the wild-type (WT) values, whereas H_2 production activity showed an increase of 60% with reduced MV as electron donor (Rousset et al., 1998). The ratios of the H_2 uptake rate to the H_2 production rate switched from 5 for the native enzyme to 2 for the P238C variant. As a result, it is possible to assume that the high potential median cluster works as an electron sink that facilitates hydrogen oxidation. The decrease of this driving force as observed with the P238C variant favors hydrogen production.

Another peculiarity in the intramolecular electron transfer chain in [NiFe] hydrogenases lies in the unusual coordination of the Fe atom from the distal [4Fe4S] center by the $N_{\delta 1}$ atom of a histidine (Color Plate 11) (Volbeda et al., 1995). This histidinyl ligation is indeed rare among the iron-sulfur proteins that commonly exhibit an all-cysteinyl ligation. Interestingly, in the native enzyme, this histidine ligand does not modify the detailed geometry of the cluster (Volbeda et al., 1995) and has no major effect on its thermodynamic properties. Remarkably, the proximal [4Fe4S] cluster that has a regular four-cysteinyl ligation presents approximately the same redox potential (~ −350 mV) as the peculiar distal cluster (Teixeira et al., 1989). This also seems to be the case in the Fe-only hydrogenases. In order to understand the role of this intriguing histidine, it was changed into a cysteine, to engineer a canonical all-cysteinyl ligation, or into a glycine, to induce the formation of a [4Fe4S] cluster with a labile and exchangeable ligand (Dementin et al., 2006). The His/Gly and His/Cys mutations dramatically decreased the H_2 oxidation activity 35- and 60-fold, respectively, while the H_2 production rates with MV as electron donor were barely affected, about 75 and 45% of that of the native enzyme. The fact that the mutations impaired only one direction of the enzyme activity but had a negligible effect on the Michaelis constants for MV indicates that the electron transfer kinetic is mainly affected (Dementin et al., 2006). In hydrogenases two electron transfer steps can be considered: the intramolecular transfer between the active site and the distal [4Fe4S] center and the intermolecular electron transfer between the distal [4Fe4S] center and the redox partner. Biochemical and electrochemical studies showed that the intermolecular electron transfer was affected for both H184C and H184G. This was quite surprising, especially for H184C, as the four-cysteinyl ligand configuration, which is observed in the vast majority of the iron-sulfur proteins, could have been assumed to be efficient in cluster coordination. But the most interesting result came from chemical rescue experiments, conducted with the H184G enzyme, monitored by cyclic voltammetry. Increasing imidazole concentration restored activity in the two directions, but the analysis of the voltammogram wave shape revealed that the intramolecular electron transfer was impaired only when the enzyme oxidized hydrogen. This unexpected result indicates that the consequence of the imidazole ligand of the distal [4Fe4S] center is to increase the rate of intramolecular electron transfer from the high potential medial center, thus facilitating the hydrogen oxidation reaction.

Respiratory [NiFe] hydrogenases from group 1 (Vignais et al., 2001; Vignais and Billoud, 2007) are involved in the cell in the uptake of hydrogen. It is remarkable that the two peculiar features of the electron transfer chain discussed above, which are the high potential of the medial cluster and the histidine ligation of the distal cluster, are crucial elements of the physiological role of the enzyme. Indeed, all modifications tested impaired H_2 oxidation and had little influence on the H_2 production reaction (Dementin et al., 2006; Rousset et al., 1998). It is therefore possible to engineer hydrogenase in order to modify the enzyme bias in such a way that it functions mainly in the direction of hydrogen production, which is the main goal of most biological hydrogen applications.

In order to investigate further how the coordination of the distal [4Fe4S] cluster affects the activity of the enzyme, different exogenous ligands were tested for their ability to restore activity of the enzyme mutated at H184 (Dementin et al., 2007). Hydrophobicity, alkalinity, and size all play a role in modulating the affinity of the ligand and the rescuing activity. Comparing the relative efficiencies of the ligands, it has been observed that among the five-atom-ring molecules, imidazole and 1-methylimidazole are more efficient than triazole. More surprising was the demonstration that the six-atom-ring pyridine can rescue hydrogenase activity; however, it does so to a lower extent. But even more surprising was the discovery that structural analogs of imidazole are not the only exogenous ligands that can bind to the empty coordination site and that ligation by pyridinic nitrogen is not a requirement for efficient catalysis. Both cyanide and thiocyanate rescued the H184G variant. The access of these molecules to the Fe atom being probably facilitated by their smaller size, their apparent affinity for the enzyme is 1 order of magnitude greater than imidazole.

Thus, the labile coordination position created in the H184G variant made possible the chemical rescue of the enzyme activity using different molecules, exhibiting different rescuing profiles (Dementin et al., 2007). It is therefore possible to control the activity of the enzyme by changing the nature and the concentration of an exogenous ligand. This type of approach is of potential interest to finely modulate the enzyme activity in biotechnological processes where hydrogenases can be used (biohydrogen production, biofuel cells, or metal reduction).

Cytoplasmic heteromultimeric bidirectional [NiFe] hydrogenases from group 3 (Vignais et al., 2001; Vignais and Billoud, 2007) include subunits able to bind a soluble cofactor, e.g., F420, NAD, or NADP, and function reversibly in vivo. The NAD(P)-reducing hydrogenases possess several subunits involved in electron transfer: one coordinates a [2Fe-2S] center, one coordinates a [4Fe-4S] center, and one coordinates a flavin mononucleotide molecule (Vignais and Colbeau, 2004). This family of enzymes is especially interesting as one member is oxygen tolerant (Burgdorf et al., 2005), the soluble hydrogenase (SH) from *Ralstonia eutropha* (Fig. 2) (see below). One other member, the bidirectional hydrogenase from the green alga *Synechocystis* (Fig. 2) (Appel et al., 2000), is involved in H_2 production from water splitting (Cournac et al., 2004) (see below). The study of the H16L mutant of the HoxH subunit of *R. eutropha* SH (Loscher et al., 2006) revealed that the enzyme was impaired in the hydrogen uptake activity while it was still able to produce hydrogen (at about 10% of the native enzyme rate). Spectroscopic studies indicated that only the [2Fe-2S] cluster was reduced and that the [4Fe4S] center remained under an oxidized state. These findings provided the first evidence for two different electron transfer pathways in the SH enzyme. The [2Fe-2S] cluster seems to be part of the electron transfer chain from NADH to the Ni-Fe site upon reductive activation of the enzyme. This pathway is functional in both WT and H16L. On the other hand, the [4Fe-4S] cluster seems to be involved predominantly in electron transfer from the Ni-Fe site to NAD^+ operating in the direction of H_2 cleavage. This pathway is active in WT but not in H16L (Loscher et al., 2006). Taking into account the similarity of this enzyme to the soluble bidirectional hydrogenase of cyanobacteria, such a finding could open a way for engineering the latter so that it might become more efficient for photosynthesis-driven H_2 production (Fig. 2).

SUBSTRATE SELECTIVITY, COMPETITION, AND LINKING

When expressed in vivo, hydrogenases interact with electron carriers which are generally at the junction of numerous redox reactions (respiration, CO_2 fix-

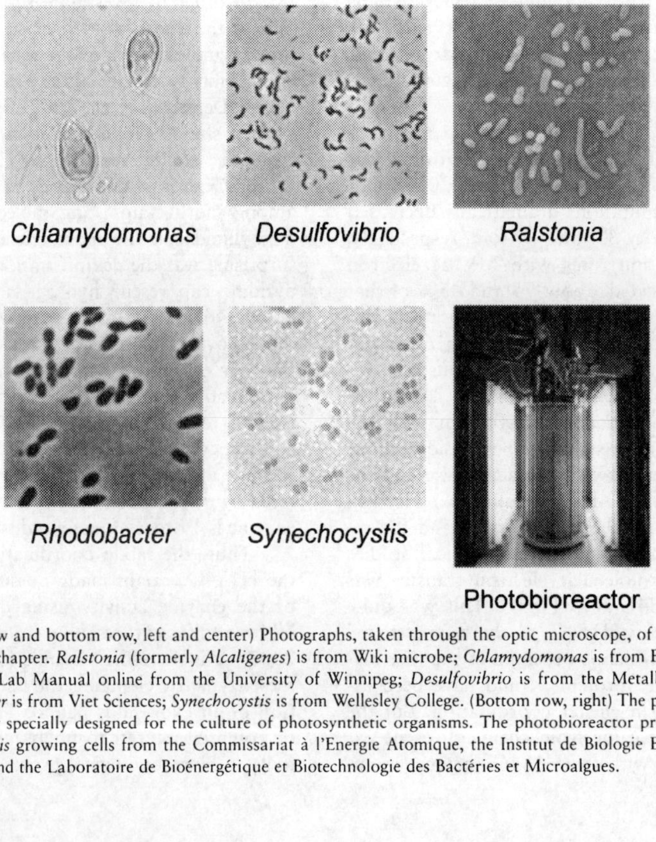

Figure 2. (Top row and bottom row, left and center) Photographs, taken through the optic microscope, of the bacterial cells presented in this chapter. *Ralstonia* (formerly *Alcaligenes*) is from Wiki microbe; *Chlamydomonas* is from Evolution Ecology and Biodiversity Lab Manual online from the University of Winnipeg; *Desulfovibrio* is from the Metalbioreduction web page; *Rhodobacter* is from Viet Sciences; *Synechocystis* is from Wellesley College. (Bottom row, right) The photobioreactor is a bacteria reactor specially designed for the culture of photosynthetic organisms. The photobioreactor presented here contains *Synechocystis* growing cells from the Commissariat à l'Energie Atomique, the Institut de Biologie Environmentale et Biotechnologie, and the Laboratoire de Bioénergétique et Biotechnologie des Bactéries et Microalgues.

ation, nitrate reduction, etc.). A key point is that these reactions that appear as competitors for biotechnological purposes are often essential for cell survival or development. In part, this explains the difficulty and the slow progress in biohydrogen research. Two research directions can be proposed to overcome this kind of limitation: improve the substrate specificity of hydrogenase or, more radically, redirect redox intermediates.

Ways to improve the substrate specificity of hydrogenase will probably arise from improvements in structural and functional knowledge. In the case of algal hydrogenases, for instance, the redox partner is ferredoxin (Fd), which is also implicated in photosynthetic carbon fixation via NADPH production by ferredoxin-$NADP^+$ reductase (FNR). No algal hydrogenase structure is available yet, but Horner et al. (2002) modeled algal hydrogenase structure and charge distribution and identified a set of amino acids likely to participate in electrostatic interaction with algal Fd. Mutagenesis at (putative) Fd binding sites in hydrogenase (and also at the Fd binding site in FNR) could be a way to modify relative affinities of these enzymes for their substrate and ultimately tune these affinities for an optimal ratio between photosynthetic and hydrogen-producing abilities. In group 3 bidirectional hydrogenases, especially those interacting with donors such as NAD(P)H ($E°' = -320$ mV), which has a higher redox potential than the H_2/H^+ couple ($E°' = -420$ mV), one of the competing reactions can be the reoxidation of hydrogen by the hydrogenase itself, which is thermodynamically more favorable. In this case, the problem of competition and the problem of directionality of the enzyme are intrinsically connected and can be circumvented by the same approach (see, for instance, the discussion on directionality of soluble hydrogenase from *R. eutropha* above, which also potentially applies to cyanobacteria). In certain cases, one could be interested in engineering the enzyme so that it might change its substrate and use a more thermodynamically favorable one for H_2 production (for instance, switching from NAD(P)H to Fd in cyanobacteria). Strategies to achieve this goal might be either (i) heterologous hydrogenase expression/replacement or (ii) engineering synthetic or chimerical enzymes based on native hydrogenase, conserving the active-site binding subunit and redesigning the electron transfer subunit(s).

Another tempting approach to favor H_2 production would be to tightly connect a specific electron carrier or a photosystem with hydrogenase, making the electrons flow directly from the photosystem to the hydrogenase or from the hydrogenase to the electron acceptor, avoiding competition with the bulk of electron carriers. In a recent study, Ihara et al. (2006b) engineered a "hard-wired" protein complex consisting of a hydrogenase and a photosystem. They designed an artificial fusion protein composed of the membrane-bound [NiFe] hydrogenase from *R. eutropha* H16 and the peripheral photosystem I (PSI) subunit PsaE of the cyanobacterium *Thermosynechococcus elongatus*. The resulting hydrogenase-PsaE fusion protein when associated with PsaE-free PSI spontaneously formed a complex which showed light-driven hydrogen production at a rate of 0.58 μmol of H_2 mg^{-1} of chlorophyll h^{-1}. The complex retained accessibility to the native electron acceptor ferredoxin, which is necessary for autotrophic growth of these cells. But unfortunately, the activity was totally suppressed in the presence of the physiological PSI partners, Fd and FNR. In an attempt to establish an H_2 photoproduction system in which the activity is not interrupted by Fd and FNR, the same group introduced a chimeric protein of PsaE and cytochrome c_3 from *Desulfovibrio vulgaris* into the cyanobacterium *Synechocystis* sp. PCC6803 (Ihara et al., 2006a). The covalent adduct of cytochrome c_3 and PsaE assembled with PsaE-free PSI and formed a complex which was still able to reduce Fd for photosynthesis (approximately 20% of the original activity). Interestingly, this complex was able to drive hydrogen production when coupled with hydrogenase from *D. vulgaris* even in the presence of Fd and FNR, although the rate was limited (around 0.30 μmol H_2/mg^{-1} of chlorophyll h^{-1}). These results suggest, however, that this type of complex may eventually be modified to produce H_2 in vivo.

UNDERSTANDING THE MOLECULAR BASIS OF OXYGEN SENSITIVITY: A KEY FOR BIOHYDROGEN APPLICATIONS

Direct biological photoproduction of hydrogen at the expense of water oxidation will unavoidably lead to a certain exposure of hydrogenase to O_2. Improving hydrogenase oxygen resistance is then a major challenge for a broad range of biotechnological applications such as hydrogen photoproduction (Fig. 2), biofuel cells, and biosensors. Biological conversion of light energy into dihydrogen by microalgae or cyanobacteria is potentially very efficient in terms of energy conservation, as 10% of the incident light energy can theoretically be recovered into hydrogen (Prince and Kheshgi, 2005). For example, an average sunlight flux of 17 GWh/ha/year can be converted with a 10% yield in 550,000 m^3 of hydrogen per ha per year, which represents 145 tons of oil equivalent per ha per year. However, photosynthetic production of dihydrogen is only a transient phenomenon under natural conditions (Cournac et al., 2004), because of hydrogenase inhibition by the dioxygen produced during water photolysis

(Leger et al., 2004; Vincent et al., 2005). As a result, actual dihydrogen production efficiencies obtained in laboratory experiments are lower than 1% (Fouchard et al., 2005; Melis et al., 2000).

Due to their high catalytic turnover and their specificity towards hydrogen, hydrogenases are also envisioned as potential catalysts to replace platinum in fuel cells, through the design of so-called biofuel cells. The development of such devices could benefit from the availability of oxygen-tolerant catalysts in order to yield more robust active surfaces and eventually to get rid of the need for a membrane separation between hydrogen-oxidizing and dioxygen-reducing electrodes (Vincent et al., 2005). Such a design might considerably alleviate the price of fuel cells in which both platinum and H^+ selective membranes represent the major costs. Oxygen-tolerant hydrogenases could also be employed in designing biosensors, to detect hydrogen presence in air, for safety applications (Bianco, 2002; Qian et al., 2002).

Green algae and cyanobacteria have very similar photosystems but differ in the type of hydrogenase they possess. The former have [FeFe] hydrogenases, which are irreversibly damaged by oxygen, whereas the latter contain [NiFe] hydrogenases, which are much more oxygen resistant. There are some examples of relatively oxygen-tolerant enzymes, such as the group 1 membrane-bound hydrogenase and the group 3 soluble hydrogenase from the Knallgas bacterium *R. eutropha* (Burgdorf et al., 2005). However, the oxygen-tolerant enzymes tend to be 2 or 3 orders of magnitude less active than the oxygen-sensitive hydrogenases. Another group of hydrogenase-like, highly oxygen-tolerant but even less active proteins is that of the group 2 hydrogen sensors from *R. eutropha* (Bernhard et al., 2001) and the HupUV proteins from *Rhodobacter capsulatus* (Elsen et al., 1996) and *Bradyrhizobium japonicum* (Black et al., 1994). As explained above, these enzymes cannot be produced directly in foreign hosts (such as cyanobacteria, for instance) because of the specific maturation machinery that prevents any hydrogenase maturation in heterologous hosts (Casalot and Rousset, 2001). Nevertheless, these enzymes represent precious inspiration sources for the study of the molecular bases of dioxygen inhibition.

Two strategies seem to have been developed through evolution to allow [NiFe] hydrogenases to be catalytically active in the presence of oxygen. In the case of the SH from *R. eutropha*, the oxygen resistance is assumed to be due to the presence of an extra CN ligand at the active site (Happe et al., 2000) that might be incorporated by a specific maturation protein, HypX (Bleijlevens et al., 2004). Interestingly, other microbes that are able to metabolize hydrogen under aerobic conditions also harbor the *hypX* gene. These species belong to various phylogenetic groups. Three of them (*R. eutropha*, *Ralstonia metallidurans*, and *Pseudomonas fluorescens*) are members of the β-proteobacteria, and three other species belong to the group of α-proteobacteria (*B. japonicum*, *Rhizobium leguminosarum*, and *Magnetospirillum magnetotacticum*). Two more species are phylogenetically distinct: *Streptomyces avermitilis*, a gram-positive bacterium, and *Aquifex aeolicus*, a hyperthermophilic bacterium. Deletion of *hypX* led to a complete knockout of hydrogenase activity in *R. leguminosarum* (Rey et al., 1996) and, in *R. eutropha*, affected the activity of the membrane-bound hydrogenase (Buhrke and Friedrich, 1998) and the oxygen resistance of the SH enzyme (Bleijlevens et al., 2004), while the hydrogen sensor remained unaffected (Buhrke et al., 2001).

The other strategy consists of reducing the gas channel size at the interface with the active site cavity. At the end of the hydrophobic channel (Color Plate 10), near the active site, two hydrophobic residues, usually a valine and a leucine that are conserved in oxygen-sensitive hydrogenases (Volbeda et al., 2002), are replaced by larger residues, isoleucine and phenylalanine, respectively, in the oxygen-tolerant hydrogen sensors (Volbeda et al., 2002). Thus, as a first approximation, increasing the bulk of residues occupying these two positions may reduce the channel diameter at that point, thereby preventing efficient dioxygen access to the active site. This hypothesis was supported by two experiments in which the bulky amino acids from *R. capsulatus* and *R. eutropha* regulatory hydrogenases were replaced by valine and leucine. In both cases, determination of the inactivation kinetics in the presence of oxygen revealed that the resulting enzymes became more sensitive to oxygen. The mutated enzymes were inactivated by oxygen and required a prolonged reductive activation to reach the maximum activity, as did the oxygen-sensitive hydrogenases (Buhrke et al., 2005; Duche et al., 2005). Interestingly, the replacement of the leucine by a phenylalanine in the SH from *R. eutropha* increased the oxygen sensitivity (Burgdorf et al., 2002). This has been assumed to be due to the bulk of phenylalanine that might block the incorporation of the extra CN ligand in the active site.

Being inspired by group 2 hydrogen-sensors, replacement of Val74 and Leu122 with isoleucine and phenylalanine, respectively, was carried out to provide oxygen tolerance to the oxygen-sensitive [NiFe] hydrogenase from *D. fructosovorans* (L. Cournac, S. Champ, A. Volbeda, C. Léger, S. Dementin, L. Martin, B. Burlat, G. Peltier, B. Guigliarelli, J. C. Fontecilla-Camps, and M. Rousset, submitted for publication). Surprisingly, these substitutions did not improve oxygen tolerance and did not significantly modify the catalytic properties of the enzyme under anoxic conditions. Consequently, it has been concluded that the residue bulkiness at these positions was not the only parameter affecting oxygen tolerance. The orientation or chemical nature of the side chain is probably also a crucial element. The next enzyme modifications tested the effects of replacing Val74 and Leu122 with methionines (Cournac et al., submitted). Methionine not only should be bulky enough to inhibit dioxygen diffusion but also is known to participate in oxidative stress responses and protection in several proteins (Stadtman, 2004, 2006; Stadtman et al., 2002). Interestingly, the V74M-L122M enzyme was catalytically active in the presence of up to 20 μM of oxygen and exhibited a slower inactivation rate in the presence of oxygen (Cournac et al., submitted). The determination of the structure of this mutant enzyme showed that the methionines remained intact and did not react with oxygen. The V74M-L122M hydrogenase oxidized by oxygen remained in the same redox state as the native enzyme oxidized anaerobically (Fernandez et al., 2005), as demonstrated by the predominance of an Ni-B EPR signal (while Ni-A is predominant in the oxygen-exposed native enzyme) and by the abundance of a hydroxyl-bridging ligand at the active site in the structure. This study was the first to show that it is possible to improve dioxygen resistance of [NiFe] hydrogenases. Extending this approach to the [NiFe] hydrogenases from the cyanobacteria might open a way to the development of a clean and sustainable hydrogen economy.

The strong interest in photoproduction of hydrogen has prompted an exploration of the green algae, in spite of the great oxygen sensitivity of their [FeFe] hydrogenases. The strategy to increase oxygen tolerance by the reduction of the hydrogen-gas channel size was applied to the HydA1 hydrogenase from the green alga *Chlamydomonas reinhardtii* (King et al., 2004). Because of the lack of structural data, the structure of HydA1 was modeled using the homologous hydrogenase from *Clostridium pasteurianum* (Peters et al., 1998), and the gas channel was modeled using the hydrophobic properties of the gas channels identified in other hydrogenases (Montet et al., 1997). The amino acids identified as lining the gas channel using this in silico work were mutated and replaced by bulkier hydrophobic residues; typically, alanines were replaced by phenylalanines or tryptophans. The resulting recombinant hydrogenases were cloned and expressed in *C. reinhardtii*. Two clones were isolated, mt18 and mt28, one of which in particular, strain mt18, retained 82 to 34% of its hydrogenase activity after exposure to 1 to 4% of O_2 concentration for 2 min. These strains are protected by the U.S. patent WO 2004/093524 A2 (King et al., 2004).

CONCLUDING REMARKS

Hydrogenase engineering is a very difficult issue not only because of the complexity of these enzymes but also because of the vital processes in which they are involved. However, in spite of the difficulties, research is very active in several countries because of the potential spin-offs that might participate in the development of a new hydrogen energy economy. The enzyme bias, substrate specificity, and oxygen resistance are the main domains in which some progress has already been made, opening the way towards future applications. But other issues, like heterologous expression of [NiFe] hydrogenases that would facilitate molecular research and organism engineering or deciphering the catalytic mechanism that would allow the development of biomimetic catalysts, are also the subjects of intensive research and will contribute to biohydrogen implementation.

Acknowledgments. We thank Maria-Luz Cardenas for critically reading the manuscript and helpful discussions.

REFERENCES

Appel, J., S. Phunpruch, K. Steinmuller, and R. Schulz. 2000. The bidirectional hydrogenase of *Synechocystis* sp. PCC 6803 works as an electron valve during photosynthesis. *Arch. Microbiol.* **173:** 333–338.

Armstrong, F. A., and S. P. Albracht. 2005. [NiFe]-hydrogenases: spectroscopic and electrochemical definition of reactions and intermediates. *Philos. Trans. R. Soc. London A* **363:**937–954.

Bernhard, M., T. Buhrke, B. Bleijlevens, A. L. De Lacey, V. M. Fernandez, S. P. Albracht, and B. Friedrich. 2001. The H_2 sensor of *Ralstonia eutropha*. Biochemical characteristics, spectroscopic properties, and its interaction with a histidine protein kinase. *J. Biol. Chem.* **276:**15592–15597.

Bianco, P. 2002. Protein modified- and membrane electrodes: strategies for the development of biomolecular sensors. *J. Biotechnol.* **82:**393–409.

Black, L. K., C. Fu, and R. J. Maier. 1994. Sequences and characterization of *hupU* and *hupV* genes of *Bradyrhizobium japonicum* encoding a possible nickel-sensing complex involved in hydrogenase expression. *J. Bacteriol.* **176:**7102–7106.

Bleijlevens, B., T. Buhrke, E. van der Linden, B. Friedrich, and S. P. Albracht. 2004. The auxiliary protein HypX provides oxygen tolerance to the soluble [NiFe]-hydrogenase of *Ralstonia eutropha* H16 by way of a cyanide ligand to nickel. *J. Biol. Chem.* **279:** 46686–46691.

Buhrke, T., B. Bleijlevens, S. P. Albracht, and B. Friedrich. 2001. Involvement of *hyp* gene products in maturation of the H(2)-sensing [NiFe] hydrogenase of *Ralstonia eutropha*. *J. Bacteriol.* **183:**7087–7093.

Buhrke, T., and B. Friedrich. 1998. *hoxX* (*hypX*) is a functional member of the *Alcaligenes eutrophus hyp* gene cluster. *Arch. Microbiol.* **170:**460–463.

Buhrke, T., O. Lenz, N. Krauss, and B. Friedrich. 2005. Oxygen tolerance of the H_2-sensing [NiFe] hydrogenase from *Ralstonia eutropha* H16 is based on limited access of oxygen to the active site. *J. Biol. Chem.* **280:**23791–23796.

Burgdorf, T., A. L. De Lacey, and B. Friedrich. 2002. Functional analysis by site-directed mutagenesis of the NAD(+)-reducing hydrogenase from *Ralstonia eutropha*. *J. Bacteriol.* **184:**6280–6288.

Burgdorf, T., O. Lenz, T. Buhrke, E. van der Linden, A. K. Jones, S. P. Albracht, and B. Friedrich. 2005. [NiFe]-hydrogenases of *Ralstonia eutropha* H16: modular enzymes for oxygen-tolerant biological hydrogen oxidation. *J. Mol. Microbiol. Biotechnol.* **10:** 181–196.

Casalot, L., and M. Rousset. 2001. Maturation of the [NiFe] hydrogenases. *Trends Microbiol.* 9:228–237.

Cournac, L., G. Guedeney, G. Peltier, and P. M. Vignais. 2004. Sustained photoevolution of molecular hydrogen in a mutant of *Synechocystis* sp. strain PCC 6803 deficient in the type I NADPH-dehydrogenase complex. *J. Bacteriol.* **186:**1737–1746.

De Lacey, A. L., V. M. Fernandez, M. Rousset, and R. Cammack. Activation and inactivation of hydrogenase function and the catalytic cycle: spectroelectrochemical studies. *Chem. Rev.* **107:** 4304–4330.

Dementin, S., V. Belle, P. Bertrand, B. Guigliarelli, G. Adryanczyk-Perrier, A. L. De Lacey, V. Fernandez, M. Rousset, and C. Léger. 2006. Changing the ligation of the distal [4Fe4S] cluster in NiFe hydrogenase impairs inter- and intramolecular electron transfers. *J. Am. Chem. Soc.* 128:5209–5218.

Dementin, S., V. Belle, S. Champ, P. Bertrand, B. Guigliarelli, A. L. De Lacey, V. M. Fernandez, C. Léger, and M. Rousset. 2007. Molecular modulation of NiFe hydrogenase activity. *Int. J. Hydrogen Energy*, in press.

Dementin, S., B. Burlat, A. L. De Lacey, A. Pardo, G. Adryanczyk-Perrier, B. Guigliarelli, V. M. Fernandez, and M. Rousset. 2004. A glutamate is the essential proton transfer gate during the catalytic cycle of the [NiFe] hydrogenase. *J. Biol. Chem.* **279:**10508–10513.

Duche, O., S. Elsen, L. Cournac, and A. Colbeau. 2005. Enlarging the gas access channel to the active site renders the regulatory hydrogenase HupUV of *Rhodobacter capsulatus* O_2 sensitive without affecting its transductory activity. *FEBS J.* 272:3899–3908.

Elsen, S., A. Colbeau, J. Chabert, and P. M. Vignais. 1996. The *hupTUV* operon is involved in negative control of hydrogenase synthesis in *Rhodobacter capsulatus*. *J. Bacteriol.* 178:5174–5181.

Fernandez, V. M., A. L. De Lacey, and M. Rousset. 2005. Native and mutant hydrogenases: unravelling structure and function. *Coordin. Chem. Rev.* **249:**1596–1608.

Fouchard, S., A. Hemschemeier, A. Caruana, J. Pruvost, J. Legrand, T. Happe, G. Peltier, and L. Cournac. 2005. Autotrophic and mixotrophic hydrogen photoproduction in sulfur-deprived *Chlamydomonas* cells. *Appl. Environ. Microbiol.* 71:6199–6205.

Happe, R. P., W. Roseboom, G. Egert, C. G. Friedrich, C. Massanz, B. Friedrich, and S. P. Albracht. 2000. Unusual FTIR and EPR properties of the H_2-activating site of the cytoplasmic NAD-reducing hydrogenase from *Ralstonia eutropha*. *FEBS Lett.* **466:**259–263.

Henstra, A. M., J. Sipma, A. Rinzema, and A. J. Stams. 2007. Microbiology of synthesis gas fermentation for biofuel production. *Curr. Opin. Biotechnol.* 18:200–206.

Higuchi, Y., T. Yagi, and N. Yasuoka. 1997. Unusual ligand structure in Ni-Fe active center and an additional Mg site in hydrogenase revealed by high resolution X-ray structure analysis. *Structure* 5:1671–1680.

Horner, D. S., B. Heil, T. Happe, and T. M. Embley. 2002. Iron hydrogenases: ancient enzymes in modern eukaryotes. *Trends Biochem. Sci.* **27:**148–153.

Ihara, M., H. Nakamoto, T. Kamachi, I. Okura, and M. Maeda. 2006a. Photoinduced hydrogen production by direct electron transfer from photosystem I cross-linked with cytochrome c3 to [NiFe]-hydrogenase. *Photochem. Photobiol.* 82:1677–1685.

Ihara, M., H. Nishihara, K. S. Yoon, O. Lenz, B. Friedrich, H. Nakamoto, K. Kojima, D. Honma, T. Kamachi, and I. Okura. 2006b. Light-driven hydrogen production by a hybrid complex of a [NiFe]-hydrogenase and the cyanobacterial photosystem I. *Photochem. Photobiol.* 82:676–682.

King, P., M. L. Ghirardi, and M. Seibert. 2004. Oxygen resistant hydrogenases and methods for designing and making same. U.S. patent WO 2004/093524 A2.

Korbas, M., S. Vogt, W. Meyer-Klaucke, E. Bill, E. J. Lyon, R. K. Thauer, and S. Shima. 2006. The iron-sulfur cluster-free hydrogenase (Hmd) is a metalloenzyme with a novel iron binding motif. *J. Biol. Chem.* **281:**30804–30813.

Kruse, O., J. Rupprecht, J. H. Mussgnug, G. C. Dismukes, and B. Hankamer. 2005. Photosynthesis: a blueprint for solar energy capture and biohydrogen production technologies. *Photochem. Photobiol. Sci.* 4:957–970.

Leach, M. R., and D. B. Zamble. 2007. Metallocenter assembly of the hydrogenase enzymes. *Curr. Opin. Chem. Biol.* **11:**159–165.

Leger, C., S. Dementin, P. Bertrand, M. Rousset, and B. Guigliarelli. 2004. Inhibition and aerobic inactivation kinetics of *Desulfovibrio fructosovorans* NiFe hydrogenase studied by protein film voltammetry. *J. Am. Chem. Soc.* **126:**12162–12172.

Loscher, S., T. Burgdorf, I. Zebger, P. Hildebrandt, H. Dau, B. Friedrich, and M. Haumann. 2006. Bias from H_2 cleavage to production and coordination changes at the Ni-Fe active site in the NAD^+-reducing hydrogenase from *Ralstonia eutropha*. *Biochemistry* 45:11658–11665.

Lyon, E. J., S. Shima, R. Boecher, R. K. Thauer, F. W. Grevels, E. Bill, W. Roseboom, and S. P. Albracht. 2004. Carbon monoxide as an intrinsic ligand to iron in the active site of the iron-sulfur-cluster-free hydrogenase H_2-forming methylenetetrahydromethanopterin dehydrogenase as revealed by infrared spectroscopy. *J. Am. Chem. Soc.* **126:**14239–14248.

Martin, W., and M. Muller. 1998. The hydrogen hypothesis for the first eukaryote. *Nature* 392:37–41.

Matias, P. M., C. M. Soares, L. M. Saraiva, R. Coelho, J. Morais, J. Le Gall, and M. A. Carrondo. 2001. [NiFe] hydrogenase from *Desulfovibrio desulfuricans* ATCC 27774: gene sequencing, three-dimensional structure determination and refinement at 1.8 Å and modelling studies of its interaction with the tetrahaem cytochrome c3. *J. Biol. Inorg. Chem.* **6:**63–81.

Melis, A., L. Zhang, M. Forestier, M. L. Ghirardi, and M. Seibert. 2000. Sustained photobiological hydrogen gas production upon reversible inactivation of oxygen evolution in the green alga *Chlamydomonas reinhardtii*. *Plant Physiol.* 122:127–136.

Montet, Y., P. Amara, A. Volbeda, X. Vernede, E. C. Hatchikian, M. J. Field, M. Frey, and J. C. Fontecilla-Camps. 1997. Gas access to the active site of Ni-Fe hydrogenases probed by X-ray crystallography and molecular dynamics. *Nat. Struct. Biol.* 4:523–526.

Pardo, A., A. L. De Lacey, V. M. Fernandez, H. J. Fan, Y. Fan, and M. B. Hall. 2006. Density functional study of the catalytic cycle of nickel-iron [NiFe] hydrogenases and the involvement of high-spin nickel(II). *J. Biol. Inorg. Chem.* **11:**286–306.

Peters, J. W., W. N. Lanzilotta, B. J. Lemon, and L. C. Seefeldt. 1998. X-ray crystal structure of the Fe-only hydrogenase (CpI) from *Clostridium pasteurianum* to 1.8 angstrom resolution. *Science* 282:1853–1858. (Erratum, 283:35, 1999; erratum, 283:2102, 1999.)

Pierik, A. J., W. Roseboom, R. P. Happe, K. A. Bagley, and S. P. Albracht. 1999. Carbon monoxide and cyanide as intrinsic ligands to iron in the active site of [NiFe]-hydrogenases. $NiFe(CN)_2CO$, Biology's way to activate H_2. *J. Biol. Chem.* **274:**3331–3337.

Pilak, O., B. Mamat, S. Vogt, C. H. Hagemeier, R. K. Thauer, S. Shima, C. Vonrhein, E. Warkentin, and U. Ermler. 2006. The crystal structure of the apoenzyme of the iron-sulphur cluster-free hydrogenase. *J. Mol. Biol.* 358:798–809.

Prince, R. C., and H. S. Kheshgi. 2005. The photobiological production of hydrogen: potential efficiency and effectiveness as a renewable fuel. *Crit. Rev. Microbiol.* 31:19–31.

Qian, D. J., C. Nakamura, S. O. Wenk, H. Ishikawa, N. Zorin, and J. Miyake. 2002. A hydrogen biosensor made of clay, poly(butylviologen), and hydrogenase sandwiched on a glass carbon electrode. *Biosens. Bioelectron.* 17:789–796.

Rey, L., D. Fernandez, B. Brito, Y. Hernando, J. M. Palacios, J. Imperial, and T. Ruiz-Argueso. 1996. The hydrogenase gene cluster of *Rhizobium leguminosarum* bv. viciae contains an additional gene (*hypX*), which encodes a protein with sequence similarity to the N10-formyltetrahydrofolate-dependent enzyme family and is required for nickel-dependent hydrogenase processing and activity. *Mol. Gen. Genet.* 252:237–248.

Rotte, C., K. Henze, M. Muller, and W. Martin. 2000. Origins of hydrogenosomes and mitochondria. *Curr. Opin. Microbiol.* 3:481–486.

Rousset, M., Y. Montet, B. Guigliarelli, N. Forget, M. Asso, P. Bertrand, J. C. Fontecilla-Camps, and E. C. Hatchikian. 1998. [3Fe-4S] to [4Fe-4S] cluster conversion in *Desulfovibrio fructosovorans* [NiFe] hydrogenase by site-directed mutagenesis. *Proc. Natl. Acad. Sci. USA* 95:11625–11630.

Rupprecht, J., B. Hankamer, J. H. Mussgnug, G. Ananyev, C. Dismukes, and O. Kruse. 2006. Perspectives and advances of biological H_2 production in microorganisms. *Appl. Microbiol. Biotechnol.* 72:442–449.

Shima, S., E. J. Lyon, M. Sordel-Klippert, M. Kauss, J. Kahnt, R. K. Thauer, K. Steinbach, X. Xie, L. Verdier, and C. Griesinger. 2004. The cofactor of the iron-sulfur cluster free hydrogenase Hmd: structure of the light-inactivation product. *Angew. Chem. Int. Ed. Engl.* 43:2547–2551.

Shima, S., and R. K. Thauer. 2007. A third type of hydrogenase catalyzing H_2 activation. *Chem. Rec.* 7:37–46.

Stadtman, E. R. 2004. Cyclic oxidation and reduction of methionine residues of proteins in antioxidant defense and cellular regulation. *Arch. Biochem. Biophys.* 423:2–5.

Stadtman, E. R. 2006. Protein oxidation and aging. *Free Radic. Res.* 40:1250–1258.

Stadtman, E. R., J. Moskovitz, B. S. Berlett, and R. L. Levine. 2002. Cyclic oxidation and reduction of protein methionine residues is an important antioxidant mechanism. *Mol. Cell. Biochem.* 234-235:3–9.

Teixeira, M., I. Moura, A. V. Xavier, J. J. Moura, J. LeGall, D. V. DerVartanian, H. D. Peck, Jr., and B. H. Huynh. 1989. Redox intermediates of *Desulfovibrio gigas* [NiFe] hydrogenase generated under hydrogen. Mossbauer and EPR characterization of the metal centers. *J. Biol. Chem.* 264:16435–16450.

Teixeira, V. H., A. M. Baptista, and C. M. Soares. 2006. Pathways of H_2 toward the active site of [NiFe]-hydrogenase. *Biophys. J.* 91:2035–2045.

Vignais, P. M., and B. Billoud. 2007. Occurrence, classification, and biological function of hydrogenases: an overview. *Chem. Rev.* 107:4206–4272.

Vignais, P. M., B. Billoud, and J. Meyer. 2001. Classification and phylogeny of hydrogenases. *FEMS Microbiol. Rev.* 25:455–501.

Vignais, P. M., and A. Colbeau. 2004. Molecular biology of microbial hydrogenases. *Curr. Issues Mol. Biol.* 6:159–188.

Vincent, K. A., A. Parkin, O. Lenz, S. P. Albracht, J. C. Fontecilla-Camps, R. Cammack, B. Friedrich, and F. A. Armstrong. 2005. Electrochemical definitions of O_2 sensitivity and oxidative inactivation in hydrogenases. *J. Am. Chem. Soc.* 127:18179–18189.

Volbeda, A., M. H. Charon, C. Piras, E. C. Hatchikian, M. Frey, and J. C. Fontecilla-Camps. 1995. Crystal structure of the nickel-iron hydrogenase from *Desulfovibrio gigas*. *Nature* 373:580–587.

Volbeda, A., E. Garcin, C. Piras, A. L. de Lacey, V. M. Fernandez, C. E. Hatchikian, M. Frey, and J. C. Fontecilla-Camps. 1996. Structure of the [NiFe] hydrogenase active site: evidence for biologically uncommon Fe ligands. *J. Am. Chem. Soc.* 118:12989–12996.

Volbeda, A., Y. Montet, X. Vernede, C. E. Hatchikian, and J. C. Fontecilla-Camps. 2002. High-resolution crystallographic analysis of *Desulfovibrio fructosovorans* [NiFe] hydrogenase. *Int. J. Hydrogen Energy* 27:1449–1461.

Bioenergy
Edited by J. Wall et al.

Chapter 21

Nitrogenase-Catalyzed Hydrogen Production by Purple Nonsulfur Photosynthetic Bacteria

CAROLINE S. HARWOOD

Hydrogen gas has received serious consideration as an alternative to petroleum because it is a clean-burning energy carrier that can be converted to electricity in hydrogen fuel cells and used to power automobiles or any other electricity-driven process. Part of the appeal of hydrogen is that it can be produced from a wide range of resources including natural gas, coal, fossil fuels, biomass, and solar energy (Turner, 2004). This chapter focuses on a biological process for hydrogen generation that depends on the environmentally benign use of biomass and solar energy.

Biological hydrogen production is catalyzed by hydrogenase or nitrogenase enzymes (Das, 2001; Prince and Kheshgi, 2005; Rupprecht et al., 2006). The principal biological function of nitrogenases is to convert nitrogen gas to ammonia, but these enzymes also generate hydrogen gas as an obligatory aspect of their catalytic cycle (Henderson, 2002). Both types of enzymes function to combine low-potential electrons with protons to produce hydrogen. Nitrogenases have an additional requirement for ATP. Isolated enzymes cannot be practically used to generate hydrogen because they require expensive cofactors in the form of ATP and reduced NADH or reduced ferredoxin. However, hydrogenases and nitrogenases will produce significant amounts of hydrogen gas in the context of living microbial cells since cells generate low-potential electrons and ATP as part of their normal metabolism. Both types of enzyme are extremely sensitive to oxygen, and the cellular milieu provides some protection against oxygen exposure. The major categories of microbes that have been investigated for hydrogen production are anaerobic bacteria, which oxidize organic compounds and produce hydrogen as a fermentation end product; cyanobacteria and green algae, which generate low-potential electrons and produce hydrogen from water through the agency of photophosphorylation; and anoxygenic photosynthetic bacteria, which generate electrons for hydrogen production from organic compounds and use energy from light to drive hydrogen generation.

The group of anoxygenic photosynthetic bacteria known as purple nonsulfur bacteria (PNSB) produce large amounts of hydrogen under normal growth conditions by using nitrogenases as opposed to hydrogenases. PNSB have advantages over fermentative bacteria in that they can potentially divert 100% of the electrons from an electron-donating substrate to hydrogen production. This is because they can utilize solar energy to drive energetically unfavorable reactions and thus are free from the thermodynamic constraints that govern the types of metabolism that can be carried out by fermentative bacteria. Fermentative bacteria are restricted in the amount of hydrogen that they can generate to about 15% of the electrons in a high-carbohydrate waste stream (Angenent et al., 2004).

PNSB have advantages over cyanobacteria and algae in that they catalyze hydrogen production under anaerobic conditions and thus do not need to protect their hydrogen-generating enzymes from inactivation by oxygen. Cyanobacteria and algae catalyze hydrogen production as a consequence of splitting water, a light-driven process that generates oxygen as well as the low-potential electrons that will be used for hydrogen production. Direct photolysis, involving the direct transfer of energized electrons from the photosynthetic apparatus to a hydrogenase, is in theory a very efficient process. However, hydrogen production by this route tends to be short-lived because of the extreme oxygen sensitivity of hydrogenases and nitrogenases. In practice, oxygenic photosynthesis needs to be temporally or spatially separated from hydrogen generation. A common

Caroline S. Harwood • Department of Microbiology, University of Washington, Seattle, WA 98195-7242.

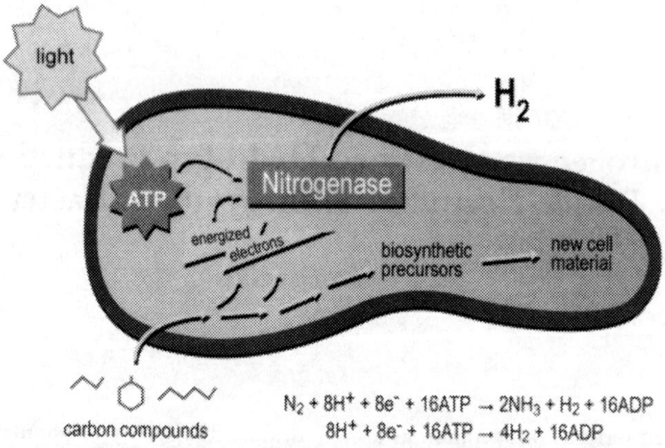

Figure 1. Nitrogenase-catalyzed hydrogen production by PNSB. PNSB can generate the electrons needed for hydrogen production by oxidizing organic compounds and selected inorganic compounds. They generate ATP by cyclic photophosphorylation. Protons derive from water or are generated along with electrons when organic compounds are oxidized. The theoretical stoichiometries for nitrogenase reactions for hydrogen production in the presence and absence of nitrogen gas are indicated. Reprinted from Rey et al. (2007) with permission of the publisher.

way to do this is referred to as indirect photolysis. First, oxygenic photosynthesis generates energized electrons, which are used to synthesize carbohydrates by carbon dioxide fixation. Later, cells are incubated in the dark or under oxygen-depleted conditions and carbohydrates are oxidized to provide electrons for hydrogen-generating enzymes (Prince and Kheshgi, 2005).

Over 20 genera of PNSB have been described (Madigan and Imhoff, 2007), and all of those tested can grow under nitrogen-fixing conditions and evolve hydrogen as a by-product of this growth mode (Madigan, 1995). PNSB use anoxygenic photosynthesis to generate the ATP required for nitrogenase-catalyzed hydrogen production. PNSB typically acquire the low-potential electrons that are needed for hydrogen production by oxidizing organic compounds (Fig. 1), but they are also capable of oxidizing selected inorganic compounds for this purpose. Most species can oxidize compounds that are present in fermentation waste streams (Barbosa et al., 2001; Brenner et al., 2001; Fißler et al., 1994; Franchi et al., 2004; Rey et al., 2007; Shi and Yu, 2006). Thus, PBSB have a potential application as dual-use organisms for wastewater treatment with accompanying production of hydrogen gas.

This chapter reviews the fundamental biology of nitrogenase-catalyzed hydrogen production by PNSB. I also describe several basic strategies to improve and stabilize hydrogen production by this group of bacteria as well as strategies to construct efficient, low-cost bioreactors for hydrogen gas production.

THE DISCOVERY OF PHOTOHYDROGEN PRODUCTION

As a graduate student, Howard Gest observed that the PNSB *Rhodospirillum rubrum* produced hydrogen when grown in a mineral salts medium containing glutamate and fumarate and incubated in light. Gest also found that photohydrogen production was inhibited by nitrogen gas. He and his advisor, Martin Kamen, interpreted this to mean that nitrogen was competing with the hydrogen for reductant and from this made an intuitive leap that led to the discovery of nitrogen fixation in photosynthetic bacteria (Gest, 1999; Gest and Kaman, 1949; Kaman and Gest, 1949). The proof of this was the demonstration that $^{15}N_2$ was incorporated into cellular nitrogenous compounds (Kamen and Gest, 1949; Lindstrom et al., 1949). For many years thereafter, it was assumed that bacterial hydrogenase enzymes, separate from nitrogenase, were responsible for hydrogen production. Gradually it became clear that hydrogen evolution and nitrogenase activity were affected by a common set of physiological conditions. Both processes were inhibited by ammonia and oxygen, and both required molybdenum (reviewed in Yoch, 1978). This led Ormerod and Gest to hypothesize that "both H_2 production and N_2 activation in *Rhodospirillum rubrum* are catalyzed by the same enzyme or enzyme complex" (Ormerod and Gest, 1962). A few years later it was shown that nitrogenase and ATP-dependent hydrogen production activities were copurified from *Azotobacter vinelandii* cell extracts (Burns and Bulen,

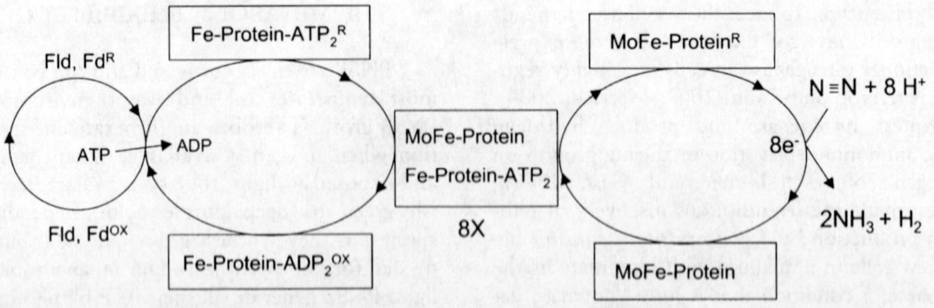

Figure 2. The nitrogenase reaction. Ferredoxins (Fd) and flavodoxins (Fld) transfer electrons generated from the oxidation of organic compounds during metabolism to the Fe protein to initiate a reaction cycle. Reprinted from Howard and Rees (1994) with permission from the *Annual Review of Biochemistry*.

1965). In the next year these studies were extended to *Rhodospirillum rubrum* (Burns and Bulen, 1966).

CHARACTERISTICS OF MOLYBDENUM NITROGENASE

Molybdenum nitrogenase catalyzes the ATP-dependent reduction of nitrogen gas to ammonia and hydrogen according to the following equation.

$$N_2 + 16ATP + 8e^- + 8H^+ \rightarrow$$
$$2NH_3 + H_2 + 16\ ADP + 16\ P_i$$

Nitrogenase is a two-component enzyme consisting of an Fe protein (dinitrogenase reductase encoded by *nifH*) of about 63 kDa and a MoFe protein (dinitrogenase, encoded by *nifDK*) of about 230 kDa. The Fe protein is a homodimer that transfers one electron at a time to the dinitrogenase in a reaction that is coupled to ATP hydrolysis. The MoFe protein is an $\alpha_2\beta_2$ heterotetramer that contains an iron-molybdenum (FeMo) cofactor, which is the substrate reduction site, and a second Fe-S metallocenter, the P cluster, which participates in electron transfer to the active site of N_2 reduction. The Fe protein undergoes a reaction cycle in which it docks with a MoFe protein, catalyzes ATP hydrolysis, transfers an electron to the dinitrogenase, and then dissociates. After each dissociation event, an electron transfer protein such as a ferredoxin or flavodoxin reduces the Fe protein to initiate another cycle (Fig. 2). Many accessory proteins encoded by various *nif* genes participate in the synthesis of the P cluster and the transition metal (MoFe) cofactor and in the assembly of the nitrogenase (Rubio and Ludden, 2005). Depending on the organism, between 15 and 20 proteins are required to assemble a functional nitrogenase enzyme. The molybdenum nitrogenase gene cluster from *Rhodopseudomonas palustris* is shown as an example in Fig. 3. Because nitrogenase assembly is such an elaborate process and because cells need to synthesize large

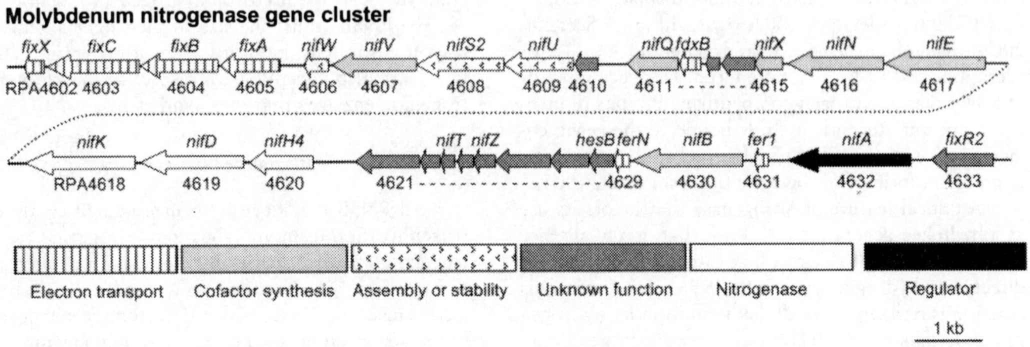

Figure 3. Organization of the molybdenum nitrogenase gene cluster in *Rhodopseudomonas palustris* strain CGA009. Gene functions are annotated according to Rubio and Ludden (2005).

amounts of this enzyme to meet the very large demands that growing cells have for fixed nitrogen, the expression of functional nitrogenase in cells is a highly regulated process (Dixon and Kahn, 2004; Merrick, 2004). Bacteria express nitrogenase and produce hydrogen only during ammonia deprivation or during growth on poor nitrogen sources (Hillmer and Gest, 1977a, 1977b). Gest made the serendipitous discovery of photohydrogen production by *Rhodospirillum rubrum* because he grew cells in a medium with glutamate as the nitrogen source, a condition that is now known to derepress nitrogenase synthesis in this bacterial species.

WHY HYDROGEN IS AN OBLIGATORY PRODUCT OF NITROGENASE

Simpson and Burris showed in 1984 that even with an external nitrogen pressure of 50 atm, approximately 25% of the electrons that flux through nitrogenase are directed to hydrogen production (Simpson and Burris, 1984). This solidified the concept that proton reduction to hydrogen is an intrinsic part of the nitrogenase catalytic mechanism. The reduction of a single N_2 molecule to ammonia is accompanied by the generation of one molecule of H_2. This requires that eight cycles of Fe protein docking and electron transfer to the MoFe protein take place. The catalytic mechanism of nitrogenase has been extraordinarily difficult to study because the MoFe protein has at least seven reduced states and it has not, in general, been possible to obtain for study homogenous enzyme preparations with just a single reduced intermediate or a single substrate-bound state. The P cluster [8Fe:7S], the FeMo cofactor [1Mo:7Fe:9S; homocitrate], bound hydrides, and partially reduced bound intermediates are all possible sites where electrons may bind or accumulate. A favored hypothesis for why hydrogen production is part of the catalytic cycle is that in order for dinitrogen to bind to the FeMo cofactor it must displace a bound dihydride (Henderson, 2002; Igarashi and Seefeldt, 2003). This displacement would generate one H_2 for every molecule of N_2 that is reduced. Evidence in favor of this is that dihydrogen competitively inhibits dinitrogen reduction. Recently hydride bound to the FeMo cofactor active site was directly observed spectroscopically in a mutant form of nitrogenase (Igarashi et al., 2005).

A critical feature of nitrogenase that favors its use as a hydrogen-generating catalyst is that, in the absence of nitrogen gas, all electron flux through the protein is directed to hydrogen gas (Eady, 1996). Moreover, this reaction is resistant to feedback inhibition by hydrogen gas and, unlike the reduction of other substrates by nitrogenase, is insensitive to inhibition by carbon monoxide (Fisher and Newton, 2002).

THE METABOLIC FLEXIBILITY OF PNSB

PNSB are easily cultivated and can be isolated from most temperate soil and water environments. Most PNSB grow as aerobes and generate energy by respiration when oxygen is available. When deprived of air and exposed to light, they turn a characteristic brownish green to deep purple color, depending on the species, as they synthesize the light-absorbing pigments needed for photosynthesis. During anaerobic growth in light, PNSB generate all their ATP by photophosphorylation. They differ from green plants and cyanobacteria in that they do not split water during photophosphorylation and thus do not produce oxygen during this process. Rather, the form of photophosphorylation carried out by the PNSB involves a cyclic flow of electrons (Fig. 4). PNSB can carry out carbon dioxide fixation to obtain carbon by photoautotrophic growth. However, most species prefer to use organic compounds for cell carbon by photoheterotrophic growth.

The PNSB as a group have great versatility in the metabolic modules that they can use for hydrogen production (Fig. 1). Collectively they have the potential to generate reductant for hydrogen from virtually every category of oxidizable compound, including a large number of structurally diverse organic compounds such as aromatic acids, fatty acids, and tricarboxylic acid cycle intermediates (Brenner et al., 2001; Harwood and Gibson, 1988). They can also oxidize inorganic compounds, including Fe(II) and reduced sulfur compounds (Croal et al., 2007; Jiao and Newman, 2007; Larimer et al., 2004). PNSB encode sophisticated light-gathering antenna systems that allow cells to take advantage of a relatively broad spectrum of light in the longer-wavelength regions (Evans et al., 2005; Hartigan et al., 2002; Roszak et al., 2003; Scheuring et al., 2006). The antennae absorb light energy and transfer it to the bacteriochlorophyll-containing reaction center, where it is converted into a proton gradient that is used to generate the ATP required for hydrogen production (Fig. 4). The PNSB adjust the size of their light antennae in response to light intensity (Scheuring et al., 2006). PNSB are also versatile with respect to the types of nitrogenase enzymes that they synthesize.

Alternative Nitrogenases

All PNSB that have been investigated catalyze nitrogen fixation using molybdenum nitrogenase (Mo nitrogenase). Several species have, in addition, alternative nitrogenases that have an iron-vanadium cofactor (V nitrogenase) or an iron-iron cofactor (Fe nitrogenase) at the active site instead of an iron-molybdenum cofactor. Two species for which nitrogen fixation has been relatively well studied, *Rhodospirillum rubrum* and

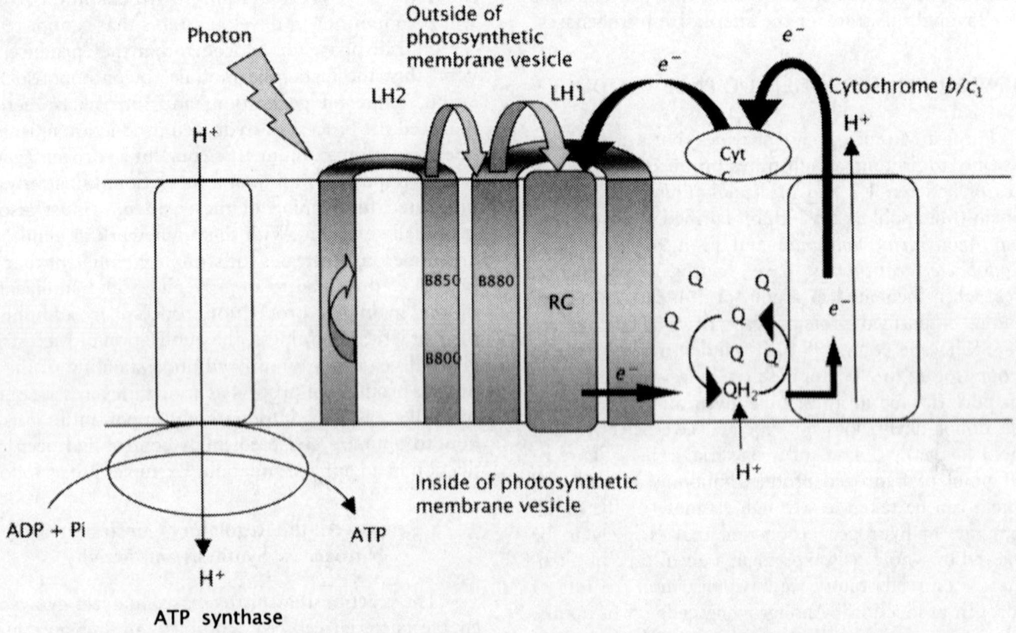

Figure 4. The photosynthetic apparatus of the PNSB, depicting ATP generation by cyclic photophosphorylation. Light energy absorbed by the peripheral light antenna (LH2) is transferred to the core light antenna (LH1) and then the reaction center (RC). These components can also absorb photons directly at the wavelengths indicated. The gray arrows indicate energy transfer reactions, and the black arrows indicate electron or proton transfer reactions. Ubiquinone molecules (Q), which are mobile in the membrane, accept two protons from the inside of membrane vesicles along with energized electrons from the reaction center. Cytochrome *b*/*c*₁ catalyzes electron transfer between ubiquinol (QH_2) and the mobile electron carrier, cytochrome *c* (cyt *c*). Electron transfer to cytochrome *c* is coupled to the translocation of protons across the membrane to create a proton gradient. The cyclic flow of electrons back to the reaction center is completed by cytochrome *c*. ATP synthase uses the proton gradient to generate ATP. Adapted from Cogdell et al. (2006).

Rhodobacter capsulatus, have Fe and Mo nitrogenases (Lehman and Roberts, 1991; Masepohl et al., 2002b). A third PNSB, *Rhodopseudomonas palustris* strain CGA009, encodes all three nitrogenase isozymes (Oda et al., 2005). Metal availability dictates the type of nitrogenase isozyme expressed by those species that encode more than one type. Generally the Mo nitrogenase is expressed in preference to the two alternative nitrogenases when Mo is present (Kutsche et al., 1996; Oda et al., 2005).

V and Fe nitrogenases are distinct enzymes that are comprised of VnfHDK and AnfHDK subunits that are homologous to the NifHDK subunits of the Mo nitrogenase. These enzymes also include VnfG and AnfG subunits as additional structural components of the dinitrogenase (Eady, 1996). Some of the cofactor synthesis and assembly proteins from the *nif* gene cluster participate in alternative nitrogenase cofactor synthesis and assembly. However, the *vnf* and *anf* gene clusters also encode proteins specific to the synthesis of the cognate nitrogenase (Masepohl et al., 2002b; Oda et al., 2005). The alternative Fe and V nitrogenases are of interest from the point of view of hydrogen production because they produce relatively more hydrogen and less ammonia than the traditional Mo nitrogenase that is synthesized by all nitrogen-fixing bacteria. The generally agreed on stoichiometries for hydrogen production by each isozyme are as follows (Masepohl et al., 2002b).

Mo nitrogenase: $N_2 + 8H^+ + 8e^- \rightarrow 2NH_3 + 1H_2$
V nitrogenase: $N_2 + 12H^+ + 12e^- \rightarrow 2NH_3 + 3H_2$
Fe nitrogenase: $N_2 + 24H^+ + 24e^- \rightarrow 2NH_3 + 9H_2$

If the hypothesis that two hydride ions must be present at the active site and displaced in order for dinitrogen to bind is correct, then it is unclear why V and Fe nitrogenases release more than one molecule of hydrogen gas for each molecule of ammonia formed. It has been suggested that V and Fe atoms affect the reactivity of their respective cofactors such that the active sites of V and Fe nitrogenases are less efficient for nitrogen fixation

(Bell et al., 2003). The effect of this is that protons are a more favored substrate for the alternative nitrogenases.

HOW MUCH HYDROGEN DO PNSB PRODUCE?

Purified Mo nitrogenase enzymes that are given access only to protons as substrates produce hydrogen at rates of between 1.2 and 2.2 μmol/min/mg of purified protein (Masepohl et al., 2002b). Purified V nitrogenase from *Azotobacter vinelandii* and Fe nitrogenase from *Rhodobacter capsulatus* have hydrogen production rates, when incubated in argon, of 1.4 and 2.4 μmol/min/mg of purified protein, respectively (Hales et al., 1986; Schneider et al., 1994). Given that nitrogenase can account for up to 2.0% of the protein in whole cells and that pure enzyme in a test tube given an artificial electron donor likely does not operate as efficiently as it would in the cytoplasm of a bacterial cell, a figure of 200 nmol of hydrogen produced/min/mg of total cell protein can be taken as a rough estimate of the maximum rate of hydrogen production that is likely to be achieved by whole cells expressing a given type of nitrogenase. In my laboratory we have measured the rate of hydrogen production by nongrowing cells of an uptake hydrogenase mutant strain (see below) of *Rhodopseudomonas palustris* expressing Mo nitrogenase and given access only to protons to be about 60 nmol/min/mg of total cell protein. This is equivalent to 0.3 mmol of H_2/liter of culture/h or 7.5 ml of H_2/liter of culture/h, units in which hydrogen production is often expressed. Specific hydrogen production rates of 10 to 80 ml of H_2/liter of culture/h have been reported for growing cultures of PNSB (Barbosa et al., 2001; Tsygankov et al., 1998).

Extrapolating from our small-scale measurements, we estimate that *Rhodopseudomonas palustris* has the potential to produce 60 g of H_2/10 kg of cells/h. This is the energy equivalent of about 1.5 gal of gasoline a day. Depending on the price of petroleum, this biological process for hydrogen production should be cost-effective without further improvement if bioreactor construction and operation costs are sufficiently low. Difficulties are to be expected in scaling up the process and in maintaining maximum rates of hydrogen production. Fortunately there are many targets for process optimization and strain improvement that have yet to be explored for hydrogen production by PNSB. Some of these are discussed in the remainder of this chapter.

STRATEGIES TO IMPROVE AND MAINTAIN HYDROGEN PRODUCTION

The development of an efficient process for hydrogen production by PNSB will require that the biocatalyst for hydrogen production, the nitrogenase, be synthesized and active under all growth conditions, and it will be important to develop strains that cannot recapture and dissipate the hydrogen that they produce. Beyond this, the metabolic modules of photophosphorylation, reductant generation, and nitrogenase activity that feed the process of hydrogen production must each operate with maximum efficiency for hydrogen generation as opposed to ammonia formation. Obviously, the integrated functioning of these modules must also be maximally efficient. With this framework in mind there are numerous strategies for strain development that can be expected to lead to improvements and stabilization of the hydrogen production process. In addition to their practical usefulness, the application of such strategies will lead to an improved understanding of the hydrogen production process as it operates in the context of whole cells. Such information is important to learning how to optimize and predictably control hydrogen production in an applied situation. Examples are as follows.

Remove All the Regulatory Constraints on Nitrogenase Synthesis and Activity

The reaction that nitrogenase catalyzes evolved to enable bacterial cells to synthesize ammonia from nitrogen gas and not for hydrogen production. Accordingly, PNSB produce hydrogen gas only when cells have a need for fixed nitrogen. With a requirement of 16 ATP per catalytic cycle, nitrogenase is an expensive enzyme to operate. It is also a rather slow enzyme with a turnover time of about 5 s^{-1} and thus needs to be synthesized in large amounts to meet the fixed-nitrogen demands of growing cells. Finally, the assembly of a mature nitrogenase is a complex process that requires the participation of ancillary proteins for the synthesis of the metal centers, the P cluster and the FeMo cofactor, that are unique to this enzyme. For all these reasons bacteria tightly repress the synthesis and activity of nitrogenases at the transcriptional level in response to fixed-nitrogen availability, most commonly ammonia (Dixon and Kahn, 2004; Merrick, 2004).

The two species of PNSB that have been best studied with respect to regulation of nitrogenase synthesis and activity are *Rhodobacter capsulatus* and *Rhodospirillum rubrum* (Masepohl et al., 2002a; Zhang et al., 2001, 2005, 2006a, 2006b). *Rhodobacter sphaeroides* and *Rhodopseudomonas palustris* have also been investigated, but to a lesser extent (Connelly et al., 2006; Joshi and Tabita, 1996; Oda et al., 2005). In each species overlapping but nonidentical sets of as many as 10 nitrogenase regulatory proteins and PII signal transduction proteins interact in a complex hierarchy to stimulate the expression and activity of the transcriptional activator, NifA, in response to fixed-nitrogen starvation. NifA directly controls the synthesis of *nif* genes in all bacteria studied (Merrick, 2004). A homologous AnfA protein controls the synthesis of *anf* genes that specify the iron-iron nitrogenase in *Rhodospirillum rubrum*, *Rhodobacter capsulatus*, and *Rhodopseudomonas palustris*, and VnfA controls the synthesis of vanadium nitrogenase in *Rhodopseudomonas palustris* (Merrick, 2004; Oda et al., 2005). The PNSB also have an additional layer of posttranslational control in which enzymatic activity is shut down upon addition of ammonia by a reversible ADP-ribosylation of the dinitrogenase reductase (Ludden and Roberts, 1995).

The development of strains in which the regulatory controls that govern nitrogenase synthesis and activity have been removed has clear advantages for a hydrogen production process, the most obvious being that such derepressed strains can in theory divert all of the electrons that they derive from oxidizing exogenous electron donors to hydrogen gas production. A second advantage is that hydrogen production by such strains will not be repressed by the presence of ammonia or any other form of fixed nitrogen that might be present in the electron-donating feedstock.

Strains of PNSB that are derepressed for nitrogenase synthesis and activity have been isolated and analyzed to various degrees. In a recent study, mutant strains of *Rhodopseudomonas palustris* were described that produce hydrogen constitutively, even under conditions (growth with ammonia) in which wild-type cells do not produce detectable amounts of hydrogen. The strains produced up to five times more hydrogen than did wild-type cells growing under nitrogen-fixing conditions. Each isolated mutant had a different single amino acid change in its NifA transcriptional regulator (Rey et al., 2007) that likely caused structural changes in the protein that allowed it to assume an active conformation proficient to stimulate *nif* gene transcription, even under normally repressing growth conditions. These mutants were obtained by applying a selection strategy which demanded that hydrogen be produced in order for cells to grow. This strategy involved inoculating *Rhodopseudomonas palustris* into an ammonium-containing medium that included a highly reduced (relatively electron-rich) carbon source. In order for cells to grow in this medium, mutations had to occur to allow the use of nitrogenase as an electron sink and hydrogen-producing enzyme rather than as a catalyst for ammonia production. A selection strategy similar in concept to this was applied to obtain *Rhodobacter sphaeroides* mutants that produce hydrogen constitutively (Joshi and Tabita, 1996). In this case the site of the mutation(s) that caused the hydrogen production phenotype was not reported. In a final example, a rational strain design was applied based on knowledge of the nitrogenase regulatory network to obtain strains of *Rhodobacter capsulatus* that produce hydrogen constitutively (Drepper et al., 2003). Strains derepressed for the expression of alternative nitrogenases have not been reported to date.

Disable the Uptake Hydrogenase

Most PNSB have uptake hydrogenases that function to consume hydrogen produced during nitrogen fixation. These are membrane-bound nickel-iron enzymes that catalyze the oxidation of hydrogen gas (Vignais and Toussaint, 1994; Vignais et al., 2001). Electrons derived from this process can be used for a variety of functions including carbon dioxide fixation and the reduction of organic compounds, such as malate, that are relatively more oxidized than cell material, to allow their complete assimilation into biomass (Rey et al., 2006). An obvious approach to improve nitrogenase-dependent hydrogen production is to disable the uptake hydrogenase by mutation or by withholding nickel from the growth medium. In fact, uptake hydrogenase mutants of *Rhodospirillum rubrum*, *Rhodobacter capsulatus*, and *Rhodopseudomonas palustris* have all been shown to produce significantly more hydrogen than the corresponding wild-type parent under nitrogen-fixing conditions (Kern et al., 1994; Ozturk et al., 2006; Rey et al., 2006; Vignais et al., 2006).

Manipulate the Photosynthetic Apparatus of PNSB

Photobiological processes for hydrogen production are often described in terms of theoretical quantum efficiencies of light conversion (Prince and Kheshgi, 2005). When evaluated in these terms, nitrogenase-catalyzed hydrogen production is less efficient than direct hydrogenase-catalyzed hydrogen production by oxygenic phototrophs, because of its additional requirement for ATP. However, nitrogenase-catalyzed photohydrogen production has about the same calculated efficiency as hydrogen production by indirect photolysis via hydrogenase by oxygenic phototrophs (Prince and Kheshgi, 2005). These efficiency calculations are based on the assumption that solar energy conversion is the rate-limiting step in photohydrogen production. This is certainly not the case for nitrogenase-catalyzed hydrogen production in the laboratory, where ample light is supplied. In this situation the generation and delivery of reductant to the nitrogenase constitute the rate-limiting step. Thus, the relative quantum inefficiency of nitrogenase-catalyzed photohydrogen production need not necessarily be viewed as an intrinsic limitation to the process.

PNSB absorb light in the near-infrared region of the spectrum of solar irradiation that reaches the surface of the Earth at wavelengths from 700 to 950 nm.

This is about 25% of the energy available from the sun. Another 45% of the available solar energy is in the visible range of 400 to 700 nm and can be captured by oxygen-evolving phototrophs. One way to augment the light-capturing abilities of each group of phototrophs is to grow them in coculture. Melis and Melnicki, for example, have designed and operated small-scale systems for biological hydrogen production in which green algae and PNSB are cocultivated in such a way that both the visible (green algae) and near-infrared (anoxygenic phototroph) parts of the solar spectrum are harvested (Melis and Melnicki, 2006).

Hydrogen-producing phototrophs are exposed to both very high and very low light intensities depending on where they are physically located in a bioreactor relative to the light source and depending on the time of day. The PNSB have evolved to thrive in anaerobic sediments and thus are not efficient at harvesting the energy that is delivered at high solar intensities. The PNSB do have the advantage that they can tolerate high light intensities without suffering the photoinhibition that is experienced by oxygenic phototrophs due to their exposure to oxygen during photophosphorylation. However, inefficiencies of light absorption are associated with increased dissipation of the light energy to heat. Because of this, large-scale processes for hydrogen production may be associated with high demands for cooling (Hoekema et al., 2006). These considerations suggest that the development of strains with increased photosynthetic efficiency would be valuable.

The PNSB have embedded in their photosynthetic membranes two major types of antennae or light-harvesting complexes referred to as "core" and "peripheral" (Zuber and Cogdell, 1995) (Fig. 4). Core antenna systems, also known as light harvesting 1, are comprised of oligomers of monomeric units of α and β polypeptides and associated bacteriochlorophylls and carotenoids. The monomeric units assemble to form a doughnut-shaped structure that surrounds an individual reaction center with a fixed stoichiometry of monomeric units to reaction center. The fundamental unit of peripheral light-harvesting systems, also known as light harvesting 2, consists of two other types of α and β polypeptides that also have noncovalently bound bacteriochlorophyll and carotenoid molecules. These units assemble to form ring-shaped structures, which in turn assemble into paracrystalline arrays around core antenna-reaction center complexes (Bahatyrova et al., 2004; Hunter et al., 2005; Scheuring and Sturgis, 2005; Scheuring et al., 2006). PNSB species vary in the number of different types of peripheral light-harvesting systems that they synthesize. Some, such as *Rhodobacter sphaeroides* and *Rhodobacter capsulatus*, have one type of peripheral light-harvesting complex (Zuber and Cogdell, 1995). Others, including *Rhodopseudomonas palustris*, can synthesize more than one type (Evans et al., 2005; Hartigan et al., 2002). Peripheral light-harvesting systems are regulated according to light intensity (Aagaard and Sistrom, 1972; Cogdell et al., 2006; Evans et al., 2005), with regulation occurring mostly at the level of transcription (Cogdell et al., 2006; Roh et al., 2004). Although PNSB have the ability to adjust the sizes of their light-harvesting antennae to maximize light absorption, the limits to this have not been systematically studied. Also worth systematic investigation is the degree to which light absorption at high and low intensities can be expanded by genetic engineering to increase or decrease the size of antennae. We also lack a detailed understanding of the molecular mechanisms that control expression of genes for light-harvesting complexes in response to light quality and intensity. Available data suggest that these mechanisms can be species specific. Thus, it will be important to elucidate their operation in the particular species that is used for a hydrogen production application.

Genetically Modify Nitrogenase for Improved Hydrogen Production

That nitrogenase has evolved for ammonia formation and not for hydrogen production suggests that there may be opportunities for improving the hydrogen production aspect of the catalytic cycle. Arguing against this is the observation that the rate-limiting step in nitrogenase catalysis is the dissociation of the Fe protein from the MoFe protein at the end of each cycle of electron delivery (Hageman and Burris, 1978). This is a complex process that may be difficult to improve upon. A more promising avenue to enzyme improvement would be to change the substrate selectivity of nitrogenase so that hydrogen is preferentially synthesized relative to ammonia, even in the presence of dinitrogen. This type of enzyme would have advantages in applied situations where nitrogen gas from the atmosphere is present. In fact, a mutant form of the *Azotobacter vinelandii* molybdenum nitrogenase with a valine-to-isoleucine substitution at amino acid 70 of the alpha subunit of the MoFe protein has been described that has a threefold-higher specific activity for hydrogen production in the presence of dinitrogen (Barney et al., 2004). The K_m of this mutant nitrogenase for nitrogen was increased over threefold relative to that of the wild-type enzyme. Because nitrogenases are highly conserved enzymes, it is likely that the same amino acid change will have the same effect on nitrogenases from species of PNSB. Other amino acid changes may also improve the relative proportion of hydrogen produced by nitrogenase.

Prevent the Diversion of Reductant to Other Metabolic Processes

Two metabolic processes that consume large quantities of reductant and thus have the potential to divert electrons away from nitrogenase-catalyzed hydrogen production by whole cells of PNSB are poly-hydroxyalkanoate (PHA) synthesis and carbon dioxide fixation. Developing strains in which these processes have been disabled by mutation can prevent diversion of reductant away from hydrogen production. PHA is a polymer assembled from 3-hydroxy fatty acids. It forms large intracellular granules in many PNSB and serves as a carbon storage polymer (Suriyamongkol et al., 2007; Vincenzini et al., 1997). The synthesis of PHAs from acetyl coenzyme A (acetyl-CoA) requires three enzymes: a β-ketothiolase, an acetoacetyl-CoA reductase, and a PHA synthase. The acetotacetyl-CoA reductase step, required for the generation of 3-hydroxybutyryl-CoA monomers, is an NADPH-dependent reaction. Since PHA granules are typically generated from the polymerization of 10,000 or more 3-hydroxybutyryl-CoA subunits, their formation can potentially have a large effect in competing with nitrogenase for electrons destined for hydrogen production. Consistent with this, PHA biosynthesis mutants of *Rhodobacter sphaeroides* and *Rhodospirillum rubrum* were found to produce more hydrogen than their wild-type parents when grown under nitrogen-fixing conditions (Hustede et al., 1993).

PNSB fix carbon dioxide into cell material via the Calvin Benson Bassham pathway as follows:

$$3CO_2 + 6NADPH + 5H_2O + 9ATP \rightarrow$$
$$C_3H_5O_3{-}PO_3 + 3H^+ + 6NADP^+ + 9ADP + 8P_i$$

Since it has a high demand for reductant, carbon dioxide fixation has the potential to be a major competitor for electrons destined for hydrogen production. One approach to prevent this from happening is to not add carbon dioxide or bicarbonate to bioreactors. However, it is impossible to prevent the formation of carbon dioxide that occurs when cells oxidize the electron-donating organic compounds that are required in the bioreactors to feed the process. Since carbon dioxide can serve as an electron sink, it would seem prudent to construct strains that are defective in carbon dioxide fixation to stabilize the hydrogen production process.

Expand the Ability of PNSB To Utilize Diverse Electron-Donating Substrates Alone and in Combination

In an ideal situation, low-cost feedstocks containing a well-defined mixture of organic compounds will serve as the electron donors for hydrogen-producing bioreactors of PNSB. A more likely scenario is that ill-defined fermentation and lignocellulosic waste streams will be the feedstocks for this process. PNSB have been reported to play a role in the remediation of a number of food industry, chemical industry, and fermentation waste streams (Kobayashi and Kobayashi, 1995). They can also predominate in swine wastewaters (Do et al., 2003; Okubo et al., 2006). In these situations, nature was allowed to take its course and the waste stream provided conditions that enriched for phototrophs. In the future one would like to treat waste streams with strains that have been optimized to produce hydrogen from the organic material that they contain. A study in which hydrogen was produced by metabolically engineered *Rhodobacter sphaeroides* strains inoculated into food waste-derived medium has demonstrated the feasibility of this approach (Franchi et al., 2004). As a group, PNSB have a well-developed ability to metabolize volatile fatty acids, compounds that are typical end products of anaerobic fermentations. Most of these types of compounds are expected to be good electron donors for hydrogen production, but this has not been carefully evaluated. Some of the protein components involved in electron transfer to nitrogenase have been identified for *Rhodospirillum rubrum* and *Rhodobacter capsulatus* (Edgren and Nordlund, 2004, 2005; Jeong and Jouanneau, 2000). However, the complete pathway of electron flow from organic compounds to nitrogenase has not been elucidated for any species. It may be that some compounds are better electron donors than others for hydrogen production. The capabilities of PNSB to use more than one carbon source simultaneously have also not been systematically evaluated. These are all areas for future study.

Another avenue for strain improvement is to expand the range of organic compounds that a particular strain can oxidize. For example, some species of PNSB have a well-developed ability to use aromatic acids and others can degrade sugars, but most cannot do both. It is likely that the substrate range of a particular species can be improved by metabolic engineering.

BIOREACTOR DESIGN

Depending on the cell density of a culture and the dimensions of the culture vessel that is used to grow cells, PNSB can become light limited, even in the research laboratory. This is because cells in the outer layers of a culture vessel shade those in the interior. Light limitation can be avoided by growing cells in small volumes of 10 to 100 ml and by controlling cell densities. As photobiological hydrogen production processes are scaled up, however, the large reduction in the ratio of surface area to volume that occurs renders the uniform illumination of reactors a technical challenge. This challenge has been addressed by the development of transparent flat-panel bioreactors, tubular bioreactors, and multilayer reactors to increase the surface area that is exposed to light (Carlozzi et al., 2006; Hoekema et al., 2002; Kondo et al., 2006). These approaches show promise, but the bioreactors can be complicated to construct and it is not clear how long they can be continuously operated. Another approach that has been successfully applied on a small scale to reduce the problem of light limitation is immobilization of cells in thin nanoporous strips of latex (Flickinger et al., 2007). Illuminated single-layer translucent latex films about 60 μm thick containing concentrated (2.0×10^8 cells/μl) nongrowing *Rhodopseudomonas palustris* cells produced hydrogen continuously at a constant rate for over 4,000 h (M. Flickinger, personal communication). The latex films were replenished with fresh acetate (the electron donor) every 5 or 6 days during this period. Latex strips with entrapped *Rhodopseudomonas palustris* cells experienced no loss in reactivity after being frozen for 1 year at −80°C. Similarly, latex strips that were illuminated after over 3 months of storage in the dark at 30°C resumed hydrogen production (Gosse et al., 2007). This suggests that latex-embedded PNSB can be prepared at a central location and transported to geographically distant sites. To maintain a sterile environment, strips of latex-entrapped cells can be prepared with a sealant top coating of a few microns in thickness that is permeable to small molecules but excludes particles the size of bacteria. The use of this sealant should make it possible to use waste streams as electron donors for hydrogen production while avoiding problems of contamination with other microbes. In future embodiments one can expect that latex strips containing different photosynthesis mutants can be prepared and stacked in layers to increase light absorption at different intensities and wavelengths (Flickinger et al., 2007; Gosse et al., 2007).

CHOICE OF SPECIES FOR HYDROGEN PRODUCTION

Most of the reported work on hydrogen production has been done with the four species mentioned in this chapter: *Rhodopseudomonas palustris*, *Rhodospirillum rubrum*, *Rhodobacter capsulatus*, and *Rhodobacter sphaeroides*. The genomes of the commonly used research laboratory strains have been sequenced for each of these species (Larimer et al., 2004; Mackenzie et al., 2006; http://img.jgi.doe.gov/cgi-bin/pub/main.cgi; http://rhodo.img.cas.cz/index.html). Each is genetically tractable. Each of the species has served as a model organism for different basic biological processes that are important for hydrogen production, and depending on the specific hydrogen production application, any one of the four species may be suitable for use in applying various of the strategies for strain development that have been mentioned in this chapter. The transcriptomes of *Rhodopseudomonas palustris* CGA009 mutant strains expressing each nitrogenase isozyme have been analyzed (Oda et al., 2005), as has the proteome of wild-type cells grown under nitrogen-fixing (and hydrogen-producing) conditions (VerBerkmoes et al., 2006). The transcriptomes of mutant strains of *Rhodopseudomonas palustris* that produce hydrogen constitutively under all growth conditions have also been analyzed (Rey et al., 2007). *Rhodopseudomonas palustris* and *Rhodobacter sphaeroides* have served as model organisms for detailed studies of the regulation and mechanism of photosynthesis (Richter et al., 2007; Scheuring et al., 2006; Tavano and Donohue, 2006). The molecular regulation of nitrogenase expression has been worked out in detail in *Rhodobacter capsulatus* and *Rhodospirillum rubrum* (Masepohl et al., 2002a; Zhang et al., 1995, 2001, 2004, 2005). Both the Mo and Fe nitrogenases have been purified and characterized from *Rhodobacter capsulatus* (Schneider et al., 1997) .

When the fundamental physiological properties of the four species are considered, *Rhodopseudomonas palustris* has attributes that make it particularly well suited for development as a platform for hydrogen production. It encodes all three nitrogenase isozymes and can access lignin monomers and acetate as sources of electrons for hydrogen. The pathways for lignin monomer degradation are quite complex (Gibson and Harwood, 2002). The lignin component of plant biomass is the second most abundant polymer on Earth and a potential renewable resource for hydrogen production. Acetate, the main compound that is generated in anaerobic food webs, is also readily available. *Rhodopseudomonas palustris* is also extraordinarily robust and can survive for months in a nongrowing "resting" state. An advantage to the use of nongrowing cells as biocatalysts is that they can potentially divert 100% of the reducing equivalents present in organic compounds to hydrogen without having to devote any of the carbon sources to biosynthesis. There are some potential disadvantages of *Rhodopseudomonas palustris* compared to the other three species. One is that it has an intrinsically slower growth rate, and this is reflected in a lower in vivo nitrogenase activity on a per cell basis (Madigan et al., 1984). Another is that the ability to degrade sugars is a strain-dependent trait for this species (Brenner et al., 2001). However, these are limitations that can likely be overcome by metabolic engineering. On balance, the extreme metabolic versatility and robustness of *Rhodopseudomonas palustris* are overarching attributes that make it an excellent biocatalyst for hydrogen production.

CONCLUSIONS

Surprisingly little research on biological hydrogen production has been done relative to the amount of print that has been devoted to this topic in reviews and in the perspective pages of scientific journals. In many ways studies of nitrogenase-catalyzed hydrogen production by anoxygenic phototrophic bacteria are still in their infancy. While we know a great deal about the processes of metabolism, nitrogenase regulation, and photophosphorylation that are needed for hydrogen to be produced, scientists have not generally investigated these topics in the context of improving our understanding of the systems biology of hydrogen evolution. A systematic approach to strain improvement and bioreactor design will be required to bring hydrogen production from the bench to large-scale bioreactors. In applying the elements of a systematic approach, it will be important to view hydrogen production in the context of the physiology of whole bacterial cells and not simply in terms of the component parts that fuel hydrogen production.

REFERENCES

Aagaard, J., and W. R. Sistrom. 1972. Control of synthesis of reaction center bacteriochlorophyll in photosynthetic bacteria. *Photochem. Photobiol.* 15:209–225.

Angenent, L. T., K. Karim, M. H. Al-Dahhan, B. A. Wrenn, and R. Domiguez-Espinosa. 2004. Production of bioenergy and biochemicals from industrial and agricultural wastewater. *Trends Biotechnol.* 22:477–485.

Bahatyrova, S., R. N. Frese, C. A. Siebert, J. D. Olsen, K. O. Van Der Werf, R. Van Grondelle, R. A. Niederman, P. A. Bullough, C. Otto, and C. N. Hunter. 2004. The native architecture of a photosynthetic membrane. *Nature* 430:1058–1062.

Barbosa, M. J., J. M. Rocha, J. Tramper, and R. H. Wijffels. 2001. Acetate as a carbon source for hydrogen production by photosynthetic bacteria. *J. Biotechnol.* 85:25–33.

Barney, B. M., R. Y. Igarashi, P. C. Dos Santos, D. R. Dean, and L. C. Seefeldt. 2004. Substrate interaction at an iron-sulfur face of the FeMo-cofactor during nitrogenase catalysis. *J. Biol. Chem.* 279: 53621–53624.

Bell, J., A. J. Dunford, E. Hollis, and R. A. Henderson. 2003. The role of Mo atoms in nitrogen fixation: balancing substrate reduction and dihydrogen production. *Angew. Chem. Int. Ed. Engl.* 42: 1149–1152.

Brenner, D. J., N. R. Krieg, and J. T. Staley. 2001. *The Apha-, Beta-, Delta-, and Epsilonproteobacteria, vol. 2, part C. In* G. M. Garrity (ed.), *Bergey's Manual of Systematic Bacteriology.* Springer, New York, NY.

Burns, R. C., and W. A. Bulen. 1965. ATP-dependent hydrogen evolution by cell-free preparations of *Azotobacter vinelandii. Biochim. Biophys. Acta* 105:437–445.

Burns, R. C., and W. A. Bulen. 1966. A procedure for the preparation of extracts from *Rhodospirillum rubrum* catalyzing N_2 reduction and ATP-dependent H_2 evolution. *Arch. Biochem. Biophys.* 113:461–463.

Carlozzi, P., B. Pushparaj, A. Degl'Innocenti, and A. Capperucci. 2006. Growth characteristics of *Rhodopseudomonas palustris* cultured outdoors, in an underwater tubular photobioreactor, and investigation on photosynthetic efficiency. *Appl. Microbiol. Biotechnol.* 73:789–795.

Cogdell, R. J., A. Gall, and J. Kohler. 2006. The architecture and function of the light-harvesting apparatus of purple bacteria: from single molecules to in vivo membranes. *Q. Rev. Biophys.* 39:227–324.

Connelly, H. M., D. A. Pelletier, T. Y Lu., P. K. Lankford, and R. L. Hettich. 2006. Characterization of PII family (GlnK1, GlnK2, and GlnB) protein uridylylation in response to nitrogen availability for *Rhodopseudomonas palustris. Anal. Biochem.* 357:93–104.

Croal, L. R., Y. Jiao, and D. K. Newman. 2007. The *fox* operon from *Rhodobacter* strain SW2 promotes phototrophic Fe(II) oxidation in *Rhodobacter capsulatus* SB1003. *J. Bacteriol.* 189:1774–1782.

Das, D., and T. N. Veziroglu. 2001. Hydrogen production by biological processes: a survey of literature. *Int. J. Hydrogen Energy* 26: 13–28.

Dixon, R., and D. Kahn. 2004. Genetic regulation of biological nitrogen fixation. *Nat. Rev. Microbiol.* 2:621–631.

Do, Y. S., T. M. Schmidt, J. A. Zahn, E. S. Boyd, A. de la Mora, and A. A. DiSpirito. 2003. Role of *Rhodobacter* sp. strain PS9, a purple non-sulfur photosynthetic bacterium isolated from an anaerobic swine waste lagoon, in odor remediation. *Appl. Environ. Microbiol.* 69:1710–1720.

Drepper, T., S. Gross, A. F. Yakunin, P. C. Hallenbeck, B. Masepohl, and W. Klipp. 2003. Role of GlnB and GlnK in ammonium control of both nitrogenase systems in the phototrophic bacterium *Rhodobacter capsulatus. Microbiology* 149:2203–2212.

Eady, R. R. 1996. Structure-function relationships of alternative nitrogenases. *Chem. Rev.* 96:3013–3030.

Edgren, T., and S. Nordlund. 2004. The *fixABCX* genes in *Rhodospirillum rubrum* encode a putative membrane complex participating in electron transfer to nitrogenase. *J. Bacteriol.* 186:2052–2060.

Edgren, T., and S. Nordlund. 2005. Electron transport to nitrogenase in *Rhodospirillum rubrum:* identification of a new *fdxN* gene encoding the primary electron donor to nitrogenase. *FEMS Microbiol. Lett.* 245:345–351.

Evans, K., A. P. Fordham-Skelton, H. Mistry, C. D. Reynolds, A. M. Lawless, and M. Z. Papiz. 2005. A bacteriophytochrome regulates the synthesis of LH4 complexes in *Rhodopseudomonas palustris. Photosynth. Res.* 85:169–180.

Fisher, K., and W. E. Newton. 2002. Nitrogen fixation—a general overview, p. 1–34. *In* G. J. Leigh (ed.), *Nitrogen Fixation at the Millennium.* Elsevier, Amsterdam, The Netherlands.

Fißler, J., C. Schirra, G.-W. Koring, and F. Giffhorn. 1994. Hydrogen production from aromatic acids by *Rhodopseudomonas palustris. Appl. Microbiol. Biotechnol.* 41:395–399.

Flickinger, M. C., J. L. Schottel, D. R. Bond, A. Aksan, and L. E. Scriven. 2007. Painting and printing living bacteria: engineering nanoporous biocatalytic coatings to preserve microbial viability and intensify reactivity. *Biotechnol. Prog.* 23:2–17.

Franchi, E., C. Tosi, G. Scolla, G. Della Penna, F. Rodriguez, and P. M. Pedroni. 2004. Metabolically engineered *Rhodobacter sphaeroides* RV strains for improved biohydrogen photoproduction combined with disposal of food wastes. *Marine Biotechnol.* 6:552–565.

Gest, H., and M. D. Kamen. 1949. Photochemical production of molecular hydrogen by *Rhodospirillum rubrum. Science* 109:558–559.

Gest, H. 1999. Memoir of a 1949 railway journey with photosynthetic bacteria. *Photosynth. Res.* 61:91–96.

Gibson, J., and C. S. Harwood. 2002. Metabolic diversity in aromatic compound utilization by anaerobic microbes. *Annu. Rev. Microbiol.* **56:**345–369.

Gosse, J. L., B. J. Engel, F. E. Rey, C. S. Harwood, L. E. Scriven, and M. C. Flickinger. 2007. Hydrogen production by photoreactive nanoporous latex coatings of nongrowing *Rhodopseudomonas palustris* CGA009. *Biotechnol. Prog.* **23:**124–130.

Hageman, R. V., and R. H. Burris. 1978. Nitrogenase and nitrogenase reductase associate and dissociate with each catalytic cycle. *Proc. Natl. Acad. Sci. USA* **75:**2699–2702.

Hales, B. J., E. E. Case, J. E. Morningstar, M. F. Dzeda, and L. A. Mauterer. 1986. Isolation of a new vanadium-containing nitrogenase from *Azotobacter vinelandii. Biochemistry* **25:**7251–7255.

Hartigan, N., H. A. Tharia, F. Sweeney, A. M. Lawless, and M. Z. Papiz. 2002. The 7.5-Å electron density and spectroscopic properties of a novel low-light B800 LH2 from *Rhodopseudomonas palustris. Biophys. J.* **82:**963–977.

Harwood, C. S., and J. Gibson. 1988. Anaerobic and aerobic metabolism of diverse aromatic compounds by the photosynthetic bacterium *Rhodopseudomonas palustris. Appl. Environ. Microbiol.* **54:**712–717.

Henderson, R. A. 2002. Advances towards the mechanism of nitrogenase, p. 223–261. *In* G. J. Leigh (ed.), *Nitrogen Fixation at the Millennium.* Elsevier, Amsterdam, The Netherlands.

Hillmer, P., and H. Gest. 1977a. H_2 metabolism in the photosynthetic bacterium *Rhodopseudomonas capsulata:* production and utilization of H_2 by resting cells. *J. Bacteriol.* **129:**732–739.

Hillmer, P., and H. Gest. 1977b. H_2 metabolism in the photosynthetic bacterium *Rhodopseudomonas capsulata:* H_2 production by growing cultures. *J. Bacteriol.* **129:**724–731.

Hoekema, S., M. Bijmans, M. Janssen, J. Tramper, and R. H. Wijffels. 2002. A pneumatically agitated flat-panel photobioreactor with gas re-circulation: anaerobic photoheterotrophic cultivation of a purple non-sulfur bacterium. *Int. J. Hydrogen Energy* **27:** 1331–1338.

Hoekema, S., R. D. Douma, M. Janssen, J. Tramper, and R. H. Wijffels. 2006. Controlling light-use by *Rhodobacter capsulatus* continuous cultures in a flat-panel photobioreactor. *Biotechnol. Bioeng.* **95:**613–626.

Howard, J. B., and D. C. Rees. 1994. Nitrogenase: a nucleotide-dependent molecular switch. *Annu. Rev. Biochem.* **63:**235–264.

Hunter, C. N., J. D. Tucker, and R. A. Niederman. 2005. The assembly and organisation of photosynthetic membranes in *Rhodobacter sphaeroides. Photochem. Photobiol. Sci.* **4:**1023–1027.

Hustede, E., A. Steinbuchel, and H. G. Schlegel. 1993. Relationship between the photoproduction of hydrogen and the accumulation of PHB in non-sulfur purple bacteria. *Appl. Microbiol. Biotechnol.* **39:**87–93.

Igarashi, R. Y., and L. C. Seefeldt. 2003. Nitrogen fixation: the mechanism of the Mo-dependent nitrogenase. *Crit. Rev. Biochem. Mol. Biol.* **38:**351–384.

Igarashi, R. Y., M. Laryukhin, P. C. Dos Santos, H. I. Lee, D. R. Dean, L. C. Seefeldt, and B. M. Hoffman. 2005. Trapping H-bound to the nitrogenase FeMo-cofactor active site during H2 evolution: characterization by ENDOR spectroscopy. *J. Am. Chem. Soc.* **127:**6231–6241.

Jeong, H. S., and Y. Jouanneau. 2000. Enhanced nitrogenase activity in strains of *Rhodobacter capsulatus* that overexpress the *rnf* genes. *J. Bacteriol.* **182:**1208–1214.

Jiao, Y., and D. K. Newman. 2007. The *pio* operon is essential for phototrophic Fe(II) oxidation in *Rhodopseudomonas palustris* TIE-1. *J. Bacteriol.* **189:**1765–1773.

Joshi, H. M., and F. R. Tabita. 1996. A global two component signal transduction system that integrates the control of photosynthesis, carbon dioxide assimilation, and nitrogen fixation. *Proc. Natl. Acad. Sci. USA* **93:**14515–14520.

Kaman, M. D., and H. Gest. 1949. Evidence for a nitrogenase system in the photosynthetic bacterium *Rhodospirillum rubrum. Science* **109:**560.

Kern, M., W. Klipp, and J. H. Klemme. 1994. Increased nitrogenase-dependent H_2 photoproduction by *hup* mutants of *Rhodospirillum rubrum. Appl. Environ. Microbiol.* **60:**1768–1774.

Kobayashi, M., and M. Kobayashi. 1995. Waste remediation and treatment using anoxygenic phototrophic bacteria, p. 1269–1282. *In* R. E. Blankenship, M. T. Madigan, and C. E. Bauer (ed.), *Anoxygenic Photosynthetic Bacteria.* Kluwer Academic Publishers, Dordrecht, The Netherlands.

Kondo, T., T. Wakayama, and J. Miyake. 2006. Efficient hydrogen production using a multi-layered photobioreactor and a photosynthetic bacterium mutant with reduced pigment. *Int. J. Hydrogen Energy* **31:**1522–1526.

Kutsche, M., S. Leimkuhler, S. Angermuller, and W. Klipp. 1996. Promoters controlling expression of the alternative nitrogenase and the molybdenum uptake system in *Rhodobacter capsulatus* are activated by NtrC, independent of sigma54, and repressed by molybdenum. *J. Bacteriol.* **178:**2010–2017.

Larimer, F. W., P. Chain, L. Hauser, J. Lamerdin, S. Malfatti, L. Do, M. L. Land, D. A. Pelletier, J. T. Beatty, A. S. Lang, F. R. Tabita, J. L. Gibson, T. E. Hanson, C. Bobst, J. L. Torres, C. Peres, F. H. Harrison, J. Gibson, and C. S. Harwood. 2004. Complete genome sequence of the metabolically versatile photosynthetic bacterium *Rhodopseudomonas palustris. Nat. Biotechnol.* **22:**55–61.

Lehman, L. J., and G. P. Roberts. 1991. Identification of an alternative nitrogenase system in *Rhodospirillum rubrum. J. Bacteriol.* **173:**5705–5711.

Lindstrom, E. S., R. H. Burris, and P. W. Wilson. 1949. Nitrogen fixation by photosynthetic bacteria. *J. Bacteriol.* **58:**313–316.

Ludden, P. W., and G. P. Roberts. 1995. The biochemistry and genetics of nitrogen fixation by photosynthetic bacteria, p. 929–947. *In* R. E. Blankenship, M. T. Madigan, and C. E. Bauer (ed.), *Anoxygenic Photosynthetic Bacteria.* Kluwer Academic Publishers, Dordrecht, The Netherlands.

Mackenzie, C., J. M. Eraso, M. Choudhary, J. H. Roh, X. Zeng, P. Bruscella, A. Puskas, and S. Kaplan. 18 June 2006. Postgenomic adventures with *Rhodobacter sphaeroides. Annu. Rev. Microbiol.* **61:**283–307.

Madigan, M., S. S. Cox, and R. A. Stegeman. 1984. Nitrogen fixation and nitrogenase activities in members of the family *Rhodospirillaceae. J. Bacteriol.* **157:**73–78.

Madigan, M. T. 1995. Microbiology of nitrogen fixation by anoxygenic photosynthetic bacteria, p. 915–928. *In* R. E. Blankenship, M. T. Madigan, and C. E. Bauer (ed.), *Anoxygenic Photosynthetic Bacteria.* Kluwer Academic Publishers, Dordrecht, The Netherlands.

Madigan, M. T., and J. F. Imhoff. 2007. International Committee on Systematics of Prokaryotes; Subcommittee on the Taxonomy of Phototrophic Bacteria: minutes of the meetings, 29 August 2006, Pau, France. *Int. J. Syst. Evol. Microbiol.* **57:**1169–1171.

Masepohl, B., T. Drepper, A. Paschen, S. Gross, A. Pawlowski, K. Raabe, K. U. Riedel, and W. Klipp. 2002a. Regulation of nitrogen fixation in the phototrophic purple bacterium *Rhodobacter capsulatus. J. Mol. Microbiol. Biotechnol.* **4:**243–248.

Masepohl, B., K. Schneider, T. Drepper, A. Müller, and W. Klipp. 2002b. Alternative nitrogenases, p. 191–222. *In* G. J. Leigh (ed.), *Nitrogen Fixation at the Millennium.* Elsevier, Amsterdam, The Netherlands.

Melis, A., and M. R. Melnicki. 2006. Integrated biological hydrogen production. *Int. J. Hydrogen Energy* **31:**1563–1573.

Merrick, M. J. 2004. Regulation of nitrogen fixation in free-living diazotrophs, p. 197–223. *In* W. Klipp, B. Mesepohl, J. R. Gallon, and W. E. Newton (ed.), *Genetics and Regulation of Nitrogen Fixation in Free-Living Bacteria.* Kluwer Academic Publishers, Dordrecht, The Netherlands.

Oda, Y., S. K. Samanta, F. E. Rey, L. Wu, X. Liu, T. Yan, J. Zhou, and C. S. Harwood. 2005. Functional genomic analysis of three nitrogenase isozymes in the photosynthetic bacterium *Rhodopseudomonas palustris. J. Bacteriol.* **187:**7784–7794.

Okubo, Y., H. Futamata, and A. Hiraishi. 2006. Characterization of phototrophic purple nonsulfur bacteria forming colored microbial mats in a swine wastewater ditch. *Appl. Environ. Microbiol.* **72:** 6225–6233.

Ormerod, J. G., and H. Gest. 1962. Symposium on metabolism of inorganic compounds. IV. Hydrogen photosynthesis and alternative metabolic pathways in photosynthetic bacteria. *Bacteriol. Rev.* **26:** 51–66.

Ozturk, Y., M. Yucel, F. Daldal, S. Mandaci, U. Gunduz, L. Turker, and I. Eroglu. 2006. Hydrogen production by using *Rhodobacter capsulatus* mutants with genetically modified electron transfer chains. *Int. J. Hydrogen Energy* **31:**1545–1552.

Prince, R. C., and H. S. Kheshgi. 2005. The photobiological production of hydrogen: potential efficiency and effectiveness as a renewable fuel. *Crit. Rev. Microbiol.* **31:**19–31.

Rey, F. E., Y. Oda, and C. S. Harwood. 2006. Regulation of uptake hydrogenase and effects of hydrogen utilization on gene expression in *Rhodopseudomonas palustris. J. Bacteriol.* **188:**6143–6152.

Rey, F. E., E. K. Heiniger, and C. S. Harwood. 2007. Redirection of metabolism for biological hydrogen production. *Appl. Environ. Microbiol.* **73:**1665–1671.

Richter, M. F., J. Baier, T. Prem, S. Oellerich, F. Francia, G. Venturoli, D. Oesterhelt, J. Southall, R. J. Cogdell, and J. Kohler. 2007. Symmetry matters for the electronic structure of core complexes from *Rhodopseudomonas palustris* and *Rhodobacter sphaeroides* PufX. *Proc. Natl. Acad. Sci. USA* **104:**6661–6665.

Roh, J. H., W. E. Smith, and S. Kaplan. 2004. Effects of oxygen and light intensity on transcriptome expression in *Rhodobacter sphaeroides* 2.4.1. Redox active gene expression profile. *J. Biol. Chem.* **279:**9146–9155.

Roszak, A. W., T. D. Howard, J. Southall, A. T. Gardiner, C. J. Law, N. W. Isaacs, and R. J. Cogdell. 2003. Crystal structure of the RC-LH1 core complex from *Rhodopseudomonas palustris. Science* **302:**1969–1972.

Rubio, L. M., and P. W. Ludden. 2005. Maturation of nitrogenase: a biochemical puzzle. *J. Bacteriol.* **187:**405–414.

Rupprecht, J., B. Hankamer, J. H. Mussgnug, G. Ananyev, C. Dismukes, and O. Kruse. 2006. Perspectives and advances of biological H2 production in microorganisms. *Appl. Microbiol. Biotechnol.* **72:**442–449.

Scheuring, S., and J. N. Sturgis. 2005. Chromatic adaptation of photosynthetic membranes. *Science* **309:**484–487.

Scheuring, S., R. P. Goncalves, V. Prima, and J. N. Sturgis. 2006. The photosynthetic apparatus of *Rhodopseudomonas palustris:* structures and organization. *J. Mol. Biol.* **358:**83–96.

Schneider, K., U. Gollan, S. Selsemeier-Voigt, W. Plass, and A. Muller. 1994. Rapid purification of the protein components of a highly active "iron only" nitrogenase. *Naturwissenschaften* **81:**405–408.

Schneider, K., U. Gollan, M. Drottboom, S. Selsemeier-Voigt, and A. Muller. 1997. Comparative biochemical characterization of the iron-only nitrogenase and the molybdenum nitrogenase from *Rhodobacter capsulatus. Eur. J. Biochem.* **244:**789–800.

Shi, X. Y., and H. Q. Yu. 2006. Continuous production of hydrogen from mixed volatile fatty acids with *Rhodopseudomonas capsulata. Int. J. Hydrogen Energy* **31:**1641–1647.

Simpson, F. B., and R. H. Burris. 1984. A nitrogen pressure of 50 atmospheres does not prevent evolution of hydrogen by nitrogenase. *Science* **224:**1095–1097.

Suriyamongkol, P., R. Weselake, S. Narine, M. Moloney, and S. Shah. 2007. Biotechnological approaches for the production of polyhydroxyalkanoates in microorganisms and plants—a review. *Biotechnol. Adv.* **25:**148–175.

Tavano, C. L., and T. J. Donohue. 2006. Development of the bacterial photosynthetic apparatus. *Curr. Opin. Microbiol.* **9:**625–631.

Tsygankov, A. A., A. S. Fedorov, T. V. Laurinavichene, I. N. Gogotov, K. K. Rao, and D. O. Hall. 1998. Actual and potential rates of hydrogen photoproduction by continuous culture of the purple non-sulphur bacterium *Rhodobacter capsulatus. Appl. Microbiol. Biotechnol.* **49:**102–107.

Turner, J. A. 2004. Sustainable hydrogen production. *Science* **305:**972–974.

VerBerkmoes, N. C., M. B. Shah, P. K. Lankford, D. A. Pelletier, M. B. Strader, D. L. Tabb, W. H. McDonald, J. W. Barton, G. B. Hurst, L. Hauser, B. H. Davison, J. T. Beatty, C. S. Harwood, F. R. Tabita, R. L. Hettich, and F. W. Larimer. 2006. Determination and comparison of the baseline proteomes of the versatile microbe *Rhodopseudomonas palustris* under its major metabolic states. *J. Proteome Res.* **5:**287–298.

Vignais, P. M., and B. Toussaint. 1994. Molecular biology of membrane-bound H_2 uptake hydrogenases. *Arch. Microbiol.* **161:**1–10.

Vignais, P. M., B. Billoud, and J. Meyer. 2001. Classification and phylogeny of hydrogenases. *FEMS Microbiol. Rev.* **25:**455–501.

Vignais, P. M., J. P. Magnin, and J. C. Willison. 2006. Increasing biohydrogen production by metabolic engineering. *Int. J. Hydrogen Energy* **31:**1478–1483.

Vincenzini, M., A. Marchini, A. Ena, and R. D. Philippis. 1997. H_2 and poly-β-hydroxybutyrate, two alternative chemicals from purple non sulfur bacteria. *Biotechnol. Lett.* **19:**759–762.

Yoch, D. C. 1978. Nitrogen fixation and hydrogen metabolism by photosynthetic bacteria, p. 657–676. *In* R. C. Clayton and W. R. Sistrom (ed.), *The Photosynthetic Bacteria.* Plenum Press, New York, NY.

Zhang, Y., A. D. Cummings, R. H. Burris, P. W. Ludden, and G. P. Roberts. 1995. Effect of an *ntrBC* mutation on the posttranslational regulation of nitrogenase activity in *Rhodospirillum rubrum. J. Bacteriol.* **177:**5322–5326.

Zhang, Y., E. L. Pohlmann, P. W. Ludden, and G. P. Roberts. 2001. Functional characterization of three GlnB homologs in the photosynthetic bacterium *Rhodospirillum rubrum:* roles in sensing ammonium and energy status. *J. Bacteriol.* **183:**6159–6168.

Zhang, Y., E. L. Pohlmann, and G. P. Roberts. 2004. Identification of critical residues in GlnB for its activation of NifA activity in the photosynthetic bacterium *Rhodospirillum rubrum. Proc. Natl. Acad. Sci. USA* **101:**2782–2787.

Zhang, Y., E. L. Pohlmann, and G. P. Roberts. 2005. GlnD is essential for NifA activation, NtrB/NtrC-regulated gene expression, and posttranslational regulation of nitrogenase activity in the photosynthetic, nitrogen-fixing bacterium *Rhodospirillum rubrum. J. Bacteriol.* **187:**1254–1265.

Zhang, Y., E. L. Pohlmann, M. C. Conrad, and G. P. Roberts. 2006a. The poor growth of *Rhodospirillum rubrum* mutants lacking PII proteins is due to an excess of glutamine synthetase activity. *Mol. Microbiol.* **61:**497–510.

Zhang, Y., D. M. Wolfe, E. L. Pohlmann, M. C. Conrad, and G. P. Roberts. 2006b. Effect of AmtB homologues on the post-translational regulation of nitrogenase activity in response to ammonium and energy signals in *Rhodospirillum rubrum. Microbiology* **152:**2075–2089.

Zuber, H., and R. J. Cogdell. 1995. Structure and organization of purple bacteria antenna complexes, p. 315–348. *In* R. E. Blankenship, M. T. Madigan, and C. E. Bauer (ed.), *Anoxygenic Photosynthetic Bacteria.* Kluwer Academic Publishers, Dordrecht, The Netherlands.

Bioenergy
Edited by J. Wall et al.

Chapter 22

Photosynthetic Water-Splitting for Hydrogen Production

MICHAEL SEIBERT, PAUL W. KING, MATTHEW C. POSEWITZ, ANASTASIOS MELIS, AND MARIA L. GHIRARDI

The world is using energy at a rate of over 4.1×10^{20} J/year (equal to over 13 TW of continuous power), but demand in 2050 could be as high as 30 TW (http://www.science.doe.gov/production/bes/reports/files/SEU_rpt.pdf). Gasoline prices at the pump are again over $3 a gallon in the United States. Levels of CO_2 could reach 600 ppm by 2035 if a "business as usual" attitude continues (http://www.sternreview.org.uk). The polar ice sheets are melting, and sea level could rise by as much as 0.88 m by the end of the century (Overpeck, et al., 2006). If all of these projections materialize, the economic consequences on the world will be devastating, and humanity will have no choice but to adapt as best it can. The consensus of the scientific community is that the release of greenhouse gases as the result of human activity is the root cause (http://www.ipcc.ch/pdf/assessment-report/ar4/wg1/ar4-wg1-spm.pdf). The challenges that we face are to act decisively and address the root cause quickly, or pay the consequences.

There are a number of options that could be implemented on a rather short timescale to partially address the challenge of global climate change, including conservation (increasing the energy efficiency of our cars, buildings, homes, and industrial processes), carbon sequestration (burying the CO_2 generated by large point sources), reforestation, nuclear power (albeit social issues are involved), and renewable energy. The last option, which currently produces about 6% of the primary energy in the United States (data from the Energy Information Administration, August 2005), includes biofuels, wind, photovoltaics, geothermal energy, ocean thermal energy conversion, wave power, and solar thermal technologies that can produce electricity, heat, and liquid fuels (e.g., ethanol and biodiesel) from biomass. Many of these same technologies can also be used to produce H_2. Currently most of the 9 to 10 million tons of H_2 used in the United States (ca. 1% of U.S. primary energy) is made by reforming natural gas, and most of this is for ammonia production and captive use in oil refineries.

BIOHYDROGEN FROM WATER

Biohydrogen production is the primary interest of this chapter with emphasis on direct solar conversion processes (sunlight plus water to H_2 by the direct linkage of photosynthesis to hydrogenase [the enzyme that releases H_2 gas] function), particularly those associated with green algae. To demonstrate the potential of a direct conversion process compared to an indirect one (e.g., biomass production in one step and conversion to a fuel such as ethanol in a second step), we show in Fig. 1 the area of the United States that would have to be covered with photobioreactors in order to displace the current amount of gasoline used in the country. The assumptions were solar-to-H_2 conversion (using the average yearly U.S. solar resource) at the current solar conversion efficiency goal of 10% (Rao and Cammack, 2001) and utilization of the H_2 in fuel-cell-powered vehicles getting 60 miles per gal of gasoline equivalent. For comparison, we have also indicated the land areas required to displace all U.S. gasoline by ethanol, if the ethanol were produced from (i) corn grain (assuming that the 23,750 mi^2 of corn devoted to ethanol production in 2006 displaces 2.4% of the gasoline used in the country) (Johnson, 2006) or (ii) cellulose (the theoretical maximum for switchgrass at 20 dry tons per acre [Kelly Ibsen, personal communication]). Realistically no one technology will be the answer, but direct conversion processes hold much promise for the future, especially in terms of efficient land utilization, if research and development (R&D) can lead to cost-competitive processes.

Michael Seibert, Paul W. King, and Maria L. Ghirardi • Chemical and Bioscience Center, National Renewable Energy Laboratory, Golden, CO 80401. **Matthew C. Posewitz** • Environmental Science and Engineering Dept., Colorado School of Mines, Golden, CO 80401. **Anastasios Melis** • Dept. of Plant & Microbial Biology, University of California, Berkeley, Berkeley, CA 94720-3102.

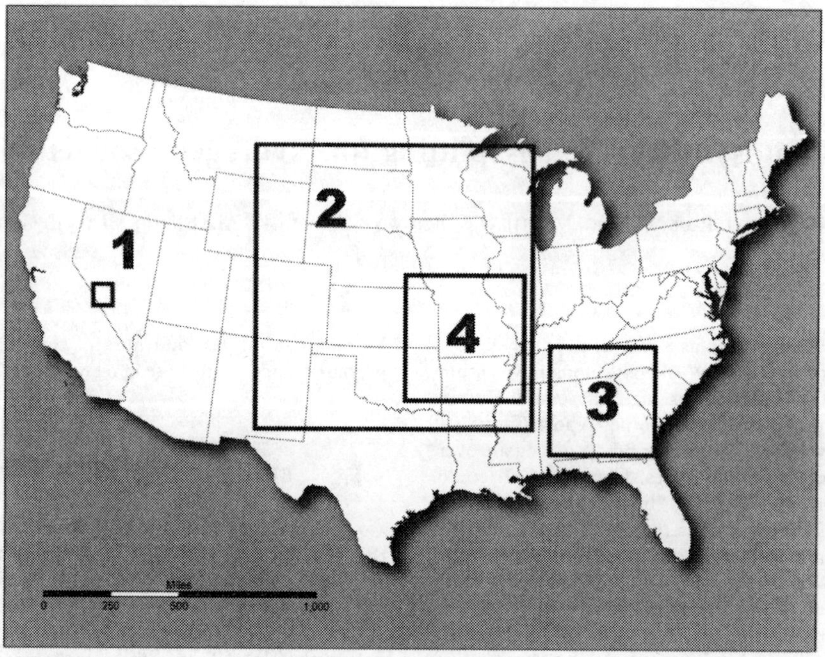

Figure 1. Approximate areas of the country required to displace all gasoline used in the United States using different technologies: algal H_2 produced from water at a future 10% solar efficiency (1), corn grain ethanol at current production yields (2), and cellulosic ethanol from switchgrass at an estimated optimal yield (3). For comparison, box 4 represents the total area of the 2006 corn crop.

A BRIEF HISTORY

The first recorded observation of a photosynthetic organism producing H_2 was reported over 100 years ago, when a natural bloom of *Anabaena*, after being placed in a glass jar, started to produce H_2 (Jackson and Ellms, 1896). Hydrogen uptake (Gaffron, 1939) and production (Gaffron and Rubin, 1942) in *Scenedesmus obliquus* were the first reports of H_2 metabolism in a green alga, and the early algal work has been reviewed recently (Homann, 2003; Melis and Happe, 2004). A complete review of the photobiological hydrogen production literature up until 1980 is also available (Weaver et al., 1980).

WHY ALGAL HYDROGEN PRODUCTION

The advantage of algae and cyanobacteria over photosynthetic bacteria for photoproducing H_2 is that water can be used as a substrate, and an organic sacrificial donor is not required. Furthermore, algae employ [FeFe]-hydrogenases (Roessler and Lien, 1984a; Happe and Naber, 1993; Happe et al., 1994), which are better H_2 production catalysts than the nitrogenases or [NiFe]-hydrogenases found in cyanobacteria and photosynthetic bacteria (Frey, 2002), and the theoretical efficiency of algal H_2 production is several times that of cyanobacterial, nitrogenase-linked systems (Bock et al., 2006). [FeFe]-hydrogenases, which are also found in many bacteria, are easier to express in recombinant or nonnative organisms than [NiFe]-hydrogenases, which often require coexpression of several organism-specific maturation genes (Boichenko et al., 2004; Prince and Kheshgi, 2005). Finally, the maximum activity of [FeFe]-hydrogenase in algae (Kosourov et al., 2002) is well matched to the maximum rate of oxygenic algal photosynthesis (Ghirardi et al., 2005).

FOCUS OF THIS REVIEW

This review emphasizes photobiological, H_2-producing organisms and processes that are able to link photosynthetic water oxidation (reductant-generation) directly to [FeFe]-hydrogenase-catalyzed H_2 production function.

BACKGROUND

Diversity of Organisms That Produce H_2

The occurrence of H_2 metabolism is widespread among microorganisms, including strictly anaerobic and facultatively aerobic bacteria, cyanobacteria, archaea, and eukaryotes such as green and red algae, diatoms, fungi, and protists (Adams et al., 1980; Weaver et al., 1980; Vignais et al., 1985; Bui and Johnson, 1996; Tamagnini et al., 2002; Voncken et al., 2002; Boichenko et al., 2004). Among green algae, hydrogenases and/or H_2 production activity has been detected in *Chlamydomonas reinhardtii* (Happe et al., 2002a; Happe and Kaminski, 2002; Forestier et al., 2003), *Chlamydomonas moewusii* (Winter et al., 2004), *Chlorococcum littorale* (Ueno et al., 1999), *Scenedesmus obliquus* (Gaffron and Rubin, 1942; Florin et al., 2001; Wunschiers et al., 2001), *Chlorella fusca* (Winkler et al., 2002), *Chlorella vulgaris* (Pow and Krasna, 1979), *Platymonas subcordiformis* (Guan et al., 2004), *Tetraselmis kochinensis* (Bhosale et al., 2007), and many others (Brand et al., 1989; Boichenko and Hoffman, 1994).

Hydrogen-Producing Enzymes

The biological catalysts involved in H_2 metabolism are either nitrogenases or hydrogenases. As discussed below, nitrogenases can produce H_2 under N_2-fixing and nonfixing conditions (Weaver et al., 1980; Houchins and Hind, 1984; Rey et al., 2007). Most N_2-fixing microorganisms (heliobacteria, photosynthetic bacteria, and some cyanobacteria) also express two physiologically different hydrogenases: an uptake enzyme (which usually oxidizes the H_2 produced by the nitrogenase, thus conserving energy) and a reversible, H_2-producing enzyme (Tamagnini et al., 2002).

Three main types of hydrogenases have been described in the literature: [NiFe], [FeFe], and FeS-cluster-free (Vignais et al., 2001). Interestingly, the three types of hydrogenases are almost completely segregated within specific groups of organisms, suggesting convergent evolution (Vignais et al., 2001; Ludwig et al., 2006). [NiFe]-hydrogenase enzymes are present in most facultative and photosynthetic eubacteria, cyanobacteria, and archaea (Wu and Mandrand, 1993; Albracht, 1994). [FeFe]-hydrogenases, on the other hand, can be found in anaerobic eubacteria, algae, fungi, and protists (Happe et al., 2002). The FeS-cluster-free Hmd enzyme (Hartmann et al., 1996; Pilak et al., 2006) found in methanogens, although catalyzing H_2 oxidation, is primarily a methylene-tetrahydromethano-pterin dehydrogenase (Lyon et al., 2004). A few organisms have been reported to express both [NiFe]- and [FeFe]-hydrogenases, including *Desulfovibrio* sp. (Fauque et al., 1988), *Thermoanaerobacter tengcongensis* (Soboh et al., 2004), and some clostridial species (Vignais et al., 2001).

Oxygenic Photosynthesis

Plant-type photosynthesis occurs on thylakoid membranes located within cyanobacteria and the chloroplasts of algae. The process involves three primary steps (Blankenship, 2002): (i) light absorption by chlorophyll, carotenoid, and/or phycobilin pigments associated with one of two photosystems (PSII or PSI), which harvest light and transfer energy to their respective reaction centers (RCs); (ii) light-induced charge separation at the RCs, where oxidants and reductants are generated; and (iii) electron transport from PSII to PSI (charge equilibration between the photosystems) through a chain of electron carriers, including plastoquinones (PQ) that couple sequential oxidation-reduction reactions to proton translocation across the thylakoid membrane. The charge-separated state generated by the PSII RC (1.8 V) is stabilized on the oxidizing side by electron donation from water through the oxygen-evolving complex, and that generated by the PSI RC (1.5 V) is stabilized on the reducing side by electron transfer to a [2Fe-2S]-containing ferredoxin (Fd). Reduced Fd is the physiological electron donor to $NADP^+$, and it is responsible for the accumulation of NADPH necessary for CO_2 fixation by the Benson-Calvin cycle through Rubisco (the first enzyme of the Calvin cycle). The other requisite for CO_2 fixation, ATP, is generated upon dissipation of the proton gradient generated by electron transport from PSII to PSI through the chloroplast ATPase enzyme.

Pathways to H_2 Production

Two distinct H_2 photoproduction pathways have been described in green algae, and there is evidence for a third, light-independent, fermentative H_2 pathway coupled to starch degradation (Gfeller and Gibbs, 1984; Kreuzberg, 1984; Ohta et al., 1987; Happe et al., 2002; Happe and Kaminski, 2002; Kosourov et al., 2003). The pathways have in common the reduction of Fd as the primary electron donor to hydrogenase. Rather than utilizing light-driven reduction of Fd, the fermentative pathway may involve a pyruvate-ferredoxin-oxidoreductase (PFOR) enzyme, similar to those found in many anaerobic systems (Gray and Gest, 1965; Ragsdale, 2003). Although PFOR-catalyzed pyruvate oxidation/Fd reduction in the *C. reinhardtii* fermentative H_2 production pathway is not proven, a PFOR gene is up-regulated in *C. reinhardtii* (Mus et al., 2007) under dark, H_2-producing conditions (Gfeller and Gibbs, 1984). This suggests that PFOR might provide a link between fermentative carbon dissimilation and H_2 production.

The two H_2 photoproduction pathways are dependent, respectively, on (i) both PSII and PSI activities and (ii) both NADP-PQ oxidoreductase (NPQR) and PSI activities (Cournac et al., 2002; Boichenko et al., 2004; Ghirardi et al., 2005; Kruse et al., 2005a). In the first of these pathways, water oxidation by PSII supplies reductants (electrons) for H_2 production through PSI and Fd (so-called direct biophotolysis). In the second pathway, NPQR transfers reductants released by the glycolytic degradation of glucose or other organic compounds in the chloroplast to the photosynthetic electron transport chain at the level of PQ. Upon illumination, the reductants are reenergized at the level of PSI and reduce hydrogenases through Fd. The relative contribution of the two pathways to overall algal H_2 photoproduction varies according to growth conditions (Laurinavichene et al., 2004) and the particular green algal species being examined. For the PSI-only route, it is around 20% in *C. reinhardtii* (Antal et al., 2003), 10% in *S. obliquus* (Randt and Senger, 1985), and as high as 73% in *C. moewusii* (Healey, 1970).

It is interesting that these pathways differ significantly in cyanobacteria, which photoproduce H_2 through an [NiFe]-hydrogenase (Appel and Schulz, 1998). Here, the direct electron donor to hydrogenase is NADPH, which is generated by electron transport from reduced Fd to $NADP^+$ and catalyzed by the Fd-NADP oxidoreductase (FNR). A second pathway, similar to the algal PSII-independent pathway described above, also exists in cyanobacteria, and the connection to the PQ pool occurs through the multisubunit, respiratory Complex I-like NDH1, which also binds the hydrogenase (Appel and Schulz, 1998; Cournac et al., 2000, 2002).

The physiological role of H_2 production in green algae is still debatable. Appel and Schulz (Appel and Schulz, 1998) proposed that hydrogenases serve as valves, helping to release the high reductant pressure that accumulates when the CO_2 fixation pathway is not functional in the light. According to their hypothesis, hydrogenases help to properly set the redox potential within the chloroplast. It was further suggested that H_2 production under anaerobic conditions, when mitochondrial electron transport is not operational, allows concomitant ATP production through photosynthesis to maintain a minimum basal metabolism (Melis et al., 2004).

Competition for Reductant at the Level of Ferredoxin

The link between photosynthesis and hydrogenase in green algae occurs at the level of reduced Fd. Fd-like proteins are present in all photosynthetic cells and play a key role in the energy transfer mechanisms of photosynthesis. As shown in Fig. 2, reduced Fd is the primary electron source for proton reduction by hydrogenases by ferrying electrons from PSI (Knaff, 1996) to the hydrogenase (Roessler and Lien, 1984b; Happe and Naber, 1993; Horner and Walinsky, 2002) as well as to numerous competing assimilatory pathways. These pathways include CO_2 fixation, nitrite reduction, glutamate synthesis, sulfite reduction, cyclic electron transport around PSI, and reduction of thioredoxin for the regulation of biosynthetic pathways. In certain physiological states, these processes compete with hydrogenase for the available pool of reduced Fd. Thus, it is of prime importance to characterize in detail the kinetics of Fd interactions with the other pathways under H_2-producing conditions in order to maximize the H_2-production yield. The primary competitor with hydrogenase for low-potential electrons under aerobic photosynthetic conditions is FNR, the protein responsible for redox coupling to CO_2 fixation (Fig. 2). For example, the Michaelis-Menten constant (K_m) of reduced Fd for each of these enzymes varies between 0.4 μM for FNR (Kurisu et al., 2005) and 25 μM in the case of sulfite reductase (Saitoh et al., 2006) and has been estimated to be 10 to 35 μM in the case of green algal hydrogenases (Roessler and Lien, 1984b; Happe and Naber, 1993; King et al., 2006a). These estimates suggest that the interaction between Fd and hydrogenase may have to be modified in order to ensure that most of the photosynthetically generated reductants will be directed to H_2 production.

In nonphotosynthetic organisms such as *Clostridium* sp., Fd was shown to be an intermediate in the nitrite and hydroxylamine reductase pathways (Valentine and Wolfe, 1963). Although nitrite reductase is well described in *C. reinhardtii* and, when up-regulated, significantly suppresses H_2 production (Aparicio et al., 1985), it is not yet known whether hydroxylamine reductase activity is present in this alga. However, the gene encoding this enzyme is one of the most up-regulated transcripts in *C. reinhardtii* during anoxia (Mus et al., 2007), suggesting that it may indeed compete with the hydrogenase pathway.

Energetics/Efficiencies of H_2 Production

H_2 production coupled to photosynthetic water oxidation can be described as follows:

$$2H_2O + 8\ \text{photons} \rightarrow 4H^+ + O_2 + 4e^- \quad \text{(photosynthesis)}\ \Delta G = 1.26\ \text{V} \quad (1)$$

$$4H^+ + 4e^- \rightarrow 2H_2\ (H_2\ \text{production})\ \Delta G = -0.03\ \text{V} \quad (2)$$

The light reactions of photosynthesis (equation 1) generate a strong oxidant (0.81 V) at PSII ($P680^+$) that is able to extract electrons from water and generate O_2 gas as a by-product. The reductant generated at PSI re-

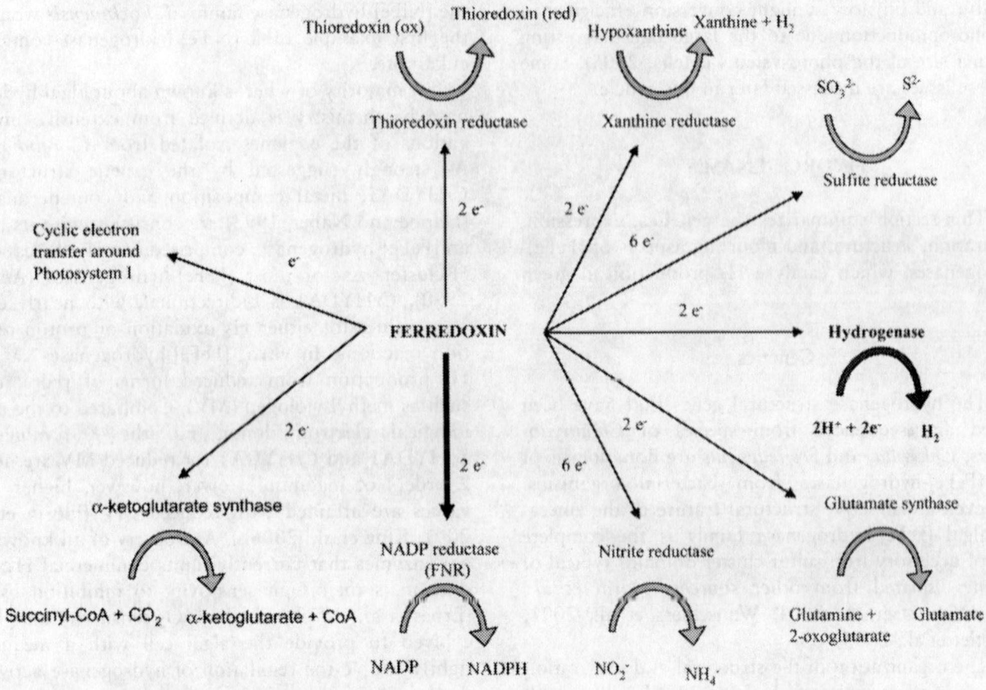

Figure 2. Enzymes that depend on electron transfer from reduced Fd in *C. reinhardtii*. The broad arrow from Fd to FNR represents the preferential flux of electrons under normal physiological conditions. Electrons are transported to Fd directly from PSI. This figure was provided by A. Dubini.

duces Fd (−0.45 V), accounting for 1.26 V of the overall reaction. The catalytic reduction of protons by reduced Fd (equation 2) is an almost isoelectric reaction ($\Delta G = -0.03$ V), since the redox potential of hydrogenases is −0.42 V. Hence, the overall process of H_2 photoproduction from water is associated with a ΔG of 1.23 V. The Gibbs free energy of H_2 is equivalent to 237×10^3 J/mol (Boichenko et al., 2004). Since the production of 2 mol of H_2 requires 8 mol of absorbed photons, with an average energy of 0.214 J/μE at 560 nm, the estimated maximum light conversion efficiency is about 28% (Boichenko et al., 2004). This estimate does not take into account the fact that green algae can absorb only about 43% of the incident light (Blankenship et al., 2002), which will lower the maximum incident light conversion efficiency to about 10 to 13% (Prince and Kheshgi, 2005; Kruse et al., 2005b; Ghirardi, 2006). The latter value represents the maximum potential percentage of the incident light energy that can be converted into H_2 energy by green algae. This value as it relates to the rate of H_2 photoproduction is important for estimating potential areas and costs of photobioreactor systems necessary to produce H_2.

Issues Affecting H_2 Production

Although promising, photobiological H_2 production systems are still far from commercial application (Melis and Happe, 2001; Levin et al., 2004; Kruse et al., 2005b; Ghirardi et al., 2007), and continued R&D will be required to solve several biochemical and engineering challenges. The most basic biochemical challenge is the inactivation of hydrogenase gene expression and activity by O_2, a by-product of oxygenic photosynthesis. Oxygen acts as a transcriptional repressor, an inhibitor of hydrogenase assembly, and an irreversible inhibitor of hydrogenase catalytic activity at the enzyme level in most but not all hydrogenases (see below).

Other biochemical issues that are known to limit algal H_2 production are (i) the existence of pathways that compete with the hydrogenase for photosynthetic reductant from Fd, such as cyclic electron transfer under anaerobic conditions and CO_2 fixation under aerobic conditions (discussed above); (ii) down-regulation of photosynthetic electron transport under conditions where the proton gradient is not dissipated (Greenbaum et al., 1995; de Vitry et al., 2004; Kruse et al.,

2005b); and (iii) low sunlight conversion efficiency of H_2 photoproduction due to the large light-harvesting antenna size of the photosystems (Melis, 2005). Some of these issues are discussed later in this article.

HYDROGENASES

This section summarizes the genetics, expression, maturation, structure, and modeling aspects of [FeFe]-hydrogenases, which catalyze H_2 production in green algae.

Genetics

The hydrogenase structural genes that have been cloned and sequenced from species of *Chlamydomonas*, *Chlorella*, and *Scenedesmus* are homologues of the [FeFe]-hydrogenases from bacterial organisms. However, a distinctive structural feature of the emerging algal [FeFe]-hydrogenase family is the complete lack of accessory iron-sulfur-cluster domains typical of enzymes isolated from other sources (Florin et al., 2001; Vignais et al., 2001; Wunschiers et al., 2001; Forestier et al., 2003).

The organization of the structural and maturation (assembly) genes in green algal genomes has been well summarized in recent reviews (Happe et al., 2002; Ghirardi et al., 2007). Unlike the organization of the [FeFe]-hydrogenase structural and maturation (*HYDE*, *HYDF*, and *HYDG*, see below) genes in some bacterial genomes, they exist as single genes in the *C. reinhardtii* genome, controlled by individual promoters (Happe et al., 2002; Posewitz et al., 2004a). So far this arrangement appears to be conserved for genes isolated from algal species. A distinctive feature of the *C. reinhardtii* maturation gene structure, which may be a common trait of algal genomes, is the fusion of *HYDE* and *HYDF* into a single gene, *HYDEF*. In all the other sequenced genomes that possess *HYDE* and *HYDF* homologues, the genes encode two distinct proteins.

Monomeric forms of [FeFe]-hydrogenases have been purified from *Chlamydomonas reinhardtii* (*Cr*HYDA1), *Chlamydomonas moewusii* (*Cm*HYDA), *Scenedesmus obliquus* (*So*HYDA1 and *So*HYDA2) and *Chlorella fusca* (*Cf*HYDA) (Roessler and Lien, 1984b; Happe and Naber, 1993; Florin et al., 2001; Wunschiers et al., 2001; Winkler et al., 2002). A monomeric hydrogenase has also been purified from the marine alga *Chlorococcum littorale* (*Cl*HYDA) (Ueno et al., 1999) and shown to catalyze H_2 evolution with the Fd isolated from the same organism. A two-subunit hydrogenase was isolated recently from *Tetraselmis kochinensis*, a species of unicellular marine alga (Bhosale et al., 2007). Provided that the protein sequence is homologous to the [NiFe]-hydrogenase family, *T. kochinensis* would be the first example of a [NiFe]-hydrogenase-containing eukaryote.

A majority of what is known about algal hydrogenase biochemistry is derived from extensive investigations of the enzymes isolated from *C. reinhardtii*. As strongly suggested by the genetic structure of *Cr*HYDA1, metal composition and content analysis (Happe and Naber, 1993) are consistent with its being an [FeFe]-hydrogenase, comprised of only the catalytic H-cluster. As for most [FeFe]-hydrogenases (Adams, 1990), *Cr*HYDA1 is bidirectional, with nearly equal V_{max} values for either H_2 oxidation or proton reduction reactions. In vitro, [FeFe]-hydrogenases catalyze H_2 production from reduced forms of redox dyes, such as methyl viologen (MV). Compared to the physiological electron donor, Fd, the K_m values of *So*HYDA1 and *Cr*HYDA1 for reduced MV are nearly 2 orders of magnitude lower; however, higher V_{max} values are attained with reduced MV (Florin et al., 2001; King et al., 2006b). A property of all known algal enzymes that currently limits commercial H_2 production is their high sensitivity to inhibition by O_2 (Erbes et al., 1979). The high O_2 sensitivity may have evolved to provide the algal cell with a means to tightly couple the regulation of hydrogenase activities to photosynthetic capacities.

Regulation of Gene Expression

As mentioned above, O_2 is a potent inactivator of [FeFe]-hydrogenases. To induce and sustain H_2 photoproduction, green algae must be cultured under conditions that maintain anaerobiosis (e.g., sulfur limitation, as discussed below). Detection of [FeFe]-hydrogenase gene expression and enzyme activities are possible only in anaerobically adapted cells. From the results of expression studies with *C. reinhardtii* mutants, it appears that both O_2 level and redox status control hydrogenase transcription, which is complex. In fact O_2 might even control transcription through effects on the levels of cellular factors (e.g., metabolites or redox mediators) that respond to changes in O_2 availability (Posewitz et al., 2004a; Ghirardi et al., 2007). Transcription of the structural and maturation genes is induced in cells at the transition from aerobic to anaerobic conditions, reaching steady-state levels under anoxia. Transcript levels of both gene sets correlate well with hydrogenase activities, suggesting that [FeFe]-hydrogenase expression and maturation are tightly coupled (Posewitz et al., 2004a). In agreement with Northern blot results, *Cr*HydA1 promoter fusions to the arylsulfatase gene expressed arylsulfatase activities only in cells that were cultured under anoxic conditions (Stirnberg and Happe, 2004). This approach was used to show that a minimal *Cr*HydA1 promoter required for anaerobic induction mapped to the region between −128 and +44 relative to the transcription start site.

In green algae, the active form of [FeFe]-hydrogenases ultimately localizes to the chloroplast stroma, the site of H_2 metabolism. Both hydrogenase structural and maturation genes, however, are nuclear encoded (Posewitz et al., 2004a). The *Cr*HYDA1, *So*HYDA1, and *Cf*HYDA1 enzymes are processed through recognition and cleavage of an N-terminal transit peptide (Happe and Naber, 1993; Happe et al., 1994; Florin et al., 2001; Winkler et al., 2002). The N-terminal peptide is missing from purified *Cr*HYDA1 (Happe et al., 1994), strong evidence that N-terminal processing functions in targeting *Cr*HYDA1 to the chloroplast. A precursor form of the hydrogenase, one that lacks an H-cluster or retains a transit peptide, has not been isolated from whole cells. Furthermore, it has not been determined in which compartment of the algal cell that hydrogenase maturation occurs (i.e., whether the apoproteins or the mature enzymes are recognized and translocated to the chloroplast stromal compartment). In the pathogenic protist *Trichomonas vaginalis*, [FeFe]-hydrogenases and maturases colocalize in the hydrogenosomes, the site of H_2 metabolism. Thus, in this eukaryote, both enzyme maturation and H_2 metabolism occur in a single compartment (Putz et al., 2006). For additional information on these topics see reviews by Horner (2002), Happe et al. (2002), and Ghirardi et al. (2007).

Enzyme Maturation and Domain Structure

In almost every instance, microbes that possess genes or predicted genes that encode proteins with signature [FeFe]-hydrogenase H-cluster motifs also contain HYDE, HYDF, and HYDG homologues (Posewitz et al., 2004a; Meyer, 2007). Among these organisms are several that are known to biosynthesize [FeFe]-hydrogenases, including *Clostridium acetobutylicum*, *Clostridium perfringens*, *Desulfovibrio desulfuricans*, *Desulfovibrio vulgaris*, *Thermotoga maritima*, and *T. tengcongensis*, among others (Vignais et al., 2001; Bock et al., 2006; Meyer, 2007). In the genome, HYD structural and maturation genes are organized into either a single (e.g., *C. acetobutylicum*), fusion (e.g., *C. reinhardtii*), or operon (e.g., *D. desulfuricans*) arrangement. An implication of single and fusion gene arrangements is that coregulated expression is required under appropriate metabolic conditions for induction of hydrogenase activity in cells. In microbial species that contain more than one [FeFe]-hydrogenase (e.g., *C. acetobutylicum* and *C. reinhardtii*), growth conditions can influence the relative activity levels of the enzymes (Gorwa et al., 1996). It remains to be determined how (e.g., gene expression or enzyme maturation) organisms mediate the response to regulate activity levels of the [FeFe]-hydrogenases.

Apohydrogenases undergo maturation by a unique and enzyme-specific set of maturases that biosynthesize the enzymes' catalytic metallocluster or H-cluster (Bock et al., 2006; Forzi and Sawers, 2007; Leach and Zamble, 2007). For [FeFe]-hydrogenases, as alluded to above, biosynthesis of the H-cluster requires the combined activities of HYDE, HYDF, and HYDG (Posewitz et al., 2004a; Rubach et al., 2005; Brazzolotto et al., 2006; King et al., 2006a). In *C. reinhardtii*, *HydEF* and *HydG* are coinduced with the structural genes in anaerobically adapted cells (Posewitz et al., 2004a), anaerobiosis being a prerequisite for enzyme maturation and activity.

HYDE and HYDG are radical *S*-adenosyl-L-methionine (Radical SAM) proteins, whereas HYDF is a GTPase (Posewitz et al., 2004a; Rubach et al., 2005; Bock et al., 2006; Brazzolotto et al., 2006). Radical SAM proteins participate in numerous biosynthetic pathways, including the production of sulfur-rich cofactors (e.g., LS, BioB) (Cicchillo and Booker, 2005; Jarrett, 2005a, 2005b), hemes (HemN) (Layer et al., 2004), and metalloclusters (NifB) (Curatti et al., 2006), and can also function in the activation of enzymes (pyruvate formate lyase activase) (Frey and Magnusson, 2003; Nicolet and Drennan, 2004) as reviewed previously (Cheek and Broderick, 2001; Sofia et al., 2001). The mechanism for Radical SAM-dependent reactions involves the reductive cleavage of SAM, in the presence of a protein-specific substrate, to generate a 5′-deoxyadenosyl radical intermediate. The radical intermediate then facilitates H-bond rearrangement on the protein-specific substrate. The identification of HYDE, HYDF, and HYDG (Posewitz et al., 2004a) has prompted intense efforts by several groups to characterize their substrates and reaction pathways in H-cluster biosynthesis required for [FeFe]-hydrogenase maturation. For a review of Radical SAM protein structures, reaction mechanisms, and pathways see Frey and Magnusson, 2003; Jarrett, 2005a, 2005b; and Fontecave, 2006. The latest research progress and state of understanding of the maturation process can be found in Peters et al. (2006), King et al. (2006a), and McGlynn et al. (2007), and are summarized in reviews by Bock et al. (2006), and Leach and Zamble (2007).

A majority of the [FeFe]-hydrogenase genes and proteins so far isolated exhibit complex structures that are organized into modular domains (Vignais et al., 2001; Horner et al., 2002; Meyer, 2007). Accessory domains contain the iron-sulfur clusters that function to electronically wire the buried catalytic sites to physiological electron carriers. Full-length and accessory-domain-only forms of [FeFe]-hydrogenases biosynthesized in the absence of the H-cluster maturases contain functional iron-sulfur clusters (Filipiak et al., 1989; Pierik et al., 1992; Atta et al., 1998; Kummerle et al., 1999), evidence that accessory-cluster maturation proceeds independent of the catalytic domain. As mentioned above, algal enzymes lack an accessory-cluster domain, yet undergo maturation by maturases homologous to those found in other hosts with more-complex enzymes (Posewitz et al., 2004a; King et al., 2006a). Moreover, truncated derivatives of the *C. acetobutylicum* [FeFe]-hydrogenase I consisting of only the catalytic domain undergo biosynthesis into active hydrogenases (King et al., 2006a). These two lines of evidence show that maturation of accessory domains and catalytic domains proceed by independent pathways. Altogether these observations support the hypothesis that structurally minimized algal enzymes might have evolved from a more structurally complex progenitor enzyme through the loss of the accessory domain (Vignais et al., 2001; Meyer, 2007). A caveat to the algal structure is that the lack of an accessory iron-sulfur cluster domain requires that reduced Fd transfer electrons directly to the H-cluster. Thus, the H-cluster in algal enzymes is likely to be more solvent accessible, or less buried, than the more-complex forms of [FeFe]-hydrogenases found in other organisms.

Catalytic Site Structure

All [FeFe]-hydrogenases consist of a highly conserved catalytic domain composed of three motifs, L1Fe, L2Fe, and L3Fe (Vignais et al., 2001; Meyer, 2007), that coordinate the H-cluster and form the catalytic site. The residues within these motifs and their function in H-cluster ligation have been elucidated from the X-ray crystal structures of bacterial [FeFe]-hydrogenases, CpI (*Clostridium pasteurianum* [FeFe]-hydrogenase I) and DdH (*D. desulfuricans* [FeFe]-hydrogenase), resolved to 1.8 and 1.6 Å, respectively (Peters et al., 1998; Nicolet et al., 1999, 2000). As predicted by extensive biochemical, analytical, and spectroscopic studies on purified [FeFe]-hydrogenases (Bennett et al., 2000; Albracht et al., 2006; Roseboom et al., 2006) and literature reviews (Adams, 1990; Evans and Pickett, 2003), the catalytic site structures of both enzymes possess the unique H-cluster comprised of a 4Fe-center bridged to an unusual 2Fe-center. The 4Fe-center is a cubane [4Fe4S] cluster, which is coordinated by three cysteines, one from each of the L1Fe, L2Fe, and L3Fe motifs. The fourth cysteine in the L3Fe motif bridges the two Fe centers and functions as the only protein ligand to the 2Fe center. Structurally the 2Fe center is a [2Fe2S] cluster with unusual Fe ligands, some of which are common to Fe ligation of the [NiFe]-hydrogenase catalytic site. Infrared (IR) and X-ray structural investigations of [NiFe]-hydrogenases first led to the discovery that the Fe atom possesses cyanyl and carbonyl ligands. A comparison of the IR spectra of [FeFe]-hydrogenase and [NiFe]-hydrogenase was the first hint that Fe ligations in the two enzymes were similar, with both exhibiting similar Fe-cyanyl and Fe-carbonyl signals (van der Spek et al., 1996; Pierik et al., 1998). Confirmation of cyanyl and carbonyl ligation of the [FeFe]-hydrogenase H-cluster came from additional IR studies (Pierik et al., 1998) combined with the announcement of the CpI and DdH X-ray structures (Peters et al., 1998; Nicolet et al., 2000). Each of the Fe atoms possesses a terminal asymmetric cyanyl and carbonyl tandem, with an additional carbonyl bridging the 2Fe center (Nicolet et al., 2000). A proposed function for these unusual ligands is the stabilization of Fe in a low-spin state (Bagley et al., 1994, 1995; Volbeda et al., 1995). The dithiolate bridge of the H-cluster 2Fe center contains an unknown organic ligand, tentatively assigned as either dithiopropane or di-(thiomethyl)amine (Peters, 1999; Nicolet et al., 2001; Fan, 2001a). Theoretical studies of 2Fe center model structures have been used to argue for the di-(thiomethyl)amine assignment, where an amine near the open coordination site of the distal Fe atom could participate in an acid-base mechanism of H_2 catalysis (Fan and Hall, 2001). In summary, an organized series of novel biochemical reactions must occur to synthesize the H-cluster including cyanyl, carbonyl, and organic thiolate ligation of the 2Fe center; ligation of the 2Fe and 4Fe centers; and finally protein coordination to achieve complete [FeFe]-hydrogenase maturation.

Molecular Dynamics Studies

Results obtained from molecular dynamics simulations and solvent accessibility mapping on molecular dynamics models of [FeFe]-hydrogenase CpI suggest that there are two pathways by which O_2 is able to reach the catalytic site (Cohen et al., 2005a). One of these pathways was identified by previous X-ray structural studies of another bacterial [FeFe]-hydrogenase (Nicolet et al., 1999). Moreover, bacterial and algal [FeFe]-hydrogenases with highly homologous catalytic domains show strikingly large variations in O_2 sensitivities (Cohen et al., 2005b). These theoretical and experimental studies have led to the hypothesis that the kinetics of O_2 diffusion, and therefore enzyme inactivation in algal enzymes, might be amenable to modification through the molecular engineering of the two gas pathways. Experimental investigations on the molecular engineering of O_2 accessibility in [FeFe]-hydrogenase are currently under way (King et al., 2006b).

Oxygen Tolerance and Photolytic H_2 Production

Finally, once a functional enzyme is assembled, it will produce H_2 only in the absence of O_2. The I_{50} for O_2 inactivation has been shown to be around 0.3% (Ghirardi et al., 2007), and exposure of recombinant [FeFe]-hydrogenases from *C. reinhardtii* to atmospheric levels of O_2 inactivates 50% of algal enzyme activities in less than 1 s (King et al., 2006b). It is important to realize, though, that some organisms have evolved mechanisms for the protection of [FeFe]-hydrogenases from inactivation by O_2. For instance, the *C. acetobutylicum* and *C. pasteurianum* enzymes are half-inactivated in, respectively, 120 to 300 and 415 seconds (King et al., 2006b). Additionally, the *D. desulfuricans* and *D. vulgaris* [FeFe]-hydrogenases have the unique property of adopting an O_2-protected, catalytically inactive state under anaerobic, oxidizing conditions (Van Dijk et al., 1983; Vincent and Armstrong, 2005). Moreover, physiological investigations of H_2 production by *Thermotoga neapolitana* and other related heterotrophic hyperthermophiles show that they can induce and sustain H_2 production in the presence of up to 8% O_2 in the culture headspace (Van Ooteghem et al., 2002, 2004). The O_2 levels that were tolerated by H_2 production in each species of *Thermotogales* varied, but they revealed this to be a property common to representative species of marine vent microbial communities (Van Ooteghem et al., 2004). Interestingly, the same level of O_2 tolerance was not observed in *T. maritima* (Van Ooteghem et al., 2004), which has a heterotrimeric [FeFe]-hydrogenase with 90 to 95% identity to the putative heterotrimeric [FeFe]-hydrogenase in *T. neapolitana*. Studies are under way to determine the unique mechanism of O_2 protection/tolerance of [FeFe]-hydrogenase in *T. neapolitana*, and how this might relate to improving O_2 tolerance of algal hydrogenases.

GENOMICS RESEARCH

All areas of biological research are being transformed by the appearance of powerful new molecular tools and the current availability of over 750 sequenced genomes. Included are sequences for phototrophic H_2-producing organisms including *Nostoc* (*Anaebana*) sp. PCC 7120, *Rhodobacter sphaeroides* ATCC 17029, *Rhodopseudomonas palustris* CGA009, *Rubrivivax gelatinosus* PM1, *Synechococcus* sp. CC9311, and *Synochocystis* sp. PCC6803, and bacterial H_2 producers *C. acetobutylicum* ATCC824D and *Escherichia coli* (http://www.genomesonline.org/). Moreover, the nearly complete *C. reinhardtii* genome sequence has provided the groundwork for experimental examination of the transcriptome, proteome, and metabolome, as well as for computational systems biology efforts, in support of algal H_2 production R&D. Seven other algal sequences are also available.

Known Genes Involved in H_2 Metabolism from a Historical Perspective

Up until the time that the first version of the *Chlamydomonas* genome was released, only eight algal genes were known to have a specific effect on H_2 photoconversion. These included the *HYDA1* and *HYDA2* [FeFe]-hydrogenase structural genes (Happe et al., 2002; Happe and Kaminski, 2002; Forestier et al., 2003), the *HYDEF* and *HYDG* hydrogenase maturation genes (Posewitz et al., 2004a), the *STA7* isoamylase gene associated with crystalline starch accumulation (Posewitz et al., 2004b), the *LHC* (Polle et al., 2000, 2001) and *TLA* (Polle et al., 2003) genes associated with the regulation of light-harvesting complexes, and the *SULP* gene (Chen et al., 2003, 2005), which controls sulfate uptake into the algal chloroplast. Of course there are homologues to these genes in many other organisms (Florin et al., 2001; Winkler et al., 2002), and the reader is referred to a recent review that provides additional information (Melis et al., 2004). It is now known that the *Chlamydomonas* genome (http://genome.jgi-psf.org/Chlre3/Chlre3.home.html; ca. 120 Mb) probably contains over 15,000 genes, with a significant number of them encoding polypeptides involved in anaerobic metabolism (Grossman et al., 2007).

Global Expression Profiling

Preliminary global gene expression studies were done with *C. reinhardtii* cultures that were sulfur deprived for up to 24 h (Zhang et al., 2004). Although useful insights were obtained regarding the early response mechanisms activated upon sulfate stress, such as increased sulfate assimilation capacity, cellular restructuring, and modulation of cell processes (Zhang et al., 2004; Pollock et al., 2005), these studies did not examine the role of long-term adaptation strategies such as H_2 production. As we will see later, anoxia required for hydrogenase induction can be achieved in the light by depriving illuminated, sealed cultures of sulfate if the incubation is long enough (Melis et al., 2000; Kosourov et al., 2002, 2003; Zhang et al., 2002; Zhang and Melis, 2002). In a more recent study, it became clear that fermentation pathways are active in the alga as the cells become anaerobic (Mus et al., 2007). Anaerobiosis was achieved in the dark by sparging cultures with an inert gas to remove O_2 (Greenbaum et al., 1983; Gfeller and Gibbs, 1984; Ghirardi et al., 1997; Posewitz et al., 2004a). In the dark, fermentation is coupled to the degradation of starch reserves (Gfeller and Gibbs, 1984; Kreuzberg, 1984; Ohta et al., 1987).

Formate, acetate, and ethanol are formed as major fermentative products, and H_2 and CO_2 gases are emitted as minor products (Gfeller and Gibss, 1984; Kreuzberg, 1984; Ohta et al., 1987). These products are also formed in the light after exposure to sulfate-deprivation stress (Kosourov et al., 2003). The formation of fermentation products is metabolically controlled at the level of pyruvate, and the ratios of the fermentative products may change as a consequence of culture conditions or the use of different laboratory strains. Pyruvate formate lyase catalyzes the oxidation of pyruvate into formate and acetyl coenzyme A (acetyl-CoA) (Wagner et al., 1992). The acetyl-CoA generated can be reduced to ethanol via the acetaldehyde dehydrogenase/alcohol dehydrogenase pathway or to acetate and ATP via the phosphotransacetylase (PTA)/acetate kinase (ACK) pathway. *Chlamydomonas* also appears to use a pyruvate decarboxylase/alcohol dehydrogenase pathway to ethanol that is putatively active in the cytoplasm (Mus et al., 2007), particularly at low pH (Kosourov et al., 2003). Interestingly, there is one report (Atteia et al., 2006) that the *Chlamydomonas* genome contains two copies each of the *PTA* and *ACK* genes, and it suggested that the conversion of acetyl-CoA to acetate occurs in both the chloroplast and mitochondrion.

In *Chlamydomonas*, pyruvate may also be oxidized to acetyl-CoA and CO_2 in the chloroplast by the PFOR, which was recently identified in the *Chlamydomonas* genome (Hemschemeier and Happe, 2005; Atteia et al., 2006). In amitochondriate eukaryotes and anaerobic microbes such as species of *Clostridia*, PFOR is known to reduce flavodoxin and/or Fd, which can provide electrons to hydrogenase for H_2 generation. The pyruvate formate lyase and PFOR enzymes of *Chlamydomonas* are likely to act in concert to balance pyruvate oxidation into acetyl-CoA with the coproduction of formate or CO_2, respectively (Mus et al., 2007). The accumulation of ethanol and formate and the acidification of the cellular compartment due to uptake of organic acids can be toxic to the cells (Kennedy et al., 1992). Thus, dynamic responses in cellular metabolism to changing growth conditions are required to balance ATP production, while limiting the accumulation of toxic metabolites.

Genomic and transcriptomic (microarray) data can provide genome-wide insights into the regulation of the complex metabolic networks utilized by *C. reinhardtii* under the anaerobic conditions associated with H_2 production. Recent work with second-generation *Chlamydomonas* microarrays confirmed that dark anoxia in the alga is characterized by the activation of an extensive set of fermentation pathways that provide cellular energy, while limiting the accumulation of potentially toxic fermentative products (Mus et al., 2007). The levels of 514 transcripts increased significantly during acclimation of the cells to dark, H_2-producing conditions. Of these transcripts, 145 encode proteins of known function, 209 encode conserved proteins of unknown function, and 160 encode putative proteins not previously identified. Elevated levels of transcripts encoding proteins associated with the production of H_2, organic acids, and ethanol were observed along with the accumulation of fermentation products. It appears that *Chlamydomonas* is capable of using a diverse set of anaerobic pathways that are unprecedented in a single organism (Mus et al., 2007). Transcripts encoding transcription/translation regulators, prolyl hydroxylases, hybrid cluster proteins, proteases, transhydrogenase, and catalase were also up-regulated. Of the 58 genes that were down-regulated under these same conditions (Mus et al., 2007), 10 encode proteins of known function, 20 encode conserved proteins of unknown function, and 28 encode putative proteins not previously identified. Of the 10 transcripts encoding proteins for which a function is known, all are putatively associated with signal transduction and metabolite transport. Curiously, the transcript level for a putative hybrid cluster protein (HCP4), also referred to as the prismane protein or hydroxylamine reductase, increases over 1,500-fold (real-time PCR results) in response to anaerobisis and is among the most significantly up-regulated genes detected under these conditions (Mus et al., 2005). The reason for this is unknown, but HCPs might oxidize reduced Fd and therefore compete with hydrogenase for electrons from reduced Fd (Valentine and Wolfe, 1963).

Metabolomics

Procedures for metabolite profiling of *C. reinhardtii* CC-125 cells, which quickly inactivate enzymatic activity, optimize extraction capacity, and are amenable to large sample sizes, were reported quite recently (Bölling and Fiehn, 2005). The study is particularly relevant to this review, because it explored profiles of Tris-acetate-phosphate-grown cells as well as cells that were deprived of sulfate under aerobic conditions. Nitrogen, phosphate, and iron deprivation profiles were also examined, and each metabolic profile was different. Sulfur depletion, as discussed extensively in "Hydrogen Production Processes," below, leads rapidly to the anaerobic conditions required for H_2 production in a sealed reactor, although the reader is cautioned that the 24-h incubation under sulfur-deprived conditions described in this paper might not have been long enough to induce hydrogenase synthesis and function (the conversion from aerobic to anoxic conditions can depend on the growth conditions). In any case, rapidly sampled cells (cell leakage controls were determined by

^{14}C-labeling techniques) were analyzed by gas chromatography coupled to time of flight mass spectrometry, and more than 100 metabolites (e.g., amino acids, carbohydrates, phosphorylated intermediates, nucleotides, and organic acids) of about 800 detected could be identified. Notably, a 50-fold increase in 4-hydroxyproline was reported (Bölling and Fiehn, 2005), consistent with significant up-regulation of a putative prolyl 4-hydroxylase (actually four prolyl hydroxylases are up-regulated in the CC-425 strain) (Mus et al., 2007). This is in contrast to an initial report that prolyl 4-hydroxylases are down-regulated under sulfur stress (Zhang et al., 2004), though first-generation *Chlamydomonas* microarrays were used in that study. Although proline hydroxylation occurs in algal cell wall proteins (Kaska et al., 1987) and 4-hydroxyproline is a prominent constituent of the glycoproteins in the cell wall (Voigt and Frank, 2003), it is unlikely that these gene products are involved in protein hydroxylation during anaerobiosis since prolyl hydroxylases require O_2 as a substrate (Hieta and Myllyharju, 2002). However, it was suggested that the 4-hydroxyproline results from enhanced cell wall degradation during sulfur deprivation (Bölling and Fiehn, 2005), where cell wall proteins are known to be rearranged (Takahashi et al., 2001). Finally, the concentration of a number of phosphorylated glycolysis intermediates increased significantly during sulfur stress (Bölling and Fiehn, 2005), consistent with the up-regulation of many genes associated with starch degradation and fermentation observed in anaerobic *Chlamydomonas* cells (Mus et al., 2007).

Proteomics

Chlamydomonas proteomics is in its infancy, but there have been a number of relevant studies as reviewed by Stauber and Hippler (2004). However, to our knowledge, there has not been any reported proteomics research in algae under H_2-producing conditions.

Global Pathway Modeling

A better understanding of anaerobic metabolism in *Chlamydomonas* and metabolic fluxes associated with diurnal periods of light and dark will facilitate the development of physiological models able to predict metabolic fluxes under various environmental conditions. However, at this point relatively little is known about (i) the regulatory mechanisms by which phototrophic microorganisms sense and acclimate to an anaerobic environment after periods of photosynthetic activity or to an aerobic environment after periods of anoxia, (ii) the complete repertoire of genes and proteins required during anaerobiosis and how changes in gene expression link to changing cellular metabolism, and (iii) changes in metabolite fluxes and the importance of these fluxes in sustaining cell viability during anoxia. Since *Chlamydomonas* can balance photosynthesis, aerobic respiration, and fermentation pathways concomitantly, at least under sulfur-deprivation conditions (Kosourov et al., 2003), it is likely that a number of the proteins that currently have no known function are involved in the regulation, metabolic partitioning, and/or function of *Chlamydomonas*' extraordinary repertoire of metabolic networks. One initial attempt to model pathways associated with biohydrogen production used a power-law sensitivity, S-system approach and correctly modeled several biochemical features of a sulfur-deprived algal culture (Horner and Wolinsky, 2002). One particularly interesting result was the placement of constraints on the amount of ATP that could be utilized by the organism in processes not involved directly in H_2 production (e.g., cell maintenance and repair; note that ATP is not used for H_2 production by sulfur-deprived algae) due to potential competition for protons. This constraint could potentially place limitations on any genetic reengineering efforts seeking to increase H_2 by the algae under anaerobic conditions. A more recent paper (Park and Moon, 2007) further demonstrated the application of S-state analysis to metabolism by studying H_2 production rates as a function of light intensity. Nevertheless, complete global models of overall cellular metabolism will have to account for the diversity of metabolic processes available in *C. reinhardtii*, as well as consider metabolite fluxes in multiple cellular compartments and cross talk between organelles (Chang et al., 2007).

HYDROGEN PRODUCTION PROCESSES

This section discusses the ways in which algae have been used to produce H_2 and some of the issues that have arisen. These, of course, must be addressed before algal H_2 production can be considered for practical purposes.

Early Studies on Green Algal H_2 Production

H_2 production activity in green algae is not observed under aerobic conditions, since O_2 is a positive suppressor of hydrogenase gene expression (Forestier et al., 2003; Stirnberg and Happe, 2004; Posewitz et al., 2004a) and a potent inhibitor of the enzyme (Erbes et al., 1979; Urbig et al., 1993). Historically, the hydrogenase pathway in green algae was induced following an obligatory anaerobic incubation of the cells in the dark followed by illumination to detect activity. Under such anaerobic conditions, the nucleus-encoded hydrogenase enzyme was expressed and targeted to the chloroplast, the photosynthetic organelle catalyzing light-mediated generation of H_2. At this point it is not known if the hydrogenase catalytic clusters are assembled in the cytosol (prior to translocation to the chloroplast) or in the chloroplast (after translocation). However, the presence of putative transit peptide sequences on the genes encoding for the maturation proteins supports the latter hypothesis. It was recognized that light absorption by the photosynthetic apparatus is essential for the generation of H_2, as light facilitates the endergonic transport of electrons in the photosynthetic apparatus via PSI to Fd. Such electrons, as discussed above, can be derived either from the oxidation of H_2O via PSII or upon the oxidation of endogenous substrate in the chloroplast (Ghirardi et al., 2000; Antal et al., 2001; Melis and Happe, 2001). In either case, reduced Fd is the electron donor to the green algal hydrogenase.

In the past, photosynthetic H_2 production activity could operate for only 30 to 90 s (Ghirardi et al., 1997), generating trace amounts of H_2 that required sophisticated instruments, such as a mass spectrometer or a Clark-type H_2 electrode to detect. The short lifetime of the process is due to the fact that O_2 is concomitantly and inevitably evolved as a by-product of photosynthetic water oxidation and electron transport in PSII of microalgae. This stringent incompatibility of the simultaneous photoevolution of O_2 by the photosynthetic apparatus and H_2 production by [FeFe]-hydrogenase has stymied development of a cost-effective, applied process for half a century.

Sulfur Nutrient Deprivation for H_2 Production

Removal of O_2 from the reaction mixture has proven to be necessary and sufficient for sustained photobiological H_2 production. For example, the addition of sodium diothionite, a powerful reductant, to the reaction mixture (Randt and Senger, 1985), or sparging of the reaction mixture with inert gases (e.g., argon or helium) (Greenbaum, 1982), which chemically or physically remove O_2, have been used to sustain the H_2 production process under conditions of continuous water oxidation and O_2 release by the photosynthetic apparatus.

A recent innovation utilized the cells' own respiratory capability to remove photosynthetically generated O_2. Sulfur nutrient deprivation in green algae causes reversible inhibition in the activity of photosynthesis (Wykoff et al., 1998). In the absence of sulfate, the rates of photosynthetic O_2 evolution fall below those of O_2 consumption by respiration (Melis et al., 2000). As a consequence, sealed cultures of the green alga *C. reinhardtii* become anaerobic in the light (Ghirardi et al., 2000), induce the hydrogenase pathway of electron transport, and continuously photoproduce H_2 gas in a sustained process that can last for several days. During H_2 gas production, the cells consume internal starch and catabolize internal protein in a highly regulated process in order to sustain mitochondrial respiration and required anaerobiosis (Ghirardi et al., 2000; Melis et al., 2000; Kosourov et al., 2002, 2003; Zhang et al., 2002). Such catabolic reactions sustain, directly and indirectly, the H_2 production process. Indeed, the pool of internal metabolites is pivotal for the duration and yield of the H_2 production process (Zhang et al., 2002; Posewitz et al., 2004b; Kruse et al., 2005b). It has been shown that the H_2 production activity of sulfur-deprived microalgae stops when the cells are depleted of internal metabolites, either soluble (Posewitz et al., 2004b) or nonsoluble (Zhang et al., 2002).

Profile analysis of selected photosynthetic proteins showed a preferential decline in the amount of Rubisco as a function of time in sulfur deprivation, a more gradual decline in the level of PSII and PSI proteins, and a change in the composition of the chlorophyll *a-b*-binding proteins (Zhang and Melis, 2002; Zhang et al., 2002). An increase in the level of the [FeFe]-hydrogenase enzyme was noted during the initial stages of sulfur deprivation (up to 72 h) followed by a decline in the level over longer periods of sulfur deprivation. These physiological measurements were augmented by microscopic observations, which showed distinct morphological changes in *C. reinhardtii* during S deprivation and H_2 production. Within 24 h of S deprivation, small, ellipsoid cells were converted into much larger, spherical cells, reflecting a 10-fold accumulation of starch. The latter originated in part from the recycling of carbon previously stored in Rubisco and in part from the residual Calvin cycle activity (White and Melis, 2006). During subsequent stages of sulfur deprivation and H_2 production, cell mass reductions were observed (24 to 120 h after sulfur deprivation), reflecting the gradual loss of endogenous starch (Zhang et al., 2002). This novel approach showed that sulfur nutrient deprivation of green algae could serve as a metabolic switch that triggers a reversible change in the metabolic flux of the cell. Compared to metabolism under sulfur-nutrient-replete conditions, it entailed a substantially altered interaction between oxygenic photosynthesis, mitochondrial respiration, catabolism of endogenous substrate, and electron transport via the hydrogenase pathway, thereby supporting light-mediated H_2 production (Ghirardi et al., 2000; Melis and Happe, 2001; Zhang et al., 2002).

Sulfur deprivation proved to be a critically successful tool in the sustained production of H_2 by green algae since, for the first time in 60 years of related research (Gaffron, 1939; Gaffron and Rubin, 1942; Melis et al., 2000), substantial amounts of pure H_2 gas could be produced continuously for about 4 days in the light and accumulated in suitable containers. These results and more recent studies (Posewitz et al., 2004b; Mus et al., 2007) suggested that genes and proteins of starch and fermentation metabolisms are relevant to the process of H_2 metabolism in unicellular green algae. Research in this direction has indicated a role of a chloroplast isoamylase enzyme (Posewitz et al., 2004b) and a chloroplast sulfate permease (Melis and Chen, 2005) in green algal H_2 metabolism. Surprisingly, not all algal species that produce H_2 under anaerobic conditions respond in the same manner to anaerobiosis induced by sulfur deprivation. For example, both *Scenedesmus obliquus* and *Scenedesmus vacuolatus* (currently known as *Chlorella fusca*) contain hydrogenases, whose activities are induced by exposure to argon gas. However, only the former photoproduced H_2 gas under sulfur deprivation (Winkler et al., 2002).

Light Utilization in H_2 Production

Photosynthesis and H_2 production in unicellular green algae can in principle operate with a nearly 100% absorbed photon utilization efficiency (Ley and Mauzeroll, 1982; Greenbaum, 1988), making them potentially efficient biocatalysts for the generation of H_2 from sunlight and water. However, under direct sunlight, as would be required in a commercial production process, green microalgal cultures show rather poor light utilization efficiency. The reason for this is that green algal photosynthesis normally saturates at about one-fifth of full sunlight intensity or less. Under bright sunlight, the rate of photon absorption by the Chl antenna arrays in PSII and PSI far exceeds the rate at which photosynthesis can utilize them. Excess photons cannot be stored in the photosynthetic apparatus but are dissipated (lost) as fluorescence or heat. Up to 80% of absorbed photons could thus be wasted (Melis et al., 1998), decreasing light utilization efficiency and compromising cellular productivity to levels far lower than the theoretical maximum. Thus, in a high-density mass culture, cells at the surface over absorb and waste sunlight, whereas cells deeper in the culture are deprived of much needed irradiance as light is strongly attenuated due to shading.

Green algae are increasingly finding a variety of applications in the areas of biomass accumulation, carbon sequestration, and energy production. The commercial use of microalgae will require utilization of strains that (i) are not subject to the above-mentioned optical pitfall and suboptimal utilization of sunlight in mass culture and (ii) operate with maximal light utilization efficiency and photosynthetic productivity under bright sunlight. To attain these performance characteristics, it is necessary to minimize the absorption of sunlight by individual cells so as to permit greater transmittance and utilization of irradiance through high-density green alga mass culture. This requirement was recognized long ago (Kok, 1953; Myers, 1957; Radmer and Kok, 1977) but could not be satisfied because algae with a "truncated light-harvesting chlorophyll antenna size" are not encountered in nature. Until recently, this problem could not be addressed and solved in the laboratory either, due to the lack of necessary tools. The advent of molecular genetics in combination with sensitive absorbance-difference kinetic spectrophotometry for the precise measurement of the chlorophyll antenna size of the photosystems in green algae has now made it possible to reduce the number of photosynthetic chlorophyll antenna molecules per reaction center.

The rationale for attempting a bioengineering approach to truncate the chlorophyll antenna size in green algae is that such modification will prevent individual cells at the surface of a high-density culture from overabsorbing sunlight and wastefully dissipating most of the energy. A truncated Chl antenna size will permit sunlight to penetrate deeper into the culture, thus engaging many more cells to contribute to useful photosynthate production and resulting in greater culture productivity and H_2 production. It has been shown in the laboratory that a truncated Chl antenna size can facilitate a three- to fourfold greater light energy conversion efficiency and photosynthetic productivity than could be achieved with fully pigmented cells (Melis et al., 1998). Such a bioengineering effort, therefore, is a potential requirement for the practical exploitation of microalgae, as it might substantially enhance photosynthetic productivity in mass culture.

A systematic approach to this problem is to identify genes that regulate the Chl antenna size of photosynthesis (Melis, 2002; Melis, 2005) and, further, to manipulate such genes so as to confer a permanently truncated Chl antenna size to the model green alga, *C. reinhardtii* (Polle et al., 2003). Identification of such genes in *Chlamydomonas* will permit the subsequent transfer of this trait to other microalgae of interest to the alga biotechnology industry. This objective has been approached upon the application of DNA insertional mutagenesis techniques (Kindle, 1990; Gumpel and Purton, 1994), screening, biochemical/molecular/genetic, and absorbance-difference kinetic spectrophotometry analyses of *C. reinhardtii* cells. Bioengineered strains of green algae, acquired during the course of such effort, may find direct application in biomass accumulation (Vazquezduhalt, 1991; Brown and Zeiler, 1993; Westermeier and Gomez, 1996), H_2 production (Zaborsky, 1998; Melis et al., 2000; Zhang et al., 2002; Zhang and Melis, 2002), and carbon sequestration (Mulloney, 1993; Nakicenovic, 1993).

In support of this effort, *TLA1*, the first "Chl antenna size regulatory gene" was cloned (Tetali et al., 2007). The partially truncated chlorophyll antenna size of the *tla1* mutant alleviates the overabsorption of incident sunlight by individual cells and the wasteful dissipation of overabsorbed irradiance (Polle et al., 2003). Results from this work might apply directly to green algal biomass accumulation, carbon sequestration, and H_2 production efforts, although the latter has not yet been demonstrated.

Rate Limitations

The rate of electron transport in the thylakoid membrane of photosynthesis is of importance for defining yield and efficiency of the overall process. The current H_2 production rate from photosynthetic microorganisms is too low for commercial viability. Low rates have been attributed to (i) the nondissipation of a proton gradient across the photosynthetic membrane, which is established during electron transport from water to the hydrogenase under anaerobic conditions (Lee and Greenbaum, 2003) and (ii) the existence of competing metabolic flux pathways for reductant (Ghirardi, 2006) (Fig. 2). Genetic means to overcome the restricting metabolic pathways, such as the insertion of a proton channel across the thylakoid membrane and the regulation of electron partitioning among reductive pathways in the chloroplast, must be attempted to significantly increase the rate of H_2 production. Finally, under aerobic conditions, with an O_2-tolerant hydrogenase catalyzing H_2 production, the competition between CO_2 fixation and hydrogenase will also have to be addressed.

Process Improvements

Up until this point, more basic aspects of biohydrogen research and the sulfur deprivation process (the only working algal H_2 production system) have been emphasized along with the major research challenges. Here, we describe briefly some of the more applied work that has sought to improve H_2 photoproduction rates and efficiencies in working systems.

Work since the discovery of sulfur-deprivation-stimulated algal H_2 photoproduction (Ghirardi et al., 2000; Melis et al., 2000) has emphasized optimization of the batch system and development of a continuous H_2 production system. Some of the advances related to batch processes included methods to replace centrifugation as a means of depriving the cultures of sulfate (Laurinavichene et al., 2002), use synchronous algal cultures (Kosourov et al., 2002; Tsygankov et al., 2002), optimize the cell density and the amount of sulfate present after sulfur deprivation (Kosourov et al., 2002, 2005), optimize the pH (Kosourov et al., 2003), optimize the light intensity (Laurinavichene et al., 2004), eliminate the acetate requirement (Fouchard et al., 2005; Tsygankov et al., 2006; Kosourov et al., 2007), and immobilize the algal cultures (Laurinavichene et al., 2006, 2008) for improving H_2 production. Finally, a continuous H_2 photoproduction system has been reported (Fedorov et al., 2005), though the rates and yields were not as good as those of batch cultures. One recent exciting report was the report of a multiple-phenotype mutant of *C. reinhardtii* that simultaneously accumulates more starch, is inhibited in cyclic electron transfer, and has higher rates of respiration (Kruse et al., 2005b). This mutant, called *stm6*, should serve as the platform upon which further improvements could be made in the future.

Photobioreactors

While discussions of applied photobioreactor systems are somewhat premature due to the relatively small amounts of H_2 that can be produced at this point, the area has been examined to a limited extent (Boichenko et al., 2004). Nevertheless, researchers have started to think about the materials (Blake et al., 2007) and cost (Ghirardi and Amos, 2004) challenges that will have to be surmounted once a commercially viable organism has been developed.

RESEARCH NEEDS AND FUTURE DEVELOPMENTS

In this review we have attempted to give the reader a sense of the current state of knowledge in the area of algal photoproduction of H_2, with an emphasis on the issues that are being considered in the field. Research needs are still significant, but the potential advantages of direct conversion processes in terms of improved land utilization are substantial. The future will see many more advances commensurate with increased interest in the area around the world and many new researchers moving into the field. Among these will be great improvements in understanding basic issues about the O_2 tolerance of hydrogenases as the result of basic structure/function studies on the protein, enzyme engineering approaches, and directed-evolution efforts. A recent successful [FeFe]-hydrogenase gene-shuffling demonstration (Nagy et al., 2007) will certainly facilitate the latter approaches. Furthermore, the diversity of known H_2-producing phototrophic organisms will be greatly expanded by a number of niche environmental discovery efforts starting up in a number of laboratories.

Acknowledgments. We thank the entire biohydrogen team at NREL for their encouragement and discussions during the writing of this review. Special thanks are due to A. Dubini and K. Ibsen for specific contributions noted in the text.

The generous support of the U.S. Department of Energy Office of Science GTL (M.S.), BES (M.L.G. and M.S.), and Hydrogen (M.S., A.M., and M.G.L.) Programs, the NREL LDRD Program (M.S.), and the AFOSR (M.C.P.) is graciously acknowledged.

REFERENCES

Adams, M. W. 1990. The structure and mechanism of iron-hydrogenases. *Biochim. Biophys. Acta* **1020**:115–145.

Adams, M. W. W., L. E. Mortenson, and J. S. Chen. 1980. Hydrogenase. *Biochim. Biophys. Acta* **594**:105–176.

Albracht, S. P. 1994. Nickel hydrogenases: in search of the active site. *Biochim. Biophys. Acta* **1188**:167–204.

Albracht, S. P., W. Roseboom, and E. C. Hatchikian. 2006. The active site of the [FeFe]-hydrogenase from *Desulfovibrio desulfuricans*. I. Light sensitivity and magnetic hyperfine interactions as observed by electron paramagnetic resonance. *J. Biol. Inorg. Chem.* **11**:88–101.

Antal, T. K., T. E. Krendeleva, T. V. Laurinavichene, V. V. Makarova, A. A. Tsygankov, M. Seibert, and A. B. Rubin. 2001. The relationship between the photosystem 2 activity and hydrogen production in sulfur deprived *Chlamydomonas reinhardtii* cells. *Dokl. Biochem. Biophys.* **381**:371–374.

Antal, T. K., T. E. Krendeleva, T. V. Laurinavichene, V. V. Makarova, M. L. Ghirardi, A. B. Rubin, A. A. Tsygankov, and M. Seibert. 2003. The dependence of algal H_2 production on Photosystem II and O_2 consumption activities in sulfur-deprived *Chlamydomonas reinhardtii* cells. *Biochim. Biophys. Acta* **1607**:153–160.

Aparicio, P. J., M. P. Azuara, A. Ballesteros, and V. M. Fernandez. 1985. Effects of light-intensity and oxidized nitrogen-sources on hydrogen-production by *Chlamydomonas reinhardtii*. *Plant Physiol.* **78**:803–806.

Appel, J., and R. Schulz. 1998. Hydrogen metabolism in organisms with oxygenic photosynthesis: hydrogenases as important regulatory devices for a proper redox poising? *J. Photochem. Photobiol. B* **47**:1–11.

Atta, M., M. E. Lafferty, M. K. Johnson, J. Gaillard, and J. Meyer. 1998. Heterologous biosynthesis and characterization of the [2Fe-2S]-containing N-terminal domain of *Clostridium pasteurianum* hydrogenase. *Biochemistry* **37**:15974–15980.

Atteia, A., R. van Lis, G. Gelius-Dietrich, A. Adrait, J. Garin, J. Joyard, N. Rolland, and W. Martin. 2006. Pyruvate formate-lyase and a novel route of eukaryotic ATP synthesis in *Chlamydomonas* mitochondria. *J. Biol. Chem.* **281**:9909–9918.

Bagley, K. A., C. J. Van Garderen, M. Chen, E. C. Duin, S. P. Albracht, and W. H. Woodruff. 1994. Infrared studies on the interaction of carbon monoxide with divalent nickel in hydrogenase from *Chromatium vinosum*. *Biochemistry* **33**:9229–9236.

Bagley, K. A., E. C. Duin, W. Roseboom, S. P. Albracht, and W. H. Woodruff. 1995. Infrared-detectable groups sense changes in charge density on the nickel center in hydrogenase from *Chromatium vinosum*. *Biochemistry* **34**:5527–5535.

Bennett, B., B. J. Lemon, and J. W. Peters. 2000. Reversible carbon monoxide binding and inhibition at the active site of the Fe-only hydrogenase. *Biochemistry* **39**:7455–7460.

Bhosale, S. H., A. Pant, and M. I. Khan. 22 February 2007. Purification and characterization of putative alkaline [Ni-Fe] hydrogenase from unicellular marine green alga, *Tetraselmis kochinensis* NCIM 1605. *Microbiol. Res.* [Epub ahead of print.]

Blake, D. M., W. Amos, M. G. Ghirardi, and M. Seibert. 2007. Materials requirements for photobiological hydrogen production. *In* R. Jones and G. Thomas (ed.), *Materials for the Hydrogen Economy*, in press. CRC Press, Boca Raton, FL.

Blankenship, R. E. 2002. *Molecular Mechanisms of Photosynthesis*. Blackwell Science, London, United Kingdom.

Bock, A., P. W. King, M. Blokesch, and M. C. Posewitz. 2006. Maturation of hydrogenases. *Adv. Microb. Physiol.* **51**:1–71.

Boichenko, V. A., and P. Hoffman. 1994. Photosynthetic hydrogen production in prokaryotes and eukaryotes: occurrence, mechanism, and functions. *Photosynthetica* **30**:527–552.

Boichenko, V. A., E. Greenbaum, and M. Seibert. 2004. Hydrogen production by photosynthetic microorganisms, p. 397–452. *In* M. D. Archer and J. Barber (ed.), *Photoconversion of Solar Energy: Molecular to Global Photosynthesis*, vol. 2. Imperial College Press, London, United Kingdom.

Bölling, C., and O. Fiehn. 2005. Metabolite profiling of *Chlamydomonas reinhardtii* under nutrient deprivation. *Plant Physiol.* **139**:1995–2005.

Brand, J. J., J. Wright, and S. Lien. 1989. Hydrogen production by eukaryotic algae. *Biotechnol. Bioeng.* 33:1482–1488.

Brazzolotto, X., J. K. Rubach, J. Gaillard, S. Gambarelli, M. Atta, and M. Fontecave. 2006. The [Fe-Fe]-hydrogenase maturation protein HydF from *Thermotoga maritima* is a GTPase with an iron-sulfur cluster. *J. Biol. Chem.* **281**:769–774.

Brown, L. M., and K. G. Zeiler. 1993. Aquatic biomass and carbon-dioxide trapping. *Energy Conversion Manage.* 34:1005–1013.

Bui, E. T., and P. J. Johnson. 1996. Identification and characterization of [Fe]-hydrogenases in the hydrogenosome of *Trichomonas vaginalis*. *Mol. Biochem. Parasitol.* 76:305–310.

Chang, C., D. Alber, P. Graf, K. Kim, and M. Seibert. 2007. Addressing unknown constants and metabolic network behaviors through petascale computing: understanding H_2 production in green algae. *J. Phys. Conf. Ser.* 38:012011.

Cheek, J., and J. B. Broderick. 2001. Adenosylmethionine-dependent iron-sulfur enzymes: versatile clusters in a radical new role. *J. Biol. Inorg. Chem.* 6:209–226.

Chen, H. C., K. Yokthongwattana, A. J. Newton, and A. Melis. 2003. SulP, a nuclear gene encoding a putative chloroplast-targeted sulfate permease in *Chlamydomonas reinhardtii*. *Planta* **218**:98–106.

Chen, H. C., A. J. Newton, and A. Melis. 2005. Role of SulP, a nuclear-encoded chloroplast sulfate permease, in sulfate transport and H_2 evolution in *Chlamydomonas reinhardtii*. *Photosynth. Res.* **84**:289–296.

Cicchillo, R. M., and S. J. Booker. 2005. Mechanistic investigations of lipoic acid biosynthesis in *Escherichia coli*: both sulfur atoms in lipoic acid are contributed by the same lipoyl synthase polypeptide. *J. Am. Chem. Soc.* 127:2860–2861.

Cohen, J., K. Kim, P. King, M. Seibert, and K. Schulten. 2005a. Finding gas diffusion pathways in proteins: application to O_2 and H_2 transport in CpI [FeFe]-hydrogenase and the role of packing defects. *Structure* **13**:1321–1329.

Cohen, J., K. Kim, M. Posewitz, M. L. Ghirardi, K. Schulten, M. Seibert, and P. King. 2005b. Molecular dynamics and experimental investigation of H(2) and O(2) diffusion in [Fe]-hydrogenase. *Biochem. Soc. Trans.* 33:80–82.

Cournac, L., K. Redding, J. Ravenel, D. Rumeau, E. M. Josse, M. Kuntz, and G. Peltier. 2000. Electron flow between photosystem II and oxygen in chloroplasts of photosystem I-deficient algae is mediated by a quinol oxidase involved in chlororespiration. *J. Biol. Chem.* **275**:17256–17262.

Cournac, L., F. Mus, L. Bernard, G. Gudeney, P. Vignais, and G. Peltier. 2002. Limiting steps of hydrogen production in *Chlamydomonas reinhardtii* and *Synechocystis* PCC 6803 as analysed by light-induced gas exchange transients. *Int. J. Hydrogen Energy* **27**:1229–1237.

Curatti, L., P. W. Ludden, and L. M. Rubio. 2006. NifB-dependent in vitro synthesis of the iron-molybdenum cofactor of nitrogenase. *Proc. Natl. Acad. Sci. USA* **103**:5297–5301.

de Vitry, C., Y. X. Ouyang, G. Finazzi, F. A. Wollman, and T. Kallas. 2004. The chloroplast Rieske iron-sulfur protein—at the crossroad of electron transport and signal transduction. *J. Biol. Chem.* **279**:44621–44627.

Erbes, D. L., D. King, and M. Gibbs. 1979. Inactivation of hydrogenase in cell-free extracts and whole cells of *Chlamydomonas reinhardtii* by oxygen. *Plant Physiol.* **63**:1138–1142.

Evans, D. J., and C. J. Pickett. 2003. Chemistry and the hydrogenases. *Chem. Soc. Rev.* **32**:268–275.

Fan, H. J., and M. B. Hall. 2001. A capable bridging ligand for Fe-only hydrogenase: density functional calculations of a low-energy route for heterolytic cleavage and formation of dihydrogen. *J. Am. Chem. Soc.* **123:**3828–3829.

Fauque, G., H. D. Peck, Jr., J. J. Moura, B. H. Huynh, Y. Berlier, D. V. DerVartanian, M. Teixeira, A. E. Przybyla, P. A. Lespinat, I. Moura, et al. 1988. The three classes of hydrogenases from sulfate-reducing bacteria of the genus *Desulfovibrio. FEMS Microbiol. Rev.* **4:**299–344.

Fedorov, A. S., S. Kosourov, M. L. Ghirardi, and M. Seibert. 2005. Continuous hydrogen photoproduction by *Chlamydomonas reinhardtii. Appl. Biochem. Biotechnol.* **121:**403–412.

Filipiak, M., W. R. Hagen, and C. Veeger. 1989. Hydrodynamic, structural and magnetic properties of *Megasphaera elsdenii* Fe hydrogenase reinvestigated. *Eur. J. Biochem.* **185:**547–553.

Florin, L., A. Tsokoglou, and T. Happe. 2001. A novel type of iron hydrogenase in the green alga *Scenedesmus obliquus* is linked to the photosynthetic electron transport chain. *J. Biol. Chem.* **276:** 6125–6132.

Fontecave, M. 2006. Iron-sulfur clusters: ever-expanding roles. *Nat. Chem. Biol.* **2:**171–174.

Forestier, M., P. King, L. Zhang, M. Posewitz, S. Schwarzer, T. Happe, M. L. Ghirardi, and M. Seibert. 2003. Expression of two [Fe]-hydrogenases in *Chlamydomonas reinhardtii* under anaerobic conditions. *Eur. J. Biochem.* **270:**2750–2758.

Forzi, L., and R. Sawers. 2007. Maturation of [NiFe]-hydrogenases in *Escherichia coli. BioMetals* **20:**565–578.

Fouchard, S., A. Hemschemeier, A. Caruana, K. Pruvost, J. Legrand, T. Happe, G. Peltier, and L. Cournac. 2005. Autotrophic and mixotrophic hydrogen photoproduction in sulfur-deprived *Chlamydomonas* cells. *Appl. Environ. Microbiol.* **71:**6199–6205.

Frey, M. 2002. Hydrogenases: hydrogen-activating enzymes. *Chembiochem* **3:**153–160.

Frey, P. A., and O. T. Magnusson. 2003. S-Adenosylmethionine: a wolf in sheep's clothing, or a rich man's adenosylcobalamin? *Chem. Rev.* **103:**2129–2148.

Gaffron, H. 1939. Reduction of CO_2 with H_2 in green plants. *Nature* **143:**204–205.

Gaffron, H., and J. Rubin. 1942. Fermentative and photochemical production of hydrogen in algae. *J. Gen. Physiol.* **26:**219–240.

Gfeller, R. P., and M. Gibbs. 1984. Fermentative metabolism of *Chlamydomonas reinhardtii.* 1. Analysis of fermentative products from starch in dark and light. *Plant Physiol.* **75:**212–218.

Ghirardi, M. L., R. K. Togasaki, and M. Seibert. 1997. Oxygen sensitivity of algal H-2-production. *Appl. Biochem. Biotechnol.* **63:** 141–151.

Ghirardi, M. L., L. Zhang, J. W. Lee, T. Flynn, M. Seibert, E. Greenbaum, and A. Melis. 2000. Microalgae: a green source of renewable H(2). *Trends Biotechnol.* **18:**506–511.

Ghirardi, M. L., and W. Amos. 2004. Renewable hydrogen from green algae. *BioCycle* **45:**59–62.

Ghirardi, M. L., P. King, S. Kosourv, M. Forestier, L. Zhang, and M. Seibert. 2005. Development of algal systems for hydrogen photoproduction—addressing the hydrogenase oxygen-sensitivity problem, p. 213. *In* C. Collings (ed.), *Artificial Photosynthesis.* Wiley-VCH Verlag, Weinheim, Germany.

Ghirardi, M. L. 2006. Hydrogen production by photosynthetic green algae. *Indian J. Biochem. Biophys.* **43:**201–210.

Ghirardi, M. L., M. C. Posewitz, P.-C. Maness, A. Dubini, J. Yu, and M. Seibert. 2007. Hydrogenases and hydrogen photoproduction in oxygenic photosynthetic organisms. *Annu. Rev. Plant Biol.* **58:** 71–91.

Gorwa, M. F., C. Croux, and P. Soucaille. 1996. Molecular characterization and transcriptional analysis of the putative hydrogenase gene of *Clostridium acetobutylicum* ATCC 824. *J. Bacteriol.* **178:** 2668–2675.

Gray, C. T., and H. Gest. 1965. Biological formation of molecular hydrogen. *Science* **148:**186–192.

Greenbaum, E. 1982. Photosynthetic hydrogen and oxygen production—kinetic-studies. *Science* **215:**291–293.

Greenbaum, E., R. R. L. Guillard, and W. G. Sunda. 1983. Hydrogen and oxygen photoproduction by marine-algae. *Photochem. Photobiol.* **37:**649–655.

Greenbaum, E. 1988. Energetic efficiency of hydrogen photoevolution by algal water splitting. *Biophys. J.* **54:**365–368.

Greenbaum, E., C. V. Tevault, and C. Y. Ma. 1995. New photosynthesis—direct photoconversion of biomass to molecular-oxygen and volatile hydrocarbons. *Energy Fuels* **9:**163–167.

Grossman, A. R., M. Croft, V. N. Gladyshev, S. S. Merchant, M. C. Posewitz, S. Prochnik, and M. H. Spalding. 2007. Novel metabolism in *Chlamydomonas* through the lens of genomics. *Curr. Opin. Plant Biol.* **10:**190–198.

Guan, Y., M. Deng, X. Yu, and W. Zhang. 2004. Two-stage photobiological production of hydrogen by marine green alga *Platymonas subcordiformis. Biochem. Eng. J.* **19:**69–73.

Gumpel, N. J., and S. Purton. 1994. Playing tag with *Chlamydomonas. Trends Cell. Biol.* **4:**299–301.

Happe, T., and J. D. Naber. 1993. Isolation, characterization and N-terminal amino acid sequence of hydrogenase from the green alga *Chlamydomonas reinhardtii. Eur. J. Biochem.* **214:**475–481.

Happe, T., B. Mosler, and J. D. Naber. 1994. Induction, localization and metal content of hydrogenase in the green alga *Chlamydomonas reinhardtii. Eur. J. Biochem.* **222:**769–774.

Happe, T., A. Hemschemeier, M. Winkler, and A. Kaminski. 2002. Hydrogenases in green algae: do they save the algae's life and solve our energy problems? *Trends Plant Sci.* **7:**246–250.

Happe, T., and A. Kaminski. 2002. Differential regulation of the Fe-hydrogenase during anaerobic adaptation in the green alga *Chlamydomonas reinhardtii. Eur. J. Biochem.* **269:**1022–1032.

Hartmann, G. C., A. R. Klein, M. Linder, and R. K. Thauer. 1996. Purification, properties and primary structure of H2-forming N5, N10-methylenetetrahydromethanopterin dehydrogenase from *Methanococcus thermolithotrophicus. Arch. Microbiol.* **165:**187–193.

Healey, F. P. 1970. The mechanism of hydrogen evolution by *Chlamydomonas moewusii. Plant Physiol.* **45:**153–159.

Hemschemeier, A., and T. Happe. 2005. The exceptional photofermentative hydrogen metabolism of the green alga *Chlamydomonas reinhardtii. Biochem. Soc. Trans.* **33:**39–41.

Hieta, R., and J. Myllyharju. 2002. Cloning and characterization of a low molecular weight prolyl 4-hydroxylase from *Arabidopsis thaliana. J. Biol. Chem.* **277:**23965–23971.

Homann, P. 2003. Hydrogen metabolism of green algae, discovery and early research—a tribute to Hans Gaffron and his coworkers. *Photosynth. Res.* **76:**93–103.

Horner, D. S., B. Heil, T. Happe, and T. M. Embley. 2002. Iron hydrogenases—ancient enzymes in modern eukaryotes. *Trends Biochem. Sci.* **27:**148–153.

Horner, J. K., and M. A. Wolinsky. 2002. A power-law sensitivity analysis of the hydrogen-producing metabolic pathway in *Chlamydomonas reinhardtii. Int. J. Hydrogen Energy* **27:**1251–1255.

Houchins, J. P., and G. Hind. 1984. Concentration and function of membrane-bound cytochromes in cyanobacterial heterocysts. *Plant Physiol.* **76:**456–460.

Jackson, D. D., and J. W. Ellms. 1896. On odors and tastes of surface waters with special reference to *Anabaena*, a microscopical organism found in certain water supplies of Massachusetts. *Rep. Mass. State Board of Health* **1897:**410–420.

Jarrett, J. 2005a. Biotin synthase: enzyme or reactant? *Chem. Biol.* **12:**409–410.

Jarrett, J. T. 2005b. The novel structure and chemistry of iron-sulfur clusters in the adenosylmethionine-dependent radical enzyme biotin synthase. *Arch. Biochem. Biophys.* **433:**312–321.

Johnson, J. 2006. Untold effects of energy farming. *Chem. Eng. News* 84:57–62.

Kaska, D. D., V. Gunzler, K. I. Kivirikko, and R. Myllyla. 1987. Characterization of a low-relative-molecular-mass prolyl 4-hydroxylase from the green-alga *Chlamydomonas-reinhardtii. Biochem. J.* 241: 483–490.

Kennedy, R. A., M. E. Rumpho, and T. C. Fox. 1992. Anaerobic metabolism in plants. *Plant Physiol.* 100:1–6.

Kindle, K. L. 1990. High-frequency nuclear transformation of *Chlamydomonas-reinhardtii. Proc. Natl. Acad. Sci. USA* 87:1228–1232.

King, P. W., M. C. Posewitz, M. L. Ghirardi, and M. Seibert. 2006a. Functional studies of [FeFe] hydrogenase maturation in an *Escherichia coli* biosynthetic system. *J. Bacteriol.* 188:2163–2172.

King, P. W., D. Svedruzic, J. Cohen, K. Schulten, M. Seibert, and M. L. Ghirardi. 2006b. Structural and functional investigations of biological catalysts for optimization of solar-driven H_2 production systems *Proc. Soc. Photo Opt. Instrum. Eng.* 6340:63400Y.

Knaff, D. B. 1996. Ferredoxin and ferredoxin dependent enzymes, p. 333–361. *In* D. R. Ort and C. F. Yocum (ed.), *Advances in Photosynthesis, vol. 4. Oxygenic Photosynthesis: The Light Reactions.* Kluwer Academic Publishers, Dordrecht, The Netherlands.

Kok, B. 1953. Experiments on photosynthesis by *Chlorella* in flashing light, p. 63–75. *In* J. S. Burlew (ed.), *Algal Culture: from Laboratory to Pilot Plant.* Carnegie Institute of Washington, Washington, DC.

Kosourov, S., A. Tsygankov, M. Seibert, and M. L. Ghirardi. 2002. Sustained hydrogen photoproduction by *Chlamydomonas reinhardtii*: effects of culture parameters. *Biotechnol. Bioeng.* 78:731–740.

Kosourov, S., M. Seibert, and M. L. Ghirardi. 2003. Effects of extracellular pH on the metabolic pathways in sulfur-deprived, H_2-producing *Chlamydomonas reinhardtii* cultures. *Plant Cell Physiol.* 44:146–155.

Kosourov, S., V. Makarova, A. S. Fedorov, A. Tsygankov, M. Seibert, and M. L. Ghirardi. 2005. The effect of sulfur re-addition on H(2) photoproduction by sulfur-deprived green algae. *Photosynth. Res.* 85:295–305.

Kosourov, S., E. Patrusheva, M. L. Ghirardi, M. Seibert, and A. Tsygankov. 2007. A comparison of hydrogen photoproduction by sulfur-deprived Chlamydomonas reinhardtii under different growth conditions. *J. Biotechnol.* 128:776–787.

Kreuzberg, K. 1984. Starch fermentation via formate producing pathway in *Chlamydomonas reinhardtii, Chlorogonium elongatum* and *Chlorella fusca. Physiol. Plant.* 61:87–94.

Kruse, O., J. Rupprecht, J. H. Mussgnugn, G. C. Dismukes, and B. Hankamer. 2005a. Photosynthesis: a blueprint for solar energy capture and biohydrogen production technologies. *Photochem. Photobiol. Sci.* 4:957–969.

Kruse, O., J. Rupprecht, K. P. Bader, S. Thomas-Hall, P. M. Schenk, G. Finazzi, and B. Hankamer. 2005b. Improved photobiological H_2 production in engineered green algal cells. *J. Biol. Chem.* 280:34170–34177.

Kummerle, R., M. Atta, J. Sculler, J. Gaillard, and J. Meyer. 1999. Structural similarities between the N-terminal domain of *Clostridium pasteurianum* hydrogenase and plant-type ferredoxins. *Biochemistry* 38:1938–1943.

Kurisu, G., D. Nishiyama, M. Kusunoki, S. Fujikawa, M. Katoh, G. T. Hanke, T. Hase, and K. Teshima. 2005. A structural basis of *Equisetum arvense* ferredoxin isoform II producing an alternative electron transfer with ferredoxin-NADP$^+$-reductase. *J. Biol. Chem.* 280:2275–2281.

Laurinavichene, T., I. Tolstygina, and A. Tsygankov. 2004. The effect of light intensity on hydrogen production by sulfur-deprived *Chlamydomonas reinhardtii. J. Biotechnol.* 114:143–151.

Laurinavichene, T. V., I. V. Tolstygina, R. R. Galiulina, M. L. Ghirardi, M. Seibert, and A. A. Tsygankov. 2002. Dilution methods to deprive *Chlamydomonas reinhardtii* cultures of sulfur for subsequent hydrogen photoproduction. *Int. J. Hydrogen Energy* 27:1245–1249.

Laurinavichene, T. V., A. S. Fedorov, M. L. Ghirardi, M. Seibert, and A. A. Tsygankov. 2006. Demonstration of sustained hydrogen photoproduction by immobilized, sulfur-deprived *Chlamydomonas reinhardtii* cells. *Int. J. Hydrogen Energy* 31:659–667.

Laurinavichene, T. V., S. N. Kosourov, M. L. Ghirardi, M. Seibert, and A. Tsyganov. 2008. Prolongation of H_2 photoproduction by sulfur-limited *Chlamydomonas reinhardtii* cultures. *J. Biotechnol.* 17:Epub ahead of print.

Layer, G., D. W. Heinz, D. Jahn, and W. D. Schubert. 2004. Structure and function of radical SAM enzymes. *Curr. Opin. Chem. Biol.* 8:468–476.

Leach, M. R., and D. B. Zamble. 2007. Metallocenter assembly of the hydrogenase enzymes. *Curr. Opin. Chem. Biol.* 11:159–165.

Lee, J. W., and E. Greenbaum. 2003. A new oxygen sensitivity and its potential application in photosynthetic H_2 production. *Appl. Biochem. Biotechnol.* 105–108:303–313.

Levin, D., L. Pitt, and M. Love. 2004. Biohydrogen production: prospects and limitations to practical application. *Int. J. Hydrogen Energy* 29:173–185.

Ley, A. C., and D. C. Mauzerall. 1982. Absolute absorption cross-sections for photosystem-II and the minimum quantum requirement for photosynthesis in *Chlorella-vulgaris. Biochim. Biophys. Acta* 680:95–106.

Ludwig, M., R. Schulz-Friedrich, and J. Appel. 2006. Occurrence of hydrogenases in cyanobacteria and anoxygenic photosynthetic bacteria: implications for the phylogenetic origin of cyanobacterial and algal hydrogenases. *J. Mol. Evol.* 63:758–768.

Lyon, E. J., S. Shima, G. Buurman, S. Chowdhuri, A. Batschauer, K. Steinbach, and R. K. Thauer. 2004. UV-A/blue-light inactivation of the 'metal-free' hydrogenase (Hmd) from methanogenic archaea. *Eur. J. Biochem.* 271:195–204.

McGlynn, S., S. Ruebush, A. Naumov, L. Nagy, A. Dubini, P. King, J. Broderick, M. Posewitz, and J. Peters. 2007. In vitro activation of [FeFe] hydrogenase: new insight into hydrogenase maturation. *J. Biol. Inorg. Chem.* 12:443–447.

Melis, A., J. Neidhardt, and J. R. Benemann. 1998. *Dunaliella salina* (Chlorophyta) with small chlorophyll antenna sizes exhibit higher photosynthetic productivities and photon use efficiencies than normally pigmented cells. *J. Appl. Phycol.* 10:515–525.

Melis, A., L. Zhang, M. Forestier, M. L. Ghirardi, and M. Seibert. 2000. Sustained photobiological hydrogen gas production upon reversible inactivation of oxygen evolution in the green alga *Chlamydomonas reinhardtii. Plant Physiol.* 122:127–136.

Melis, A., and T. Happe. 2001. Hydrogen production. Green algae as a source of energy. *Plant Physiol.* 127:740–748.

Melis, A. 2002. Green alga hydrogen production: progress, challenges and prospects. *Int. J. Hydrogen Energy* 27:1217–1228.

Melis, A., and T. Happe. 2004. Trails of green alga hydrogen research—from Hans Gaffron to new frontiers. *Photosynth. Res.* 80: 401–409.

Melis, A., M. Seibert, and T. Happe. 2004. Genomics of green algal hydrogen research. *Photosynth. Res.* 82:277–288.

Melis, A. 2005. Bioengineering of green algae to enhance photosynthesis and hydrogen production, p. 229–240. *In* A. F. Collins and C. Critchley (ed.), *Artificial Photosynthesis: From Basic Biology to Industrial Application.* Wiley-VCH Verlag, Weinheim, Germany.

Melis, A., and H. C. Chen. 2005. Chloroplast sulfate transport in green algae—genes, proteins and effects. *Photosynth. Res.* 86:299–307.

Meyer, J. 2007. [FeFe] hydrogenases and their evolution: a genomic perspective. *Cell. Mol. Life Sci.* 64:1063–1084.

Mulloney, J. A. 1993. Mitigation of carbon-dioxide releases from power production via sustainable agri-power—the synergistic combination of controlled environmental agriculture (large commercial greenhouses) and disbursed fuel-cell power-plants. *Energy Conversion Manage.* 34:913–920.

Mus, F., L. Cournac, V. Cardettini, A. Caruana, and G. Peltier. 2005. Inhibitor studies on non-photochemical plastoquinone reduction and H(2) photoproduction in *Chlamydomonas reinhardtii. Biochim. Biophys. Acta* 1708:322–332.

Mus, F., A. Dubini, M. Seibert, M. C. Posewitz, and A. R. Grossman. Anaerobic acclimation in *Chlamydomonas reinhardtii*: anoxic gene expression, hydrogenase induction, and metabolic pathways. *J. Biol. Chem.* 282:25475–25486.

Myers, J. 1957. Algal culture, p. 649–668. *In* R. E. Kirk and D. E. Othmer (ed.), *Encyclopedia of Chemical Technology*. Interscience, New York, NY.

Nagy, L. E., J. E. Meuser, S. Plummer, M. Seibert, M. L. Ghirardi, P. W. King, D. Ahmann, and M. C. Posewitz. 2007. Application of gene-shuffling for the rapid generation of novel [FeFe]-hydrogenase libraries. *Biotechnol. Lett.* 29:421–430.

Nakicenovic, N. 1993. Carbon-dioxide mitigation measures and options. *Environ. Sci. Technol.* 27:1986–1989.

Nicolet, Y., C. Piras, P. Legrand, C. E. Hatchikian, and J. C. Fontecilla-Camps. 1999. *Desulfovibrio desulfuricans* iron hydrogenase: the structure shows unusual coordination to an active site Fe binuclear center. *Structure* 7:13–23.

Nicolet, Y., B. J. Lemon, J. C. Fontecilla-Camps, and J. W. Peters. 2000. A novel FeS cluster in Fe-only hydrogenases. *Trends Biochem. Sci.* 25:138–143.

Nicolet, Y., A. L. de Lacey, X. Vernede, V. M. Fernandez, E. C. Hatchikian, and J. C. Fontecilla-Camps. 2001. Crystallographic and FTIR spectroscopic evidence of changes in Fe coordination upon reduction of the active site of the Fe-only hydrogenase from *Desulfovibrio desulfuricans. J. Am. Chem. Soc.* 123:1596–1601.

Nicolet, Y., and C. L. Drennan. 2004. AdoMet radical proteins—from structure to evolution—alignment of divergent protein sequences reveals strong secondary structure element conservation. *Nucleic. Acids Res.* 32:4015–4025.

Ohta, S., K. Miyamoto, and Y. Miure. 1987. Hydrogen evolution as a consumption mode of reducing equivalents in green algal fermentation. *Plant Physiol.* 83:1022–1026.

Overpeck, J. T., B. L. Otto-Bliesner, G. H. Miller, D. R. Muhs, R. B. Alley, and J. T. Kiehl. 2006. Paleoclimatic evidence for future ice-sheet instability and rapid sea-level rise. *Science* 311:1747–1750.

Park, W., and I. Moon. 2007. A discrete multi states model for the biological production of hydrogen by phototrophic microalga. *Biochem. Eng. J.* 36:19–27.

Peters, J. W., W. N. Lanzilotta, B. J. Lemon, and L. C. Seefeldt. 1998. X-ray crystal structure of the Fe-only hydrogenase (CpI) from *Clostridium pasteurianum* to 1.8 angstrom resolution. *Science* 282: 1853–1858.

Peters, J. W. 1999. Structure and mechanism of iron-only hydrogenases. *Curr. Opin. Struct. Biol.* 9:670–676.

Peters, J. W., R. K. Szilagyi, A. Naumov, and T. Douglas. 2006. A radical solution for the biosynthesis of the H-cluster of hydrogenase. *FEBS Lett.* 580:363–367.

Pierik, A. J., W. R. Hagen, J. S. Redeker, R. B. Wolbert, M. Boersma, M. F. Verhagen, H. J. Grande, C. Veeger, P. H. Mutsaers, R. H. Sands, et al. 1992. Redox properties of the iron-sulfur clusters in activated Fe-hydrogenase from *Desulfovibrio vulgaris* (Hildenborough). *Eur. J. Biochem.* 209:63–72.

Pierik, A. J., M. Hulstein, W. R. Hagen, and S. P. Albracht. 1998. A low-spin iron with CN and CO as intrinsic ligands forms the core of the active site in [Fe]-hydrogenases. *Eur. J. Biochem.* 258:572–578.

Pilak, O., B. Mamat, S. Vogt, C. H. Hagemeier, R. K. Thauer, S. Shima, C. Vonrhein, E. Warkentin, and U. Ermler. 2006. The crystal structure of the apoenzyme of the iron-sulphur cluster-free hydrogenase. *J. Mol. Biol.* 358:798–809.

Polle, J. E. W., J. R. Benemann, A. Tanaka, and A. Melis. 2000. Photosynthetic apparatus organization and function in the wild type and a chlorophyll b-less mutant of *Chlamydomonas reinhardtii*. Dependence on carbon source. *Planta* 211:335–344.

Polle, J. E. W., K. K. Niyogi, and A. Melis. 2001. Absence of lutein, violaxanthin and neoxanthin affects the functional chlorophyll antenna size of photosystem-II but not that of photosystem-I in the green alga *Chlamydomonas reinhardtii. Plant Cell Physiol.* 42: 482–491.

Polle, J. E. W., S. D. Kanakagiri, and A. Melis. 2003. tla1, a DNA insertional transformant of the green alga *Chlamydomonas reinhardtii* with a truncated light-harvesting chlorophyll antenna size. *Planta* 217:49–59.

Pollock, S. V., W. Pootakham, N. Shibagaki, J. L. Moseley, and A. R. Grossman. 2005. Insights into the acclimation of *Chlamydomonas reinhardtii* to sulfur deprivation. *Photosynth. Res.* 86:475–489.

Posewitz, M. C., P. W. King, S. L. Smolinski, L. Zhang, M. Seibert, and M. L. Ghirardi. 2004a. Discovery of two novel radical S-adenosylmethionine proteins required for the assembly of an active [Fe] hydrogenase. *J. Biol. Chem.* 279:25711–25720.

Posewitz, M. C., S. L. Smolinski, S. Kanakagiri, A. Melis, M. Seibert, and M. L. Ghirardi. 2004b. Hydrogen photoproduction is attenuated by disruption of an isoamylase gene in *Chlamydomonas reinhardtii. Plant Cell* 16:2151–2163.

Pow, T., and A. I. Krasna. 1979. Photoproduction of hydrogen from water in hydrogenase-containing algae. *Arch. Biochem. Biophys.* 194:413–421.

Prince, R. C., and H. S. Kheshgi. 2005. The photobiological production of hydrogen: potential efficiency and effectiveness as a renewable fuel. *Crit. Rev. Microbiol.* 31:19–31.

Putz, S., P. Dolezal, G. Gelius-Dietrich, L. Bohacova, J. Tachezy, and K. Henze. 2006. Fe-hydrogenase maturases in the hydrogenosomes of *Trichomonas vaginalis. Eukaryot. Cell* 5:579–586.

Radmer, R., and B. Kok. 1977. Photosynthesis—limited yields, unlimited dreams. *BioScience* 27:599–605.

Ragsdale, S. W. 2003. Pyruvate ferredoxin oxidoreductase and its radical intermediate. *Chem. Rev.* 103:2333–2346.

Randt, C., and H. Senger. 1985. Participation of the two photosystems in light dependent hydrogen evolution in *Scenedesmus obliquus. Photochem. Photobiol. Sci.* 42:553–557.

Rao, K. K., and R. Cammack. 2001. Producing hydrogen as a fuel, p. 201–230. *In* R. Cammack, M. Frey, and R. Robson (ed.), *Hydrogen as a Fuel.* Taylor and Francis, London, United Kingdom.

Rey, F. E., E. K. Heiniger, and C. S. Harwood. 2007. Redirection of metabolism for biological hydrogen production. *Appl. Environ. Microbiol.* 73:1665–1671.

Roessler, P. G., and S. Lien. 1984a. Activation and de novo synthesis of hydrogenase in *Chlamydomonas. Plant Physiol.* 76:1086–1089.

Roessler, P. G., and S. Lien. 1984b. Purification of hydrogenase from *Chlamydomonas reinhardtii. Plant Physiol.* 75:705–709.

Roseboom, W., A. L. De Lacey, V. M. Fernandez, E. C. Hatchikian, and S. P. Albracht. 2006. The active site of the [FeFe]-hydrogenase from *Desulfovibrio desulfuricans*. II. Redox properties, light sensitivity and CO-ligand exchange as observed by infrared spectroscopy. *J. Biol. Inorg. Chem.* 11:102–118.

Rubach, J. K., X. Brazzolotto, J. Gaillard, and M. Fontecave. 2005. Biochemical characterization of the HydE and HydG iron-only hydrogenase maturation enzymes from *Thermatoga maritima. FEBS Lett.* 579:5055–5060.

Saitoh, T., T. Ikegami, M. Nakayama, K. Teshima, H. Akutsu, and T. Hase. 2006. NMR study of the electron transfer complex of plant ferredoxin and sulfite reductase. *J. Biol. Chem.* 281:10482–10488.

Soboh, B., D. Linder, and R. Hedderich. 2004. A multisubunit membrane-bound [NiFe] hydrogenase and an NADH-dependent Fe-only hydrogenase in the fermenting bacterium *Thermoanaerobacter tengcongensis. Microbiology* 150:2451–2463.

Sofia, H. J., G. Chen, B. G. Hetzler, J. F. Reyes-Spindola, and N. E. Miller. 2001. Radical SAM, a novel protein superfamily linking unresolved steps in familiar biosynthetic pathways with radical mechanisms: functional characterization using new analysis and information visualization methods. *Nucleic. Acids Res.* **29**:1097–1106.

Stauber, E. J., and M. Hippler. 2004. *Chlamydomonas reinhardtii* proteomics. *Plant Physiol. Biochem.* **42**:989–1001.

Stirnberg, M., and T. Happe. 2004. Identification of a *cis*-acting element controlling anaerobic expression of the HYDA gene from *Chlamydomonas reinhardtii*, p. 117–127. *In* J. Miyake, Y. Igarashi, and M. Rogner (ed.), *Biohydrogen III*. Elsevier, New York, NY.

Takahashi, H., C. E. Braby, and A. R. Grossman. 2001. Sulfur economy and cell wall biosynthesis during sulfur limitation of *Chlamydomonas reinhardtii. Plant Physiol.* **127**:665–673.

Tamagnini, P., R. Axelsson, P. Lindberg, F. Oxelfelt, R. Wunschiers, and P. Lindblad. 2002. Hydrogenases and hydrogen metabolism of cyanobacteria. *Microbiol. Mol. Biol. Rev.* **66**:1–20.

Tetali, S. D., M. Mitra, and A. Melis. 2007. Development of the light-harvesting chlorophyll antenna in the green alga *Chlamydomonas reinhardtii* is regulated by the novel Tla1 gene. *Planta* **225**:813–829.

Tsygankov, A., S. Kosourov, M. Seibert, and M. L. Ghirardi. 2002. Hydrogen photoproduction under continuous illumination by sulfur-deprived, synchronous *Chlamydomonas reinhardtii* cultures. *Int. J. Hydrogen Energy* **27**:1239–1244.

Tsygankov, A. A., S. N. Kosourov, I. V. Tolstygina, M. L. Ghirardi, and M. Seibert. 2006. Hydrogen production by sulfur-deprived *Chlamydomonas reinhardtii* under photoautotrophic conditions. *Int. J. Hydrogen Energy* **31**:1574–1584.

Ueno, Y., N. Kurano, and S. Miyachi. 1999. Purification and characterization of hydrogenase from the marine green alga, *Chlorococcum littorale. FEBS Lett.* **443**:144–148.

Urbig, T., R. Schulz, and H. Senger. 1993. Inactivation and reactivation of the hydrogenases of the green algae *Scenedesmus obliquus* and *Chlamydomonas reinhardtii. Z. Naturforsch. Sect. C* **48**:41–45.

Valentine, R. C., and R. S. Wolfe. 1963. Role of ferredoxin in the metabolism of molecular hydrogen. *J. Bacteriol.* **85**:1114–1120.

van der Spek, T. M., A. F. Arendsen, R. P. Happe, S. Yun, K. A. Bagley, D. J. Stufkens, W. R. Hagen, and S. P. Albracht. 1996. Similarities in the architecture of the active sites of Ni-hydrogenases and Fe-hydrogenases detected by means of infrared spectroscopy. *Eur. J. Biochem.* **237**:629–634.

Van Dijk, C., A. Van Berkel-Arts, and C. Veeger. 1983. The effect of re-oxidation on the reduced hydrogenase of *Desulfovibrio vulgaris* strain Hildenborough and its oxygen stability. *FEBS Lett.* **156**:340–344.

Van Ooteghem, S. A., S. K. Beer, and P. C. Yue. 2002. Hydrogen production by the thermophilic bacterium *Thermotoga neapolitana. Appl. Biochem. Biotechnol.* **98**:177–189.

Van Ooteghem, S. A., A. Jones, D. van der Lelie, B. Dong, and D. Mahajan. 2004. H-2 production and carbon utilization by *Thermotoga neapolitana* under anaerobic and microaerobic growth conditions. *Biotechnol. Lett.* **26**:1223–1232.

Vazquezduhalt, R. 1991. Light effect on neutral lipids accumulation and biomass composition of *Botryococcus sudeticus* (Chlorophyceae). *Cryptogam. Algol.* **12**:109–119.

Vignais, P. M., A. Colbeau, J. C. Willison, and Y. Jouanneau. 1985. Hydrogenase, nitrogenase, and hydrogen metabolism in the photosynthetic bacteria. *Adv. Microb. Physiol.* **26**:155–234.

Vignais, P. M., B. Billoud, and J. Meyer. 2001. Classification and phylogeny of hydrogenases. *FEMS Microbiol. Rev.* **25**:455–501.

Vincent, K. A., and F. A. Armstrong. 2005. Investigating metalloenzyme reactions using electrochemical sweeps and steps: fine control and measurements with reactants ranging from ions to gases. *Inorg. Chem.* **44**:798–809.

Voigt, J., and R. Frank. 2003. 14-3-3 proteins are constituents of the insoluble glycoprotein framework of the *Chlamydomonas* cell wall. *Plant Cell* **15**:1399–1413.

Volbeda, A., M. H. Charon, C. Piras, E. C. Hatchikian, M. Frey, and J. C. Fontecilla-Camps. 1995. Crystal structure of the nickel-iron hydrogenase from *Desulfovibrio gigas. Nature* **373**:580–587.

Voncken, F. G., B. Boxma, A. H. van Hoek, A. S. Akhmanova, G. D. Vogels, M. Huynen, M. Veenhuis, and J. H. Hackstein. 2002. A hydrogenosomal [Fe]-hydrogenase from the anaerobic chytrid *Neocallimastix* sp. L2. *Gene* **284**:103–112.

Wagner, A. F. V., M. Frey, F. A. Neugebauer, W. Schafer, and J. Knappe. 1992. The free-radical in pyruvate formate-lyase is located on glycine-734. *Proc. Natl. Acad. Sci. USA* **89**:996–1000.

Weaver, P. F., S. Lien, and M. Seibert. 1980. Photobiological production of hydrogen. *Solar Energy* **24**:3–45.

Westermeier, R., and I. Gomez. 1996. Biomass, energy contents and major organic compounds in the brown alga *Lessonia nigrescens* (Laminariales, Phaeophyceae) from Mehuin, south Chile. *Botanica Marina* **39**:553–559.

White, A. L., and A. Melis. 2006. Biochemistry of hydrogen metabolism in *Chlamydomonas reinhardtii* wild type and a Rubisco-less mutant. *Int. J. Hydrogen Energy* **31**:455–464.

Winkler, M., B. Heil, B. Heil, and T. Happe. 2002. Isolation and molecular characterization of the [Fe]-hydrogenase from the unicellular green alga *Chlorella fusca. Biochim. Biophys. Acta* **1576**:330–334.

Winter, G., T. Buhrke, A. K. Jones, and B. Friedrich. 2004. The role of the active site-coordinating cysteine residues in the maturation of the H_2-sensing [NiFe] hydrogenase from *Ralstonia eutropha* H16. *Arch. Microbiol.* **182**:138–146.

Wu, L. F., and M. A. Mandrand. 1993. Microbial hydrogenases: primary structure, classification, signatures and phylogeny. *FEMS Microbiol. Rev.* **10**:243–269.

Wunschiers, R., K. Stangier, H. Senger, and R. Schulz. 2001. Molecular evidence for a Fe-hydrogenase in the green alga *Scenedesmus obliquus. Curr. Microbiol.* **42**:353–360.

Wykoff, D. D., J. P. Davies, A. Melis, and A. R. Grossman. 1998. The regulation of photosynthetic electron transport during nutrient deprivation in *Chlamydomonas reinhardtii. Plant Physiol.* **117**: 129–139.

Zaborsky, O. R. 1998. *BioHydrogen.* Plenum Publishing Corporation, New York, NY.

Zhang, L., T. Happe, and A. Melis. 2002. Biochemical and morphological characterization of sulfur-deprived and H_2-producing *Chlamydomonas reinhardtii* (green alga). *Planta* **214**:552–561.

Zhang, L. P., and A. Melis. 2002. Probing green algal hydrogen production. *Philos. Trans. R. Soc. Lond. B* **357**:1499–1507.

Zhang, Z. D., J. Shrager, M. Jain, C. W. Chang, O. Vallon, and A. R. Grossman. 2004. Insights into the survival of *Chlamydomonas reinhardtii* during sulfur starvation based on microarray analysis of gene expression. *Eukaryot. Cell* **3**:1331–1348.

5. FUEL CELLS: ELECTRICITY

Bioenergy
Edited by J. Wall et al.

Chapter 23

Electricity Production with Electricigens

DEREK R. LOVLEY AND KELLY P. NEVIN

The term "microbial fuel cells" has been used to describe a rather wide diversity of systems in which electricity is produced with the help of microorganisms. As previously reviewed in detail (Shukla et al., 2004; Lovley, 2006a; see also chapter 24 of this volume), the concept of microbial fuel cells dates back to at least 1910 when early studies by Potter (Potter, 1910, 1911) revealed that small amounts of electricity could be harvested from microbial cultures. Since that time there has been a cyclic waxing and waning of interest in microbial fuel cells that can be attributed to the periodic emergence of new concepts for improving the electrical connection between microbes and electrodes as well as sporadic inspiration for possible new applications of microbial fuel cells. There currently is renewed interest in microbial fuel cells, at least in part because improvements in the engineering of microbial fuel cells are beginning to increase the power outputs (Logan, 2005; Rabaey and Verstraete, 2005; Cheng et al., 2006; Logan et al., 2007). Furthermore, as summarized in this chapter, the discovery of microorganisms known as electricigens has recently made it clear that highly efficient and sustainable microbial fuel cells that may be used in a diversity of applications are on the horizon.

The purpose of this chapter is to review the microbiology of microbial fuel cells with emphasis on fuel cells powered by electricigens. Practical applications of microbial fuel cells are not discussed in detail. The primary reason for this is that the application of microbial fuel cells is in its infancy. Although there is often much emphasis on improved power outputs in the recent microbial fuel cell literature, in general these research-grade fuel cells have little practical value. In general, if the energy required to pump fuel/media into the systems and to stir the systems was included in the calculations, most of these systems would be net consumers of energy. Furthermore, many of these fuel cells use nonsustainable systems, such as cathodes that use ferric cyanide as the oxidant. Therefore, even though there are research programs funded to develop microbial fuel cells for a variety of applications, ranging from running small electronic devices to powering automobiles, as of now there are few current practical applications. The only immediate use of microbial fuel cells appears to be sediment microbial fuel cells that can power electronic monitoring devices with electricity extracted from marine sediments (Tender et al., 2002). This application is feasible because the power demand of many monitoring devices is relatively low and intermittent, making it feasible to slowly charge capacitors during periods of low demand and then use the stored power during peak power demand. For almost all other applications significant increases in the power output of microbial fuel cells will be required. Possibilities for increasing the power output of electricigen-powered microbial fuel cells with engineering and microbiological approaches are discussed in this chapter.

DISTINCTIONS BETWEEN ELECTRICIGENS AND OTHER MICROBES EMPLOYED IN MICROBIAL FUEL CELLS

The term electricigen was coined to make a clear distinction in the mechanisms of power production in microbial fuel cells (Lovley, 2006a). Electricigens are microorganisms that conserve energy to support growth by completely oxidizing organic compounds to carbon dioxide with direct electron transfer to the anodes of microbial fuel cells. Electricity production with electricigens is significantly different from that of other types of microorganisms (Table 1).

For example, early microbial fuel cells employed fermentable substrates as fuels and were powered by fermentative microorganisms (Katz et al., 2003; Shukla et al., 2004; Lovley, 2006a). Fermentative microorganisms typically convert a fermentable fuel, such as glucose, to small-chain organic acids, hydrogen, and

Derek R. Lovley and Kelly P. Nevin • Department of Microbiology, University of Massachusetts, Amherst, MA 01003.

Table1. Comparison of benefits associated with production of electricity with different types of microorganisms

Organism	Coulombic efficiency (%)	Self-sustaining	Exogenous mediator required	Employed in open environments
Fermentative organisms	<10	No	Not applicable	Yes
Fermentative with added mediator	<10	No	Yes	No
Shewanella-based system	<33	Yes	No	No
Electricigen	>90	Yes	No	Yes

carbon dioxide. Electricity production results from the interaction of poorly defined reduced compounds that are produced under the low redox conditions generated during fermentation or possibly some direct electron transfer between the fermentative microorganisms and the anode surface. However, most of the electrons that were in the initial fuel are found in fermentation products that are not electrochemically active. These certainly include the organic acids, which are the primary products of fermentation, but even hydrogen does not react well with the anode materials employed in most microbial fuel cells. Therefore, recovery of the electrons present in the fuel as electricity (i.e., coulombic efficiency) is very low. Furthermore, at high rates of metabolism the accumulation of fermentation acids in the system can inhibit the growth of the fermentative microorganisms. Thus, these fuel cells cannot be operated for long periods of time.

Addition of mediators, compounds that can enter the cell, accept electrons from various intracellular electron carriers, exit the cell in the reduced state, and then donate electrons to the anode, can significantly increase rates of current production in microbial fuel cells powered by fermentative microorganisms (Bennetto, 1990; Lovley, 2006a). The drawback of adding mediators is that these compounds are often toxic to humans, preventing the use of such mediators in microbial fuel cells that interact with the environment, such as when electricity is harvested from sewage, other organic wastes, or aquatic sediments. Mediators often do not have long-term stability. This limits the lifetime of the microbial fuel cells. Furthermore, even in the presence of mediators the fermentative microorganisms produce fermentation acids, which will eventually pickle the system. Most of the electrons initially present in the fuel are recovered in the fermentation acids rather than as electricity, and thus, coulombic efficiency remains low in these systems.

The requirement for an exogenous mediator is alleviated in microbial fuel cells powered by *Shewanella* species (Kim et al., 2002; Lanthier et al., 2007). However, the only organic fuels that *Shewanella* species have been reported to convert to electricity are a few short-chain fatty acids, such as lactate and pyruvate. These are not practical fuels for most applications. Furthermore, *Shewanella* species only incompletely oxidize these fuels to acetate. This greatly limits the coulombic efficiency because, as with the fermentative systems, most of the electrons in the fuel are recovered in waste products rather than electricity. Another factor limiting current production with *Shewanella* species appears to be an inability to produce thick biofilms on the anode surface (Kim et al., 2002; Lanthier et al., 2007). This contrasts with the substantial biofilms that some electricigens, such as *Geobacter sulfurreducens*, can produce, which can lead to much higher power densities (discussed in detail below). *Shewanella* species are not typically found on the anodes of microbial fuel cells harvesting electricity from complex organic wastes, such as aquatic sediments, suggesting that they are not competitive with electricigens in open environments.

Therefore, electricity production with electricigens has a number of advantages. Of great significance is the high coulombic efficiency that results from these microorganisms being able to completely oxidize organic fuels to carbon dioxide with an electrode serving as the sole electron acceptor. In fact, until an anode material that can effectively and sustainably catalyze the abiotic oxidation of fermentation products is developed, it will be impossible to have a microbial fuel cell that can have high coulombic efficiency without employing an electricigen. This is because even if nonelectricigens, such as fermentative microorganisms, carry out the initial metabolism of the organic fuel, at least one electricigen that can effectively recover the electrons from the metabolic products of the nonelectricigen will be required in order to achieve high electron recoveries as electricity.

Another advantage to electricigen-powered microbial fuel cells is their long-term sustainability. This results from the fact that electricigens conserve energy for maintenance and growth from electron transfer to anodes. Electricigen-based microbial fuel cells have been run for more than 2 years without a decline in power output.

The ability of electricigens to directly transfer electrons to the anode surface also alleviates the need for unstable, and potentially toxic, mediators. This simplifies the design of the microbial fuel cells and lowers their costs. Furthermore, this makes it possible to employ electricigen-based fuel cells in open environments.

In fact, as described in more detail below, just placing a bare graphite anode in sediments or soils will naturally enrich a biofilm of electricigens on the anode surface.

ELECTRICIGENS AVAILABLE IN PURE CULTURE

A number of electricigens are available in pure culture. The most heavily studied are electricigens in the family *Geobacteraceae*, especially *G. sulfurreducens*. The possibility that members of the *Geobacteraceae* might oxidize organic compounds with an electron acceptor was first realized when it was discovered that *Geobacteraceae* were specifically enriched on anodes harvesting electricity from aquatic sediments (Bond et al., 2002; Tender et al., 2002; Holmes et al., 2004b). In marine sediments, microorganisms most closely related to *Desulfuromonas* species predominated, whereas in freshwater sediments *Geobacter* species were most prevalent. These two genera have similar physiologies, oxidizing short-chain fatty acids to carbon dioxide with Fe(III) oxides serving as the electron acceptor (Lovley et al., 2004). Both *Desulfuromonas acetoxidans* and *Geobacter metallireducens* were able to grow via acetate oxidation with an electrode serving as the sole electron acceptor (Bond et al., 2002). However, subsequent studies have primarily focused on *G. sulfurreducens* (Bond and Lovley, 2003; Holmes et al., 2006; Reguera et al., 2006; Richter et al., 2007; K. P. Nevin, H. Richter, S. F. Covalla, J. P. Johnson, T. L. Woodard, H. Jia, M. Zhang, and D. R. Lovley, submitted for publication) because (i) *G. sulfurreducens* produces power densities (i.e., 2 W/m^2 of anode surface; 2 kW/m^3 of fuel cell volume) as high as have been reported for any pure culture or microbial consortia (Nevin et al., submitted); (ii) *G. sulfurreducens* is closely related to the *Geobacteraceae* that predominate on the surface of anodes harvesting electricity from aquatic sediments (Holmes et al., 2004b); (iii) a pure culture is necessary in order to evaluate mechanisms of extracellular electron transfer to anodes in a systematic manner; (iv) the complete genome sequence is available (Methé et al., 2003); (v) a genetic system is available (Coppi et al., 2001); (vi) whole-genome DNA microarrays for analyzing gene expression are available (Methé et al., 2005; Postier et al., 2007); (vii) databases for proteomic studies have already been developed (Ding et al., 2006; Khare et al., 2006); and (viii) detailed information on the physiology of this organism, including a genome-based in silico model (Mahadevan et al., 2006), is available (Lovley et al., 2004).

Like *G. metallireducens*, *G. sulfurreducens* can oxidize acetate with an electrode serving as the sole electron acceptor. It can also produce electricity with hydrogen as the electron donor. *G. metallireducens* can also use monoaromatic compounds (Bond et al., 2002) and ethanol (K. P. Nevin, unpublished data) as electron donors for electricity production. It seems likely that other electron donors, such as longer-chain fatty acids, that serve as electron donors for Fe(III) reduction by *G. metallireducens* will also yield electricity, but these have not yet been tested. The known *Geobacter* species do not oxidize sugars and must rely on fermentative microorganisms to convert carbohydrates to fermentation products that they can then oxidize.

The ability of *Geobacter* and *Desulfuromonas* species to donate electrons to electrodes is assumed to be a fortuitous result of their ability to transfer electrons to other extracellular electron acceptors, such as Fe(III) oxides (Lovley et al., 1987; Lovley and Phillips, 1988) and humic substances (Lovley et al., 1996a), because as far as is known there has not been an opportunity for these organisms to produce electricity until the first sediment microbial fuel cells were deployed. Using Fe(III) as an electron acceptor has proven to be a good strategy for isolating electricigens from the anode surfaces of sediment microbial fuel cells (Holmes et al., 2004c). Furthermore, screening culture collections of Fe(III)-reducing microorganisms has uncovered some interesting electricigens.

For example, *Rhodoferax ferrireducens* was isolated from subsurface sediments as an Fe(III) reducer (Finneran et al., 2003). Not only is it phylogenetically distinct from the *Geobacteraceae*, but also it has the unique capacity to oxidize various mono- and disaccharides to carbon dioxide with Fe(III) serving as the electron acceptor (Chaudhuri and Lovley, 2003). *R. ferrireducens* is also capable of producing electricity with these sugars as fuels (Chaudhuri and Lovley, 2003). Other Fe(III) reducers that have subsequently been found to produce electricity include *Geothrix fermentans* (Bond and Lovley, 2005) and *Desulfobulbus propionicus* (Holmes et al., 2004a).

However, the ability to use Fe(III) oxide as an electron acceptor does not necessarily confer the capacity for electron transfer to electrodes. For example, *Pelobacter carbinolicus*, a member of the *Geobacteraceae*, can use Fe(III) oxide as an electron acceptor but did not produce current in a microbial fuel cell (Richter et al., 2007). This was despite the fact that *P. carbinolicus* could grow in the fuel cell chamber if given a substrate, such as acetoin, that could be fermented, or, as detailed below, when grown syntrophically with *G. sulfurreducens*.

It seems likely that, just as a wide phylogenetic and physiological diversity of microorganisms have the capacity for Fe(III) reduction (Lovley et al., 2004), many types of microorganisms may be able to function as electricigens. It is possible that some of these might be able to produce electricity even if they are not effective Fe(III) reducers. Thus, although enriching and isolating electricigens with Fe(III) as a surrogate electron acceptor is a good strategy for recovering some electricigens, direct isolation on electrodes may be a productive alternative strategy if the appropriate isolation techniques can be developed.

INTERACTIONS BETWEEN ELECTRICIGENS AND OTHER MICROORGANISMS TO PRODUCE ELECTRICITY

As stated above, the activity of at least one electricigen is an absolute requirement for microbial fuel cells to effectively convert organic fuels to electricity. However, a diversity of microorganisms that are not electricigens can also contribute to current production and, in some instances, may be a necessity. For example, as noted above, *Geobacteraceae* are specifically enriched on the anodes of sediment microbial fuel cells, but it is unlikely that they are directly oxidizing the complex organic matter in sediments with electron transfer to electrodes. The same is true in systems converting swine waste to electricity in which *Geobacter* species account for 70% of the microorganisms on the anodes (K. B. Gregory, S. A. Sullivan, and D. R. Lovley, submitted for publication). Studies with sediments in which *Geobacter* species are using their natural electron acceptor, Fe(III) oxides, rather than electrodes, have suggested that complex organic matter is degraded via the cooperation of fermentative microorganisms and *Geobacter* species with the fermenters converting fermentable substrates to the fermentation products that *Geobacter* species can oxidize (Lovley and Phillips, 1989; Lovley, 1991). A similar cooperative processing of complex organic matter to electricity is likely.

In instances in which other microorganisms are converting organic fuels into simpler fuels that electricigens can oxidize, the microorganisms carrying out the initial step(s) in the metabolism of the fuel need not be specifically attached to the anode. The products of fermentation may diffuse from their site of generation at some distance from the anode to the anode surface where the electricigens convert them to electricity. A simple example of this was observed in cocultures of *P. carbinolicus* and *G. sulfurreducens* (Richter et al., 2007). As noted above, *P. carbinolicus* is not capable of electricity production. However, cocultures of *P. carbinolicus* and *G. sulfurreducens* produced electricity from ethanol, a fuel which *G. sulfurreducens* cannot oxidize (Caccavo et al., 1992). Further investigation revealed that *P. carbinolicus* was converting ethanol to acetate and hydrogen, which *G. sulfurreducens* converted to electricity (Richter et al., 2007). Although *G. sulfurreducens* was found almost exclusively on the anode surface, as many *P. carbinolicus* cells were planktonic as were attached to the anode.

TYPES OF ELECTRICIGEN-BASED MICROBIAL FUEL CELLS

A wide diversity of microbial fuel cell configurations have been described (Katz et al., 2003; Cheng et al., 2006; Lovley, 2006a). Studies with electricigens have utilized modifications of several of these designs (Fig. 1).

Sediment microbial fuel cells, which are typically colonized predominately by electricigens, consist of a graphite plate buried in the anoxic zone of the sediments connected electrically to a graphite plate in the overlying, aerobic water (Reimers et al., 2001; Tender et al., 2002). For practical applications electrical devices are placed within this circuit. For research purposes the circuit typically contains a fixed resistance.

For pure culture investigations the basic principles of the sediment microbial fuel cell can be replicated in an "H-cell" which was adapted from previous studies by Zeikus and colleagues (Park et al., 1999). The H-cell consists of an anode chamber that is maintained under anaerobic conditions. The anode is typically a graphite stick. The cathode chamber contains a graphite stick cathode. The cathode chamber can be bubbled with air to provide oxygen as the electron acceptor. Alternatively, Fe(III) cyanide, typically 50 mM, is added as the oxidant. The two chambers are separated with a DuPont Nafion membrane, which permits protons, produced during organic matter oxidation in the anode chamber, to diffuse through to the cathode chamber, while restricting passage of oxygen or Fe(III) cyanide from the cathode chamber to the anode chamber. A fixed resistor is placed in the circuit between the anode and cathode. The level of this external resistance is generally determined from an analysis of power output at different resistances and then picking the resistance that yields maximum power. Although these H-cells are convenient for microbiological investigations, their power output (14 mW/m^2) is low. At or near maximal power production, current levels are also rather low, ca. 0.4 mA, in a standard H-cell. The low rate of consumption of fuel means that relatively low concentrations of fuel (10 mM acetate, for example) can power the system for days. Therefore, H-cells are typically run in batch mode without a continuous fuel feed.

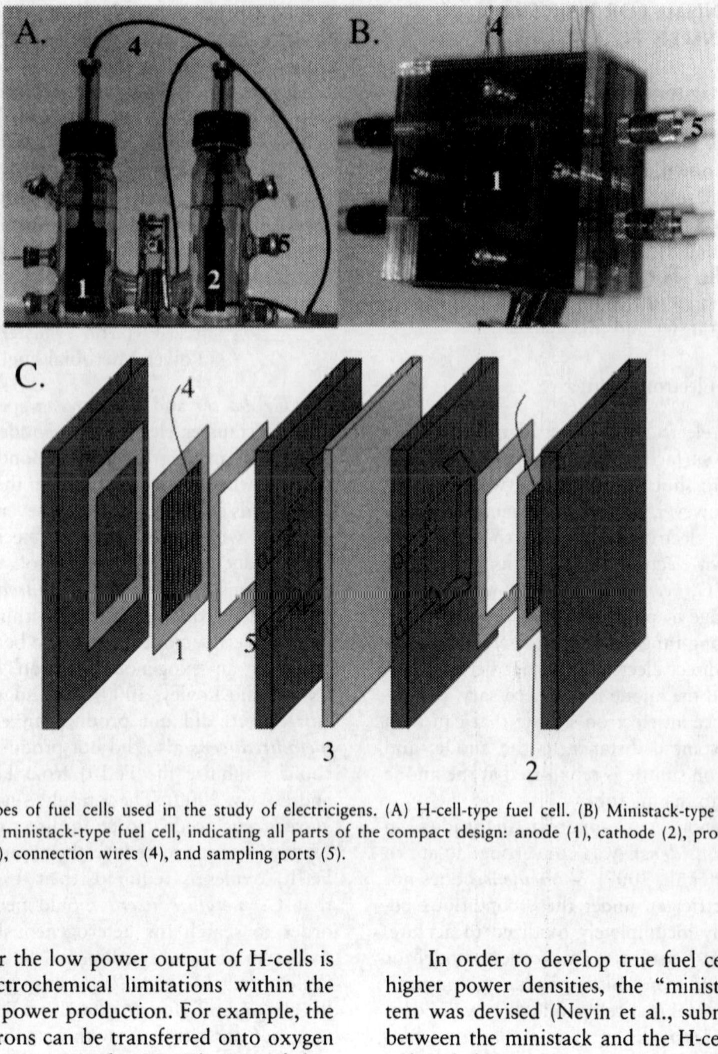

Figure 1. Types of fuel cells used in the study of electricigens. (A) H-cell-type fuel cell. (B) Ministack-type fuel cell. (C) Schematic of ministack-type fuel cell, indicating all parts of the compact design: anode (1), cathode (2), proton exchange membrane (3), connection wires (4), and sampling ports (5).

One reason for the low power output of H-cells is that there are electrochemical limitations within the H-cells that hinder power production. For example, the rate at which electrons can be transferred onto oxygen in H-cells can limit power production. The rate of electron transfer can be increased somewhat if Fe(III) cyanide is added as the oxidant, but these systems are still cathode limited. This limitation can be overcome with a potentiostat, which poises the anode at a fixed potential and ensures that transfer of electrons is not limited by cathodic reactions. This benefit comes at the cost of an input of energy, so a potentiostat is not a true fuel cell, but rather an electrochemically controlled system appropriate for mimicking anode reactions under ideal conditions. With a potentiostat, current levels in H-cells can reach much higher levels (ca. 14 mA versus 0.4 mA), and thus, potentiostat-controlled H-cells are typically run with a continuous input of medium in order to provide sufficient fuel.

In order to develop true fuel cells that can run at higher power densities, the "ministack" fuel cell system was devised (Nevin et al., submitted). Differences between the ministack and the H-cell include a greatly reduced distance between the anode and cathode and a large membrane surface area relative to the anode and cathode chamber volumes. In order to overcome the limitation of the cathode, ministack anodes were made substantially smaller than the cathode. When the current density of the ministack fuel cell (with the anode one-eighth the size of the cathode) was as great as that of an anode poised by the potentiostat, the limiting factor in the ministack became bacterial electron production. After these design and cathode electrochemical limitations were removed, current and power densities of 4.56 A/m^2 and 1.88 W/m^2, respectively, at a nominal resistance of 560 Ω were obtained. Further reduction of the anode volume resulted in power densities of 2.15 kW/m^3 at a nominal resistance of 200 Ω.

MECHANISMS FOR ELECTRON TRANSFER TO ANODES

Optimizing the output of electricigen-based microbial fuel cells will require an understanding of how electricigens transfer electrons to the anode surface. If this mechanism is known, then it may aid in the design of materials that will interact better with the appropriate electrical contacts between the cells and the anodes than the graphite that is presently employed in most microbial fuel cells. Furthermore, such knowledge could help in the design of the anode surface to ensure optimum electron transfer and fuel delivery.

Electron Shuttles

By definition, electricigens directly transfer electrons to the anode surface without the need of an exogenous mediator to shuttle electrons between the cell and the anode. However, in some instances electricigens may facilitate electron transfer to the anode by producing their own electron shuttle. This appears to be the case with *G. fermentans* (Bond and Lovley, 2005). The advantage to producing an electron shuttle is that it permits long-range electrical interaction with an anode because direct electrical contact between the microorganisms and the anode is not necessary. Microorganisms can reduce an electron shuttle that can then diffuse over a substantial distance to the anode, and then once the electron shuttle is reoxidized at the anode it can diffuse back to the microbe.

This phenomenon was apparent in studies in which *Shewanella oneidensis* was converting lactate to electricity (Lanthier et al., 2007). *S. oneidensis* does not function as an electricigen under these conditions because lactate is only incompletely oxidized to acetate. However, *S. oneidensis* serves as an excellent model for interaction of an electron-shuttle-producing microorganism with a microbial fuel cell anode. It is well documented that *S. oneidensis* can produce an electron shuttle to promote electron transfer between the cell and insoluble electron acceptors (Newman and Kolter, 2000; Nevin and Lovley, 2002b; Rosso et al., 2003; Lies et al., 2005). In microbial fuel cells at least as many cells of *S. oneidensis* were planktonic as were attached to the anode surface (Lanthier et al., 2007). Presumably, this was possible because the planktonic cells electrically interacted with the anode surface via electron shuttling.

However, producing an electron shuttle is energetically expensive (Mahadevan et al., 2006). Therefore, in open systems a microorganism can invest substantial energy into the synthesis of an electron shuttle and the electron shuttle may not return to that microbe but be lost in the environment. This may be one reason that *Geobacter* species predominate over *Geothrix* species in open environments whether Fe(III) oxides or electrodes are serving as the electron acceptor (Nevin and Lovley, 2002a; Holmes et al., 2004b). In microbial fuel cells *G. fermentans* may diminish loss of the electron shuttle with the production of a thick extracellular matrix (Bond and Lovley, 2005). However, production of the extracellular matrix also has an energetic cost that would likely lower competitiveness with electricigens, such as *Geobacter* species, that, as discussed below, do not need to make such a matrix.

Direct Electron Transfer—Low-Power Microbial Fuel Cells

Geobacter and *Rhodoferax* species are thought to directly transfer electrons to anodes without the production of an electron shuttle (Bond and Lovley, 2003; Chaudhuri and Lovley, 2003). In the case of *Geobacter* species this concept is based on several lines of evidence, some of which are linked to mechanisms for Fe(III) oxide reduction. Initial studies on Fe(III) oxide reduction demonstrated that *G. metallireducens* could not reduce Fe(III) oxide sequestered within porous beads, but that the Fe(III) oxides within the beads were readily reduced if an exogenous electron shuttle was added (Nevin and Lovley, 2000). This suggested that *G. metallireducens* did not produce an electron shuttle. *G. metallireducens* also did not produce a chelator, which could solubilize the Fe(III) from Fe(III) oxide (Nevin and Lovley, 2000). These results suggested that *G. metallireducens* needed to be in direct contact with Fe(III) oxides in order to reduce them. If direct contact with Fe(III) oxides is required, then it would be expected that *G. metallireducens* would need to be motile in order to search for heterogeneously dispersed Fe(III) oxides in soils and sediments. Subsequent studies revealed that *G. metallireducens* specifically produced flagella and pili during growth on Fe(III) or Mn(IV) oxides, but not during growth on soluble electron acceptors, including chelated Fe(III) (Childers et al., 2002).

Direct electron transfer to Fe(III) oxides would also require electrical contacts between the outer surface of the cell and the mineral surface. The development of a genetic system for *G. sulfurreducens* (Coppi et al., 2001) made it feasible to evaluate the possible role of proteins in extracellular electron transfer. Deleting genes for several outer membrane *c*-type cytochromes (Leang et al., 2003; Mehta et al., 2005) or a putative multicopper protein with possible redox properties (Mehta et al., 2006) inhibited Fe(III) oxide reduction.

Early studies with the H-cell systems, which as described above have low power levels, suggested that *G. sulfurreducens* also required close cell-electron acceptor association for electron transfer to fuel cell anodes.

Completely exchanging the fuel cell medium to remove any planktonic cells and electron shuttles that had been released into the medium did not diminish power production in *G. sulfurreducens* fuel cells (Bond and Lovley, 2003). Scanning electron micrographs revealed that the anodes were covered with cells in a near monolayer, suggesting intimate cell-anode contact, consistent with direct electron transfer from the cells to the anode. Similar results were observed with *R. ferrireducens* (Chaudhuri and Lovley, 2003).

In order to learn more about the electrical contacts between *G. sulfurreducens* and anodes in these early, relatively low-power fuel cell systems, gene transcript levels of *G. sulfurreducens* growing on anode surfaces were compared, using a whole-genome DNA microarray, with transcript levels in *G. sulfurreducens* growing with Fe(III) citrate as the electron acceptor (Holmes et al., 2006). There was a significant increase in transcript levels for a number of genes. The greatest increase in transcript levels was for *omcS*, which encodes an outer membrane *c*-type cytochrome that had previously been shown to be required for reduction of Fe(III) oxide, but not Fe(III) citrate (Mehta et al., 2005). Quantitative PCR demonstrated that transcript levels for *omcS* increased as current increased. When *omcS* was deleted, power production was greatly diminished (Holmes et al., 2006). Expressing *omcS* in *trans* on a plasmid restored power production.

Transcript levels for another outer membrane *c*-type cytochrome, OmcE, were also higher during growth on an anode, and expression of *omcE* also increased with increasing current (Holmes et al., 2006). Current production was initially inhibited in a strain in which *omcE* was deleted, but this strain adapted to produce more current over time. This phenotype of inhibition followed by adaptation is similar to what had previously been reported for growth of this mutant strain on Fe(III) oxide (Mehta et al., 2005).

These results suggested that OmcS and OmcE are involved in electron transfer to the anodes of this first generation of *G. sulfurreducens* microbial fuel cells. Both of these cytochromes are located on the outer surface of the cell, are easily sheared off the cells, and are predicted to be highly hydrophilic. These considerations suggest that they could serve as electrical contacts between *G. sulfurreducens* and fuel cell anodes. This conclusion was supported by the finding that, although pili have been proposed to be the final conduit for electron transfer between *G. sulfurreducens* and Fe(III) oxides (Reguera et al., 2005), a mutant deficient in pilin production produced current as well as wild-type cells in these studies. Electron transfer from cells to anodes should be feasible in such systems, in which most of the cells appear to be in close physical contact with the anode surface.

Electron Transfer via Microbial Nanowires in High-Power Density Fuel Cells

Although electron transfer to anodes via outer surface *c*-type cytochromes is a feasible explanation for current production in low-power density H-cell systems, this cannot be the sole mechanism for electron transfer in systems with higher rates of current flow. This is because biofilms, multiple cell layers thick, form as current levels increase in less electrochemically limited potentiostat or fuel cell systems. This was first noted in H-cells in which the anode potential was poised with a potentiostat. Relatively thick (>40 μm) biofilms of *G. sulfurreducens* formed on the anodes, even when the H-cells were operated in batch mode with a one-time addition of 10 mM acetate (Reguera et al., 2006). Surprisingly, there was a direct correlation between current production and the amount of biomass on the anode or biofilm thickness, even though cells that were not in intimate contact with the anode surface accounted for most of the biomass. These results suggested that cells at substantial distance from the anode were contributing as effectively to current production as were cells closer to the anode surface. Electron transfer from outer surface *c*-type cytochromes to the anode could not account for this phenomenon.

One possibility for long-range electron transfer from *G. sulfurreducens* to the anode is through the specialized pili, known as "microbial nanowires." Previous studies have suggested that these pili are the final conduit between the cell and Fe(III) oxides (Reguera et al., 2005). This concept is based on the following observations: (i) at optimal growth temperatures pili are specifically expressed during growth on Fe(III) oxides, but not on soluble electron acceptors, including Fe(III) citrate; (ii) Fe(III) oxides appear to specifically associate with the pili; (iii) deleting the gene for PilA, the structural pilin protein, eliminates the capacity for Fe(III) oxide reduction; and (iv) measurements with an atomic-force microscope fitted with a conductive tip suggest that the pili are electrically conductive. If microbial nanowires are capable of conducting electrons to Fe(III) oxides, then it might not be surprising if they could also conduct electrons to the anodes of microbial fuel cells.

A mutant in which *pilA* had been deleted produced much less power than wild-type cells in the potentiostat-poised system (Reguera et al., 2006). Maximum current in the mutant strain was ca. 1 mA, and the mutant cells did not stack to form a thick biofilm. Instead, most of the cells were in close contact with the anode. These results demonstrate that the pili are not required for low rates of electron transfer to the anode and explain why in previous studies with H-cell fuel cells deleting *pilA* had no impact on current production.

However, for higher rates of current production the pili are required (Reguera et al., 2006). Pilin production is also required for high-density current levels in the mini-stack fuel cell system (H. Richter, unpublished results).

It is conceivable that the pilin requirement could be due to some feature other than their conductivity. For example, when the *pilA*-deficient mutant was grown on glass surfaces with the soluble electron acceptor, fumarate, the thickness of the biofilm was approximately one-half the thickness of the wild-type cells (Reguera et al., 2007). It is unlikely that pilin would play a role in fumarate reduction, which is catalyzed at the inner membrane (Butler et al., 2006). Thus, these results suggested that the pilin might be important in biofilm structure (Reguera et al., 2007). However, subsequent studies revealed that when the *pilA*-deficient mutant was inoculated into an anode chamber in which the anode was not connected to the cathode and in which fumarate was added to the continuous feed, the mutant formed biofilms as thick as the wild type (Nevin, unpublished results). These results suggest that pili are not necessary for the development of thick biofilms on graphite surface unless that graphite surface is also serving as the electron acceptor.

From these results it seems possible that the pili are responsible for the long-range electron transfer through the biofilms of high-power density of *G. sulfurreducens* microbial fuel cells. However, further investigation of this phenomenon is required. For example, it remains to be determined whether the pili must directly contact the anode surface or whether there is cell-to-cell electron transfer through the biofilm. The longest pili yet observed in cultures of *G. sulfurreducens* have been ca. 20 μm, too short to reach from the outer surface of biofilms over 50 μm thick to the anode surface. It may be that, as previously suggested (Reguera et al., 2005), intertwined pili could produce a conductive network. Another possibility is that pili may connect to cells deeper in the biofilm and that electrons passed to these cells are then transferred toward the anode via the pili of those cells. It is also possible that the pili are just one component of a more complex conductive matrix in the anode biofilm.

The electron transfer via pili to electrodes proposed for *G. sulfurreducens* must be significantly different from the electron transfer to electrodes recently proposed for *S. oneidensis* (Gorby et al., 2006). Although the structures associated with *S. oneidensis* were also suggested to be conductive pili, the structures are ca. 100 nm in diameter, much wider than the 3- to 5-nm-diameter pili of *G. sulfurreducens*. The simple, yet definitive experiment of deleting gene(s) necessary for pilin production and determining if this eliminated formation of the structures was not conducted. Furthermore, although it is suggested that the structures were required for electron transfer to electrodes, this conclusion was not based on any direct evidence. The only related experiment was one in which genes for outer membrane *c*-type cytochromes were deleted. Deletion of these cytochromes resulted in diminished power production. It was suggested that the cytochromes were associated with the pilin, making the pilin conductive. However, an alternative explanation is that these outer membrane cytochromes were associated with other portions of the outer cell surface and that this was the point of cell-electrode contact. This is an important point because unlike *G. sulfurreducens*, *S. oneidensis* does not make thick biofilms on the surfaces of fuel cell anodes. Therefore, long-range electron transfer mechanisms are unlikely to be required, especially when it is considered that *S. oneidensis* also is capable of producing a soluble electron shuttle that sup-

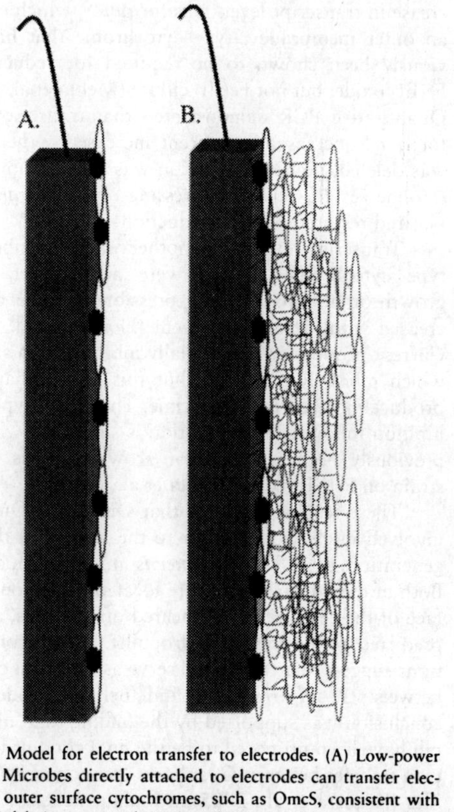

Figure 2. Model for electron transfer to electrodes. (A) Low-power fuel cell. Microbes directly attached to electrodes can transfer electrons via outer surface cytochromes, such as OmcS, consistent with the *omcS* deletion mutant being greatly impaired in power production, but deletion of *pilA* having no impact. (B) High-power density fuel cell. Microbes directly attached to the anode can transfer electrons via outer surface cytochromes. Microbes not attached to the anode can transfer electrons to the electrode and/or other cells closer to the electrodes via conductive pili.

ports substantial growth of planktonic cells in microbial fuel cells.

To date, *G. sulfurreducens* is the only electricigen documented to form thick biofilms on anodes. Studies to determine if this is a widespread capability among electricigens are warranted. However, with the data at hand it appears that *G. sulfurreducens*, and by analogy possibly other electricigens, may have at least two mechanisms for electron transfer to anodes (Fig. 2). Cells in close association with the anode may transfer electrons to the anode via outer surface *c*-type cytochromes because in these systems *G. sulfurreducens* can readily make power without the ability to make pilin, but the *c*-type cytochromes are required. Cells at a distance from the anode too great for electron transfer via outer surface *c*-type cytochromes to be feasible may accomplish electron transfer to the anode via the electrically conductive pili. This concept, though still rudimentary, serves as the present basis for strategies to design fuel cells and/or organisms to increase power production.

ENGINEERING ELECTRICIGEN FUEL CELLS

There are many possible options for enhanced design of microbial fuel cells. Many of these, such as approaches to increase reaction rates at the cathode surface or to overcome potential limitations due to the rate of proton exchange between the anode and cathode compartments, probably do not have a significant microbiological component. However, information on the mechanisms of electron transfer between microorganisms and anodes may lead to other improvements, most notably in the design and/or patterning of anode materials and genetic engineering to enhance electrical connections between electricigens and anodes.

Development of Novel Anode Materials and/or Patterns

The finding that electricigens attached to the anode surface are responsible for power production suggests that power output can be increased by increasing the amount of surface area that the electricigens can colonize while still having ready access to fuel. Simple changes in anode surface area have increased power production (Chaudhuri and Lovley, 2003). It seems likely that much more sophisticated patterning of anode materials is on the horizon.

Evaluation of a wide diversity of materials has demonstrated that *G. sulfurreducens* is capable of transferring electrons to many conductive metals and polymers (Richter et al., 2008; C. Xu and K. P. Nevin, unpublished data). Incorporation of moieties known to be electron acceptors for *Geobacter* species, such as quinones, Fe(III), or Mn(IV), influenced the redox potential at which electrons were transferred to the anodes of sediment fuel cells and temporarily increased power production, but long-term power production was not enhanced because of other limitations in the system (Lowy et al., 2006). In pure-culture studies with *G. sulfurreducens* no anode materials have proven to provide more long-term maximum power than plain graphite electrodes (Nevin, unpublished data). This suggests either that graphite interacts with electron transfer components as well as any other conductive material yet evaluated or that factors other than the rate at which *G. sulfurreducens* can transfer electrons to the anode limit current production. Detailed electrochemical analyses to discriminate between these two possibilities are under way.

Genetic Engineering

The fact that, as far as is known, there has been no evolutionary pressure on microorganisms to produce electricity suggests that electricigens may not be optimized for electricity production. Obvious possible routes to optimization include increasing the number of electrical contacts between electricigens and the anode. However, to date this approach has not been successful. Introducing genes to increase production of the outer surface cytochrome, OmcS, or pilin did not increase power production (M. Izallalen, unpublished data). It is likely that extracellular electron transfer is a complex and highly regulated process that involves the interaction of multiple redox proteins. If so, then it is not surprising that increasing production of just the proteins thought to carry out the terminal electron transfer step may not significantly increase power production.

Adaptive Evolution

The likely complexity of extracellular electron transfer and previous lack of selective pressure on electricigens to produce electricity has suggested that it may be possible to evolve electricigens for enhanced current production (Lovley, 2006a, 2006b). Preliminary studies have demonstrated that *G. sulfurreducens* can be adapted for faster extracellular electron transfer, and this is associated with the accumulation of a number of mutations in the adapted strain (Z. Summers and H. Yi, unpublished data). Which of these mutations confers the capacity for faster extracellular electron transfer is currently under investigation. It has also been possible to adapt *G. sulfurreducens* to transfer electrons at lower potentials, but the nature of this adaptation is not yet known. Not only may adapted strains prove to be power producers superior to the currently available electricigens, but elucidating which mutations lead to increased electricity production should also yield significant insights into the mechanisms for electron transfer to anodes.

ENVIRONMENTAL RESTORATION WITH ELECTRICIGENS

There are several strategies by which electricigen-electrode interactions may be exploited in the bioremediation of organic and/or metal contaminants in sediments and the subsurface. For example, *Geobacter* species have been shown to be important in the degradation of petroleum and landfill leachate contaminants in groundwater (Lovley et al., 1989; Anderson et al., 1998; Rooney-Varga et al., 1999; Roling et al., 2001; Lin et al., 2005). This oxidation of contaminants coupled to the reduction of Fe(III) can be accelerated with the addition of Fe(III) chelators or electron-shuttling compounds (Lovley et al., 1994, 1996a, 1996b). *G. metallireducens*, the pure-culture isolate frequently used as a model for this metabolism, is capable of oxidizing aromatic compounds with an electrode serving as the sole electron acceptor (Bond and Lovley, 2002). Preliminary studies with flowthrough columns of subsurface sediments suggested that providing electrodes as electron acceptors might increase aromatic hydrocarbon degradation (R. T. Anderson, personal communication).

Geobacter species are capable of accepting electrons from electrodes poised at low potential for the reduction of various electron acceptors (Gregory et al., 2004; Gregory and Lovley, 2005). *Geobacter*-catalyzed reduction of U(VI) to U(IV) at electrode surfaces can precipitate uranium contamination from groundwater, precipitating U(IV) on the electrode (Gregory and Lovley, 2005). This potentially attractive approach to extracting uranium contamination from the subsurface needs to be further evaluated in field trials. *G. lovleyi* can use an electrode poised at low potential as the electron donor for the reductive dechlorination of perchloroethylene and trichloroethylene (S. M. Strycharz, K. P. Nevin, R. A. Sanford, F. E. Löffler, and D. R. Lovley, submitted for publication). Thus, electrodes might represent an alternative strategy for introducing electrons into the subsurface to support microbial dechlorination of chlorinated contaminants.

CONCLUSIONS

Of the microorganisms known to contribute to electricity production in microbial fuel cells, only electricigens offer the possibility of highly efficient, self-sustaining conversion of waste organic matter and renewal biomass to electricity. However, the study of electricity production with electricigens is clearly in its infancy. Substantial increases in power output will be required in order for the applications of electricigen-driven microbial fuel cells to expand beyond powering monitoring devices via sediment microbial fuel cells. This will require more information on the mechanisms for extracellular electron transfer in electricigens. The lack of previous evolutionary pressure on microorganisms to produce current suggests that the process can be improved. The coupling of optimized electricigens with advanced microbial fuel cell designs may greatly expand the applications of microbial fuel cells.

Acknowledgments. Research from our laboratory summarized here was supported by the Dept. of Energy's Genomics: GTL and environmental Remediation and Sciences Programs and the Office of Naval Research.

REFERENCES

Anderson, R. T., J. Rooney-Varga, C. V. Gaw, and D. R. Lovley. 1998. Anaerobic benzene oxidation in the Fe(III)-reduction zone of petroleum-contaminated aquifers. *Environ. Sci. Technol.* **32:**1222–1229.

Bennetto, H. P. 1990. 'Bugpower'—electricity from microbes, p. 66–82. *In* A. Scott (ed.), *Frontiers of Science*. Blackwell Publishing, Cambridge, MA.

Bond, D. R., D. E. Holmes, L. M. Tender, and D. R. Lovley. 2002. Electrode-reducing microorganisms that harvest energy from marine sediments. *Science* **295:**483–485.

Bond, D. R., and D. R. Lovley. 2003. Electricity production by *Geobacter sulfurreducens* attached to electrodes. *Appl. Environ. Microbiol.* **69:**1548–1555.

Bond, D. R., and D. R. Lovley. 2005. Evidence for involvement of an electron shuttle in electricity generation by *Geothrix fermentans*. *Appl. Environ. Microbiol.* **71:**2186–2189.

Butler, J. E., R. H. Glaven, A. Esteve-Nunez, C. Nunez, E. S. Shelobolina, D. R. Bond, and D. R. Lovley. 2006. Genetic characterization of a single bifunctional enzyme for fumarate reduction and succinate oxidation in *Geobacter sulfurreducens* and engineering of fumarate reduction in *Geobacter metallireducens*. *J. Bacteriol.* **188:**450–455.

Caccavo, F., Jr., R. P. Blakemore, and D. R. Lovley. 1992. A hydrogen-oxidizing, Fe(III)-reducing microorganism from the great bay estuary, New Hampshire. *Appl. Environ. Microbiol.* **58:**3211–3216.

Chaudhuri, S. K., and D. R. Lovley. 2003. Electricity generation by direct oxidation of glucose in mediatorless microbial fuel cells. *Nat. Biotechnol.* **21:**1229–1232.

Cheng, S., H. Liu, and B. E. Logan. 2006. Increased power generation in a continuous flow MFC with advective flow through the porous anode and reduced electrode spacing. *Environ. Sci. Technol.* **40:**2426–2432.

Childers, S. E., S. Ciufo, and D. R. Lovley. 2002. *Geobacter metallireducens* accesses insoluble Fe(III) oxide by chemotaxis. *Nature* **416:**767–769.

Coppi, M. V., C. Leang, S. J. Sandler, and D. R. Lovley. 2001. Development of a genetic system for *Geobacter sulfurreducens*. *Appl. Environ. Microbiol.* **67:**3180–3187.

Ding, Y. H. R., K. K. Hixson, C. S. Giometti, A. Stanley, A. Esteve-Nunez, T. Khare, S. L. Tollaksen, W. H. Zhu, J. N. Adkins, M. S. Lipton, R. D. Smith, T. Mester, and D. R. Lovley. 2006. The proteome of dissimilatory metal-reducing microorganism *Geobacter sulfurreducens* under various growth conditions. *Biochim. Biophys. Acta* **1764:**1198–1206.

Finneran, K., C. V. Johnsen, and D. R. Lovley. 2003. *Rhodoferax ferrireducens* sp. nov., a psychrotolerant, facultatively anaerobic bacterium that oxidizes acetate with the reduction of Fe(III). *Int. J. Syst. Evol. Microbiol.* 53:669–673.

Gorby, Y. A., S. Yanina, J. S. Mclean, K. M. Rosso, D. Moyles, A. Dohnalkova, T. J. Beveridge, I. S. Chang, B. H. Kim, K. S. Kim, D. E. Culley, S. B. Reed, M. F. Romine, D. A. Saffarini, E. A. Hill, L. Shi, D. A. Elias, D. W. Kennedy, G. Pinchuk, K. Watanabe, S. Ishii, B. E. Logan, and K. H. Nealson. 2006. Electrically conductive bacterial nanowires produced by *Shewanella oneidensis* strain MR-1 and other microorganisms. *Proc. Natl. Acad Sci. USA* 103:11358–11363.

Gregory, K. B., D. R. Bond, and D. R. Lovley. 2004. Graphite electrodes as electron donors for anaerobic respiration. *Environ. Microbiol.* 6:596–604.

Gregory, K. B., and D. R. Lovley. 2005. Remediation and recovery of uranium from contaminated subsurface environments with electrodes. *Environ. Sci. Technol.* 39:8943–8947.

Holmes, D. E., D. R. Bond, and D. R. Lovley. 2004a. Electron transfer by *Desulfobulbus propionicus* to Fe(III) and graphite electrodes. *Appl. Environ. Microbiol.* 70:1234–1237.

Holmes, D. E., D. R. Bond, R. A. O'Neil, C. E. Reimers, L. R. Tender, and D. R. Lovley. 2004b. Microbial communities associated with electrodes harvesting electricity from a variety of aquatic sediments. *Microb. Ecol.* 48:178–190.

Holmes, D. E., S. K. Chaudhuri, K. P. Nevin, T. Mehta, B. A. Methe, A. Liu, J. E. Ward, T. L. Woodard, J. Webster, and D. R. Lovley. 2006. Microarray and genetic analysis of electron transfer to electrodes in *Geobacter sulfurreducens*. *Environ. Microbiol.* 8:1805–1815.

Holmes, D. E., J. S. Nicoll, D. R. Bond, and D. R. Lovley. 2004c. Potential role of a novel psychrotolerant *Geobacteraceae*, *Geopsychrobacter electrodiphilus* gen. nov., sp. nov., in electricity production by the marine sediment fuel cell. *Appl. Environ. Microbiol.* 70:6023–6030.

Katz, E., A. N. Shipway, and I. Wilner. 2003. Biochemical fuel cells, p. 355–381. *In* W. Vielstich, A. Lamm, and H. A. Gasteiger (ed.), *Handbook of Fuel Cells: Fundamentals, Technology, and Applications.* John Wiley & Sons, Ltd., Chichester, United Kingdom.

Khare, T., A. Esteve-Nunez, K. P. Nevin, W. H. Zhu, J. R. Yates, D. R. Lovley, and C. S. Giometti. 2006. Differential protein expression in the metal-reducing bacterium *Geobacter sulfurreducens* strain PCA grown with fumarate or ferric citrate. *Proteomics* 6:632–640.

Kim, H. J., H. S. Park, M. S. Hyun, I. S. Chang, M. Kim, and B. H. Kim. 2002. A mediator-less microbial fuel cell using a metal reducing bacterium, *Shewanella putrefaciens*. *Enzyme Microb. Technol.* 30:145–152.

Lanthier, M., K. B. Gregory, and D. R. Lovley. 2007. Growth with high planctonic biomass in Shewanella oneidensis fuel cells. *FEMS Microbiol. Lett.* 278:29–35.

Leang, C., M. V. Coppi, and D. R. Lovley. 2003. Omcb, a *c*-type polyheme cytochrome, involved in Fe(III) reduction in *Geobacter sulfurreducens*. *J. Bacteriol.* 185:2096–2103.

Lies, D. P., M. E. Hernandez, A. Kappler, R. E. Mielke, J. A. Gralnick, and D. K. Newman. 2005. *Shewanella oneidensis* MR-1 uses overlapping pathways for iron reduction at a distance and by direct contact under conditions relevant for biofilms. *Appl. Environ. Microbiol.* 71:4414–4426.

Lin, B., M. Braster, B. M. Van Breukelen, H. W. Van Verseveld, H. V. Westerhoff, and W. F. M. Roling. 2005. *Geobacteraceae* community composition is related to hydrochemistry and biodegradation in an iron-reducing aquifer polluted by a neighboring landfill. *Appl. Environ. Microbiol.* 71:5983–5991.

Logan, B. E. 2005. Simultaneous wastewater treatment and biological electricity generation. *Water Sci. Technol.* 52:31–37.

Logan, B. E., S. Cheng, V. Watson, and G. Estadt. 2007. Graphite fiber brush anodes for increased power production in air-cathode microbial fuel cells. *Environ. Sci. Technol.* 41:3341–3346.

Lovley, D. R. 1991. Dissimilatory Fe(III) and Mn(IV) reduction. *Microbiol. Rev.* 55:259–287.

Lovley, D. R. 2006a. Bug juice: harvesting electricity with microorganisms. *Nat. Rev. Microbiol.* 4:497–508.

Lovley, D. R. 2006b. Taming electricigens: how electricity-generating microbes can keep going, and going—faster. *Scientist* 20:46.

Lovley, D. R., M. J. Baedecker, D. J. Lonergan, I. M. Cozzarelli, E. J. P. Phillips, and D. I. Siegel. 1989. Oxidation of aromatic contaminants coupled to microbial iron reduction. *Nature* 339:297–299.

Lovley, D. R., J. D. Coates, E. L. Blunt-Harris, E. J. P. Phillips, and J. C. Woodward. 1996a. Humic substances as electron acceptors for microbial respiration. *Nature* 382:445–448.

Lovley, D. R., D. E. Holmes, and K. P. Nevin. 2004. Dissimilatory Fe(III) and Mn(IV) reduction. *Adv. Microbiol. Phys.* 49:219–286.

Lovley, D. R., and E. J. Phillips. 1989. Requirement for a microbial consortium to completely oxidize glucose in Fe(III)-reducing sediments. *Appl. Environ. Microbiol.* 55:3234–3236.

Lovley, D. R., and E. J. P. Phillips. 1988. Novel mode of microbial energy metabolism: organic carbon oxidation coupled to dissimilatory reduction of iron or manganese. *Appl. Environ. Microbiol.* 54:1472–1480.

Lovley, D. R., J. F. Stolz, G. L. Nord, and E. J. P. Phillips. 1987. Anaerobic production of magnetite by a dissimilatory iron-reducing microorganism. *Nature* 330:252–254.

Lovley, D. R., J. C. Woodward, and F. H. Chapelle. 1994. Stimulated anoxic biodegradation of aromatic hydrocarbons using Fe(III) ligands. *Nature* 370:128–131.

Lovley, D. R., J. C. Woodward, and F. H. Chapelle. 1996b. Rapid anaerobic benzene oxidation with a variety of chelated Fe(III) forms. *Appl. Environ. Microbiol.* 62:288–291.

Lowy, D., L. Tender, J. Zeikus, D. Park, and D. R. Lovley. 2006. Harvesting energy from the marine sediment-water interface II—kinetic activity of anode materials. *Biosens. Bioelectron.* 21: 2058–2063.

Mahadevan, R., D. R. Bond, J. E. Butler, A. Esteve-Nunez, M. V. Coppi, B. O. Palsson, C. H. Schilling, and D. R. Lovley. 2006. Characterization of metabolism in the Fe(III)-reducing organism *Geobacter sulfurreducens* by constraint-based modeling. *Appl. Environ. Microbiol.* 72:1558–1568.

Mehta, T., S. E. Childers, R. Glaven, D. R. Lovley, and T. Mester. 2006. A putative multicopper protein secreted by an atypical type II secretion system involved in the reduction of insoluble electron acceptors in *Geobacter sulfurreducens*. *Microbiology* 152:2257–2264.

Mehta, T., M. V. Coppi, S. E. Childers, and D. R. Lovley. 2005. Outer membrane *c*-type cytochromes required for Fe(III) and Mn(IV) oxide reduction in *Geobacter sulfurreducens*. *Appl. Environ. Microbiol.* 71:8634–8641.

Methé, B. A., K. E. Nelson, J. A. Eisen, I. T. Paulsen, W. Nelson, J. F. Heidelberg, D. Wu, M. Wu, N. Ward, M. J. Beanan, R. J. Dodson, R. Madupu, L. M. Brinkac, S. C. Daugherty, R. T. Deboy, A. S. Durkin, M. Gwinn, J. F. Kolonay, S. A. Sullivan, D. H. Haft, J. Selengut, T. M. Davidsen, N. Zafar, O. White, B. Tran, C. Romero, H. A. Forberger, J. Weidman, H. Khouri, T. V. Feldblyum, T. R. Utterback, S. E. Van Aken, D. R. Lovley, and C. M. Fraser. 2003. The genome of *Geobacter sulfurreducens*: insights into metal reduction in subsurface environments. *Science* 302:1967–1969.

Methé, B. A., J. Webster, K. P. Nevin, J. Butler, and D. R. Lovley. 2005. DNA microarray analysis of nitrogen fixation and Fe(III) reduction in *Geobacter sulfurreducens*. *Appl. Environ. Microbiol.* 71:2530–2538.

Nevin, K. P., and D. R. Lovley. 2000. Lack of production of electron-shuttling compounds or solubilization of Fe(III) during reduction of insoluble Fe(III) oxide by *Geobacter metallireducens*. *Appl. Environ. Microbiol.* 66:2248–2251.

Nevin, K. P., and D. R. Lovley. 2002a. Mechanisms for Fe(III) oxide reduction in sedimentary environments. *Geomicrobiol. J.* 19:141–159.

Nevin, K. P., and D. R. Lovley. 2002b. Novel mechanisms for accessing insoluble Fe(III) oxide during dissimilatory Fe(III) reduction by *Geothrix fermentans*. *Appl. Environ. Microbiol.* 68:2294–2299.

Newman, D. K., and R. Kolter. 2000. A role for excreted quinones in extracellular electron transfer. *Nature* 405:93–97.

Park, D. H., M. Laivenieks, M. V. Guettler, M. K. Jain, and J. G. Zeikus. 1999. Microbial utilization of electrically reduced neutral red as the sole electron donor for growth and metabolite production. *Appl. Environ. Microbiol.* 65:2912–2917.

Postier, B. L., R. Didonato, Jr., K. P. Nevin, A. Liu, B. Frank, D. R. Lovley, and B. A. Methé. 2007. Benefits of in-situ synthesized microarrays for analysis of gene expression in understudied microorganisms. *J. Microbiol. Methods* [Epub ahead of print.]

Potter, M. C. 1910. On the difference of potential due to the vital activity of microorganisms. *Proc. Univ. Durham Phil. Soc.* 3:245–249.

Potter, M. C. 1911. Electrical effects accompanying the decomposition of organic compounds. *Proc. R. Soc. Lond. B* 84:260–276.

Rabaey, K., and W. Verstraete. 2005. Microbial fuel cells: novel biotechnology for energy generation. *Trends Biotechnol.* 23:291–298.

Reguera, G., K. D. McCarthy, T. Mehta, J. S. Nicoll, M. T. Tuominen, and D. R. Lovley. 2005. Extracellular electron transfer via microbial nanowires. *Nature* 435:1098–1101.

Reguera, G., K. P. Nevin, J. S. Nicoll, S. F. Covalla, T. L. Woodard, and D. R. Lovley. 2006. Biofilm and nanowire production leads to increased current in *Geobacter sulfurreducens* fuel cells. *Appl. Environ. Microbiol.* 72:7345–7348.

Reguera, G., R. B. Pollina, J. S. Nicoll, and D. R. Lovley. 2007. Possible non-conductive role of *Geobacter sulfurreducens* pili nanowires in biofilm formation. *J. Bacteriol.* 189:2125–2127.

Reimers, C. E., L. M. Tender, S. Fertig, and W. Wang. 2001. Harvesting energy from the marine sediment-water interface. *Environ. Sci. Technol.* 35:192–195.

Richter, H., M. Lanthier, K. P. Nevin, and D. R. Lovley. 2007. Lack of electricity production by *Pelobacter carbinolicus* indicates that the capacity for Fe(III) oxide reduction does not necessarily confer electron transfer ability to fuel cell anodes. *Appl. Environ. Microbiol.* 73:5347–5353.

Richter, H., K. McCarthy, K. P. Nevin, J. P. Johnson, V. M. Rotello, and D. R. Lovley. 2008. Electricity generation by *Geobacter sulfurreducens* attached to gold electrodes. *Langmuir* [Epub ahead of print.].

Roling, W. F. M., B. M. Van Breukelen, B. L. Braster, and H. W. Van Verseveld. 2001. Relationships between microbial community structure and hydrochemistry in a landfill leachate-polluted aquifer. *Appl. Environ. Microbiol.* 67:4619–4629.

Rooney-Varga, J. N., R. T. Anderson, J. L. Fraga, D. Ringelberg, and D. R. Lovley. 1999. Microbial communities associated with anaerobic benzene degradation in a petroleum-contaminated aquifer. *Appl. Environ. Microbiol.* 65:3056–3064.

Rosso, K. M., J. M. Zachara, J. K. Fredrickson, Y. A. Gorby, and S. C. Smith. 2003. Nonlocal bacterial electron transfer to hematite surfaces. *Geochim. Cosmochim. Acta* 67:1081–1087.

Shukla, A. K., P. Suresh, S. Berchmans, and A. Rajendran. 2004. Biological fuel cells and their applications. *Curr. Sci.* 87:455–468.

Tender, L. M., C. E. Reimers, H. A. Stecher, D. E. Holmes, D. R. Bond, D. A. Lowy, K. Pilobello, S. J. Fertig, and D. R. Lovley. 2002. Harnessing microbially generated power on the seafloor. *Nat. Biotechnol.* 20:821–825.

Bioenergy
Edited by J. Wall et al.

Chapter 24

Microbial Fuel Cells as an Engineered Ecosystem

PETER AELTERMAN, KORNEEL RABAEY, LIESJE DE SCHAMPHELAIRE, PETER CLAUWAERT, NICO BOON, AND WILLY VERSTRAETE

The pursuit of the highest energetic gain shapes microbial communities as they develop in all kinds of ecosystems. The ability of microorganisms to maximize their energy gain during the conversion of substrate determines their chances to survive, grow, and become dominant within a microbial community. In this context, not only the availability of substrate and nutrients is a necessity but also the presence of an appropriate electron acceptor. The anode and cathode compartments of a microbial fuel cell (MFC) are engineered ecosystems in which microbial communities function in close interaction with a solid but conductive surface. As both anode and cathode electrodes interfere with the energy metabolism of the microorganisms, they impose a selective pressure which is normally not present in natural environmental ecosystems. For communities growing in MFCs, the availability of both substrate and electron acceptors becomes a spatially dependent issue. Moreover, the solid nature of the electrode forces bacteria to develop specialized strategies in order to transfer electrons to, or derive electrons from, the electrode. These factors have major consequences for the bacteria developing within the biofilm attached to the surface of the electrodes. Within this chapter we discuss the driving force for bacteria to generate and consume electricity. We aim to provide an overview of the microbial communities found in MFCs, the interactions that drive the community structure, the processes performed by the communities, and how engineering affects the microbial resources within MFCs.

THE GENERAL ENERGY METABOLISM OF MICROORGANISMS

Respiring Bacteria Gain Energy by the Transfer of Electrons to External Acceptors

During respiration, microorganisms liberate electrons from an electron-rich substrate at a low redox potential and transfer these electrons through a number of electron transport complexes through the cell membrane, where a final electron acceptor is reduced (Schlegel, 1992). Microorganisms do not use the energy produced by the flow of electrons in a direct way; instead, the flow of electrons is used to create a proton gradient across the cell membrane as described by Mitchell (1961). The energy released by the inward flux of the protons through a membrane complex (ATP synthase) is used to regenerate energy carrier molecules, such as ATP. By creating this proton gradient, the potential difference between the electron donor (i.e., the substrate at low potential) and the electron acceptor at a high potential is translated into a process for the generation of energy. The higher the potential difference between the electron donor and electron acceptor, the higher the proton-driven potential difference and the higher the potential amount of ATP which can be refueled. Respiring microorganisms can use a large variety of different electron acceptors, ranging from oxygen, nitrate, iron, and manganese oxides to sulfate, but their ability to use the acceptor with the highest redox potential will increase their energy for growth

Peter Aelterman, Liesje De Schamphelaire, Peter Clauwaert, Nico Boon, and Willy Verstraete • Laboratory of Microbial Ecology and Technology (LabMET), Ghent University, Coupure Links 653, B-9000 Ghent, Belgium. **Korneel Rabaey** • Advanced Water Management Centre, University of Queensland, Brisbane, QLD 4072 Australia.

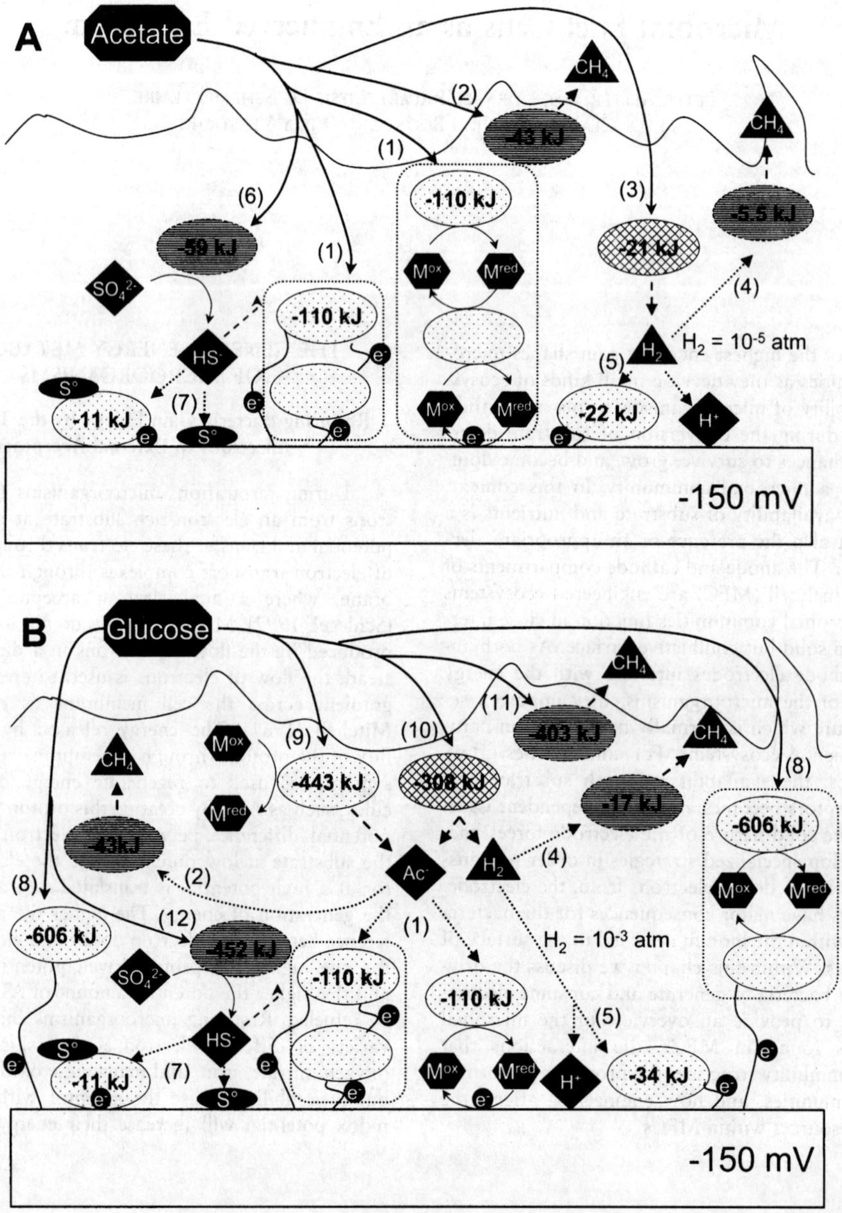
A
Acetate
CH4
(2)
-43 kJ
CH4
(1)
(6)
-110 kJ
(3)
-5.5 kJ
-59 kJ
(1)
-21 kJ
SO4 2-
-110 kJ
Mox
Mred
(4)
HS-
H2 = 10-5 atm
S°
(7)
(5)
-11 kJ
-22 kJ
S°
H+
-150 mV
B
Glucose
CH4
(11)
Mox
(9)
(10)
-403 kJ
CH4
CH4
-443 kJ
-308 kJ
(8)
Mred
-17 kJ
-606 kJ
-43kJ
Ac-
H2
(4)
Mox
Mred
(8)
(2)
(1)
-606 kJ
(12)
-452 kJ
H2 = 10-3 atm
SO4 2-
-110 kJ
-110 kJ
(5)
HS-
S°
(7)
-11 kJ
H+
-34 kJ
S°
-150 mV

(Madigan et al., 2000) and is their incentive to explore alternative electron acceptors.

Fermenting Bacteria Generate Energy by the Internal Recirculation of Electrons

In many environments, the availability of electron acceptors is limited, which impedes microorganisms from using the respiratory pathway. In these cases, which are abundant in many environmental conditions, fermenting organisms are likely to establish themselves. Fermentation is an ATP-regenerating metabolic process in which degradation products or organic substrates serve as electron donors as well as electron acceptors (Schlegel, 1992). The advantage of this pathway is that fermenting organisms are able to grow in numerous environments which are nonsupportive for organisms that use the respiratory pathway only because suitable electron acceptors are lacking. Fermenting organisms are important within the overall microbial processes in nature for their ability to degrade polymeric compounds into readily degradable monomers. However, fermentation is energetically far less efficient than respiration as only 1 to 4 mol of ATP are formed during the fermentation of glucose whereas 26 to 38 mol of ATP are formed during the aerobic degradation of glucose (Schlegel, 1992). This is also reflected in the Gibbs free energy value, which for the fermentation of glucose is lower than for aerobic respiration by a factor of 7. The remainder of the Gibbs free energy is not lost but is contained within the excreted fermentation products such as volatile fatty acids, hydrogen, alcohols, and many more. The trade-off between their low energetic yield and their ability to colonize niches devoid of readily available electron donors and acceptors determine the success of fermenting organisms in many ecosystems.

THE GENERATION OF ELECTRICITY USING MICROORGANISMS

The Biocatalyst Liberates Electrons from Substrates

Microbial electricity generation in MFCs relies on the drive of bacteria to acquire maximum energy. Unlike natural environmental systems, the anode compartment of an MFC is an engineered environment in which the availability of soluble electron acceptors is limited. The main electron acceptor present, enabling bacteria to use respiratory processes, is a solid conductive electrode. The higher amount of metabolic energy released by transferring electrons to the electrode rather than other electron acceptors drives bacteria to colonize the electrode and develop electron transfer strategies. The result is a process in which bacteria serve as a biocatalyst to transform an electron-rich substrate into electrons, which are transferred to the electrode; into protons, which migrate to the cathode; and into oxidized products, which leave the reactor. The electrons flow through an external electrical circuit to the cathode electrode, where a final electron acceptor is reduced by a chemical (Zhao et al., 2006) or microbial catalyst (Clauwaert et al., 2007a, 2007b). Due to the potential difference between the anoxic anode and the cathode compartment, the flow of electrons generates electrical power.

The Electrode Potential Controls the Energy Gain of the Biocatalyst

The potential difference between the electron donor and acceptor couple, together with the amount of electrons transferred, determines the energy available for bacteria. The redox potential of the soluble electron donor and electron acceptor can be derived from their standard reduction potential (Thauer et al.,

Figure 1. Overview of the possible microbial degradation pathways of acetate (A) and glucose (B) within the biofilm on the anode of an MFC. Fermenting cells are crosshatched. Cells using alternative electron acceptors are shaded. Microorganisms transferring electrons to the electrode by using mediators (M) or nanowires are dotted. Full lines indicate the use of acetate (A) or glucose (B) as the electron donor, while dotted lines indicate the use of intermediate electron donors. Striped lines represent possible losses of electron donors. Triangles represent gaseous components, and diamonds are aqueous components. Gibbs free values are expressed per mole of (intermediate) substrate as shown in reactions 1 to 12 and are calculated based on Thauer et al. (1977) and assuming substrate and intermediary product concentrations of 0.01 M, an anode potential of −150 mV versus standard hydrogen for both direct and mediated electron transfer, a pH of 7, and hydrogen partial pressures as indicated in panels A and B. Interspecies electron transfer is grouped by a frame; the division of the available Gibbs free energy is not taken into account. Reactions are as follows: (1) $CH_3COO^- + 4H_2O \rightarrow 2HCO_3^- + 9H^+ + 8e^-$; (2) $CH_3COO^- + H_2O \rightarrow CH_4 + HCO_3^-$; (3) $CH_3COO^- + 4H_2O \rightarrow 4H_2 + 2HCO_3^- + H^+$; (4) $H_2 + 1/4HCO_3^- + 1/4H^+ \rightarrow 1/4CH_4 + 3/4H_2O$; (5) $H_2 \rightarrow 2H^+ + 2e^-$; (6) $CH_3COO^- + SO_4^{2-} \rightarrow 2HCO_3 + HS^-$; (7) $HS^- \rightarrow S^0 + H^+ + 2e^-$; (8) $C_6H_{12}O_6 + 12H_2O \rightarrow 6HCO_3^- + 30H^+ + 24e^-$; (9) $C_6H_{12}O_6 + 4H_2O \rightarrow 2CH_3COO^- + 2HCO_3^- + 12H^+ + 8e^-$; (10) $C_6H_{12}O_6 + 4H_2O \rightarrow 2CH_3COO^- + 2HCO_3^- + 4H_2 + 4H^+$; (11) $C_6H_{12}O_6 + 3H_2O \rightarrow 3CH_4 + 3HCO_3^- + 3H^+$; (12) $C_6H_{12}O_6 + 3SO_4^{2-} \rightarrow 6HCO_3^- + 3HS^- + 3H^+$.

1977). However, this potential can rise or decrease depending on a number of factors including the ratio of the amount of oxidized and reduced species, the pH, and the temperature of the surrounding medium as indicated by the Nernst equation (Thauer et al., 1977). The potential of the graphite electrodes commonly used in MFCs, however, depends not only on the parameters of the surrounding medium, but also on the activity of the (bio)catalyst and the external resistance applied. The latter represents an important tool to change the potential of the electrodes. Figure 1 shows the Gibbs free energy values for several electron donor and acceptor couples within an MFC ecosystem. The value of the Gibbs free energy reflects the energy potentially available for the bacteria and is calculated as $\Delta G = -n.F.\Delta E$, where n is the number of electrons transferred within the reaction, F is the Faraday constant (96,485 C/mol e^-), and ΔE is the electrochemical potential difference (in volts) between the electron donor and acceptor. The more negative the Gibbs free energy value, the higher the potential energy gain for the bacteria and the higher the potential growth yield (Heijnen, 1999).

Microorganisms strive to gain as much energy as possible; the use of the donor/acceptor couple yielding the highest energy gain enables a microorganism to grow faster and to outcompete other species, which increases their opportunities to survive. Freguia et al. (2007) found that the microbial growth yield in MFCs depended on the type of substrate and the external resistance applied to the MFC and ranged from 0 to 0.54 g of biomass C formed per g of substrate C used. Glucose has a lower standard potential than acetate and should enable the bacteria to generate more energy and to increase their growth yield. However, the data of Freguia et al. (2007) did not support this, probably due to a lower adaptation of the microbial community to glucose as electron donor. A lowering of the external resistance resulted in an increase of the microbial growth yield. When the external resistance (R) is decreased, the ratio between the MFC voltage (U) and the current (I) generated must decrease according to Ohm's law (R = U/I). This results in a decrease of the MFC voltage, which is translated in an increase and decrease of the potential of the anode and cathode, respectively, and/or an increase of the produced current. From a thermodynamic point of view, an increase of the anode potential results in a more negative Gibbs free energy value for the bacteria. An increase of the current for a set anode potential will not change the Gibbs free energy value but will affect the electron transfer kinetics by allowing a higher electron transfer rate and subsequent increase of the rate of the metabolic processes. From an engineering point of view, one wants to gain as many of the electrons at the highest potential difference, i.e., at the lowest anode potential and at the highest cathode potential. The electrode potential represents an important tool to control and increase the biocatalyst activity in relation to electricity generation. Moreover, the electrode potential will, as the key factor in the energy metabolism, determine the trade-off between fermenting and respiring organisms, thereby influencing the microbial composition.

MICROBIAL DIVERSITY IN MFCs

Several studies have described the microbial composition in MFCs. When comparing these data, several conclusions regarding the microbial community composition can be derived.

First, various inocula can be used to successfully enrich electron-transferring organisms in an MFC. Activated sludge (Lee et al., 2003), anaerobic sludge (Rabaey et al., 2003), domestic wastewater (Liu et al., 2004), industrial wastewater (Prasad et al., 2006), marine sediment (Bond et al., 2002), and aquatic sediments (Holmes et al., 2004b) have been reported as sources of anodophilic microorganisms. Although the best results in terms of power output thus far have been described when using activated or anaerobic sludge, most environments contain sufficient anodophilic microorganisms that can serve as an appropriate inoculum.

Second, several authors have concluded that MFCs strongly enrich organisms that utilize the electrode as final electron acceptor (Kim et al., 2006), both in a direct and in an indirect way. In all studies investigating the microbial composition, the dominant species present in the enriched cultures were profoundly different from the original dominant species within the inoculum. For an evolving microbial community, it was found by recording polarization curves, monitoring the internal resistance, and analyzing the microbial community that both the electrochemical and the microbial features evolved to a higher level, resulting in a tripling of the power output during a 3-month period (Aelterman et al., 2006b). The significant increase of the power output and maximum current generation could likely be attributed to an evolution of the community, which was dominated by a gram-positive species identified as *Brevibacillus* sp. During the evolution of the community, a transient change of the voltage curve at high currents was noted (Fig. 2). The steep decrease of the voltage at high currents (Time 2) was attributed to an insufficient transfer of fuel or removal of waste products to or from the biocatalytic sites within the community. However, when the MFC performance was maximal (Time 3) the steep voltage decrease at increasing currents disappeared. Clearly, the selective pressure of the MFC biases the community composition and biofilm structure towards electrochemical competence.

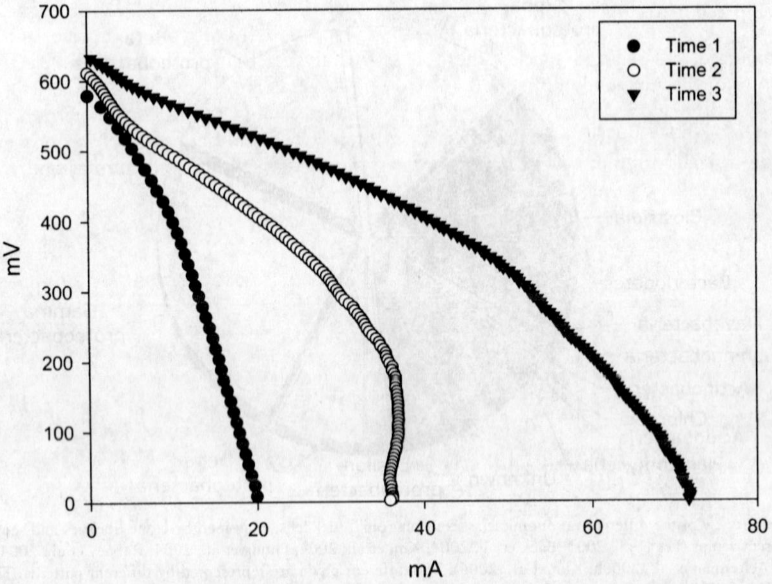

Figure 2. Evolution of the polarization curves of an acetate-fed MFC during a 3-month period (adapted from Aelterman et al., 2006b). While at Time 1 (152 days after start-up) a linear decrease of the voltage is noted at increasing currents, the polarization curve at Time 2 (175 days after start-up) is dominated by a sharp decrease of the voltage at maximum current production. At Time 3 (201 days after start-up) the overall performance of the MFC had increased and maximum current production had tripled. In addition, the sharp decrease of the voltage at high currents, attributed to mass transfer losses, was not noted any more.

Third, although the MFC is an appropriate device to enrich electricity-producing communities, a typical electricity-generating microbial community has not been established yet. Figure 3 gives an overview of the different taxonomic classes in microbial fuel cells, examined by clone libraries and sequencing in eight different studies (Lee et al., 2003; Back et al., 2004; Kim et al., 2004, 2006; Phung et al., 2004; Rabaey et al., 2004; Logan et al., 2005; Aelterman et al., 2006b). The majority of reported taxonomic classes are *Proteobacteria* (64%) followed by *Firmicutes* (13%) and nonclassified sequences (13%). Although the gram-negative bacteria represent almost two-thirds of the reported organisms, the presence of gram-positive *Firmicutes* seems to be a functional necessity. There is a distinct difference between the cell wall of gram-positive and gram-negative bacteria: gram-positive bacteria have, in contrast to gram-negative bacteria, a thick peptidoglycan layer surrounding the cell; moreover, they lack a periplasmatic envelope, which is reported to contain electron transfer functionalities (Madigan et al., 2000). Because of this, gram-positive bacteria were expected to participate less within the direct electron transport. Yet, Fig. 3 shows that gram-positive bacteria encompass 13% of the sequenced species in the selected research. Kim et al. (2006) have shown, by staining the gram-positive and gram-negative species, that gram-positive organisms were positioned on top of an apparent (mono)layer of gram-negative microorganisms. Besides, several researchers found a community in which gram-positive microorganisms were present such as *Brevibacillus* species (Aelterman et al., 2006a), *Clostridium butyricum* (Park et al., 2001), and *Enterococcus* sp. (Rabaey et al., 2004). Nevertheless, initial MFC research focused on the use of gram-negative metal-reducing organisms such as *Geobacter* spp. and *Shewanella* spp., which have been investigated intensively for their electron-transferring properties (Kim et al., 1999; Newman and Kolter, 2000; Bond and Lovley, 2003; Reguera et al., 2005; Gorby et al., 2006). It was found that *Shewanella* spp. were able to generate high-power outputs in a miniaturized MFC (Ringeisen et al., 2006), axenic *Geobacter* cultures are known to generate electricity in MFCs (Bond and Lovley, 2003), and both *Geobacter* spp. and *Shewanella* spp. have been reported in enriched MFC microbial communities (Logan et al., 2005; Aelterman et al., 2006b; Kim et al., 2006). However, the dominance of *Geobacter* spp. and *Shewanella* spp. has not been found to be obligatory to support high electrical power outputs using mixed microbial communities (Rabaey et al., 2004; Aelterman et al., 2006b).

A final remark about the community composition concerns the function of the MFC reactor itself. Kim et

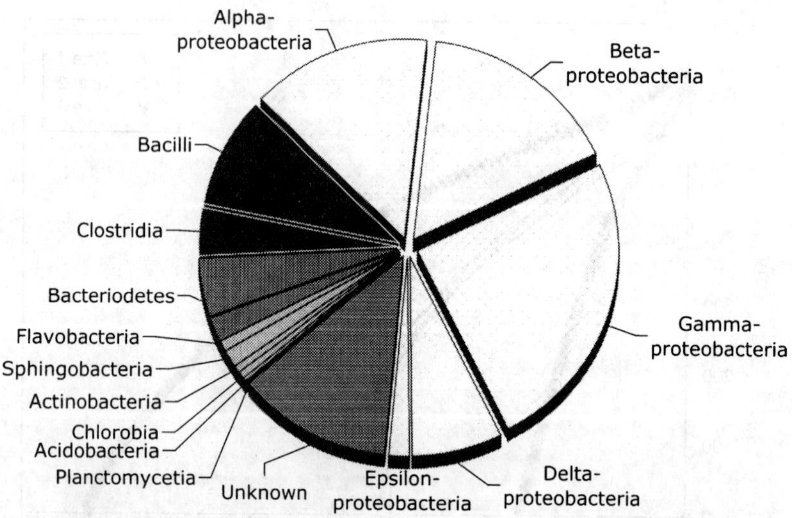

Figure 3. Overview of the different taxonomic classes in microbial fuel cells, examined by clone libraries and sequencing in eight different setups (Lee et al., 2003; Back et al., 2004; Kim et al., 2004; Phung et al., 2004; Rabaey et al., 2004; Logan et al., 2005; Aelterman et al., 2006b; Kim et al., 2006). The different phyla are represented by different patterns. The relative amount of the different phyla is given between brackets: *Proteobacteria*, dotted pattern (64%); *Firmicutes*, black (13%); *Bacteroides*, vertical pattern (7%); other phyla, crosshatched (3%); nonclassified sequences, horizontal pattern (13%).

al. (2006) have investigated the community structure, compartmentalization, and activity in an MFC. The MFCs were inoculated with activated sludge and fed continuously with artificial wastewater containing glucose and glutamate. The differences between the bacterial compositions of the communities located in the anodic biofilm, the bacterial clumps, and the planktonic and membrane biofilm were investigated and revealed that *Gammaproteobacteria* phylotypes were present at higher numbers within libraries from the bacterial clumps and electrode biofilm than in other parts of the fuel cell. This might be caused by the biofilm-forming capacities of several *Gammaproteobacteria* (Sutherland, 2001).

Although several genera are frequently reported, no typical MFC community composition is found or selected for. From the large diversity and significant number of uncultured species, it is likely that the capacity to use an electrode—and insoluble electron acceptors in general—as the final electron acceptor is ubiquitous in nature. This is very promising, both for finding new anodophilic microorganisms and for the robustness of the electricity-generating processes. The more numerous the microorganisms capable of growing within the anode compartment, the higher the likelihood that all niches within the ecosystems will be occupied, guaranteeing a stable and effective operation of the microbial electricity-generating process.

MICROBIAL INTERACTIONS IN MFCs

The large variety of microorganisms found in MFCs suggests that many organisms can interact within the electricity-generating process. Unlike the development of biofilms on nonconductive carrier materials, the biofilms growing on the electrodes face a duality between the availability of the electron donor (maximum at the top of the biofilm) on the one hand and the proximity of the electron acceptor on the other hand (closest at the bottom of the biofilm). The biofilm structure needs to be optimized for a sufficient influx of substrate and efflux of waste products while maximizing the transfer of electrons to the electrode. The influences of both the electron transfer interactions and the substrate transport within the biofilm on the development of the microbial community are discussed below.

Electron Transfer Interactions within the Biofilm

Several electron transfer mechanisms have been studied and unraveled thus far, but the complex electron transfer interactions of the bacteria within the biofilm remain unknown. The electron transfer mechanisms thus far found in MFCs are closely related to the mechanisms investigated for dissimilatory metal-reducing microorganisms. The main electron transfer pathways are the direct electron transfer by cell contact (Bond and Lovley, 2003) or nanowires (Reguera et al., 2005) and the electron transfer using mediating molecules (Newman and Kolter, 2000). Both pathways enable microorganisms to occupy different niches within the community.

Direct electron transfer

The direct membrane-associated electron transfer by the involvement of cytochromes has been described for *Shewanella* and *Geobacter* species (Kim et al., 2002; Bond and Lovley, 2003). Lee et al. (2003) conclude that direct cell contact is the main route of electron transfer in an acetate-enriched MFC. They suggest that the electrons are probably transferred to the electrode at an early stage in the electron transfer chain originating from coenzyme Q and that the electrons are transferred to the cell surface through unknown mediators. Recently, nanowires have been described as a more effective way to directly transfer electrons to the electrode within a biofilm structure (Reguera et al., 2005; Gorby et al., 2006). Nanowires (20-μm length) are pili or pilus-like structures which are conductive and are found to attach to the electrode or possibly also to other cells. They have been reported for *Geobacter* and *Shewanella* species, and possibly a number of other species possess them as well, but their exact role within the transport of electrons to the surface of graphite electrodes is not yet fully understood. It seems likely that pili increase the power output because they promote long-range electron transfer across the multilayer biofilms on anodes (Reguera et al., 2005). Research has shown that *Geobacter sulfurreducens* nanowires also have a (nonconductive) role in the biofilm formation (Reguera et al., 2007). Furthermore, the nanowires in *Geobacter*, interacting in a different way with the surface of metal oxides compared to the surface of graphite electrodes, are not an absolute necessity to generate current (Holmes et al., 2006).

Within the bacterium itself, electrons can be transported over considerable distances and at high rates using tunnelling and redox complexes (Page et al., 1999). However, tunnelling through vacuum is clearly much slower than electron tunnelling between covalently bridged redox centers in synthetic systems (Beratan et al., 1992). Based on Page et al. (1999), the distance between the exterior of the cell wall and the electrode must be in the order of 0 to 2 nm to effectively and rapidly transfer electrons from the cell to the electrode. As a consequence, bacteria using the direct electron transfer pathway are required to be, at least with the electron-transferring complex, in very close contact with the electrode surface. However, in order to reach within 0 to 2 nm of a dolomite surface, *Pseudomonas aeruginosa* has to overcome an energy barrier of at least 170 kT (Grasso et al., 1996). Although the microbial surface thermodynamics in MFCs have not been investigated yet and are dependent on the charge and size of the bacteria, the microbial exopolysaccharide production, the ionic concentration of the medium, the properties and charge of the support material, and many other factors, these initial theoretical observations transform the direct cell contact mechanism into a pathway with thermodynamic constraints. However, the development of nanowires enables bacteria to touch the electrode with an extremity without the necessity to fully or partially cover the electrode surface. This greatly lowers the energy required to reach the electrode within the 0- to 2-nm zone and improves the chance of microorganisms, using conductive extremities, to participate within the electricity generation. Pili as long as 20 μm have been observed on *G. sulfurreducens*. Moreover, cell-to-cell electron transfer via intertwined pili might establish a "nano power grid" in which the pilus network transfers electrons through the 40- to 50-μm-thick anode biofilm (Reguera et al., 2006). Figure 4 visualizes the distance dependency of the several electron transfer mechanisms within a 100 μm biofilm.

Mediators

Mediators or electron-shuttling molecules are soluble molecules which can be reversibly oxidized and reduced. They function as a shuttle for electrons from the microorganism to the electrode (Allen and Bennetto, 1993). In the early studies of MFCs, exogenous mediators like neutral red, hexacyanoferrate, thionin, or 2-hydroxy-1,4-naphthoquinone were generally added to the systems in order to increase the electron transfer rate of nonenriched microbial communities (Kim et al., 2000; Park and Zeikus, 2000). Rabaey et al. (2004, 2005a) have reported that the production of endogenous mediators produced by the microorganisms within the microbial community can improve the electricity generation of MFCs. *Pseudomonas* spp. and *Geothrix fermentans* have been described as mediator-producing organisms within a microbial community (Rabaey et al., 2004, 2005a; Bond and Lovley, 2005). Likely, this feature is not limited to these species alone, but thus far endogenous production has not been studied for other organisms in MFCs. The presence of environmental molecules with electron-mediating properties such as humic acids in the reactor medium itself may prove sufficient to establish the electrical link, as shown for *Desulfitobacterium hafniense* (Milliken and May, 2007).

To support this type of electron transfer, a sufficiently high concentration of oxidized redox shuttles in the bacterial microenvironment is necessary, both from a thermodynamic and from a kinetic point of view. The

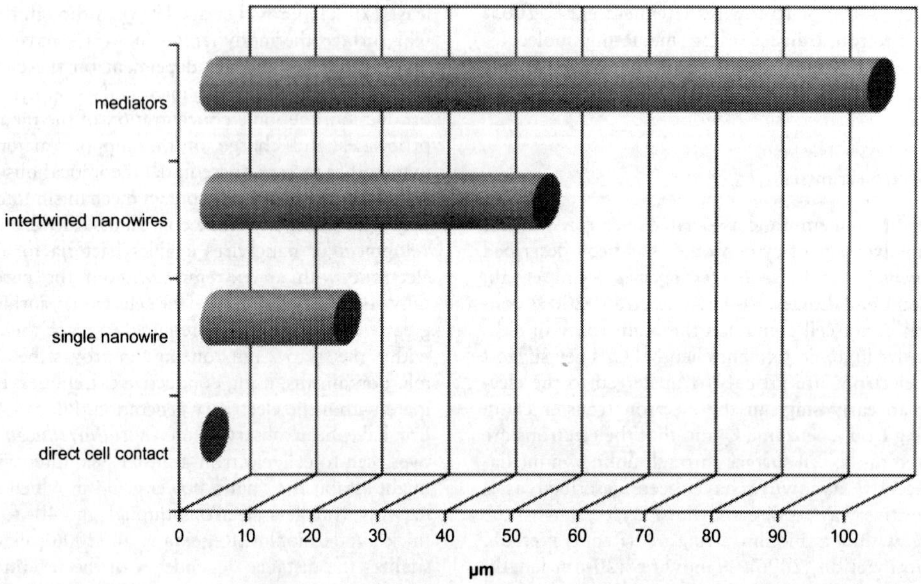

Figure 4. Overview of the potential working range of the different direct and indirect electron transfer mechanisms (direct cell contact, single nanowires, intertwined nanowires, and mediators), assuming a biofilm thickness of 100 µm.

level of oxidized mediators can be kept high by a rapid electron transfer to the electrode or by the use of reduced mediators as an electron donor by other microorganisms within the community (Hernandez and Newman, 2001). Although the addition of mediators to planktonic cells has resulted in an increase of the current (Park and Zeikus, 2000), the use of mediators in biofilm systems appears more advantageous. The production of an exopolysaccharide matrix in biofilms slows down the diffusion of small molecules between the biofilm and the external electrolyte and could therefore support an effective recycling of mediators within the biofilm (Costerton, 1995). Calculations by Rabaey et al. (2007) have indicated that, influenced by electrostatic interactions, redox shuttles can be attracted by electron acceptors such as an electrode. We hypothesize that mediators are potentially available within the complete depth of the biofilm. When the supply of oxidized mediators, resulting in a redox buffer, is ensured, microorganisms reducing mediators within the whole depth of the biofilm are able to participate in the electron transfer, as indicated in Fig. 4.

Transferring electrons by cell bridging

The presence of flagellum-like filaments between the propionate fermenter *Pelotomaculum thermopropionicum* and a methanogen, *Methanothermobacter thermautotrophicus*, which morphologically resembled the previously described nanowires for *Shewanella* sp. (Gorby et al., 2006) has been visualized (Ishii et al., 2005). On the microscopic pictures of *Shewanella* sp. biofilms on electrodes, the direct contact of cells using nanowires was also visualized (Gorby et al., 2006). These data suggest the possibility that the microorganisms themselves may function as a docking station to transfer electrons from one organism to the other or to the electrode (Logan and Regan, 2006). The interspecies exchange of electrons or reduced and oxidized mediators may further extend the range of microorganisms able to transfer electrons to the electrode within the biofilm. This form of synthrophy can be sustained due to redox gradients which may evolve within the biofilm. These gradients affect the potential of the mediator and its function to serve as an electron acceptor or a donor. The outcome is a division of the available energy between the synthrophic microorganisms interacting as shown in Fig. 1.

Parallel Metabolic Processes

Quenching of electrons

Beside the microorganisms which are able to use the electrode as a final electron acceptor, microorganisms using alternative energy-generating pathways may evolve. These processes may result in a decrease of the performance of MFCs due to a lowering of the coulombic efficiency, i.e., the ratio of the electrons generated during current production and the electrons present in the supplied substrate. In particular, methanogenesis occurring in mixed communities has been found by several authors to quench the MFC power output (He et al., 2005; Kim et al., 2005). Moreover, methanogenesis accounted for 35 to 58% of the soluble chemical oxygen demand removed at a loading rate of 1.0 g of chemical oxygen demand/liter/day (He et al., 2005). An overloading of the substrate, possibly in combination with a limited capacity of the MFC to generate current, creates a niche for methanogens. The production of excess hydrogen and sulfide, the leaching of mediators out of the biofilm, and the use of reduced electron shuttles by organisms not involved in the electron transfer can also be considered to be additional forms of quenching.

Empowerment of the metabolic network

Fermenting microorganisms convert the substrate into partly oxidized products but are not considered to take part in the electricity-generating process in a direct way. However, the use of a monoculture of *Escherichia coli* K-12 producing fermentation products which are directly converted in an MFC using a polymer-modified Pt-anode has been shown to yield high current densities of 1.5 mA/cm^2 (Schroder et al., 2003). We hypothesize that fermenting organisms, able to colonize specific niches within the microbial community of an MFC, can be advantageous for the MFC operation. Kim et al. (2006) showed by confocal laser scanning microscopy the presence of gram-negative and gram-positive bacteria forming microcolonies and microbial clumps throughout the electrode surface of glucose- and glutamate-fed MFCs. Electrodes enriched with acetate did not show microbial clumps. It might be plausible that the microbes in the clumps ferment the complex substrate and the electrochemically active microbes in the biofilm oxidize the fermentation products. Figure 1B indicates that the energy remaining in the fermentation products (acetate and H_2) is sufficient to sustain the electron-transferring microorganisms, resulting in an electron recovery from the fermentation products and empowerment of the metabolic and electron transfer network. Although fermenting organisms are able to empower the metabolic network, their presence results in the loss of a limited amount of the energy and electrons embodied in the substrate for the growth and maintenance of these organisms.

Sulfurous compounds acting as inorganic shuttles

Alternative electron acceptors such as carbonate, sulfate, and nitrate are potential scavengers for electrons. However, when the reduced products are possible electron donors for the microbial community or when the reduced products can react electrochemically with the electrode, some of the electrons can be recovered. While the production of methane and nitrogen gas are inevitable losses, the simultaneous removal of sulfate and sulfide by the microbial reduction of sulfate to sulfide and the subsequent oxidation of sulfide at the anode of an MFC have been described (Habermann and Pommer, 1991; Rabaey et al., 2006). Although sulfide can be oxidized electrochemically at the electrode, Fig. 1 indicates that bacteria are able to gain a limited amount of energy during the oxidation of sulfide to sulfur. Rabaey et al. (2006) have isolated sulfide-oxidizing bacteria from a sulfide-fed MFC and noted the precipitation of possibly biogenic sulfur compounds on the electrode surface. However, the oxidation of 1 mol of sulfide to sulfur releases only 2 of the 8 mol of electrons present. In addition, the equilibrium between the aqueous HS^- and the volatile H_2S is pH dependent with an increasing loss of H_2S at a lower pH. To lower these losses and to increase the electron recovery, a further oxidation of sulfur to sulfate, recovering an additional six electrons, is beneficial. The additional oxidation of sulfur to sulfate has been shown in a study using *Desulfobulbus propionicus* with an electrode poised at +520 mV versus standard hydrogen serving as the sole electron acceptor (Holmes et al., 2004a). However, this electrode potential is exceptionally high, disabling a notable energy recovery. In addition to the direct recovery of electrons from sulfide, the presence of sulfur (elemental or polysulfides) and/or thiosulfate is described to complete a sulfur-mediated reduction of ferrihydrite by *Sulfurospirillum deleyianum* under environmental conditions (Straub and Schink, 2004). In sediment-based MFCs containing high levels of sulfate, the cycling of sulfur components represents a key process in the generation of electricity, indicating the role of sulfur components as a mediator (Holmes et al., 2004a; Ryckelynck et al., 2005).

ENGINEERING OF THE MICROBIAL RESOURCES WITHIN THE MFC

Anode Operation

The microbial oxidation of organic molecules in MFCs is being investigated extensively; examples are acetate, butyrate, glucose, cysteine, proteins, sucrose, starch, ethanol, methanol (Rabaey et al., 2003; Min and Logan, 2004; He et al., 2005; Liu et al., 2005; Logan et al., 2005; Heilmann and Logan, 2006), and more complex organics, containing a large variety of different readily and nonreadily degradable molecules such as domestic wastewater (Liu et al., 2004) or the effluent of anaerobic digesters (Aelterman et al., 2006a). The composition of the wastewater strongly affects the power output of MFCs, with reported power outputs for a digester effluent and a domestic wastewater of, respectively, 59 and 48 W/m^3 (Rabaey et al., 2005b). Spiking the wastewater with acetate resulted in an increase of the power output, indicating that the higher the rapidly biodegradable fraction within the wastewater is, the higher the power output will be (Rabaey et al., 2005b). Increasing the volumetric loading rate from 2.5 to 10 g biochemical oxygen demand/liter/day resulted in a doubling of the power production (Moon et al., 2006). The effect of intermittent feeding on the microbial electricity generation has only recently been investigated by Freguia et al. (2007). They examined the electron and carbon balances in MFCs to identify the inefficiencies of the electricity-generating process. By monitoring and analyzing the off gases, the produced current, the substrate removal, and the biomass production, they have shown that upon supplying the bacteria with acetate after a period of starvation, the microbial cells were able to store carbon as polymeric material inside their cell. These substances were incorporated during the feeding phase and were utilized by the bacteria to generate current after the acetate was depleted. Up to 57% of the total current generated occurred after the depletion of the external carbon source. While batch feeding is unlikely to occur in practice, this capacity for storage indicates that bacterial communities can generate reasonably constant power with fluctuating influent strengths, as prevailing during wastewater treatment.

The type of substrate fed to an MFC potentially has an impact on the structure and composition of the microbial community. As seen in Fig. 1, the more reduced the substrate, the more energy is available to divide across the community. This may lead to an increase of the possible interactions and niches. Up till now, no clear image of the effect of the type of substrate on the microbial community has been available. Liu et al. (2005) examined the electricity generation using a single-chambered MFC having acetate or butyrate in the influent. Power generated with acetate (800 mg/liter; 12.7 W/m^3) was up to 66% higher than that fed with butyrate (1,000 mg/liter; 7.6 W/m^3), demonstrating that acetate is a preferred aqueous substrate for electricity generation in MFCs. The reported coulombic efficiencies and overall energy recovery were 10 to 31% and 3 to 7% for acetate and 8 to 15% and 2 to 5% for butyrate, indicating substantial electron and energy losses to processes other than electricity generation. These results show that although the Gibbs free energy for the complete oxidation at pH 7 of 1 mol of butyrate (−524 kJ) is higher than for acetate (−215 kJ), the overall performance of a butyrate-fed MFC was lower. A possible reason for this discrepancy is the limited period of enrichment (140 h) during the described batch experiments, resulting in an unstable microbial community not able to fully metabolize the substrate. Freguia et al. (2007) observed that a change of the substrate from acetate to glucose applied to an acetate-adapted community resulted in an increase of the liberation of hydrogen and methane gas, indicating that the acetate-enriched community was not able to fully convert the energy released within the fermentation products into electricity.

Cathode Operation

One of the major issues in MFC technology is the search for a good functioning cathode. A fast and spontaneous reduction reaction at fairly high redox potentials (around +300 mV versus SHE [standard hydrogen electrode] or higher) is required to sustain a considerable electricity production.

Chemical cathodes

Conventionally, the cathodic process of a microbial fuel cell is depicted as an oxygen reduction reaction. This type of reduction reaction on a graphite electrode has slow kinetics, and a catalyst is therefore required. A platinum coating is an obvious but expensive and unsustainable solution many researchers adapted. Alternative and cheaper catalysts such as pyrolyzed phthalocyanines have already been tested successfully (Zhao et al., 2006). MFCs reducing oxygen at the cathode can have their solid cathode in an aqueous solution but can also have a permeable cathode directly exposed to the air (open-air cathode) (Liu and Logan, 2004). Chemical solutions containing an alternative for oxygen as electron acceptor allow higher mass transfer efficiencies and thus higher power outputs. Typically, a ferricyanide solution is used (Park and Zeikus, 2000). This chemical solution is, however, highly unsustainable and should be avoided in the future.

Biological cathodes

A promising and rapidly expanding class of cathodic reactions is the microbially mediated cathodic process. This comprises processes in which the cathode chemically reduces an electron acceptor and bacteria regenerate the pool of electron acceptors by reoxidizing them as well as processes in which bacteria directly accept electrons from the solid cathode for the reduction of a dissolved electron acceptor. In any case, bacteria mediate the flow of electrons from an electron donor at low redox potential to an electron acceptor at higher redox potential and by doing so, they are able to gain energy, as shown in Fig. 5.

Cycles of iron and manganese at the cathodic surface allow the generation of an inexpensive and sustainable cathode. Ter Heijne et al. (2006) described a cathodic process, requiring a bipolar membrane to maintain a pH of 2.5 in the cathode, in which soluble ferric iron was the electron acceptor. It is possible to regenerate the ferric iron by using iron oxidizing microorganisms such as *Thiobacillus ferrooxidans* as shown in Fig. 5 (Ter Heijne et al., 2007). Rhoads et al. (2005) and Shantaram et al. (2005) reported systems in which the electrons resulting from the cathode were chemically reducing manganese oxides deposited by manganese oxidizing bacteria such as *Leptothrix discophora*. However, these studies could not yet prove the actual cycling of manganese.

Several research groups studied the direct transfer of electrons from the cathode to the microorganisms resulting in a cathode-driven denitrification (Gregory et al., 2004; Clauwaert et al., 2007a). In order to gain energy from this reaction, the redox potential of the cathode should be sufficiently lower than that of the nitrate reduction step (+0.433 mV versus standard hydrogen). This lowers the attainable cathodic potential and limits the power output which can be reached with this type of cathode. On the other hand, the process implies the removal of a wastewater component from an aqueous stream which makes it suitable for wastewater treatment. The development of an oxygen-driven microbial cathode is the real quest to maximize the power output of MFCs. Bergel et al. (2005) demonstrated that it is possible to let a biofilm catalyze the oxygen reduction reaction. This was shown with a marine biofilm on stainless steel in salt water. Clauwaert et al. (2007b) have found a process in which microbial catalysis of oxygen reduction occurs at graphite electrodes in mineral medium. The omnipresence of oxygen as the electron acceptor and the absence of an expensive catalyst (the microbiota function as catalysts) render this process into a cathodic reaction well suited for many configurations.

The use of microbial cathodes opens perspectives for inexpensive and sustainable MFCs. Furthermore, some specific wastewater treatment processes including the degradation of chlorinated compounds such as trichloroethene (Aulenta et al., 2007) can be performed at the cathode, while generating electrical current.

The MFC Design

In order to get a beneficial microbial community structure, mainly constituted of electron-transferring organisms, a good MFC design is necessary. We suggest that a good reactor design will enable a sufficient and uniform substrate distribution and will have a large electrode surface (>100 m^2/m^3) with good attachment properties and an internal resistance lower than 2 $\Omega \cdot$ liter. The first parameter maximizes the conversion of fuel to electrons and will support a good proton transport to the cathode and sustain a proper redox ratio between the substrate and waste products (Cheng et al., 2006; Logan et al., 2006). The last two parameters influence the electrochemical environment of the bacteria. Increasing the electrode surface and lowering the internal resistance will decrease the current density-dependent losses. As a result, the microorganisms can generate a higher current for a set potential enabling

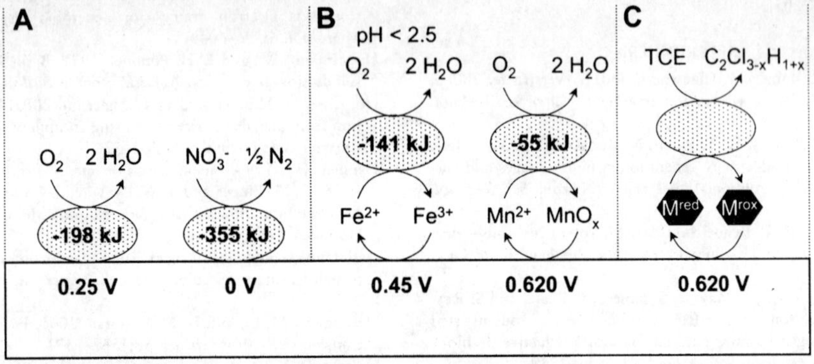

Figure 5. Schematic overview of the cathode reactions catalyzed by microorganisms (represented by dotted ovals). (A) Direct transfer of electrons from the cathode to the microorganisms (Gregory et al., 2004; Clauwaert et al., 2007a, 2007b); (B) metal-mediated electron transfer (Rhoads et al., 2005; Terheijne et al., 2006); and (C) bioelectrochemically driven dechlorination at the cathode (Aulenta et al., 2007). The free Gibbs energy values are expressed per mole of electron acceptor and are calculated using the represented electrode potentials expressed versus standard hydrogen, a saturated oxygen concentration of 8 mg/liter and a NO_3^- concentration of 0.01 M, both at pH 7 unless noted otherwise.

higher power outputs. A good reactor design, resulting in an increase of the drive for bacteria to transfer electrons to the electrode, will select for the bacteria best able to survive in MFCs. As a result, a highly anodophilic community will evolve, biomass concentrations will increase, and the overall current generation will rise, resulting in a win-win situation for both the bacteria and the electricity consumers.

CONCLUSIONS

Microbial communities in fuel cells are not more complex than those in other technical systems such as activated-sludge treatment tanks or anaerobic digesters. Just as for these systems, we must learn to identify the engineering parameters which should be imposed in order to render them efficient in terms of rate of conversion and effective in terms of power production.

At present, the factors governing MFC output are the presence of a sufficient microbial diversity at the onset, the setting of the electrode potentials so that there is a proper trade-off for the bacteria in terms of gain for their metabolism relative to the power delivered to the system, the avoidance of quenching reactions trapping electrons for other reactions (such as, for example, methane production), and finally the provision of sufficient fluxes of electrons by either imposing small distances or facilitating electron conductive matrixes.

Acknowledgments. This research was funded by a Ph.D. grant (IWT grant 41294 to P.A. and IWT grant 51305 to P.C.) of the Institute for the Promotion of Innovation through Science and Technology in Flanders (IWT-Vlaanderen), a Ph.D. grant from the Bijzonder Onderzoeksfonds of the Ghent University (BOF grant 01D24405 to L.D.S.), the Postdoctoral Fellow Scheme at the University of Queensland (to K.R.), and a grant from the Research Foundation—Flanders (FWO grant G017205).

REFERENCES

Aelterman, P., K. Rabaey, P. Clauwaert, and W. Verstraete. 2006a. Microbial fuel cells for wastewater treatment. *Water Sci. Technol.* 54:9–15.

Aelterman, P., K. Rabaey, H. T. Pham, N. Boon, and W. Verstraete. 2006b. Continuous electricity generation at high voltages and currents using stacked microbial fuel cells. *Environ. Sci. Technol.* 40:3388–3394.

Allen, R. M., and H. P. Bennetto. 1993. Microbial fuel-cells—electricity production from carbohydrates. *Appl. Biochem. Biotechnol.* 39:27–40.

Aulenta, F., A. Catervi, M. Majone, S. Panero, P. Reale, and S. Rossetti. 2007. Electron transfer from a solid-state electrode assisted by methyl viologen sustains efficient microbial reductive dechlorination of TCE. *Environ. Sci. Technol.* 41:2554–2559.

Back, J. H., M. S. Kim, H. Cho, I. S. Chang, J. Y. Lee, K. S. Kim, B. H. Kim, Y. I. Park, and Y. S. Han. 2004. Construction of bacterial artificial chromosome library from electrochemical microorganisms. *FEMS Microbiol. Lett.* 238:65–70.

Beratan, D. N., J. N. Onuchic, J. R. Winkler, and H. B. Gray. 1992. Electron-tunneling pathways in proteins. *Science* 258:1740–1741.

Bergel, A., D. Feron, and A. Mollica. 2005. Catalysis of oxygen reduction in PEM fuel cell by seawater biofilm. *Electrochem. Commun.* 7:900–904.

Bond, D. R., D. E. Holmes, L. M. Tender, and D. R. Lovley. 2002. Electrode-reducing microorganisms that harvest energy from marine sediments. *Science* 295:483–485.

Bond, D. R., and D. R. Lovley. 2003. Electricity production by *Geobacter sulfurreducens* attached to electrodes. *Appl. Environ. Microbiol.* 69:1548–1555.

Bond, D. R., and D. R. Lovley. 2005. Evidence for involvement of an electron shuttle in electricity generation by *Geothrix fermentans*. *Appl. Environ. Microbiol.* 71:2186–2189.

Cheng, S., H. Liu, and B. E. Logan. 2006. Increased power generation in a continuous flow MFC with advective flow through the porous anode and reduced electrode spacing. *Environ. Sci. Technol.* 40:2426–2432.

Clauwaert, P., K. Rabaey, P. Aelterman, L. De Schamphelaire, H. T. Pham, P. Boeckx, N. Boon, and W. Verstraete. 2007a. Biological denitrification in microbial fuel cells. *Environ. Sci. Technol.* 41: 3354–3360.

Clauwaert, P., D. Van der Ha, N. Boon, K. Verbeken, M. Verhaege, K. Rabaey, and W. Verstraete. 2007b. Open air biocathode enables effective electricity generation with microbial fuel cells. *Environ. Sci. Technol.* 41:7564–7569.

Costerton, J. W. 1995. Overview of microbial biofilms. *J. Ind. Microbiol.* 15:137–140.

Freguia, S., K. Rabaey, S. Yuan, and J. Keller. 2007. Electron and carbon balances in microbial fuel cells reveal temporary bacterial storage behavior during electricity generation. *Environ. Sci. Technol.* 41:2915–2921.

Gorby, Y. A., S. Yanina, J. S. McLean, K. M. Rosso, D. Moyles, A. Dohnalkova, T. J. Beveridge, I. S. Chang, B. H. Kim, K. S. Kim, D. E. Culley, S. B. Reed, M. F. Romine, D. A. Saffarini, E. A. Hill, L. Shi, D. A. Elias, D. W. Kennedy, G. Pinchuk, K. Watanabe, S. Ishii, B. Logan, K. H. Nealson, and J. K. Fredrickson. 2006. Electrically conductive bacterial nanowires produced by *Shewanella oneidensis* strain MR-1 and other microorganisms. *Proc. Natl. Acad. Sci. USA* 103:11358–11363.

Grasso, D., B. F. Smets, K. A. Strevett, B. D. Machinist, C. J. Van-Oss, R. F. Giese, and W. Wu. 1996. Impact of physiological state on surface thermodynamics and adhesion of *Pseudomonas aeruginosa*. *Environ. Sci. Technol.* 30:3604–3608.

Gregory, K. B., D. R. Bond, and D. R. Lovley. 2004. Graphite electrodes as electron donors for anaerobic respiration. *Environ. Microbiol.* 6:596–604.

Habermann, W., and E. H. Pommer. 1991. Biological fuel-cells with sulfide storage capacity. *Appl. Microbiol. Biotechnol.* 35:128–133.

He, Z., S. D. Minteer, and L. T. Angenent. 2005. Electricity generation from artificial wastewater using an upflow microbial fuel cell. *Environ. Sci. Technol.* 39:5262–5267.

Heijnen, J. J. 1999. Bioenergetics of microbial growth, p. 267–291. *In* M. C. Flickinger and S. W. Drew (ed.), *Bioprocess Technology: Fermentation, Biocatalysis, Bioseparation.* John Wiley and Sons, Hoboken, NJ.

Heilmann, J., and B. E. Logan. 2006. Production of electricity from proteins using a microbial fuel cell. *Water Environ. Res.* 78:531–537.

Hernandez, M. E., and D. K. Newman. 2001. Extracellular electron transfer. *Cell. Mol. Life Sci.* 58:1562–1571.

Holmes, D. E., D. R. Bond, and D. R. Lovley. 2004a. Electron transfer by *Desulfobulbus propionicus* to Fe(III) and graphite electrodes. *Appl. Environ. Microbiol.* 70:1234–1237.

Holmes, D. E., D. R. Bond, R. A. O'Neill, C. E. Reimers, L. R. Tender, and D. R. Lovley. 2004b. Microbial communities associated with electrodes harvesting electricity from a variety of aquatic sediments. *Microb. Ecol.* 48:178–190.

Holmes, D. E., S. K. Chaudhuri, K. P. Nevin, T. Mehta, B. A. Methe, A. Liu, J. E. Ward, T. L. Woodard, J. Webster, and D. R. Lovley. 2006. Microarray and genetic analysis of electron transfer to electrodes in *Geobacter sulfurreducens*. *Environ. Microbiol.* 8:1805–1815.

Ishii, S., T. Kosaka, K. Hori, Y. Hotta, and K. Watanabe. 2005. Coaggregation facilitates interspecies hydrogen transfer between *Pelotomaculum thermopropionicum* and *Methanothermobacter thermautotrophicus*. *Appl. Environ. Microbiol.* 71:7838–7845.

Kim, B. H., H. J. Kim, M. S. Hyun, and D. H. Park. 1999. Direct electrode reaction of Fe(III)-reducing bacterium, *Shewanella putrefaciens*. *J. Microbiol. Biotechnol.* 9:127–131.

Kim, B. H., H. S. Park, H. J. Kim, G. T. Kim, I. S. Chang, J. Lee, and N. T. Phung. 2004. Enrichment of microbial community generating electricity using a fuel-cell-type electrochemical cell. *Appl. Microbiol. Biotechnol.* 63:672–681.

Kim, G. T., G. Webster, J. W. T. Wimpenny, B. H. Kim, H. J. Kim, and A. J. Weightman. 2006. Bacterial community structure, compartmentalization and activity in a microbial fuel cell. *J. Appl. Microbiol.* 101:698–710.

Kim, H. J., H. S. Park, M. S. Hyun, I. S. Chang, M. Kim, and B. H. Kim. 2002. A mediator-less microbial fuel cell using a metal reducing bacterium, *Shewanella putrefaciens*. *Enzyme Microb. Technol.* 30:145–152.

Kim, J. R., B. Min, and B. E. Logan. 2005. Evaluation of procedures to acclimate a microbial fuel cell for electricity production. *Appl. Microbiol. Biotechnol.* 68:23–30.

Kim, N., Y. Choi, S. Jung, and S. Kim. 2000. Effect of initial carbon sources on the performance of microbial fuel cells containing *Proteus vulgaris*. *Biotechnol. Bioeng.* 70:109–114.

Lee, J., N. T. Phung, I. S. Chang, B. H. Kim, and H. C. Sung. 2003. Use of acetate for enrichment of electrochemically active microorganisms and their 16s rDNA analyses. *FEMS Microbiol. Lett.* 223:185–191.

Liu, H., S. A. Cheng, and B. E. Logan. 2005. Production of electricity from acetate or butyrate using a single-chamber microbial fuel cell. *Environ. Sci. Technol.* 39:658–662.

Liu, H., and B. E. Logan. 2004. Electricity generation using an aircathode single chamber microbial fuel cell in the presence and absence of a proton exchange membrane. *Environ. Sci. Technol.* 38: 4040–4046.

Liu, H., R. Ramnarayanan, and B. E. Logan. 2004. Production of electricity during wastewater treatment using a single chamber microbial fuel cell. *Environ. Sci. Technol.* 38:2281–2285.

Logan, B. E., B. Hamelers, R. Rozendal, U. Schrorder, J. Keller, S. Freguia, P. Aelterman, W. Verstraete, and K. Rabaey. 2006. Microbial fuel cells: methodology and technology. *Environ. Sci. Technol.* 40:5181–5192.

Logan, B. E., C. Murano, K. Scott, N. D. Gray, and I. M. Head. 2005. Electricity generation from cysteine in a microbial fuel cell. *Water Res.* 39:942–952.

Logan, B. E., and J. M. Regan. 2006. Electricity-producing bacterial communities in microbial fuel cells. *Trends Microbiol.* 14:512–518.

Madigan, M. T., J. M. Martinko, and J. Parker. 2000. *Brock Biology of Microorganisms*. Prentice-Hall, Upper Saddle River, NJ.

Milliken, C. E., and H. D. May. 2007. Sustained generation of electricity by the spore-forming, gram-positive, *Desulfitobacterium hafniense* strain DCB2. *Appl. Microbiol. Biotechnol.* 73:1180–1189.

Min, B., and B. E. Logan. 2004. Continuous electricity generation from domestic wastewater and organic substrates in a flat plate microbial fuel cell. *Environ. Sci. Technol.* 38:5809–5814.

Mitchell, P. 1961. Coupling of phosphorylation to electron and hydrogen trnasfer by a chemiosmotic type of mechanism. *Nature* (London) 191:144–148.

Moon, H., I. S. Chang, and B. H. Kim. 2006. Continuous electricity production from artificial wastewater using a mediator-less microbial fuel cell. *Bioresour. Technol.* 97:621–627.

Newman, D. K., and R. Kolter. 2000. A role for excreted quinones in extracellular electron transfer. *Nature* 405:94–97.

Page, C. C., C. C. Moser, X. X. Chen, and P. L. Dutton. 1999. Natural engineering principles of electron tunnelling in biological oxidation-reduction. *Nature* 402:47–52.

Park, D. H., and J. G. Zeikus. 2000. Electricity generation in microbial fuel cells using neutral red as an electronophore. *Appl. Environ. Microbiol.* 66:1292–1297.

Park, H. S., B. H. Kim, H. S. Kim, H. J. Kim, G. T. Kim, M. Kim, I. S. Chang, Y. K. Park, and H. I. Chang. 2001. A novel electrochemically active and Fe(III)-reducing bacterium phylogenetically related to *Clostridium butyricum* isolated from a microbial fuel cell. *Anaerobe* 7:297–306.

Phung, N. T., J. Lee, K. H. Kang, I. S. Chang, G. M. Gadd, and B. H. Kim. 2004. Analysis of microbial diversity in oligotrophic microbial fuel cells using 16s rDNA sequences. *FEMS Microbiol. Lett.* 233:77–82.

Prasad, D., T. K. Sivaram, S. Berchmans, and V. Yegnaraman. 2006. Microbial fuel cell constructed with a microorganism isolated from sugar industry effluent. *J. Power Sources* 160:991–996.

Rabaey, K., N. Boon, M. Hofte, and W. Verstraete. 2005a. Microbial phenazine production enhances electron transfer in biofuel cells. *Environ. Sci. Technol.* 39:3401–3408.

Rabaey, K., N. Boon, S. D. Siciliano, M. Verhaege, and W. Verstraete. 2004. Biofuel cells select for microbial consortia that self-mediate electron transfer. *Appl. Environ. Microbiol.* 70:5373–5382.

Rabaey, K., P. Clauwaert, P. Aelterman, and W. Verstraete. 2005b. Tubular microbial fuel cells for effeient electricity generation. *Environ. Sci. Technol.* 39:8077–8082.

Rabaey, K., G. Lissens, S. D. Siciliano, and W. Verstraete. 2003. A microbial fuel cell capable of converting glucose to electricity at high rate and efficiency. *Biotechnol. Lett.* 25:1531–1535.

Rabaey, K., J. Rodriguez, L. Blackall, J. Keller, D. Batstone, W. Verstraete, and K. H. Nealson. 2007. Microbial ecology meets electrochemistry: electricity driven and driving communities. *ISME J.* 1:9–18.

Rabaey, K., K. Van de Sompel, L. Maignien, N. Boon, P. Aelterman, P. Clauwaert, L. De Schamphelaire, H. T. Pham, J. Vermeulen, M. Verhaege, P. Lens, and W. Verstraete. 2006. Microbial fuel cells for sulfide removal. *Environ. Sci. Technol.* 40:5218–5224.

Reguera, G., K. D. McCarthy, T. Mehta, J. S. Nicoll, M. T. Tuominen, and D. R. Lovley. 2005. Extracellular electron transfer via microbial nanowires. *Nature* 435:1098–1101.

Reguera, G., K. P. Nevin, J. S. Nicoll, S. F. Covalla, T. L. Woodard, and D. R. Lovley. 2006. Biofilm and nanowire production leads to increased current in *Geobacter sulfurreducens* fuel cells. *Appl. Environ. Microbiol.* 72:7345–7348.

Reguera, G., R. B. Pollina, J. S. Nicoll, and D. R. Lovley. 2007. Possible nonconductive role of *Geobacter sulfurreducens* pilus nanowires in biofilm formation. *J. Bacteriol.* 189:2125–2127.

Rhoads, A., H. Beyenal, and Z. Lewandowski. 2005. Microbial fuel cell using anaerobic respiration as an anodic reaction and biomineralized manganese as a cathodic reactant. *Environ. Sci. Technol.* 39:4666–4671.

Ringeisen, B. R., E. Henderson, P. K. Wu, J. Pietron, R. Ray, B. Little, J. C. Biffinger, and J. M. Jones-Meehan. 2006. High power density from a miniature microbial fuel cell using *Shewanella oneidensis* DSP10. *Environ. Sci. Technol.* 40:2629–2634.

Ryckelynck, N., H. A. Stecher, and C. E. Reimers. 2005. Understanding the anodic mechanism of a seafloor fuel cell: interactions between geochemistry and microbial activity. *Biogeochemistry* 76: 113–139.

Schlegel, H. 1992. *General Microbiology*, 7th ed. Cambridge University Press, Cambridge, United Kingdom.

Schroder, U., J. Niessen, and F. Scholz. 2003. A generation of microbial fuel cells with current outputs boosted by more than one order of magnitude. *Angew. Chem.-Int. Ed.* 42:2880–2883.

Shantaram, A., H. Beyenal, R. Raajan, A. Veluchamy, and Z. Lewandowski. 2005. Wireless sensors powered by microbial fuel cells. *Environ. Sci. Technol.* 39:5037–5042.

Straub, K. L., and B. Schink. 2004. Ferrihydrite-dependent growth of *Sulfurospirillum deleyianum* through electron transfer via sulfur cycling. *Appl. Environ. Microbiol.* 70:5744–5749.

Sutherland, I. W. 2001. The biofilm matrix—an immobilized but dynamic microbial environment. *Trends Microbiol.* 9:222–227.

Ter Heijne, A., H. V. M. Hamelers, and C. J. N. Buisman. 2007. Microbial fuel cell operation with continuous biological ferrous iron oxidation of the catholyte. *Environ. Sci. Technol.* 41:4130–4134.

Ter Heijne, A., H. V. M. Hamelers, V. De Wilde, R. A. Rozendal, and C. J. N. Buisman. 2006. A bipolar membrane combined with ferric iron reduction as an efficient cathode system in microbial fuel cells. *Environ. Sci. Technol.* 40:5200–5205.

Thauer, R. K., K. Jungermann, and K. Decker. 1977. Energy-conservation in chemotropic anaerobic bacteria. *Bacteriol. Rev.* 41:100–180.

Zhao, F., F. Harnisch, U. Schroder, F. Scholz, P. Bogdanoff, and I. Herrmann. 2006. Challenges and constraints of using oxygen cathodes in microbial fuel cells. *Environ. Sci. Technol.* 40:5193–5199.

6. ORGANIC SOLVENTS

Bioenergy
Edited by J. Wall et al.

Chapter 25

Molecular Aspects of Butanol Fermentation

Carlos J. Paredes, Shawn W. Jones, Ryan S. Senger, Jacob R. Borden, Ryan Sillers, and Eleftherios T. Papoutsakis

Worldwide energy consumption has doubled in the past 30 years and is expected to increase by 50% by the year 2025 (Ragauskas et al., 2006), driven by the growth of the world economy and population. For now, energy supplies rely heavily on finite, nonrenewable fossil fuels whose combustion has increasingly apparent detrimental impact on the climate of our planet. One of the possible alternatives to lessen and/or eliminate our dependence on fossil fuels is the use of renewable resources. Along these lines, the use of biomass for the production of carbon-neutral biofuels and biologically produced chemicals and materials has led to the concept of biorefinery (Ragauskas et al., 2006), covered elsewhere in this book.

Among the biologically produced compounds, butanol is emerging as an important product with a promising role as a biofuel given its superior chemical properties when compared to ethanol: lower volatility and hydrophilicity, higher miscibility with other hydrocarbons, and higher energy content per unit mass. Butanol is not a newcomer to the world of biologically produced chemicals or fuels. In fact, the anaerobic acetone-butanol-ethanol (ABE) clostridial fermentation (Jones and Woods, 1986; Lesnik et al., 2001; Rogers, 1986) was the primary source of butanol and acetone for over 40 years until the early 1950s, when the petrochemical process prevailed for economic reasons.

This chapter focuses on the regulation of solvent formation in solventogenic clostridia and in particular in *Clostridium acetobutylicum* ATCC 824, the most widely studied solventogenic *Clostridium* at a genetic level, and until recently the only sequenced one. *Clostridium beijerinckii* has been recently sequenced, but its sequence and annotation have not yet been formally published. *C. beijerinckii* and other clostridia are also discussed to the extent that relevant molecular details are known and pertinent to the subject matter.

GROWTH, SPORULATION, AND SOLVENT PRODUCTION: AN OVERVIEW OF THE KEY FUNDAMENTAL AND PRACTICAL ISSUES

In a typical batch culture, *C. acetobutylicum* and all solventogenic clostridia undergo a phase of vegetative growth whereby the primary metabolic products of carbohydrate metabolism are the organic acids acetate and butyrate; this is known as acidogenesis (Jones and Woods, 1986). Such an acid production causes a drop in the pH of the culture, and at some point, cells start sporulating (Dürre et al., 1995; Long et al., 1983, 1984a, 1984b; Paredes et al., 2005). At the same time, they transform the previously produced acids into the (temporarily at least) less toxic solvents acetone (or, in some clostridia, isopropanol), butanol, and ethanol; this is referred to as solventogenesis (Jones and Woods, 1986). The first part of the culture is widely known as acidogenesis, whereas the second part is usually referred to as solventogenesis. Culture and macroscopic parameters affecting the initiation of solventogenesis have been extensively reviewed (Girbal and Soucaille, 1998; Hüsemann and Papoutsakis, 1988; Jones and Woods, 1986; Meyer and Papoutsakis, 1989; Woods, 1995), and key ones include the culture pH and the levels of butyrate and possibly acetate. However, the metabolic landscape leading to solvent production is considerably more complicated. Continuous cultures of *C. acetobutylicum* have shown (Girbal et al., 1995a) the existence of three metabolic states capable of producing acids, alcohols, and acetone in different proportions. The key parameters leading to each of these states are the pH of the culture and the NAD(P)H availability. When *C. acetobutylicum* is grown on glucose at neutral pH, the result is an acidogenic culture with production of acetic and butyric acids. When grown at low pH on glucose, the culture becomes solventogenic as it produces acetone, butanol,

Carlos J. Paredes, Shawn W. Jones, Ryan S. Senger, Jacob R. Borden, Ryan Sillers, and Eleftherios T. Papoutsakis • Department of Chemical and Biological Engineering, Northwestern University, Evanston, IL 60208.

and ethanol. Finally, when grown at neutral pH under conditions of high NAD(P)H availability, the culture is termed alcohologenic as only butanol and ethanol (but not acetone) are produced (Girbal and Soucaille, 1994).

Over the last 2 to 3 decades, some of the questions or issues that have been extensively researched in the context of solvent production by solventogenic clostridia include the link between solventogenesis and sporulation; the molecular basis of solvent formation; the mechanisms for acid and solvent tolerance; how we can manipulate the ratio between the different fermentation products; and how different substrates affect the amounts of alcohols, acetone, and acids produced. This chapter focuses on these and related questions with emphasis on more-recent and genomically based work that has not been previously reviewed. Key older and recent reviews (Dürre, 2005; Dürre et al., 1995; Girbal and Soucaille, 1998; Jones and Woods, 1986; Paredes et al., 2005; Woods, 1995) should also be consulted on this topic to add to the comprehension of this chapter.

PRIMARY METABOLISM, SOLVENT PRODUCTION, AND GENE EXPRESSION: THE MAIN PLAYERS AND THEIR EXPRESSION PATTERNS

The first step in the conversion of carbohydrates into solvents is their uptake. Like in most organisms, in clostridia, this is accomplished mostly via the phosphotransferase systems, which combine the uptake and phosphorylation of the carbohydrate into a single step using phosphoenolpyruvate as phosphodonor. Phosphotransferase systems in clostridia have been recently reviewed by Mitchell and Tangney (Mitchell and Tangney, 2005), and there is experimental evidence that such systems are responsible for the uptake of cellobiose, fructose, glucose, lactose, maltose, mannitol, and sucrose in *C. acetobutylicum* and fructose, glucitol, glucose, lactose, mannitol, and sucrose in *C. beijerinckii*.

Alsaker and Papoutsakis (Alsaker and Papoutsakis, 2005) studied the *C. acetobutylicum* transcriptional program during the exponential and early stationary phases when using glucose as a substrate. Their results show that, despite particular variations between the eleven glycolytic genes responsible for the conversion of glucose to acetyl-coenzyme A (acetyl-CoA), the overall pathway shows very little temporal regulation. The conversion of glucose to acetyl-CoA generates an excess of NADH which is reduced by ferredoxin (Fig. 1). During acidogenesis, ferredoxin is in turn regenerated by the hydrogenase (*hydA*) producing H_2 in the process (Jones and Woods, 1986). Acetyl-CoA is then converted to either acetate or butyryl-CoA (Jones and Woods, 1986). The genes needed to convert acetyl-CoA to acetate, phosphate acetyltransferase (*pta*) and acetate kinase (*ack*) genes, show little change in expression with a slight increase just after the initiation of solventogenesis that returns to baseline expression values (Alsaker and Papoutsakis, 2005; Winzer et al., 1997). The conversion of acetyl-CoA to butyryl-CoA involves six genes that show few temporal expression patterns for batch cultures (Alsaker and Papoutsakis, 2005; Harris et al., 2002; Tummala et al., 1999). However, a study using continuous culture found a temporal expression pattern for the thiolase (*thl*) gene (Winzer et al., 2000). From butyryl-CoA, butyrate is formed by phosphate butyryltransferase (*ptb*) and butyrate kinase (*buk*) (Jones and Woods, 1986); both of these genes show increased expression immediately before initiation of solventogenesis but mostly remain at a constant expression level (Alsaker and Papoutsakis, 2005; Harris et al., 2002; Tummala et al., 1999).

As the concentrations of butyrate and acetate increase, the pH begins to drop and a yet-unknown signal (Dürre and Hollergschwandner, 2004; Paredes et al., 2005) triggers the sporulation cascade via the phosphorylation of the transcription factor Spo0A that in turn causes the switch from acidogenesis to solventogenesis (Dürre and Hollergschwandner, 2004; Harris et al., 2002; Paredes et al., 2005). Upon activation, Spo0A upregulates the key solventogenic genes *aad* (*adhe1*), *ctfA*, *ctfB*, *adc*, and *bdhB* (Alsaker and Papoutsakis, 2005; Feustel et al., 2004; Gerischer and Durre, 1992; Harris et al., 2002; Sauer and Durre, 1995; Walter et al., 1992). Regulation of *bdhA* by Spo0A is unclear (Dürre, 2005): although two putative Spo0A binding sites have been found in its promoter region (Ravagnani et al., 2000), under some conditions it seems to be constitutively expressed (Sauer and Durre, 1995). The genes *aad* (*adhE1*), *ctfA*, and *ctfB* form the *sol* operon and thus show a similar expression pattern (Alsaker and Papoutsakis, 2005). The remaining genes are all transcribed separately, with *adc* and *bdhB* showing a slight increase in expression before the activation of Spo0A (Alsaker and Papoutsakis, 2005). Reassimilation of the acids is achieved by butyrate-acetoacetate-CoA transferase, coded by *ctfA* and *ctfB*, converting the acids into their respective CoA derivatives (Andersch et al., 1983; Dürre et al., 1995; Hartmanis et al., 1984). These reassimilation reactions are coupled with the formation of acetone, whereby the product is converted to acetone by acetoacetate decarboxylase (coded by *adc*) (Dürre et al., 1995; Laursen and Westheimer, 1966). Butyryl-CoA is converted to butanol via one of the butyraldehyde dehydrogenases (which are part of the bifunctional, aldehyde/alcohol dehydrogenase proteins [AdhE1 and AdhE2] encoded, respectively, by *aad* [*adhe1*] and *adhe2*) and a butanol dehydrogenase (encoded by one or several of the following

Chromosomal genes

hydA	CAC0028
ldh1	CAC0267
ldh2	CAC1543
buk2	CAC1660
pta	CAC1742
ack/askA	CAC1743
ldh3	CAC2691
hbd	CAC2708
etfA	CAC2709
etfB	CAC2710
bcd	CAC2711
crt	CAC2712
thl	CAC2873
buk	CAC3075
ptb	CAC3076
bdhB	CAC3298
bdhA	CAC3299
ldh4	CAC3552

pSOL1 megaplasmid genes

pdc	CAP0025
adhE/adhE2	CAP0035
thil	CAP0078
aad/adhE1	CAP0162
ctfA	CAP0163
ctfB	CAP0164
adc	CAP0165

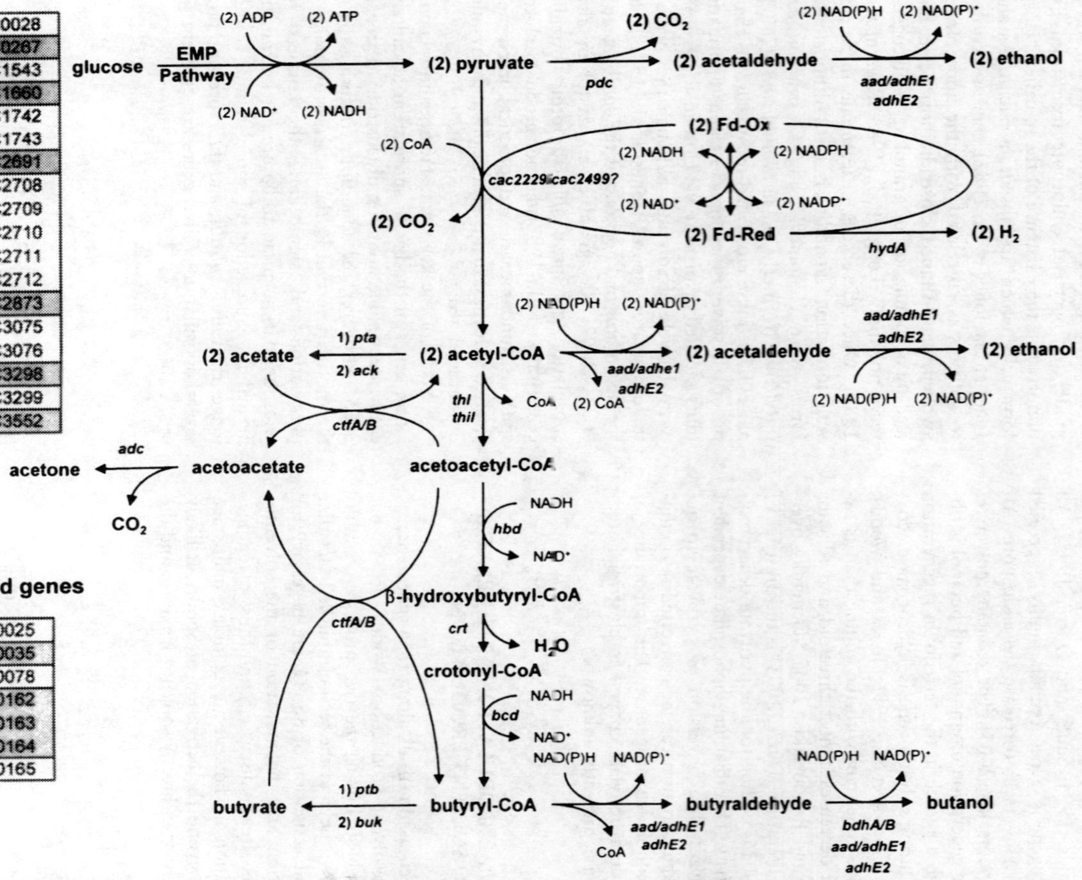

Figure 1. Metabolic pathways leading to acid and solvent formation in clostridia. The gene names shown correspond to those in *C. acetobutylicum* ATCC 824. Genes belonging to the same transcriptional unit share the same background (i.e., shaded or unshaded) on the left-hand table. The two boxes correspond to the genes encoding the proteins shown in the graph. Systematic gene names are shown according to the genome annotation of *C. acetobutylicum* ATCC 824 (Nölling et al., 2001) whereby CAC is used for genes located on the chromosome and CAP is used for genes located on the pSOL1 megaplasmid.

genes: *aad* [*adhe1*], *adhe2*, *bdhA*, and *bdhB*), whereas ethanol is formed likely by the same enzymes as butanol (Bertram et al., 1990; Dürre et al., 1995; Nair and Papoutsakis, 1994; Sauer and Durre, 1995). All these solventogenesis genes remain highly expressed until the cell lyses itself to release the endospore (Alsaker and Papoutsakis, 2005). The reducing power required for the generation of ethanol and butanol is obtained through a switch in the ferredoxin regeneration mechanism. During solventogenesis, hydrogenase activity decreases (Gorwa et al., 1996; Jones and Woods, 1986), which is consistent with the finding that its expression level decreases almost threefold and remains low (Alsaker and Papoutsakis, 2005). Generally overlooked in the literature (Dürre, 2005) is the presence of a gene (CAP0025) coding for the pyruvate decarboxylase (*pdc*) activity. This thiamine-diphosphate-dependent enzyme catalyzes the conversion of pyruvate to acetaldehyde, which in turn can be converted into ethanol by an alcohol dehydrogenase. The importance of the pyruvate decarboxylase enzyme in clostridial ethanol production is presently unknown.

CLOSTRIDIAL SPORULATION AS IT RELATES TO SOLVENT PRODUCTION

It has been assumed all along that sporulation is necessary for solvent production (Woods, 1995), but the molecular basis linking the two processes remains largely unknown beyond the well-established fact that expression and activation of Spo0A are necessary for both sporulation and the induction of the solventogenic genes (as already discussed) and thus solventogenesis. Sporulation is, however, a complex multistage process, and beyond the activation of Spo0A, it likely impacts directly or indirectly solvent formation and the associated bioprocessing. Figure 2 presents the different stages in clostridial sporulation named after the corresponding stages in *Bacillus subtilis*, the de facto model for bacterial endospore formation. Especially important is the existence of the so-called "clostridial"-form cells which are absent in bacilli. As shown by Jones et al. (Jones et al., 1982), the conversion of acids to solvents is correlated with the appearance of the swollen, cigar-shaped clostridial-form cells characterized by accumulation of granulose (a glycogen-like reserve polymer) (Long et al., 1983; Reysenbach et al., 1986). Once these cells differentiate further, the solvent formation capability is lost. Although these cell forms were examined using *Clostridium saccharobutylicum* (formerly *C. acetobutylicum* P262 [Keis et al., 2001]), these clostridial-form cells are characteristic of all solvent-forming clostridia. Sporulation mutants blocked at or near the clostridial stage produced normal levels of solvents, although they were unable to form mature spores (Jones et al., 1982). However, the disruption or knockout of the *spo0A* gene (Harris et al., 2002; Ravagnani et al., 2000) or any other sporulation mutant unable to form clostridial-form cells (Jones et al., 1982) did not produce solvents. Despite its fundamental and practical importance, the clostridial-form remains genetically and physiologically uncharacterized.

Many of the key sigma factors and related sporulation genes of the bacilli sporulation cascade have an easily identifiable homolog in clostridia (Paredes et al., 2005). However, the conditions triggering the start of the sporulation, that is, the Spo0A activation by phosphorylation and, significantly, the kinase(s) for this phosphorylation in clostridia are not known, the time needed for completion of sporulation is considerably longer than in *B. subtilis*, and the expression profiles of several sporulation-related genes are different (Paredes et al., 2005).

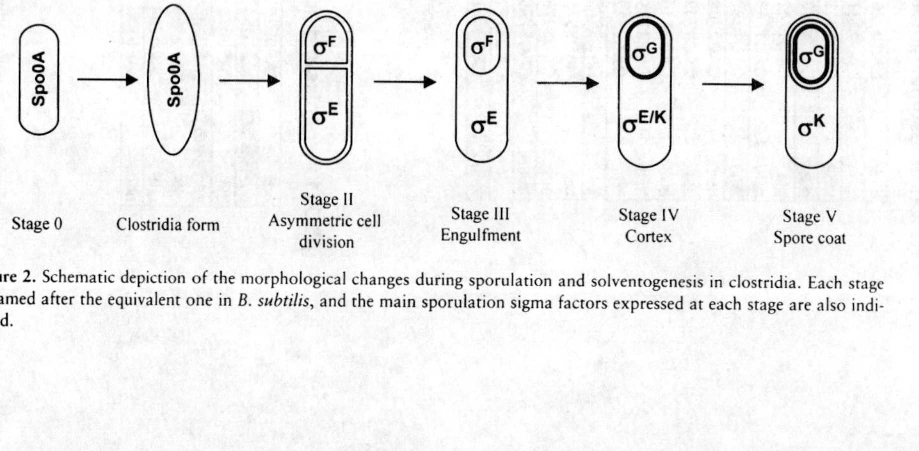

Figure 2. Schematic depiction of the morphological changes during sporulation and solventogenesis in clostridia. Each stage is named after the equivalent one in *B. subtilis*, and the main sporulation sigma factors expressed at each stage are also indicated.

Briefly, sporulation is a tightly regulated, both spatially and temporally, differentiation program used as a defense mechanism by some organisms in response to unfavorable environmental conditions. The end result of this process is a dormant, tough, nonreproductive structure that is highly resistant to harsh physical and chemical conditions. Under favorable conditions, these spores are able to germinate and the new cells are capable of vegetative growth and (if needed) sporulation again. In the case of the *Firmicutes*, the sporulation involves the orderly expression of several sigma factors (σ^H, σ^F, σ^E, σ^G, and σ^K) upon activation of the master regulation of sporulation Spo0A. This expression is also compartmentalized, with σ^F and σ^G being expressed in the forespore whereas σ^E and σ^K are expressed in the mother cell. In bacilli, *sigH* expression is induced before sporulation and upregulates *spo0A* and the *sigF* operon; however, in *C. acetobutylicum* the expression level of *sigH* does not undergo much temporal change (Alsaker and Papoutsakis, 2005; Dürre and Hollergschwandner, 2004; Predich et al., 1992; Wu et al., 1991). Analysis of the possible σ^H regulon in *C. acetobutylicum* (based on the orthologous genes from *B. subtilis*) revealed that though the expression of *sigH* may not vary, the sigma factor is expressed and active (Alsaker and Papoutsakis, 2005). After Spo0A becomes activated (phosphorylated) by an as-yet-uncharacterized mechanism (Dürre and Hollergschwandner, 2004; Paredes et al., 2005), upregulation of the operons encoding *sigF*, *sigE*, *sigG*, and *spoIIE* is observed (Alsaker and Papoutsakis, 2005; Harris et al., 2002). The operon encoding *sigF*, the *spoIIA* operon, also contains *spoIIAA*, the anti-anti-σ^F factor, and *spoIIAB*, the anti-σ^F factor (Dürre and Hollergschwandner, 2004; Paredes et al., 2005). Upon translation, the anti-σ^F factor SpoIIAB binds to σ^F and prevents it from becoming active; when the anti-anti-σ^F factor SpoIIAA is dephosphorylated by SpoIIE, it will bind to SpoIIAB and release σ^F (Dürre and Hollergschwandner, 2004; Paredes et al., 2005). However, SpoIIAB can phosphorylate SpoIIAA, preventing the SpoIIAA-SpoIIAB complex, and thus, a threshold of dephosphorylated SpoIIAA must be produced before σ^F becomes active (Dürre and Hollergschwandner, 2004; Paredes et al., 2005).

In bacilli, *sigE* and *sigG* are transcribed separately, with active σ^F and σ^E needed for the transcription of *sigG*, but in *C. acetobutylicum*, *sigE* and *sigG* are computationally predicted to be on the same operon and their similar expression patterns seem to confirm this prediction (Harris et al., 2002; Paredes et al., 2004; Stragier and Losick, 1996). σ^K is the final sigma factor in the sporulation cascade in bacilli. Although it has been annotated in *C. acetobutylicum*, its expression levels are very low, and thus, there is serious doubt that it has been correctly identified (or that it is even present) in *C. acetobutylicum* ATCC 824.

Spo0A ACTIVATION: BUTYRYL-P AND SENSOR KINASES

As stated above, the mechanism of Spo0A activation in clostridia remains unknown. Unlike what is observed for the related and widely studied spore-forming *B. subtilis*, carbon and nitrogen starvation does not induce sporulation (Long et al., 1984a) and many of the genetic elements leading to Spo0A activation in *B. subtilis* are missing in clostridia (Nölling et al., 2001; Paredes et al., 2005). Proposed mechanisms for Spo0A activation include phosphorylation by a sensor kinase or kinases, phosphorylation by butyryl-phosphate, or a modified phosphorelay system (Paredes et al., 2005). In *B. subtilis*, Spo0A is phosphorylated by five different sensor kinases (Jiang et al., 2000). All five kinases are "orphan" kinases that are not directly associated with a response regulator in the same operon (Fabret et al., 1999). In *C. acetobutylicum*, there are five orphan kinases, and two of these exhibit a significant induction just prior to the onset of sporulation (when gene targets, such as the *sol* operon and *adc*, of activated Spo0A are induced) (Paredes et al., 2005). Butyryl-phosphate has also been proposed as a signal molecule inducing solvent formation and sporulation (Harris et al., 2000). A butyrate kinase deletion mutant was found to produce lower levels of butyrate, resulting in increased solvent production and earlier sporulation (Green et al., 1996; Harris et al., 2000). Later studies confirmed the increased concentration and earlier accumulation of butyryl-phosphate in the butyrate kinase mutant, suggesting a possible role as a phosphate donor in cellular signaling or perhaps as the molecule phosphorylating Spo0A (Zhao et al., 2005).

ORGANIZATION AND REGULATION OF SOLVENT GENES

As shown in Fig. 1, the genes responsible for solvent production in *C. acetobutylicum* include *ctfA/B* and *adc* (acetone formation), *bdhA/B*, and two *adhE* genes: *aad* (*adhE1*) and *adhE2*. All but the *bdhA/B* genes are located on a large megaplasmid, pSOL1, whose loss not only leads to the virtual elimination of solvent production but also abolishes sporulation (Cornillot et al., 1997). The five solventogenic genes on pSOL1 are organized into three operons (Paredes et al., 2004). The *sol* operon contains *aad* (*adhE1*)-*ctfA*-*ctfB*, whereas the *adc* and *adhE2* genes are organized in separate monocistronic operons. The *sol* operon and *adc* are located adjacent to each other but are transcribed from opposite strands. The *adhE2* gene is separated by about 50 kbp from the *sol* operon. The chromosomal *bdhA* and *bdhB* genes are organized into two consecutive monocistronic operons (Paredes et al., 2004).

As stated previously, both the *adc* gene and the *sol* operon are directly controlled by Spo0A (Ravagnani et al., 2000; Thormann et al., 2002) and the disruption of the Spo0A motif upstream of the *sol* operon results in the elimination of promoter activity. Interestingly, overexpression of the promoter element alone causes a decrease in solvent production. It has been suggested (Thormann et al., 2002) that such a decrease could be due to the titration of an as-yet-unknown activator.

The occurrence of the key solventogenic genes in the pSOL1 megaplasmid provides a unique opportunity for their in vivo study by complementing one of the two degenerate strains (M5 and DG1) which have lost pSOL1. The complementation of a degenerate strain either with the primary *adhE* gene (*aad* or *adhE1*) or with the secondary *adhE2* gene effectively restored butanol production to near one-half the level of the wild-type (WT) strain (Fontaine et al., 2002; Nair and Papoutsakis, 1994). It has been suggested that the lower levels of butanol produced by the pSOL1-deficient strains complemented with either *aad* or *adhE2* are due to either their inability to uptake acid products (due to the lack of *ctfA/B* and *adc*) or to the fact that these nonsporulating strains may have greater sensitivity to solvent products than the WT strain (Fontaine et al., 2002).

SOLVENT PRODUCTION IS AFFECTED BY THE CARBON SUBSTRATE AND SEVERAL ENVIRONMENTAL CONDITIONS

As stated above, NAD(P)H availability plays a significant role in the distribution of fermentation end products. Continuous cultures of *C. acetobutylicum* grown at low pH on glucose produce acetone, butanol, and ethanol, whereas when grown at neutral pH under conditions of high NAD(P)H availability (see below), an alcohologenic metabolism producing butanol and ethanol but not acetone is observed. Increases of the NAD(P)H availability have been obtained by CO sparging (Husemann and Papoutsakis, 1989; Meyer and Papoutsakis, 1989), using artificial electron carriers like neutral red (Girbal et al., 1995b), providing alternatives to the use of the hydrogenase (Girbal et al., 1995b; Grupe and Gottschalk, 1992), or using mixtures of substrates (like glycerol and glucose) with different degrees of reduction (Girbal and Soucaille, 1994; Vasconcelos et al., 1994).

The possibility of influencing the NAD(P)H availability (and thus product selectivity in favor of butanol production, the most desirable product) by using substrates with different degrees of reduction (Tangney et al., 1998, 2001, 2003; Tangney and Mitchell, 2000, 2007) is a strategy of potentially large practical importance. To do so, it is important that the influence of all components used in the typical and practically important complex media (such as those based on molasses, starch, corn mash, corn fiber, agricultural waste streams, whey, etc. [Bahl et al., 1986; Jesse et al., 2002; Qureshi et al., 2001; Zverlov et al., 2006]) be studied and understood. Table 1 presents data on how the butanol-to-acetone ratio changes with different medium compositions. The use of whey as a substrate results in high butanol levels with nearly no acetone production (Bahl et al., 1986). However, cultures grown on lactose alone (the main carbohydrate present in whey) could not replicate these results and citrate seems to play a very important role. The proportions between the different solvents can also be manipulated by regulating the amount of oligoelements present on the culture medium. For instance, cultures growing under iron limitation show an increased butanol-to-acetone ratio (Bahl et al., 1986), probably due to the iron requirement of acetate decarboxylase, the final enzyme in acetone formation.

Production of butanol from starch is very attractive economically. Numerous clostridial species express amylolytic enzymes capable of starch degradation, although there has been relatively little characterization of these enzymes. Solvent production using starch as a substrate has been most extensively studied for *C. beijerinckii*, where mutant strains have been developed that produce nearly 25 g of acetone-butanol-ethanol solvents/liter of medium (Annous and Blaschek, 1990; Jesse et al., 2002). Examination of the genome of *C. acetobutylicum* yields six proteins annotated as amylases or possible amylases (Nölling et al., 2001). Three of these are located on the chromosome, and three are on the pSOL1 megaplasmid. Screening for amylase activity between the WT and the DG1 (Nair, 1995) pSOL1-deficient strains demonstrated amylolytic activity for both strains, but it was higher in the WT under starch-only conditions (Sabathe et al., 2002). Growing on a medium containing both starch and glucose, only the WT strain maintained noticeable amylolytic activity, suggesting that the chromosomal amylases are catabolite repressed. Initial isolation of an extracellular amylase from *C. acetobutylicum* did not show catabolite repression, as it was produced at similar levels whether grown on starch or glucose (Paquet et al., 1991). Other studies, however, reported increased in-

Table 1. Effects of substrates and medium composition on solvent production

Strain (reference)	Medium	Addition	Concn (mM) of: Butanol	Acetone	Ethanol	Butyrate	Acetate	Butanol-to-acetone ratio
C. acetobutylicum DSM 792, 1731, 1732 (Bahl et al., 1986)	Whey		122	1.2	37	14	13	101.6
		Minerals	75	1.3	17	23	19	57.8
	Whey, citrate removed		65	4.3	17	36	23	15.1
		Citrate (6 mM)	99	1.8	22	43	19	55.0
		Minerals	77	7	17	54	31	11.0
	Synthetic	Glucose	154	75	22	1.3	8.2	2.1
		Lactose	156	60	30	NA[a]	NA	2.6
		Glucose + galactose	87	39	12	26	14	2.2
		Glucose + lactate	132	34	14	23	23	3.9
		Lactose + lactate	114	30	10	45	32	3.8
C. beijerinckii BA101 (Jesse et al., 2002)	Starch + yeast extract (96 h)		202	117	22	11	NA	1.7
	Starch-based packing peanuts + yeast extract (110 h, 80 g/liter)		192	77	NA	6	NA	2.5
	Agricultural wastes + yeast extract (72 h, 96.4 g/liter)		183	115	NA	8	32	1.6
C. beijerinckii BA101 (Qureshi et al., 2001)	Spray-dried soy molasses (120 h)	Glucose (25.3 g/liter)	258	72	10	10	15	3.6
	Glucose (120 h)		271	74	9	7	14	3.7
C. acetobutylicum (Zverlov et al., 2006)	Molasses		135	110	33	NA	NA	1.2

[a]NA, not available.

duction of amylolytic enzymes in cultures grown on starch compared to cultures grown on glucose (Annous and Blaschek, 1991).

MECHANISMS OF PRODUCT (INCLUDING SOLVENT) TOLERANCE AND GENERATION OF TOLERANT STRAINS

Microbial susceptibility to the accumulation of toxic metabolites (e.g., solvents such as ethanol, acetone, butanol, and isopropanol and acids such as acetate, butyrate, and lactate) has been the subject of considerable study (Baer et al., 1987; Bowles and Ellefson, 1985; Ingram, 1976; Weber and de Bont, 1996). Solvents have a varying ability to intercalate into the cell membrane, altering fluidity and compromising membrane protein functions (Sardessai and Bhosle, 2002). Solvent accumulation in the cytosol also impacts protein folding and sugar transport and metabolism (Ounine et al., 1985). Acid metabolites, which freely diffuse across the cell membrane when protonated, affect cellular physiology through both free-proton and anion interactions. Excess free protons tend to equilibrate the energetically valuable proton gradient (Baronofsky et al., 1984; Kashket, 1987) while also potentially impacting purine bases of genomic DNA (Choi et al., 2000). Mechanisms for solvent and acid tolerance are directly related to their toxic effects. Active adjustment of the relative composition of membrane lipids counteracts increased fluidity imposed by solvents (Baer et al., 1987, 1989; Vollherbst-Schneck et al., 1984). Efflux pumps transport protons and solvents out of the cytosol, while stress genes/proteins (e.g., *groESL*, *dnaK*, *hsp18*, and *lonA*) necessary for protein folding and degradation are upregulated, thus mitigating cellular damage (Tomas et al., 2003). Biofilm formation, production of emulsifying enzymes, and transport and generation of osmolytes (e.g., proline, glycine, and betaine) may also protect cells from toxic acids and solvents (Sardessai and Bhosle, 2002).

Microbiological and, more recently, molecular biology techniques have been used to create strains with improved solvent or acid tolerance. In the most widely applied strategy, cells are grown in either batch or continuous culture and are exposed to increasing amounts of a toxic compound (Soucaille et al., 1987; Yomano et al., 1998). Over time and increased exposure, genetic mutations accumulate, leading to a more tolerant phenotype, and these have resulted in 40% improvement in solvent tolerance (Soucaille et al., 1987). An adaptation of this technique involves the use of chemical agents (Wong and Bennett, 1996) and mutagens (Clark et al., 1989; Junelles et al., 1987; Murray et al., 1983) that increase the likelihood of genetic mutation. Metabolic engineering has also been useful for generation of tolerant phenotypes (Harris et al., 2000, 2001; Tomas et al., 2003). As an extension of directed metabolic engineering, libraries of genetic variants (e.g., genomic, cDNA, and insertional inactivation libraries) can be enriched in the batch and continuous cultures described above for selection of genetic variants with improved tolerance (Borden and Papoutsakis, 2007).

Clearly, solvent and acid tolerance can be classified as a complex genetic trait, and its characterization requires not only the study of the "key players" but also the study of the organism response as a whole. Application of DNA microarray technology is a first step towards this holistic understanding of this phenomenon (Alsaker, 2006; Alsaker et al., 2004; Tomas et al., 2004). Recent work by Alsaker et al. (Alsaker, 2006) used microarrays to detail the transcript level changes resulting from sublethal levels of acetate, butyrate, and butanol added to a growing culture. These results showed that several stress protein transcripts and the *sol* operon (*aad-ctfA-ctfB*) were differentially upregulated by acetate, butyrate, and butanol stresses, including *lonA* (CAC0456), *hrcA-grpE-dnaK* (CAC1280 through CAC1282), *groESL* (CAC2704 and CAC2703), *ctsR-yacH-yacI-clpC* (CAC3192 through CAC3189), *hsp90* (*htpG*, CAC3315), *hsp18* (CAC3714), and an HtrA-like serine protease (CAC2433), expanding the previous reports (Alsaker et al., 2004; Tomas et al., 2004) about the butanol stress response in *C. acetobutylicum*. Many of the genes related to branched-chain amino acid biosynthesis (leucine, isoleucine, and valine) were differentially upregulated following butanol and butyrate stress but were expressed at a lower level following acetate stress. Also upregulated was the *aro* locus (CAC0892 through CAC0899) encoding the genes for the biosynthesis of chorismate, a branch point and key intermediate for phenylalanine, tyrosine, and tryptophan biosynthesis. The general downregulation of the fatty acid biosynthesis locus genes (CAC3568 through CAC3579) after butanol stress (Alsaker, 2006) is consistent with the quick adjustment of the saturated and unsaturated fatty acid content previously reported (Tomas et al., 2004). Butanol stress also upregulated the glycerol metabolism genes *glpA* and *glpF*. However, although glycerol metabolism genes have been implicated in hexane tolerance (Tomas et al., 2004) and xylene and cyclohexane (Shimizu et al., 2005) tolerance in *Escherichia coli*, the expression of *glpF* and *glpA* may not necessarily enhance triacyl-glycerol metabolism, and the role of these genes in the solvent stress response is not obvious. CAP0102, a membrane protein, was strongly induced by all three metabolite stresses, and previous reports have shown that this gene is induced by butanol stress (Alsaker et al., 2004; Tomas et al., 2004) and the onset of solventogenesis and stationary phase (Alsaker and Papoutsakis, 2005). CAP0102 is structurally similar to the MMPL (mycobacterial membrane protein large) class of proteins. In mycobacteria these genes are often located in the same loci as fatty acid and polyketide synthesis genes, suggesting that MMPLs export these products (Minnikin et al., 2002).

Given the large number of genes involved in complex phenotypes such as chemical tolerance, it seems only logical to use high-throughput techniques for elucidating the main contributors to this phenotype (Borden and Papoutsakis, 2007; Chiu et al., 2007; Gill et al., 2002; Kim et al., 2007). Such techniques have also been extended for the first time to gram-positive organisms such as *C. acetobutylicum* (Borden and Papoutsakis, 2007). Using a genomic library constructed from randomly sheared *C. acetobutylicum* 824 genomic DNA, single-batch and serial transfer enrichment techniques were used to identify fragments conferring resistance to the toxic metabolites butanol and butyrate, and DNA microarrays were used to monitor the process of library enrichment through serial transfers (Borden and Papoutsakis, 2007). Notable among the methodological improvements in this work were the improved performances of gene inserts from serial enrichment as opposed to a single round of enrichment in batch culture (Borden and Papoutsakis, 2007). Also, genomic DNA restriction (by membrane-bound endonucleases common among clostridia [Klapatch et al., 1996; Mermelstein et al., 1992; Ozkan et al., 2001]) that complicated DNA microarray hybridization was circumvented by PCR amplification of plasmid-inserted genomic fragments (Borden and Papoutsakis, 2007).

Among the library inserts enriched through serial transfer were a set of transcriptional regulators—CAC0977, CAC1463, CAC1869, and CAC2495 (Borden and Papoutsakis, 2007). Transcriptional regulators potentially control a much larger gene set; hence, they might impact multiple tolerance mechanisms, and their use has previously been exploited to enhance tolerance (Alper et al., 2006). In particular, CAC0977 is annotated as an Lrp (leucine-responsive regulatory protein)

(Nölling et al., 2001), which has homologs in 45% of sequenced bacterial genomes (Thaw et al., 2006) and impacts upwards of 10% of the *E. coli* genome (Tani et al., 2002). The most enriched gene fragment, containing CAC1869, was able to confer a 90% increase in tolerance to butanol, as determined by measuring growth (i.e., optical density) after 12 h of exposure to increasing butanol concentrations (Borden and Papoutsakis, 2007). CAC1869 is homologous to xenobiotic responsive element transcriptional regulators, which contribute to an array of tolerance mechanisms in other microorganisms (Azcarate-Peril et al., 2004; Labie et al., 1989; Liu et al., 2003). Borden et al. concluded that the process of serial enrichment in the presence of butanol exacerbates the process of strain degeneration (Borden and Papoutsakis, 2007) (loss of the pSOL1 megaplasmid necessary for both sporulation and solvent production), but also that degenerate *C. acetobutylicum* strains have lower overall solvent tolerance than WT cells (Borden and Papoutsakis, 2007). The latter conclusion was tested, and as expected WT *C. acetobutylicum* ATCC 824 showed a significantly higher tolerance to butanol than the degenerate M5 strain. It therefore seems likely that one or more of the 178 pSOL1 genes is responsible for increased tolerance to butanol.

METABOLIC FLUX ANALYSIS AND GENOME-SCALE MODELS FOR UNDERSTANDING AND EXPLOITING GENOME-SCALE REGULATION FOR SOLVENT PRODUCTION

The combination of biochemical pathways and mass balances as a tool to quantify the primary metabolism of *C. acetobutylicum* ATCC 824 has a long history. Early attempts relied on the first systematic use of metabolic stoichiometry in order to calculate metabolic fluxes and carry out a large-scale metabolic flux analysis on the overall primary metabolism of *C. acetobutylicum* (Papoutsakis, 1984). An improved model was developed later (Desai et al., 1999) and successfully applied to the calculation of the metabolic fluxes for several genetically engineered strains (Harris et al., 2000, 2001; Tummala et al., 2003a, 2003b), including a pioneering attempt combining transcriptional and metabolic flux data (Tummala et al., 2003b). Nevertheless, the widespread use of complex media and the link between sporulation and solvent production certainly have played against the generalized use of such a tool to characterize the metabolic state of solvent producing clostridia.

These first models were used to study the fluxes in WT and genetically altered strains to not only identify bottlenecks of primary metabolism but also gain insight into the effects of governing regulatory mechanisms. However, a new paradigm is evolving whereby the combination of the genome annotation (Nölling et al., 2001) with publicly available reaction (Kanehisa et al., 2006), enzymatic (Schomburg et al., 2004), and transporter (Ren et al., 2007; Saier et al., 2006) databases is used to construct a genome-scale reaction network (Edwards et al., 1999, 2001; Heinemann et al., 2005). In the case of *C. acetobutylicum*, this network comprises over 500 reactions involving more than 400 metabolites describing (i) primary metabolism, (ii) starch and hexose sugar metabolism, (iii) amino acid biosynthesis, (iv) nucleotide biosynthesis, (v) nitrogen fixation, (vi) cell wall and lipid biosyntheses, (vii) metabolism of vitamins and cofactors, and (viii) a network of transporter proteins governing metabolite exchange across the cell membrane. The use of such a comprehensive model not only to illuminate the metabolic fluxes associated with specific genetic alterations or mutant strains but also to simulate the most plausible response of the culture under a variety of physical environments and genotypes presents new opportunities as well as challenges for the study of bacterial responses to external stimuli (e.g., acid or solvent response). Certainly these regulated metabolic shifts not only impact the expression of hundreds of genes but also result in changes in the bacterial biomass composition itself. Elusive aspects like the altered metabolic demands of a changed lipid profile and its impact on the growth rate and the redirection of metabolic fluxes through primary metabolism could be computationally analyzed by this new generation of genome-scale models. However, to become a central piece for rational design of new and improved solventogenic strains, new approaches should be sought to accommodate the necessary layers of regulatory controls that convert the potentials of a genotype into several possible phenotypes.

Acknowledgments. This work was supported by NSF grants BES-0331402 and BES-0418157 and the U.S. Department of Energy grant DE-FG36-03GO13160. J.R.B. was supported by an NIH/NIGMS Biotechnology Training grant (T32-GM08449-11).

REFERENCES

Alper, H., J. Moxley, E. Nevoigt, G. R. Fink, and G. Stephanopoulos. 2006. Engineering yeast transcription machinery for improved ethanol tolerance and production. *Science* **314:**1565–1568.

Alsaker, K. V. 2006. Microarray-based transcriptional analyses of stationary phase phenomena and stress responses in *Clostridium acetobutylicum*. Ph.D. thesis. Northwestern University, Evanston, IL.

Alsaker, K. V., and E. T. Papoutsakis. 2005. Transcriptional program of early sporulation and stationary-phase events in *Clostridium acetobutylicum*. *J. Bacteriol.* **187:**7103–7118.

Alsaker, K. V., T. R. Spitzer, and E. T. Papoutsakis. 2004. Transcriptional analysis of *spo0A* overexpression in *Clostridium acetobutylicum* and its effect on the cell's response to butanol stress. *J. Bacteriol.* **186:**1959–1971.

Andersch, W., H. Bahl, and G. Gottschalk. 1983. Level of enzymes involved in acetate, butyrate, acetone and butanol formation by

Clostridium acetobutylicum. *Eur. J. Appl. Microbiol. Biotechnol.* 18:327–332.

Annous, B. A., and H. P. Blaschek. 1991. Isolation and characterization of *Clostridium acetobutylicum* mutants with enhanced amylolytic activity. *Appl. Environ. Microbiol.* 57:2544–2548.

Annous, B. A., and H. P. Blaschek. 1990. Regulation and localization of amylolytic enzymes in *Clostridium acetobutylicum* ATCC-824. *Appl. Environ. Microbiol.* 56:2559–2561.

Azcarate-Peril, M. A., E. Altermann, R. L. Hoover-Fitzula, R. J. Cano, and T. R. Klaenhammer. 2004. Identification and inactivation of genetic loci involved with *Lactobacillus acidophilus* acid tolerance. *Appl. Environ. Microbiol.* 70:5315–5322.

Baer, S. H., H. P. Blaschek, and T. L. Smith. 1987. Effect of butanol challenge and temperature on lipid composition and membrane fluidity of butanol-tolerant *Clostridium acetobutylicum*. *Appl. Environ. Microbiol.* 53:2854–2861.

Baer, S. H., D. L. Bryant, and H. P. Blaschek. 1989. Electron spin resonance analysis of the effect of butanol on the membrane fluidity of intact cells of *Clostridium acetobutylicum*. *Appl. Environ. Microbiol.* 55:2729–2731.

Bahl, H., M. Gottwald, A. Kuhn, V. Rale, W. Andersch, and G. Gottschalk. 1986. Nutritional factors affecting the ratio of solvents produced by *Clostridium acetobutylicum*. *Appl. Environ. Microbiol.* 52:169–172.

Baronofsky, J. J., W. J. Schreurs, and E. R. Kashket. 1984. Uncoupling by acetic acid limits growth of and acetogenesis by *Clostridium thermoaceticum*. *Appl. Environ. Microbiol.* 48:1134–1139.

Bertram, J., A. Kuhn, and P. Durre. 1990. Tn916-induced mutants of *Clostridium acetobutylicum* defective in regulation of solvent formation. *Arch. Microbiol.* 153:373–377.

Borden, J. R., and E. T. Papoutsakis. 2007. Dynamics of genomic-library enrichment and identification of solvent tolerance genes in *Clostridium acetobutylicum*. *Appl. Environ. Microbiol.* 73:3061–3068.

Bowles, L. K., and W. L. Ellefson. 1985. Effects of butanol on *Clostridium acetobutylicum*. *Appl. Environ. Microbiol.* 50:1165–1170.

Chiu, H. C., T. L. Lin, and J. T. Wang. 2007. Identification and characterization of an organic solvent tolerance gene in *Helicobacter pylori*. *Helicobacter* 12:74–81.

Choi, S. H., D. J. Baumler, and C. W. Kaspar. 2000. Contribution of *dps* to acid stress tolerance and oxidative stress tolerance in *Escherichia coli* O157:H7. *Appl. Environ. Microbiol.* 66:3911–3916.

Clark, S. W., G. N. Bennett, and F. B. Rudolph. 1989. Isolation and characterization of mutants of *Clostridium acetobutylicum* ATCC-824 deficient in acetoacetyl-coenzyme A:acetate/butyrate:coenzyme A-transferase (EC 2.8.3.9) and in other solvent pathway enzymes. *Appl. Environ. Microbiol.* 55:970–976.

Cornillot, E., R. V. Nair, E. T. Papoutsakis, and P. Soucaille. 1997. The genes for butanol and acetone formation in *Clostridium acetobutylicum* ATCC 824 reside on a large plasmid whose loss leads to degeneration of the strain. *J. Bacteriol.* 179:5442–5447.

Desai, R. P., L. K. Nielsen, and E. T. Papoutsakis. 1999. Stoichiometric modeling of *Clostridium acetobutylicum* fermentations with non-linear constraints. *J. Biotechnol.* 71:191–205.

Dürre, P. 2005. Formation of solvents in clostridia, p. 671–693. *In* P. Dürre (ed.), *Handbook on Clostridia*. CRC Press, Boca Raton, FL.

Dürre, P., R. J. Fischer, A. Kuhn, K. Lorenz, W. Schreiber, B. Sturzenhofecker, S. Ullmann, K. Winzer, and U. Sauer. 1995. Solventogenic enzymes of *Clostridium acetobutylicum*—catalytic properties, genetic organization, and transcriptional regulation. *FEMS Microbiol. Rev.* 17:251–262.

Dürre, P., and C. Hollergschwandner. 2004. Initiation of endospore formation in *Clostridium acetobutylicum*. *Anaerobe* 10:69–74.

Edwards, E. S., R. Ramakrishna, C. H. Schilling, and B. O. Palsson. 1999. Metabolic flux analysis, p. 13–57. *In* S. Y. Lee and E. T. Papoutsakis (ed.), *Metabolic Engineering*. Marcel Dekker, New York, NY.

Edwards, J. S., R. U. Ibarra, and B. O. Palsson. 2001. In silico predictions of *Escherichia coli* metabolic capabilities are consistent with experimental data. *Nat. Biotechnol.* 19:125–130.

Fabret, C., V. A. Feher, and J. A. Hoch. 1999. Two-component signal transduction in *Bacillus subtilis*: how one organism sees its world. *J. Bacteriol.* 181:1975–1983.

Feustel, L., S. Nakotte, and P. Dürre. 2004. Characterization and development of two reporter gene systems for *Clostridium acetobutylicum*. *Appl. Environ. Microbiol.* 70:798–803.

Fontaine, L., I. Meynial-Salles, L. Girbal, X. Yang, C. Croux, and P. Soucaille. 2002. Molecular characterization and transcriptional analysis of *adhE2*, the gene encoding the NADH-dependent aldehyde/alcohol dehydrogenase responsible for butanol production in alcohologenic cultures of *Clostridium acetobutylicum* ATCC 824. *J. Bacteriol.* 184:821–830.

Gerischer, U., and P. Durre. 1992. mRNA analysis of the *adc* gene region of *Clostridium acetobutylicum* during the shift to solventogenesis. *J. Bacteriol.* 174:426–433.

Gill, R. T., S. Wildt, Y. T. Yang, S. Ziesman, and G. Stephanopoulos. 2002. Genome-wide screening for trait conferring genes using DNA microarrays. *Proc. Natl. Acad. Sci. USA* 99:7033–7038.

Girbal, L., C. Croux, I. Vasconcelos, and P. Soucaille. 1995a. Regulation of metabolic shifts in *Clostridium acetobutylicum* ATCC 824. *FEMS Microbiol. Rev.* 17:287–297.

Girbal, L., and P. Soucaille. 1994. Regulation of *Clostridium acetobutylicum* metabolism as revealed by mixed-substrate steady-state continuous cultures: role of NADH/NAD ratio and ATP pool. *J. Bacteriol.* 176:6433–6438.

Girbal, L., and P. Soucaille. 1998. Regulation of solvent production in *Clostridium acetobutylicum*. *Trends Biotechnol.* 16:11–16.

Girbal, L., I. Vasconcelos, S. Saintamans, and P. Soucaille. 1995b. How neutral red modified carbon and electron flow in *Clostridium acetobutylicum* grown in chemostat culture at neutral pH. *FEMS Microbiol. Rev.* 16:151–162.

Gorwa, M. F., C. Croux, and P. Soucaille. 1996. Molecular characterization and transcriptional analysis of the putative hydrogenase gene of *Clostridium acetobutylicum* ATCC 824. *J. Bacteriol.* 178:2668–2675.

Green, E. M., Z. L. Boynton, L. M. Harris, F. B. Rudolph, E. T. Papoutsakis, and G. N. Bennett. 1996. Genetic manipulation of acid formation pathways by gene inactivation in *Clostridium acetobutylicum* ATCC 824. *Microbiology* 142:2079–2086.

Grupe, H., and G. Gottschalk. 1992. Physiological events in *Clostridium acetobutylicum* during the shift from acidogenesis to solventogenesis in continuous culture and presentation of a model for shift induction. *Appl. Environ. Microbiol.* 58:3896–3902.

Harris, L. M., L. Blank, R. P. Desai, N. E. Welker, and E. T. Papoutsakis. 2001. Fermentation characterization and flux analysis of recombinant strains of *Clostridium acetobutylicum* with an inactivated *solR* gene. *J. Ind. Microbiol. Biotechnol.* 27:322–328.

Harris, L. M., R. P. Desai, N. E. Welker, and E. T. Papoutsakis. 2000. Characterization of recombinant strains of the *Clostridium acetobutylicum* butyrate kinase inactivation mutant: need for new phenomenological models for solventogenesis and butanol inhibition? *Biotechnol. Bioeng.* 67:1–11.

Harris, L. M., N. E. Welker, and E. T. Papoutsakis. 2002. Northern, morphological, and fermentation analysis of *spo0A* inactivation and overexpression in *Clostridium acetobutylicum* ATCC 824. *J. Bacteriol.* 184:3586–3597.

Hartmanis, M. G. N., T. Klason, and S. Gatenbeck. 1984. Uptake and activation of acetate and butyrate in *Clostridium acetobutylicum*. *Appl. Microbiol. Biotechnol.* 20:66–71.

Heinemann, M., A. Kummel, R. Ruinatscha, and S. Panke. 2005. *In silico* genome-scale reconstruction and validation of the *Staphylococcus aureus* metabolic network. *Biotechnol. Bioeng.* 92:850–864.

Husemann, M. H. W., and E. T. Papoutsakis. 1989. Comparison between in vivo and in vitro enzyme activities in continuous and batch fermentations of *Clostridium acetobutylicum*. *Appl. Microbiol. Biotechnol.* 30:585–595.

Hüsemann, M. H. W., and E. T. Papoutsakis. 1988. Solventogenesis in *Clostridium acetobutylicum* fermentations related to carboxylic-acid and proton concentrations. *Biotechnol. Bioeng.* 32:843–852.

Ingram, L. O. 1976. Adaptation of membrane lipids to alcohols. *J. Bacteriol.* 125:670–678.

Jesse, T. W., T. C. Ezeji, N. Qureshi, and H. P. Blaschek. 2002. Production of butanol from starch-based waste packing peanuts and agricultural waste. *J. Ind. Microbiol. Biotechnol.* 29:117–123.

Jiang, M., W. Shao, M. Perego, and J. A. Hoch. 2000. Multiple histidine kinases regulate entry into stationary phase and sporulation in *Bacillus subtilis*. *Mol. Microbiol.* 38:535–542.

Jones, D. T., A. van der Weshuizen, S. Long, E. R. Allcock, S. J. Reid, and D. R. Woods. 1982. Solvent production and morphological changes in *Clostridium acetobutylicum*. *Appl. Environ. Microbiol.* 43:1434–1439.

Jones, D. T., and D. R. Woods. 1986. Acetone-butanol fermentation revisited. *Microbiol. Rev.* 50:484–524.

Junelles, A. M., R. Janatiidrissi, A. Elkanouni, H. Petitdemange, and R. Gay. 1987. Acetone-butanol fermentation by mutants selected for resistance to acetate and butyrate halogen analogs. *Biotechnol. Lett.* 9:175–178.

Kanehisa, M., S. Goto, M. Hattori, K. F. Aoki-Kinoshita, M. Itoh, S. Kawashima, T. Katayama, M. Araki, and M. Hirakawa. 2006. From genomics to chemical genomics: new developments in KEGG. *Nucleic Acids Res.* 34:D354–D357.

Kashket, E. R. 1987. Bioenergetics of lactic-acid bacteria: cytoplasmic pH and osmotolerance. *FEMS Microbiol. Rev.* 46:233–244.

Keis, S., R. Shaheen, and D. T. Jones. 2001. Emended descriptions of *Clostridium acetobutylicum* and *Clostridium beijerinckii*, and descriptions of *Clostridium saccharoperbutylacetonicum* sp. nov. and *Clostridium saccharobutylicum* sp. nov. *Int. J. Syst. Evol. Microbiol.* 51:2095–2103.

Kim, M. J., G. H. Lim, E. S. Kim, C. B. Ko, K. Y. Yang, J. A. Jeong, M. C. Lee, and C. S. Kim. 2007. Abiotic and biotic stress tolerance in *Arabidopsis* overexpressing the multiprotein bridging factor 1a (MBF1a) transcriptional coactivator gene. *Biochem. Biophys. Res. Commun.* 354:440–446.

Klapatch, T. R., A. L. Demain, and L. R. Lynd. 1996. Restriction endonuclease activity in *Clostridium thermocellum* and *Clostridium thermosaccharolyticum*. *Appl. Microbiol. Biotechnol.* 45:127– 131.

Labie, C., F. Bouche, and J. P. Bouche. 1989. Isolation and mapping of *Escherichia coli* mutations conferring resistance to division inhibition protein DicB. *J. Bacteriol.* 171:4315–4319.

Laursen, R. A., and F. H. Westheimer. 1966. Active site of acetoacetate decarboxylase. *J. Am. Chem. Soc.* 88:3426–3430.

Lesnik, E. A., R. Sampath, H. B. Levene, T. J. Henderson, J. A. McNeill, and D. J. Ecker. 2001. Prediction of *rho*-independent transcriptional terminators in *Escherichia coli*. *Nucleic Acids Res.* 29:3583–3594.

Liu, Y., J. Zhou, M. V. Omelchenko, A. S. Beliaev, A. Venkateswaran, J. Stair, L. Wu, D. K. Thompson, D. Xu, I. B. Rogozin, E. K. Gaidamakova, M. Zhai, K. S. Makarova, E. V. Koonin, and M. J. Daly. 2003. Transcriptome dynamics of *Deinococcus radiodurans* recovering from ionizing radiation. *Proc. Natl. Acad. Sci. USA* 100:4191–4196.

Long, S., D. T. Jones, and D. R. Woods. 1984a. Initiation of solvent production, clostridial stage and endospore formation in *Clostridium acetobutylicum* P262. *Appl. Microbiol. Biotechnol.* 20:256–261.

Long, S., D. T. Jones, and D. R. Woods. 1984b. The relationship between sporulation and solvent production in *Clostridium acetobutylicum*-P262. *Biotechnol. Lett.* 6:529–534.

Long, S., D. T. Jones, and D. R. Woods. 1983. Sporulation of *Clostridium acetobutylicum* P262 in a defined medium. *Appl. Environ. Microbiol.* 45:1389–1393.

Mermelstein, L. D., N. E. Welker, G. N. Bennett, and E. T. Papoutsakis. 1992. Expression of cloned homologous fermentative genes in *Clostridium acetobutylicum* ATCC 824. *Biotechnology* (New York) 10:190–195.

Meyer, C. L., and E. T. Papoutsakis. 1989. Increased levels of ATP and NADH are associated with increased solvent production in continuous cultures of *Clostridium acetobutylicum*. *Appl. Microbiol. Biotechnol.* 30:450–459.

Minnikin, D. E., L. Kremer, L. G. Dover, and G. S. Besra. 2002. The methyl-branched fortifications of *Mycobacterium tuberculosis*. *Chem. Biol.* 9:545–553.

Mitchell, W. J., and M. Tangney. 2005. Carbohydrate uptake by the phosphotransferase system and other mechanisms, p. 155–175. *In* P. Dürre (ed.), *Handbook on Clostridia*. CRC Press, Boca Raton, FL.

Murray, W. D., K. B. Wemyss, and A. W. Khan. 1983. Increased ethanol-production and tolerance by a pyruvate-negative mutant of *Clostridium saccharolyticum*. *Eur. J. Appl. Microbiol. Biotechnol.* 18:71–74.

Nair, R. 1995. Molecular characterization and regulation of a multifunctional aldehyde/alcohol dehydrogenase gene and its use for metabolic engineering of *Clostridium acetobutylicum* ATCC 824. Ph.D. thesis. Northwestern University, Evanston, IL.

Nair, R. V., and E. T. Papoutsakis. 1994. Expression of plasmid-encoded *aad* in *Clostridium acetobutylicum* M5 restores vigorous butanol production. *J. Bacteriol.* 176:5843–5846.

Nölling, J., G. Breton, M. V. Omelchenko, K. S. Makarova, Q. Zeng, R. Gibson, H. M. Lee, J. Dubois, D. Qiu, J. Hitti, Y. Wolf, R. L. Tatusov, F. Sabathe, L. Doucette-Stamm, P. Soucaille, M. J. Daly, G. N. Bennett, E. V. Koonin, and D. R. Smith. 2001. Genome sequence and comparative analysis of the solvent-producing bacterium *Clostridium acetobutylicum*. *J. Bacteriol.* 183:4823–4838.

Ounine, K., H. Petitdemange, G. Raval, and R. Gay. 1985. Regulation and butanol inhibition of D-xylose and D-glucose uptake in *Clostridium acetobutylicum*. *Appl. Environ. Microbiol.* 49:874–878.

Ozkan, M., S. G. Desai, Y. Zhang, D. M. Stevenson, J. Beane, E. A. White, M. L. Guerinot, and L. R. Lynd. 2001. Characterization of 13 newly isolated strains of anaerobic, cellulolytic, thermophilic bacteria. *J. Ind. Microbiol. Biotechnol.* 27:275–280.

Papoutsakis, E. T. 1984. Equations and calculations for fermentations of butyric-acid bacteria. *Biotechnol. Bioeng.* 26:174–187.

Paquet, V., C. Croux, G. Goma, and P. Soucaille. 1991. Purification and characterization of the extracellular alpha-amylase from *Clostridium acetobutylicum* ATCC-824. *Appl. Environ. Microbiol.* 57: 212–218.

Paredes, C. J., K. V. Alsaker, and E. T. Papoutsakis. 2005. A comparative genomic view of clostridial sporulation and physiology. *Nat. Rev. Microbiol.* 3:969–978.

Paredes, C. J., I. Rigoutsos, and E. T. Papoutsakis. 2004. Transcriptional organization of the *Clostridium acetobutylicum* genome. *Nucleic Acids Res.* 32:1973–1981.

Predich, M., G. Nair, and I. Smith. 1992. *Bacillus subtilis* early sporulation genes *kinA*, *spo0F*, and *spo0A* are transcribed by the RNA polymerase containing sigma-H. *J. Bacteriol.* 174:2771–2778.

Qureshi, N., A. Lolas, and H. P. Blaschek. 2001. Soy molasses as fermentation substrate for production of butanol using *Clostridium beijerinckii* BA101. *J. Ind. Microbiol. Biotechnol.* 26:290–295.

Ragauskas, A. J., C. K. Williams, B. H. Davison, G. Britovsek, J. Cairney, C. A. Eckert, W. J. Frederick, J. P. Hallett, D. J. Leak, C. L. Liotta, J. R. Mielenz, R. Murphy, R. Templer, and T. Tschaplinski. 2006. The path forward for biofuels and biomaterials. *Science* 311:484–489.

Ravagnani, A., K. C. Jennert, E. Steiner, R. Grunberg, J. R. Jefferies, S. R. Wilkinson, D. I. Young, E. C. Tidswell, D. P. Brown, P. Youngman, J. G. Morris, and M. Young. 2000. Spo0A directly controls the switch from acid to solvent production in solvent-forming clostridia. *Mol. Microbiol.* 37:1172–1185.

Ren, Q., K. Chen, and I. T. Paulsen. 2007. TransportDB: a comprehensive database resource for cytoplasmic membrane transport systems and outer membrane channels. *Nucleic Acids Res.* 35: D274–D279.

Reysenbach, A. L., N. Ravenscroft, S. Long, D. T. Jones, and D. R. Woods. 1986. Characterization, biosynthesis, and regulation of granulose in *Clostridium acetobutylicum*. *Appl. Environ. Microbiol.* 52:185–190.

Rogers, P. 1986. Genetics and biochemistry of *Clostridium* relevant to development of fermentation processes. *Adv. Appl. Microbiol.* 31:1–60.

Sabathe, F., C. Croux, E. Cornillot, and P. Soucaille. 2002. amyP, a reporter gene to study strain degeneration in *Clostridium acetobutylicum* ATCC 824. *FEMS Microbiol. Lett.* 210:93–98.

Saier, M. H., Jr., C. V. Tran, and R. D. Barabote. 2006. TCDB: the Transporter Classification Database for membrane transport protein analyses and information. *Nucleic Acids Res.* 34:D181–D186.

Sardessai, Y., and S. Bhosle. 2002. Tolerance of bacteria to organic solvents. *Res. Microbiol.* 153:263–268.

Sauer, U., and P. Durre. 1995. Differential induction of genes related to solvent formation during the shift from acidogenesis to solventogenesis in continuous culture of *Clostridium acetobutylicum*. *FEMS Microbiol. Lett.* 125:115–120.

Schomburg, I., A. Chang, C. Ebeling, M. Gremse, C. Heldt, G. Huhn, and D. Schomburg. 2004. BRENDA, the enzyme database: updates and major new developments. *Nucleic Acids Res.* 32:D431–D433.

Shimizu, K., S. Hayashi, T. Kako, M. Suzuki, N. Tsukagoshi, N. Doukyu, T. Kobayashi, and H. Honda. 2005. Discovery of *glpC*, an organic solvent tolerance-related gene in *Escherichia coli*, using gene expression profiles from DNA microarrays. *Appl. Environ. Microbiol.* 71:1093–1096.

Soucaille, P., G. Joliff, A. Izard, and G. Goma. 1987. Butanol tolerance and autobacteriocin production by *Clostridium acetobutylicum*. *Curr. Microbiol.* 14:295–299.

Stragier, P., and R. Losick. 1996. Molecular genetics of sporulation in *Bacillus subtilis*. *Annu. Rev. Genet.* 30:297–341.

Tangney, M., J. K. Brehm, N. P. Minton, and W. J. Mitchell. 1998. A gene system for glucitol transport and metabolism in *Clostridium beijerinckii* NCIMB 8052. *Appl. Environ. Microbiol.* 64:1612–1619.

Tangney, M., A. Galinier, J. Deutscher, and W. J. Mitchell. 2003. Analysis of the elements of catabolite repression in *Clostridium acetobutylicum* ATCC 824. *J. Mol. Microbiol. Biotechnol.* 6:6–11.

Tangney, M., and W. J. Mitchell. 2000. Analysis of a catabolic operon for sucrose transport and metabolism in *Clostridium acetobutylicum* ATCC 824. *J. Mol. Microbiol. Biotechnol.* 2:71–80.

Tangney, M., and W. J. Mitchell. 2007. Characterisation of a glucose phosphotransferase system in *Clostridium acetobutylicum* ATCC 824. *Appl. Microbiol. Biotechnol.* 74:398–405.

Tangney, M., G. T. Winters, and W. J. Mitchell. 2001. Characterization of a maltose transport system in *Clostridium acetobutylicum* ATCC 824. *J. Ind. Microbiol. Biotechnol.* 27:298–306.

Tani, T. H., A. Khodursky, R. M. Blumenthal, P. O. Brown, and R. G. Matthews. 2002. Adaptation to famine: a family of stationary-phase genes revealed by microarray analysis. *Proc. Natl. Acad. Sci. USA* 99:13471–13476.

Thaw, P., S. E. Sedelnikova, T. Muranova, S. Wiese, S. Ayora, J. C. Alonso, A. B. Brinkman, J. Akerboom, J. van der Oost, and J. B. Rafferty. 2006. Structural insight into gene transcriptional regulation and effector binding by the Lrp/AsnC family. *Nucleic Acids Res.*. 34:1439–1449.

Thormann, K., L. Feustel, K. Lorenz, S. Nakotte, and P. Dürre. 2002. Control of butanol formation in *Clostridium acetobutylicum* by transcriptional activation. *J. Bacteriol.* 184:1966-1973.

Tomas, C. A., J. Beamish, and E. T. Papoutsakis. 2004. Transcriptional analysis of butanol stress and tolerance in *Clostridium acetobutylicum*. *J. Bacteriol.* 186:2006–2018.

Tomas, C. A., N. E. Welker, and E. T. Papoutsakis. 2003. Overexpression of *groESL* in *Clostridium acetobutylicum* results in increased solvent production and tolerance, prolonged metabolism, and changes in the cell's transcriptional program. *Appl. Environ. Microbiol.* 69:4951–4965.

Tummala, S. B., S. G. Junne, and E. T. Papoutsakis. 2003a. Antisense RNA downregulation of coenzyme A transferase combined with alcohol-aldehyde dehydrogenase overexpression leads to predominantly alcohologenic *Clostridium acetobutylicum* fermentations. *J. Bacteriol.* 185:3644–3653.

Tummala, S. B., S. G. Junne, C. J. Paredes, and E. T. Papoutsakis. 2003b. Transcriptional analysis of product-concentration driven changes in cellular programs of recombinant *Clostridium acetobutylicum* strains. *Biotechnol. Bioeng.* 84:842–854.

Tummala, S. B., N. E. Welker, and E. T. Papoutsakis. 1999. Development and characterization of a gene expression reporter system for *Clostridium acetobutylicum* ATCC 824. *Appl. Environ. Microbiol.* 65:3793–3799.

Vasconcelos, I., L. Girbal, and P. Soucaille. 1994. Regulation of carbon and electron flow in *Clostridium acetobutylicum* grown in chemostat culture at neutral pH on mixtures of glucose and glycerol. *J. Bacteriol.* 176:1443–1450.

Vollherbst-Schneck, K., J. A. Sands, and B. S. Montenecourt. 1984. Effect of butanol on lipid composition and fluidity of *Clostridium acetobutylicum* ATCC 824. *Appl. Environ. Microbiol.* 47:193–194.

Walter, K. A., G. N. Bennett, and E. T. Papoutsakis. 1992. Molecular characterization of two *Clostridium acetobutylicum* ATCC 824 butanol dehydrogenase isozyme genes. *J. Bacteriol.* 174:7149–7158.

Weber, F. J., and J. A. de Bont. 1996. Adaptation mechanisms of microorganisms to the toxic effects of organic solvents on membranes. *Biochim. Biophys. Acta* 1286:225–245.

Winzer, K., K. Lorenz, and P. Durre. 1997. Acetate kinase from *Clostridium acetobutylicum*: a highly specific enzyme that is actively transcribed during acidogenesis and solventogenesis. *Microbiology* 143:3279–3286.

Winzer, K., K. Lorenz, B. Zickner, and P. Durre. 2000. Differential regulation of two thiolase genes from *Clostridium acetobutylicum* DSM 792. *J. Mol. Microbiol. Biotechnol.* 2:531–541.

Wong, J., and G. N. Bennett. 1996. The effect of novobiocin on solvent production by *Clostridium acetobutylicum*. *J. Ind. Microbiol.* 16:354–359.

Woods, D. R. 1995. The genetic engineering of microbial solvent production. *Trends Biotechnol.* 13:259–264.

Wu, J. J., P. J. Piggot, K. M. Tatti, and C. P. Moran. 1991. Transcription of the *Bacillus subtilis* SpoIIA locus. *Gene* 101:113–116.

Yomano, L. P., S. W. York, and L. O. Ingram. 1998. Isolation and characterization of ethanol-tolerant mutants of *Escherichia coli* KO11 for fuel ethanol production. *J. Ind. Microbiol. Biotechnol.* 20:132–138.

Zhao, Y. S., C. A. Tomas, F. B. Rudolph, E. T. Papoutsakis, and G. N. Bennett. 2005. Intracellular butyryl phosphate and acetyl phosphate concentrations in *Clostridium acetobutylicum* and their implications for solvent formation. *Appl. Environ. Microbiol.* 71: 530–537.

Zverlov, V. V., O. Berezina, G. A. Velikodvorskaya, and W. H. Schwarz. 2006. Bacterial acetone and butanol production by industrial fermentation in the Soviet Union: use of hydrolyzed agricultural waste for biorefinery. *Appl. Microbiol. Biotechnol.* 71: 587–597.

Bioenergy
Edited by J. Wall et al.
© 2008 ASM Press, Washington, DC

Chapter 26

Practical Aspects of Butanol Production

THADDEUS C. EZEJI AND HANS P. BLASCHEK

Butanol is an industrially important chemical produced in substantial amounts during fermentation of carbohydrates by solventogenic clostridia. Butanol is a four-carbon alcohol which can be used as a direct solvent in the production of plasticizers, butyl acetate, butylamines, amino resins, glycol ethers, butyl acrylate, and methacrylate. Butanol fermentation has an interesting history that goes back to Louis Pasteur. Pasteur made the first observation in 1861 that bacteria produce butanol. In 1912, Chaim Weizmann discovered a microorganism called *Clostridium acetobutylicum*, which was able to ferment starch to acetone, butanol, and ethanol (ABE). The first commercial ABE fermentation plant was built in Terre Haute, IN, in 1918 by Commercial Solvents Corporation. This venture supplied butanol for conversion to butyl acetate, which was the primary component for paint lacquers.

In the early 1960s, ABE fermentation was discontinued in the United States due to unfavorable economic conditions brought about by competition with the petrochemical industry. Fermentation butanol (biobutanol) is currently interesting for its potential use as fuel, fuel extender, or chemical. Butanol has research and motor octane numbers of 113 and 94, respectively, compared to 111 and 92 for ethanol (Ladisch, 1991), which makes it a better fuel extender than ethanol presently used in gasoline. It also has a high cetane number (CN25; diesel averages CN45 and ethanol averages CN9) which allows butanol to be blended with petrodiesel (http://peswiki.com/index.php/Directory:Butanol). Biobutanol shares many of the benefits of other biofuels like ethanol, including reducing carbon dioxide emissions by closing the carbon cycle and providing markets to U.S. farmers. In addition, due to the current high prices of gasoline and given the probability of future increases in petroleum prices, butanol fermentation may again be carried out at an industrial scale.

However, the butanol fermentation process suffers from several limitations such as the high cost of substrate and butanol toxicity/inhibition to the fermenting microorganisms, resulting in low concentrations of butanol in the fermentation broth. Despite renewed interest in biobutanol and higher oil prices, biobutanol is not economically competitive when produced using traditional batch fermentation, a technology that was successfully employed during the prepetrochemical era. Biomass, which represents both cellulosic and noncellulosic materials, contains the most abundant source of fermentable carbohydrates that can be fermented into butanol. According to the Department of Energy (DOE) "Roadmap for Biomass Technology in the United States" (http://www.bioenergyupdate.com/magazine/security/Bioenergy%20Update%2001-03/bioenergy_update_January_2003.htm), biobased transportation fuels are projected to increase from 0.5% of U.S. consumption in 2001 to 4% in 2010, 10% in 2020, and further to 20 to 30% in 2030, or about 60 billion gal of gasoline equivalent per year. The production of butanol from low-cost lignocellulosic biomass that does not compete with food crops may be the key to meeting the DOE biofuel target. This approach would be consistent with using cheap, readily available, and sustainable substrate sources for biobutanol production.

To solve the problem of product toxicity, various alternative in situ butanol removal systems, such as pervaporation (Groot et al., 1984), perstraction (Qureshi et al., 1992), and reverse osmosis (Garcia et al., 1986), as well as adsorption (Nielson et al., 1988), liquid-liquid extraction (Evans and Wang, 1988), and gas stripping (Groot et al., 1989; Ezeji et al., 2003) have been studied by many investigators. This chapter reviews practical approaches for biobutanol production, including strategies for reducing or eliminating butanol toxicity to the culture, as well as exploiting both physiological and nutritional aspects of the fermenting microorganisms in order to achieve better product specificity and yield.

Thaddeus C. Ezeji • Department of Animal Sciences and Ohio Agricultural Research and Development Center, The Ohio State University, Wooster, OH 44691. **Hans P. Blaschek** • Center for Advanced BioEnergy Research and Department of Food Science & Human Nutrition, University of Illinois, Urbana, IL 61801.

BUTANOL FERMENTATION COPRODUCTS

Valuable fermentation coproducts that can be manufactured together with biobutanol include acetone and ethanol. Acetone, a coproduct of butanol fermentation, is a feedstock chemical for the manufacture of methyl isobutyl ketone, methacrylates, methyl butanol, bisphenyl A, methyl isobutylcarbinol, isophorone, and diacetone alcohol, as well as a solvent for paints, lacquers, resins, nitrocellulose, varnishes, and cellulose acetate. Acetone is produced chemically by the cumene hydroperoxide process or the catalytic dehydrogenation of isopropanol (Nelson and Webb, 1979). Presently, ethanol is the most common renewable fuel. The role of ethanol as fuel extender and oxygenate cannot be overemphasized. Therefore, the flooding of the market with butanol fermentation coproducts, should butanol fermentation be commercialized, is not envisaged.

ACETONE-BUTANOL PRODUCTION

There are a number of clostridial cultures capable of producing butanol in significant concentrations from different carbohydrate sources. The *Clostridium* spp. and strains that have been studied in the past and are still being studied include *C. beijerinckii* 8052 and its hyper-butanol-producing mutant derivative strain *C. beijerinckii* BA101, *C. acetobutylicum* 260, *C. acetobutylicum* P262, *C. butylicum* 592, and *C. acetobutylicum* 824. It is believed that *C. acetobutylicum* was the culture used in the first commercial ABE plant that was built in 1918 by Commercial Solvents Corporation in Terre Haute, IN. Sentrachem in South Africa did employ ABE fermentation for commercial production of butanol through the early 1980s, and the strain used was *C. acetobutylicum* P262. Unfortunately, the plant was shut down due to unfavorable economic conditions brought about by competition with the petrochemical industry.

Recent developments in liquid biofuel technology, the uncertainty of petroleum supplies, the finite nature of fossil fuels, and environmental concerns have revived research efforts aimed at obtaining butanol from renewable resources. During the past 15 years, advances have been made in strain development combined with new isolations, and these have, at least partially, overcome some of the problems associated with biobutanol production (Annous and Blaschek, 1991; Dürre, 1998; Qureshi and Blaschek, 2001a, 2001b). Montoya et al. (2000) isolated clostridial strains that produced up to 25.2 g of total ABE/liter of medium. The peculiar nature of the solventogenic clostridia makes it necessary to carry out upstream manipulations in order to increase butanol titer, yield, and productivity during fermentation. *C. beijerinckii* BA101, a genetically modified strain of *C. beijerinckii* 8052, has been reported to accumulate up to 33 g of total ABE/liter of medium under optimized conditions (Chen and Blaschek, 1999).

Factors Affecting Butanol Fermentation

Substrate and nutrient requirements

Solventogenic clostridia can utilize a wide range of carbohydrates and can be grown on simple media such as ground whole corn, grains, potatoes, molasses, whey permeate, or on semidefined and defined media with glucose as the carbon source. Due to the saccharolytic property of the solventogenic clostridia, starchy carbohydrates can be fermented to butanol without prior saccharification of the starch (Ezeji et al., 2005a). When semidefined and defined media are used, a wide array of vitamins and minerals are required in addition to a carbohydrate source. Generally, a mixture of organic and inorganic nitrogen sources is required for good growth and solvent production by *C. acetobutylicum* (Madihah et al., 2001) and *C. beijerinckii* (Ezeji et al., 2003). *C. acetobutylicum* 824 can grow on defined media, while *C. beijerinckii* 8052 requires at least semidefined media for growth and butanol production. To date, the best medium for culturing *C. beijerinckii* 8052 or the *C. beijerinckii* BA101 mutant strain is a semidefined medium called P2, which in addition to containing defined ingredients, also contains 1 g of yeast extract per liter of medium (Qureshi and Blaschek, 1999). Organic nitrogen sources such as yeast extract provide essential amino acids and growth factors that promote good *C. beijerinckii* growth. Corn steep water, a by-product of the corn wet-milling industry, which contains a rich complement of important nutrients such as amino acids, vitamins, nitrogen, and minerals (Hull et al., 1996), was determined to be a good substitute for the more expensive organic-nitrogen-source yeast extract (Parekh et al., 1998, 1999; Ezeji et al., 2005b). This finding is important as it impacts the economics of the AB fermentation. While corn steep water is a rich source of growth promoters, it may contain growth inhibitors such as lactic acid, sulfites, phytic acid, and heavy metals (Zn and Cu) and their salts, which must be diluted to reduce their inhibitory effect before corn steep water can be used as a nutrient source for fermentation (Parekh et al., 1998).

The effect of carbon limitation on the onset and maintenance of solvent production in batch, fed-batch, and continuous fermentations is well known. The presence of a limited amount of carbon results in predominantly acids being produced (Ezeji et al., 2005a; 2007c). It has been our experience that the shift to solvent production can occur in cultures in which a large excess of sugar is present (Campos et al., 2002; Ezeji et al., 2005a; 2007c). The solventogenic phase is characterized by the reassimilation of acetate and butyrate and the conversion to the corresponding solvent, namely, ABE. A schematic illustration of the pathway and the enzymes involved can be seen in Fig. 1. There is strong evidence that phosphate limitation is an inducer of butanol formation and strain stability in *C. acetobutylicum* 824 (Bahl et al., 1982). As a result, phosphate has an influence in directing the biochemical pathway towards butyrate or butanol formation (Fig. 1).

Furthermore, enzyme synthesis and control of electron flow in the glycolytic pathway are vital to the regulation of the butanol fermentation pathways. Electron flow can be reversed, and butanol yield usually responds to factors that influence the direction of the electron flow (Mitchell, 1998). This observation has caused researchers to test the effect of numerous reducing compounds such as carbon monoxide gassing, addition of methyl viologen, and neutral red addition into the fermentation medium during the ABE fermentation. In the presence of electron carriers, butanol and ethanol formation may be stimulated at the expense of acetone synthesis (Mitchell, 1998). In addition, a completely different product, 1,3-propanediol, is synthesized by *C. acetobutylicum* 824 and *C. butyricum* VPI 3266 when glycerol, which is a more reduced substrate than glucose, is used as the carbon source (González-Pajuelo et al., 2006). Therefore, the development of an optimal

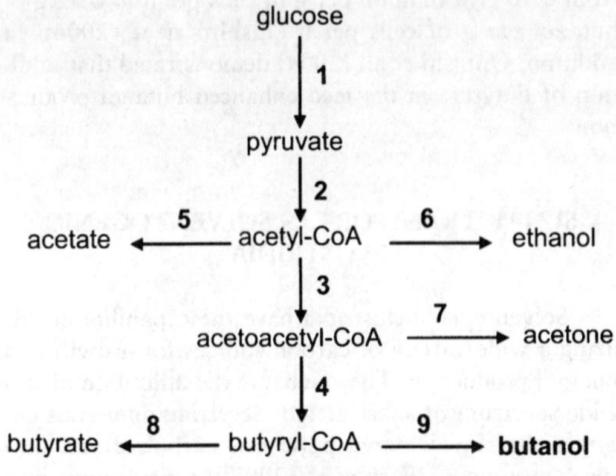

Figure 1. Simplified metabolism of glucose by solventogenic clostridia. Symbols: 1, glucose uptake by the PTS and conversion to pyruvate by the Embden-Meyerhof-Parnas pathway; 2, pyruvate-ferredoxin oxidoreductase; 3, thiolase; 4, 3-hydroxybutyryl-CoA dehydrogenase, crotonase, and butyryl-CoA dehydrogenase; 5, phosphate acetyltransferase and acetate kinase; 6, acetaldehyde dehydrogenase and ethanol dehydrogenase; 7, acetoacetyl-CoA:acetate/butyrate:CoA transferase and acetoacetate decarboxylase; 8, phosphate butyltransferase and butyrate kinase; 9, butyraldehyde dehydrogenase and butanol dehydrogenase.

fermentation medium and process for biobutanol production lies in our understanding of the physiology of the bacteria and associated critical interactions between carbon pathways and electron flow (Ezeji et al., 2007b).

Inoculum

In order to prepare an inoculum, laboratory spore stocks of solventogenic clostridia such as *C. beijerinckii* and *C. acetobutylicum* are heat shocked at 80°C for 10 min, and vegetative cells are grown anaerobically in cooked meat medium (Difco Laboratories, Detroit, MI) containing 20 g of glucose per liter at 35°C, followed by transferring 5 ml of actively growing culture (16 to 18 h old) to 100 ml of tryptone-glucose-yeast extract medium. Cells are usually grown anaerobically in this medium for 3 to 6 h at 35°C until the optical density at 600 nm reaches 0.8 to 1.0. The solvent production medium containing glucose, liquefied corn starch, or lignocellulosic hydrolysate as the carbon source is then inoculated at a concentration of 5 to 7% (vol/vol). An increase or decrease of the inoculum percentage may affect butanol production. A profound effect of inoculum size on the degeneration of *C. acetobutylicum* strain 824 was demonstrated by Hartmanis et al. (1986). They observed that when *C. acetobutylicum* 824 was subcultured on a daily basis, the production of butanol steadily decreased; and by 150 transfers the culture had degenerated. When the inoculum size was doubled from 3.3 to 6.7%, the cells rapidly reverted to the production of butanol (Hartmanis et al., 1986; Kashket and Cao, 1995).

Temperature

Solventogenic clostridia used in the bioproduction of butanol from defined or semidefined media under submerged fermentation have an optimum temperature between 30 and 35°C. The effect of temperature on solvent production by *C. acetobutylicum* was studied in the range of 25 to 40°C by McNeil and Kristiahsen (1985). They found that the solvent yield decreased with increasing temperature, seemingly because of a reduction in acetone production. In addition, butanol toxicity to the clostridial cells tends to increase with an increase in temperature, which results in lower accumulation of butanol in the bioreactor during fermentation. Considering total solvent yield and productivity only, the optimum fermentation temperature is ca. 35°C (McNeil and Kristiahsen, 1985). Similar results had been obtained previously by McCutchan and Hickey (1954), who had shown that solvent yields remained fairly stable around 31% at a temperature range of 30 to 33°C. When the fermentation temperature was increased to 37°C, the solvent yield decreased to 23 to 25%.

pH

Butanol fermentation is characterized by two distinct phases. The first, or acidogenic phase, involves the rapid formation of acetic and butyric acids. These acids are secreted into the medium, causing the pH of the medium to decrease to a low value. The second phase (solventogenic) commences after the critical pH "break" point has been reached, during which acids are reassimilated for their subsequent conversion to butanol and acetone.

Culture pH, internal and external concentrations of acetic and butyric acids, and residual sugar concentration have all been implicated in the switch from acid to solvent production. Long et al. (1984) observed that in butanol fermentation by *C. acetobutylicum* P262, a total acid concentration in excess of 4 g/liter is associated with inhibition of cell division and the triggering of solvent production. Specifically, solvent production starts at an undissociated butyric acid concentration between 1.6 and 1.9 g/liter (Monot et al., 1983). The fact that high concentrations of acetate and butyrate are required to trigger solventogenesis at pH 7 focuses attention on the effect of the undissociated forms of these acids on the cell culture. Several related parameters such as acidic pH, high weak-acid concentrations, and high cell densities appear to be involved in triggering solvent production (George and Chen, 1983). In industrial AB production, pH has been shown to play an important role in solvent production. During the AB fermentation by *C. beijerinckii* BA101 in P2 medium (Ezeji et al., 2005a), the fermentation was initiated by growing the culture in P2 medium with a pH value near neutrality (pH 6.8) and then allowing the pH to fall to the optimum value (pH 5.0 to 5.5), which triggers solventogenesis. During batch fermentation in P2 medium by *C. beijerinckii* BA101, the culture is able to maintain the optimum pH value for butanol production after the onset of solventogenesis and requires no external pH adjustment (Ezeji et al., 2003). However, the pH is continuously adjusted when the continuous fermentation is run in a chemostat bioreactor system.

Acetate and butyrate in feed

The tendency for solvent-producing clostridia to lose the ability to produce solvents following repeated subculture or continuous cultivation, a phenomenon known as degeneration, is well known. Previous investigations carried out by Chen and Blaschek (1999) did show that the addition of sodium acetate to the fermentation medium prevented degeneration in *C. beijerinckii* NCIMB 8052 and *C. beijerinckii* BA101. In this study, it was shown that the addition of 20 mM sodium acetate was able to stabilize solvent production by *C. beijerinckii* NCIMB 8052 and maintained the cell density, whereas growth decreased rapidly in the absence of additional acetate. RNA and enzyme analyses indicated that coenzyme A (CoA) transferase was highly expressed, and the presence of a *sol* operon (the *sol* operon encodes aldehyde/alcohol dehydrogenase [*aad*] and two CoA transferases [acetoacetyl-CoA, *ctfA*, and acetate-butyrate-CoA, *ctfB*, the key enzymes involved in ABE formation]) and the presence of a *ptb-buk* operon (the *ptb-buk* operon encodes the enzymes [e.g., butyrate kinase] responsible for the conversion of butyryl-CoA to butyrate) in *C. beijerinckii* NCIMB 8052 were confirmed. They concluded that the addition of acetate to MP2 medium may induce the expression of the *sol* operon, which ensures solvent production and prevents strain degeneration in *C. beijerinckii* NCIMB 8052. In related studies carried out by Hüsemann and Papoutsakis (1990), it was observed that the addition of acetate or propionate to uncontrolled-pH batch *C. acetobutylicum* cultures did not affect the initiation of solventogenesis but did enhance final solvent concentrations compared with those of unchallenged cultures. This observation was explained in terms of the increased buffering capacity of the medium brought about by the added acids, which resulted in the protection of the culture against premature growth inhibition due to low culture pH values.

Addition of butyric acid in the fermentation medium during batch cultivation of *C. saccharoperbutylacetonicum* has been shown to induce solvent production, and a specific butanol production rate increased from 0.10 g of butanol per g of cells per h to 0.42 g of butanol per g of cells per h (Tashiro et al., 2004). In addition, Qureshi et al. (2004) demonstrated that addition of butyrate in the feed enhanced butanol production.

SUGAR TRANSPORT IN SOLVENTOGENIC CLOSTRIDIA

Solventogenic clostridia have the capability of utilizing a wide variety of carbon sources for growth and butanol production. They facilitate the utilization of this wide spectrum of substrates by secreting numerous enzymes that break down polymeric carbohydrates into sugar monomers (Ezeji et al., 2007b) for transport into the cells via specific membrane-bound transport systems.

Solventogenic clostridia and many other anaerobes transport sugars into cell membranes through a phosphoenolpyruvate (PEP)-dependent phosphotransferase system (PTS) (Mitchell, 1998; Lee and Blaschek, 2001), and this PTS is involved in the transfer of a phosphate group from PEP to the substrate sugar. While glucose and fructose phosphorylation were supported by PEP, indicating the involvement of a PTS in uptake of these sugars, galactose appears to be transported by a nonphosphotransferase mechanism because a significant rate of phosphorylation of this sugar was supported by ATP rather than PEP (Mitchell, 1998). However, *C. beijerinckii* BA101 and *C. beijerinckii* 8052 have been shown to involve both transport systems simultaneously in the uptake of glucose during fermentation with the nonphosphotransferase mechanism (ATP-dependent glucokinase) predominant during the solventogenic phase when the PTS was repressed (Lee and Blaschek, 2001; Lee et al., 2005; Ezeji et al., 2007a). Phosphorylation of galactose has been suggested to be catalyzed by galactokinase (Mitchell, 1998; Ezeji et al., 2007a). Invariably an efficient galactose uptake depends on the expression of galactokinase genes and the activity of galactokinase enzymes such as ATP-α-D-galactose-1-phosphate transferase-galactokinase, uridine diphosphoglucose-D-galactose-1-phosphate uridylyltransferase, and uridine diphosphogalactose-4-epimerase during the solventogenic growth phase, when PTS is repressed (Daldal and Applebaum, 1985; Ezeji et al., 2007a).

LIMITATIONS OF BUTANOL FERMENTATION

Butanol Toxicity

A major problem associated with the typical butanol fermentation is product inhibition due to butanol toxicity, and although butyric acid is more toxic than butanol, its concentration during a normal butanol fermentation remains low (Ezeji et al., 2005b). During the AB fermentation, solventogenic clostridia are able to carry out cellular functions until the concentration of the butanol reaches inhibitory levels at around 12 to 13 g/liter. Above this level, further cellular metabolism ceases (Jones and Woods, 1986). At toxic levels, butanol has been found to inhibit nutrient transport, the membrane-bound ATPase, and sugar uptake (Bowles and Ellefson, 1985). Butanol has also been found to affect the lipid composition and fluidity of the cell membrane (Baer et al., 1987). Solventogenic clostridia respond to high levels of butanol by modifying the fatty acid (saturation) composition of their lipids, and this increase in the degree of saturation of cellular lipids likely occurs in order to compensate for the increase in fluidity of the cell membrane induced by the butanol (Baer et al., 1987).

Since butanol is highly toxic to solventogenic clostridia, metabolic engineering of various microorganisms not typically associated with butanol production but resistant to butanol toxicity has been suggested. In addition, simultaneous fermentation and recovery to relieve butanol toxicity (discussed in detail in "Advanced AB Fermentation: Simultaneous Fermentation and Recovery" below) may be the best solution given the currently available solventogenic clostridial strains. These approaches may be useful for increasing butanol concentration in the fermentation broth.

Culture Degeneration

Degeneration is the process by which solventogenic clostridia lose their ability to produce AB and/or sporulate. Typically, solventogenic clostridia exist in a mixed culture of solvent-producing and degenerate cells. Repeated subculturing of the vegetative cells of solventogenic clostridia, as opposed to being grown from heat shocked spores, leads to strain-specific degeneration (T. C. Ezeji and H. P. Blaschek, unpublished data). Kashket and Cao (1995), during their examination on clostridial strain degeneration, demonstrated that degeneration can be caused by excessive acidification of the culture during exponential growth phase. They presented data which demonstrated that during uncontrolled fermentation of glucose to acetic and butyric acids by *C. beijerinckii* 8052, there was rapid growth which led to the rate of acid production being disproportionately greater than the rate of induction of the solventogenic pathway enzymes (conversion of these acids to solvents). As a result, the pH of the fermentation medium drops to bactericidal levels and the cells are unable to switch to solventogenesis and sporulation. Results obtained from recent investigations on generation times and stability of solventogenic clostridia carried out in our laboratory are consistent with Kashket and Cao's observations. Therefore, the exogenous addition of acetate or butyrate during butanol fermentation should be carefully regulated to avoid the degeneration of the fermenting solventogenic clostridia cells.

The degeneration phenomenon appears to be strain dependent. While degenerate cells of *C. beijerinckii* 8052 can lose the ability to both produce solvent and sporulate, *C. acetobutylicum* 824 can lose solvent-producing ability during long-term continuous cultivation while retaining the abilty to sporulate (Stevens et al., 1985). In batch cultures, it is clear that the degenerate cells easily outgrow the nondegenerate cells even though the nondegenerate cells differentiate more rapidly (Kutzenok and Aschner, 1952; Kashket and Cao, 1995). However, the degenerate cells have advantages over the nondegenerate cells due to their rapid cell division. In continuous cultures the degenerate cells predominate because they tend to accumulate as the culture is washed out (Kashket and Cao, 1995).

Changes in colony morphology have been associated with the degeneration of solvent-producing strains of *C. acetobutylicum*. Adler and Crow (1987) demonstrated that the most efficient solvent-producing strains gave rise exclusively to colonies with dense centers containing large numbers of spores and many outgrowths of various morphologies developed from the perimeter of such colonies after several days of incubation. Conversely, the most degenerate cultures gave rise to large diffuse colonies that did not contain spores, did not produce solvents, and formed no outgrowths after several days of incubation. It was concluded that the colony type associated with vigorous, solvent-producing, spore-forming cultures was designated type I, and the colonies representing different stages of the degeneration process were designated types II through IV (Adler and Crow, 1987). These observations may provide a way of monitoring and predicting the degeneration and butanol-producing capability of a given *C. acetobutylicum* culture.

ADVANCED AB FERMENTATION: SIMULTANEOUS FERMENTATION AND PRODUCT RECOVERY

During the past 2 decades, significant amounts of research have been performed on the use of alternative fermentation and product recovery techniques for biobutanol production (Ezeji et al., 2007b). The simultaneous (integrated) fermentation and product recovery approach appears to have solved most of the early problems associated with the butanol fermentation such as substrate and product inhibition, poor reactor productivity, and high-volume process streams encountered in the traditional ABE batch fermentation process. These techniques have involved the use of immobilized and cell recycle continuous bioreactors and product recovery employing alternative techniques such as adsorption, gas stripping, ionic liquids, liquid-liquid extraction, pervaporation, aqueous two-phase separation, supercritical extraction, and perstraction, etc. The application of some of these techniques to the ABE fermentation process is described below.

Batch Fermentation Culture Systems

Traditionally, batch fermentation was the only type of fermentation system employed for the commercial production of biobutanol due to lack of superior or alternative technologies. During the 1940s and 1950s, butanol production on an industrial scale (Terre Haute, IN, and Peoria, IL) was carried out using large batch fermentors ranging in capacity from 200,000 to 800,000 liters. The industrial process used 8 to 10% corn mash, which was cooked for 90 min at 130 to 133°C. Cane molasses was also used to produce butanol in a commercial plant in South Africa until the early 1980s. In industrial processes, beet and blackstrap molasses were diluted to give a fermentation sugar concentration of 50 to 75 g/liter, usually 60 g/liter. The molasses solution was sterilized at 107 to 120°C for 15 to 60 min followed by adding organic and inorganic nitrogen, phosphorus, and buffering chemicals. This fermentation was characterized by a low cell concentration in the reactor, poor yield, and reactor productivity due to product (butanol) toxicity problems associated with this fermentation. Distillation was the method of choice to recover the butanol, which is carried out at the end of the fermentation process.

To improve the product yield and reactor productivity, bring about a reduction in process streams, and improve the economics of biobutanol production, it may be necessary to use integrated butanol fermentation technologies. Gas stripping is one of the most important techniques for removing butanol from the fermentation broth during batch butanol fermentation. Gas stripping allows for selective removal of volatiles from fermentation broth and uses no membrane (Ezeji et al., 2003). During gas stripping, stringent anaerobic conditions can be maintained using a makeup gas usually consisting of oxygen-free nitrogen. It should be noted that the production of biobutanol is associated with the generation of gases (CO_2 and H_2). In an attempt to make the process of butanol recovery from the fermentation broth simpler and more economical, fermentation gases (CO_2 and H_2) have been used to recover ABE products (Ezeji et al., 2003, 2004). The gases are bubbled through the fermentation broth followed by cooling of the gas (or gases) in a condenser. As the gases are bubbled through the bioreactor, the gas mixture captures butanol including other volatiles such as acetone and ethanol, and they are condensed in the condenser followed by collection in a solvent receiver. Once the butanol is condensed, the gases are recycled back to the bioreactor to capture more butanol. This process continues until all the substrate in the bioreactor is utilized by the culture. Gas stripping has been successfully applied to remove solvents from batch fermentations (Ezeji et al., 2003). In addition to the removal of butanol, gas stripping technology makes it possible for a concentrated sugar solution (150 g/liter) to be fed to the reactors in order to reduce process volume streams and economize the butanol production process. The reader is referred to Ezeji et al., 2003, for a schematic diagram and detailed description of this process. In some cases, a separate gas stripper can be used in which the fermentation broth is circulated over a stripping column. The stripping gas is introduced at the bottom of the stripping column, and the stripped butanol is condensed and recovered as shown in Fig. 2.

Pervaporation is a membrane-based technique that has been applied successfully in integrated batch fermentation and butanol recovery. In this system, the

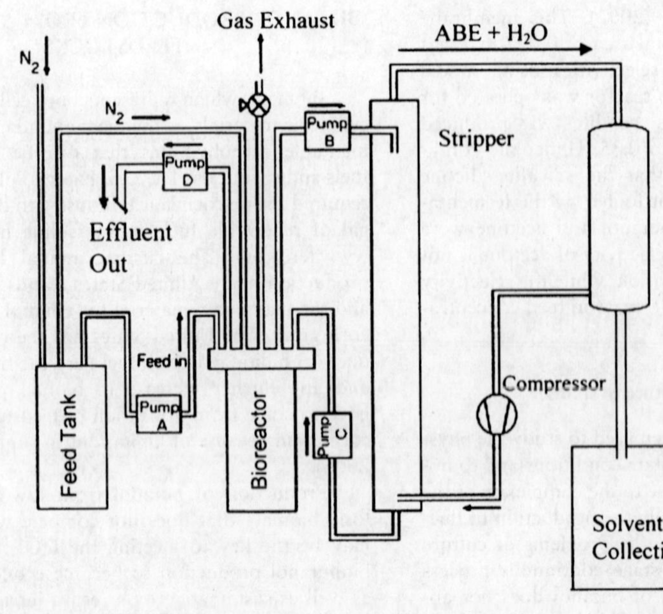

Figure 2. Schematic diagram of ABE production by *C. beijerinckii* BA101 and recovery by gas stripping process with a separate stripping column.

volatiles such as butanol, acetone, and ethanol diffuse through a solid membrane, leaving behind dissolved solids (nutrients and sugars) and microbial cells, and solvent productivities can be improved up to 200% over control batch fermentations. When the *C. beijerinckii* BA101 hyper-butanol-producing mutant was used in a pervaporative butanol fermentation, it produced 51.5 g of total ABE/liter of medium compared to 24.2 g of total ABE/liter of medium produced in the control batch reactor (Qureshi and Blaschek, 1999). Since the permeate is a mixture of acetone, butanol, and ethanol (and small concentrations of acids), distillation is required to separate butanol from the mixture.

Fed-Batch Culture Systems

Traditionally, fed-batch fermentation is a technique in which the bioreactor is initiated with a relatively low substrate concentration in order to reduce substrate inhibition, and as the substrate is consumed, it is replaced by adding a concentrated substrate solution at a low rate, while keeping the substrate concentration in the reactor below the toxic level (Ezeji et al., 2005b). This may or may not be a growth-limiting substrate. Since butanol fermentation is limited by both substrate and product inhibition due to osmotic pressure and butanol toxicity, respectively, it is difficult to carry out fed-batch fermentation without integration of the fermentation with simultaneous product recovery. In such a case, the reactor is initiated in a batch mode with a low substrate concentration (usually 60 to 100 g/liter) and low fermentation medium volume, usually about one-half the volume of the bioreactor. The bioreactor is inoculated with the culture, and the fermentation process is initiated. Gas stripping has been successfully employed to remove solvents from fed-batch fermentations (Ezeji et al., 2004). By feeding the reactor at a slow and controlled rate, substrate toxicity is kept below inhibitory levels, while the solvent removal technique is applied simultaneously to reduce butanol toxicity. Application of these two techniques takes care of two toxicity problems: one for substrate inhibition and another for butanol inhibition. Using this approach, a total of 500 g of glucose was used to produce 232.8 g of solvents (77.7 g of acetone, 151.7 g of butanol, and 3.4 g of ethanol) in 1 liter of culture broth, and solvent productivities were improved by 400% over the control batch fermentation productivities.

Pervaporation has also been applied for simultaneous fed-batch fermentation and butanol removal (Qureshi and Blaschek, 2000; Qureshi et al., 2002). Membrane stability, high selectivity, and high flux are the desired factors when selecting membranes for fed-batch simultaneous butanol fermentation and recovery by pervaporation. The membrane, a silicalite-silicone composite used in a butanol fed-batch pervaporative process, was produced by Qureshi et al. (2002). This membrane (306 μm thick and with a total area of 0.022 m^2) was characterized for flux and selectivities using model ABE solutions. The fed-batch reactor was operated for 870 h, and 154.97 g of ABE per liter was produced with an ABE yield of 0.31 to 0.35. Under the conditions tested, it was found that the silicalite-silicone composite membrane was not fouled by the fermentation broth. Selectivities of butanol and acetone were unaffected. In addition, the selectivity of acetone is not sensitive to the feed concentration, while the selectivity of butanol increased with a decrease in feed concentration.

Continuous Culture Systems

Continuous culture is often used to study the physiology of a culture in steady-state conditions and to improve reactor productivity. Due to the complexity of the butanol fermentation, especially the production of fluctuating levels of butanol and the problems of culture stability, the use of a single-stage continuous process for the industrial production of butanol does not appear to be feasible at this time. This has been demonstrated in laboratory scale bioreactors. In a single-stage continuous system, high reactor productivity may be obtained, but it is at the expense of low product concentration compared to that achieved in a batch process. It is well known that the final concentration of solvents in a batch culture system rarely exceeds 20 g/liter (Ezeji et al., 2005a). The final concentration in a typical continuous-culture system is even lower when compared to that achieved in a batch culture system. As a means of increasing product concentration in the effluent and reducing fluctuations in butanol concentration, two or more multistage continuous-fermentation systems have been investigated (Bahl et al., 1982). Often, this is done by allowing cell growth, acid production, and butanol production to occur in separate bioreactors. While downtime and lag phase can be eliminated using a continuous culture, the problem of product inhibition remains. This problem can be eliminated by the application of novel product removal techniques.

An integrated continuous butanol fermentation and recovery approach was investigated in our laboratory (T. C. Ezeji, N. Qureshi, and H. P. Blaschek, U.S. provisional patent 60/504,280). In this case, the solvents were removed by gas stripping. In this system, 1,163 g of glucose was utilized per liter of culture broth. The total amount of ABE produced per liter of reaction volume was 460 g. This suggests that gas stripping is a novel technique, which can be successfully applied to continuous butanol fermentation processes.

BUTANOL PRODUCTION FROM ALTERNATIVE FEEDSTOCKS

Biomass, which represents both cellulosic and noncellulosic materials, is the most abundant source of fermentable carbohydrates that can be fermented into fuels and chemicals. The U.S. Energy Policy Act of 2005 required the petrochemical industry to blend 7.5 billion gal of renewable fuels into gasoline by 2012 (http://www.ferc.gov). The current annual biofuel ethanol production in the United States stands at 5 billion gal and the major raw material for ethanol fermentation is corn grain. The United States has the capacity to produce 13 billion gal of biofuel per year from corn alone, and any further increase in biofuel production will have to come from utilization of feedstocks other than corn grain because of limitations in supply (Gray et al., 2006).

Production of butanol from low-cost lignocellulosic biomass that does not compete with food crops may be the key to meeting the DOE target, and for biobutanol production to become economically viable as well as sustainable. As a result, interest in the cultivation of lignocellulosic crops such as switchgrass, *Miscanthus*, etc., for subsequent conversion into fermentable sugars is receiving more attention. In addition, industrial and agricultural coproducts such as corn fiber, corn stover, dried distillers' grains with solubles (DDGS), wheat straw, rice straw, soybean residues, as well as various types of agricultural and industrial wastes are presently considered potential renewable feedstocks for the production of fermentable sugars.

Whereas the major carbohydrate in corn grain is starch, lignocellulosic biomass is composed of 56 to 72% fermentable carbohydrates (cellulose and hemicellulose) (dry weight) (Table 1). In addition, lignocellulosic biomass contains lignin, which provides structural characteristics to the feedstock and encapsulates the cellulose component. The encapsulation of the cellulose by lignin makes the lignocellulosic biomass almost inaccessible to hydrolytic enzymes, and as a result more difficult to hydrolyze than starchy materials. Due to the recalcitrant nature of these lignocellulosic feedstocks, their pretreatment often requires a combination of physical, chemical, and heat treatments to disrupt the structure and convert it into a more hydrolyzable form. In addition to pretreatment, enzymes such as xylanase, β-xylosidase, and several other complementary enzymes such as acetylxylan esterase, α-arabinofuranosidase, α-glucuronidase, α-galactosidase, and ferulic and/or *p*-coumaric acid esterase are required for complete lignocellulosic deconstruction (Ezeji et al., 2007b). During acid hydrolysis, a complex mixture of microbial inhibitors is generated. Lignins are oxidized or degraded to form phenolic compounds, and parts of the

Table 1. Composition of representative potential lignocellulosic raw materials for biobutanol production[a]

Compound	Content (% dry wt)							
	Corn	Corn fiber	Corn stover	DDGS	Wheat straw	Sugarcane bagasse	Switchgrass	Poplar
Hexan								
Glucan	2.4[b]	36.5	36.1	22	36.6	38.1	32.2	39.8
Galactan		2.9	2.5	0.3	2.4	1.1	0	0
Mannan		NA	1.8	NA	0.8	NA	0.4	2.4
Total	71.7[c]	39.4	40.4	22.3	39.8	39.2	32.6	42.2
Pentan	5.5[d]							
Xylan		18.4	21.4	9.5	19.2	23.3	20.3	14.8
Arabinan		13.3	3.5	5.5	2.4	2.5	24	16
Total	5.5	31.7	24.9	15	21.6	25.8	24	16
Total fermentable	79.6	71.1	65.3	37.3	61.4	65	56.6	58.2
Lignin	0.2	6.9	17.2	3.1	14.5	18.4	23.2	29.1

[a]NA, not available. Data on switchgrass and poplar from Chung et al. (2005). Data on wheat straw and sugarcane bagasse from Lee (1997). Data on corn stover from Laureano-Perez et al. (2005). Data on corn fiber and DDGS from our laboratory (our unpublished data). Data on corn from Gulati et al. (1996).
[b]Cellulose.
[c]Starch.
[d]Xylan + arabinan.

sugars that are released during hydrolysis are also degraded into products that may inhibit cell growth and fermentation. Examples of the inhibitory compounds include furfural, hydroxymethyl furfural, and acetic, ferulic, glucuronic, and *p*-coumaric acids, etc. (Zaldivar et al., 1999; Ezeji et al., 2007a). For complete depolymerization of lignocellulosic biomass, it is difficult to totally avoid the generation of inhibitory compounds irrespective of the pretreatment and hydrolysis method utilized.

The fermentation of lignocellulosic hydrolysate by alcohol-producing microorganisms remains a challenge not only from the standpoint of potential inhibitory compounds, but also with respect to the types of sugars present. Most bacteria utilize glucose as the preferred carbon source, and only when glucose is limiting are the pentose sugars utilized. However, solventogenic clostridia have been shown to utilize other sugars such as cellobiose, galactose, mannose, arabinose, and xylose (Ezeji et al., 2007a; Ezeji and Blaschek, in press). Although the solventogenic clostridia have the ability to utilize a wide variety of carbohydrates, the genes which make up the carbohydrate catabolic operons are usually expressed when the respective substrates are present in the fermentation medium or when the preferred carbon sources are depleted (Ezeji and Blaschek, unpublished). Therefore, for efficient fermentation of lignocellulosic hydrolysates containing a wide variety of sugar monomers, it is imperative to heat shock solventogenic clostridia in media containing the representative carbohydrates that are present in the fermentation medium (Ezeji et al., 2007a).

BIOREACTOR MONITORING SYSTEMS

Generally, process controls during fermentations may be classified as physical (agitation speed, temperature, pressure, and aeration rate), chemical (pH, redox potential, dissolved oxygen, and dissolved CO_2), and biological (biomass concentration, oxygen uptake, and production rates of H_2, CO_2, CH_4, etc.) variables. The biological components as well as the physical and chemical environments of fermentation are vital in bioreactor monitoring. Online monitoring can be performed using several sensors or combinations of sensors, in which case they are brought into direct contact with the bioreactor and the fermentation broth/products. Usually, the physical variables are measured by in-line sensors in the bioreactor. While physical sensors are always an integral part of the bioreactor unit, the chemical sensors are not and do not usually come with the bioreactor. For the measurement or control of chemical variables, you can buy the individual sensors (usually small in size but in a complete unit) and couple them to the bioreactor unit.

During fermentation, cell mass is one of the parameters that is routinely monitored. Monitoring cell mass will indicate the growth status of the fermenting microorganism as well as the progress of the fermentation. The standard off-line method for monitoring cell growth of solventogenic clostridia during butanol fermentation is indirect measurement of cell mass via measuring the optical density of the growing cells at a 600-nm wavelength by using a spectrophotometer. The measured values can then be used to estimate cell concentrations by using a predetermined correlation between optical density at 600 nm and cell dry weight. The cell populations in bioreactors can be characterized online in terms of morphological changes in order to examine the vitality of the fermentation. In practice, there is no ideal in situ online cell concentration sensor available today, but the common method for this kind of monitoring is in situ microscopy equipped with an image analyzer (Sonnleitner et al., 1992). Online monitoring of cell concentration is expected to be further improved in the future, and we contend that using different sensor types in parallel may be a promising future solution to obtaining good online monitoring of biomass.

Optimal cell growth of many microorganisms depends heavily on tight pH control because the cells produce acids which sometimes make up the major component of the metabolic product. Therefore, pH is an important parameter commonly measured during fermentation. Steam-sterilizable glass pH electrodes are still the best for use in bioreactors. On the other hand, the solventogenic clostridia have the ability to reassimilate the fermentation intermediates (acids) for solvent or butanol production and stabilize the pH in the process. pH measurement during butanol fermentation is important in order to accurately monitor the fermentation progress and in extreme cases prevent "acid crash." Accurate online pH monitoring is important for early detection of poorly buffered pH media. In poorly buffered media, the pH can decrease to below 4.5 during the early part of fermentation due to sudden termination of solventogenesis, and a phenomenon known as acid crash may occur. The acid crash is associated with a rapid termination of solventogenesis when the combined concentrations of undissociated acetic and butyric acids exceed a critical threshold value in the fermentation broth (Maddox et al., 2000). The acid crash should not be confused with culture degeneration, which occurs as a result of an apparent failure in switching from the acidogenic to the solventogenic phase and takes place over a period of time as might occur during continuous fermentation.

Substrate consumption is another indicator of the metabolic status of the fermenting cells. The exact concentration of substrate during butanol fermentation is critical for the maintenance of optimum feed concentrations required to sustain solventogenesis in the bioreactor during butanol fermentation. Substrate consumption (glucose) can be determined using a hexokinase and glucose-6-phosphate dehydrogenase (Sigma Chemicals, St. Louis, MO)-coupled enzymatic assay (Ezeji et al., 2003), which requires about 5 min to analyze, or by membrane-bound enzyme biosensors, which requires only 1 min. Substrate consumption (hexose and pentose sugars) can be monitored by high-pressure liquid chromatography (HPLC), which requires about 30 min to run. HPLC has the advantage of analyzing multiple metabolites concurrently. However, HPLC analysis is time-consuming and does not provide real-time feedback on the state of the fermentation. Therefore, quicker methods should be developed.

The exit gas composition from the bioreactor, especially CO_2, can be analyzed by a gas analyzer using infrared detectors. Carbon dioxide is of importance because its concentration in the exit gas can be indicative of the fermentation health or contamination. Mass spectroscopy may be used to analyze all gaseous components including H_2.

Redox potential, which is related to the overall availability of free electrons in the fermentation broth, can be measured during the butanol fermentation. The value of the potential is the sum total of the oxidizing or reducing agents' concentrations. The metabolic activity of solventogenic clostridia, including the carbon flux in the metabolic pathways, is influenced by the redox potential of the culture environment. Therefore, monitoring as well as controlling the redox potential of the fermentation broth during butanol fermentation may be a vital tool in achieving optimal butanol production.

Agitation and sparging of bioreactors always generate foam during butanol fermentation. Foam is the entrapment of gas in a lipid, polysaccharide, or protein matrix and if not controlled, can empty the bioreactor. Mechanical foam breakers can be used to physically break and blow the foam back down during butanol fermentation. For pilot or industrial-scale butanol fermentation, a combination of nontoxic antifoams and mechanical foam breakers is recommended.

CONCLUSION

Biobutanol production is a relatively complicated process because the solventogenic clostridia undergo a biphasic type of fermentation and the fermentation product (butanol) is toxic to the producing cultures. The fermentation suffers from a number of limitations such as low titer, yield, and productivity. Improvements in the performance of the solventogenic clostridia and the utilization of cheap alternative feedstocks are strategies for moving biobutanol fermentation research to a competitive economical industrial process. Nevertheless, simultaneous butanol fermentation and in-line product recovery processes involving membranes, liquid-liquid extraction, perstraction, pervaporation, gas stripping, etc., have dramatically reduced butanol toxicity to the fermenting microorganisms and improved butanol productivity. It is our opinion that the integrated approach involving simultaneous production and removal of butanol in the fermentation reactor can be carried out economically on an industrial scale.

Acknowledgments. This work is supported by funding from the U.S. Department of Agriculture (AG2001-35504-10668 and AG2006-35504-17419) and the Department of Energy (DOE PU 541-0338-01).

REFERENCES

Adler, H. I., and W. Crow. 1987. A technique for predicting the solvent-producing ability of *Clostridium acetobutylicum*. *Appl. Environ. Microbiol.* 53:2496–2499.

Annous, B. A., and H. P. Blaschek. 1991. Isolation and characterization of *Clostridium acetobutylicum* mutants with enhanced amylolytic activity. *Appl. Environ. Microbiol.* 57:2544–2548.

Baer, S. H., H. P. Blaschek, and T. L. Smith. 1987. Effect of butanol challenge and temperature on lipid composition and membrane fluidity of butanol-tolerant *Clostridium acetobutylicum*. *Appl. Environ. Microbiol.* 53:2854–2861.

Bahl, H., W. Andersch, and G. Gottschalk. 1982. Continuous production of acetone and butanol by *Clostridium acetobutylicum* in a two-stage phosphate limited chemostat. *Eur. J. Appl. Microbiol. Biotechnol.* 15:201–205.

Bowles, L. K., and W. L. Ellefson. 1985. Effects of butanol on *Clostridium acetobutylicum*. *Appl. Environ. Microbiol.* 50:1165–1170.

Campos, E. J., N. Qureshi, and H. P. Blaschek. 2002. Production of acetone butanol ethanol from degermed corn using *Clostridium beijerinckii* BA101. *Appl. Biochem. Biotechnol.* 98:552–561.

Chen, C. K., and H. P. Blaschek. 1999. Acetate enhances solvent production and prevents degeneration in *Clostridium beijerinckii* BA101. *Appl. Microbiol. Biotechnol.* 52:170–173.

Chung, Y.-C., A. Bakalinsky, and M. H. Penner. 2005. Enzymatic saccharification and fermentation of xylose-optimized dilute acid-treated lignocellulosics. *Appl. Biochem. Biotechnol.* 121-124:947–961.

Daldal, F., and J. Applebaum. 1985. Cloning and expression of *Clostridium pasteurianum* galactokinase gene in *Escherichia coli* K-12 and nucleotide sequence analysis of a region affecting the amount of the enzyme. *J. Mol. Biol.* 186:533–545.

Dürre, P. 1998. New insights and novel developments in clostridial acetone/butanol/isopropanol fermentation. *Appl. Microbiol. Biotechnol.* 49:639–648.

Evans, P. J., and H. W. Wang. 1988. Enhancement of butanol fermentation by *Clostridium acetobutylicum* in the presence of decanol-oleyl alcohol mixed extractants. *Appl. Environ. Microbiol.* 54:1662–1667.

Ezeji, T. C., and H. P. Blaschek. Fermentation of dried distillers' grains and solubles (DDGS) hydrolysates to solvents and value-added products by solventogenic clostridia. *Bioresour. Technol.* (Epub ahead of print.)

Ezeji, T. C., N. Qureshi, and H. P. Blaschek. 2007a. Butanol production from agricultural residues: impact of degradation products on *Clostridium beijerinckii* growth and butanol fermentation. *Biotechnol. Bioeng.* 97:1460–1469.

Ezeji, T. C., N. Qureshi, and H. P. Blaschek. 2007b. Bioproduction of butanol from biomass: from genes to bioreactors. *Curr. Opin. Biotechnol.* 18:220–227.

Ezeji, T. C., N. Qureshi, and H. P. Blaschek. 2007c. Production of acetone butanol ethanol (ABE) in a continuous flow bioreactor using degermed corn and *Clostridium beijerinckii*. *Process Biochem.* 42:34–39.

Ezeji, T. C., N. Qureshi, and H. P. Blaschek. 2005a. Continuous butanol fermentation and feed starch retrogradation: butanol fermentation sustainability using *Clostridium beijerinckii* BA101. *J. Biotechnol.* 115:179–187.

Ezeji, T. C., N. Qureshi, and H. P. Blaschek. 2005b. Industrially relevant fermentations, p. 797–812. *In* P. Durre, (ed.), *Handbook on Clostridia*. Taylor and Francis, New York, NY.

Ezeji, T. C., N. Qureshi, and H. P. Blaschek. 2004. Acetone-butanol-ethanol production from concentrated substrate: reduction in substrate inhibition by fed-batch technique and product inhibition by gas stripping. *Appl. Microbiol. Biotechnol.* 63:653–658.

Ezeji, T. C., N. Qureshi, and H. P. Blaschek. 2003. Production of butanol by *Clostridium beijerinckii* BA101 and in-situ recovery by gas stripping. *World J. Microbiol. Biotechnol.* 19:595–603.

Garcia, A., E. L. Lanotti, and J. L. Fischer. 1986. Butanol fermentation liquor production and separation by reverse osmosis. *Biotechnol. Bioeng.* 28:785–791.

George, H. A., and J.-S. Chen. 1983. Acidic conditions are not obligatory for onset on butanol formation by *Clostridium beijerinckii* (synonym, *C. butylicum*). *Appl. Environ. Microbiol.* 46:321–327.

González-Pajuelo, M., I. Meynial-Salles, M. F. Filipa, P. Soucaille, and I. Vasconcelos. 2006. Microbial conversion of glycerol to 1,3-propanediol: physiological comparison of a natural producer, *Clostridium butyricum* VPI 3266, and an engineered strain, *Clostridium acetobutylicum* DG1(pSPD5). *Appl. Environ. Microbiol.* 72:96–101.

Gray, K. A., L. Zhao, and M. Emptage. 2006. Bioethanol. *Curr. Opinion. Chem. Biol.* 10:141–146.

Groot, W. J., R. G. J. M. van der Lans, and K. C. A. M. Luyben. 1989. Batch and continuous butanol fermentation with free cells: integration with product recovery by gas stripping. *Appl. Microbiol. Biotechnol.* 32:305–308.

Groot, W. J., C. E. van den Oever, and N. W. F. Kossen. 1984. Pervaporation for simultaneous product recovery in the butanol/isopropanol batch fermentation. *Biotechnol. Lett.* 6:709–714.

Gulati, M., K. Kohlmann, M. R. Ladisch, R. Hespell, and R. J. Bothast. 1996. Assessment of ethanol production options for corn products. *Bioresour. Technol.* 58:253–264.

Hartmanis, M. G. N., H. Ahlman, and S. Gatenbeck. 1986. Stability of solvent formation in *Clostridium acetobutylicum* during repeated subculturing. *Appl. Microbiol. Biotechnol.* 23:369–371.

Hull, S., B. Y. Yang, D. Venzke, K. Kulhavy, and R. Montgomery. 1996. Composition of corn steep water during steeping. *J. Agric. Food Chem.* 44:1857–1863.

Hüsemann, M. H., and E. T. Papoutsakis. 1990. Effects of propionate and acetate additions on solvent production in batch cultures of *Clostridium acetobutylicum*. *Appl. Environ. Microbiol.* 56:1497–1500.

Jones, D. T., and D. R. Woods. 1986. Acetone-butanol fermentation revisited. *Microbiol. Rev.* 50:484–524.

Kashket, E. R., and Z.-Y. Cao. 1995. Clostridial strain degeneration. *FEMS Microbiol. Rev.* 17:307–315.

Kutzenok, A., and M. Aschner. 1952. Degenerative processes in a strain of *Clostridium butylicum*. *J. Bacteriol.* 64:829–838.

Ladisch, M. R. 1991. Fermentation-derived butanol and scenarios for its uses in energy-related applications. *Enzyme Microb. Technol.* 13:280–283.

Laureano-Perez, L., F. Teymouri, H. Alizadeh, and B. E. Dale. 2005. Understanding factors that limit enzymatic hydrolysis of biomass. *Appl. Biochem. Biotechnol.* 121–124:1081–1099.

Lee, J. 1997. Biological conversion of lignocellulosic biomass to ethanol. *J. Biotechnol.* 56:1–24.

Lee, J., and H. P. Blaschek. 2001. Glucose uptake in *Clostridium beijerinckii* NCIMB 8052 and the solvent-hyperproducing mutant BA101. *Appl. Environ. Microbiol.* 67:5025–5031.

Lee, J., W. J. Mitchell, M. Tangney, and H. P. Blaschek. 2005. Evidence for the presence of alternative glucose transport system in *Clostridium beijerinckii* NCIMB 8052 and the solvent-hyperproducing mutant BA101. *Appl. Environ. Microbiol.* 71:3384–3387.

Long, S., D. T. Jones, and D. R. Woods. 1984. The relationship between sporulation and solvent production in *Clostridium acetobutylicum* P262. *Biotechnol. Lett.* 6:529–534.

Maddox, I. S., E. Steiner, S. Hirsch, S. Wessner, N. A. Gutierrez, J. R. Gapes, and K. C. Schuster. 2000. The cause of "acid crash" and "acidogenic fermentations" during the batch acetone-butanol-ethanol (ABE-) fermentation process. *J. Mol. Microbiol. Biotechnol.* 2:95–100.

Madihah, M. S., A. B. Ariff, K. M. Sahaid, A. A. Suraini, and M. I. A. Karim. 2001. Direct fermentation of gelatinized sago starch to acetone-butanol-ethanol by *Clostridium acetobutylicum. World J. Microbiol. Biotechnol.* 17:567–576.

McCutchan, W. N., and R. J. Hickey. 1954. The butanol-acetone fermentations. *Ind. Ferment.* 1:347–388.

McNeil, B., and B. Kristiahsen. 1985. Effect of temperature upon growth rate and solvent production in batch cultures of *Clostridium acetobutylicum. Biotechnol. Lett.* 7:499–502.

Mitchell, W. J. 1998. Physiology of carbohydrates to solvent conversion by clostridia. *Adv. Microb. Physiol.* 39:31–130.

Monot, F., J. M. Engasser, and H. Petitdemange. 1983. Regulation of acetone butanol production in batch and continuous cultures of *Clostridium acetobutylicum. Biotechnol. Bioeng. Symp.* 13:207–216.

Montoya, D., S. Sandra, S. Edelberto, and W. H. Schwarz. 2000. Isolation of mesophilic solvent-producing clostridia from Colombian sources: physiological characterization, solvent production and polysaccharide hydrolysis. *J. Biotechnol.* 79:117–126.

Nelson, D. L., and B. P. Webb. 1979. *Kirk-Othmer Encyclopedia of Chemical Technology*, 3rd ed., vol. IV, p. 179–190. John Wiley, New York, NY.

Nielson, L., M. Larsson, O. Holst, and B. Mattiasson. 1988. Adsorbents for extractive bioconversion applied to the acetone butanol fermentation. *Appl. Microbiol. Biotechnol.* 28:335–339.

Parekh, M., J. Formanek, and H. P. Blaschek. 1999. Pilot-scale production of butanol by *Clostridium beijerinckii* BA101 using a low-cost fermentation medium based on corn steep water. *Appl. Microbiol. Biotechnol.* 51:152–157.

Parekh, M., J. Formanek, and H. P. Blaschek. 1998. Development of a cost-effective glucose-corn steep medium for production of butanol by *Clostridium beijerinckii. J. Ind. Microbiol. Biotechnol.* 21:187–191.

Qureshi, N., and H. P. Blaschek. 2001a. ABE production from corn: a recent economic evaluation. *J. Ind. Microbiol. Biotechnol.* 27: 292–297.

Qureshi, N., and H. P. Blaschek. 2001b. Evaluation of recent advances in butanol fermentation, upstream and downstream processing. *Bioproc. Biosyst. Eng.* 24:219–226.

Qureshi, N., and H. P. Blaschek 2000. Butanol production using hyper-butanol producing mutant strain of *Clostridium beijerinckii* BA101 and recovery by pervaporation. *Appl. Biochem. Biotechnol.* 84:225–235.

Qureshi, N., and H. P. Blaschek. 1999. Production of acetone butanol ethanol (ABE) by a hyper-producing mutant strain of *Clostridium beijerinckii* BA101 and recovery by pervaporation. *Biotechnol. Prog.* 15:594–602.

Qureshi, N., P. Karcher, M. Cotta, and H. P. Blaschek. 2004. High-productivity continuous biofilm reactor for butanol production. *Appl. Biochem. Biotechnol.* 113–116:713–721.

Qureshi, N., I. S. Maddox, and A. Friedl. 1992. Application of continuous substrate feeding to the ABE fermentation: relief of product inhibition using extraction, perstraction, stripping, and pervaporation. *Biotechnol. Prog.* 8:382–390.

Qureshi, N., M. M., Meagher, J. Huang, and H. P. Hutkins. 2002. Acetone butanol ethanol (ABE) recovery by pervaporation using silicalite-silicone composite membrane from fed-batch reactor of *Clostridium acetobutylicum. J. Membr. Sci.* 187:93–102.

Sonnleitner, B., G. Locher, and A. Fiechter. 1992. Biomass determination. *J. Biotechnol.* 25:5–22.

Stevens, G. M., R. A. Holt, J. C. Gotschal, and J. G. Morris. 1985. Studies on the stability of solvent production by *Clostridium acetobutylicum* in continuous culture. *J. Appl. Bacteriol.* 59:597–605.

Tashiro, Y., K. Takeda, G. Kobayashi, K. Sonomoto, A. Ishizaki, and S. Yoshino. 2004. High butanol production by *Clostridium saccharoperbutylacetonicum* N1-4 in fed-batch culture with pH-stat continuous butyric acid and glucose feeding method, *J. Biosci. Bioeng.* 98:263–268.

Zaldivar, J., A. Martinez, and L. O. Ingram. 1999. Effect of selected aldehydes on the growth and fermentation of ethanologenic *Escherichia coli. Biotechnol. Bioeng.* 65:24–33.

Bioenergy
Edited by J. Wall et al.

Chapter 27

Biorefineries for Solvents: the MixAlco Process

Cesar B. Granda and Mark T. Holtzapple

From ethanol to vegetable oils, solvents have always been extremely important to the chemical industry from its inception. Until the advent of the petrochemical industry, solvents were biologically derived. Acetone, for instance, before World War I, was produced from dry distillation of calcium acetate obtained from the neutralization of vinegar with calcium carbonate or lime. This process, in turn, was replaced by the ABE (acetone/butanol/ethanol) fermentation. However, because agriculture and processing were inefficient, production of solvents was limited. With the petrochemical revolution, more extensive production, more variety, and more efficient solvents became economically feasible.

Today, improvements in agriculture and bioprocessing, in addition to high petroleum prices, come together to make biologically derived solvents economically and logistically viable. Bioprocessing, however, must allow for the production of a larger variety of solvents than before because new solvents have become available due to the chemical flexibility that the petrochemical industry has provided, and such demand must be supplied. Most bioprocesses, however, are limited to only a few products. ABE fermentation by *Clostridium acetobutylicum*, for instance, which is one of the most productively diverse fermentations, allows for mainly four solvents, namely, ethanol, acetone, butanol, and acetic acid (other minor products being acetoin and butyric acid) (Klass, 1998). Most other fermentation processes are designed to produce only one product (e.g., ethanol or acetic acid). In addition, free sugars are needed for the fermentation to occur. Such sugars must be produced directly from the cultivated crop, as in the case of sugarcane or sugar beets, or they have to be obtained by acid or enzymatic hydrolysis, as in the case of starch-bearing crops (e.g., corn), or, although with more difficulty, from lignocellulosic biomass. In spite of the difficulty and higher complexity, because of the higher productivities and abundance, lignocellulosic biomass would be the preferred feedstock.

One of the main barriers that the biochemical industry, which uses sugars as intermediates (i.e., sugar platform), must overcome to be able to compete with the petrochemical industry is their high processing costs. In addition to the need to decrease pretreatment costs prior to the fermentation, such as the enzymes or chemicals needed for hydrolysis, the fermentation costs must also be diminished. Because the fermentation normally employs pure cultures, the need to maintain sterility and to monitor viability and integrity of such cultures is essential. Such techniques, which involve using expensive vessels and equipment, sterilizing the feedstock, monitoring and treating infections, wasting batches that may be ruined by infecting bacteria among other things, resemble pharmaceutical applications, where the main product is generally a highly priced specialty compound. Neither the fuel nor the solvent market can comfortably afford such high-cost processing, especially because, for a larger variety of solvents to be manufactured, further chemistry would be necessary downstream.

Another alternative for the production of alcohol solvents is the thermochemical platform, which converts syngas (carbon monoxide and hydrogen) produced from the gasification of biomass, into Fischer-Tropsch (F-T) fuels and mixed alcohols through catalysis. The F-T fuels produced can be diesel or other hydrocarbons, and the mixed alcohols may range from methanol (C_1) all the way to octanol (C_8). Such a process has the disadvantage of needing expensive gasifiers, which are difficult to control, and extensive gas cleanup to avoid catalyst poisoning. In addition, at the moment, for mixed alcohol production, only modest catalyst yields (<40%) are attained (National Renewable Energy Laboratory, 2006). Such a process could become advantageous in the long term, with continuing research that will allow costs and yields to improve. In such a case, the F-T fuels and the mixed alcohols can be the base compounds for further chemistry that

Cesar B. Granda and Mark T. Holtzapple • Department of Chemical Engineering, Texas A&M University, College Station, TX 77843-3122.

would allow other solvents to be manufactured, such as esters and ethers.

An alternative to the sugar and thermochemical platforms is a promising technology for the production of chemicals, such as fuels and solvents, known as the MixAlco process. The MixAlco process, invented and patented by Mark Holtzapple and colleagues at Texas A&M University, makes use of a mixed culture of anaerobic microorganisms; therefore, no sterility is necessary. These microorganisms produce a mixture of carboxylic acids ranging from acetic acid (C_2) to heptanoic acid (C_7); such carboxylic acids are neutralized to control pH with a buffering agent, such as calcium carbonate or ammonium bicarbonate, thus producing carboxylate salts. From these carboxylate salts, further chemistry can be performed to obtain a wide variety of chemicals, which include many different types of fuels and solvents.

THE TECHNOLOGY

Figure 1 shows some of the different paths that may be taken to produce chemicals from carboxylate salts or carboxylic acids. Such chemistry may be afforded because of the low production costs of the carboxylate salts. The heart of this low-cost processing is the fermentation. The fermentation by the mixed culture of microorganisms is no different from conventional anaerobic digestion for the production of methanogenic biogas.

Anaerobic Digestion

In anaerobic fermentation, the mixed culture is a stable ecosystem with the various microorganisms playing essential roles necessary for the common benefit of such community. This culture is therefore very efficient and robust, not needing sterility or periodic revitalization. With pure cultures, on the other hand, sterility is needed to avoid contamination of foreign microorganisms and periodic revitalization might be necessary due to loss of viability in the culture. In addition to the robustness and efficiency of such an ecosystem, the wide variety of microorganisms allows for flexibility in the feedstocks that may be broken down and digested. The anaerobic mixed culture breaks down and digests virtually most biodegradable substances. Starch, cellulose, hemicellulose, fats, and proteins are converted into simpler molecules and finally digested and converted into carboxylic acids without the need of supplementation with chemicals or special enzymes, as the microorganisms are able to make their own. Anaerobic digestion has four major stages as follows.

1. Hydrolysis, whereby complex molecules such as starch, cellulose, hemicellulose, fats, and proteins are broken down into smaller molecules such as free sugars, fatty acids, and amino acids.
2. Acidogenesis, whereby these smaller molecules are converted into carboxylic acids, such as acetic, propionic, and butyric acids (from C_2 to C_7), and by-products such as carbon dioxide (CO_2), ammonia (NH_3), and hydrogen sulfide (H_2S) are produced.
3. Acetogenesis, whereby some of these higher acids are converted into acetic acid, hydrogen (H_2), and CO_2.
4. Methanogenesis, whereby the carboxylic acids (acetic and the higher acids) and H_2 may be converted into methane (CH_4) and CO_2.

Because in the case of the MixAlco process the desired product is the carboxylic acids, and CH_4 is an undesired side product derived from further degradation of these acids, methanogenesis must be inhibited. Methanogenesis can be inhibited by maintaining a low pH; however, low pH also inhibits the production of

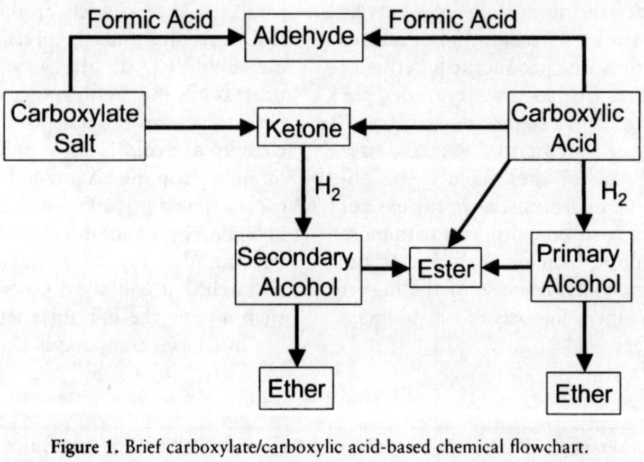

Figure 1. Brief carboxylate/carboxylic acid-based chemical flowchart.

the desired carboxylic acids. The literature suggests that the addition of compounds such as monensin, piromiellitic diimide, and 2-bromoethanesulfonic acid can inhibit methanogenesis (Martin and Macy, 1985). For the fermentation in the MixAlco process, methane analogs, such as bromoform or iodoform, are preferred as methane inhibitors. The dose of such compounds for effective inhibition of methane is very small (only a few parts per million) (Agbogbo, 2005). When ammonium bicarbonate (NH_4HCO_3) is used as the buffering agent, however, the ammonium ions (NH_4^+), when present in high enough concentrations, also act as methane analogs, thus inhibiting methane and avoiding the use of other inhibiting compounds.

Process Description

Besides depending on the type of product being made, the process configuration also depends upon the type of buffering agent used in the fermentation to control pH. The buffering agent determines the cation for the salts. Two buffering agents have been mainly investigated due to their convenience of use and ease of recoverability and regeneration: calcium carbonate ($CaCO_3$) and ammonium bicarbonate (NH_4HCO_3). Figure 2 shows a flow diagram of the MixAlco process when $CaCO_3$ is used as the buffering agent, and Fig. 3 shows the process flow diagram when NH_4HCO_3 is employed. Both configurations assume that the feedstock is a lignocellulosic material; therefore, it requires pretreatment to increase biodigestibility. Pretreatment and fermentation are the first steps in the process. Biomass is pretreated with lime under aeration to increase digestibility and fermented anaerobically using carboxylic acid-forming microorganisms. Because calcium carbonate or ammonium bicarbonate is added to control pH, carboxylate salts of calcium or ammonium are obtained (i.e., calcium or ammonium acetate, propionate, butyrate, valerate, caproate, and heptanoate). The salt solution is concentrated in an evaporator. Water is recycled to the pretreatment and fermentation, and the concentrated salt solution continues to downstream conversion.

The calcium salts may be dried and then thermally converted by means of dry distillation into ketones (e.g., acetone, methyl ethyl ketone, diethyl ketone, dipropyl ketone, methyl propyl ketone, and ethyl propyl ketone), and those ketones may be hydrogenated into secondary alcohols (e.g., isopropanol, 2-butanol, 3-pentanol, 2-pentanol, 3-hexanol, and 4-heptanol). Alternatively, the calcium salts may undergo acid springing (see "Acid springing" below for details of this process), whereby carboxylic acids (i.e., acetic, propionic, butyric, valeric, caproic, and heptanoic acid) are produced. The carboxylic acids may then be esterified, and the resulting esters can undergo hydrogenolysis to produce primary alcohols (i.e., ethanol, propanol, butanol, pentanol, hexanol, and heptanol).

The ammonium salts can be directly esterified, and the resulting esters can be hydrogenolyzed to yield the same primary alcohols. Alternatively, the salts may undergo acid springing to produce carboxylic acids. The carboxylic acids can be converted into ketones through catalysis, and the ketones can be hydrogenated to produce secondary alcohols.

Pretreatment

In lignocellulosic fiber, cellulose is encapsulated by lignin, which keeps the enzymes secreted by the microorganisms from reaching it; therefore, pretreatment is necessary to remove lignin. It has been found (Agbogbo,

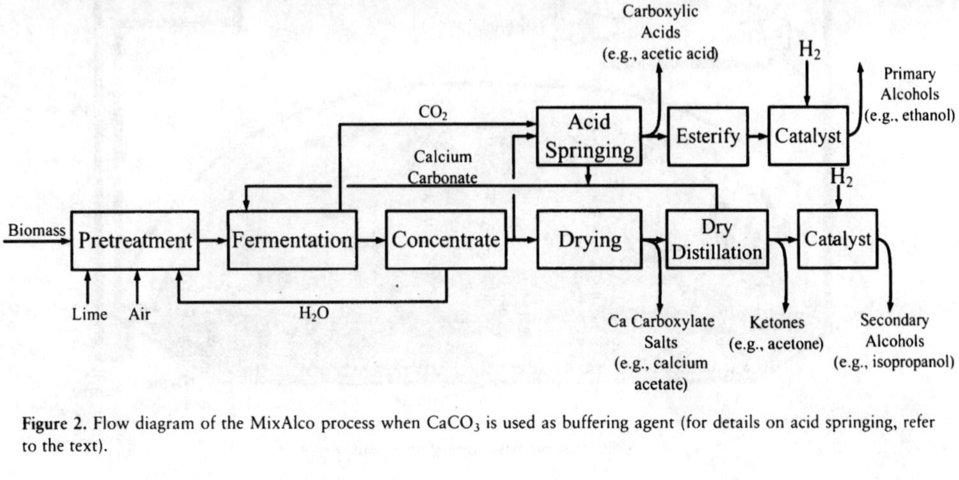

Figure 2. Flow diagram of the MixAlco process when $CaCO_3$ is used as buffering agent (for details on acid springing, refer to the text).

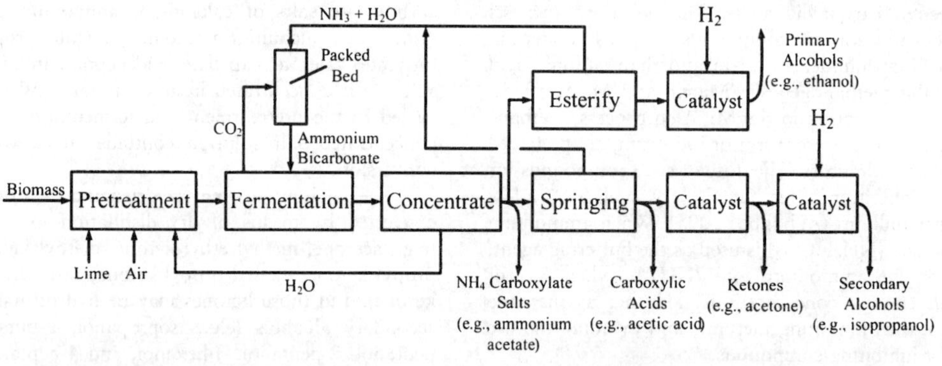

Figure 3. Flow diagram of the MixAlco process when NH_4HCO_3 is used as buffering agent (for details on acid springing, refer to the text).

2005; Granda Cotlear, 2004; Kim, 2004; Sierra Ramirez, 2005; Thanakoses, 2002) that pretreatment with lime and air efficiently delignifies lignocellulosic materials, increasing digestibility significantly. For the process, ground biomass is mixed intimately with about 100 kg of quicklime/dry ton of biomass and diluted with water to 3% solids to form a pumpable slurry. A pile (Fig. 4) is formed by directing the flow of this pumpable slurry on a gravel bed, which has been lined with a resistant geomembrane. The pile is formed in a way similar to that employed in the Ritter process for sugarcane bagasse wet storage (Atchison and Hettenhaus, 2004). The liquid that drains from the pile is returned to the slurrying tank.

Once the pile is built, liquid is recirculated on a regular basis from the gravel bed, where it collects as it trickles through, to the top of the pile, thus keeping it wet. The preferred temperature for this pretreatment is about 50°C, which should not be a problem to maintain as the reaction is exothermic. The cover shown in Fig. 4 is optional for large piles during pretreatment, which should not be affected by weather.

Lime alone is able to remove about one-third of the lignin, and in low-lignin lignocellulosic biomass, such as corn stover (lignin content, ~18%), it might be sufficient. However, with biomass such as sugarcane bagasse (lignin content ~22%), oxidation using air is necessary to bring the lignin content to about 10 to

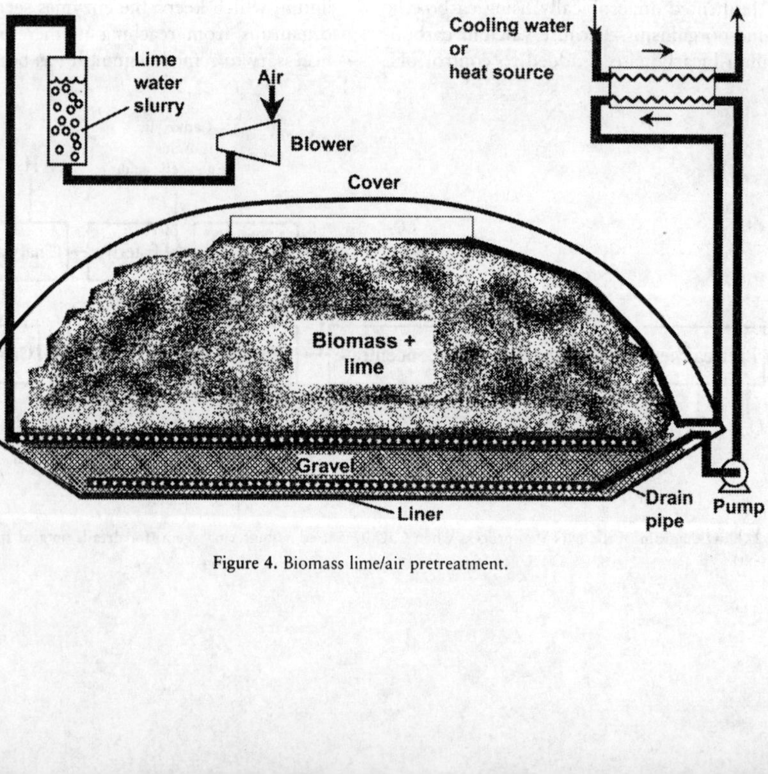

Figure 4. Biomass lime/air pretreatment.

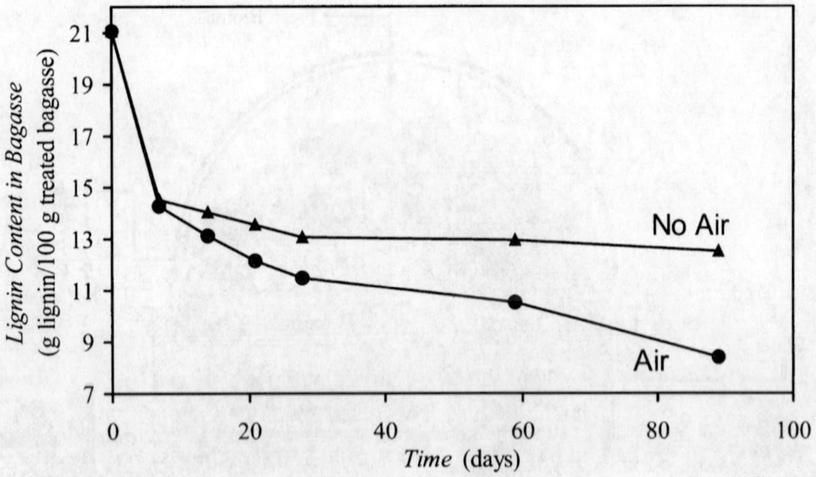

Figure 5. Delignification during lime pretreatment of sugarcane bagasse with and without air at 50°C (Granda Cotlear, 2004).

12% (Fig. 5), which is where maximum digestibility is attained. Embedded in the gravel bed there are air distributors, which are used to aerate the pile and allow oxidation to occur. This pretreatment takes about 4 to 6 weeks, depending on the lignin content.

Anaerobic fermentation

The next step in the process is the fermentation, which can take place in a fermentor or may occur in the pretreatment pile itself. The pretreated biomass pile can be dismantled and the material sent to a fermentor, which, more than likely, will be made of concrete. Although many fermentor configurations may be envisioned, the preferred option is a countercurrent laminar-flow fermentor (Fig. 6), where the solids are loaded at the top, while fresh water is injected at the bottom. The liquid product, on the other hand, is recovered at the top, while digested solids leave the fermentor at the bottom. This countercurrent operation offsets inhibitions caused by high product concentrations and digested biomass, whereby the fresh liquid contacts digested biomass, while concentrated liquid contacts fresh biomass.

The fermentor is initially inoculated with microorganisms. It has been found (Thanakoses, 2002) that, because the product from the fermentation consists of salts, microorganisms from marine swamps can better withstand the stresses imposed by the high salt concentrations; therefore, this is the inoculum of choice. The operating temperature in the fermentor should be between 40 and 55°C. A nutrient source is needed in this fermentation; therefore, the preferred feedstock is 80% carbohydrates (i.e., lignocellulosic residues) and 20% nutrients (e.g., chicken manure, swine manure, and sewage sludge). If, on the other hand, the cells leaving with the digested biomass are recycled, then the nutrient requirement decreases. To achieve 80% conversion, with a salt concentration of about 60 g/liter, about 1-month residence time is necessary (to achieve 70% conversion, only about 2 weeks are required) (Thanakoses, 2002). One month might seem long when compared with the shorter residence times in enzymatic hydrolysis/ethanol fermentation (~1 week); however, these fermentors (made of concrete) are about six times cheaper than the fermentor needed in enzymatic hydrolysis/ethanol fermentation (made of stainless steel). In theory, in-ground plastic fermentors can also be employed, which would be 30 times cheaper.

If the fermentation is to occur in piles, when the pH has decreased after pretreatment is done, the blowers are stopped and the piles are also initially inoculated with acid-forming microorganisms from marine environments to start the fermentation. The advantage of the use of piles is that no containers are needed as the piles do not require containment except at the base, which translates into significant capital savings. Any biorefinery would need to have an inventory of biomass of several months, especially in regions where the harvest occurs only once or twice every year. The biomass must be maintained and stored more than likely in piles. So, it can be said that in this process configuration biomass is being pretreated and fermented while it is stored. For fermentation, the piles would need to be covered to maintain anaerobiosis and so that gases may be directed to a scrubber for odor control and for recovery of volatilized acids. Several piles would be needed so that a round-robin system (Agbogbo, 2005; Agbogbo and Holtzapple, 2007) may be set up (Fig. 7), where some piles are being built or taken apart, others

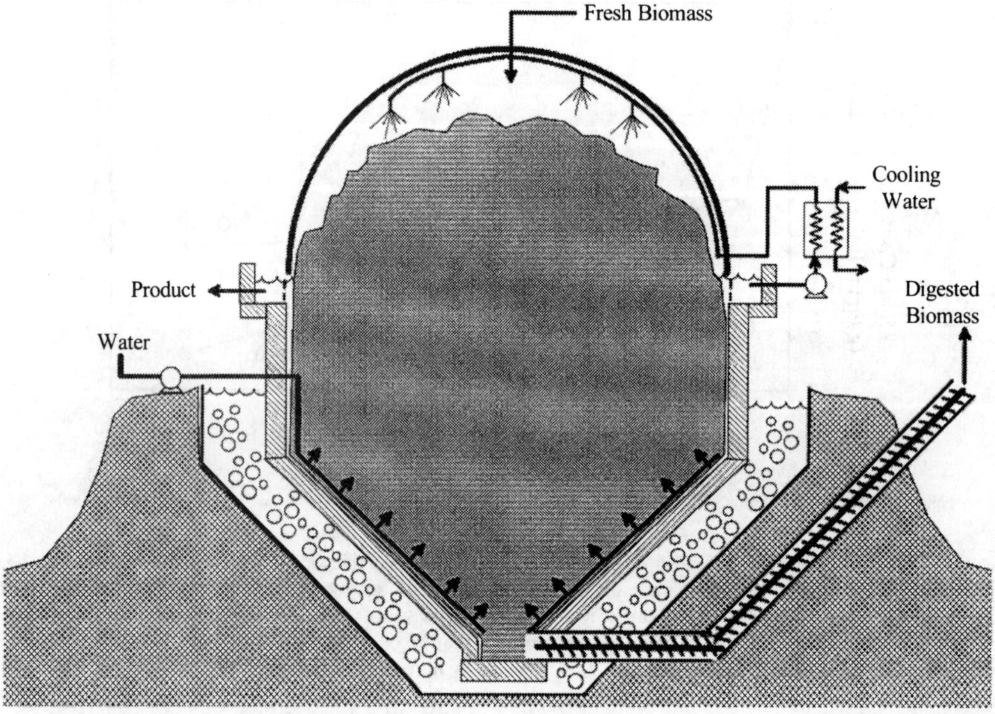

Figure 6. Countercurrent laminar-flow fermentor.

are being pretreated, and others are fermenting. The piles in the fermentation mode would see liquid transferred from one pile to another, where the freshest liquid will enter the most digested pile, and the liquid product with the highest salt concentration will exit the freshest pile. Transferring liquid in this fashion will allow for a higher concentration of carboxylate salts in the product and for higher conversions, as it will offset inhibitions caused by the high salt concentrations and the recalcitrance of digested biomass. The round-robin system is set up so that only fluids are moved and redirected accordingly but the piles do not move, thus minimizing solids handling, which would occur only during pile assembly and disassembly.

When $CaCO_3$ is used as the buffering agent, because it is fairly insoluble, it can be added in excess at the beginning of the fermentation and it would be consumed as the acids are produced. With $CaCO_3$ the pH is controlled at about 5.8 to 6. However, the microorganisms thrive at pH 7, and their metabolism slows as pH starts to drop below 6. Ammonium bicarbonate, on the other hand, maintains the pH at about 7.2; therefore, when this buffer was employed, the fermentation rates improved. Thermophilic fermentation (~55°) also shows higher rates than mesophilic fermentation (~40°).

The acid profile is affected by both the buffer and the operating temperature. Table 1 shows typical acid profiles when a given buffering agent and a given operating temperature are used. Mesophilic conditions (40°) and $CaCO_3$ tend to favor the higher acids, whereas thermophilic conditions (55°) and NH_4HCO_3 favor acetate formation. The given profile is important because it reflects also the profile for the final products (e.g., if the primary alcohol route is chosen, at 40° and using NH_4HCO_3, the final product will be about 69% ethanol, 3% propanol, 26% butanol, and 2% other higher alcohols).

When calcium carbonate is used as the buffer, iodoform, which is added on a regular basis, is needed to inhibit methanogenesis. The typical iodoform dosage is about 15 to 20 ppm/day (Agbogbo, 2005). On the other hand, when ammonium bicarbonate is used, the ammonium ions present act as the inhibitor, so iodoform addition is necessary only during start-up.

The digested biomass from the fermentation may be burned in the boilers to provide energy to run the process. Calculations show that less than 50% of the total digested biomass is needed to provide all the en-

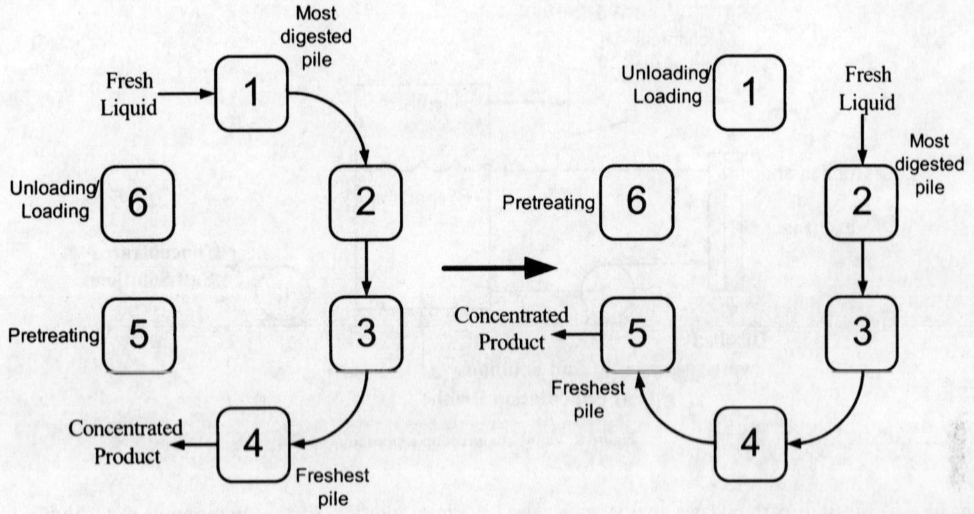

Figure 7. Sketch of the round-robin system operation.

ergy needs of a MixAlco plant. In addition, the rest of the undigested residue may be gasified to produce hydrogen for the process.

Salt solution dewatering

The acid concentration in the fermentation broth coming out of the fermentor is about 50 g/liter (60 g of salts/liter); therefore, dewatering of this broth is necessary. After fermentation, the fermentation broth undergoes clarification or descuming using flocculants followed by filtration, and then it is sent to a vapor compression evaporator, where most of the water is removed (Fig. 8). If the broth contains calcium carboxylate salts, crystallization takes place in the evaporator, and the salts are removed by filtration or centrifugation. On the other hand, if the broth contains ammonium carboxylate salts, the broth is concentrated to only about 40% (wt/vol), which removes ~90% of the water.

Vapor compression evaporators make use of mechanical power (i.e., a compressor or a blower) to pressurize the evaporated steam. Then, this pressurized steam is sent to a heat exchanger where it provides the latent heat of vaporization for more water to be evaporated. Although vapor compression evaporators are known to the industry, certain innovations make this evaporator very efficient. The efficiency of this vapor compression evaporator is equivalent to a 40–80-effect evaporator (Lara-Ruiz, 2005). The innovations consist of the use of a very efficient compressor and a high overall heat transfer coefficient attained in the heat exchanger. As a comparison, ethanol distillation and subsequent drying employ no less than 3 kg of steam per liter of ethanol produced (Maiorella, 1985), which is equivalent to about 29% of the combustion heat of ethanol. Using this efficient vapor compression evaporator, the energetics to dewater the broth from 6% salt to 99% are equivalent to only 6% of the combustion heat of the salts (Lara-Ruiz, 2005); thus, the efficiency of dewatering these salts is higher than ethanol distillation.

Table 1. Typical carboxylic acid profiles when using different buffers and temperatures

	Profile (% by wt)			
Acid	$CaCO_3$		NH_4HCO_3	
	40°C	55°C	40°C	55°C
Acetic	41	80	71	84
Propionic	15	4	3	3
Isobutyric	<1	<1	<1	1
Butyric	21	15	24	9
Isovaleric	<1	<1	<1	2
Valeric	8	<1	<1	<1
Caproic	12	<1	<1	<1
Heptanoic	3	<1	<1	<1

Ketone production by dry distillation of calcium carboxylate salts

Dry distillation is an old technique, which, as mentioned, was used before World War I to produce acetone from calcium acetate. It consists of thermally converting the calcium carboxylate salts into ketones by

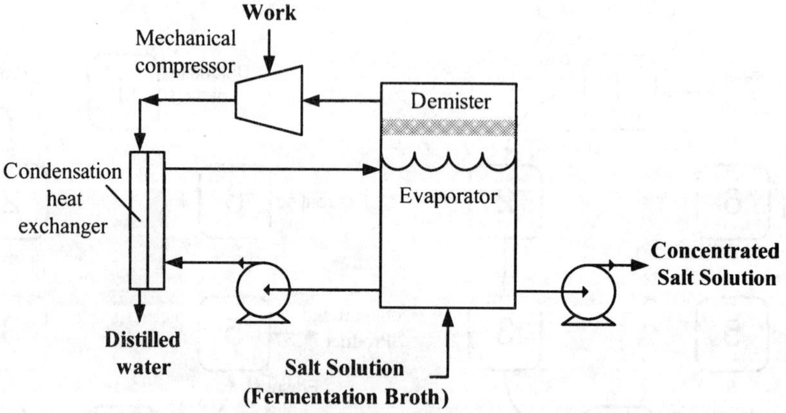

Figure 8. Vapor compression evaporator.

making use of the two carboxylates in the molecule, which, after giving off CO_2 and the calcium, join together to form the corresponding ketone.

When a pure calcium carboxylate salt is used, the salt molecules can react only with themselves; thus, only a pure symmetric ketone is formed.

$$(RCOO)_2Ca \rightarrow RCOR + CaCO_3$$

When a mixture of carboxylate salts is used, the salt molecules react both with each other and with the other carboxylate salts present; thus, a mixture of ketones are formed. For instance, when two different calcium carboxylate salts [e.g., $(RCOO)_2Ca$ and $(R'COO)_2Ca$] undergo dry distillation, the possible reactions are as follows:

$$(RCOO)_2Ca \rightarrow RCOR + CaCO_3$$

$$(RCOO)_2Ca + (R'COO)_2Ca \rightarrow 2\ R'COR + 2CaCO_3$$

$$(R'COO)_2Ca \rightarrow R'COR' + CaCO_3$$

A mixture of three different ketones (i.e., RCOR, R'COR, and R'COR') is thus expected.

The crystallized calcium carboxylate salts from the evaporator are sent to a dryer to decrease their moisture content. The dry salts are then sent to the dry distillation unit. As seen in Table 1, when calcium carbonate is used as the buffering agent in the fermentation, a mixture of basically six calcium carboxylate salts is produced, which could range from calcium acetate (C_2) to calcium heptanoate (C_7). These salts will combine with each other and, in turn, yield a rich mixture of 21 ketones ranging from acetone (C_3) all the way to diheptyl ketone (7- tridecanone) (C_{13}). When NH_4HCO_3 is used as the buffering agent, calcium salts can be obtained by exchanging the NH_4^+ ion with Ca^{2+} through lime addition to the ammonium salt solution, or by adding lime or calcium carbonate to the acids generated from acid springing (Fig. 3) (see "Acid springing" for details of this process).

Yeh (2002) demonstrated that very high yields can be attained during dry distillation of actual fermentation calcium carboxylate salts. Figure 9 shows the conversion that can be attained for the fermentation calcium salts at different temperatures and times. For instance, at 440° about 25 min is needed to attain >99% conversion. Because of the higher percentage of acetate, most of the ketone mixture will be acetone. The mixture of ketones can then be further separated by distillation into the different ketones as needed.

Ketone hydrogenation

The ketones obtained from dry distillation of the calcium carboxylate salts may be sold as such, or they

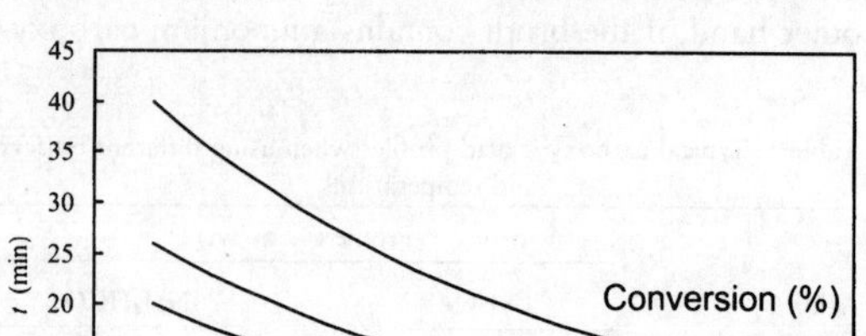

Figure 9. Thermal conversion to ketones of calcium carboxylate salts from MixAlco fermentation (Yeh, 2002).

may be further converted by hydrogenation into secondary alcohols (e.g., isopropanol, 2-butanol, and 3-pentanol). A catalyst, such as Raney nickel or platinum, may be employed. The reaction with Raney nickel was found to yield optimal results (>99% conversion) at 130°, 200 g per liter of Raney nickel catalyst, 35 min of residence time, and 15 atm of hydrogen (Aldrett-Lee, 2000). The reaction is as follows:

$$RCOR' + H_2 \rightarrow RCHOHR'$$

where R can be equal to R′, or they may be different.

Optimization of the process showed that three stages are needed (Fig. 10). This reaction is exothermic; therefore, some steam is generated, which can be used in other parts of the MixAlco plant. From this unit a mixture of 21 different secondary alcohols may be obtained, ranging from isopropanol (C_3) all the way to 7-tridecanol (C_{13}). As with the ketones, the alcohols may be separated by distillation as desired.

Acid springing

From the carboxylate salts, whether they are ammonium or calcium salts, the acids can also be recovered. In industrial carboxylic acid fermentations (e.g., citric acid), because of the need to control pH, salts of the buffering agent used (commonly calcium carbonate or lime) are also obtained. To convert the salts into the acids then, traditionally, sulfuric acid has been used. When calcium carbonate or lime buffering agents are used, calcium sulfate (gypsum) is produced. When the carboxylic acids are manufactured as specialty chemicals, disposal of gypsum is feasible, because, although limited, gypsum has its uses. However, when the goal is to produce commodity chemicals or fuels, disposal of gypsum becomes a problem because of the huge quantities that would be produced; therefore, in the MixAlco process, sulfuric acid is not preferred for acid regeneration from the salts.

The method used in the MixAlco process to recover the acids from the salts has been dubbed "acid springing," and, in theory, it does not consume any chemicals, nor does it produce by-products. For acid springing, the salts do not need to be dry but rather a concentrated solution or a slurry of the salts is needed. In the case of calcium salts, the calcium is first switched with a low-molecular-weight (LMW) tertiary amine (e.g., triethylamine). The calcium is removed from the solution by bubbling CO_2, which causes it to precipitate as calcium carbonate. The calcium carbonate may be recycled to the fermentation. The LMW amine carboxylate (e.g., triethylammonium acetate) is then switched with a high-molecular-weight (HMW) tertiary amine (e.g., trioctylamine), and the resulting HMW amine carboxylate (e.g., trioctylammonium acetate) is subsequently heated to >220° to "spring" or crack the acids off it. The acid product leaves the column at the top, and the HMW amine, which leaves at the bottom, is recycled to repeat the procedure (Williamson, 2000). Figure 11 illustrates this process. If ammonium carboxylate salts are to undergo acid springing, the process is greatly simplified (Fig. 12), as the LMW amine is not needed. The concentrated ammonium salt solution from the evaporator enters the switching column, where it is contacted with the HMW tertiary amine (e.g., trioctylamine). The remaining water and

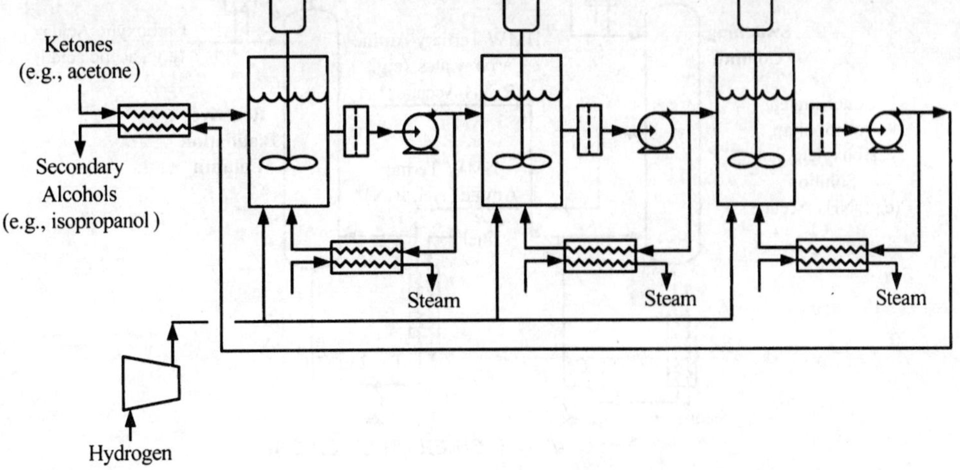

Figure 10. Secondary alcohols production by ketone hydrogenation.

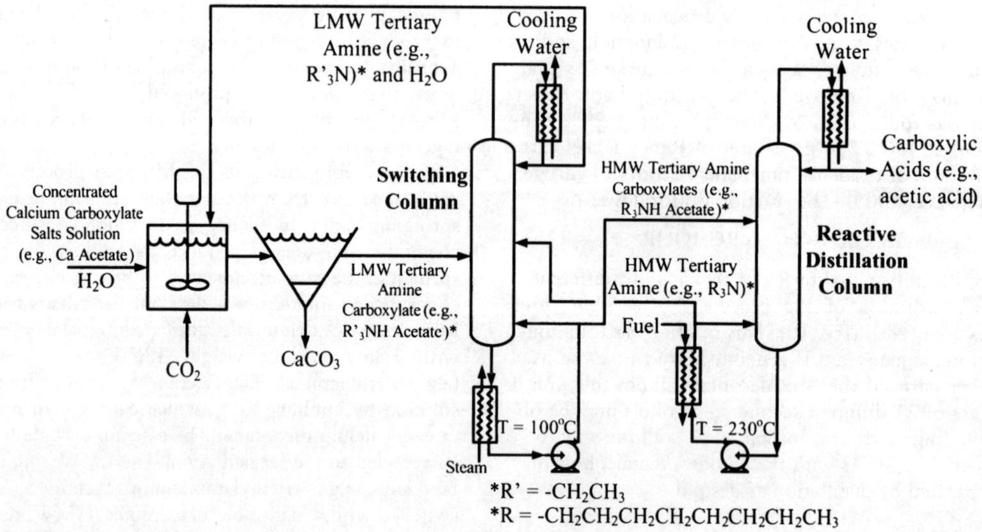

Figure 11. Acid springing system for calcium carboxylate salts.

the ammonia are driven off and sent to the packed bed to be contacted with carbon dioxide to regenerate the ammonium bicarbonate needed in the fermentation. From this switching, the HMW tertiary amine carboxylates are produced (e.g., trioctylammonium acetate), which are sent to the reactive distillation column, where they are heated to a temperature above 220° to thermally crack or decompose the acids from the amine. As with calcium carboxylate salt acid springing, the acids leave at the top of the reactive distillation column,

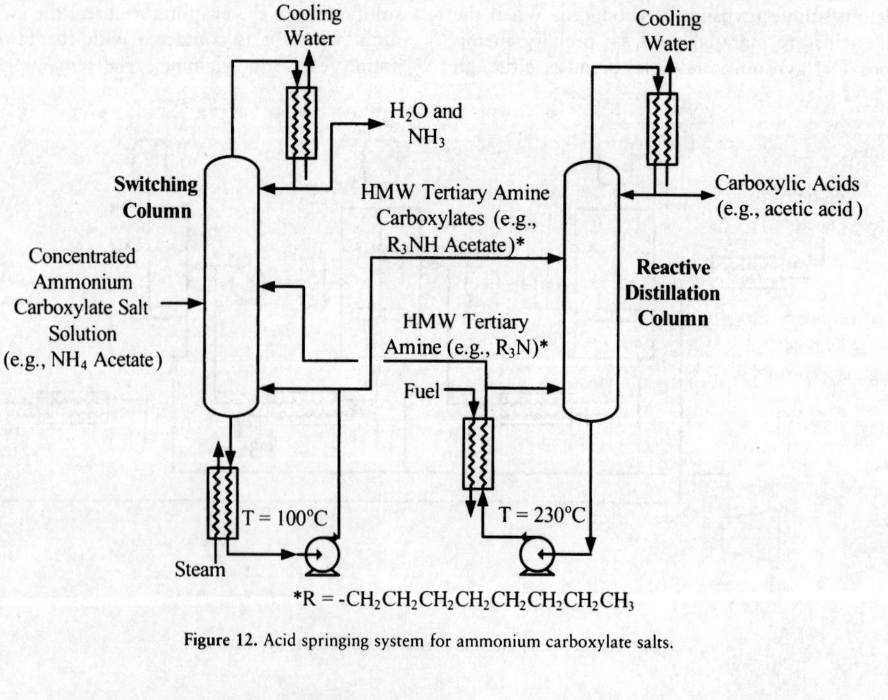

Figure 12. Acid springing system for ammonium carboxylate salts.

whereas the HMW tertiary amine is recovered at the bottom and recycled to the switching column to repeat the process. From acid springing, the corresponding eight carboxylic acids may be obtained, as shown in Table 1, which range from acetic acid (C_2) to heptanoic acid (C_7).

Acid gas phase catalytic ketonization

The resulting acids from acid springing can be sold into the chemical market as such, or they may be further converted into ketones. As mentioned, it would be possible to accomplish this by reacting the acids with calcium carbonate, which would yield the calcium salts of the acids and water. The water is driven off, and the resulting salts are dried. These dry salts are then conveyed to a reactor where they can be thermally decomposed into ketones and calcium carbonate by dry distillation as described in "Ketone production by dry distillation of calcium carboxylate salts" above. Nonetheless, the inherent inconvenience with this method is solids handling.

An alternative method to convert the acids into ketones, whereby no solids handling is needed, is using gas phase catalytic conversion. The acids are vaporized and passed through a catalytic bed of zirconium oxide at about 400° (Fig. 13). The products from this reaction would be ketones, water, and carbon dioxide. The reaction is as follows:

$$2RCOOH \rightarrow RCOR + CO_2 + H_2O$$

If two different acids react with each other, the reaction is as follows:

$$RCOOH + R'COOH \rightarrow RCOR' + CO_2 + H_2O$$

Because we have a mixture of acids ranging from acetic acid to heptanoic acid, just as with dry distillation, there will be many different types of ketones, ranging from acetone to 3-tridecanone, formed as the same type of acid reacts with itself and with the other acids present. Preliminary tests in an unoptimized reactor showed that this reaction is very efficient, attaining conversions higher than 95% with no observable side reactions (D. Ingram, unpublished data).

Salt esterification/ester hydrogenolysis

If producing primary alcohols (e.g., ethanol) is desired, either the carboxylic acids obtained from acid springing, or the LMW amine carboxylate complex (e.g., triethylammonium acetate) from the first step in the acid springing of calcium carboxylate salts, or the concentrated ammonium carboxylate salt solution directly from the evaporator can be sent to an esterification column. In this esterification column, the carboxylic acids, the LMW amine carboxylate complex, or the ammonium carboxylate salt solution is contacted with an HMW alcohol (e.g., heptanol) in the presence of an acid catalyst (e.g., H-β zeolites) to yield esters (e.g., heptyl acetate or heptyl propionate). The esterification of carboxylic acids is a common practice in industry. For instance, in the production of

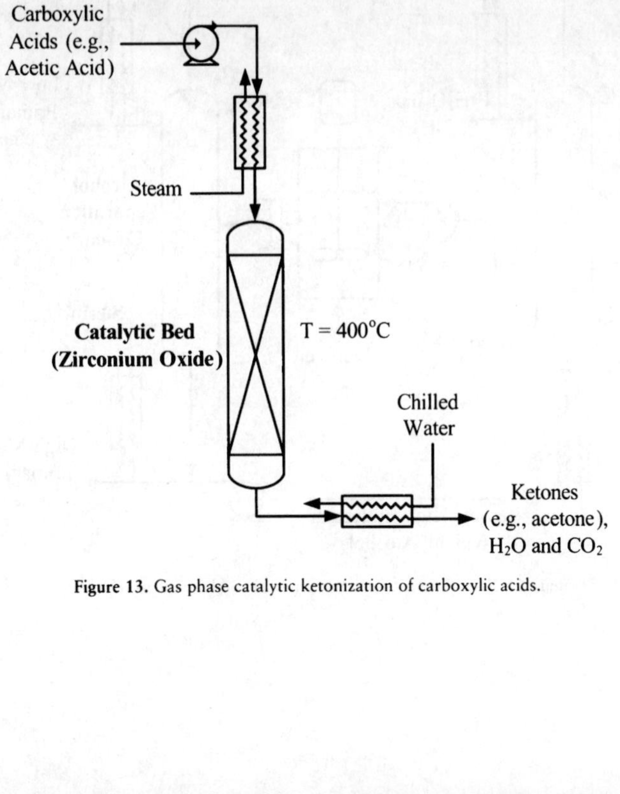

Figure 13. Gas phase catalytic ketonization of carboxylic acids.

biodiesel, free fatty acids may be converted to methyl esters (biodiesel) by means of this reaction. However, more convenient is the fact that the LMW amine carboxylate, from the first step in the acid springing of calcium carboxylate salts, can also undergo esterification without the need to go all the way to the carboxylic acid. Even more convenient yet is the fact that the ammonium carboxylate salts can be esterified directly after the evaporator without the need of any extra step (Filachione et al., 1951). In the esterification column, only water is produced if the carboxylic acid is employed. If the LMW amine carboxylate complex is used, water and the LMW amine leave through the top of the esterification column and return to the carbonation reactor to repeat the procedure. If the ammonium carboxylate salts are esterified, water and ammonia leave the system in the esterification column distillate and are sent back to the packed bed for regeneration of the NH_4HCO_3 buffer needed in fermentation, while the resulting esters and the unreacted alcohol leave at the bottom (Fig. 14).

The reaction for the esterification is as follows:

$$RCOONH_4 + R'OH \leftrightarrows RCOOR' + H_2O + NH_3$$

Although the reaction is reversible, it is driven to completion by constantly removing water and ammonia (or the water and the LMW amine if the LMW amine carboxylate is used). As mentioned, the esterification is acid catalyzed.

The resulting esters are then hydrogenolyzed in the presence of a catalyst. Copper chromite is a commercially available hydrogenolysis catalyst that is widely used in industry, for instance, for the production of detergent alcohols from fatty esters. However, it employs high pressures and temperatures (>200° and >40 atm). Reduced CuO/ZnO, on the other hand, employs milder conditions (~150° at <25 atm) (Bradley et al., 1982); therefore, this catalyst is preferred. After hydrogenolysis, the resulting alcohols are then sent to a distillation column where the LMW alcohols (e.g., ethanol, propanol, and butanol), which exit at the top (distillate), are separated from the HMW alcohols, which exit at the bottom. Some of the HMW alcohols are recycled back to the esterification column (Fig. 14).

The following is the reaction for the hydrogenolysis of the esters:

$$RCOOR' + 2H_2 \rightarrow R\text{-}CH_2OH + R'OH$$

The hydrogen can be obtained from many sources, such as gasification of the undigested residue from the fermentation. From the same gasification process, steam and electricity can be generated to provide energy for the plant.

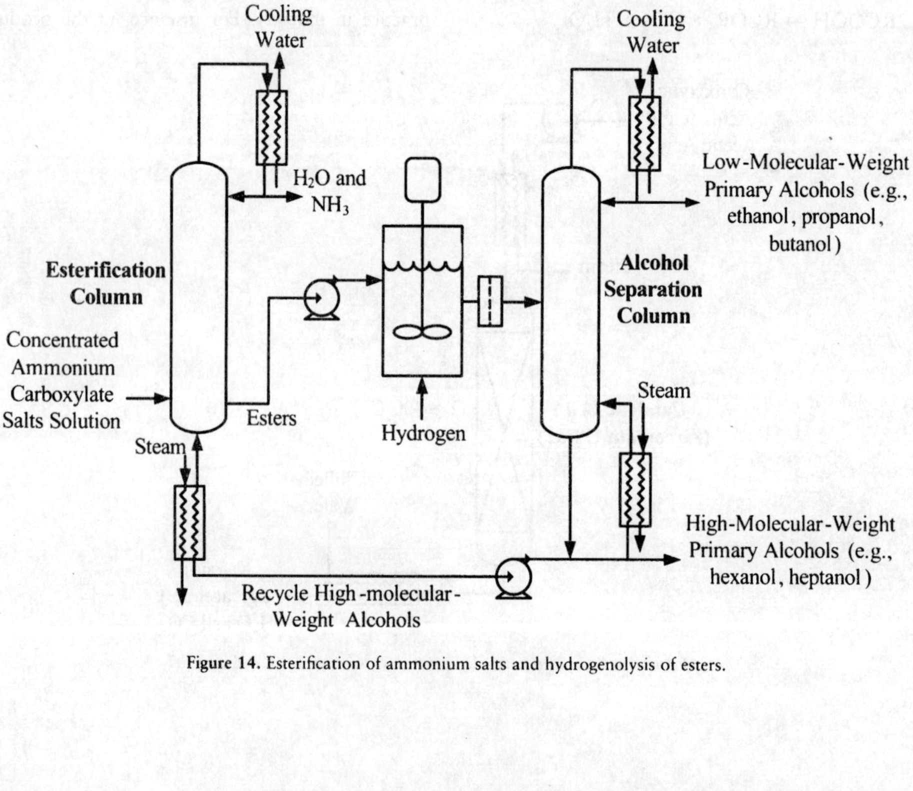

Figure 14. Esterification of ammonium salts and hydrogenolysis of esters.

DISCUSSION AND CONCLUSIONS

The MixAlco process offers an economical and robust biomass bioconversion method. Savings in this fermentation allow the use of well-established chemistry to produce a wide array of fuels and chemicals, including many types of solvents, and still be economically practical and profitable. The MixAlco process can generate from biomass over 100 chemicals of many types (Table 2) such as carboxylic acids, ketones, esters, aldehydes, ethers, and primary and secondary alcohols. Many of those chemicals, such as acetic acid, acetone, methyl ethyl ketone, diethylketone, ethanol, propanol, butanol, isopropanol, and sec-butanol, to name a few, are important and valuable industrial solvents. In this treatise, we have reviewed a few of the processes downstream of the fermentation that may be employed to produce some of those chemicals. The wide diversity of products that the MixAlco process provides allows meeting the also diverse solvent demand that the petrochemical industry has created.

As mentioned above, unlike specialty biochemicals manufacturing, where due to the large profit margins one can employ biopharmaceutical techniques, the biomass-based fuel and commodity chemical industry cannot comfortably afford this. The MixAlco process avoids the need for biopharmaceutical techniques by using a mixed culture of naturally occurring microorganisms.

Conversion of lignocellulose directly to alcohols through hydrolysis and fermentation, although straightforward, is a little difficult. From the pretreatment/enzymatic hydrolysis of lignocellulosic biomass, xylose and other C_5 sugars from hemicellulose are also produced. However, typical ethanologenic yeasts, for instance, *Saccharomyces cerevisiae*, cannot readily ingest C_5 sugars. Although progress has been made with other types of yeasts that can ingest C_5 sugars, such as *Pichia stipitis*, and genetically modified organisms that can ingest both C_5 and C_6 sugars (e.g., glucose), such organisms are more difficult to grow and, in the case of genetically modified organisms, likely will encounter public opposition in addition to disposal difficulties. The naturally occurring mixed culture employed in the MixAlco process, on the other hand, can readily ingest both C_5 and C_6 sugars.

The theoretical mass yield of acetic acid in the fermentation should be 100%, because, unlike ethanol or butanol fermentation, acetic acid is mostly produced without the production of CO_2; therefore, mass is not lost. The reaction for acetic acid fermentation is as follows:

$$C_6H_{12}O_6 \rightarrow 3CH_3COOH$$

In the MixAlco fermentation, some carbon dioxide is released, but this is minor; thus, the mass yields are still high. If the undigested residue (mostly lignin) is then used in gasification to provide the hydrogen to make the alcohols, it is ensured that most of the energy present in the biomass will end up in the product. In addition, the hydrogen may be produced from other renewable sources, such as wind or solar energy, by electrolysis; thus, energy from these sources can end up in the product as well. These facts have driven some to consider switching from ethanol to acetic acid fermentation even when free sugars are used as the feedstock (Verser and Eggeman, 2005).

The fact that intermediate products in the MixAlco process (e.g., acids and ketones) are now more valuable

Table 2. Some of the chemicals that can be produced using the MixAlco process

Type	Chemicals
Carboxylic acids	Acetic acid, propionic acid, butyric acid, valeric acid, caproic acid, enanthic acid
Ketones	Acetone, methyl ethyl ketone, diethyl ketone, 2-hexanone, 2-heptanone, 2-octanone, 3-pentanone, 3-hexanone, 3-heptanone, 3-octanone, 3-nonanone, 4-heptanone, 4-octanone, 4-nonanone, 4-decanone, 5-nonanone, 5-decanone, 5-undecanone, 6-undecanone, 6-dodecanone, 7-tridecanone
Primary alcohols	Ethanol, 1-propanol, 1-butanol, 1-pentanol, 1-hexanol, 1-heptanol
Secondary alcohols	Isopropanol, sec-butanol, 2-pentanol, 2-hexanol, 2-heptanol, 2-octanol, 3-pentanol, 3-hexanol, 3-heptanol, 3-octanol, 3-nonanol, 4-heptanol, 4-octanol, 4-nonanol, 4-decanol, 5-nonanol, 5-decanol, 5-undecanol, 6-undecanol, 6-dodecanol, 7-tridecanol
Esters	Ethyl acetate, ethyl propionate, ethyl butyrate, ethyl pentanate, ethyl hexanate, ethyl heptanate, propyl acetate, propyl propionate, propyl butyrate, propyl pentanate, propyl hexanate, propyl heptanate, butyl acetate, butyl propionate, butyl butyrate, butyl pentanate, butyl hexanate, butyl heptanate, pentyl acetate, pentyl propionate, pentyl butyrate, pentyl pentanate, pentyl hexanate, pentyl heptanate, hexyl acetate, hexyl propionate, hexyl butyrate, hexyl pentanate, hexyl hexanate, hexyl heptanate, heptyl acetate, heptyl propionate, heptyl butyrate, heptyl pentanate, heptyl hexanate, heptyl heptanate
Ethers	Diethyl ether, ethyl propyl ether, ethoxy butane, ethoxy pentane, ethoxy hexane, ethoxy heptane, dipropyl ether, propoxy butane, propoxy pentane, propoxy hexane, propoxy heptane, dibutyl ether, butoxy pentane, butoxy hexane, butoxy heptane, dipentyl ether, pentoxy hexane, pentoxy heptane, dihexyl ether, hexoxy heptane, diheptyl ether
Aldehydes	Acetaldehyde, propanal, butanal, pentanal, hexanal, heptanal

than some of the final products that have bigger markets (e.g., alcohols) is very advantageous because it ensures that the technology is profitable at every stage of its development. The near-term approach would be to process waste biomass into acids and ketones. Then, as the technology grows and becomes more widespread and the market for these chemicals becomes more saturated, it might be convenient to process cheap biomass into alcohols. Finally, when the technology is fully established and very large plants are built, it will be possible to pay more for the biomass and grow crops specifically for energy and chemical production, which will be mostly processed into alcohols for the fuel market.

REFERENCES

Agbogbo, F. K. 2005. Anaerobic fermentation of rice straw and chicken manure to carboxylic acids. Ph.D. dissertation. Texas A&M University, College Station.

Agbogbo, F. K., and M. T. Holtzapple. 2007. Fixed-bed fermentation of rice straw and chicken manure using a mixed culture of marine mesophilic microorganisms. *Bioresour. Technol.* 98:1586–1595.

Aldrett-Lee, S. 2000. Catalytic hydrogenation of liquid ketones with emphasis on gas-liquid mass transfer. Ph.D. dissertation. Texas A&M University, College Station.

Atchison, J. E., and J. R. Hettenhaus. 2004. *Innovative Methods for Corn Stover Collecting, Handling, Storing and Transporting (NREL/SR-510-33893).* National Renewable Energy Laboratory (NREL), Golden, CO. http://www.nrel.gov/docs/fy04osti/33893.pdf.

Bradley, M. W., N. Harris, and K. Turner. November 1982. Process for Hydrogenolysis of Carboxylic Acid Esters. Patent WO 82/03854.

Filachione, E. M., E. J. Costello, and C. H. Fisher. 1951. Preparation of esters by reaction of ammonium salts with alcohols. *J. Am. Chem. Soc.* 73:5265–5267.

Granda Cotlear, C. B. 2004. Sugarcane juice extraction and preservation, and long-term lime pretreatment of bagasse. Ph.D. dissertation. Texas A&M University, College Station.

Kim, S. 2004. Lime pretreatment and enzymatic hydrolysis of corn stover. Ph.D. dissertation. Texas A&M University, College Station.

Klass, D. L. 1998. *Biomass for Renewable Energy, Fuels, and Chemicals*, p. 431. Academic Press, San Diego, CA.

Lara-Ruiz, J. H. J. 2005. An advanced vapor-compression desalination system. Ph.D. dissertation. Texas A&M University, College Station.

Maiorella, B. L. 1985. Ethanol. *In* M. Young (ed.), *Comprehensive Biotechnology*, vol. 3. Pergamon Press, Oxford, United Kingdom.

Martin, S. A., and J. M. Macy. 1985. Effects of monensin, pyromellitic diimide and 2-bromoethanesulfonic acid on rumen fermentation in vitro. *J. Anim. Sci.* 60:544–550.

National Renewable Energy Laboratory (NREL). 2006. *Equipment Design and Cost Estimation for Small Modular Biomass Systems, Synthesis Gas Cleanup, and Oxygen Separation Equipment—Task 9: Mixed Alcohols from Syngas—State of Technology.* Contract No. DE-AC36-99-GO10337, subcontract report NREL/SR-510-39947. National Renewable Energy Laboratory, Golden, CO. http://www.nrel.gov/docs/fy06osti/39947.pdf.

Sierra Ramirez, R. 2005. Long-term lime pretreatment of poplar wood. M.S. thesis. Texas A&M University, College Station.

Thanakoses, P. 2002. Conversion of bagasse and corn stover to mixed carboxylic acids using a mixed culture of mesophilic microorganisms. Ph.D. dissertation. Texas A&M University, College Station.

Verser, D., and T. Eggeman. 2005. Process for producing ethanol. U.S. patent 6,927,048. http://www.zeachem.com.

Williamson, S.A. 2000. Conversion of carboxylate salts to carboxylic acids via reactive distillation. M.S. thesis. Texas A&M University, College Station.

Yeh, H. 2002. Pyrolytic decomposition of carboxylate salts. M.S. thesis. Texas A&M University, College Station.

Bioenergy
Edited by J. Wall et al.

Chapter 28

Increased Biofuel Production by Metabolic Engineering of *Clostridium acetobutylicum*

LEIGHANN SULLIVAN, MILES C. SCOTCHER, AND GEORGE N. BENNETT

Microbially based conversion technologies are potential contributors to fuel and chemical industries because microorganisms ferment, ideally, inexpensive feedstocks to industrially important products. With regard to the fuel industry, a handful of microorganisms convert substrates into organic solvents, for example, butanol or ethanol, which can be used as gasoline fuel extenders. The organic solvent butanol is arguably a better fuel extender than ethanol (Schwarz and Gapes, 2006); hence, this review focuses primarily on butanol. In order to increase microbial butanol yields and tolerance to high butanol concentrations, the pathways of the host microorganism need to be globally manipulated. Microbial metabolic engineering has already proven to be a successful strategy in producing products for many chemical industries from various organisms: amino acids (Aiba et al., 1980; Shio, 1986; Ikeda and Katsumata, 1992; Colon et al., 1995a, 1995b; Kumagai, 2000); vitamins and precursors (Shimada et al., 1998; Alper et al., 2005); flavanoids (Yan et al., 2005); chemical precursors (Nakamura and Whited, 2003; Hong et al., 2004); cofactors (Lopez de Felipe et al., 1998); biosurfactants (Koch et al., 1991); antibiotics (Malpartida and Hopwood, 1984; Bartel et al., 1990; Isogai et al., 1991; Stanzak et al., 1986; Smith et al., 1990); polymers (Peoples and Sinskey, 1989; Tong et al., 1991; Tong and Cameron, 1992; Cameron et al., 1998); and chemotherapeutics (Madduri et al., 1998) related to human immunodeficiency virus (Buckland et al., 1999) or antimalaria drugs (Ro et al., 2006).

Metabolic engineering employs a systemic alteration of biosynthetic pathways leading to metabolite production levels different from those that the host organism would otherwise manufacture. These cellular changes ought not to detract from normal physiological characteristics too drastically. Overproduction of a particular metabolite should not be at the expense of the ability of the microorganism to grow well. Maintaining a robust growth profile is important because certain metabolites are synthesized only at a particular time during growth. Growth-dependent metabolite production is particularly valid in *Clostridium acetobutylicum* fermentation. *C. acetobutylicum* initially ferments sugar into the organic acids acetate and butyrate during exponential phase. At the transition to stationary phase, the acids begin to reassimilate into the organic solvents acetone and butanol, and then solvents are generally produced in earnest during stationary phase. Figure 1 shows the metabolic pathways of *C. acetobutylicum*.

METABOLIC ENGINEERING STRATEGIES FOR *CLOSTRIDIUM ACETOBUTYLICUM*

The engineering strategies to identify which gene(s) to modulate are varied; they range from basic characterization to correlation with known genes, to conditional screens, and to alteration of a key modulator affecting many downstream players. Strategies to increase *C. acetobutylicum* metabolite production have historically focused on single gene perturbations. The gene to be perturbed was determined by knowledge of the pathways and the enzymes involved; however, this strategy is not practical when little is known about the cellular networks of newly sequenced microorganisms. Comparative analysis may identify genes to manipulate that will drastically affect cell-wide activities such as genes for transcriptional regulators and stress proteins. One strategy to identify genes responsible for a global physiological change is to screen a genomic library under a particular test condition. Undiscovered genes contributing to complex phenotypes may correlate transcriptionally or proteomically with

Leighann Sullivan • Department of Biochemistry & Cell Biology, Rice University, Houston, TX 77251-1892. **Miles C. Scotcher** • Foodborne Contaminants Research Unit, Western Regional Research Center, USDA Agricultural Research Service, 800 Buchanan St., Albany, CA 94710. **George N. Bennett** • Department of Biochemistry & Cell Biology, Rice University, Houston, TX 77251-1892.

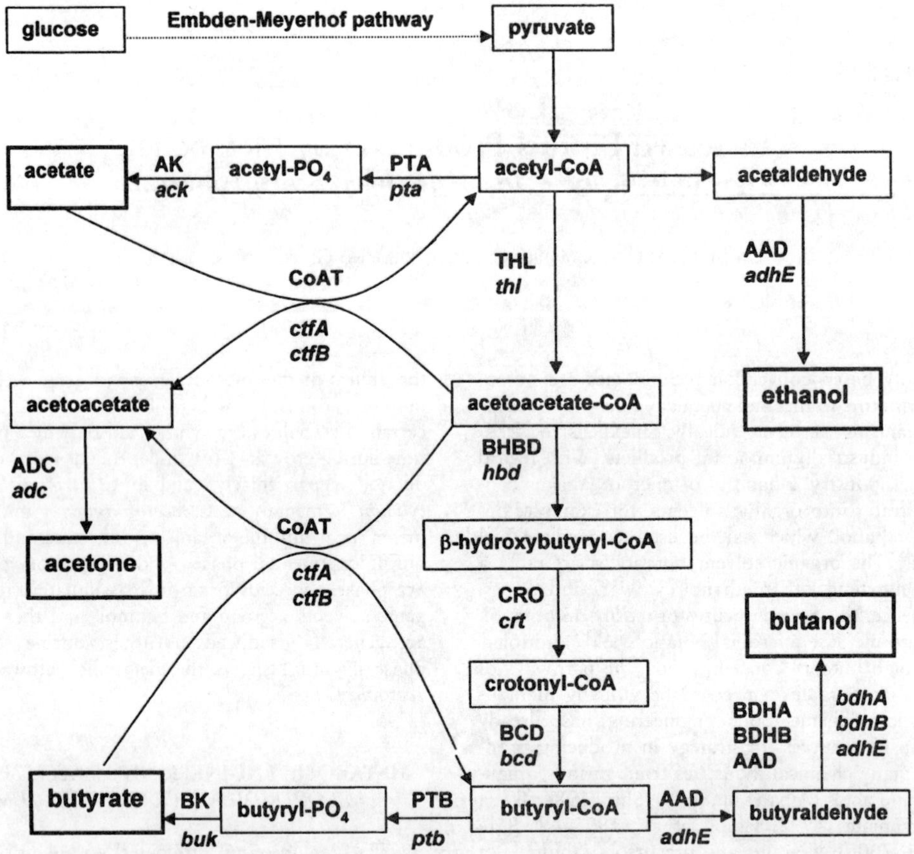

Figure 1. Metabolic pathways of *C. acetobutylicum*. The acid and solvent products are in bold boxes: acetate, butyrate, ethanol, acetone, and butanol. Genes are shown in italics, with those underlined residing on the pSOL1 plasmid. Enzyme names are capitalized: AK, acetate kinase; PTA, phosphotransacetylase; THL, thiolase; CoAT, CoA transferase; AAD, alcohol/aldehyde dehydrogenase; ADC, acetoacetate decarboxylase; BHBD, β-hydroxylbutyryl dehydrogenase; CRO, crotonase; BCD, butyryl-CoA dehydrogenase; BK, butyrate kinase; PTB, phosphotransbutyrylase; BDHA and BDHB, butanol dehydrogenase isozymes A and B (adapted from Bahl and Durre, 2001, and Mitchell, 1998).

genes or gene products known to be involved. Therefore, the most beneficial metabolic engineering strategies will be those that integrate complementary fields: "omics" technologies, computational systems biology, protein engineering, and synthetic biology as described previously (Tyo et al., 2007) and shown in Fig. 2. However, before such higher-level integration could be undertaken, the foundation of metabolic engineering was laid by experimentally modifying one gene at a time and by basic biochemical analysis of the organism and pathways.

The pioneering molecular analyses to elucidate the genes encoding the metabolic pathway enzymes of solvent production involved gene identification, sequencing and promoter characterization, and later more global theoretical analysis of operon structure or genomic arrangements. The sequence and arrangement of the genes involved in the acetone production (*adc*) (Cary et al., 1990; Petersen et al., 1993) and the butyrate synthesis (*ptb-buk*) pathways (Cary et al., 1988; Walter et al., 1993) were determined almost 15 years ago. Likewise, the genes for butanol production, namely the *sol* operon (*adhE* [also known as *aad*], *ctfA*, and *ctfB*), have been cloned and sequenced, and their promoters have been characterized (Fischer et al., 1993) during the same time period.

In addition to describing the genes of the pathway enzymes, the enzymes themselves were initially measured during fermentations and later isolated and characterized. The fermentation-phase-dependent enzyme

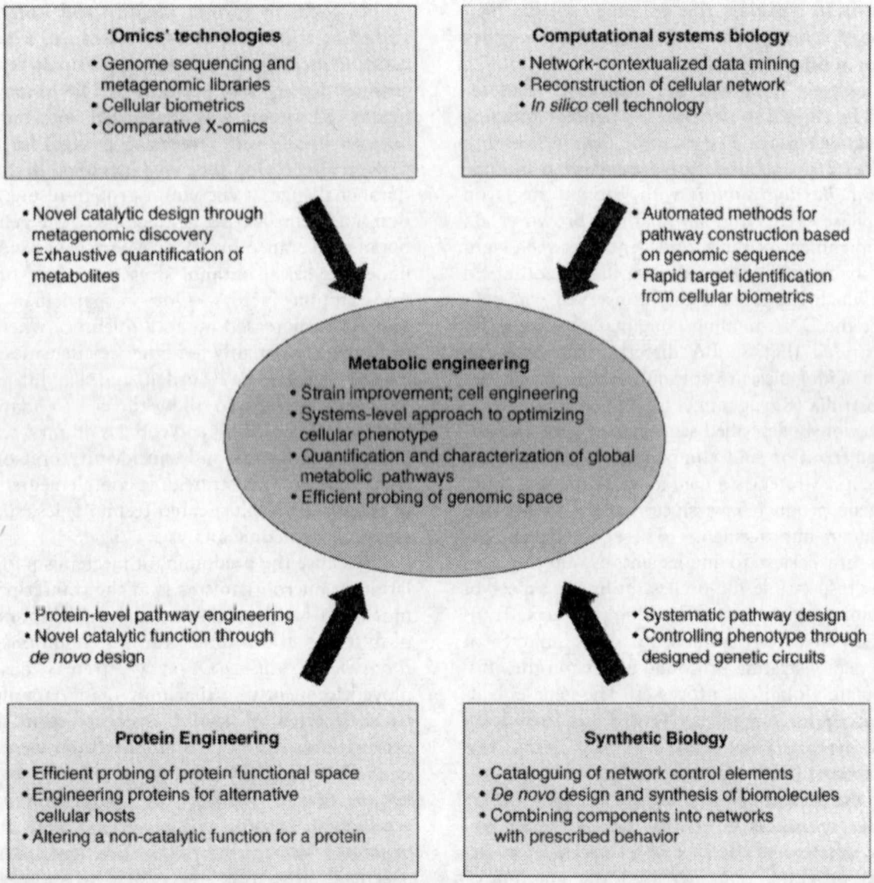

Figure 2. Metabolic engineering stands to gain significantly from advances in complementary biological fields. Omics technologies and computational systems biology can provide large amounts of data about a cellular state and the means by which to analyze it, whereas protein engineering and synthetic biology can provide tool sets for new ways to manipulate a cell to improve the cellular properties. These four fields have a unique set of expertise that could be applied to further metabolic engineering analysis and implementation. Reprinted from *Trends in Biotechnology* (Tyo et al., 2007) with permission of the publisher.

levels involved in acid formation were characterized. Phosphotransacetylase (Pta) and acetate kinase (Ack) decrease on the onset of solventogenesis. Three consecutively acting enzymes involved in the conversion of acetyl-coenzyme A (CoA) to butyryl-CoA, thiolase, beta-hydroxybutyryl-CoA dehydrogenase, and crotonase were coordinated in their expression and maximal activity, reaching a peak at the end of exponential phase. Butyryl-CoA dehydrogenase activity was low and difficult to analyze. Phosphotransbutyrylase (Ptb) activity rapidly decreased at the onset of solventogenesis, while butyrate kinase (Buk) activity increased dramatically (Hartmanis and Gatenbeck, 1984). Thiolase (Wiesenborn et al., 1988), Ptb (Wiesenborn et al., 1989b), and CoA transferase (Wiesenborn et al., 1989a) were purified, and their properties were determined. Intracellular levels of CoA and its derivatives were analyzed (Grupe and Gottschalk, 1992; Boynton et al., 1994) as well as the corresponding acyl-phosphates (Zhao et al., 2005), and their possible role in regulating flux through the pathway was considered. After the basic knowledge of genes, enzymes, and pathways is discovered for one organism, then by comparative biology a putative function can be ascribed to a similar gene or protein in a related organism. While the assigned function may or may not be the same for the two organisms in question, this putative function provides an initial direction for testing homologous genes and proteins in other organisms.

After multiple homologs are identified, then sequences can be aligned to illustrate the protein domains and conservative regions. For example, Spo0A was first recognized as a phosphorylation-dependent transcription factor in *Bacillus subtilis* with known effects on stationary-phase events and sporulation (Brown et al., 1994). An alignment of the *spo0A* genes between eight bacilli and six clostridia showed three highly conserved regions of which the most highly conserved was suggested to be the DNA-binding domain. Subsequent research illustrated that Spo0A directly controlled the switch from acidogenesis to solventogenesis in solvent-forming clostridia (Ravagnani et al., 2000).

The previously described strategies of gene and enzyme characterization and comparative analysis focus on single genes. Strategies aimed at manipulating more than one gene product have since evolved to provide further cellular improvements. These multigene improvements are easiest to implement by altering one gene that is responsible for orchestrating an aspect of global physiology involving many gene products. Transcription factors directly influence a great number of genes, thus demonstrating profound effects on the ability to modulate global cell processes. The gene encoding the transcriptional regulator Spo0A has been inactivated and overexpressed (Harris et al., 2002). The *spo0A*-inactivated strain was severely limited in the solvents it produced and showed significant sporulation defects. Overexpression of *spo0A* caused higher butanol titers, accelerated the rate of its sporulation developmental program, and increased the amount of spores produced. Importantly, the *spo0A* overexpression strain demonstrated increased tolerance and prolonged metabolism in response to butanol stress.

Transcription factors are not the only orchestrators of many gene products. Another elicitor for the synthesis of many related genes is the basic phenomenon of stress. A common stress that many clostridia must cope with is that posed by acids and solvents, especially alcohols, which are particularly detrimental to membrane integrity. Thus, increased expression of genes encoding heat shock proteins would be expected to help restructure the cellular milieu as well as adapt the cell to withstand increased solvent levels. Indeed, overexpression of *groESL* decreased the inhibitory effects of butanol challenge by 85% on *C. acetobutylicum* growth. Final solvent concentrations in cultures of the modified strain were significantly higher, likely because active metabolism lasted more than twice as long as the control (Tomas et al., 2003b). Therefore, increased accumulation of stress proteins, specifically GroEL and GroES, allowed *C. acetobutylicum* to tolerate higher solvent levels.

In order to further identify and isolate genes involved in regulating solvent tolerance, a screen of *C. acetobutylicum* cells containing plasmids representing a genomic library was performed. The library of recombinant cell strains was challenged with butanol either once or serially with increasing dosages for an acclimatization effect. One gene was identified in the single butanol challenge; it encodes a protein of unknown function, and cells bearing a subclone of the gene increased butanol tolerance by 13%. Many genes were enriched under the serial butanol stress test, four of which are transcription factors. One in particular, CAC1869, gave 81% increased butanol tolerance when subcloned and grew consistently to higher cell densities with a prolonged metabolism (Borden and Papoutsakis, 2007). This gene appears to allow the cell to adapt to longer production cycles of solvents. A library screen is exploratory in nature and can identify previously unsuspected genes. This strategy is complemented by the use of technically sophisticated technologies, such as DNA arrays or proteome analysis.

Because the predominant mechanism of gene regulation in microorganisms is at the transcriptional level, much can be gleaned from the transcriptional profiles of different strains under various conditions. In *C. acetobutylicum*, the DNA array strategy has been employed frequently so that now these data abound. The transcriptomes of *spo0A* overexpression and *spo0A* gene inactivation in *C. acetobutylicum* were contrasted in order to identify new genes affected by Spo0A, a known master regulator of sporulation and solvent production. Most of the genes that were differentially expressed between the two strains appear to be part of a general stress response related to spore and solvent production; they were genes of fatty acid metabolism, motility and chemotaxis, heat shock response, and cell division. The genes not part of a general stress response may have a role in increasing butanol tolerance. The putative butanol tolerance genes were expressed at higher levels immediately after butanol challenge in the *spo0A* overexpression strain. These genes are involved in butyryl-CoA synthesis, butyrate formation and assimilation, cell division, gyrase activity, DNA synthesis and repair, and fatty acid synthesis. Some gene families were primarily affected in one strain or the other. Only the *spo0A* overexpression strain, but not the *spo0A*-null strain, showed drastic decreases in glycolytic pathway genes. Likewise, in the *spo0A*-inactivated strain two Spo0A-dependent regulators were significantly changed, i.e., *abrB* and *sigF*, while these two genes showed little change when *spo0A* was overexpressed (Alsaker et al., 2004).

Transcriptional analysis of the *groESL* overexpression strain, which is more butanol tolerant, showed elevated expression of motility and chemotaxis genes and decreased expression of other stress response genes (Tomas et al., 2003b). The solvent tolerance and stress responses were then evaluated further by challenging the *C. acetobutylicum* strain overexpressing *groESL*, *C. acetobutylicum* 824(pGROE1), with low or high concentrations of butanol. The transcriptional profile showed many genes with similar expression patterns for *C. acetobutylicum* 824(pGROE1) and the control, indicating that these genes make up a general butanol stress response. Genes that increased in expression included those involved in stress responses, solvent formation, butyrate formation, butyryl-CoA biosynthesis, and specific genes encoding fructose bisphosphate aldolase, and a *kinA* homolog of *B. subtilis*. Downregulated genes were for fatty acid synthesis and glycolysis, and a few were for sporulation. However, a small sample of genes had expression patterns suggesting that these genes are associated with increased tolerance to butanol. These genes encode rare lipoprotein A, arginine transporter, and hemin permease. These expression data suggest that stress, even from the solvent itself, can further induce solventogenesis and related processes (Tomas et al., 2004).

C. acetobutylicum genome-wide gene expression patterns at the transition from exponential to stationary phase were examined by DNA arrays for novel genes with expression patterns that correlate with genes known to be up- or downregulated at this shift. In addition to the solvent-forming genes on pSOL1, the majority of other genes on the megaplasmid also showed an upregulation at the onset of solventogenesis. These data suggest that these additional genes may be involved in stationary-phase phenomena. Most sporulation genes show patterns similar to those in *B. subtilis*, with the exception of *sigE*. The expression of *sigE* in *C. acetobutylicum* occurs 7.5 h after detectable expression of *spo0A*, consistent with the lag time between the appearance of the swollen clostridial form and the detection of spores (Alsaker and Papoutsakis, 2005).

DNA array analysis has also implicated butyryl phosphate as a regulatory molecule of solventogenesis in *C. acetobutylicum*. In a butyrate kinase mutant, butyryl phosphate increases during exponential phase as do solvent formation and stress genes. This increase in butyryl phosphate corresponds to a downregulation of flagellar genes. Butyryl phosphate may regulate transcription factor activity by acting as a phosphodonor (Zhao et al., 2005).

In addition to the plentiful transcriptional data, a few proteomic studies have been performed with *C. acetobutylicum* as a strategy of determining which proteins are altered under various conditions. For a summary of the proteomic data see Table 1. Dramatic changes in protein accumulation patterns in *C. acetobutylicum* between acid production and solvent production phases were observed when samples of cells from the different growth phases were compared by proteomics. During solventogenesis, almost one-half of the proteins analyzed increased more than twofold while about one-quarter decreased to less than half, compared to the acidogenic proteome. Nine proteins induced during solventogenesis were identified, and one of those was previously known to be a solvent formation protein, acetoacetate decarboxylase. Other proteins synthesized in higher amounts during solventogenesis were three general stress proteins, two serine biosynthesis enzymes, and a seryl-tRNA synthetase (Schaffer et al., 2002).

A comparison of proteomes from wild-type, *spo0A* gene inactivation, and *spo0A* overexpression strains of *C. acetobutylicum* over various fermentation time points revealed protein changes, and among the 23 proteins identified were heat shock stress response, acid and solvent formation, or transcription and translation proteins. Some of the proteins that increased during wild-type fermentations were stress proteins, an acetone pathway enzyme, a glycogen regulator, and two uncharacterized proteins with a novel repetitive domain. Some proteins decreased during fermentation. Analysis of the *spo0A* modulated strains showed that six proteins were not detected in the *spo0A*-null strain, suggesting that these proteins require Spo0A, while *spo0A* overexpression affected the abundance of certain proteins involved in glycolysis, translation, heat shock stress response, and energy production (Sullivan and Bennett, 2006).

Changes in proteins in other clostridia have been studied and may have relevance to the physiology of the cell during solvent formation. For example, membrane protein accumulation in ethanol-adapted strains of *Clostridium thermocellum* was evaluated. More than one-half of the membrane proteins identified were differentially expressed, with most of those being downregulated in the ethanol-adapted strain compared to the control. A significant portion of the downregulated proteins were involved with carbohydrate transport and metabolism, whereas one-third of the upregulated proteins were chemotaxis and signal transduction proteins. These data suggest that membrane-associated proteins in the ethanol-adapted strain are either synthesized in lesser quantities or not properly incorporated into the cell membrane (Williams et al., 2007). *Clostridium difficile* produces many cell wall-associated virulence factors including cell wall protein Cwp66, flagella, and a high-molecular-weight surface layer protein. Its genome predicts many more cell wall-associated virulence factors, so proteomic analysis was undertaken in order to identify more virulence factors. In this study, 49 different proteins were identified and among them were two highly abundant surface layer proteins, flagellum

Table 1. Proteomic analyses summarized[a]

Species	Proteomes compared	No. identified[b]	Diff[c]	Inc[d], classification	Dec[d], classification	Reference
C. thermocellum	WT vs. ethanol-adapted membrane proteins	81	46/81	12/46 Chemotaxis Signal transduction Metal ion transport	34/46 10/34 Carbohydrate metabolism Carbohydrate transport	Williams et al., 2007
C. acetobutylicum ATCC 824	WT exponential vs. WT stationary	23	13/23	8/23 Stress Solvent induction Novel extracellular proteins Granulose metabolism	5/23 Glucose metabolism Energy production Acid/solvent core enzymes Translation	Sullivan and Bennett, 2006
	WT control vs. *spo0A* overexpression	23	7/23	4/23 Solvent induction Stress Energy production Acid/solvent core enzyme	3/23 Glucose metabolism Translation	
	WT control vs. *spo0A* null	23		17 proteins detected	6/23 proteins absent in *spo0A* null Novel extracellular proteins Solvent induction Stress Granulose metabolism	
C. acetobutylicum DSM 792	WT acidogenic vs. WT solventogenic	10	86/130	9/10 Stress Serine biosynthesis Serine transfer to tRNA Solvent induction Leucine biosynthesis Vitamin B6 biosynthesis	1/10 Glucose metabolism	Schaffer et al., 2002
C. difficile	Glycine cell wall protein extraction	50	12/50	HMW-surface layer protein 6 different paralogs of HMW-surface layer protein LMW-surface layer protein flagellum components		Wright et al., 2005
	Lysozyme cell wall protein extraction	69	45/69	HMW-surface layer protein 4 different paralogs of HMW-surface layer protein LMW-surface layer protein Cytoplasmic proteins		

[a]Abbreviations: WT, wild type; HMW, high molecular weight; LMW, low molecular weight.
[b] Number of proteins identified in *C. thermocellum* and *C. acetobutylicum* studies; number of protein spots identified in *C. difficile* study.
[c]Diff, differentially accumulated proteins between proteomes compared in *C. thermocellum* and *C. acetobutylicum* studies; number of unique proteins identified among the protein spots examined in *C. difficile* study.
[d]Inc or Dec, increased or decreased in relevant strains compared to wild-type or wild-type control strains; in *spo0A*-null strain, Inc refers to whether proteins were detected and Dec refers to whether proteins were absent at any time; in *C. difficile* study, only the classification of proteins is indicated, not whether there was an increase or a decrease.

components, and paralogs of the high-molecular-weight surface layer protein. These identified proteins may represent new virulence factors (Wright et al., 2005). While gene identification strategies can determine which candidate genes to manipulate for strain improvement, the methods needed to employ these genetic improvements are under development and are described below.

CURRENT METHODS FOR METABOLIC ENGINEERING IN *C. ACETOBUTYLICUM*

Great strides have been recently made toward manipulating the metabolic pathways of *C. acetobutylicum*. Strain characteristics have been improved by virtue of the genetic engineering and gene inactivation tools designed specifically for clostridia. The genetic engineering of clostridia required the development of basic molecular biological methods for gene cloning, expression, and transfer. Gene inactivation can be either random or targeted. Random mutagenesis with transposons has been used successfully. Replicative as well as nonreplicative plasmids have been used for targeted gene inactivation. The current status for reduction of expression of multiple genes is antisense RNA technology. Generally up to two gene products can be targeted. However, facile methods for the complete elimination of multiple gene products have been slow to develop because of the technical difficulty involved. For the next generation of metabolically engineered strains, methods to accumulate multiple genetic alterations will be necessary.

Some basic genetic tools for *C. acetobutylicum* have been developed. The *C. acetobutylicum* genome has been sequenced (Nolling et al., 2001). Cloning of *C. acetobutylicum* genes is fairly routine based on PCR methodology, and expression of these genes in *Escherichia coli* is often successful. Shuttle vectors and a reliable genetic transfer system, i.e., electroporation of methylated DNA, have been developed. Cloned genes of *C. acetobutylicum* have also been overexpressed on multicopy plasmids under either their native promoter or a strong promoter. For a general background on the molecular biology related to metabolic engineering of *C. acetobutylicum*, the reader is referred to a recent book covering many aspects of these organisms (Durre, 2005) and earlier articles by Woods (1995) and Mermelstein et al. (1994).

Single-gene inactivation may be random or targeted. Random mutagenesis has been shown in *C. acetobutylicum* and *C. difficile* with the use of a transposon. Transposon *916*, Tn*916*, was transferred from *E. coli* via conjugation to *C. acetobutylicum*, thus demonstrating transfer between a gram-negative and a gram-positive organism (Bertram et al., 1991). Tn*916* was further used to create *C. acetobutylicum* mutants deficient in protease activity (Sass et al., 1993) and granulose accumulation among other phenotypes related to solvent formation (Mattsson and Rogers, 1994). In *C. difficile*, Tn*916* shows a preference for particular sites in its genome which seem to be the native locations for a Tn*916*-like element (Wang et al., 2000). This transposon has been used for gene cloning and for targeted downregulation in *C. difficile*. A 1.1-kb fragment of toxin B was cloned into a toxin-deficient strain of *C. difficile* by mating with *B. subtilis* (Mullany et al., 1994). Tn*916* was used to downregulate an adhesion gene, *cwp66*, by antisense RNA within the DNA region of Tn*916*. The resultant *cwp66* downregulation Tn*916* construct was inserted in the *B. subtilis* genome such that it was transferred when *B. subtilis* was mated, with the modified Tn*916* being then integrated into the recipient *C. difficile* genome (Roberts et al., 2003).

In many microorganisms, such as *E. coli*, gene inactivations are conducted using nonreplicative plasmids or suicide plasmids so that there is no need to cure the plasmid from the transformant strain afterwards. That method has found some limited use in *C. acetobutylicum*, and four genes were inactivated using such a method: the primary *adhE* gene (Green and Bennett, 1996), the primary *buk* gene (Green et al., 1996), *pta* gene (Green et al., 1996), and *solR* genes (Nair et al., 1999). Because the transformation efficiency is significantly lower in *C. acetobutylicum* than in *E. coli*, targeted gene inactivation using replicative plasmids with the requisite curing has been shown to be an alternative strategy. The gene encoding the master transcriptional regulator Spo0A was inactivated in such a fashion (Harris et al., 2002).

Antisense RNA technology has been successfully employed in *C. acetobutylicum* for directed gene expression downregulation. The antisense gene fragment is cloned onto a replicative plasmid under control of a strong clostridial promoter. This technology was first used in *C. acetobutylicum* to individually downregulate the butyrate synthesis pathway genes *ptb* and *buk* (Desai and Papoutsakis, 1999). Antisense RNA has since been used to downregulate the acetone pathway genes *adc*, *ctfA*, and *ctfB* (Tummala et al., 2003a, 2003b), the regulator *abrB310* (Scotcher et al., 2005), the sporulation gene *spoIIE* (Scotcher and Bennett, 2005), and the hydrogenase gene *hydA* (Watrous et al., 2003). These fundamental gene manipulation techniques have been applied to altering the pattern of metabolites, especially focusing on making butanol represent a larger fraction of the solvents, because this change would be an economically useful goal of metabolic engineering of *C. acetobutylicum*.

SPECIFIC APPLICATIONS OF METABOLIC ENGINEERING TO SOLVENT PRODUCTION AND TOLERANCE

Acid and Solvent Enzymes

Strategies to increase the butanol fraction of solvents in *C. acetobutylicum* have addressed manipulating expression levels of genes for acid and solvent production, regulators, heat shock proteins, sporulation, and cell membrane synthesis. Early metabolic engineering efforts capitalized on the discovery of solvent-related genes to manipulate *C. acetobutylicum* pathways. These genes were introduced in either wild-type or solvent-deficient strains, and the effect on acid and solvent production was assayed. When the genes for acid or solvent formation were overexpressed in the wild type, little effect was noted, apparently due to the high natural amount of these activities present in the cell or regulatory processes.

The introduction of plasmids bearing genes of acid pathways showed only a modest effect on the acid content (Walter et al., 1994). This study suggested that increasing the levels of certain enzymes may not affect the overall metabolite profile if they are already in adequate amount and the network is limited by other factors (Walter et al., 1994). In situations where new genes were introduced into a cell not already having the gene, an unusual pattern of solvents was obtained. For example, *adhE* was expressed from a plasmid, pCAAD, in *C. acetobutylicum* M5 (Clark et al., 1989), which is not capable of solvent production because it has lost the pSOL1 plasmid. *C. acetobutylicum* M5(pCAAD) cultures grown at pH 5.0 produced nearly identical levels of ethanol and acetate but much less butyrate (99 mM versus 170 mM) and much more butanol (84 mM versus 0 mM) than the control strain *C. acetobutylicum* M5 (pCCL), in which pCCL is the control plasmid (Nair and Papoutsakis, 1994). This experiment showed that butanol could be produced without acetone formation in a suitable strain.

A combinatorial approach modified the expression of a gene essential for solvent formation (*adhE*) as well as another metabolic gene to alter the pathway fluxes and product formations. This experiment overexpressed the major alcohol-forming gene, *adhE*, while simultaneously reducing the level of expression of the CoA transferase gene with antisense RNA. This strain was designated 824(pAADB1). In a strain containing a plasmid bearing the antisense construct alone, 824 (pCTFB1AS), acetone and butanol levels were reduced considerably. The addition of *adhE* overexpression plas-

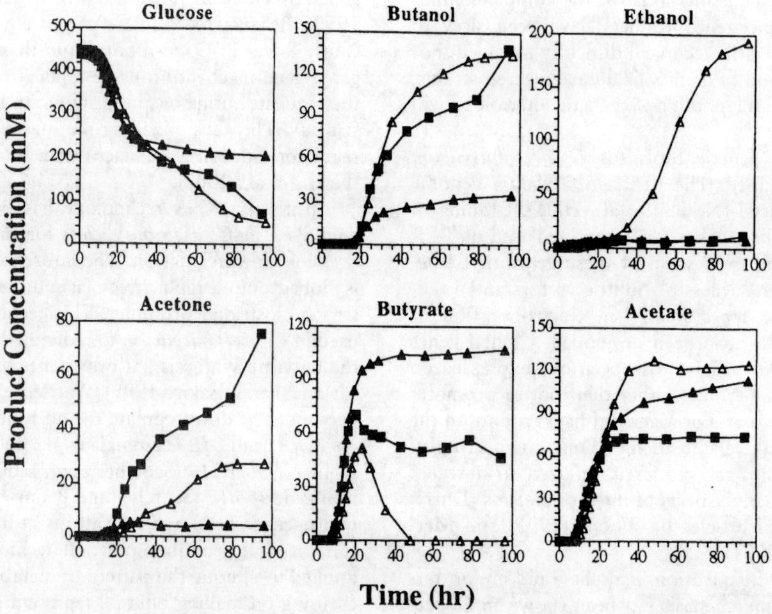

Figure 3. Fermentation kinetics of 824(pSOS95del), 824(pCTFB1AS), and 824(pAADB1). (A) Glucose, acid, and solvent profiles. The name of each profile in 824(pSOS95del) (■), 824(pCTFB1AS) (▲), and 824(pAADB1) (△) is indicated above each graph. Reprinted from the *Journal of Bacteriology* (Tummala et al., 2003a) with permission of the publisher.

mid showed high production of butanol and ethanol with lower formation of acetone (Fig. 3). Furthermore, it was considered that in the 824(pAADB1) strain, butanol production likely would have been higher had flux to butyryl-CoA not been limiting due to butyrate depletion during fermentation. This result shows that a considerably modified pattern of metabolites can be achieved with an overall high maintenance of alcohol-based products (Tummala et al., 2003a).

The inactivation of the gene encoding a core acid production enzyme has demonstrated the highly encouraging solvent production capabilities of such strains. Butyrate kinase, encoded by *buk*, catalyzes the formation of butyrate from butyryl-phosphate via the transfer of the phosphate moiety to ADP. *buk* is found downstream of *ptb* in a bicistronic operon under transcriptional control of the *ptb* promoter (Walter et al., 1993). In strain PJC4BK, where *buk* had been inactivated, butyrate kinase activity decreased by approximately 80%, resulting in a corresponding decrease in maximum butyrate concentrations by 66% relative to the wild type (Green et al., 1996). Butyrate kinase activity and butyrate production were restored when PJC4BK was complemented with a plasmid-borne copy of *buk* (Green and Bennett, 1998).

Further fermentations showed that strain PJC4BK can produce extremely high solvent concentrations, specifically ethanol (57 mM; 260% increase over the wild type), and butanol (225 mM; 42% increase over the wild type) (Fig. 4). When transformed with a vector overexpressing *adhE*, maximum ethanol production was further elevated to 98 mM, with butanol unchanged. Butanol concentrations of 225 mM equate to 16.7 g/liter, an increase of 3.7 g/liter more than the previously established limit for butanol tolerance (Jones and Woods, 1986; Harris et al., 2000).

In addition to improved final solvent concentrations, these studies reveal that metabolic flux analysis is critical to understanding solvent production in *C. acetobutylicum*. In strain PJC4BK, butanol production is initiated when low butyrate concentrations (1 mM) are detected, suggesting that high levels of undissociated butyrate are not needed for solventogensis (Harris et al., 2000). In strain PJC4BK, a shift in acid formation fluxes towards acetate production was observed, whereas in strain 824(pRD4), where *buk* expression is repressed using a vector expressing antisense RNA to *buk*, a shift in acid flux was not observed, yet both strains ultimately exhibited a similar increase in butanol production (Desai et al., 1999).

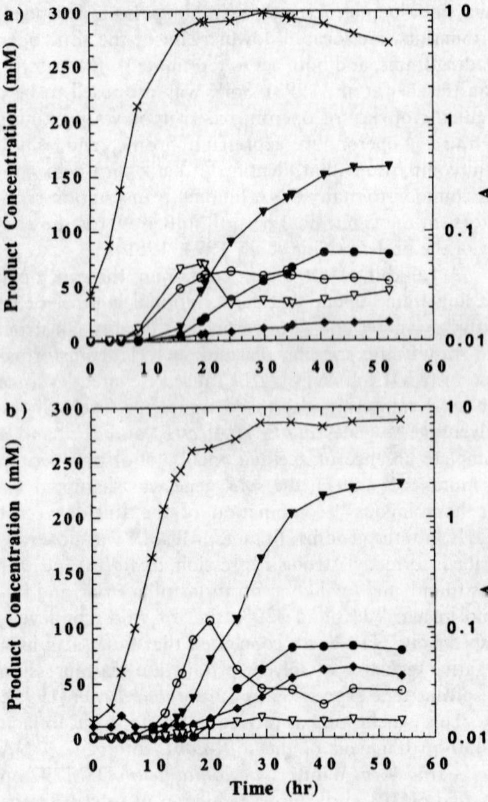

Figure 4. Fermentation profiles of WT (a) and PJC4BK (b) at pH 5.0. Shown are optical density A_{600} (–X–) and product concentrations of acetate (–○–), acetone (–●–), butyrate (–△–), butanol (–▲–), and ethanol (–◆–). The data at pH 5.0 show a more dramatic effect than those published earlier for pH 5.5 cultures by Green et al. (1996). Reprinted from *Biotechnology and Bioengineering* (Harris et al., 2000) with permission of the publisher.

Regulators

In addition to increasing or decreasing gene expression levels of specific acid and solvent genes, the manipulation of the regulators SolR and Spo0A have also altered the solvent profiles. The function of SolR is disputed, SolR having been originally described as a transcriptional repressor (Nair et al., 1999) and further studies suggesting it is a glycosylating modulating enzyme (Thormann and Durre, 2001). Spo0A is a fairly well studied master transcription factor known to activate or repress solvent-related genes. The 957-bp open reading frame, designated *solR* (originally known as ORF1 in strain ATCC 824 and ORF5 in strain DSM792), is located 663 bp upstream of *adhE* and encodes a 319-amino-acid, 37-kDa protein (Fischer et al., 1993; Nair et al., 1994). The gene *solR* is expressed from a single transcription start site, with elevated levels of the *solR* transcript observed in cultures prior to the transition from acidogenic to solventogenic growth.

Two stem-loop structures, resembling rho-independent terminators, are located downstream of the *solR* open reading frame, and both act to terminate *solR* transcription (Fischer et al., 1993). SolR was proposed to be a regulator of the *sol* operon, based upon its proximity to the *sol* operon, its expression profile, and amino acid comparisons that identified a helix-turn-helix motif, common to many DNA-binding transcription regulators, in the C-terminal half of SolR towards the center of the protein (Nair et al., 1994, 1999).

A 1.6-kb DNA fragment containing the *solR* open reading frame under the transcriptional control of its native promoter and terminators was used to construct the *solR* overexpression plasmid, pCO1. Transformation of pCO1 into ATCC 824 caused elevated levels of the *solR* transcript, decreased transcript levels of the solventogenic genes *adhE*, *ctfA*, *ctfB*, and *adc*, and a complete absence of acetone and butanol production. In mutants B and H, the *solR* gene was disrupted via the homologous recombination of the suicide vector pO1X into the genome. In mutant B *solR* was not transcribed correctly, strong expression of *adhE* and *adc* was found, and final concentrations of acetone and butanol reached 210 and 320% that of wild-type levels, respectively. Nair et al. concluded that SolR acts as a negative regulator of solvent production via repression of solvent gene expression in *C. acetobutylicum* (1999).

This conclusion has proven to be incorrect. In fact, a 430-bp fragment of the *solR-adhE* intergenic DNA can be transformed into *C. acetobutylicum* DSM792 on plasmid pIMP1 and cause a reduction in solvent levels, possibly by the titrating out of an unknown factor essential for the expression of solvent genes (Thormann et al., 2002). When either wild-type strains or mutants B and H of *C. acetobutylicum* were complemented with a plasmid harboring either an intact or partially deleted copy of *solR*, solvent production was eliminated in the wild-type background and severely decreased in the mutant backgrounds with both the full and partial *solR* gene complementation constructs. These data indicate that the *solR* open reading frame is not required to elicit this phenotype (Scotcher et al., 2003). The molecular mechanisms responsible for high solvent phenotype in mutants B and H remain unclear (Scotcher et al., 2003).

When overexpressed on plasmid pORF5H in DSM792, SolR was found to localize to the extracellular membrane of the cell and the glycosylation state of several exoproteins was significantly altered, yet little effect on solvent production was observed. Purified SolR exhibited no DNA binding activity in vitro, and supported by amino acid comparisons showing significant sequence similarity between SolR and O-glycosylating enzymes, it was concluded that SolR plays a role in protein glycosylation/deglycosylation (Thormann and Durre, 2001).

Irrespective of the role of SolR, mutants B and H have proven to be valuable strains producing high levels of solvent. In fermentation studies, mutant H was shown to produce higher levels of solvents, utilizing glucose at a higher rate as a consequence. When transformed with a plasmid expressing *adhE*, solvent levels were even higher than in mutant H alone (Harris et al., 2001). If strains of *C. acetobutylicum* are to be developed for industrial use, these mutants could form a foundation upon which further metabolic engineering might be performed.

The master transcriptional regulator Spo0A is another regulator that when perturbed alters solvent profiles. The *spo0A* locus was originally described in 1978 in *B. subtilis* (Trowsdale et al., 1978), and the critical role of Spo0A in the initiation of the sporulation cascade in gram-positive bacteria has been well elucidated (Stephenson and Lewis, 2005). A homolog to *spo0A* is present at position 2071 on the chromosome of *C. acetobutylicum* (Nolling et al., 2001) and has been shown to be required for both the initiation of sporulation and the production of solvents (Brown et al., 1994; Harris et al., 2002; Durre and Hollergschwandner, 2004).

The Spo0A binding motif, known as the 0A box, is found upstream of several solventogenic genes, including *adhE* and *adc*, in *C. acetobutylicum* and *Clostridium beijerinckii* (Ravagnani et al., 2000; Thormann et al., 2002). In *C. beijerinckii* it was shown that Spo0A binds to, and regulates transcription from, the promoter regions of *adc* and *ptb* via the 0A box (Ravagnani et al., 2000).

In *C. acetobutylicum*, a *spo0A*-disrupted strain designated SKO1 was formed. Analysis of metabolites of cultures of SKO1 showed that acetone production was almost eliminated, with butanol levels decreased to only 8% of wild-type levels. SKO1 cultures grew 31% slower than wild-type cultures, and when entering sporulation, SKO1 cells formed long filaments of connected rods and failed to sporulate. A *spo0A*-overexpresser strain designated 824(pMSPOA) exhibited 20% higher final solvent concentrations, grew faster and to a 50% greater cell density, and transitioned to sporulation earlier than the control strain. It was concluded that *spo0A* overexpression actually accelerates sporulation at the expense of solvent production (Harris et al., 2002).

Large-scale proteome and transcriptome analyses have revealed that Spo0A also regulates the expression of genes involved in motility and chemotaxis, electron transport, sugar metabolism, butanol tolerance, and heat shock protein induction (Tomas et al., 2003a; Alsaker et al., 2004; Sullivan and Bennett, 2006). Any changes that occur in these processes may in turn have an effect on cellular metabolism including energetics and carbon flux, which in turn could influence solvent production.

The dogmatic model of a fermentation culture of *C. acetobutylicum* had been of a linear process, with acidogenic exponential growth followed by stationary solventogenic growth, which finally transitioned into spore production. It is now apparent that sporulation and solventogenesis are simultaneously initiated in response to Spo0A activation and that any separation of the processes, which might be desirable in the development of an industrially useful strain of *C. acetobutylicum*, must occur genetically downstream of *spo0A*. Experiments with SKO1 and 824(pMSPOA) have allowed many other important areas of the *C. acetobutylicum* life cycle to be investigated and may suggest themes to be manipulated to improve strain development and solvent production.

Heat Shock Proteins

In addition to altering the level of expression for genes encoding acid and solvent enzymes as well as regulators, increasing expression of heat shock genes has been explored as a means to maintain protein stability when cultures are faced with high butanol titers. The rationale is that if the metabolic and solventogenic enzymes could be stabilized, then perhaps butanol production could be extended past its normal inhibitory point. The bicistronic *groESL* operon, comprised of *groES* and *groEL*, was originally identified in bacteriophage-transfected *E. coli* mutants (Tilly et al., 1981). The operon *groESL* is expressed under normal physiological conditions but is strongly upregulated in response to heat shock and other stress-related stimuli to the cell (Narberhaus and Bahl, 1992). GroEL and GroES form an essential chaperonin that assists in ATP-dependent protein folding events (Xu and Sigler, 1998).

In *C. acetobutylicum*, at least 15 heat shock proteins, including GroEL and GroES, have been identified (Bahl et al., 1995). The genes *groESL* can be induced by a heat shock of 42°C, and elevated levels of several heat shock proteins including GroEL can be detected during late exponential growth at the transition from acidogenic to solventogenic growth (Pich et al., 1990; Bahl et al., 1995). In strain M5, which does not undergo the transition to solventogenesis, *groEL* and *groES* expression is significantly reduced during late exponential growth (Tomas et al., 2003a).

As solventogenic growth progresses, *C. acetobutylicum* fermentations are subjected to increasing chemical stress, particularly by the accumulating butanol concentrations. A butanol concentration of approximately 13 g/liter is normally sufficient to cause a complete cessation of growth and solvent production (Bowles and Ellefson, 1985; Jones and Woods, 1986).

In fermentations of strain 824(pGROE1), in which the *groESL* operon is overexpressed, final acetone and butanol concentrations were 54 and 32% greater, respectively, than in wild-type fermentations. When challenged by the exogenous addition of butanol during mid-exponential growth, the growth of strain 824 (pGROE1) was inhibited up to 85% less than the growth of the corresponding control strain. The active metabolism of 824(pGROE1) lasted 2.5 times longer than that of the control strain (Tomas et al., 2003b). Transcriptome analysis revealed elevated expression of solventogenic and stress-protein genes, in both 824(pGROE1) and the control strain, in a butanol dose-dependent manner (Tomas et al., 2004).

A similar response of extended active metabolism and increased butanol tolerance was observed in the *spo0A* overexpression strain 824(pMSPOA) when subjected to butanol challenge (Alsaker et al., 2004). It is clear that improved protection from the damaging effects of butanol, as seen with manipulating heat shock genes and transcriptional regulators, is critical to the development of an industrially viable strain of *C. acetobutylicum*.

Sporulation

The strategy to increase butanol production has been addressed by separating sporulation from solvent production. In *B. subtilis*, the expression of *spoIIE* is sporulation specific and is upregulated by activated Spo0A (York et al., 1992). SpoIIE is localized to the septum between mother cell and forespore and dephosphorylates the anti-anti-sigma factor SpoIIAA to facilitate the derepression of σ^F in the forespore. The correct localization and phosphatase activity of SpoIIE are both required to allow sporulation to continue through Stage II (Arigoni et al., 1996; Feucht et al., 1999; Lucet et al., 2000).

A homolog to *spoIIE* has been identified at position 3205 on the chromosome of *C. acetobutylicum* (Nolling et al., 2001). Expression of *spoIIE* was shown to be restricted to mid-to-late solventogenesis and required an active copy of *spo0A*. In small-batch cultures of the *spoIIE*-knockdown strain 824(pASspo), maximum acetone and butanol levels were shown to be 50 and 132% greater, respectively, than in the control strain. After 140 h, a significantly reduced number of sporulating cells were found in cultures of strain 824(pASspo), with a significant proportion exhibiting morphological defects. It was concluded that decreased expression of *spoIIE* did not directly affect solventogenesis but caused a delay in sporulation, allowing the window for solvent production to be extended (Scotcher and Bennett, 2005).

In more optimally pH-controlled batch fermentations of strain 824(pASspo), solvent production did not differ from that in control strains, although the morphological defects were still observed (E. T. Papoutsakis, personal communication). This suggests that the extension of solventogenesis via sporulation delay is notably effective under nonoptimal fermentation conditions. Nevertheless, strains in which sporulation is delayed, thus prolonging the solventogenic phase, may be considered to confer significant economic advantages to industrial-scale fermentations and may widen the operational window for high productivity.

Cell Membrane Synthesis Genes

Among the strategies discussed to increase the butanol fraction of solvents in *C. acetobutylicum*, that is, by manipulating expression levels of genes for acid and solvent production, regulators, heat shock proteins, and sporulation, altering gene expression for cell membrane synthesis would be expected to directly address solvent tolerance as well as production capabilities. Membranes of cells grown under solvent stress conditions showed an increased amount of fatty acids containing a cyclopropane modification and some increase in chain length (Baer and Blaschek, 1987; MacDonald and Goldfine, 1991). These changes would be expected to make the membrane more rigid and less perturbed by organic solvents. Zhao et al. (2003) cloned and overexpressed the cyclopropane fatty acid synthase gene (*cfa*) of *C. acetobutylicum*. The expression of the clostridial *cfa* gene in *E. coli* complemented a *cfa*-deficient strain in terms of fatty acid composition and growth rate under solvent stress. Overexpression of the *cfa* gene in the wild-type *C. acetobutylicum* and in a butyrate kinase-deficient strain showed that the construct was functional in *C. acetobutylicum* and increased the cyclopropane fatty acid content of early-log-phase cells. The acid and butanol resistance of these early-stage cells was also increased based on growth measurements. Solvent production in the *cfa*-overexpressing strain was decreased, suggesting that too early expression or overexpression of *cfa* results in changes in membrane properties that dampen the full induction of solventogenesis.

FUTURE TARGETS FOR METABOLIC ENGINEERING

A number of targets for future metabolic engineering have been found by the careful examination of transcriptional programs (Cheng et al., submitted for publication). Among interesting genes and patterns are several iron transport and regulatory genes that show higher expression during the transition to the solvent production stage. It is noted from earlier work (Peguin and Soucaille, 1995) that iron limitation increased solventogenesis. Of possible interest is the expression of genes of the biosynthetic pathways for hydrophobic amino acids and glycosyltransferases during the solvent phase. As is the case with *B. subtilis*, a number of sigma factors are found in *C. acetobutylicum*, and analysis of the expression and antisense experiments with some of these has indicated that reduced expression of certain sigma-factor genes can modulate sporulation and perhaps be useful in manipulating the solvent-producing state. Other regulators can potentially be identified by this approach, enabling a more global approach to metabolic engineering of *C. acetobutylicum*.

METABOLIC ENGINEERING OF OTHER FUEL- AND CHEMICAL-RELATED PRODUCTS

1,3-Propanediol

Among other compounds that have been produced by clostridia and are of biotechnological interest are specifically 1,3-propanediol (1,3-PD) and hydrogen. The compound 1,3-PD can be formed from glycerol by a variety of organisms, and engineering of microbes for its production has been the subject of work by DuPont over the last decade. The study of the genes encoding the 1,3-PD operon of *Clostridium butyricum* VPI1718 (Raynaud et al., 2003) revealed three genes, *dhaB1*, *dhaB2*, and *dhaT*. When cloned and expressed in *E. coli* these genes enable the cells to produce 1,3-PD. *dhaB1* and *dhaB2* encode a new type of glycerol dehydratase and its activator protein. This enzyme, unlike most glycerol dehydratases, is not vitamin B_{12} dependent and thus could reduce the cost of a biological process for the production of 1,3-PD. Further work involved the metabolic engineering of *C. acetobutylicum*. The 1,3-PD pathway from *C. butyricum* was introduced on a plasmid (pSPD5) into a mutant of *C. acetobutylicum*, DG1, that was no longer capable of solvent formation (Gonzalez-Pajuelos et al., 2005). The recombinant strain grew well, and chemostat cultures of this strain grown on glucose alone produced acids (acetate, butyrate, and lactate) and a high level of hydrogen. However, when it was grown on glycerol, 1,3-PD became the major product. In a fed-batch culture, the engineered *C. acetobutylicum* DG1(pSPD5) strain was able to produce 1,3-PD at a higher concentration (1,104 mM) and productivity than the natural producer *C. butyricum* VPI 3266. This strain could be used in long-term continuous production of 1,3-PD with a high volumetric productivity (3 g liter^{-1} h^{-1}). The physiological behavior of the recombinant *C. acetobutylicum* DG1(pSPD5) strain was also compared with that of the natural 1,3-PD producer *C. butyricum* (Gonzalez-Pajuelo et al., 2006).

Hydrogen

Efforts have been made to alter the production of hydrogen in clostridium species for biotechnology purposes. For example, it would be useful to direct hydrogen production from biomass, for fuel cell applications, to maintain increased reductant capacity for the improved production of solvents, and to improve the capacity of cells to reduce exogenous molecules. Examples of genetic engineering directed toward affecting the level of functional hydrogenase include the efforts to overexpress hydrogenase from various species in clostridium. The gene for the Fe-hydrogenase has been studied (Gorwa et al., 1996), and this and other hydrogenase genes have been overexpressed in *C. acetobutylicum* (Girbal et al., 2005). Overexpression of hydrogenase genes in other species of *Clostridium* has also been reported (Morimoto et al., 2005). The [Fe]-hydrogenase gene (*hydA*) was cloned from *Clostridium paraputrificum* M-21 and placed into a shuttle vector, pJIR751, used with *Clostridium perfringens*, and expressed in *C. paraputrificum*. Hydrogen gas productivity of the recombinant strain increased up to 1.7-fold compared with the parental *C. paraputrificum*. In the recombinant, overexpression of *hydA* abolished lactic acid production and increased acetic acid production by consumption of NADH, which then was apparently routed through the ferredoxin-NADH oxidoreductase and made available for hydrogen formation rather than for reduction of pyruvate to lactate.

In an opposite strategy, the function of hydrogenase was decreased in *C. acetobutylicum* by using an antisense approach (Watrous et al., 2003). The role of Fe-hydrogenase on the reduction of 2,4,6-trinitrotoluene (TNT) in *C. acetobutylicum* was evaluated. Hydrogenase expression levels were altered in various strains, and the specific TNT reduction rates were examined. The level of hydrogenase correlated with the organism's ability to reduce TNT. Strains exhibiting underexpression of hydrogenase produced slower rates of reduction of TNT.

In addition to controlling the levels of hydrogenase itself, hydrogenase activity has been modulated by altering another metabolic gene. It was observed that hydrogenase activity increased by 50% in a mutant altered by disruption of the acetate kinase gene (*ack*) of *Clostridium tyrobutyricum* (Liu et al., 2006).

CONCLUSION

The recent history of metabolic engineering in *C. acetobutylicum* serves as an excellent case study for microbial metabolic engineering in general. Traditional methods involving the knockdown or overexpression of core metabolic genes can achieve phenotypic improvement of solvent production.

However, the studies detailed in this review reflect an increased level of complexity in the process of metabolic engineering and recognize that attempts to improve the desired phenotype must be accompanied by a deeper understanding of the organism as a whole if success in strain engineering is to be achieved.

This is illustrated by findings that key transcription factors such as Spo0A, sporulation, heat shock proteins, fatty acid synthesis, metabolic flux, and even undetermined factors as exemplified by the *solR* studies can all influence solvent production. We anticipate that an engineered strain of *C. acetobutylicum*, combining aspects of all factors that contribute towards elevated solvent production, could contribute to the economic viability of *C. acetobutylicum* fermentations on an industrial scale.

Acknowledgments. We recognize the support of the U.S. Department of Agriculture grants 00-35504-9269 and 2006-35504-17294 and National Science Foundation grant BES-0418289.

REFERENCES

Aiba, S., T. Imanaka, and H. Tsunekawa. 1980. Enhancement of tryptophan production by *Escherichia coli* as an application of genetic-engineering. *Biotechnol. Lett.* 2:525–530.

Alper, H., K. Miyaoku, and G. Stephanopoulos. 2005. Construction of lycopene-overproducing *E. coli* strains by combining systematic and combinatorial gene knockout targets. *Nat. Biotechnol.* 23:612–616.

Alsaker, K. V., and E. T. Papoutsakis. 2005. Transcriptional program of early sporulation and stationary-phase events in *Clostridium acetobutylicum*. *J. Bacteriol.* 187:7103–7118.

Alsaker, K. V., T. R. Spitzer, and E. T. Papoutsakis. 2004. Transcriptional analysis of spo0A overexpression in *Clostridium acetobutylicum* and its effect on the cell's response to butanol stress. *J. Bacteriol.* 186:1959–1971.

Arigoni, F., L. Duncan, S. Alper, R. Losick, and P. Stragier. 1996. SpoIIE governs the phosphorylation state of a protein regulating transcription factor sigma F during sporulation in *Bacillus subtilis*. *Proc. Natl. Acad. Sci. USA* 93:3238–3242.

Baer, S. H., and H. P. Blaschek. 1987. Effect of butanol challenge and temperature on lipid composition and membrane fluidity of butanol-tolerant *Clostridium acetobutylicum*. *Appl. Environ. Microbiol.* 53:2854–2861.

Bahl, H., H. Muller, S. Behrens, H. Joseph, and F. Narberhaus. 1995. Expression of heat shock genes in *Clostridium acetobutylicum*. *FEMS Microbiol. Rev.* 17:341–348.

Bahl, H., and P. Durre. 2001. *Clostridia—Biotechnology and Medical Applications*. Wiley-VCH, New York, NY.

Bartel, P. L., C. B. Zhu, J. S. Lampel, D. C. Dosch, N. C. Connors, W. R. Strohl, J. M. Beale, Jr., and H. G. Floss. 1990. Biosynthesis of anthraquinones by interspecies cloning of actinorhodin biosynthesis genes in streptomycetes: clarification of actinorhodin gene functions. *J. Bacteriol.* 172:4816–4826.

Bertram, J., M. Stratz, and P. Durre. 1991. Natural transfer of conjugative transposon Tn*916* between gram-positive and gram-negative bacteria. *J. Bacteriol.* 173:443–448.

Borden, J. R., and E. T. Papoutsakis. 2007. Dynamics of genomic-library enrichment and identification of solvent-tolerance genes in *Clostridium acetobutylicum*. *Appl. Environ. Microbiol.* 73:3061–3068.

Bowles, L. K., and W. L. Ellefson. 1985. Effects of butanol on *Clostridium acetobutylicum*. *Appl. Environ. Microbiol.* 50:1165–1170.

Boynton, Z. L., G. N. Bennett, and F. B. Rudolph. 1994. Intracellular concentrations of coenzyme A and its derivatives from *Clostridium acetobutylicum* ATCC 824 and their roles in enzyme regulation. *Appl. Environ. Microbiol.* 60:39–44.

Brown, D. P., L. Ganova-Raeva, B. D. Green, S. R. Wilkinson, M. Young, and P. Youngman. 1994. Characterization of spo0A homologues in diverse *Bacillus* and *Clostridium* species identifies a probable DNA-binding domain. *Mol. Microbiol.* 14:411–426.

Buckland, B. C., S. W. Drew, N. C. Connors, M. M. Chartrain, C. Lee, P. M. Salmon, K. Gbewonyo, W. Zhou, P. Gailliot, R. Singhvi, R. C. Olewinski, Jr., W. J. Sun, J. Reddy, J. Zhang, B. A. Jackey, C. Taylor, K. E. Goklen, B. Junker, and R. L. Greasham. 1999. Microbial conversion of indene to indandiol: a key intermediate in the synthesis of CRIXIVAN. *Metab. Eng.* 1:63–74.

Cameron, D. C., N. E. Altaras, M. L. Hoffman, and A. J. Shaw. 1998. Metabolic engineering of propanediol pathways. *Biotechnol. Prog.* 14:116–125.

Cary, J. W., D. J. Petersen, E. T. Papoutsakis, and G. N. Bennett. 1988. Cloning and expression of *Clostridium acetobutylicum* phosphotransbutyrylase and butyrate kinase genes in *Escherichia coli*. *J. Bacteriol.* 170:4613–4618.

Cary, J. W., D. J. Petersen, E. T. Papoutsakis, and G. N. Bennett. 1990. Cloning and expression of *Clostridium acetobutylicum* ATCC 824 acetoacetyl-coenzyme A:acetate/butyrate:coenzyme A transferase in *Escherichia coli*. *Appl. Environ. Microbiol.* 56: 1576–1583.

Cheng, N., S. W. Jones, B. Tracey, C. J. Paredes, R. Sillers, and E. T. Papoutsakis. The transcriptional program of clostridial sporulation. Submitted for publication.

Clark, S. W., G. N. Bennett, and F. B. Rudolph. 1989. Isolation and characterization of mutants of *Clostridium acetobutylicum* ATCC 824 deficient in acetoacetyl-coenzyme A:acetate/butyrate:coenzyme A transferase (EC 2.8.3.9) and in other solvent pathway enzymes. *Appl. Environ. Microbiol.* 55:970–976.

Colon, G. E., M. S. Jetten, T. T. Nguyen, M. E. Gubler, M. T. Follettie, A. J. Sinskey, and G. Stephanopoulos. 1995a. Effect of inducible *thrB* expression on amino acid production in *Corynebacterium lactofermentum* ATCC 21799. *Appl. Environ. Microbiol.* 61: 74–78.

Colon, G. E., T. T. Nguyen, M. S. Jetten, A. J. Sinskey, and G. Stephanopoulos. 1995b. Production of isoleucine by overexpression of ilvA in a *Corynebacterium lactofermentum* threonine producer. *Appl. Microbiol. Biotechnol.* 43:482–488.

Desai, R. P., L. M. Harris, N. E. Welker, and E. T. Papoutsakis. 1999. Metabolic flux analysis elucidates the importance of the acid-formation pathways in regulating solvent production by *Clostridium acetobutylicum*. *Metab. Eng.* 1:206–213.

Desai, R. P., and E. T. Papoutsakis. 1999. Antisense RNA strategies for metabolic engineering of *Clostridium acetobutylicum*. *Appl. Environ. Microbiol.* 65:936–945.

Durre, P., and C. Hollergschwandner. 2004. Initiation of endospore formation in *Clostridium acetobutylicum*. *Anaerobe* 10:69–74.

Durre, P. 2005. *CRC Handbook of Clostridia*. CRC Press, Taylor & Francis, Boca Raton, FL.

Feucht, A., R. A. Daniel, and J. Errington. 1999. Characterization of a morphological checkpoint coupling cell-specific transcription to septation in *Bacillus subtilis*. *Mol. Microbiol.* 33:1015–1026.

Fischer, R. J., J. Helms, and P. Durre. 1993. Cloning, sequencing, and molecular analysis of the *sol* operon of *Clostridium acetobutylicum*, a chromosomal locus involved in solventogenesis. *J. Bacteriol.* 175:6959–6969.

Girbal, L., G. von Abendroth, M. Winkler, P. M. Benton, I. Meynial-Salles, C. Croux, J. W. Peters, T. Happe, and P. Soucaille. 2005. Homologous and heterologous overexpression in *Clostridium acetobutylicum* and characterization of purified clostridial and algal Fe-only hydrogenases with high specific activities. *Appl. Environ. Microbiol.* 71:2777–2781.

Gonzalez-Pajuelo, M., I. Meynial-Salles, F. Mendes, J. C. Andrade, I. Vasconcelos, and P. Soucaille. 2005. Metabolic engineering of *Clostridium acetobutylicum* for the industrial production of 1,3-propanediol from glycerol. *Metab. Eng.* 7:329–336.

Gonzalez-Pajuelo, M., I. Meynial-Salles, F. Mendes, P. Soucaille, and I. Vasconcelos. 2006. Microbial conversion of glycerol to 1,3-propanediol: physiological comparison of a natural producer, *Clostridium butyricum* VPI 3266, and an engineered strain, *Clostridium acetobutylicum* DG1(pSPD5). *Appl. Environ. Microbiol.* 72:96–101.

Gorwa, M. F., C. Croux, and P. Soucaille. 1996. Molecular characterization and transcriptional analysis of the putative hydrogenase gene of *Clostridium acetobutylicum* ATCC 824. *J. Bacteriol.* 178:2668–2675.

Green, E. M., and G. N. Bennett. 1996. Inactivation of an aldehyde/alcohol dehydrogenase gene from *Clostridium acetobutylicum* ATCC 824. *Appl. Biochem. Biotechnol.* 57-58:213–221.

Green, E. M., and G. N. Bennett. 1998. Genetic manipulation of acid and solvent formation in *Clostridium acetobutylicum* ATCC 824. *Biotechnol. Bioeng.* 58:215–221.

Green, E. M., Z. L. Boynton, L. M. Harris, F. B. Rudolph, E. T. Papoutsakis, and G. N. Bennett. 1996. Genetic manipulation of acid formation pathways by gene inactivation in *Clostridium acetobutylicum* ATCC 824. *Microbiology* 142:2079–2086.

Grupe, H., and G. Gottschalk. 1992. Physiological events in *Clostridium acetobutylicum* during the shift from acidogenesis to solventogenesis in continuous culture and presentation of a model for shift induction. *Appl. Environ. Microbiol.* 58:3896–3902.

Harris, L. M., L. Blank, R. P. Desai, N. E. Welker, and E. T. Papoutsakis. 2001. Fermentation characterization and flux analysis of recombinant strains of *Clostridium acetobutylicum* with an inactivated solR gene. *J. Ind. Microbiol. Biotechnol.* 27:322–328.

Harris, L. M., R. P. Desai, N. E. Welker, and E. T. Papoutsakis. 2000. Characterization of recombinant strains of the *Clostridium acetobutylicum* butyrate kinase inactivation mutant: need for new phenomenological models for solventogenesis and butanol inhibition? *Biotechnol. Bioeng.* 67:1–11.

Harris, L. M., N. E. Welker, and E. T. Papoutsakis. 2002. Northern, morphological, and fermentation analysis of *spo0A* inactivation and overexpression in *Clostridium acetobutylicum* ATCC 824. *J. Bacteriol.* 184:3586–3597.

Hartmanis, M. G., and S. Gatenbeck. 1984. Intermediary metabolism in *Clostridium acetobutylicum*: levels of enzymes involved in the formation of acetate and butyrate. *Appl. Environ. Microbiol.* 47:1277–1283.

Hong, S. H., J. S. Kim, S. Y. Lee, Y. H. In, S. S. Choi, J. K. Rih, C. H. Kim, H. Jeong, C. G. Hur, and J. J. Kim. 2004. The genome sequence of the capnophilic rumen bacterium *Mannheimia succiniciproducens*. *Nat. Biotechnol.* 22:1275–1281.

Ikeda, M., and R. Katsumata. 1992. Metabolic engineering to produce tyrosine or phenylalanine in a tryptophan-producing *Corynebacterium glutamicum* strain. *Appl. Environ. Microbiol.* 58:781–785.

Isogai, T., M. Fukagawa, I. Aramori, M. Iwami, H. Kojo, T. Ono, Y. Ueda, M. Kohsaka, and H. Imanaka. 1991. Construction of a 7-aminocephalosporanic acid (7ACA) biosynthetic operon and direct production of 7ACA in *Acremonium chrysogenum*. *Bio/Technology* 9:188–191.

Jones, D. T., and D. R. Woods. 1986. Acetone-butanol fermentation revisited. *Microbiol. Rev.* 50:484–524.

Koch, A. K., O. Kappeli, A. Fiechter, and J. Reiser. 1991. Hydrocarbon assimilation and biosurfactant production in *Pseudomonas aeruginosa* mutants. *J. Bacteriol.* 173:4212–4219.

Kumagai, H. 2000. Microbial production of amino acids in Japan. *Adv. Biochem. Eng. Biotechnol.* **69:**71–85.

Liu, X., Y. Zhu, and S. T. Yang. 2006. Construction and characterization of ack deleted mutant of *Clostridium tyrobutyricum* for enhanced butyric acid and hydrogen production. *Biotechnol. Prog.* **22:**1265–1275.

Lopez de Felipe, F., M. Kleerebezem, W. M. de Vos, and J. Hugenholtz. 1998. Cofactor engineering: a novel approach to metabolic engineering in *Lactococcus lactis* by controlled expression of NADH oxidase. *J. Bacteriol.* **180:**3804–3808.

Lucet, I., A. Feucht, M. D. Yudkin, and J. Errington. 2000. Direct interaction between the cell division protein FtsZ and the cell differentiation protein SpoIIE. *EMBO J.* **19:**1467–1475.

MacDonald, D. L., and H. Goldfine. 1991. Effects of solvents and alcohols on the polar lipid composition of *Clostridium butyricum* under conditions of controlled lipid chain composition. *Appl. Environ. Microbiol.* **57:**3517–3521.

Madduri, K., J. Kennedy, G. Rivola, A. Inventi-Solari, S. Filippini, G. Zanuso, A. L. Colombo, K. M. Gewain, J. L. Occi, D. J. MacNeil, and C. R. Hutchinson. 1998. Production of the antitumor drug epirubicin (4′-epidoxorubicin) and its precursor by a genetically engineered strain of *Streptomyces peucetius*. *Nat. Biotechnol.* **16:**69–74.

Malpartida, F., and D. A. Hopwood. 1984. Molecular cloning of the whole biosynthetic pathway of a *Streptomyces* antibiotic and its expression in a heterologous host. *Nature* **309:**462–464.

Mattsson, D. M., and P. Rogers. 1994. Analysis of Tn*916*-induced mutants of *Clostridium acetobutylicum* altered in solventogenesis and sporulation. *J. Ind. Microbiol.* **13:**258–268.

Mermelstein, L. D., N. E. Welker, D. J. Petersen, G. N. Bennett, and G. N. Bennett. 1994. Genetic and metabolic engineering of *Clostridium acetobutylicum* ATCC 824. *Ann. N. Y. Acad. Sci.* **721:**54–68.

Mitchell, W. J. 1998. Physiology of carbohydrate to solvent conversion by clostridia. *Adv. Microb. Physiol.* **39:**31–130.

Morimoto, K., T. Kimura, K. Sakka, and K. Ohmiya. 2005. Overexpression of a hydrogenase gene in *Clostridium paraputrificum* to enhance hydrogen gas production. *FEMS Microbiol. Lett.* **246:**229–234.

Mullany, P., M. Wilks, L. Puckey, and S. Tabaqchali. 1994. Gene cloning in *Clostridium difficile* using Tn916 as a shuttle conjugative transposon. *Plasmid* **31:**320–323.

Nair, R. V., G. N. Bennett, and E. T. Papoutsakis. 1994. Molecular characterization of an aldehyde/alcohol dehydrogenase gene from *Clostridium acetobutylicum* ATCC 824. *J. Bacteriol.* **176:**871–885.

Nair, R. V., and E. T. Papoutsakis. 1994. Expression of plasmid-encoded aad in *Clostridium acetobutylicum* M5 restores vigorous butanol production. *J. Bacteriol.* **176:**5843–5846.

Nair, R. V., E. M. Green, D. E. Watson, G. N. Bennett, and E. T. Papoutsakis. 1999. Regulation of the *sol* locus genes for butanol and acetone formation in *Clostridium acetobutylicum* ATCC 824 by a putative transcriptional repressor. *J. Bacteriol.* **181:**319–330.

Nakamura, C. E., and G. M. Whited. 2003. Metabolic engineering for the microbial production of 1,3-propanediol. *Curr. Opin. Biotechnol.* **14:**454–459.

Narberhaus, F., and H. Bahl. 1992. Cloning, sequencing, and molecular analysis of the *groESL* operon of *Clostridium acetobutylicum*. *J. Bacteriol.* **174:**3282–3289.

Nolling, J., G. Breton, M. V. Omelchenko, K. S. Makarova, Q. Zeng, R. Gibson, H. M. Lee, J. Dubois, D. Qiu, J. Hitti, Y. I. Wolf, R. L. Tatusov, F. Sabathe, L. Doucette-Stamm, P. Soucaille, M. J. Daly, G. N. Bennett, E. V. Koonin, and D. R. Smith. 2001. Genome sequence and comparative analysis of the solvent-producing bacterium *Clostridium acetobutylicum*. *J. Bacteriol.* **183:**4823–4838.

Peguin, S., and P. Soucaille. 1995. Modulation of carbon and electron flow in *Clostridium acetobutylicum* by iron limitation and methyl viologen addition. *Appl. Environ. Microbiol.* **61:**403–405.

Peoples, O. P., and A. J. Sinskey. 1989. Poly-beta-hydroxybutyrate (PHB) biosynthesis in *Alcaligenes eutrophus* H16. Identification and characterization of the PHB polymerase gene (phbC). *J. Biol. Chem.* **264:**15298–15303.

Petersen, D. J., J. W. Cary, J. Vanderleyden, and G. N. Bennett. 1993. Sequence and arrangement of genes encoding enzymes of the acetone-production pathway of *Clostridium acetobutylicum* ATCC824. *Gene* **123:**93–97.

Pich, A., F. Narberhaus, and H. Bahl. 1990. Induction of heat shock proteins during initiation of solvent formation in *Clostridium acetobutylicum*. *Appl. Microbiol. Biotechnol.* **33:**697–704.

Ravagnani, A., K. C. Jennert, E. Steiner, R. Grunberg, J. R. Jefferies, S. R. Wilkinson, D. I. Young, E. C. Tidswell, D. P. Brown, P. Youngman, J. G. Morris, and M. Young. 2000. Spo0A directly controls the switch from acid to solvent production in solvent-forming clostridia. *Mol. Microbiol.* **37:**1172–1185.

Raynaud, C., P. Sarcabal, I. Meynial-Salles, C. Croux, and P. Soucaille. 2003. Molecular characterization of the 1,3-propanediol (1,3-PD) operon of *Clostridium butyricum*. *Proc. Natl. Acad. Sci. USA* **100:**5010–5015.

Ro, D. K., E. M. Paradise, M. Oullet, K. J. Fisher, K. L. Newman, J. M. Ndungu, K. A. Ho, R. A. Eachus, T. S. Ham, J. Kirby, M. C. Chang, S. T. Withers, Y. Shiba, R. Sarpong, and J. D. Keasling. 2006. Production of the antimalarial drug precursor artemisinic acid in engineered yeast. *Nature* **440:**940–943.

Roberts, A. P., C. Hennequin, M. Elmore, A. Collignon, T. Karjalainen, N. Minton, and P. Mullany. 2003. Development of an integrative vector for the expression of antisense RNA in *Clostridium difficile*. *J. Microbiol. Methods* **55:**617–624.

Sass, C., J. Walter, and G. N. Bennett. 1993. Isolation of mutants of *Clostridium acetobutylicum* ATCC-824 deficient in protease activity. *Curr. Microbiol.* **26:**151–154.

Schaffer, S., N. Isci, B. Zickner, and P. Durre. 2002. Changes in protein synthesis and identification of proteins specifically induced during solventogenesis in *Clostridium acetobutylicum*. *Electrophoresis* **23:**110–121.

Schwarz, W. H., and R. Gapes. 2006. Butanol—rediscovering a renewable fuel. *BioWorld Europe* **1:**16–19.

Scotcher, M. C., and G. N. Bennett. 2005. SpoIIE regulates sporulation but does not directly affect solventogenesis in *Clostridium acetobutylicum* ATCC 824. *J. Bacteriol.* **187:**1930–1936.

Scotcher, M. C., K. X. Huang, M. L. Harrison, F. B. Rudolph, and G. N. Bennett. 2003. Sequences affecting the regulation of solvent production in *Clostridium acetobutylicum*. *J. Ind. Microbiol. Biotechnol.* **30:**414–420.

Scotcher, M. C., F. B. Rudolph, and G. N. Bennett. 2005. Expression of *abrB310* and *sinR*, and effects of decreased *abrB310* expression on the transition from acidogenesis to solventogenesis, in *Clostridium acetobutylicum* ATCC 824. *Appl. Environ. Microbiol.* **71:**1987–1995.

Shimada, H., K. Kondo, P. D. Fraser, Y. Miura, T. Saito, and N. Misawa. 1998. Increased carotenoid production by the food yeast *Candida utilis* through metabolic engineering of the isoprenoid pathway. *Appl. Environ. Microbiol.* **64:**2676–2680.

Shio, I. 1986. Production of individual amino acids: tryptophan, phenylalanine, and tyrosine, p. 188–206. *In* I. Aida, K. Nakayama, K. Takinami, and H. Yamada (ed.), *Biotechnology of Amino Acid Production*. Elsevier, Tokyo, Japan.

Smith, D. J., M. K. Burnham, J. Edwards, A. J. Earl, and G. Turner. 1990. Cloning and heterologous expression of the penicillin biosynthetic gene cluster from *Penicillum* [sic] *chrysogenum*. *Bio/Technology* **8:**39–41.

Stanzak, R., P. Matsushima, R. H. Baltz, and R. N. Rao. 1986. Cloning and expression in *Streptomyces lividans* of clustered erythromycin biosynthesis genes from *Streptomyces erythreus*. *Bio/Technology* **4:**229–232.

Stephenson, K., and R. J. Lewis. 2005. Molecular insights into the initiation of sporulation in gram-positive bacteria: new technologies for an old phenomenon. *FEMS Microbiol. Rev.* **29**:281–301.

Sullivan, L., and G. N. Bennett. 2006. Proteome analysis and comparison of *Clostridium acetobutylicum* ATCC 824 and Spo0A strain variants. *J. Ind. Microbiol. Biotechnol.* **33**:298–308.

Thormann, K., and P. Durre. 2001. Orf5/SolR: a transcriptional repressor of the sol operon of *Clostridium acetobutylicum*? *J. Ind. Microbiol. Biotechnol.* **27**:307–313.

Thormann, K., L. Feustel, K. Lorenz, S. Nakotte, and P. Durre. 2002. Control of butanol formation in *Clostridium acetobutylicum* by transcriptional activation. *J. Bacteriol.* **184**:1966–1973.

Tilly, K., N. McKittrick, C. Georgopoulos, and H. Murialdo. 1981. Studies on *Escherichia coli* mutants which block bacteriophage morphogenesis. *Prog. Clin. Biol. Res.* **64**:35–45.

Tomas, C. A., K. V. Alsaker, H. P. Bonarius, W. T. Hendriksen, H. Yang, J. A. Beamish, C. J. Paredes, and E. T. Papoutsakis. 2003a. DNA array-based transcriptional analysis of asporogenous, nonsolventogenic *Clostridium acetobutylicum* strains SKO1 and M5. *J. Bacteriol.* **185**:4539–4547.

Tomas, C. A., J. Beamish, and E. T. Papoutsakis. 2004. Transcriptional analysis of butanol stress and tolerance in *Clostridium acetobutylicum*. *J. Bacteriol.* **186**:2006–2018.

Tomas, C. A., N. E. Welker, and E. T. Papoutsakis. 2003b. Overexpression of groESL in *Clostridium acetobutylicum* results in increased solvent production and tolerance, prolonged metabolism, and changes in the cell's transcriptional program. *Appl. Environ. Microbiol.* **69**:4951–4965.

Tong, I. T., and D. C. Cameron. 1992. Enhancement of 1,3-propanediol production by cofermentation in *Escherichia coli* expressing *Klebsiella pneumoniae* dha regulon genes. *Appl. Biochem. Biotechnol.* **34-35**:149–159.

Tong, I. T., H. H. Liao, and D. C. Cameron. 1991. 1,3-Propanediol production by *Escherichia coli* expressing genes from the *Klebsiella pneumoniae dha* regulon. *Appl. Environ. Microbiol.* **57**:3541–3546.

Trowsdale, J., S. M. Chen, and J. A. Hoch. 1978. Evidence that *spo0A* mutations are recessive in *spo0A*$^-$/*spo0A*$^+$ merodiploid strains of *Bacillus subtilis*. *J. Bacteriol.* **135**:99–113.

Tummala, S. B., S. G. Junne, and E. T. Papoutsakis. 2003a. Antisense RNA downregulation of coenzyme A transferase combined with alcohol-aldehyde dehydrogenase overexpression leads to predominantly alcohologenic *Clostridium acetobutylicum* fermentations. *J. Bacteriol.* **185**:3644–3653.

Tummala, S. B., N. E. Welker, and E. T. Papoutsakis. 2003b. Design of antisense RNA constructs for downregulation of the acetone formation pathway of *Clostridium acetobutylicum*. *J. Bacteriol.* **185**:1923–1934.

Tyo, K. E., H. S. Alper, and G. N. Stephanopoulos. 2007. Expanding the metabolic engineering toolbox: more options to engineer cells. *Trends Biotechnol.* **25**:132–137.

Walter, K. A., R. V. Nair, J. W. Cary, G. N. Bennett, and E. T. Papoutsakis. 1993. Sequence and arrangement of two genes of the butyrate-synthesis pathway of *Clostridium acetobutylicum* ATCC 824. *Gene* **134**:107–111.

Walter, K. A., L. D. Mermelstein, and E. T. Papoutsakis. 1994. Studies of recombinant *Clostridium acetobutylicum* with increased dosages of butyrate formation genes. *Ann. N. Y. Acad. Sci.* **2**:69–72.

Wang, H., A. P. Roberts, and P. Mullany. 2000. DNA sequence of the insertional hot spot of Tn916 in the *Clostridium difficile* genome and discovery of a Tn916-like element in an environmental isolate integrated in the same hot spot. *FEMS Microbiol. Lett.* **192**:15–20.

Watrous, M. M., S. Clark, R. Kutty, S. Huang, F. B. Rudolph, J. B. Hughes, and G. N. Bennett. 2003. 2,4,6-Trinitrotoluene reduction by an Fe-only hydrogenase in *Clostridium acetobutylicum*. *Appl. Environ. Microbiol.* **69**:1542–1547.

Wiesenborn, D. P., F. B. Rudolph, and E. T. Papoutsakis. 1988. Thiolase from *Clostridium acetobutylicum* ATCC 824 and its role in the synthesis of acids and solvents. *Appl. Environ. Microbiol.* **54**: 2717–2722.

Wiesenborn, D. P., F. B. Rudolph, and E. T. Papoutsakis. 1989a. Coenzyme A transferase from *Clostridium acetobutylicum* ATCC 824 and its role in the uptake of acids. *Appl. Environ. Microbiol.* **55**:323–329.

Wiesenborn, D. P., F. B. Rudolph, and E. T. Papoutsakis. 1989b. Phosphotransbutyrylase from *Clostridium acetobutylicum* ATCC 824 and its role in acidogenesis. *Appl. Environ. Microbiol.* **55**: 317–322.

Williams, T. I., J. C. Combs, B. C. Lynn, and H. J. Strobel. 2007. Proteomic profile changes in membranes of ethanol-tolerant *Clostridium thermocellum*. *Appl. Microbiol. Biotechnol.* **74**:422–432.

Woods, D. R. 1995. The genetic engineering of microbial solvent production. *Trends Biotechnol.* **13**:259–264.

Wright, A., R. Wait, S. Begum, B. Crossett, J. Nagy, K. Brown, and N. Fairweather. 2005. Proteomic analysis of cell surface proteins from *Clostridium difficile*. *Proteomics* **5**:2443–2452.

Xu, Z., and P. B. Sigler. 1998. GroEL/GroES: structure and function of a two-stroke folding machine. *J. Struct. Biol.* **124**:129–141.

Yan, Y., A. Kohli, and M. A. Koffas. 2005. Biosynthesis of natural flavanones in *Saccharomyces cerevisiae*. *Appl. Environ. Microbiol.* **71**:5610–5613.

York, K., T. J. Kenney, S. Satola, C. P. Moran, Jr., H. Poth, and P. Youngman. 1992. Spo0A controls the sigma A-dependent activation of *Bacillus subtilis* sporulation-specific transcription unit *spoIIE1*. *J. Bacteriol.* **174**:2648–2658.

Zhao, Y., L. A. Hindorff, A. Chuang, M. Monroe-Augustus, M. Lyristis, M. L. Harrison, F. B. Rudolph, and G. N. Bennett. 2003. Expression of a cloned cyclopropane fatty acid synthase gene reduces solvent formation in *Clostridium acetobutylicum* ATCC 824. *Appl. Environ. Microbiol.* **69**:2831–2841.

Zhao, Y., C. A. Tomas, F. B. Rudolph, E. T. Papoutsakis, and G. N. Bennett. 2005. Intracellular butyryl phosphate and acetyl phosphate concentrations in *Clostridium acetobutylicum* and their implications for solvent formation. *Appl. Environ. Microbiol.* **71**: 530–537.

7. MICROBIAL OIL RECOVERY

Bioenergy
Edited by J. Wall et al.
© 2008 ASM Press, Washington, DC

Chapter 29

Emerging Oil Field Biotechnologies: Prevention of Oil Field Souring by Nitrate Injection

GERRIT VOORDOUW

One of the biggest challenges facing society today is how to chart the best course towards our energy future (Hall et al., 2003). Our society currently depends greatly on fossil fuels (40% oil, 24% gas, and 22% coal) with smaller contributions from nuclear energy (6%) and renewables (8%). The environmental consequences of this heavy reliance on fossil fuels are well known and include significant increases of CO_2 in the Earth's atmosphere contributing to global warming. Hence, governments are under considerable pressure to act to halt or reverse environmental consequences of our reliance on fossil fuels. Key steps towards a sustainable energy future will be to (i) reduce per capita energy consumption, (ii) have renewables contribute a larger fraction of our energy supply, and (iii) make extraction and use of fossil fuels as efficient and "green" as possible. Much of the focus in this book and in recent announcements on planned research investments is on the second step. Indeed, one may claim that research on the third step is somewhat ignored. Although perhaps less trendy, improving efficiency in the use and extraction of fossil fuels is very important, because it is expected that these will remain the main component of our energy supply for many years to come. The problems associated with the production of sufficient fossil fuel to meet the world's growing demand should not be underestimated. New technologies are required to reduce the environmental impact of production and use, to extend the life span of existing conventional oil and gas reservoirs, and to improve recovery of heavy oil and bitumen. Microbiology is one of the disciplines that can offer novel solutions. Although its potential has been recognized for over 50 years, the role that microorganisms have played in shaping the oil component of our fossil fuel energy resource is only now becoming understood and reliable technologies involving microbes are only now being implemented. These technologies allow reduction of souring, the in situ production of hydrogen sulfide by sulfate-reducing bacteria (SRB) and associated corrosion. Research-driven, improved applications of these biotechnologies will reduce input of water and energy associated with fossil fuel production. Together with other emerging technologies (e.g., geological storage of produced CO_2), these biotechnologies will thus contribute to the greening of our fossil fuel use, a strategy that is key to solving problems associated with our continued energy supply.

OIL PRODUCTION BY WATER INJECTION

Much of the world's oil is produced by water injection. Water injected through injection wells helps to maintain the reservoir pressure required to sweep the oil to the surface through production wells. When injected water breaks through, an oil-water mixture is produced. After separation from produced oil, the produced water can be reinjected, a production strategy referred to as PWRI (produced water reinjection). PWRI is commonly practiced in landlocked reservoirs where access to water may be limited (Fig. 1). Because not all injected water is recovered in the producing wells, additional water (makeup water) needs to be drawn from a nearby water supply (lake, river, or aquifer). PWRI is rarely practiced when seawater is injected, as in offshore oil recovery operations, because the water supply is abundant and there is little competition for alternative use (domestic, agricultural, or recreational). Hence, in many offshore operations the produced water is discharged and fresh seawater is continually injected. A desire to reduce environmental impact is driving adoption of PWRI in seawater injections. However, such a strategy is not yet common. The fraction of water produced from oil fields subjected to water injection (the water cut) generally increases with time. In older reservoirs, as found in the Western Canadian Sedimentary

Gerrit Voordouw • Department of Biological Sciences, University of Calgary, Calgary, Alberta, T2N 1N4, Canada.

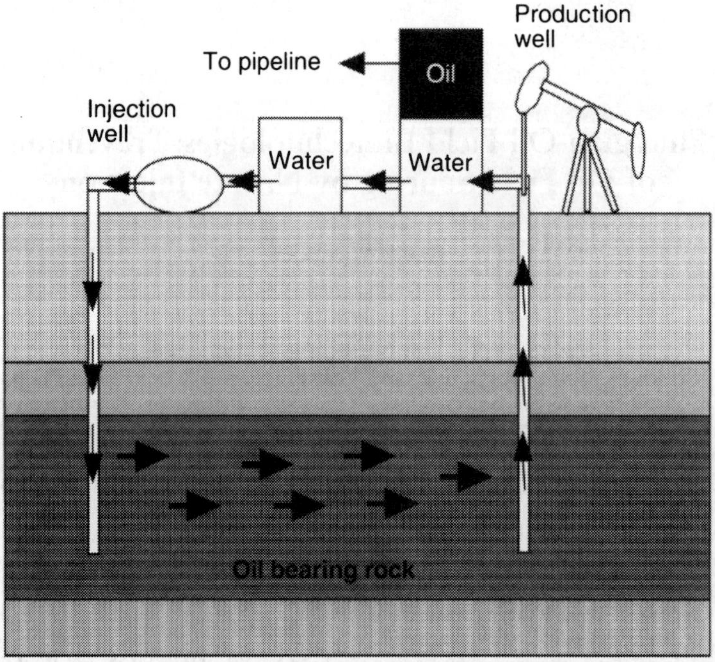

Figure 1. Schematic of oil production by water injection under a PWRI regime. Reproduced from Voordouw (1999).

Basin, which have been water injected for over 50 years, the water cut can be 90% or higher. High water cuts do not reflect the absence of oil in the oil-bearing subsurface, but rather our inability to produce the fraction of oil left behind (typically 40 to 70%). Economic production of the remaining oil from aging reservoirs requires technologies to decrease the water cut by making the water sweep areas other than the path of least resistance and to decrease negative impacts such as souring and associated corrosion to reduce production costs.

IMPACT OF NITRATE ON THE OIL FIELD SULFUR CYCLE

Oil production by water injection often results in increased sulfide levels (souring), because SRB couple the oxidation of degradable oil organics to the reduction of sulfate to sulfide (Fig. 2A). High levels of sulfide represent increased toxicity and corrosion risk, and mitigation strategies are generally considered necessary. The problem can be especially severe when seawater, which has a high sulfate concentration (~30 mM), is injected. An example is provided by seawater flooding of the Skjold field. Total daily production of sulfide increased from 100 kg/day initially to up to 1,100 kg/day after 5 years of seawater injection (Larsen, 2002). These high amounts of sulfide can be removed by chemical treatment of the produced oil/water/gas, but an alternative strategy is to prevent formation of the sulfide in the first place.

There is now a general consensus that the sulfide in these souring reservoirs is produced by SRB, presumably in the zone where sulfate-containing injection water mixes with oil organics-containing formation water (Sunde and Torsvik, 2005). The content of degradable oil organics present in the formation water may be a key determinant of the level of souring that can occur. Availability of other required nutrients (e.g., phosphate) may be another factor. As an example the composition of produced water (possibly a mixture of formation water and injected seawater) from Ekofisk, an oil field in the North Sea, is given in Table 1 (Kaster et al., 2007) and compared with that of seawater. As seen from Table 1, the produced water provides potentially a more favorable environment for microbial growth with significant concentrations of volatile fatty acids (VFA) (acetate, propionate, and butyrate), sulfate, ammonium, and some phosphate, which are mostly absent from seawater. The fact that both sulfate and VFA

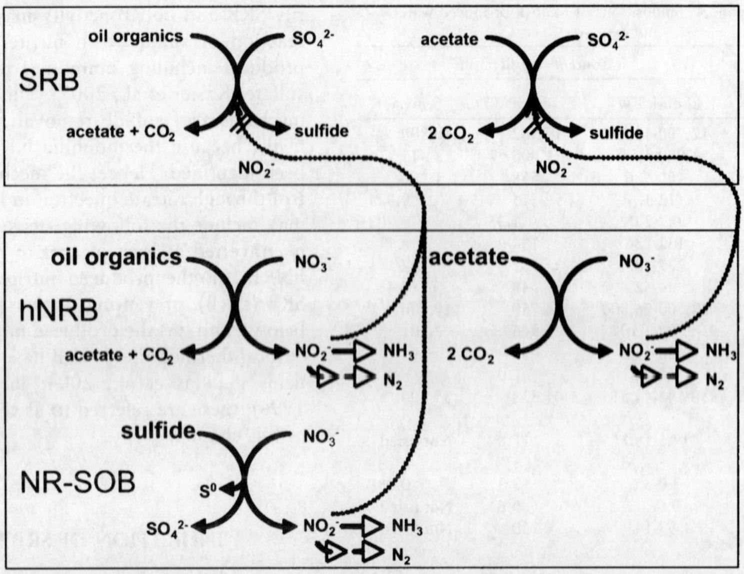

Figure 2. Survey of microbial groups impacting the sulfur cycle in oil fields. (A) SRB couple incomplete oxidation of oil organics to acetate and CO_2, or complete oxidation of acetate to CO_2 to the reduction of sulfate to sulfide. (B) hNRB couple incomplete oxidation of oil organics or complete oxidation of acetate to CO_2 to reduction of nitrate to nitrite and then to either nitrogen or ammonia. (C) NR-SOB oxidize sulfide to sulfur or sulfate, with nitrate being reduced to nitrite and then to either nitrogen (with NO and N_2O as intermediates) or to ammonia (without intermediates). Note that some NR-SOB/hNRB do not reduce nitrate beyond nitrite. Also nitrite is a powerful SRB inhibitor as indicated by the dotted lines.

are present at substantial concentrations potentially indicates the absence of high SRB activity. This may be caused by the absence of another limiting nutrient or due to the high resident temperature (80-90°C) at Ekofisk, where souring is not considered a significant problem.

In situ reservoir temperature depends on reservoir depth. At depths of up to 1 km moderate temperatures (20 to 40°C) favor growth of mesophiles, whereas at depths exceeding 2 to 3 km, higher resident temperatures (50 to 80°C) restrict growth to thermophiles. Even in fields with resident temperatures too high for microbial growth, growth may occur near the injection wells, where cold injection water and hot formation water mix, creating a thermal viability zone, where sulfide is produced, which is subsequently transferred through the reservoir to the production wells. Depending on their geological composition, fields may have significant sulfide binding capacity through reaction with iron minerals, e.g., with siderite ($FeCO_3$): $FeCO_3$ + HS^- FeS + HCO_3^- (Vance and Thrasher, 2005). This would delay emergence of sulfide in produced water and oil.

Field-wide prevention of production of sulfide may be achieved by injection of nitrate, which serves as an alternative electron acceptor for the microbial community downhole. Nitrate stimulates nitrate-reducing, sulfide-oxidizing bacteria (NR-SOB), which remove sulfide as indicated in Fig. 2C. The presence of this bacterial group and its stimulation with injected nitrate were first demonstrated in the Coleville field in Kindersley (Saskatchewan, Canada), a field with a resident temperature of 30°C, harboring a mesophilic microbial population. Injection of 300-ppm nitrate for 50 days gave on average a 70% reduction of sulfide concentrations in injection and production wells and stimulated growth of *Thiomicrospira* sp. strain CVO (Telang et al., 1997), an autotrophic NR-SOB belonging to the epsilon division of the proteobacteria (Gevertz et al., 2000) that uses CO_2 as the sole carbon source while coupling oxidation of sulfide to sulfur and sulfate with reduction of nitrate to nitrite, nitric oxide, nitrous oxide, and finally nitrogen (Fig. 2C). Instead of multistep reduction of nitrate to nitrite and N_2 (represented by the multiple arrows in Fig. 2C), NR-SOB can also reduce nitrate to nitrite and then in a single step to ammonia, a process referred to as dissimilatory nitrate reduction to ammonia. Nitrate injection also stimulates heterotrophic nitrate-reducing bacteria (hNRB), which couple either incomplete oxidation of degradable oil

Table 1. Chemical compositions of Ekofisk-produced water and seawater

Analyte	Content (mg/liter) in: Ekofisk PW[a]	Ekofisk PW[b]	Seawater[c]
Sodium	12,700-14,424	13,562	10,800
Calcium	1,090-1,510	1,300	411
Magnesium	490-626	558	1,290
Barium	12-18	15	0.021
Iron	0.7-2.0	1.35	0.0034
Strontium	132-236	184	8.1
Potassium	277-287	282	392
Boron	44-52	48	4.45
Chloride	22,700-26,800	24,750	19,400
Sulfate	782-1,101	946	2,701
Bicarbonate	436-438	437	145
pH	6.8-7.3	7.0	8.0
Total dissolved solids	38,541-44,515	41,528	35,000
Ammonia nitrogen	17.5-18.1	17.8	Not listed
Phosphorus	3.0-3.6	3.3	0.088
Phosphate	0.6	0.6	Not listed
Dissolved organic carbon	99-141	120	Not listed
Acetate	136-152	144	0[d]
Propionate	<5	<5	0[d]
Butyrate	27	27	0[d]

[a]Range of values observed for five separate produced water samples (Kaster et al., 2007).
[b]Mid-range values.
[c]Obtained from www.seafriends.org.nz/oceano/seawater.htm; composition for 3.5% salinity (35 g of salts/liter).
[d]Presumed values.

organics (to acetate and CO_2) or complete oxidation of oil organics (to CO_2 only) to the reduction of nitrate to nitrite, nitrogen, or ammonia (Fig. 2B). For example, *Thauera* sp. strain N2 produces N_2, with little production of nitrite, whereas *Sulfurospirillum* spp. producing nitrite and ammonia have been isolated from oil fields which couple oxidation of oil organics or sulfide to dissimilatory nitrate reduction to ammonia. Interestingly, these were shown to have both hNRB and NR-SOB activity (Hubert and Voordouw, 2007).

The discovery that NR-SOB participate in souring control at Coleville challenged earlier theories that nitrate eliminated SRB activity primarily by competitive exclusion, a mechanism in which hNRB use the same oil organics as SRB (Fig. 2A and B) eventually outcompeting SRB, when nitrate is injected continuously, because of the 10- to 20-fold-higher energy yield when oil organics are oxidized with nitrate than with sulfate (Hitzman et al., 1995). Surveys of other fields in the Western Canadian Sedimentary Basin have since shown that all three groups are usually present, with hNRB often dominating NR-SOB, except at Coleville (Eckford and Fedorak, 2002; Voordouw et al., 2007). Nitrate-dependent sulfide removal in systems containing primarily SRB and hNRB activity may result from chemical reaction of sulfide with nitrite, yielding a variety of products including ammonia, polysulfide, sulfur, and sulfate (Kaster et al., 2007). This may well be the main mechanism of sulfide removal from high-temperature fields, because thermophilic NR-SOB have so far not been identified. Hence, the mechanism of souring control through nitrate injection in high-temperature fields may include the following steps: (i) the injected nitrate is converted at least in part to nitrite by thermophilic hNRB; (ii) the produced nitrite inhibits thermophilic SRB (tSRB), preventing further sulfide production (see below); and (iii) the produced nitrite reacts with sulfide. Several thermophilic hNRB have been isolated from oil fields (Salinas et al., 2004; Jackson and McInerney, 1996); these are referred to as tNRB for the remainder of this chapter.

INHIBITION OF SRB BY NITRITE

Nitrite is formed by partial reduction of nitrate by NR-SOB or hNRB. Indeed, some of these (e.g., *Arcobacter* sp. strain FWKO-B [Gevertz et al., 2000]) form nitrite as the respiration end product and thus convert nitrate quantitatively into nitrite. Nitrite has also been added directly to oil and gas fields to inhibit SRB and remediate sulfide (Sturman et al., 1999). Although nitrite is known to have general bacteriostatic properties, it has, in addition, a very specific inhibitory action on SRB. Nitrite is a strong inhibitor of dissimilatory sulfite reductase (Dsr), the SRB enzyme responsible for sulfide production. Wolfe et al. (1994) showed that Dsr, which normally catalyzes the reaction $HSO_3^- + 6H^+ + 6e \rightarrow HS^- + 3H_2O$, can also catalyze the six-electron reduction of nitrite to ammonia ($NO_2^- + 7H^+ + 6e \rightarrow NH_3 + 2H_2O$). Tighter binding and slower turnover of nitrite by the enzyme causes it to act as an effective competitive inhibitor of sulfite reduction. Many mesophilic SRB have a periplasmic nitrite reductase (NrfHA), which reduces nitrite to ammonia, to prevent this inhibition. Both subunits contain *c*-type heme, and the enzyme is associated with the periplasmic face of the cytoplasmic membrane (Simon, 2002). Inhibition of Dsr by nitrite causes a strong gene expression response in the SRB *Desulfovibrio vulgaris* Hildenborough, in which genes for sulfate reduction enzymes are downregulated, whereas those for NrfHA are upregulated (Haveman et al., 2004; He et al., 2006).

The presence or absence of NrfHA significantly alters the dynamics of interaction of SRB with NRB, as shown in Fig. 3. Two mesophilic SRB strains oxidizing lactate with sulfate were incubated with nitrate, which is not reduced until NR-SOB *Thiomicrospira* sp. strain

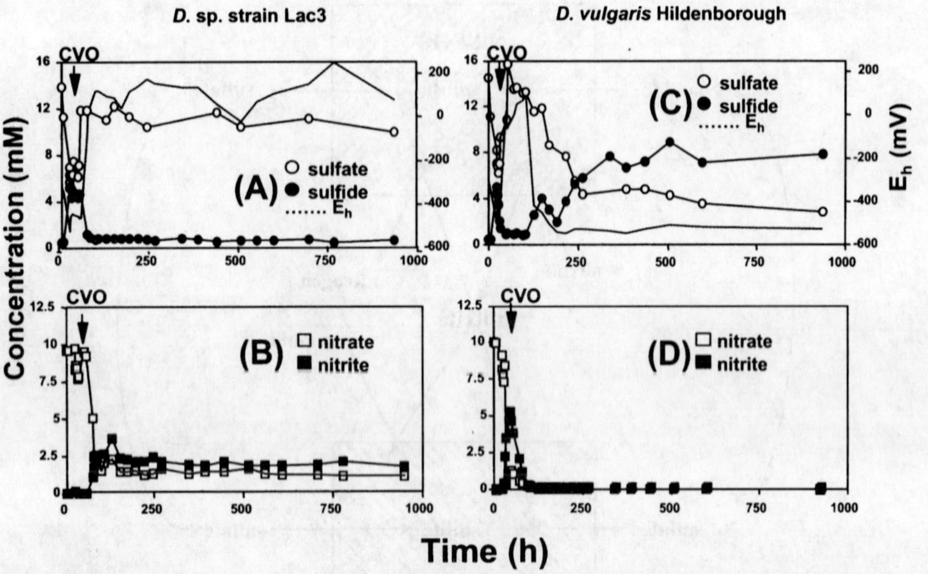

Figure 3. Effect of strain CVO and nitrate on sulfide production by SRB. (A and B) Sulfide production by *Desulfovibrio* sp strain Lac3 is permanently inhibited because the organism lacks nitrite reductase. (C and D) Sulfide production by *D. vulgaris* Hildenborough is transiently inhibited because this organism contains nitrite reductase. Adapted from Greene et al. (2003).

CVO is added. Addition of strain CVO to *Desulfovibrio* sp. strain Lac3, which lacks *nrfHA* genes, gives oxidation of sulfide to sulfate (Fig. 3A), coupled to reduction of nitrate to nitrite (Fig. 3B) and nitrogen (not shown). Because insufficient sulfide was present to reduce all nitrate and nitrite to nitrogen and because NrfHA is absent from the SRB, this leads to permanent inhibition of sulfide production and stalling of the pseudosymbiosis. When strain CVO is added to the *nrfHA*-containing *D. vulgaris* Hildenborough, inhibition is only transient. Once all nitrate and nitrite have been reduced by the NR-SOB using sulfide as the electron donor and by the SRB using lactate as the electron donor to reduce nitrite to ammonia, sulfate reduction starts again (Fig. 3C). The interaction between the SRB and NR-SOB is a pseudosymbiosis, outlined in Fig. 4. The two types of cells do not catalyze their usual metabolic function at the same time. Sulfide oxidation by the NR-SOB raises the redox potential (Fig. 3A and C), and SRB only catalyze lactate-dependent nitrite reduction under these conditions. Sulfate reduction requires a low redox potential (Fig. 3C). A variety of mesophilic SRB have been shown to have nitrite reductase (Greene et al., 2003). Addition of strain CVO and nitrate gave a gene expression response in *D. vulgaris* Hildenborough similar to that which occurred when nitrite was added directly (Haveman et al., 2005), indicating that the transient inhibition by this organism (Fig. 3C) is caused by produced nitrite.

DETERMINANTS OF NITRATE OR NITRITE DOSE

Whether the SRB community does or does not have nitrite reductase is an important determinant of the required nitrate dose (Fig. 2). In the absence of SRB nitrite reductase the added nitrate dose (10 mM) was adequate for removal of all sulfide (Fig. 2A and B), whereas in the presence of SRB nitrite reductase a higher nitrate dose, capable of oxidizing all degradable organic carbon present, must be added. Although this has not been researched in great detail, it appears that thermophilic SRB consortia lack nitrite reductase. If tNRB also lack nitrite reductase, then these reduce nitrate to nitrite but not further. A very low nitrate dose can be sufficient to stop further sulfide production in such systems. Examples include studies with laboratory upflow bioreactors, inoculated with produced water from the Ninian North Sea and Kuparuk North Slope fields and operated at 60°C. In these bioreactors 0.71 mM nitrate was converted to 0.51 to 0.71 mM nitrite (Reinsel et al., 1996), preventing sulfide production by tSRB. The effectiveness of nitrate in reducing souring in

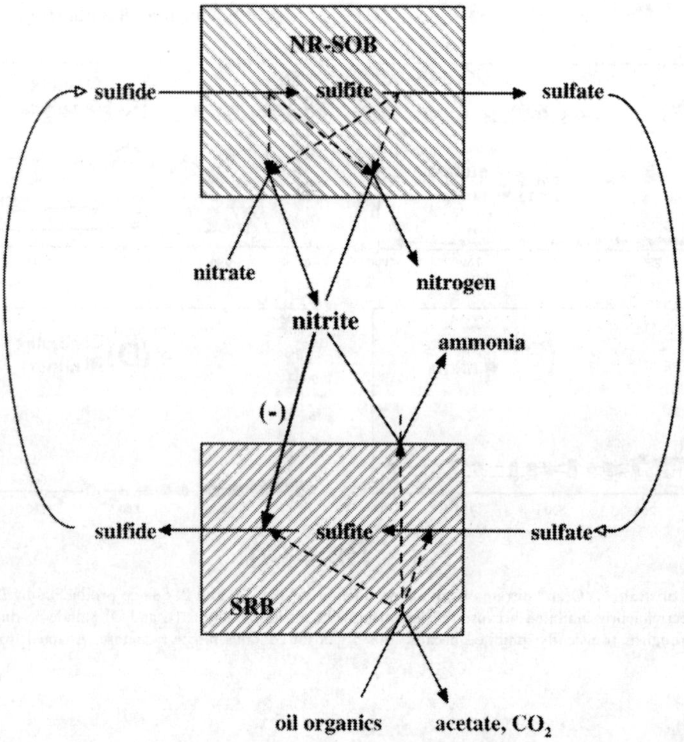

Figure 4. Role of nitrite reductase in the pseudosymbiotic relationship between SRB and NR-SOB. Adapted from Greene et al. (2003).

high-temperature North Sea reservoirs injected with seawater has been well documented (Larsen et al., 2004; Thorstenson et al., 2002; Sunde et al., 2004). As an example, Sunde and Torsvik (2005) demonstrated that field-wide, long-term injection of 35-ppm (0.28 mM) nitrate into the Gullfaks field permanently lowered sulfide concentrations from the trend line of increasing sulfide concentrations that would be expected if no nitrate had been injected.

However, the situation may be less favorable in shallow, landlocked reservoirs with a moderate downhole temperature in which oil is produced through a PWRI regime. Because the SRB population in such reservoirs has nitrite reductase, the effective dose is dictated by the total electron donor concentration (i.e., sulfide plus degradable organic carbon). In model upflow bioreactors operated at room temperature (22°C), Hubert et al. (2003) showed that the required nitrate dose needed to contain souring in such systems is proportional to the electron donor (lactate) concentration. This held true irrespective of whether nitrate or nitrite was used to reduce souring. In fact the nitrate dose needed to prevent formation of sulfide was lower than the nitrite dose, because nitrate has more oxidative power. Myhr et al. (2002) reported that a very low nitrate dose (0.5 mM) sufficed to prevent H_2S production in an oil-saturated column operated at room temperature. Apparently, the oil provided little degradable organic carbon that could be used by the SRB and hNRB present on the time scale of the experiment. This presents a conundrum in trying to estimate the required dose under these conditions: which oil organics should be counted as degradable and be used to calculate the dose? It may not suffice to count only those that are water soluble like the VFA (acetate, propionate, and butyrate). Other hydrocarbons, including low-molecular-mass aliphatics and aromatics such as hexadecane and toluene, are known to be degradable by both SRB and hNRB. We may also need to account for the different distributions of these organics among aqueous and oil phases. Hence, the effective concentration of degradable oil organics is not easily estimated and the required nitrate dose needs to be established by trial and error. A potential problem in such trial and error attempts is that when the dose is set too low, only partial oxidation of sulfide to sulfur occurs, coupled to complete reduction of nitrate, which allows nitrate to be completely reduced to nitrogen and ammonia. In contrast, when a sufficiently high dose is chosen, the sulfide will be completely oxidized to sulfate with only partial reduction of nitrate to nitrite, which inhibits further SRB activity (Greene et al., 2003). The latter scenario is preferable, especially because sulfur is considered to be quite corrosive.

Hence, it appears that nitrate injection into oil fields has the potential to reduce souring and associated corrosion, especially in high-temperature fields with limited sulfide concentrations where the injected nitrate is effectively converted into nitrite without further reduction to nitrogen and ammonia. Its successful use in onshore low-temperature reservoirs subjected to PWRI requires more study and may require high doses in cases where high concentrations of sulfide and/or of degradable oil organics are present. Regardless, because nitrate is eventually converted to chemically inert nitrogen or ammonia, nitrate injection is an environmentally friendly, green method of sulfide removal, preferable over other methods of chemical sulfide removal. Its continuous, field-wide use is now commonly practiced in offshore operations in the North Sea and is steeply on the rise elsewhere.

CONTROL OF CORROSION

The role of SRB in microbially influenced corrosion has been studied since the early 1930s, when SRB-mediated removal of cathodic hydrogen present on the surface of metallic iron (Fe^0) in contact with water was suggested as a possible corrosion mechanism (von Wolzogen Kuehr and van der Vlugt, 1934). More recently it has been demonstrated that SRB can oxidize Fe^0 cometabolically (Cord-Ruwisch and Widdel, 1986). Strains using Fe^0 efficiently as the sole electron donor for sulfate reduction have also been isolated (Dinh et al., 2004). It was suggested that these efficiently corroding SRB have electron transport complexes in their outer membrane that directly accept electrons from the steel surface, similar to those found in the outer membrane of Fe(III)-reducing bacteria (Dinh et al., 2004). Earlier papers proposed that in *D. vulgaris* the high-molecular-weight cytochrome (Hmc) complex fulfilled that role (Van Ommen et al., 1995). Electron transport through the Hmc complex connects periplasmic hydrogen oxidation to cytoplasmic sulfate reduction (Dolla et al., 2000), and the Hmc complex resides, therefore, in the inner, not in the outer membrane (Rossi et al., 1993; Heidelberg et al., 2004). Hence, the mechanism by which SRB extract electrons from the metal surface is still unsolved. Microbial corrosion can contribute to pipeline failures, which can in turn reduce production for prolonged periods of time, as was demonstrated by the failure of a British Petroleum pipeline and the associated oil spill of the Alaska North Slope region in the summer of 2006. Rapid corrosion may occur in systems subject to intermittent aeration, especially if high sulfide concentrations are present. Chemical reaction of H_2S and oxygen can give high local sulfur concentrations, which contribute to rapid pitting corrosion (Fig. 5). SRB can contribute to rapid corrosion in such partially

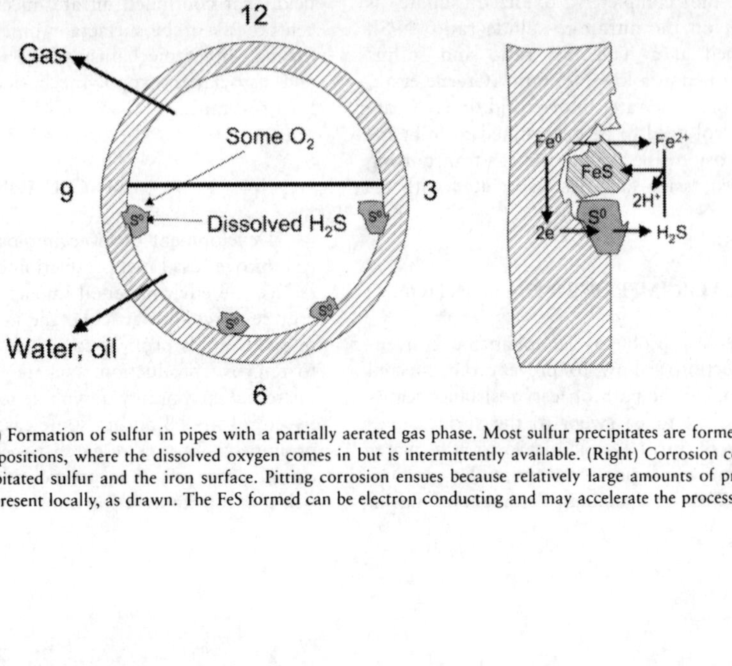

Figure 5. (Left) Formation of sulfur in pipes with a partially aerated gas phase. Most sulfur precipitates are formed at the 4 and 8 o'clock positions, where the dissolved oxygen comes in but is intermittently available. (Right) Corrosion cell formed between precipitated sulfur and the iron surface. Pitting corrosion ensues because relatively large amounts of precipitated sulfur can be present locally, as drawn. The FeS formed can be electron conducting and may accelerate the process.

aerated systems because of their ability to reduce sulfate to sulfide, which is then chemically converted to sulfur and/or polysulfide upon air exposure (Hamilton, 2003). Rapid corrosion of an oil-transporting sea line in the Congo correlated with the presence of high numbers of thiosulfate-reducing bacteria, which may in turn be indicative of air exposure as thiosulfate forms by the following reaction: $2H_2S + 2O_2 \rightarrow S_2O_3^{2-} + 2H^+ + H_2O$ (Crolet, 2005).

Hence, when managing the oil field sulfur cycle, formation of sulfur/polysulfide by partial oxidation of sulfide with oxygen or with nitrate should be avoided. Studies in upflow bioreactors with buried metal coupons indicated that under conditions of complete souring control, corrosion rates were on average lower in the presence of nitrate and much lower in the presence of nitrite (Hubert et al., 2005). On the other hand, Vik et al. (2007) found that nitrate increased corrosion rates in a field study, indicating that careful evaluation of this technology is required to achieve reduced sulfide levels, as well as corrosion rates. Instead of nitrite, biocides are also frequently used to limit SRB action in aboveground facilities (oil-water separators, pipelines, etc.). It was found recently that, because of its SRB-specific action, nitrite is highly synergistic with commonly used biocides such as glutaraldehyde, benzalkonium chloride, and bronopol, which affect microbial functions more broadly (Greene et al., 2006). Inclusion of some nitrite allows the significantly reduced use of costly and toxic biocides. This lowers costs, reduces environmental impact, and increases worker safety. In summary, it appears that when sulfide is being remediated by injection of nitrate, partial oxidation of sulfide to sulfur/polysulfide may have to be avoided to limit corrosion risk. Whether sulfur/polysulfide or the completely oxidized sulfate is formed depends on the nitrate-to-sulfide ratio (N/S). Sulfate is formed at a high N/S ratio and sulfur/polysulfide is formed at a low N/S ratio (Greene et al., 2003). Hence, corrosion rates under conditions of partial souring control need to be determined to fully appreciate the corrosion risk for oil- and gas-transporting pipelines and processing facilities, associated with nitrate injection.

POTENTIAL FOR IMPROVED PRODUCTION

One of the key problems in enhancing conventional oil production is how to prevent the injected water from traveling the path of least resistance, causing less and less oil to be swept to the surface over time. Production of 5 to 10% oil and 95 to 90% water is not uncommon in marginal conventional reservoirs, such as in the Western Canadian Sedimentary Basin, which have been water-injected for a considerable length of time. Enhanced recovery methods can include microbially enhanced oil recovery, which involves injection of bulk carbohydrates together with nutrients (e.g., phosphate) and a microbial inoculum. This yields large amounts of fermentation products and some biomass. The former include organic acids, dissolving carbonates and altering the micropore structure of the oil-holding rock; alcohols and other organics, dissolving into the oil and reducing viscosity; and gases, increasing reservoir pressure (McInerney et al., 2005). Production of biosurfactants (e.g., lipopeptides or rhamnolipids), which reduce the capillary pressure that holds the oil, is also often intended. A fermentation balance for products, including biosurfactants, formed in situ in an oil field injected with glucose, nutrients, and a fermenting *Bacillus* strain has recently been provided (Youssef et al., 2007).

The need to provide both a hydrocarbon substrate and a fermenting inoculum designed for high yields of biosurfactant adds to the complexity of traditional microbially enhanced oil recovery. A simpler procedure would be to stimulate microbial growth downhole by injection of electron acceptors (nitrate or oxygen) and nutrients (e.g., phosphate) only. Indeed, continuous injection of nitrate into oil fields with the initial objective of reducing souring has been reported to result in increased oil production (Dennis and Hitzman, 2007). Enhanced oil recovery resulting from nitrate or oxygen injection, possibly due to biosurfactant production by bacteria downhole reducing the injected electron acceptors, has been patented (Sunde, 1992; Sunde and Torsvik, 2001). At present the mechanism underlying this process is unclear: it has not been shown experimentally, either in laboratory bioreactors or in the field, that continued nitrate injection yields high concentrations of biosurfactant or enhances oil recovery through other mechanisms, like formation of biomass and associated exopolysaccharide blocking paths of least resistance.

CONCLUSIONS

Development of biotechnologies has the potential to improve fossil fuel production efficiency and thereby reduce the environmental impact of this dominant energy resource. In particular the injection of nitrate into oil fields holds promise to reduce souring and possibly to improve production. Nitrate injection changes the microbial community downhole to one in which the activities of hNRB and in some cases NR-SOB are more prominent. Further research is needed to define application protocols that prevent increased corrosion, a risk that nitrate injection aims to prevent.

Acknowledgment. Research in my laboratory was supported by Discovery and Strategic Grants from the Natural Science and Engineering Research Council of Canada (NSERC).

REFERENCES

Cord-Ruwisch, R. and F. Widdel. 1986. Corroding iron as a hydrogen source for sulphate-reducing bacteria. *Appl. Microbiol. Biotechnol.* 25:169–174.

Crolet, J.-L. 2005. Microbial corrosion in the oil industry: a corrosionist's view, p. 143–169. *In* B. Ollivier and M. Magot (ed.), *Petroleum Microbiology*. ASM Press, Washington, DC.

Dennis, D. M., and D. O. Hitzman. 2007. Advanced nitrate-based technology for sulfide control and improved oil recovery, Paper 106154. *In SPE International Symposium on Oilfield Chemistry.* Society of Petroleum Engineers, Houston, TX.

Dinh, H. T., J. Kuever, M. Mussmann, A. W. Hassel, M. Stratmann, and F. Widdel. 2004. Iron corrosion by novel anaerobic microorganisms. *Nature* 427:829–832.

Dolla, A., B. K. J. Pohorelic, J. K. Voordouw, and G. Voordouw. 2000. Deletion of the *hmc* operon of *Desulfovibrio vulgaris* subsp. *vulgaris* Hildenborough hampers hydrogen metabolism and low-redox-potential niche establishment. *Arch. Microbiol.* 174:143–151.

Eckford, R. E., and P. M. Fedorak. 2002. Planktonic nitrate-reducing bacteria and sulfate-reducing bacteria in some western Canadian oil field waters. *J. Ind. Microbiol. Biotechnol.* 29:83–92.

Gevertz, D., A. J. Telang, G. Voordouw, and G. E. Jenneman. 2000. Isolation and characterization of strains CVO and FWKO B: two novel nitrate-reducing, sulfide-oxidizing bacteria isolated from oil field brine. *Appl. Environ. Microbiol.* 66:2491–2501.

Greene, E. A., V. Brunelle, G. E. Jenneman, and G. Voordouw. 2006. Synergistic inhibition of biogenic sulfide production by combinations of the metabolic inhibitor nitrite and biocides. *Appl. Environ. Microbiol.* 72:7897–7901.

Greene, E. A., C. Hubert, M. Nemati, G. E. Jenneman, and G. Voordouw. 2003. Nitrite reductase activity of sulfate-reducing bacteria prevents their inhibition by nitrate-reducing, sulfide-oxidizing bacteria. *Environ. Microbiol.* 5:607–617.

Hall, C., P. Tharakan, J. Hallock, C. Cleveland, and M. Jefferson. 2003. Hydrocarbons and the evolution of human culture. *Nature* 426:318–322.

Hamilton, W. A. 2003. Microbially influenced corrosion as a model system for the study of metal microbe interactions: a unifying electron transfer hypothesis. *Biofouling* 19:65–76.

Haveman, S. A., E. A. Greene, C. P. Stilwell, J. K. Voordouw, and G. Voordouw. 2004. Physiological and gene expression analysis of inhibition of *Desulfovibrio vulgaris* Hildenborough by nitrite. *J. Bacteriol.* 186:7944–7950.

Haveman, S. A., E. A. Greene, and G. Voordouw. 2005. Gene expression analysis of the mechanism of inhibition of *Desulfovibrio vulgaris* Hildenborough by nitrate-reducing, sulfide-oxidizing bacteria. *Environ. Microbiol.* 7:1461–1465.

He, Q., K. H. Huang, Z. He, E. J. Alm, M. W. Fields, T. C. Hazen, A. P. Arkin, J. D. Wall, and J. Zhou. 2006. Energetic consequences of nitrite stress in *Desulfovibrio vulgaris* Hildenborough, inferred from global transcriptional analysis. *Appl. Environ. Microbiol.* 72:4370–4381.

Heidelberg, J. F., R. Seshadri, S. A. Haveman, C. J. Hemme, I. A. Paulsen, J. F. Kolonay, J. A. Eisen, N. Ward, B. Methe, L. M. Brinkac, S. C. Daugherty, R. T. Deboy, R. J. Dodson, A. S. Durkin, R. Madupu, W. C. Nelson, S. A. Sullivan, D. Fouts, D. H. Haft, J. Selengut, J. D. Peterson, T. M. Davidsen, N. Zafar, L. Zhou, R. Radune, G. Dimitrov, M. Hance, K. Tran, H. Khouri, J. Gill, T. R. Utterback, T. V. Feldblyum, J. D. Wall, G. Voordouw, and C. M. Fraser. 2004. The genome sequence of the anaerobic, sulfate-reducing bacterium *Desulfovibrio vulgaris* Hildenborough: consequences for its energy metabolism, biocorrosion and reductive metal bioremediation. *Nat. Biotechnol.* 22:554–559.

Hitzman, D. O., G. T. Sperl, and K. A. Sandbeck. 1995. Method for reducing the amount of and preventing the formation of hydrogen sulfide in aqueous system. U.S. patent 5405531.

Hubert, C., M. Nemati, G. E. Jenneman, and G. Voordouw. 2003. Containment of biogenic sulfide production in continuous up-flow, packed-bed bioreactors with nitrate or nitrite. *Biotechnol. Prog.* 19:338–345.

Hubert, C., M. Nemati, G. Jenneman, and G. Voordouw. 2005. Corrosion risk associated with microbial souring control using nitrate or nitrite. *Appl. Microbiol. Biotechnol.* 68:272–282.

Hubert, C., and G. Voordouw. 2007. Oil field souring control by nitrate-reducing *Sulfurospirillum* spp. that outcompete sulfate-reducing bacteria for organic electron donors. *Appl. Environ. Microbiol.* 73:2644–2652.

Jackson, B. E., and M. J. McInerney. 1996. Thermophilic denitrifying bacteria isolated from petroleum reservoirs and the environmental factors that influence their metabolism, p. 751–761. *In 3rd International Petroleum Environmental Conference, Albuquerque, NM.*

Kaster, K. M., A. Grigoriyan, G. Jenneman, and G. Voordouw. 2007. Effect of nitrate and nitrite on two thermophilic, sulfate-reducing enrichments from an oil field in the North Sea. *Appl. Microbiol. Biotechnol.* 75:195–203.

Larsen, J. 2002. Downhole nitrate applications to control sulfate reducing bacteria activity and reservoir souring. Paper 02025. *In Corrosion 2002.* NACE International, Houston, TX.

Larsen, J., M. H. Rod and S. Zwolle. 2004. Prevention of reservoir souring in the Halfdan field by nitrate injection. Paper 04761. *In Corrosion 2004.* NACE International, Houston, TX.

McInerney, M. J., D. P. Nagle, and R. M. Knapp. 2005. Microbially enhanced oil recovery: past present and future, p. 215–237. *In* B. Ollivier and M. Magot (ed.), *Petroleum Microbiology*. ASM Press, Washington, DC.

Myhr, S., B. L. P. Lillebo, E. Sunde, J. Beeder, and T. Torsvik. 2002. Inhibition of microbial H_2S production in an oil reservoir model column by nitrate injection. *Appl. Microbiol. Biotechnol.* 58:400–408.

Reinsel, M. A., J. T. Sears, P. S. Steward, and M. J. McInerney. 1996. Control of microbial souring by nitrate, nitrite or glutaraldehyde injection in a sandstone column. *J. Ind. Microbiol.* 17:128–136.

Rossi, M., W. B. R. Pollock, M. Reij, R. Fu, R. Keon, and G. Voordouw. 1993. The *hmc* operon of *Desulfovibrio vulgaris* Hildenborough encodes a potential transmembrane redox protein complex. *J. Bacteriol.* 175:4699–4711.

Salinas, M. B., M. L. Fardeau, J. L. Cayol, L. Casalot, B. K. Patel, P. Thomas, J. L. Garcia, and B. Ollivier. 2004. *Petrobacter succinatimandens* gen. nov., sp. nov., a moderately thermophilic, nitrate-reducing bacterium isolated from an Australian oil well. *Int. J. Syst. Evol. Microbiol.* 54:645–649.

Simon, J. 2002. Enzymology and bioenergetics of respiratory nitrite ammonification. *FEMS Microbiol. Rev.* 24:285–309.

Sturman P. J., D. M. Goeres, and M. A. Winters. 1999. Control of hydrogen sulfide in oil and gas wells with nitrite injection, SPE 56772. *Proceedings of the SPE Annual Technical Conference.* Society of Petroleum Engineers, Richardson, TX.

Sunde, E. 1992. Method of microbial enhanced oil recovery. Patent WO 92/13172.

Sunde, E., B.-L. P. Lillebo, G. Bodtker, T. Torsvik, and T. Thorstenson. 2004. H_2S inhibition by nitrate injection on the Gullfaks field, Paper 04760. *In Corrosion 2004.* NACE International, Houston, TX.

Sunde, E., and T. Torsvik. 2001. Method of microbial enhanced oil recovery. Patent WO 01/33040.

Sunde, E., and T. Torsvik. 2005. Microbial control of hydrogen sulfide production in oil reservoirs, p. 201–213. *In* B. Ollivier and M. Magot (ed.), *Petroleum Microbiology*. ASM Press, Washington, DC.

Telang, A.J., S. Ebert, J. M. Foght, D. W. S. Westlake, G. E. Jenneman, D. Gevertz, and G. Voordouw. 1997. The effect of nitrate injection on the microbial community in an oil field as monitored by reverse sample genome probing. *Appl. Environ. Microbiol.* **63:** 1785–1793.

Thorstenson, T., G. Bodtker, B. P. Lillebo, T. Torsvik, E. Sunde, and J. Beeder. 2002. Biocide replacement by nitrate in seawater injection systems. Paper 02025. *In Corrosion 2002.* NACE International, Houston, TX.

Van Ommen Kloeke, F., R. D. Bryant, and E. J. Laishley. 1995. Localization of cytochromes in the outer membrane of *Desulfovibrio vulgaris* (Hildenborough) and their role in anaerobic corrosion. *Anaerobe* **1:**351–358.

Vance, I., and D. R. Thrasher. 2005. Reservoir souring: mechanisms and prevention, p. 123–142. *In* B. Ollivier and M. Magot (ed.), *Petroleum Microbiology*. ASM Press, Washington, DC.

Vik, E. A., A. O. Janbu, F. Garshol, L. B. Henninge, S. Engebretsen, C. Kuijvenhoven, D. Oliphant, and W. P. Hendriks. 2007. Nitrate-based souring mitigation of produced water: side effects and challenges from the Draugen produced-water reinjection pilot. Paper 106178. *In SPE International Symposium on Oilfield Chemistry.* Society of Petroleum Engineers, Houston, TX.

Von Wolzogen Kuehr, C. A. H., and I. S. van der Vlugt. 1934. The graphitization of cast iron as an electrobiochemical process in anaerobic soil. *Water* **18:**147–165.

Voordouw, G. 1999. Microbial communities in oil fields, p. 313–330. *In* F. G. Priest and M. Goodfellow (ed.), *Applied Microbial Systematics*. Kluwer Academic Publishers, Dordrecht, The Netherlands.

Voordouw, G., B. Buziak, S. Lin, A. Grigoriyan, K. M. Kaster, G. E. Jenneman, and J. J. Arensdorf. 2007. Use of nitrate or nitrite for the management of the sulfur cycle in oil and gas fields. Paper 106288. *SPE International Symposium on Oil Field Chemistry.* Society of Petroleum Engineers, Houston, TX.

Wolfe, B. M., S. M. Lui, and J. A. Cowan. 1994. Desulfoviridin, a multimeric-dissimilatory sulfite reductase from *Desulfovibrio vulgaris* (Hildenborough) purification, characterization, kinetics and EPR studies. *Eur. J. Biochem.* **223:**79–89.

Youssef, N., D. R. Simpson, K. E. Duncan, M. J. McInerney, M. Folmsbee, T. Fincher, and R. M. Knapp. 2007. In situ biosurfactant production by *Bacillus* strains into a limestone petroleum reservoir. *Appl. Environ. Microbiol.* **73:**1239–1247.

Bioenergy
Edited by J. Wall et al.

Chapter 30

Microbial Approaches for the Enhanced Recovery of Methane and Oil from Mature Reservoirs

JOSEPH M. SUFLITA AND MICHAEL J. MCINERNEY

INTRODUCTION AND PERSPECTIVE

Energy is essential to all life on the planet, but humans differentially require more of this resource as they employ machines to do work to raise their standard of living. The demand for energy is unlikely to wane, since world population growth is projected to increase by nearly 45% over the next 4 decades to more than 9 billion people by mid-century (U.S. Census Bureau, International Data Base, August 2006; http://www.census.gov/ipc/www/world.html). The addition of roughly 75 million people each year places added stress on all Earth resources. The demand for energy will increase even more rapidly than the population as world living standards trend upwards. The numbers are truly staggering. In about 3 decades, energy demand will increase 71%, from 421 to 722 quadrillion Btu by the year 2030 (Energy Information Administration, 2006). Economic prosperity and national security for the foreseeable future depend on how societies manage several energy-related grand challenges. These include meeting the increased needs for energy, increasing the efficiency of energy use, developing new supplies, and doing so in an environmentally responsible and sustainable fashion.

Where will energy for an ever-growing population come from? To answer, we must first consider how it is currently supplied. Worldwide demand for energy is now met largely by the combustion of fossil fuels—oil, coal, and natural gas. Collectively, these sources supply $\geq$85% of world energy needs (Energy Information Administration, 2006). This degree of reliance on finite fossil fuel reserves and concern over the impact of combustion-associated CO_2 emissions have prompted greater interest in diversifying the global energy portfolio with more carbon-free or carbon-neutral fuels. However, even the most optimistic projections suggest that renewable energy sources will meet less than 10% of world requirements through 2030 (Energy Information Administration, 2006). So fossil fuels are very likely to retain their position as the dominant form of energy on the planet over this same time period. In fact, we will likely always need fossil fuels since we do much more with them than simply burn them. For example, we use many petroleum components as raw materials for manufacturing processes.

Therefore, even if all the biotechnological energy sources mentioned in the other chapters of this book are enormously successful, fossil fuels will remain absolutely vital to the economic and social well-being of all modern societies. However, environmental sensitivity over global climate change could influence the patterns of fossil energy use. In this regard, natural gas (>95% methane) is likely to become increasingly important, as it is a cleaner-burning energy source resulting in less CO_2, NO_x, SO_2, and particulates per Btu generated than other fossil fuels. Methane is an integral part of the global carbon cycle and, as such, is quite a valuable biofuel that is also sustainable, being both produced and consumed by microorganisms. It is indeed a more potent greenhouse gas than carbon dioxide when released to the atmosphere, but anthropogenic emissions can be captured or converted to useful forms, transported, and used to do work.

Eventually, the world will transition to altered energy use patterns, but fossil fuels will remain a part of the overall energy equation. Given our current dependency on fossil fuels and the dwindling supplies thereof, efforts must be made to ensure availability of these resources and to minimize the energy-associated environmental footprint. To be certain, significant advances in new technology will be required to address the grand challenges mentioned above. This chapter tries to indicate how biotechnology might assist in achieving these goals with specific respect to fossil energy reserves.

Joseph M. Suflita • Institute for Energy and the Environment and Department of Botany and Microbiology, Sarkeys Energy Center, Boyd St., University of Oklahoma, Norman, OK 73019. **Michael J. McInerney** • Department of Botany and Microbiology, 770 Van Vleet Oval, University of Oklahoma, Norman, OK 73019.

THE U.S. DOMESTIC ENERGY NEED

There is an ever-widening gulf between U.S. energy needs and the nation's ability to meet the demand. Projections for the immediate future indicate that this situation will be exacerbated, as domestic energy production remains stable while consumption continues to grow despite improvements in energy efficiency, conservation measures, and renewable energy usage. U.S. energy consumption will increase from roughly 100 to 131 quadrillion Btu, or 31%, by 2030 (Energy Information Administration, 2007). Like any other country, the United States will attempt to ensure adequate energy supplies via a myriad of mechanisms, but this diversification has meant greater reliance on external suppliers and complex foreign entanglements. Any serious attempt to reverse this trend will need to consider improved mechanisms for tapping unconventional energy supplies as well as the development of new technology to more fully exploit domestic reserves.

Meeting the demand for energy in the transportation sector will continue to be a critical domestic energy need. Domestic liquid fuel consumption is projected to increase by >25% by 2030, with the use of liquid fuel for transportation accounting for the vast majority of this demand (Energy Information Administration, 2007). The use of nonpetroleum sources such as ethanol and of unconventional oil sources (shale oil, gas-to-liquids, and coal-to-liquids) will increase substantially, but each will only account for <10% of the demand by 2030. Thus, the dependence on imports to meet >60% of the demand for liquid fuels will aggravate trade deficits and intensify reliance on external suppliers.

While the likelihood of discovering large, new oil reserves is low, a large untapped resource of currently unrecoverable oil exists if technologies can be developed to exploit this resource. Current oil production technologies recover only about one-third to one-half of the oil contained in reservoirs (Planckaert, 2005). It is estimated that more than 300 billion barrels (47.6 Gm^3) of oil remain in U.S. reservoirs after conventional technologies reach their economic limit (Lundquist et al., 2001). Globally, only about 1 trillion barrels (0.16 Tm^3) of oil have been recovered from a total world supply of about 2 to 4 trillion barrels (0.3 to 0.6 Tm^3) (Hall et al., 2003). Although many natural gas reserves also remain unexploited, there are substantive obstacles to the exploitation of these resources. A critical feature of the U.S. energy sector is the importance of marginal wells whose production is <10 barrels of oil or <60,000 ft^3 of natural gas per day. Currently, 27% of the oil (about the same amount imported from Saudi Arabia) and 8% of the natural gas produced in the United States onshore (excluding Alaska) are produced from marginal wells. These are wells that are marginal in both production and profitability and are at risk of being prematurely abandoned. It is estimated that about 110 million barrels of oil was lost due to the plugging and abandonment of marginal wells between 1994 and 2003. Increasing energy recovery from these domestic resources would seemingly be a critical element of any overarching national policy. However, improving production from marginal wells is of little concern to the major oil companies.

FACTORS THAT LIMIT OIL PRODUCTION

The accumulation of small particles of clay, scale, paraffins, and asphaltenes in the region near the well bore can block the main drainage routes and prevent or reduce the migration of oil into the well (Whillhite, 1986; Planckaert, 2005). Thus, recoverable oil may be available only a short distance from the well if the drainage patterns can be reestablished. Technologies that repair or prevent such depositions can lead to marked stimulations in oil production. It is important to note that such technologies improve oil production rate but may not increase the ultimate amount of oil that is recovered from a reservoir. However, increasing oil production rates or reducing the need for well servicing makes a well more profitable to operate and thus extends the economic life of marginal fields.

There are several main reasons why conventional waterflooding technology (e.g., the injection of brine in the reservoir) recovers only a fraction of the oil in place (Whillhite, 1986; Planckaert, 2005). The waterflood may not uniformly contact the entire reservoir due to permeability variations. In effect, the brine can preferentially flow through high-permeability regions and avoid low-permeability regions. The viscosity of the brine, being much less than that of oil, moves more rapidly in the reservoir, pushing through the oil and reaching the production well first. Lastly, large amounts of oil are entrapped in small pores by capillary pressure, and large reductions (>100-fold) in the interfacial tension between the oil and aqueous phases are needed to mobilize this oil (Sabatini et al., 2000). This chapter focuses on how microbial technologies can be used to overcome these limitations and increase production and recovery of energy.

MICROBIAL METHANE RECOVERY

Of course, one approach to increasing energy recovery from oil reservoirs is to convert the available carbon to a form that is more readily recoverable. Such is the logic of converting oil to methane and carbon dioxide (Fig. 1A). This scenario is predicated on the susceptibility of various hydrocarbons to anaerobic biodegradation. The subject has been reviewed repeatedly in recent years (Heider et al., 1999; Beller, 2000; Sporman and Widdel, 2000; Zwolinski et al., 2000; Widdel and Rabus, 2001; Boll et al., 2002; Chakraborty and Coates, 2004; Meckenstock et al., 2004; Suflita et al., 2004; Rabus, 2005; Heider, 2007), and no attempt is made to do so again here. It is very clear that the oxidation of numerous hydrocarbons can be coupled with the reduction of various electron acceptors and initial reactions involved are often novel and relatively few. This no doubt reflects the fact that hydrocarbons are ancient forms of organic matter and there was selection and evolutionary conservation for preferred catalytic pathways. Several of the requisite microorganisms have

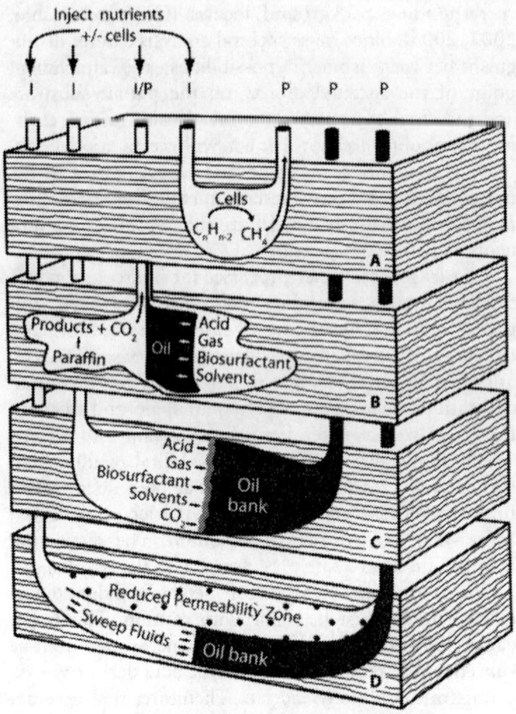

Figure 1. Biotechnological approaches for the recovery of energy from residual oil deposits. Panels A through D represent areas of the terrestrial subsurface amenable to the different processes described in the text and in Table 1. (A) Hydrocarbon metabolism results in methane production. (B) Microbial well stimulation removes paraffins and/or creates useful products that increase oil drainage into the well. (C) Microbially enhanced waterflooding where microbial metabolism creates useful products to mobilize entrapped oil. (D) Microbially selective plugging blocks high-permeability channels and redirects the recovery fluid into previously bypassed regions of the reservoir to recover additional oil.

been identified, the predominant pathways have been clarified, and the biochemical details as well as the genetics continue to emerge. What has become clearer in recent years is that some of the hydrocarbonoclastic anaerobes can couple their metabolism in syntrophic associations that ultimately allow for the conversion of the petroleum components to the smallest hydrocarbon, i.e., methane (Edwards and Grbic-Galic, 1994; Kazumi et al., 1997; Weiner and Lovley, 1998; Beller and Edwards, 2000; Ulrich and Edwards, 2003; Townsend et al., 2003, 2004; Zengler et al., 1999a; Anderson and Lovley, 2000; Siddique et al., 2006, 2007). The amount of methane recoverable from a particular hydrocarbon can be estimated based on the Buswell equation (Symons and Buswell, 1933). For hexadecane, 3.4 mol of methane/mol of carbon dioxide are predicted for every mole of parent substrate utilized:

$$C_{16}H_{34} + 7.5\ H_2O \longrightarrow 3.75\ CO_2 + 12.25\ CH_4$$

Therefore, anaerobic metabolism has the prospect of resulting in a methane gas stream that will require minimal purification. There is also no doubt that methane can be used as a transportation fuel or readily converted to a transportation fuel (van Wechem and Senden, 1994).

Many hydrocarbons are initially oxidized by covalent linkage across the double bond of fumarate to form succinyl derivatives. The reaction was initially established with studies designed to elucidate the anaerobic metabolic fate of toluene. Since denitrifying bacteria grow relatively rapidly at the expense of hydrocarbons, many of the biochemical and genetic details on anaerobic toluene metabolism were established with various strains of *Azoarcus* and *Thauera*. These *Betaproteobacteria* catalyze fumarate addition reactions by a novel glycyl radical-containing enzyme, benzylsuccinate synthase. Other organisms belonging to the *Alpha*- and *Deltaproteobacteria* catalyze toluene decay in an analogous fashion, attesting to the widespread distribution of this capability. In addition, other alkylated mono- and polyaromatic hydrocarbons undergo similar activation reactions when degraded by pure and mixed anaerobic cultures. Strains of pure denitrifiers, sulfate reducers, and methanogenic enrichments are able to form methyl-, dimethyl- and naphthylsuccinic acid intermediates during the metabolism of dimethyl- or trimethylbenzenes as well as methylnaphthalene (Annweiler et al., 2000; Beller and Edwards, 2000; Beller and Spormann, 1997a, 1997b; Beller and Spormann, 1998, 1999; Coschigano et al., 1998; Kane et al., 2002; Krieger et al., 1999; Kühner et al., 2005; Leuthner et al., 1998; Morasch et al., 2004; Rabus and Heider, 1998; Shinoda et al., 2005; Shinoda et al., 2004; Verfurth et al., 2004; Zengler et al., 1999b).

This metabolic strategy can also be extrapolated to the anaerobic biodegradation of normal and cyclic alkanes (Callaghan et al., 2006; Cravo-Laureau et al., 2005; Kropp et al., 2000; Rabus et al., 2001; Wilkes et al., 2002; Rios-Hernandez et al., 2003). Pure cultures of *Beta*-, *Gamma*-, and *Deltaproteobacteria* are known to catalyze the anaerobic decomposition of various *n*-alkanes, but the attachment point of the succinyl moiety is not the terminal methyl group of the parent substrate. Rather, fumarate is added to a subterminal (C-2 or C-3) carbon atom to form an alkylsuccinate intermediate. The substrate specificity associated with pure *n*-alkane-degrading bacteria is rather narrow. Typically the difference between the smallest and largest *n*-alkane attacked by pure cultures varies by only 6 to 10 carbons in chain length. Nevertheless, the stoichiometry of alkane mineralization, the metabolites produced in hydrocarbon-degrading cultures, and in situ detection and production of alkylsuccinates in petroleum-contaminated field locales show that the diversity of *n*-alkanes known to be amenable to this bioconversion, while quite wide (at least from C_3 to C_{16}), is not fully appreciated.

Collectively, anaerobes exhibit an impressive array of *n*-alkanes that can be metabolized, but the extremes deserve special attention. The first step in anaerobic oxidation of methane is believed to be a reversal of the final step in methanogenesis (Boetius et al., 2000; Orphan et al., 2001; Raghoebarsing et al., 2006). The catalytic entity involved is minimally a consortium of nitrate- or sulfate-reducing bacteria and archaea. However, an alternate pathway is worth suggesting. While investigating corrosion processes in oil pipelines on the north slope of Alaska, we detected the corresponding alkylsuccinate metabolites that would be expected if methane, ethane, propane, and butane were metabolized by fumarate addition reactions (our unpublished observation). This finding is at least consistent with the long-term recycling of these light hydrocarbons to help pressurize subsurface formations during oil extraction procedures. Since the putative intermediates can be formed in other ways that do not necessarily involve hydrocarbon metabolism (Kawamura and Kaplan, 1987), such findings, while intriguing, must be considered speculative and need confirmation through the isolation of the requisite organisms. Comparable investigations on the metabolism of high-molecular-weight paraffins reveal that *n*-alkanes up to 50 carbons in chain length can be biodegraded in the absence of oxygen (L. M. Gieg, K. E. Duncan, and J. M. Suflita, presented at the 11th International Symposium on Microbial Ecology, Vienna, Austria, August 2006). To our knowledge, this represents the highest-molecular-weight *n*-alkane subject to either aerobic or anaerobic decay. Paraffinic substrates are solid at room temperature and associated with a host of problems in the oil industry (Mistra et al., 1995). However, the mechanism(s) for the anaerobic bioconversion of solid paraffins is not known with certainty.

Alicyclic compounds comprise a major fraction (up to 12% [wt/wt]) of the organic molecules in hydrocarbon mixtures depending on the nature of the originating petroleum formation (Malins, 1977). There is direct evidence that these molecules can also be activated by fumarate addition (Rios-Hernandez et al., 2003) and mineralized by anaerobic microorganisms. However, many times the evidence attesting to the metabolism of these compounds is only inferential. Mass spectral detection of the resulting succinate derivatives are two mass units less than the corresponding *n*-alkane metabolites of comparable carbon number (C_{n-2}). These derivatives are regularly found as putative intermediates in hydrocarbon-contaminated areas but not in corresponding background locales (Gieg and Suflita, 2002, 2005). Since mass spectral analysis cannot distinguish between isomeric possibilities, the attachment point of the succinyl moiety on the parent substrate and therefore the exact chemical identity of the resulting metabolite are not known with certainty. Metabolites with identical mass spectral features might be expected if alkenes were similarly metabolized, but these molecules are rarely quantitatively important components in petroleum mixtures.

At least three other processes for hydrocarbon metabolism under anaerobic conditions have been suggested including direct carboxylation, methylation, and oxygen-independent hydroxylation reactions. Carboxylation has been reported as important for unsubstituted aromatic hydrocarbons like naphthalene and phenanthrene, as well as several *n*-alkanes (Zhang and Young, 1997; Annweiler et al., 2002; So et al., 2003; Callaghan et al., 2006). Methylation has been documented for both benzene and naphthalene (Ulrich et al., 2005; Safinowski and Meckenstock, 2006). The distinction between these two fate processes is not always straightforward, and interpretation of metabolites detected in culture fluids must be made with caution. This is because substrates that are initially methylated can be subsequently oxidized to carboxylic acid derivatives via a fumarate addition reaction. Therefore, the mere detection of a carboxylated metabolite does not convincingly argue that direct carboxylation from inorganic carbon is operative.

The direct carboxylation of hydrocarbons has been most convincingly demonstrated with $^{13}CO_2$ incorporation studies into isotopically labeled metabolites. In cases where parent hydrocarbons were activated by other mechanisms and subsequently metabolized, no label would be expected in carboxylated metabolites after incubation with ^{13}C-bicarbonate. For instance,

2-methylnaphthalene-degrading enrichments produced 2-naphthoic acid but did not incorporate ^{13}C in the carboxyl group of the metabolite when incubated with labeled bicarbonate (Sullivan et al., 2001). Similarly, a sulfidogenic consortium capable of benzene mineralization did not incorporate ^{13}C carbon in the carboxyl group of benzoate when the cells were supplemented with ^{13}C-bicarbonate, suggesting that the carboxyl group in the metabolite was not a product of direct carboxylation (Phelps et al., 2001). In contrast, studies attesting to the direct incorporation of ^{13}C-bicarbonate strongly suggest that more-elaborate explanations for the resulting carboxylated metabolites are simply unnecessary.

The formation of hydroxylated metabolites is also one of the more prominent bioconversion processes for anaerobic hydrocarbon decay. In fact, early reports show that the oxygen in the resulting metabolite originates from water. Hydroxylation has been suggested for the anaerobic biodegradation of benzene, toluene, ethylbenzene, and propylbenzene (Vogel and Grbic-Galic, 1986; Caldwell and Suflita, 2000; Eriksson et al., 2005); Weiner and Lovley, 1998; Chakraborty and Coates, 2005; Ulrich et al., 2005, Knicmeyer et al., 2003). Alkylphenols have been suggested as indicators of microbial hydrocarbon metabolism in petroliferous deposits (Lucach et al., 2002). However, this reaction has not been reported for normal or cyclic alkane hydrocarbons.

Of course, the bacteria responsible for the initiation of petroleum biodegradation by the aforementioned mechanisms are not capable of directly converting these substrates into methane. They must work in concert with methanogens, which are notoriously limited to simple electron donors (predominantly hydrogen or acetate) to support their metabolism. As is typical in methanogenic environments, the catalytic entity involved in the mineralization of complex forms of organic matter is not an individual microorganism. Rather, several organisms enter into thermodynamically based syntrophic associations. Complex forms of organic matter are broken down into simpler components and eventually into volatile fatty acids or alcohols. These components, in turn, are converted to acetate, hydrogen, and carbon dioxide by syntrophic bacteria that rely on the removal of acetate and hydrogen by methanogens or other organisms to render the energetics of this bioconversion favorable. Hydrogenotrophic methanogens produce methane from hydrogen and carbon dioxide, whereas aceticlastic methanogens produce methane and carbon dioxide from acetate. Of course, homoacetogens can produce acetate from hydrogen and carbon dioxide, and this end product is then a suitable substrate for methanogenesis.

It is in this sense that anaerobic hydrocarbon biodegradation can best be viewed, but the syntrophic association with methanogens need not be too elaborate. For instance, a completely oxidizing sulfate-reducing isolate, *Desulfoglaeba alkanexedens* (Davidova et al., 2006) was capable of anaerobic dodecane utilization in the absence of sulfate but in a coculture with *Methanospirillum hungatei* JF1 (I. Davidova and J. M. Suflita, unpublished observation). This isolate is a deltaproteobacterium belonging to the order *Syntrophobacterales*. These bacteria are known to couple with methanogens to form syntrophic associations. In effect, the methanogen serves as the electron acceptor for the sulfate reducer that can then catalyze transformations that the methanogen is incapable of doing. It is therefore not unreasonable to presume that similar microbial associations can exist in petroliferous subsurface formations and catalyze hydrocarbon conversions to methane and carbon dioxide. In their study of methanogenic hexadecane utilization, Zengler et al. (1999a) postulated on a consortium that included bacteria to convert alkanes to acetate and H_2 and both types of methanogens to convert the latter molecules into methane. Indeed, most bacterial clones sequenced from this consortium were *Deltaproteobacteria* related to the genus *Syntrophus*, while archaeal clones included both acetoclastic (*Methanosaeta*) and hydrogenotrophic (*Methanospirillum* and *Methanoculleus*) methanogens. Similarly, both types of methanogens have been implicated in the conversion of hydrocarbons to methane (Kleikemper et al., 2005; Dojka et al., 1998; Struchtemeyer et al., 2005; Watanabe et al., 2002).

Ecologically, one would expect methanogenesis as a predominant process in organic-matter-rich, but electron-acceptor (e.g., oxygen, nitrate, iron, sulfate, etc.)-depleted, environments that contain the requisite microorganisms and appropriate environmental conditions. Geological evidence suggests that the inherent ability of anaerobes to consume hydrocarbons is a phenomenon that has occurred for millennia depending on nutrient availability, water, temperature, and the presence of the requisite microorganisms (Head et al., 2003; Larter et al. 2006). Aitken et al. (2004) detected metabolites characteristic of anaerobic hydrocarbon decay in oil samples from across the globe. Magot (2005) recently reviewed evidence for the detection of a variety of anaerobic microorganisms including acetogens and methanogens in petroleum reservoir fluids. Methane is now believed to be the primary by-product of anaerobic oil biodegradation in many petroliferous deposits (Sakata et al., 1997; Pallasser, 2000; Head et al., 2003). The preferential microbial utilization of light hydrocarbons in reservoirs is believed to be responsible for the formation of massive heavy-oil deposits on the planet. Heavy oils differ from others by their relatively high viscosity, density (low API gravity), acidity, and concentration of resin and asphaltic components (Huang and Larter, 2005).

Most likely, such alterations occurred over geologic time in anaerobic petroleum reservoirs at temperatures less than 80°C (Roling et al., 2003; Wilhelms et al., 2001). This would appear to be at odds with the observation that nondegraded oils are found in shallow basins below this temperature. However, many of these reservoirs have been uplifted from deeper, hotter regions of the Earth where critical microbial components were likely inactivated or eliminated by the ambient temperatures. When such reservoirs were subsequently uplifted to cooler regions where conditions are more favorable for microbial proliferations, biodegradation was limited. One of the more obvious reasons for the lack of metabolic activity is the absence of the requisite hydrocarbon-degrading bacteria. Therefore, the prospect of inoculating such reservoirs with organisms capable of converting hydrocarbons to methane seems entirely feasible. However, given the rather narrow substrate specificities associated with the few hydrocarbonclastic anaerobes known to date, a complex microbial assemblage is likely to be necessary to attack the wide range of petroleum components. We have enriched such an inoculum from a gas condensate-contaminated aquifer for its ability to completely mineralize the *n*-alkane fraction (C_{13} to C_{34} range) of a model petroleum to methane and carbon dioxide (Townsend et al., 2003). This inoculum was cultivated for over 2 years on oil and successfully used to attack residual oil in crushed sandstone cores from a marginal well field in Oklahoma. Only incubations receiving the inoculum were able to convert the residual oil to methane (Suflita et al., 2004). Such findings help lay a foundation for the inoculation of marginal reservoirs with the goal of stimulating anaerobic oil decomposition and the enhanced recovery of energy.

It is clear that the aforementioned reactions represent important anaerobic activation mechanisms for structurally diverse hydrocarbons, and one can envision that the same type of transformations might occur in oil reservoirs (Head et al., 2003; Roling et al., 2003; Larter et al., 2003). Gases of biological origin often accompany the biodegradation of oils in deep anoxic horizons (James and Burns, 1984; Pallasser, 2000). Isotopically light methane ($\delta^{13}C$ from −45 to −59%), consistent with a biological origin, and in situ rates of methane production ranging from 1.3 to 80 nmol of methane liter $^{-1}$ day^{-1} were observed in oil fields under various environmental conditions (Ivanov et al., 1983; Borzenkov et al., 1997; Nazina et al., 1995; Bonch-Osmolovskaya et al., 2003). In fact, relatively high rates of methanogenesis were estimated in formation waters of the Jurassic horizon despite comparatively high temperatures (2,299 m deep; 84°C). Hybridization of 16S rRNA obtained from formation water with group-specific phylogenetic probes revealed the presence of thermophilic methanogens and heterotrophs (Bonch-Osmolovskaya et al., 2003). Laboratory incubations of formation waters and raw production fluids from two deep high-temperature petroleum reservoirs in California demonstrated active methane production at in situ temperatures (70 to 83°C). Total community DNA analysis revealed archaeal phylotypes closely related to thermophilic methanogens and sulfidogenic archaea as well as bacterial thermophiles such as *Thermatoga* spp., *Thermococcus* spp., *Thermoanaerobacter* spp., and *Desulfothiovibrio* spp. (Orphan et al., 2003). These findings suggest that while "paleosterilization" may help explain the absence of critical microbial components and associated metabolic activity in shallow oil formations that at one time experienced temperatures of >80°C, it does not preclude the selection of organisms able to occupy comparable niches and form metabolically analogous syntrophic associations that result in high-temperature methane formation.

MICROBIAL OIL RECOVERY

In contrast to the bioconversion of fossil fuels to recover energy in the form of methane, microorganisms can also be used to increase petroleum production from reservoirs. There are a number of microbial processes that are potentially useful, either to improve oil production rates from individual wells or to increase the ultimate amount of oil recovered from a reservoir (Fig. 1; Table 1) (ZoBell, 1947a, 1947b). It is important to note that for most microbial oil recovery technologies, multiple products and activities are involved and it is likely that these act synergistically. For example, very little entrapped oil was mobilized from sand-packed columns and sandstone cores when the injected brine contained only a biosurfactant, but >40% of the entrapped oil was recovered when the brine contained the biosurfactant, a polymer, and 2,3-butanediol (Maugdalya et al., 2004).

Hydrocarbon Metabolism

The incomplete oxidation of hydrocarbons could generate alcohols or fatty acids from the hydrocarbon molecule or stimulate the production of biosurfactants and bioemulsifiers derived from microbial cells (Trebbeau de Acevedo and McInerney, 1996). The beneficial effects of these microbial products would be to reduce viscosity and increase the mobility of the oil in the reservoir (Fig. 1B and C; Table 1).

Table 1. Products and activities of microorganisms potentially useful for enhanced oil recovery or improved oil production[a]

Product or activity	Examples	Effect on oil recovery
Gases	CO_2 and CH_4	Reservoir repressurization, oil viscosity reduction
Acids	Acetic, butyric, lactic	Scale removal, improved drainage, porosity increase in carbonates
Solvents	Ethanol, butanol, acetone, 2,3-butanediol	Oil viscosity reduction, wettability alteration, enhancement of interfacial tension reduction
Polymers	Polysaccharides, proteins	Mobility control, permeability rectification
Bioemulsifiers	Heteropolysaccharides, proteins	Oil emulsification, wax and paraffin control, wettability alteration
Biosurfactants	Glycolipids, lipopeptides	Interfacial tension reduction, emulsification, wettability alteration
Hydrocarbon metabolism	Aerobic	Paraffin control, viscosity alteration, improved drainage
	Anaerobic	Methane production and above problems
Biomass formation	Microbial cells	Selective plugging

[a]Adapted from ZoBell (1947a, 1947b).

A number of studies report the use of proprietary mixtures of hydrocarbon-degrading bacteria and nutrients to remove or prevent paraffin deposition and/or to stimulate oil production (Table 2) (Giangiacomo, 1997; Nelson and Schneider, 1993; Pelger, 1991; Streeb and Brown, 1992; Portwood, 1995). Experimental evidence to support hydrocarbon metabolism is often not strong and limited to alterations of the physical properties of crude oil such as viscosity, pour point (the lowest temperature at which oil flows when cooled), and cloud point (the temperature at which paraffins begin to precipitate from a liquid state) (Brown, 1992; Nelson and Schneider, 1993). Other studies found that microbial treatments were beneficial by reducing the frequency of hot-oil treatments (Santamaria and George, 1991). In a pilot-scale study, three facultatively anaerobic microbes isolated from waxy oil production wells in the Liaohe oil field in China were used to treat paraffin-laden wells (He et al., 2003). The authors also reported success in reduction in the frequency of thermal and chemical treatments as well as a substantial increase in crude oil production. In contrast, other investigators found that the use of microbial formulations to treat paraffin deposition was ineffective (Lazar et al., 1999) or inconclusive (Wilson et al., 1993; Ferguson et al., 1996). Stoichiometric data that specifically link hydrocarbon alteration to electron acceptor use are lacking. Gieg et al. (L. Gieg, M. McInerney, J. Suflita, and G. Jenneman, presented at the 11th Annual International Petroleum Environmental Conference, Albuquerque, NM, October 2004) found slight emulsification but no evidence of hydrocarbon metabolism when several crude oils were incubated under a variety of conditions with a proprietary mixture of hydrocarbon-degrading bacteria according to the manufacturer's guidelines. The injection of a commercial formulation of bacteria and nutrients into sandstone cores followed by incubation times of 1 to 8 weeks to allow for in situ microbial growth and activity did not mobilize entrapped oil from cores (Rouse et al., 1992).

Due to the proprietary nature of inoculants, they are typically used under strict manufacturer supervision and have generally not been available for objective scientific scrutiny. While the mechanism of action of commercial hydrocarbon-degrading microbial formulations is at least debatable, the technology has survived in the marketplace for many years and reports attesting to the effectiveness of such products have appeared (Table 2) (Nelson and Schneider, 1993; Pelger, 1991; Portwood, 1995). Several reports indicate that wells treated with the inoculants have less paraffin deposition on production equipment (Pelger, 1991) and require less-frequent servicing than other well treatment technologies (Brown, 1992; Santamaria and George, 1991; Streeb and Brown, 1992). Reductions in operating costs resulting from effective paraffin control would extend the economic lifetime of marginal wells regardless of whether oil production increased. A number of reports indicate the effectiveness of commercial hydrocarbon-degrading consortia in arresting the decline in oil production rates of individual wells (well stimulation) or on a reservoir-wide basis (Table 2) (Brown, 1992; Nelson and Schneider, 1993; Oppenheimer and Hiebert, 1989; Pelger, 1991; Portwood, 1995; Streeb and Brown, 1992). A success rate of microbial treatments of about 78% of 322 projects based on the aforementioned criteria and no reports of a detrimental impact of oil production were noted (Portwood, 1995). Paraffin utilization by anaerobic microorganisms has now been documented (see above) and may ultimately yield an inoculum with a known mechanistic basis.

Stimulation of in situ hydrocarbon metabolism by indigenous microbes by the injection of oxygenated brine with and without the addition of inorganic nutrients is another approach to mobilize entrapped oil (Table 2). In laboratory studies, the continued injection of oxygenated brine and inorganic nutrients decreased the residual oil saturation in sandstone cores (Sunde et al., 1992), and with sufficient oxygen supplies, aerobic hydrocarbon-degrading cultures formed metabolites that

Table 2. Survey of microbial processes to improve or enhance oil recovery

Process	Method	Results	References
Inocula for paraffin removal and/or hydrocarbon metabolism	Periodic injection of a proprietary mixture of microbes, nutrients, and "biocatalyst"	Less-frequent servicing of equipment; arrested or slowed the decline in oil production rate; oil production rates increased for some wells	Nelson and Schneider, 1993; Streeb and Brown, 1992; Pelger, 1991; Portwood, 1995; Maure et al., 2005
In situ hydrocarbon metabolism	Cyclic injection of aerated water with nitrogen and phosphorus sources during waterflood	Oil production rate increased (2 to 200%); in situ microbial activity detected	Ivanov et al., 1993; Matz et al., 1992; Murygina et al., 1995
In situ growth of indigenous bacteria	Injection of inorganic nutrients with or without a carbohydrate during waterflood	Arrested or slowed the decline in oil production rate; oil production increased in one study	Brown et al., 2002; Hitzman et al., 2004; Sheehy, 1990
In situ gas, acid, and solvent production by inoculum	Injection of a mixture of facultative and anaerobic microbes with molasses with or without inorganic nutrients into individual production wells or during waterflood	Oil production rate increased (7–733%); microbial products detected; later technologies with adapted microbial consortia more effective	Hitzman, 1983, 1988; Lazar et al., 1993; Wagner et al., 1995; Nazina et al., 1999
In situ biosurfactant production by inoculum	Injection of a mixture of microbes containing a biosurfactant producer and periodic injection of carbohydrates with or without inorganic nutrients	Oil production rate increased (13–50%); water production rate decreased; large amounts of biosurfactant detected	Bryant et al., 1990, 1993; Youssef et al., 2007
Permeability modification	Injection of carbohydrates with or without other nutrients; inoculum used in some cases	Interwell flow pattern altered; brine injected into new zones; oil production rate increased	Knapp et al., 1992; Gullapalli et al., 2000; Nagase et al., 2002

decreased oil/aqueous phase interfacial tension by 4 orders of magnitude (Kowalewski et al., 2005). In field studies, the injection of oxygen with or without inorganic nutrients increased oil recovery by 15 to 45% in four different oil formations (Ivanov et al. 1993; Matz et al., 1992; Murygina et al., 1995). The average daily increase in oil production reported in one study was over 300 barrels per day per field compared to pretreatment levels (Matz et al., 1992). However, it is not possible to determine from the information provided in these reports how oil recoveries were calculated, e.g., net increase in production above an established baseline or a slower rate in the decline in production.

Several studies report the beneficial effects of the injection of inorganic nutrients on oil production (Table 2). However, the underlying mechanism(s) for these observations is far from clear. Two studies report that the addition of nutrients to the injected brine slowed or arrested the decline in oil production in the treated portion of the reservoir (Brown et al., 2002; Hitzman et al., 2004). Brown et al. (2002) estimate that the treatment extended the economic life of the reservoir by 5 to 12 years and recovered an additional 11,000 m^3 of oil during the first 42 months of the project (Brown et al., 2002). In another study, the injection of inorganic nutrients increased oil production by 40%, a rate that was maintained for more than 1 year, compared to the oil production rate prior to treatment (Sheehy, 1990).

In Situ Fermentative Metabolism

In situ organic acid production can dissolve carbonate minerals and thus could remove the buildup of scale in the vicinity of the well and increase the porosity of carbonate reservoirs (Adkins et al., 1992; Udegbunam et al., 1991) (Fig. 1B and C; Table 1). Solvents could alter the wettability of the oil-rock interface, releasing oil from the porous matrix as well as reducing oil viscosity. Due to the solubility of gases in oil, CO_2 and CH_4 production may swell the oil and reduce its viscosity, which would make the oil more mobile. Laboratory core experiments found that in situ growth of fermentative bacteria recovered from 16 to 32% of the entrapped oil after the sandstone cores had been extensively flooded with brine (Bryant and Douglas, 1988; Marsh et al., 1995). Large amounts of acetate, butyrate, butanol, ethanol, and CO_2 were made, and their production coincided with oil recovery (Marsh et al., 1995). However, the injection of cell-free culture fluids that contained the acids and solvents did not recover entrapped oil, suggesting that in situ gas production, most likely CO_2, or some other mechanism such as permeability reduction was the reason for the enhanced oil recovery (Marsh et al., 1995). Other studies have implicated gas production in the oil recovery process more directly (Behlulgil and Mehmetoglu, 2002; Desouky et al., 1996). In situ CO_2 production could lead to the formation of calcite ($CaCO_3$) precipitates that reduce the porosity and permeability in preferred reservoir pore throats (Chapelle and Bradley, 1997). This could be an effective mechanism to block preferential flow paths and may help explain the flow alterations observed in a hypersaline oil reservoir after molasses addition (Knapp et al., 1992).

In well stimulation approaches (Fig. 1B), a single microorganism or a mixture of microorganisms and a fermentable carbohydrate, usually molasses, is injected into a production well. The process appears to be most effective in carbonate wells with an oil gravity of 875 to 965 kg/m^3, salinity of <100 g/liter, and a temperature around 35 to 40°C (Hitzman, 1983). In many cases, evidence for microbial activity was obtained, including an increase in cell numbers, a decrease in pH and wellhead pressure, or an increase in oil production (Hitzman, 1983, 1988; Lazar, 1991). In one instance, oil production increased by 700% (Wang et al., 1995). However, inconsistency of performance may be a reason why this technology is not more widely used (Hitzman, 1983, 1988; Lazar, 1991).

Analogous to well stimulation technologies, the injection of fermentative bacteria and carbohydrate-based nutrients into a reservoir is used to increase the ultimate amount of oil recovered (Fig. 1C). Oil production increased after microbial activity was detected, suggesting a causal relationship (Hitzman, 1983, 1988; Lazar and Constantinescu, 1985). The effect on oil production has been reported to last for periods of months to years (Hitzman, 1983, 1988; Lazar and Constantinescu, 1985). The stimulation of in situ fermentative metabolism by *Clostridium tyrobutyricum*, which was used as the inoculum, increased oil production compared to control wells by 29% in one case (Nazina et al., 1999) and by 100% in another case (Wagner et al., 1995). Oil production of carbonaceous sandstone reservoirs in Romania increased by 30 to 40 barrels per day when treated with a mixed microbial population adapted for rapid growth under reservoir conditions with molasses as the substrate (Lazar et al., 1988, 1991, 1993).

A decrease in the pH of the reservoir brine by 1 to 2 units (Lazar, 1987; Lazar et al., 1988; Yarbrough and Coty, 1983), and the production of acids (Ohna, 1999), carbon dioxide (Nazina et al., 1999), and butanol (Davidson and Russell, 1988) provide evidence of in situ microbial activity. In some cases, the lack of detailed information on the operation of the oil field makes it difficult to determine whether factors other than the microbial treatments influenced oil recovery. It is also not clear whether the lost revenue due to the cessation of oil production was considered in such assessments. The use of microbial populations adapted to reservoir conditions and large amounts of nutrients may explain the enhanced performance of field trials conducted in the 1990s (Lazar et al., 1993; Wagner et al., 1995; Wang et al., 1995).

Biosurfactants

Lipopeptide biosurfactants made by various *Bacillus* species (Lin et al., 1994; Javaheri et al., 1985; McInerney et al., 1990; Desai and Banat, 1997; Yakimov et al., 1997; Yonebayashi et al., 1997) and the rhamnolipid produced by *Pseudomonas* species (Maier, 2003) reduce the interfacial tension between the hydrocarbon and aqueous phases to very low levels (<0.01 mN/m). In addition, the critical micelle concentrations are low (20 to 50 mg/liter), indicating that the biosurfactants may be effective at low concentrations (Lin et al., 1994). Changes in biosurfactant activity have been correlated to specific changes in the fatty acid composition of lipopeptide biosurfactants (Youssef et al., 2005).

The use of biosurfactants to mobilize entrapped hydrocarbon from model porous systems has met with mixed success (Fig. 1C). The injection of rhamnolipid biosurfactant into sand-packed columns mobilized up to 75% of the residual hexadecane, but over 40 pore volumes of the biosurfactant solutions was needed, which would be impractical for commercial applications (Herman et al., 1997). A number of studies report the recovery of entrapped oil from sandstone cores by in situ biosurfactant production, but oil recoveries were low (5 to 20%) and required multiple cycles of nutrient injection (Marsh et al., 1995; Thomas et al., 1993; Sugihardjo and Pratomo, 1999; Yakimov et al., 1997; Yonebayashi et al., 1997). The injection of only 1 to 2 pore volumes of the lipopeptide biosurfactant (about 50 to 100 mg/liter) made by *Bacillus mojavensis* mobilized about 50% of the oil entrapped in sandstone cores so long as a polymer and an alcohol were present (Maugdalya et al., 2004).

There are few trials that involved the use of biosurfactant-producing bacteria (Bryant et al., 1990, 1993; Youssef et al., 2007). In two studies, the biosurfactant producer, *Bacillus* strain JF-2, and molasses were added to the injected brine (Bryant et al., 1990, 1993). The amounts of inoculum and molasses used were small relative to the reservoir volume that was accessed by the waterflood: 100 liters of inoculum and molasses followed by weekly treatments with 6 liters of molasses per well (Bryant et al., 1990), and about 100 barrels each of the inoculum and molasses followed by daily injection of about 175 liters of molasses into an injection manifold that serviced 19 injection wells (Bryant et al., 1993). In both cases, a very moderate 14% increase in oil production and improvements in the oil-to-water ratio were noted. The increase in oil productivity for the Mink field was less than one barrel per day (15 production wells) (Bryant et al., 1990) and about three barrels per day for the Phoenix field (21 production wells) (Bryant et al., 1993). The small increase in oil productivity may be why this process has not been implemented on a commercial scale.

An important question is whether biosurfactant concentrations sufficient to mobilize entrapped oil can be produced in situ. Youssef et al. (2007) showed that about 90 mg/liter with peak concentrations of up to 350 mg of lipopeptide biosurfactants per liter were made when two biosurfactant-producing bacillus strains and nutrients were injected into production wells. Biosurfactant production was not observed in control wells that did not receive the inocula. Laboratory experiments show that more than 50% of the residual oil is mobilized from sandstone cores if the biosurfactant concentration exceeds 50 mg/liter (Maudgalya et al., 2004). In addition, cultivation-dependent and cultivation-independent approaches showed that the two strains used as the inoculum were maintained and could be retrieved from production fluids of the inoculated wells (Youssef et al., 2007). This work shows that in situ microbial processes do generate quantities of metabolites sufficient for oil recovery and that the use of exogenously added organisms is feasible.

Permeability Modifications

The stimulation of indigenous microbial growth (Fig. 1D) by the addition of nutrients reduced the permeability of Berea sandstone cores (Jenneman et al., 1984; Raiders et al., 1986) and improved the sweep efficiency of various laboratory model systems (Raiders et al., 1986, 1989). In situ microbial growth also mobilized small amounts (4 to 8%) of the oil entrapped from sandstone cores with each nutrient injection (Raiders et al., 1989). The injection of nutrients and *Leuconostoc* species greatly reduced the permeability of fused glass columns (initial permeabilities of 4.9 to 6.9 μm^2) (Kim and Fogler, 2000; Shaw et al., 1985).

Several field trials demonstrate the efficacy of microbial permeability modification (Fig. 1D; Table 2). The addition of molasses and nitrate blocked a major channel and reduced interwell permeability variation in a hypersaline oil reservoir (Knapp et al., 1992). Evidence that in situ microbial metabolism occurred included an increase in alkalinity and microbial numbers in the production fluids from several wells (Knapp et al., 1992). The injection of a spore preparation of a bacterium similar to *Bacillus licheniformis* with molasses and nitrate blocked one highly transmissive zone and drastically reduced fluid intake in another zone (Gullapalli et al., 2000). The injection of molasses and a polymer-producing *Enterobacter* sp. altered the interwell flow pattern and increased oil production rates (Nagase et al., 2002). It was estimated that about 13,400 barrels of incremental oil were produced in 7 months.

CONCLUSIONS

It is not likely that carbon-neutral and/or carbon-free energy sources will be available at the scale needed to meet increasing energy demands in the near future. By all indications, fossil fuels will continue to be a dominant form of energy for the next several decades. Of the available options, methane is cleaner burning, abundant, sustainable, and available at the requisite scale. For these reasons, the use of natural gas may become more and more important. Large amounts of petroleum still remain in U.S. domestic reservoirs and can be tapped if technologies can be developed to recover or transform this vital resource. One approach is the in situ biodegradation of entrapped petroleum to methane, as the combustion of this energy form is associated with fewer environmental consequences. Diverse kinds of hydrocarbons can be anaerobically degraded including normal and cyclic alkanes, as well as mono- and polynuclear aromatic hydrocarbons. These are some of the most quantitatively important hydrocarbon components in oil. However, there are relatively few mechanisms that anaerobic microorganisms employ to activate hydrocarbons including fumarate addition, oxygen-independent hydroxylation, carboxylation, and methylation reactions.

Biotechnology can also be used to mobilize entrapped petroleum to increase access to domestic petroleum reserves. However, microbial oil recovery technologies have not gained widespread acceptance in the industry. Given the diversity of studies that report beneficial responses associated with microbial technologies, it is fair to ask why such processes are not used more extensively. The answer is likely to be rooted in the interplay of factors that enter into business decisions in an inherently conservative industry. However, a few remarks are worth noting. While the economics of microbial oil recovery efforts appear reasonably attractive (Portwood, 1995; Hitzman et al., 2004; Brown et al., 2002; Maure et al., 2005), success is often defined as the slowing or arresting of the rate of oil production decline. If success were based on other criteria, such as substantive growth in the rate of oil production or increases in the ultimate oil recovery factor, the efficacy of many microbial processes would be viewed very differently. With marginal oil fields, the risk of implementing microbial oil recovery technologies is low since these fields are often close to the end of their economic life span. With more-productive oil fields, the goal tends to be an increase in the ultimate recovery factor. To determine if this goal can be achieved, extensive reservoir characterization and modeling are generally needed. This is often where microbial technologies fall short. The needed mechanistic and stoichiometric information is not available to determine the efficacy for most of the varied microbial processes and approaches. Microbial oil recovery processes will gain more widespread acceptance and application only when quantitative measures of performance can be reliably obtained. Given the obvious upsurge in energy use for the foreseeable future and the dependency on nondomestic supplies, it is most prudent to develop the needed technologies and assessment tools to reliably transition microbial energy recovery technologies to the field.

Acknowledgments. Partial support for this work was obtained from Department of Energy and National Science Foundation grants.

We thank V. Parisi, L. Gieg, and I. Davidova for critical evaluation of the manuscript and the sharing of preliminary findings. We also thank Neil Wofford for the development of Fig. 1.

REFERENCES

Adkins, J. P., R. S. Tanner, E. O. Udegbunam, M. J. McInerney, and R. M. Knapp. 1992. Microbially enhanced oil recovery from unconsolidated limestone cores. *Geomicrobiol. J.* 10:77–86.

Aitken, C. M., D. M. Jones, and S. R. Larter. 2004. Evidence for anaerobic hydrocarbon biodegradation in deep sub-surface oil reservoirs. *Nature* 431:291–294.

Anderson, R. T., and D. R. Lovley. 2000. Hexadecane decay by methanogenesis. *Nature* 404:722–723.

Annweiler, E., A. Materna, M. Safinowski, A. Kappler, H. H. Richnow, W. Michaelis, and R. U. Meckenstock. 2000. Anaerobic degradation of 2-methylnaphthalene by a sulfate-reducing enrichment culture. *Appl. Environ. Microbiol.* 66:5329–5333.

Annweiler, E., W. Michaelis, and R. U. Meckenstock. 2002. Identical ring cleavage products during anaerobic degradation of naphthalene, 2-methylnaphthalene, and tetralin indicate a new metabolic pathway. *Appl. Environ. Microbiol.* 68:852–858.

Behlulgil, K., and M. T. Mehmetoglu. 2002. Bacteria for improvement of oil recovery: a laboratory study. *Energy Sources* 24:413–421.

Beller, H. R. 2000. Metabolic indicators for detecting in situ anaerobic alkylbenzene degradation. *Biodegradation* 11:125–139.

Beller, H. R., and A. M. Spormann. 1997a. Benzylsuccinate formation as a means of anaerobic toluene activation by sulfate-reducing strain PRTOL1. *Appl. Environ. Microbiol.* 63:3729–3731.

Beller, H. R., and A. M. Spormann. 1997b. Anaerobic activation of toluene and o-xylene by addition to fumarate in denitrifying strain T. *J. Bacteriol.* 179:670–676.

Beller, H. R., and A. M. Spormann. 1998. Analysis of the novel benzylsuccinate synthase reaction for anaerobic toluene activation based on structural studies of the product. *J. Bacteriol.* 180:5454-5457.

Beller, H. R., and A. M. Spormann. 1999. Substrate range of benzylsuccinate synthase from *Azoarcus* sp. strain T. *FEMS Microbiol. Lett.* 178:147–153.

Beller, H. R., and E. A. Edwards. 2000. Anaerobic toluene activation by benzylsuccinate synthase in a highly enriched methanogenic culture. *Appl. Environ. Microbiol.* 66:5503–5505.

Boetius A., K. Ravenschlag, C. J. Schubert, D. Rickert, F. Widdel, A. Gieseke, R. Amann, B. B. Jørgensen, U. Witte, and O. Pfannkuche. 2000. A marine microbial consortium apparently mediating anaerobic oxidation of methane. *Nature* 407:623–626.

Boll, M., G. Fuchs, and J. Heider. 2002. Anaerobic oxidation of aromatic compounds and hydrocarbons. *Curr. Opin. Chem. Biol.* 6: 604–611.

Bonch-Osmolovskaya, E. A., M. L. Miroshnichenko, A. V. Lebedinsky, N. A. Chernyh, T. N. Nazina, V. S. Ivoilov, S. S. Belyaev, E. S. Boulygina, Y. P. Lysov, A. N. Perov, A. D. Mirzabekov, H. Hippe, E. Stackebrandt, S. L'Haridon, and C. Jeanthon. 2003. Radioisotopic, culture-based, and oligonucleotide microchip analyses of thermophilic microbial communities in a continental high-temperature petroleum reservoir. *Appl. Environ. Microbiol.* 69:6143–6151.

Borzenkov, I. A., S. S. Belyaev, Y. M. Miller, I. A. Davydova, and M. V. Ivanov. 1997. Methanogenesis in the highly mineralized stratal waters of the Bonduzhskoe oil field. *Microbiology* 66:104–110.

Brown, F. G. 1992. Microbes: the practical and environmental safe solution to production problems, enhanced production, and enhanced oil recovery. SPE 23955. *In Proceedings of the 1992 Permian Basin Oil and Gas Recovery Conference.* Society of Petroleum Engineers, Richardson, TX.

Brown, L. R., A. A. Vadie, and J. O. Stephens. 2002. Slowing production decline and extending the economic life of an oil field: new MEOR technology. *SPE Reservoir Eval. Eng.* 5:33–41.

Bryant, R. S., and J. Douglas. 1988. Evaluation of microbial systems in porous media for EOR. *SPE Reservoir Eng.* 3:489–495.

Bryant, R. S., A. K. Stepp, K. M. Bertus, T. E. Burchfield, and M. Dennis. 1993. Microbial-enhanced waterflooding field pilots. *Dev. Petrol. Sci.* 39:289–306.

Bryant, R. S., T. E. Burchfield, D. M. Dennis, and D. O. Hitzman. 1990. Microbial-enhanced waterflooding: Mink Unit project. *SPE Reservoir Eng.* 5:9–13.

Caldwell, M. E., and J. M. Suflita. 2000. Detection of phenol and benzoate as intermediates of anaerobic benzene biodegradation under different terminal electron-accepting conditions. *Environ. Sci. Technol.* 34:1216–1220.

Callaghan, A. V., L. M. Gieg, K. G. Kropp, J. M. Suflita, and L. Y. Young. 2006. A comparison of alkane metabolism under sulfate-reducing conditions among two isolates and a bacterial consortium. *Appl. Environ. Microbiol.* 72:2896–2905.

Chakraborty R., and J. D. Coates. 2005. Hydroxylation and carboxylation—two crucial steps of anaerobic benzene degradation by *Dechloromonas* strain RCB. *Appl. Environ. Microbiol.* 71: 5427–5432.

Chakraborty, R., and J. D. Coates. 2004. Anaerobic degradation of monoaromatic compounds. *Appl. Environ. Biotechnol.* 64:437–446.

Chapelle, F. H., and P. M. Bradley. 1997. Alteration of aquifer geochemistry by microorganisms, p. 558–564. *In* C. J. Hurst, G. R. Knudsen, M. J. McInerney, L. D. Stetzenbach, and M. V. Walter. (ed.), *Manual of Environmental Microbiology*. ASM Press, Washington, DC.

Coschigano, P. W., T. S. Wehrman, and L. Y. Young. 1998. Identification and analysis of genes involved in anaerobic toluene metabolism by strain T1: putative role of a glycine free radical. *Appl. Environ. Microbiol.* 64:1650–1656.

Cravo-Laureau, C., V. Grossi, D. Raphel, R. Matheron, and A. Hirschler-RŽa. 2005. Anaerobic *n*-alkane metabolism by a sulfate-reducing bacterium, *Desulfatibacillum aliphaticivorans* strain CV2803. *Appl. Environ. Microbiol.* 71:3458–3467.

Davidova, I. A., K. E. Duncan, O. K. Choi, and J. M. Suflita. 2006. *Desulfoglaeba alkanexedens* gen. nov., sp. nov., an n-alkane-degrading, sulfate-reducing bacterium. *Int. J. Syst. Evol. Microbiol.* 56:2737–2742.

Davidson, S. W., and H. H. Russell. 1988. A MEOR pilot test in the Loco field, p. VII 1–VII 12. *In* T. E. Burchfield and R. S. Bryant (ed.), *Proceedings of the Symposium on the Application of Microorganisms to Petroleum Technology.* National Technical Information Service, Springfield, VA.

Desai, J., and I. M. Banat. 1997. Microbial production of surfactants and their commercial potential. *Microbiol. Mol. Biol. Rev.* 61:47–64.

Desouky, S. M., M. M. Abdel-Daim, M. H. Sayyouh, and A. S. Dahab. 1996. Modeling and laboratory investigation of microbial enhanced oil recovery. *J. Petrol. Sci. Eng.* 15:309–320.

Dojka, M. A., P. Hugenholtz, S. K. Haack, and N. R. Pace. 1998. Microbial diversity in a hydrocarbon- and chlorinated-solvent-contaminated aquifer undergoing intrinsic bioremediation. *Appl. Environ. Microbiol.* 64:3869–3877.

Edwards, E. A., and D. Grbic-Galic. 1994. Anaerobic degradation of toluene and *o*-xylene by a methanogenic consortium. *Appl. Environ. Microbiol.* 60:313–322.

Energy Information Administration. 2007. *Annual Energy Outlook DOE/EIA-0383(2007)*. U.S. Department of Energy, Washington, DC.

Energy Information Administration. 2006. *International Energy Outlook*. U.S. Department of Energy, Washington, DC.

Eriksson, S., T. Ankner, K. Abrahamsson, and L. Hallbeck. 2005. Propylphenols are metabolites in the anaerobic biodegradation of propylbenzene under iron-reducing conditions. *Biodegradation* 16:253–263.

Ferguson, K. R., C. T. Lloyd, D. Spencer, and J. Hoeltgen. 1996. *Microbial Pilot Test for the Control of Paraffin and Asphaltenes at Prudhoe Bay.* SPE Paper 36630. Society of Petroleum Engineers, Richardson, TX.

Giangiacomo, L. 1997. *Paraffin Control Project.* FC9544/96PT12. Rocky Mountain Oil Field Testing Center, BDM Oklahoma/NIPER, Bartlesville, OK.

Gieg, L. M., and J. M. Suflita. 2002. Detection of anaerobic metabolites of saturated and aromatic hydrocarbons in petroleum-contaminated aquifers. *Environ. Sci. Technol.* 36:3755–3762.

Gieg, L. M., and J. M. Suflita. 2005. Metabolic indicators of anaerobic hydrocarbon biodegradation in petroleum-laden environments, p. 337–356. *In* B. Ollivier and M. Magot Michel (ed.), *Petroleum Microbiology.* ASM Press, Washington, DC.

Gullapalli, I. L., J. H. Bae, K. Heji, and A. Edwards. 2000. Laboratory design and field implementation of microbial profile modification process. *SPE Reservoir Eval. Eng.* 3:42–49.

Hall, C., P. Tharakan, J. Hallock, C. Cleveland, and M. Jefferson. 2003. Hydrocarbons and the evolution of human culture. *Nature* 426:318–322.

He, Z., B. Mei, W. Wang, J. Sheng, S. Zhu, L. Wang, and T. F. Yen. 2003. A pilot test using microbial paraffin-removal technology in Liaohe oilfield. *Petrol. Sci. Technol.* 21:201–210.

Head, I. M., D. M. Jones, and S. R. Larter. 2003. Biological activity in the deep subsurface and the origin of heavy oil. *Nature* 246:344–352.

Heider, J. 2007. Adding handles to unhandy substrates: anaerobic hydrocarbon activation mechanisms. *Curr. Opin. Chem. Biol.* 11:188–194.

Heider, J., A. M. Sporman, H. R. Beller, and F. Widdel. 1999. Anaerobic bacterial metabolism of hydrocarbons. *FEMS Microbiol. Rev.* 22:459–473.

Herman, D. C., R. J. Lenhard, and R. J. Miller. 1997. Formation and removal of hydrocarbon residual in porous media: effects of attached bacteria and biosurfactants. *Environ. Sci. Technol.* 31:1290–1294.

Hitzman, D. O. 1983. Petroleum microbiology and the history of its role in enhanced oil recovery, p. 162–218. *In* E. C. Donaldson and J. B. Clarke (ed.), *Proceedings of the 1982 International Conference on Microbial Enhancement of Oil Recovery.* CONF-8205140. U.S. Department of Energy, Bartlesville, OK.

Hitzman, D. O. 1988. Review of microbial enhanced oil recovery field tests, p. VI 1–VI41. *In* T. E. Burchfield and R. S. Bryant (ed.), *Proceedings of the Symposium on Applications of Microorganisms to Petroleum Technology.* National Technical Information Service, Springfield, VA.

Hitzman, D. O., M. Dennis, and D. C. Hitzman. 2004. Recent successes: MEOR using synergistic H_2S prevention and increased oil recovery systems. SPE 89453. In *Fourteenth Symposium on Improved Oil Recovery.* Society of Petroleum Engineers, Richardson, TX.

Huang, H., and S. Larter. 2005. Biodegradation of petroleum in subsurface geological reservoirs, p. 91–121. *In* B. Ollivier and M. Magot (ed.), *Petroleum Microbiology.* ASM Press, Washington, DC.

Ivanov, M. V., S. S. Belyaev, I. A. Borzenkov, I. F. Glumov, and R. R. Ibatullin. 1993. Additional oil production during field trials in Russia. *Dev. Petrol. Sci.* 39:373–381.

Ivanov, M. V., S. S. Belyaev, A. M. Zyakun, V. A. Bondar, and K. K. Laurinavichus. 1983. Microbial methanogenesis in an exploited oil field. *Geokhimiya* 11:1647–1654.

James, A. T., and B. J. Burns. 1984. Microbial alteration of subsurface natural gas accumulations. *Bull. Am. Assoc. Petrol. Geol.* 68:957–960.

Javaheri, M., G. E. Jenneman, M. J. McInerney, and R. M. Knapp. 1985. Anaerobic production of a biosurfactant by *Bacillus licheniformis* JF-2. *Appl. Environ. Microbiol.* 50:698–700.

Jenneman, G. E., R. M. Knapp, M. J. McInerney, D. E. Menzie, and D. E. Revus. 1984. Experimental studies of in-situ microbial enhanced oil recovery. *Soc. Petrol. Eng. J.* 24:33–37.

Kane, S. R., H. R. Beller, T. C. Legler, and R. T. Anderson. 2002. Biochemical and genetic evidence of benzylsuccinate synthase in toluene-degrading, ferric iron-reducing *Geobacter metallireducens. Biodegradation* 13:149–154.

Kawamura, K., and I. R. Kaplan. 1987. Dicarboxylic acids generated by thermal alteration of kerogen and humic acids. *Geochim. Cosmochim. Acta* 51:3201–3207.

Kazumi, J., M. E. Caldwell, J. M. Suflita, D. R. Lovley, and L. Y. Young. 1997. Anaerobic degradation of benzene in diverse anoxic environments. *Environ. Sci. Technol.* 31:813–818.

Kim, D.-S., and H. S. Fogler. 2000. Biomass evolution in porous media and its effects on permeability under starvation conditions. *Biotechnol. Bioeng.* 69:47–56.

Kleikemper, J., S. A. Pombo, M. H. Schroth, W. V. Sigler, M. Pesaro, and J. Zeyer. 2005. Activity and diversity of methanogens in a petroleum hydrocarbon-contaminated aquifer. *Appl. Environ. Microbiol.* 71:149–158.

Knapp, R. M., M. J. McInerney, J. D. Coates, J. L. Chisholm, D. E. Menzie, and V. K. Bhupathiraju. 1992. Design and implementation of a microbially enhanced oil recovery field pilot, Payne County, Oklahoma. SPE 24818. *In SPE Annual Technical Conference*, Society of Petroleum Engineers, Richardson, TX.

Kniemeyer, O., T. Fischer, H. Wilkes, F. O. Glöckner, and F. Widdel. 2003. Anaerobic degradation of ethylbenzene by a new type of marine sulfate-reducing bacterium. *Appl. Environ. Microbiol.* 69:760–768.

Kowalewski, E., I. Rueslatten, T. Boassen, E. Sunde, J. A. Stensen, B. L. P. Lillbeo, G. Bodtker, and T. Torsvik. 2005. Analyzing microbial improved oil recovery production from core floods. IPTC 10924. *In The International Petroleum Technology Conference.* Society of Petroleum Engineers, Richadson, TX.

Krieger, C. J., H. R. Beller, M. Reinhard, and A. M. Spormann. 1999. Initial reactions in anaerobic oxidation of *m*-xylene by the denitrifying bacterium *Azoarcus* sp. strain T. *J. Bacteriol.* 181:6403–6410.

Kropp, K. G., I. A. Davidova, and J. M. Suflita. 2000. Anaerobic oxidation of *n*-dodecane by an addition reaction in a sulfate-reducing bacterial enrichment culture. *Appl. Environ. Microbiol.* 66:5393–5398.

Kühner, S., L. Wöhlbrand, I. Fritz, W. Wruck, C. Hultschig, P. Hufnagel, M. Kube, R. Reinhardt, and R. Rabus. 2005. Substrate-dependent regulation of anaerobic degradation pathways for toluene and ethylbenzene in a denitrifying bacterium, strain EbN1. *J. Bacteriol.* 187:1493–1503.

Larter, S., A. Wilhelms, I. Head, M. Koopmans, A. Aplin, R. DiPrimo, C. Zwach, M. Erdmann, and N. Telnaes. 2003. The controls on the composition of biodegraded oils in the deep subsurface-part I. Biodegradation rates in petroleum reservoirs. *Org. Geochem.* 34:601–613.

Larter, S., H. Huang, J. Adams, B. Bennett, O. Jokanola, T. Oldenburg, M. Jones, I. Head, C. Riediger, and M. Fowler. 2006. The controls on the composition of biodegraded oils in the deep subsurface. Part II. Geological controls on subsurface biodegradation fluxes and constraints on reservoir-fluid property prediction. *AAPG Bull.* 90:921–938.

Lazar, I. 1987. Research on the microbiology of MEOR in Romania, p. 124–151. *In* J. W. King and D. A. Stevens (ed.), *Proceedings of the First International MEOR Workshop.* National Technical Information Service, Springfield, VA.

Lazar, I. 1991. MEOR field trials carried out over the world during the last 35 years. *Dev. Petrol. Sci.* 31:485–530.

Lazar, I., and P. Constantinescu. 1985. Field trial results of microbial enhanced oil recovery, p. 122–143. *In* J. E. Zajic and E. C. Donaldson (ed.), *Microbes and Oil Recovery.* Bioresources Publications, Inc., El Paso, TX.

Lazar, I., A. Voicy, C. Nicolescu, D. Mucenica, S. Dobrota, I. G. Petrisor, M. Stefanescu, and L. Sandulescu. 1999. The use of naturally occurring selectively isolated bacteria for inhibiting paraffin deposition. *J. Petrol. Sci. Eng.* 22:161–169.

Lazar, I., S. Dobrota, and M. Stefanescu. 1988. Some considerations concerning nutrient support injected into reservoirs subjected to microbiological treatment, p. XIV 1–XIV 6. *In* T. E. Burchfield and R. S. Bryant (ed.), *Proceedings of the Symposium on Applications of Microorganisms to Petroleum Technology.* National Technical Information Service, Springfield, VA.

Lazar, I., S. Dobrota, M. C. Stefanescu, L. Sandulescu, R. Paduraru, and M. Stefanescu. 1993. MEOR, recent field trials in Romania: reservoir selection, type of inoculum, protocol for well treatment and line monitoring. *Dev. Petrol. Sci.* 39:265–288.

Lazar, I., S. Dorota, M. Stefanescu, L. Sandulescu, P. Constantinescu, C. Morosanu, N. Botea, and O. Iliescu. 1991. Preliminary results of some recent MEOR field trials in Romania. *Dev. Petrol. Sci.* 33:365–386.

Leuthner, B., C. Leutwein, H. Schulz, P. Horth, W. Haehnel, E. Schiltz, H. Schagger, and J. Heider. 1998. Biochemical and genetic characterization of benzylsuccinate synthase from *Thauera aromatica:* a new glycyl radical enzyme catalyzing the first step in anaerobic toluene metabolism. *Mol. Microbiol.* 28:615–628.

Lin, S.-C., M. A. Minton, M. M. Sharma, and G. Georgiou. 1994. Structural and immunological characterization of a biosurfactant produced by *Bacillus licheniformis* JF-2. *Appl. Environ. Microbiol.* 60:31–38.

Lucach, S. O., B. F. J. Bowler, N. Frewin, and S. R. Larter. 2002. Variation in alkylphenol distributions in a homogeneous oil suite from the Dhahaban petroleum system of Oman. *Org. Geochem.* 33:581–594.

Lundquist, A., D. Cheney, C. L. Powell, P. O'Niell, G. Norton, A. M. Veneman, D. L. Evans, N. Y. Minda, S. Abraham, J. M. Allbaugh, C. T. Whitman, J. A. Bolten, M. E. Daniels, L. B. Lindsey, and R. Barrales. 2001. *Energy for a New Century: Increasing Domestic Energy Production. National Energy Policy Report of the National Energy Policy Development Group.* U.S. Government Printing Office, Washington, DC.

Magot, M. 2005. Indigenous microbial communities in oil fields, p. 21–33. *In* B. Ollivier and M. Magot (ed.), *Petroleum Microbiology.* ASM Press, Washington, DC.

Maier, R. M. 2003. Biosurfactants: evolution and diversity in bacteria. *Adv. Appl. Microbiol.* 52:101–121.

Malins, D. C. 1977. *Effects of Petroleum on Arctic and Subarctic Marine Environments and Organics*, vol. 1, p. 24. Academic Press, New York, NY.

Marsh, T. L., X. Zhang, R. M. Knapp, M. J. McInerney, P. K. Sharma, and B. E. Jackson. 1995. Mechanisms of microbial oil recovery by *Clostridium acetobutylicum* and *Bacillus* strain JF-2, p. 593–610. *In* R. S. Bryant and K. L. Sublette (ed.), *The Fifth International Conference on Microbial Enhanced Oil Recovery and Related Problems for Solving Environmental Problems.* CONF-9509173. National Technical Information Service, Springfield, VA.

Matz, A. A., A. Y. Borisov, Y. G. Mamedov, and R. R. Ibatulin. 1992. Commercial (pilot) test of microbial enhanced oil recovery methods. *In Proceedings of the Eighth Symposium on Enhanced Oil Recovery, Part 2 (of 2), 22 to 24 April 1992, Tulsa, OK.* Society of Petroleum Engineers, Richardson, TX.

Maudgalya, S., R. M. Knapp, M. J. McInerney, D. P. Nagle, and M. J. Folsmbee. 2004. Development of a bio-surfactant-based enhanced oil recovery procedure. SPE 89473. *In Proceedings of the SPE/DOE Improved Oil Recovery Symposium.* Society of Petroleum Engineers, Richardson, TX.

Maure, A., A. A. Saldana, and A. R. Juarez. 2005. Biotechnology application to EOR in Talara off-shore oil fields, northwest Peru. SPE 94934. *In SPE Latin American and Caribbean Petroleum Engineering Conference.* Society of Petroleum Engineers, Richardson, TX.

McInerney, M. J., M. Javaheri and D. P. Nagle. 1990. Properties of the biosurfactant produced by *Bacillus licheniformis* strain JF-2. *J. Ind. Microbiol.* 5:95–102.

Meckenstock, R. U., M. Safinowski, and C. Griebler. 2004. Anaerobic degradation of polycyclic aromatic hydrocarbons. *FEMS Microbiol. Ecol.* 49:27–36.

Mistra, S., S. Baruah, and K. Singh. 1995. Paraffin problems in crude oil production and transportation: a review. *SPE Product. Facil.* **February**:50–54.

Morasch, B., B. Schink, C. C. Tebbe, and R. U. Meckenstock. 2004. Degradation of o-xylene and *m*-xylene by a novel sulfate-reducer belonging to the genus *Desulfotomaculum. Arch. Microbiol.* **181:** 407–417.

Murygina, V. P., A. A. Mats, M. U. Arinbassrov, Z. Z. Salamov, and A. B. Cherkasov. 1995. Oil field experiments of microbial improved oil recovery in Vyngapour, West Siberia, Russia, p. 87–94. *In* R. S. Bryant and K. L. Sublette (ed.), *The Fifth International Conference on Microbial Enhanced Oil Recovery and Related Problems for Solving Environmental Problems.* CONF-9509173. National Technical Information Service, Springfield, VA.

Nagase, K., S. T. Zhang, H. Asami, N. Yazawa, K. Fujiwara, H. Enomoto, C. X. Hong, and C. X. Laing. 2002. A successful test of microbial EOP process in Fuyu Oilfield, China. SPE 75238. *In SPE/DOE Improved Oil Recovery Symposium.* Society of Petroleum Engineers, Richardson, TX.

Nazina, T. N., A. E. Ivanova, I. A. Borzenkov, S. S. Belyaev, and M. V. Ivanov. 1995. Occurrence and geochemical activity of microorganisms in high temperature, water-flooded oil fields of Kazakhstan and Western Siberia. *Geomicrobiol. J.* 13:181–192.

Nazina, T. N., A. E. Ivanova, V. S. Ivoilov, Y. M. Miller, G. F. Kandaurova, R. R. Ibatullin, S. S. Belyaev, and M. V. Ivanov. 1999. Results of the trial of the microbiological method for the enhancement of oil recovery at the carbonate collector of the Romashkinskoe oil field: biogeochemical and productional characteristics. *Microbiology* (Moscow) 68:222–226.

Nelson, L., and D. R. Schneider. 1993. Six years of paraffin control and enhanced oil recovery with the microbial product, Para-Bac. *Dev. Petrol. Sci.* 39:355–362.

Ohna, K. 1999. Implementation and performance of a microbial enhanced oil recovery field pilot in Fuyu oilfield, China. SPE 54328. *In Proceedings of the SPE Asia Pacific Oil and Gas Conference, Jakarta, Indonesia.* Society of Petroleum Engineers, Richardson, TX.

Oppenheimer, C. H., and F. Hiebert. 1989. *Microbial Techniques for Paraffin Reduction in Producing Oil Wells.* Final Report. U.S.

Department of Energy contract FG19086BC 14014. DOE/BC/ 14014-9, (DE 89000741). U. S. Department of Energy, Bartlesville, OK.

Orphan, V. J., C. H. House, K.-U. Hinrichs, K. D. McKeegan, and E. F. DeLong. 2001. Methane-consuming archaea revealed by directly coupled isotopic and phylogenetic analysis. *Science* **293:** 484–487.

Orphan, V. J., S. K. Goffredi, E. F. Delong, and J. R. Boles. 2003. Geochemical influence on diversity and microbial processes in high temperature oil reservoirs. *Geomicrobiol. J.* 20:295–311.

Pallasser, R. J. 2000. Recognising biodegradation in gas/oil accumulations through the $\delta^{13}C$ composition of gas components. *Org. Geochem.* 31:1363–1373.

Pelger, J. W. 1991. Microbial enhanced oil recovery treatments and wellbore stimulation using microorganisms to control paraffin, emulsion, corrosion, and scale formation. *Dev. Petrol. Sci.* **31:**451–466.

Phelps, C. D., X. Zhang, and L. Y. Young. 2001. Use of stable isotopes to identify benzoate as a metabolite of benzene degradation in a sulfidogenic consortium. *Environ. Microbiol.* 3:600–603.

Planckaert, M. 2005. Oil reservoirs and oil production, p. 3–19. *In* B. Ollivier and M. Magot (ed.), *Petroleum Microbiology.* ASM Press, Washington, DC.

Portwood, J. T. 1995. A commercial microbial enhanced oil recovery process: statistical evaluation of a multi-project database, p. 51–76. *In* R. S. Bryant and K. L. Sublette (ed.), *The Fifth International Conference on Microbial Enhanced Oil Recovery and Related Problems for Solving Environmental Problems.* CONF-9509173. National Technical Information Service, Springfield, VA.

Rabus, R. 2005. Biodegradation of hydrocarbons under anoxic conditions, p. 277–299. *In* B. Ollivier and M. Magot (ed.), *Petroleum Microbiology.* ASM Press, Washington, DC.

Rabus, R., and J. Heider. 1998. Initial reactions of anaerobic metabolism of alkylbenzenes in denitrifying and sulfate-reducing bacteria. *Arch. Microbiol.* 170:377–384.

Rabus, R., H. Wilkes, A. Behrends, A. Armstroff, T. Fischer, A. J. Pierik, and F. Widdel. 2001. Anaerobic initial reaction of *n*-alkanes in a denitrifying bacterium: evidence for (1-methylpentyl)succinate as initial product and for involvement of an organic radical in *n*-hexane metabolism. *J. Bacteriol.* 183:1707–1715.

Raghoebarsing, A. A., A. Pol, K. T. van de Pas-Schoonen, A. J. Smolders, K. F. Ettwig, W. I. Rijpstra, S. Schouten, J. S. Damste, H. J. Op den Camp, M. S. M. Jetten, and M. Strous. 2006. A microbial consortium couples anaerobic methane oxidation to denitrification. *Nature* 440:918–921.

Raiders, R. A., M. J. McInerney, D. E. Revus, H. M. Torbati, R. M. Knapp, and G. E. Jenneman. 1986. Selectivity and depth of microbial plugging in Berea sandstone cores. *J. Ind. Microbiol.* **1:**195–203.

Raiders, R. A., R. M. Knapp, and M. J. McInerney. 1989. Microbial selective plugging and enhanced oil recovery. *J. Ind. Microbiol.* 4:215–230.

Rios-Hernandez, L. A., L. M. Gieg, and J. M. Suflita. 2003. Biodegradation of an alicyclic hydrocarbon by a sulfate-reducing enrichment from a gas condensate-contaminated aquifer. *Appl. Environ. Microbiol.* 69:434–443.

Roling, W., F. M. Wilfred, I. M. Head, and S. R. Larter. 2003. The microbiology of hydrocarbon degradation in subsurface petroleum reservoirs: perspectives and prospects. *Res. Microbiol.* **154:**321–328.

Rouse, B., F. Hiebert, and L. W. Lake. 1992. Laboratory testing of a microbial enhanced oil recovery process under anaerobic conditions. SPE 24819. *In SPE Annual Technical Conference.* Society of Petroleum Engineers, Richardson, TX.

Sabatini, D., R. Knox, J. Harwell, and B. Wu. 2000. Integrated design of surfactant enhanced DNAPL remediation. Efficient super-solubilization and gradient systems. *J. Contam. Hydrol.* 45:99–121.

Safinowski, M., and R. U. Meckenstock. 2006. Methylation is the initial reaction in anaerobic naphthalene degradation by a sulfate-reducing enrichment culture. *Environ. Microbiol.* 8:347–352.

Sakata, S., Y. Sano, T. Maekawa, and S.-I. Igari. 1997. Hydrogen and carbon isotopic composition of methane as evidence for biogenic origin of natural gases from the Green Tuff Basin, Japan. *Org. Geochem.* 26:399–407.

Santamaria, M. M., and R. E. George. 1991. Controlling paraffin-deposition-related problems by the use of bacteria treatments. SPE 22851. *In Proceedings of the SPE Annual Technical Conference and Exhibition.* Society of Petroleum Engineers, Richardson, TX.

Shaw, J. C., B. Bramhill, N. C. Wardlaw, and J. W. Costerton. 1985. Bacterial fouling in a model core system. *Appl. Environ. Microbiol.* 50:693–701.

Sheehy, A. J. 1990. Field studies of microbial EOR. SPE/DOE 20254. *In Proceedings of the SPE/DOE Seventh Symposium on Enhanced Oil Recovery.* Society of Petroleum Engineers, Richardson, TX.

Shinoda, Y., J. Akagi, Y. Uchihashi, A. Hiraishi, H. Yukawa, H. Yurimoto, Y. Sakai, and N. Kato. 2005. Anaerobic degradation of aromatic compounds by magnetospirillum strains: isolation and degradation genes. *Biosci. Biotechnol. Biochem.* 69:1483–1491.

Shinoda, Y., Y. Sakai, H. Uenishi, Y. Uchihashi, A. Hiraishi, H. Yukawa, H. Yurimoto, and N. Kato. 2004. Aerobic and anaerobic toluene degradation by a newly isolated denitrifying bacterium, *Thauera* sp. strain DNT-1. *Appl. Environ. Microbiol.* 70:1385–1392.

Siddique, T., P. M. Fedorak, and J. M. Foght. 2006. Biodegradation of short-chain n-alkanes in oil sands tailings under methanogenic conditions. *Environ. Sci. Technol.* 40:5459–5464.

Siddique, T., P. M. Fedorak, M. D. MacKinnon, and J. M. Foght. 2007. Metabolism of BTEX and naphtha compounds to methane in oil sands tailings. *Environ. Sci. Technol.* 41:2350–2356.

So, C. M., C. D. Phelps, and L. Y. Young. 2003. Anaerobic transformation of alkanes to fatty acids by a sulfate-reducing bacterium, strain Hxd3. *Appl. Environ. Microbiol.* 69:3892–3900.

Sporman, A. M., and F. Widdel. 2000. Metabolism of alkylbenzenes, alkanes and other hydrocarbons in anaerobic bacteria. *Biodegradation* 11:85–105.

Streeb, L. P., and F. G. Brown. 1992. MEOR-altamont/bluebell field project. SPE 24334. *In Proceedings of the Rocky Mountain Regional Meeting and Exhibition.* Society of Petroleum Engineers, Richardson, TX.

Struchtemeyer, C. G., M. S. Elshahed, K. E. Duncan, and M. J. McInerney. 2005. Evidence for aceticlastic methanogenesis in the presence of sulfate in a gas condensate-contaminated aquifer. *Appl. Environ. Microbiol.* 71:5348–5353.

Suflita, J. M., I. A. Davidova, L. M. Gieg, M. Nanny, and R. C. Prince. 2004. Anaerobic hydrocarbon biodegradation and the prospects for microbial enhanced energy production. *In* R. Vazquez-Duhalt and R. Quintero-Ramirez (ed.), *Petroleum Biotechnology. Developments and Perspectives*, vol. 151. Elsevier Science, Amsterdam, The Netherlands.

Sugihardjo, E. H., and S. W. Pratomo. 1999. Microbial core flooding experiments using indigenous microbes. SPE 57306. *In Proceedings of the SPE Asia Pacific Improved Oil Recovery Conference.* Society of Petroleum Engineers, Richardson, TX.

Sullivan, E. R., X. Zhang, C. Phelps, and L. Y. Young. 2001. Anaerobic mineralization of stable-isotope-labeled 2-methylnaphthalene. *Appl. Environ. Microbiol.* **67:**4353—4357.

Sunde, E., J. Beeder, R. K. Nilsen, and T. Torsvik. 1992. Aerobic microbial enhanced oil recovery for offshore use. *In Proceedings of the Eighth Symposium on Enhanced Oil Recovery.* Society of Petroleum Engineers, Richardson, TX.

Symons, G. E., and A. M. Buswell. 1933. The methane fermentation of carbohydrates. *J. Am. Chem. Soc.* 55:2028–2036.

Thomas, C. P., G. A. Bala, and M. L. Duvall. 1993. Surfactant-based enhanced oil recovery mediated by naturally occurring microorganisms. *Soc. Petrol. Eng. Reservoir Eng.* 11:285–291.

Townsend, G. T., R. C. Prince, and J. M. Suflita. 2003. Anaerobic oxidation of crude oil hydrocarbons by the resident microorganisms of a contaminated anoxic aquifer. *Environ. Sci. Technol.* 37:5213–5218.

Townsend, G. T., R. C. Prince, and J. M. Suflita. 2004. Anaerobic biodegradation of alicyclic constituents of gasoline and natural gas condensate by bacteria from an anoxic aquifer. *FEMS Microbiol. Ecol.* 49:129–135.

Trebbeau de Acevedo, G., and M. J. McInerney. 1996. Emulsifying activity in thermophilic and extremely thermophilic microorganisms. *J. Ind. Microbiol.* 16:1–7.

Udegbunam, E. O., J. P. Adkins, R. M. Knapp, M. J. McInerney, and R. S. Tanner. 1991. Assessing the effects of microbial metabolism and metabolites on reservoir pore structure. SPE 22846. *In Proceedings of the SPE Annual Technical Conference.* Society of Petroleum Engineers, Richardson, TX.

Ulrich, A. C., and E. A. Edwards. 2003. Physiological and molecular characterization of anaerobic benzene-degrading mixed cultures. *Environ. Microbiol.* 5:92–102.

Ulrich, A. C., H. R. Beller, and E. A. Edwards. 2005. Metabolites detected during biodegradation of 13C6-benzene in nitrate-reducing and methanogenic enrichment cultures. *Environ. Sci. Technol.* 39:6681–6691.

van Wechem, V. M. H., and M. M. G. Senden. 1994. Conversion of natural gas to transportation fuels via the Shell Middle Distillate Process, p. 43–71. *In* H. E. Curry-Hyde and R. F. Howe (ed.), *Natural Gas Conversion. II. Proceedings of the 3rd Natural Gas Conversion Symposium, Sydney, Australia, 4 to 9 July 1993. Studies in Surface Science and Catalysis,* 81. Elsevier Science B.V., Amsterdam, The Netherlands.

Verfurth, K., A. J. Pierik, C. Leutwein, S. Zorn, and J. Heider. 2004. Substrate specificities and electron paramagnetic resonance properties of benzylsuccinate synthases in anaerobic toluene and *m*-xylene metabolism. *Arch. Microbiol.* 181:155–162.

Vogel, T. M., and D. Grbic-Galic. 1986. Incorporation of oxygen from water into toluene and benzene during anaerobic fermentative transformation. *Appl. Environ. Microbiol.* 52:200–202.

Wagner, M., D. Lungerhausen, H. Murtada, and G. Rosenthal. 1995. Development and application of a new biotechnology of molasses in-situ method: detailed evaluation for selected wells in the Romashkino carbonate reservoir, p. 153–174. *In* R. S. Bryant and K. L. Sublette (ed.), *The Fifth International Conference on Microbial Enhanced Oil Recovery and Related Problems for Solving Environmental Problems.* CONF-9509173. National Technical Information Service, Springfield, VA.

Wang, X.-Y., Y.-F. Xue, G. Dai, L. Zhao, Z.-S. Wang, J.-L. Wang, T.-H. Sun, and X.-J. Li. 1995. Application of bio-huff-"n"-puff technology at Jilin oil field, p. 115–128. *In* R. S. Bryant and K. L. Sublette (ed.), *The Fifth International Conference on Microbial Enhanced Oil Recovery and Related Problems for Solving Environmental Problems.* CONF-9509173. National Technical Information Service, Springfield, VA.

Watanabe, K., Y. Kodama, N. Hamamura, and N. Kaku. 2002. Diversity, abundance, and activity of archaeal populations in oil-contaminated groundwater accumulated at the bottom of an underground crude oil storage cavity. *Appl. Environ. Microbiol.* **68:**3899–3907.

Weiner, J. M., and D. R. Lovley. 1998. Rapid benzene degradation in methanogenic sediments from a petroleum-contaminated aquifer. *Appl. Environ. Microbiol.* **64:**1937–1939.

Whillhite, G. P. 1986. *Waterflooding.* Society of Petroleum Engineers, Richardson, TX.

Widdel, F., and R. Rabus. 2001. Anaerobic biodegradation of saturated and aromatic hydrocarbons. *Curr. Opin. Biotechnol.* **12:** 259–276.

Wilhelms, A., S. R. Larter, I. Head, P. Farrimond, R. di-Primio, and C. Zwach. 2001. Biodegradation of oil in uplifted basins prevented by deep-burial sterilization. *Nature* **411:**1034–1037.

Wilkes, H., R. Rabus, T. Fischer, A. Armstroff, A. Behrends, and F. Widdel. 2002. Anaerobic degradation of *n*-hexane in a denitrifying bacterium: further degradation of the initial intermediate (1-methylpentyl)succinate via C-skeleton rearrangement. *Arch. Microbiol.* 177:235–243.

Wilson, J. J., W. Chee, C. O'Grady, and M. D. Bishop. 1993. Field study of downhole microbial paraffin control, p. 346-357. *In Proceedings of the 40th Annual Southwestern Petroleum Short Course, Lubbock, TX.*

Yakimov, M. M., M. M. Amro, M. Bock, K. Boseker, H. L. Fredrickson, D. G. Kessel, and K. N. Timmis. 1997. Potential of *Bacillus licheniformis* strains for in situ enhanced oil recovery. *J. Petrol. Sci. Eng.* **18:**147–160.

Yarbrough, H. F., and V. F. Coty. 1983. Microbially enhanced oil recovery from the Upper Cretaceous Nacatoch formation, Union County, Arkansas, p. 162–218. *In* E. C. Donaldson and J. B. Clarke (ed.), *Proceedings of the 1982 International Conference on Microbial Enhancement of Oil Recovery.* CONF-8205140. Technology Transfer Branch, U.S. Department of Energy, Bartlesville, OK.

Yonebayashi, H., K. Ono, H. Enomoto, T. Chida, C.-X. Hong, and K. Fujiwara. 1997. Microbial enhanced oil recovery field pilot in a waterflooded reservoir. SPE 38070. *In 1997 Asia Pacific Oil and Gas Conference and Exhibition.* Society of Petroleum Engineers, Richardson, TX.

Youssef, N., K. E. Duncan, and M. J. McInerney. 2005. Importance of the 3-hydoxy fatty acid composition of lipopeptides for biosurfactant activity. *Appl. Environ. Microbiol.* 71:7690–7695.

Youssef, N. H., D. R. Simpson, K. E. Duncan, M. J. McInerney, M. J. Folmsbee, T. Fincher, and R. M. Knapp. 2007. In situ biosurfactant production by *Bacillus* strains injected into a limestone petroleum reservoir. *Appl. Environ. Microbiol.* 73:1239–1247.

Zengler, K., H. H. Richnow, R. Rosselló-Mora, W. Michaelis, and F. Widdel. 1999a. Methane formation from long-chain alkanes by anaerobic microorganisms. *Nature* 401:266–269.

Zengler, K., J. Heider, R. Rossello-Mora, and F. Widdel. 1999b. Phototrophic utilization of toluene under anoxic conditions by a new strain of *Blastochloris sulfoviridis. Arch. Microbiol.* 172:204–212.

Zhang, X., and L. Y. Young. 1997. Carboxylation as an initial reaction in the anaerobic metabolism of naphthalene and phenanthrene by sulfidogenic consortia. *Appl. Environ. Microbiol.* **63:** 4759–4764.

ZoBell, C. E. 1947a. Bacterial release of oil from oil-bearing materials, part I. *World Oil* **126:**36–47.

ZoBell, C. E. 1947b. Bacterial release of oil from oil-bearing materials, part II. *World Oil* **127:**35–41.

Zwolinski, M. D., R. F. Harris, and W. J. Hickey. 2000. Microbial consortia involved in the anaerobic degradation of hydrocarbons. *Biodegradation* **11:**141–158.

8. EXPLOITING MICROBIAL GENOMES FOR ENERGY PRODUCTION

Bioenergy
Edited by J. Wall et al.

Chapter 31

Genomic Prospecting for Microbial Biodiesel Production

ATHANASIOS LYKIDIS AND NATALIA IVANOVA

Biodiesel is defined as the monoalkyl esters of long-chain fatty acids and constitutes an alternative fuel for compression-ignition (diesel) engines. The most common biodiesel constituent used today is fatty acid methyl esters. Currently, the origin of fatty acids is triacylglycerols (TAGs) recovered from vegetable oils and animal fats. Biodiesel is a nontoxic, completely biodegradable fuel with reduced sulfur emissions. It is produced in various countries on a large scale. The main source of biodiesel in the United States is soybean oil and yellow grease, whereas in Europe and Asia it is rapeseed and palm oil, respectively. In all cases, the current and projected production cost of biodiesel is two to three times higher than the petroleum-based product. Based on these projections it has been assumed that biodiesel cannot be a competitive alternative to petroleum-based fuel. However, microbial metabolic engineering presents a unique opportunity to lower the costs associated with the raw materials used in biodiesel production.

Biodiesel production from plant oils encounters certain limitations regarding mainly the availability of oil-seed supplies in suitable quantities and at competitive prices. On the contrary, the use of microbial systems for biodiesel production, although not exploited industrially until now, holds the promise to overcome these limitations. In addition, microbes can be tailored to utilize various carbon sources as feedstock for the production of oils, such as waste or agricultural byproducts. This chapter focuses on the structure and regulation of the pathways utilized by various microbes (bacteria, algae, and yeasts) for the production of fatty acids and TAGs (Fig. 1).

From 1978 to 1996 the U.S. Department of Energy supported a program entitled "Aquatic Species Program: Biodiesel from Algae" (http://www1.eere.energy.gov/biomass/pdfs/biodiesel_from_algae.pdf), which focused on the production of biodiesel from algae grown in ponds, utilizing waste CO_2 from coal-fired power plants. The program generated a collection of approximately 300 oil-producing species, mostly green algae and diatoms. It also explored the commercial utilization of these organisms in open-pond production systems. Funding for this program ceased in 1995 due to budget limitations. More than 1 decade later we can see that the major limitation of the program has been the absence of genetic and molecular tools for exploration and manipulation of the genes, pathways, and regulatory elements responsible for oil accumulation in these species. The program ended at the period during which major inroads in understanding the genes and regulatory mechanisms operating in lipid metabolism were being made in studies of other organisms. Therefore, improvement of the performance of these algal cultures by manipulating the key enzymes and regulators and/or knocking out catabolic pathways could not be approached at the time of the Aquatic Species Program. At present, we already have whole-genome sequences of seven algal species and more genomes are expected to be sequenced in the next few years. These data should provide the foundation for a better understanding of the biosynthetic pathways and regulatory networks of lipid metabolism in these organisms.

LIPID-ACCUMULATING MICROBES

Biosynthesis and accumulation of TAGs have been reported in a number of microbes including bacteria, algae, and yeasts. The buildup of TAG in bacteria has been observed mainly in organisms belonging to the actinomycetes group, such as *Mycobacterium*, *Streptomyces*, *Rhodococcus*, and *Nocardia* (Alvarez and

Athanasios Lykidis and Natalia Ivanova • Genome Biology Program, DOE-Joint Genome Institute, 2800 Mitchell Dr., Walnut Creek, CA 94598.

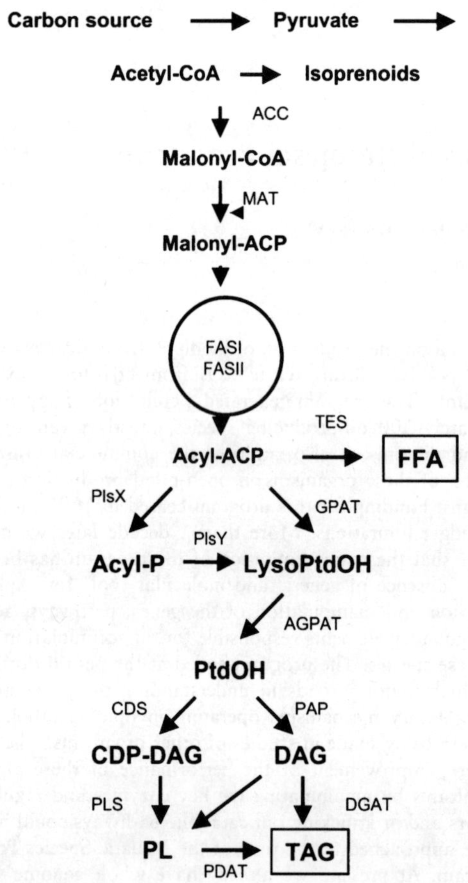

Figure 1. Overview of lipid biosynthetic pathways. Carbon sources are converted to pyruvate, which is subsequently converted to acetyl-CoA. Acetyl-CoA enters lipid biosynthesis via the action of acetyl-CoA carboxylase (ACC). Fatty acid biosynthesis proceeds through the action of either type I (FASI) or dissociated type II (FASII) systems. The resulting acyl groups are subsequently attached on the glycerol-3-phosphate backbone either via an acyl phosphate intermediate or through the action of specific acyltransferases. PtdOH is distributed between CDP-DAG and DAG. CDP-DAG and DAG are converted mainly to phospholipids via the action of phospholipid synthases. Excess DAG is diverted towards TAG. TAGs can also be derived from phospholipids via the action of PDAT. Abbreviations: MAT; malonyl-CoA:acyl-carrier-protein transacylase; TES, acyl-acyl carrier protein thioesterase; CDS, CDP-diacylglycerol synthetase; PLS, phospholipid synthases (referring to a variety of enzymes utilizing CDP-DAG and/or DAG for phospholipid synthesis); FFA, free fatty acids; PL, phospholipid. See the text for other abbreviations.

Steinbuchel, 2002). *Streptomyces* species accumulate TAGs during the postexponential phase to concentrations ranging between 50 and 150 mg/liter medium (Olukoshi and Packter, 1994). Other actinomycetes also accumulate TAGs and may contain up to 8 to 20% of cellular mass (dry weight) in fatty acids (Alvarez and Steinbuchel, 2002). A separate group of gram-negative bacteria has been shown to accumulate TAG and wax esters (esters of long-chain fatty alcohols and fatty acids). For instance, cells of *Acinetobacter* species accumulate large amounts of wax esters when they grow in a nitrogen-limited medium in the presence of alkanes or alkanols (Ishige et al., 2002).

In addition to bacteria, two classes of eukaryotic microbes contain organisms with remarkable abilities to accumulate lipids. Certain fungi, such as *Mortierella ramanniana* and species of *Lipomyces* and *Rhodotorula*, can accumulate substantial amounts of TAGs above 50% of their dry weight (Ratledge and Wynn, 2002). Three genomes of oleaginous fungi have been completely sequenced, *Aspergillus nidulans* (Galagan et al., 2005), *Debaryomyces hansenii* (Dujon et al., 2004), and *Yarrowia lipolytica* (Dujon et al., 2004), and they will serve as our paradigms in this review. These organisms accumulate approximately 20 to 30% of their cellular mass as lipids.

Besides fungi, algae have been long known to accumulate lipids and particularly TAGs (Guschina and Harwood, 2006). Algae have attracted particular attention for biotechnological applications, and several efforts are under way to exploit their utility in biodiesel production. The effects of various growth parameters (nutrient limitation, CO_2, and light intensity) on the lipid composition of several algal species have been studied. Factors that favor increased intracellular TAG levels include nutrient limitation, increased CO_2 concentration, and increased light density. Silicon, nitrogen, or phosphorus limitation caused an increase in TAG content and a concomitant decrease of polar lipids in the freshwater diatom *Stephanodiscus minutulus* (Lynn et al., 2000). Growth of algae (*Tichocarpus crinitus*) at high light densities resulted in a 50% increase in the levels of intracellular TAGs (Khotimchenko and Yakovleva, 2005). The same correlation between light intensity and TAG accumulation was observed in the green filamentous alga *Cladophora* spp. (Napolitano, 1994). TAGs were also accumulated at higher levels in *Chlamydomonas* sp. grown in medium at pH 1 than in cells grown at higher pHs (Tatsuzawa et al., 1996). Data from seven algal genomes are currently available: the red algae *Cyanidioschyzon merolae* (Matsuzaki et al., 2004) and *Galdieria sulphuraria* (Barbier et al., 2005), the diatoms *Thalassiosira pseudonana* (Armbrust et al., 2004) and *Phaeodactylum tricornutum* (unpublished; sequence data available at http://www.jgi.doe.gov), the chlorophytes *Ostreococcus tauri* and *Ostreococcus lucimarinus* (Derelle et al., 2006; Palenik et al., 2007), and *Chlamydomonas reinhardtii* (unpublished; sequence data available at http://www.jgi.doe.gov).

FATTY ACID SYNTHESIS

Initiation of Fatty Acid Synthesis

Acetyl coenzyme A (CoA) is a central metabolite synthesized from pyruvate. Acetyl-CoA carboxylase (ACC) is the committed step in fatty acid biosynthesis and constitutes the entrance gate of carbon towards cellular lipids. It catalyzes the ATP-dependent conversion of acetyl-CoA to malonyl-CoA. Acetyl-CoA is a key metabolite and a branching point in intracellular carbon distribution. Bacterial ACC is heteromeric (type II ACC) and is composed of four subunits encoded by distinct genes: the biotin carboxyl carrier protein (*accB*); biotin carboxylase (*accC*); and two proteins catalyzing the carboxyltransferase reaction (*accA* and *accD*) (Cronan and Waldrop, 2002).

The ACC subunits described above are fused into a single polypeptide encoded by one gene (type I ACC) in higher eukaryotes, resulting in a homomeric enzyme form. The *Saccharomyces cerevisiae* genome encodes two ACC enzymes. The first (ACC1) is a 2,233-amino-acid-long protein, and it is located in the cytosol (Al-Feel et al., 1992; Hasslacher et al., 1993), whereas the second (Hfa1) is targeted to mitochondria and is approximately 2,120 amino acids long (Hoja et al., 2004). The two yeast ACC isoforms appear to have unique functions in yeast physiology since mutations in the two genes have different phenotypes and the genes do not complement each other. *ACC1* is involved in cytoplasmic fatty acid biosynthesis, and its disruption is lethal and cannot be compensated by external fatty acids. In contrast, *HFA1* mutants are viable but fail to grow on lactate and glycerol and have reduced intracellular levels of lipoic acid (Tehlivets et al., 2007). Analysis of the genomes of *A. nidulans*, *D. hansenii*, and *Y. lipolytica* demonstrates that each of them contains only one ACC isoform more closely related to the cytosolic yeast ACC1 (locus tags AN6126.2, DEHA0B05632g, and YALI0C11407g, respectively).

In contrast to yeasts, many plants contain both type I and type II ACC forms, the heteromeric form in the plastid and the homomeric form in the cytosol (Sasaki and Nagano, 2004). The only isolated algal ACC gene, that of *Cyclotella cryptica* (Roessler and Ohlrogge, 1993), codes for a single polypeptide of 2,089 amino acids. However, ACC homologs are easily identifiable in all available algal genomes: *T. pseudonana*, *P. tricornutum*, and *O. tauri* genomes encode two large ACC polypeptides, whereas *C. merolae* and *G. sulphuraria* both have only one type I ACC homolog in the available data. Notably, *C. reinhardtii* is missing type I ACC homologs, but it has a type II heteromeric ACC. The ACC1 isoform of *T. pseudonana* contains a chloroplast-targeting sequence at the amino terminus of the predicted coding sequence model, whereas the ACC2 predicted protein sequence is not complete at the respective amino-terminal region.

Several lines of evidence from different organisms suggest that the ACC step is a key regulator of carbon flow towards fatty acids and lipids. In bacteria, overexpression of ACC activity increases the flow of carbon through the fatty acid-lipid biosynthesis branch of cellular metabolism (Davis et al., 2000). Overexpression of *Arabidopsis* ACC in the amyloplasts of potato tubers increased fatty acid synthesis and led to a fivefold increase in the amount of accumulated TAGs (Klaus et al., 2004). Conversely, down-regulation of the ACC1 expression in *Brassica napus* resulted in lower lipid content (Slabas et al., 2002). Although the effect was not substantial (lipid levels in mutant strains were only 10 to 30% lower than in the wild-type plants), the results were consistent with the proposed role of this enzymatic step in regulating carbon flux towards fatty acids.

The regulation of bacterial ACC has been studied in *Escherichia coli* and *Bacillus subtilis*. Transcription levels of the four *acc* genes in *E. coli* and *B. subtilis* are directly correlated with the growth rate (Li and Cronan, 1993; Marini et al., 2001). However, *accBC* expression from the native promoter did not depend on gene copy number, suggesting a transcriptional regulation mechanism. Subsequent work demonstrated that the AccB protein functions as an autoregulator of the *accBC* operon in *E. coli*, suppressing its expression (James and Cronan, 2004).

In yeast ACC activity is regulated at multiple levels. Acc1 gene expression is controlled in coordination with membrane phospholipid biosynthesis by the Ino2/Ino4 and Opi1 transcription factors (Hasslacher et al., 1993). In addition, posttranslational modification by phosphorylation appears to inactivate yeast ACC in vivo (Witters and Watts, 1990). The *S. cerevisiae* Ino2/Ino4 and Opi1 transcription factors are major regulators of lipid metabolism and control the expression of multiple genes such as fatty acid synthases and phospholipid biosynthesis genes (Carman and Henry, 1999). They exert their function by binding to Upstream Activating Sequences INO (UAS_{INO}) elements. Ino2 and Ino4 form a heterodimeric activator complex that binds the 10-bp UAS_{INO} element. Opi1 is a negative regulator required for the repression of UAS_{INO}-associated genes. Opi1 is retained in the endoplasmic reticulum (ER) by forming a complex with the membrane protein Scs2p. Intracellular levels of phosphatidic acid (PtdOH) act as a signal regulating the translocation of Opi1 to the nucleus. Opi1 binds PtdOH, and high concentrations of PtdOH in the ER membrane are required for its retention in the ER. However, when PtdOH concentrations drop, Opi1 is released and translocated to the nucleus, where it represses the transcription of lipid biosynthetic genes.

A. nidulans, *D. hansenii*, and *Y. lipolytica* contain proteins (AN8817.2, DEHA0G03894g, and YALI0C14784g, respectively) with approximately 25% identity to *S. cerevisiae* Opi1. The relatively low similarity score does not allow confident prediction that these proteins are involved in the regulation of lipid metabolism in a manner similar to Opi1. Furthermore, there are no proteins with significant similarity to Ino2p and Ino4p in *A. nidulans*, *D. hansenii*, and *Y. lipolytica*, suggesting the existence of a divergent set of regulators controlling fatty acid and lipid metabolism in these oleaginous fungi.

Elongation Steps of Fatty Acid Synthesis

Two distinct enzyme systems catalyzing the synthesis of fatty acids have been described. The type I systems (FASI), typically found in yeasts and mammals, include large multifunctional proteins harboring all the necessary enzymatic activities in one polypeptide. In contrast, the type II systems are composed of monofunctional proteins. Fatty acid biosynthesis in most bacteria proceeds with the type II, or dissociated, system (FASII). In both systems fatty acid biosynthesis involves five distinct enzymatic steps: (i) the first step, catalyzed by malonyl-CoA:ACP transacylase, transfers the malonyl group from CoA to acyl carrier protein (ACP) to form malonyl-ACP, which serves as an immediate donor of the two-carbon acetyl units used in fatty acid elongation; (ii) the condensing step, catalyzed by β-ketoacyl-ACP synthases, adds the two-carbon unit to the growing acyl-ACP; (iii) the NADPH-dependent reduction step, catalyzed by β-ketoacyl-ACP reductases, yields β-hydroxyacyl-ACP; (iv) the dehydration step, catalyzed by β-hydroxyacyl-ACP dehydrases, yields *trans*-2-enoyl-ACP; and (v) the last reductase step, catalyzed by enoyl-ACP reductases, forms a saturated acyl-ACP serving, in turn, as the substrate for another condensation reaction (Marrakchi et al., 2002). A repetitive series of reactions follows adding one two-carbon unit per cycle, until a saturated fatty acid of 16 to 18 carbons is formed. Some *Actinobacteria*, e.g., *Mycobacterium*, *Corynebacterium*, and *Nocardia* species utilize both FASI and FASII systems. Work with *Mycobacterium* species has shown that the FASI system is utilized for the production of fatty acids up to 18 carbons long and FASII elongates them to 26 to 30 carbons (Schweizer and Hofmann, 2004). However, several *Actinobacteria* genera of industrial interest, e.g., *Streptomyces*, do not have a FASI system and utilize exclusively the dissociated pathway. Several recent reviews summarize the current understanding of the genes that participate in bacterial fatty acid synthesis (Campbell and Cronan, 2001b; Zhang et al., 2003).

The type II system is also present in plants, and it is localized in the plastid, although all of its components are encoded by the nuclear genome (Ohlrogge and Jaworski, 1997). Similar to plant genomes, all available algal genomes (*C. merolae*, *T. pseudonana*, *P. tricornutum*, *O. tauri*, *O. lucimarinus*, and *C. reinhardtii*) encode homologs of the type II components, indicating the presence of the dissociated system for fatty acid biosynthesis in these organisms and its probable localization to the chloroplast.

While most higher eukaryotes utilize a FASI containing all five activities in a single large polypeptide, fungi have these activities located on two polypeptides, as exemplified by the FAS1 (β subunit) and FAS2 (α subunit) enzymes of *S. cerevisiae* (Tehlivets et al., 2007). The yeast FAS holoenzyme is a hexameric complex, $\alpha6\beta6$, located in the cytosol. Homologs of the *S. cerevisiae* FAS1 and FAS2 subunits are easily recognized in the genomes of *D. hansenii*, *Y. lipolytica*, and *A. nidulans*. In addition, *S. cerevisiae* contains a complete type II fatty acid synthesis system localized in mitochondria; products of this fatty acid synthase are utilized for lipoic acid formation, and they do not appear to be significant contributors to phospholipid and TAG formation.

Termination Steps of Fatty Acid Biosynthesis

Fatty acid biosynthesis systems have evolved to produce acyl chains approximately 16 to 18 carbon atoms long; longer fatty acids are produced in many organisms by specific systems (fatty acid elongases). In bacteria there is no specific mechanism for terminating acyl chain elongation; when acyl-ACP reaches 16 or 18 carbons, it becomes a substrate for the acyltransferases that will attach the fatty acyl chain onto the glycerol backbone to produce phospholipids. However, plants have a special class of enzymes that terminate the acyl chain elongation by hydrolyzing the thioester bond of acyl-ACP, thus releasing free fatty acids and ACP. This enzyme is known as fatty acyl-ACP thioesterase (FAT) and is localized in the plastid. Free fatty acids produced by FAT exit the plastid and are reesterified with CoA. The resulting acyl-CoAs are utilized for glycerolipid biosynthesis in the ER. Based on sequence similarity and substrate specificity, two classes of thioesterases have been described: the FATA class, which is active on unsaturated acyl-ACPs, and the FATB class, which prefers saturated acyl groups (Salas and Ohlrogge, 2002). Unlike higher plants that contain multiple FAT genes, the genomes of algae *C. reinhardtii*, *O. tauri*, and *O. lucimarinus* contain only one FAT homolog, while *C. merolae* and the diatoms *T. pseudonana* and *P. tricornutum* have no FAT homologs in the current assemblies, so the mechanism of acyl chain termination in these algae is unknown.

An important observation with regard to the possibility of microbial biodiesel production was made when FAT enzymes were overproduced in *E. coli*. Expression of a plant (*Umbellularia californica*) medium-chain FAT in an *E. coli* strain deficient in fatty acid oxidation resulted in a fourfold increase of the fatty acid output of the bacterial culture (Voelker and Davies, 1994). Thus far it represents the most efficient way to uncouple fatty acid formation from phospholipid and membrane biosynthesis in *E. coli*. The resulting fatty acids were enriched in acyl chains of 14 carbon atoms, indicating that the production of fatty acids with specific acyl length may be feasible using enzymes with the appropriate specificity. Lowering the acyl chain length of the fatty acid esters may have substantial effects on the physical properties and overall performance of the resulting fuels.

Regulation of Fatty Acid Biosynthesis

Fatty acid synthesis is regulated at multiple points, and different mechanisms operate to provide transcriptional and biochemical control of the carbon flow towards lipids. The regulation of fatty acid biosynthesis has been studied with *E. coli* and *B. subtilis*, which serve as models for gram-negative and gram-positive bacteria, respectively. Extensive work with *E. coli* demonstrated that fatty acid biosynthesis is regulated coordinately with phospholipid biosynthesis and growth in response to changes in the environment. Fatty acid biosynthesis ceases when bacteria enter the stationary phase. This response is mediated by the intracellular levels of ppGpp (guanosine 5-diphosphate-3-diphosphate). Elevated levels of ppGpp in vivo inhibit PlsB, the first acyltransferase catalyzing the attachment of acyl groups to glycerol-3-phosphate (Heath et al., 1994). The inhibition of this first step of PtdOH formation leads to elevated levels of acyl-ACPs (Heath and Rock, 1996a; Rock and Jackowski, 1982), which, in turn, serve as feedback inhibitors of three key enzymes in the pathway. The ACC reaction is inhibited by long-chain acyl-ACPs (Davis and Cronan, 2001), resulting in a decreased flux of malonate groups into the pathway. The first enzyme in fatty acid biosynthesis, FabH, is also inhibited by acyl-ACPs (Heath and Rock, 1996a, 1996b) resulting in the attenuation of fatty acid biosynthesis initiation. Finally, acyl-ACPs inhibit the last enzyme in fatty acid elongation, FabI, slowing the overall rate of fatty acid formation (Heath and Rock, 1996a). Fatty acid biosynthesis can be uncoupled from phospholipid synthesis and growth stage by overexpression of thioesterases (Cho and Cronan, 1995; Dormann et al., 1995; Jiang and Cronan, 1994; Ohlrogge et al., 1995; Voelker and Davies, 1994). Thioesterases hydrolyze the acyl-ACP thioester releasing free fatty acid and ACP, terminating the acyl elongation cycle. The resulting free fatty acids are secreted en masse to the medium, and these strains are very good candidates for further exploration with regard to biodiesel production.

Transcriptional regulation of bacterial fatty acid biosynthesis is, in general, poorly understood. There are two well-documented examples of transcriptional regulation of bacterial fatty acid biosynthesis, one in gram-negative and one in gram-positive bacteria. In *E. coli*, two TetR-type transcriptional regulators, the activator FadR and the repressor FabR, have been shown to regulate the expression of *fabA* and *fabB* genes. The FadR was initially discovered as a repressor of fatty acid degradation genes (DiRusso and Nunn, 1985; Overath et al., 1969) and was subsequently shown to be an activator of *fabA* (Henry and Cronan, 1992) and *fabB* (Campbell and Cronan, 2001a). Long-chain acyl-CoAs regulate the DNA binding activity of FadR (DiRusso et al., 1992, 1998; Raman and DiRusso, 1995). The second *E. coli* regulator, FabR, acts as a repressor of the *fabA* and *fabB* genes (Zhang et al., 2002), but it is currently unclear which ligand(s) modulates its activity. FadR and FabR have similar phylogenetic distribution, and their homologs can be found in the species of *Escherichia*, *Salmonella*, *Yersinia*, *Vibrio*, *Haemophilus*, and *Shewanella*.

A separate transcriptional system regulating the expression of FASII enzymes has been described in *B. subtilis* and is mediated by FapR (Schujman et al., 2003). Subsequent work demonstrated that malonyl-CoA functions as the direct and specific signal for the control of FapR activity (Schujman et al., 2006). Binding of malonyl-CoA, the key intermediate in fatty acid formation, promotes dissociation of the FapR-DNA complex, resulting in transcription of FASII genes. FapR homologs can be identified in various gram-positive bacteria, such as bacilli, listerias, and staphylococci. There is no substantial information on the regulation of fatty acid biosynthesis in most bacteria including cyanobacteria, actinobacteria, etc.

Most of the information regarding the regulation of FAS1 and FAS2 gene expression in fungi comes from the studies with *S. cerevisiae*. In this organism FAS1 and FAS2 expression is controlled by the lipid precursors choline and inositol through UAS_{INO} sequences (Schuller et al., 1992). In addition, *FAS1* and *FAS2* genes are activated by the general transcription factors Gcr1, Abf1p, Rap1p, and Reb1 and contain the respective binding sites in their promoter regions (Greenberg and Lopes, 1996; Schuller et al., 1994). Analysis of *A. nidulans*, *D. hansenii*, and *Y. lipolytica* genomes indicates that these oleaginous fungi do not harbor proteins with significant similarity to the *Saccharomyces*

Gcr1p and Abf1p transcription factors. However, they contain homologs of the Reb1 regulator (AN4618.2, DEHA0E13266g, and YALI0F20460g, respectively) and *D. hansenii* contains a protein (DEHA0C13959g) with a relatively low similarity score (expectation value of −30) to Rap1p. There is no experimental information on the regulators of fatty acid synthesis in these oleaginous fungi, and the above genomic observations are consistent with the hypothesis that, despite the high similarity in the structure and organization of the FAS genes, their regulation may vary significantly among organisms. An interesting observation regarding the interrelationship between the *Saccharomyces FAS1* and *FAS2* genes refers to the dependence of *FAS2* mRNA abundance on *FAS1* expression (Wenz et al., 2001). Overexpression of FAS1p increases *FAS2* levels, whereas *FAS1* mutants have decreased levels of *FAS2* mRNA. In contrast, *FAS1* expression is independent of *FAS2* expression. The responsible regulatory site is located within the coding sequence of *FAS2*. Whether this autoregulatory mechanism is conserved in other fungi remains to be established.

Unsaturated Fatty Acid Biosynthesis

Besides acyl chain length, the saturation degree of the fatty acids plays an important role in the physical properties of the resulting biodiesel. Particularly, it influences the viscosity at low temperatures. *cis*-Unsaturated and polyunsaturated fatty acids (UFAs and PUFAs) play vital roles in membrane biology, and their relative content influences the physical properties of most membrane systems regulating the fluidity of cellular membranes. Two major mechanisms have been described that synthesize UFAs: the oxygen-dependent fatty acid desaturation pathway, which has representatives in both bacteria and eukaryotes, and the anaerobic pathway, which inserts the double bond concomitantly with the elongation of the acyl length as exemplified by the *E. coli* system. The oxygen-dependent dehydrogenation at a specific position of the acyl chain is catalyzed by fatty acid desaturases and results in a *cis*-double bond. Fatty acid desaturases with specificity for different positions of the acyl chain have been described in a variety of organisms; for instance, *B. subtilis* has a $\Delta 5$ (acting on the fifth carbon of the acyl chain) desaturase (Aguilar et al., 1998). The biotechnological potential of fatty acid desaturases for the synthesis of UFAs and PUFAs has been studied extensively (Huang et al., 2004; Qi et al., 2004). Fatty acid desaturases are particularly abundant in soil and aquatic bacteria. All cyanobacteria that have been sequenced to date have multiple desaturases in their genomes, which have presumably distinct specificities, potentially resulting in synthesis of polyunsaturated fatty acids. A two-component regulatory system that controls expression of the *B. subtilis* desaturase at low temperatures has been described (Aguilar et al., 2001). The histidine kinase DesK senses changes in membrane fluidity caused by low temperatures or other factors and activates the response regulator DesR, which induces transcription of the desaturase (Aguilar and de Mendoza, 2006). Close homologs of the DesK-DesR system can be identified in other bacilli but not in other organisms, suggesting the existence of divergent (at least in terms of sequence similarity) regulatory systems.

S. cerevisiae has a single fatty acid desaturase, Ole1p, generating a double bond in the Δ_9 position of fatty acids (Stukey et al., 1989). *A. nidulans* and *D. hansenii* each have two fatty acid desaturase genes in their genomes (AN6731.2 and AN4135.2 for the former and DEHAOF04268g and DEHAOF25432g for the latter), whereas *Y. lipolytica* has one (YALI0C05951g). Transcription of *Ole1* in *S. cerevisiae* depends on two transcription factors, SPT23 and MGA2. Disruption of either gene has little effect on Ole1, whereas disruption of both genes results in auxotrophy for UFA (Zhang et al., 1999). *A. nidulans*, *D. hansenii*, and *Y. lipolytica* genomes encode proteins with relatively good similarity to the SPT23 and MGA2 proteins (20 to 30% amino acid identity); these homologs may be involved in regulation of UFA synthesis.

Multiple desaturases have been also characterized from plants, and much work has been focused on their action and regulation because of the significant dietary value of polyunsaturated fatty acids. Multiple desaturases are easily recognized in the genomes of all algae sequenced to date: *C. reinhardtii* has seven potential fatty acid desaturases, *T. pseudonana* has thirteen, *O. tauri* has eight, and *C. merolae* has four.

TRANSFER OF ACYL GROUPS TO THE MEMBRANE; SYNTHESIS OF PtdOH

After they are formed, acyl groups are sequentially anchored onto the glycerol-3-phosphate backbone to form PtdOH. Two enzyme systems have been described that mediate the attachment of the first acyl group to generate 1-acylglycerol-3-phosphate (LysoPtdOH): the universal glycerol-3-phosphate acyltransferase (GPAT) system and the PlsX-PlsY acyl-phosphate-mediated system, which appears to be confined to bacteria. GPAT catalyzes the initial acyl transfer from acyl-ACP onto glycerol-3-phosphate yielding LysoPtdOH. Recent work has uncovered an alternative route for the synthesis of LysoPtdOH in bacteria via the PlsX-PlsY pathway (Lu et al., 2006). PlsX catalyzes the reaction between acyl-ACP and phosphate, resulting in the formation of a novel intermediate, acyl-phosphate. PlsY subsequently transfers the acyl group from acyl-phosphate to glycerol-3-phosphate, yielding LysoPtdOH. The PlsX-PlsY system for lipid biosynthesis is abundant in bacteria, whereas the GPAT reaction appears to be limited to a few gram-negative genera and *Actinobacteria*. The absence of PlsX-PlsY homologs in *Actinobacteria* suggests that many oleaginous bacteria utilize exclusively the GPAT reaction for the initiation of TAG buildup.

The GPAT reaction is the only route described in eukaryotes for the association of the acyl groups with the glycerol backbone. LysoPtdOH is subsequently acylated to yield PtdOH in a step catalyzed by 1-acylglycerol-3-phosphate acyltransferase (AGPAT). *S. cerevisiae* contains two genes encoding GPAT activities, YKR067w and YBL011w. *D. hansenii* has also two homologous genes, whereas *Y. lipolytica* and *A. nidulans* have only one GPAT gene. In *S. cerevisiae* the second acyltransferase (AGPAT) is encoded by the *SLC1* gene, and single copies of this gene can be found in the other fungal genomes.

Two classes of GPATs have been described in *Arabidopsis thaliana*: one isoform is located in the plastid (Nishida et al., 1993), whereas a separate gene family encompassing six members encodes extraplastidial GPATs (Zheng et al., 2003) presumably residing on the ER. Five *Arabidopsis* genes code for AGPAT enzymes, one of which is targeted to the plastid (Kim and Huang, 2004). The compartmentalization of PtdOH synthesis in two separate organelles (plastid and ER) is a well-developed theme in plant lipid biology. Examination of the algal genomes indicates that they contain only the plastid versions of the two acyltransferases involved in PtdOH formation. *T. pseudonana*, *P. tricornutum*, *C. merolae*, *C. reinhardtii*, and *O. tauri* are all missing recognizable extraplastidial GPAT homologs, suggesting either that PtdOH synthesis occurs exclusively in the plastid or that the enzymes have dual localization both in the plastid and in the ER. If the former hypothesis is true, then lipid trafficking between the plastid and the ER would be important since the ER is the organelle where phospholipid and TAG synthesis occurs. Overexpression of plant and bacterial GPAT activities in *A. thaliana* resulted in increased seed oil content (Jain et al., 2000). Also, overexpression of the yeast AGPAT in plants leads to a 50% increase in TAG levels (Zou et al., 1997). The above observations are consistent with the general theme that overexpression of any enzymatic step in the pathways leading to TAG increases the flow of carbon through the lipid pathway and causes accumulation of TAGs.

TAG BIOSYNTHESIS

Phosphatidate Phosphatases

PtdOH is a key branching point in de novo lipid metabolism, and it is converted either to CDP-diacylglycerol (DAG) or DAG depending on the organism. CDP-DAG and DAG serve as intermediates in membrane phospholipid biosynthesis and, in addition, DAG is converted to TAG. Phosphatidate phosphatases (PAPs) in coordination with phospholipid-producing enzymes are key regulators of the flux of carbon towards TAGs. PAPs catalyze the conversion of PtdOH to DAG; the primary destination of DAG is the synthesis of membrane phospholipids, whereas excess DAG is directed towards TAG. Recent work with *S. cerevisiae* has identified the PAP1 enzyme involved in de novo lipid biosynthesis as the yeast homolog of lipin (Han et al., 2006). Lipin was first identified in mammalian cells as a regulator of lipid metabolism exerting a major effect on fat accumulation (Peterfy et al., 2001; Phan and Reue, 2005). Lipin homologs can be identified in all eukaryotic genomes sequenced to date including algae and plants. *C. merolae*, *T. pseudonana*, and *C. reinhardtii* have one copy of the lipin gene in their genomes, whereas *A. thaliana* has two copies.

The existence of PAP enzymes in certain bacteria is inferred by the structure of the biochemical pathways (Dowhan, 1997). For example, cyanobacteria synthesize DAG as an obligate intermediate in the formation of glycolipids and sulfolipids, which are important molecules for the organization of the thylakoid membranes. Certain gram-positive bacteria also use DAG in the synthesis of glycolipids. However, a bacterial PAP has not been described and further research is needed to identify the genes coding for this activity.

Acyltransferases

Two types of reactions have been described that lead to formation of TAG: the first is acyl-CoA dependent, utilizes DAG and acyl-CoA, and is catalyzed by diacylglycerol acyltransferase (DGAT) (Buhman et al., 2001), whereas the second is acyl-CoA independent and uses phospholipid molecules as donors of the acyl group. The latter reaction is catalyzed by phospholipid: diacylglycerol acyltransferase (PDAT) (Dahlqvist et al., 2000; Stahl et al., 2004). The presence of DGAT and PDAT homologs in genomic sequences can serve as a useful marker for the ability of various organisms to synthesize TAGs. DGATs have been characterized from various organisms including mammals (Buhman et al., 2001), plants (Zou et al., 1999), yeast (Sorger and Daum, 2002), and bacteria (Daniel et al., 2004;

Kalscheuer and Steinbuchel, 2003). In contrast, PDAT enzymes have been characterized in plants and yeast only (Dahlqvist et al., 2000; Stahl et al., 2004). In yeast the PDAT reaction is catalyzed by *LRO1* (LCAT Related Open reading frame) gene product. The relative contribution of the fungal DGAT and PDAT pathways to TAG accumulation was studied with *S. cerevisiae* utilizing *DGA1* and *LRO1* mutants. Overall, the results suggest that Lro1 contributes to TAG formation during exponential growth whereas Dga1 has significant participation in the stationary phase and during sporulation (Oelkers et al., 2002). DGAT and PDAT homologs can be identified in *D. hansenii* and *Y. lipolytica* genomes, suggesting the presence of both mechanisms for TAG formation in these organisms. In contrast, characterization of an *A. thaliana* knockout line indicated that the PDAT enzyme is not a major contributor to seed TAG content (Mhaske et al., 2005). DGAT enzyme levels are important regulators of the cellular TAG content: overexpression of *Arabidopsis* DGAT1 in tobacco leaves and yeast increases TAG levels (Bouvier-Nave et al., 2000). Overexpression in *Arabidopsis* seeds also results in TAG accumulation (Jako et al., 2001).

DGAT and PDAT homologs can be easily identified in all algal genomes sequenced to date, although they exhibit distinct distribution patterns. Examination of the algal genomic sequences reveals that *C. reinhardtii* has three genes encoding DGATs but it is missing a PDAT homolog. The diatoms *T. pseudonana* and *P. tricornutum* have both DGAT and PDAT homologs. On the other hand, *O. tauri* lacks recognizable DGAT homologs but it does have a PDAT enzyme, suggesting that TAGs are derived from phospholipids in this organism. *C. merolae* has DGAT but not a PDAT.

The first bacterial DGAT enzyme was identified and cloned in *Acinetobacter baylyi* strain ADP1 (Kalscheuer and Steinbuchel, 2003). In addition, it was shown that this enzyme is in fact bifunctional and it also catalyzes the reaction between acyl-CoA and fatty alcohol leading to the formation of wax esters (wax ester synthase [WS]). DGAT/WS homologs can be identified in multiple bacteria from diverse phylogenetic origins. In addition to many strains of mycobacteria and other soil *Actinobacteria* (*Nocardioides*, *Rhodococcus*, and *Streptomyces*), DGAT homologs can be recognized in soil *Alphaproteobacteria* (*Bradyrhizobium japonicum*), aquatic and sediment *Betaproteobacteria* (*Polaromonas* sp. JS666, *Rhodoferax ferrireducens* DSM 15236), aquatic *Gammaproteobacteria* (*Photobacterium profundum*, *Psychrobacter arcticus*, *Psychrobacter cryohalolentis* K5, *Hahella chejuensis* KCTC 2396, *Alcanivorax borkumensis*), aquatic bacteroidetes (*Salinibacter ruber* DMS 13855), and acidobacteria (*Solibacter usitatus* Ellin6076). Biochemical studies indicated that the *Acinetobacter* DGAT/WS can utilize acyl-CoAs of variable length, and its utility in producing fatty acid esters was exploited by overexpression in different host backgrounds (Kalscheuer et al., 2004; Stoveken et al., 2005; Uthoff et al., 2005). Coexpression of the *Acinetobacter* WS/DGAT with plant acyl-CoA reductase (which catalyzes the conversion of acyl-CoAs to fatty alcohols) in *E. coli* caused the production of wax esters of up to 1% of the cellular weight (dry weight) (Kalscheuer et al., 2006b). Expression of the above WS/DGAT in an *S. cerevisiae* strain lacking all four acyltransferases responsible for neutral lipid biosynthesis (TAGs and sterol esters) restored TAG formation but not sterol ester synthesis (Kalscheuer et al., 2004). It also resulted in the formation of fatty acid ethyl esters and fatty acid isoamyl esters, demonstrating the ability of this enzyme to utilize short-chain alcohols as substrates. This remarkable property is directly applicable to biodiesel production. In a recent study, Kalscheuer et al. (Kalscheuer et al., 2006a) exploited *E. coli* as a host for the production of biodiesel. They coupled the expression of the *Acinetobacter* WS/DGAT with the expression of pyruvate decarboxylase and alcohol dehydrogenase from the anaerobic ethanologenic bacterium *Zymomonas mobilis*. Coexpression of the three genes resulted in intracellular accumulation of fatty acid ethyl esters consisting mainly of ethyl oleate in concentrations up to 26% of bacterial dry biomass. However, the formation of fatty acid ethyl esters was strictly dependent on the presence of sodium oleate in the medium, underlining the inability of the WS/DGAT system to utilize de novo-formed fatty acids in *E. coli*. This result is another manifestation of the tight control of fatty acids biosynthesis and their destination in *E. coli*, which has to be uncoupled in order to channel fatty acids towards production of biodiesel. However, it also opens the door to the exploration of other bacterial species, particularly those able to amass fatty acids, as potential hosts.

The *Arabidopsis* AP2/EREBP transcription factor WRINKLED1 (WRI1) has been identified as a regulator of seed TAG accumulation (Focks and Benning, 1998), and it was named after the wrinkled appearance of the seed coat. Subsequent analysis indicated that *wri1* mutants are defective in a regulatory factor that controls the glycolytic breakdown of sugars and the availability of precursors for fatty acid biosynthesis and ultimately TAG accumulation (Cernac and Benning, 2004; Ruuska et al., 2002), highlighting the interrelationship between carbon flow through glycolysis and lipid metabolism. WRI1 homologs can be identified in the genomes of algae *O. tauri*, *O. lucimarinus*, and *C. reinhardtii* but not in *G. sulphuraria*, *C. merolae*, *T. pseudonana*, or *P. tricornutum*.

CORRELATION OF LIPID ACCUMULATION WITH THE CELLULAR REDOX STATUS

In addition to acetyl-CoA, fatty acid biosynthesis needs substantial amounts of NADPH to proceed. For the synthesis of every mole of C_{18} fatty acid there is a need for 18 mol of NADPH. Biochemical studies with oleaginous fungi suggested that malic enzyme is a critical step in supplying the necessary amount of NADPH for fatty acid synthesis occurring under conditions of nitrogen starvation, which induces lipid accumulation (Ratledge and Wynn, 2002; Ratledge, 2004). It has been suggested that instead of the traditional pyruvate-dependent mechanism, a distinct route for converting glucose to acetyl-CoA operates in oleaginous fungi. Under nitrogen-limiting conditions, cells activate AMP deaminase in an effort to scavenge additional nitrogen from intracellular resources. AMP deaminase converts AMP to inosine-5-monophosphate and ammonia. The decrease in AMP concentration has an inhibitory effect on isocitrate dehydrogenase, a tricarboxylic acid cycle enzyme, which requires AMP for its activity. The accumulating isocitrate is readily converted to citrate via aconitase, and citrate is subsequently effluxed from the mitochondrion to the cytoplasm. Cytosolic citrate is converted to acetyl-CoA and oxaloacetate via the action of ATP-citrate lyase, which catalyzes the following reaction:

$$\text{citrate} + \text{CoA} + \text{ATP} \rightarrow \text{acetyl-CoA} + \text{oxaloacetate} + \text{ADP} + \text{Pi}$$

The resulting acetyl-CoA is used for fatty acid biosynthesis, whereas the oxaloacetate is converted to malate (via malate dehydrogenase), which enters the mitochondria as the counterion in the citrate efflux. The above route provides additional acetyl-CoA units for lipid buildup; in addition, malic enzyme supplies NADPH by catalyzing the following reaction:

$$\text{malate} + \text{NADP}^+ \rightarrow \text{pyruvate} + CO_2 + \text{NADPH}$$

The above schema is supported by observations in studies of *A. nidulans* mutants lacking malic enzyme activity (Wynn et al., 1999). The mutant cells accumulated only one-half of the amount of lipids that wild-type cells did. Further research is necessary to acquire a complete picture of the interrelationships and dependence of lipid biosynthesis on the overall redox state of the cell, especially in bacteria and algae.

CONCLUSIONS AND PERSPECTIVE

Our understanding of the biochemical and genetic steps that constitute and regulate lipid metabolism and oil accumulation has increased significantly during the last few years. The genes and pathways responsible for fatty acid and TAG biosynthesis have been identified in multiple model organisms. However, most of the work was performed with a mindset of either identifying potential drug targets in animal and plant pathogens or improving the nutritional value of agricultural products. A new end point of lipid research has to be introduced, and this is the massive generation of fatty acids and/or TAGs that will serve as the raw material for biodiesel production.

Although the basic biochemical reactions utilized in lipid biosynthesis are essentially identical among the major domains of life, the multiple versions of enzymes, pathways, and regulatory networks that participate indicate the fine-tuning of these pathways to the environment. The emerging theme from genome comparisons underlines the evolution of distinct regulatory mechanisms in various phylogenetic groups. Carbon flow towards fatty acids occurs in order for the cell to synthesize membrane lipids. In general, overexpression of specific enzymes does not have dramatic effects in overall lipid metabolism because other steps in the pathway become limiting. Therefore, genetic engineering of specific steps probably will have limited impact in overall lipid metabolism. However, genetic modification of a combination of anabolic, catabolic, and regulatory steps may lead to profound increases in lipid accumulation. Further research into the regulatory mechanisms that govern various branches of lipid metabolism is necessary in order to obtain a complete description of the molecular mechanisms that determine fatty acid and lipid accumulation.

All free-living organisms have the machinery to synthesize fatty acids, and conceptually, they could be exploited for biodiesel production. However, the photosynthetic organisms provide the unique opportunity to couple CO_2 sequestration to lipid accumulation and subsequent biodiesel production. In light of the recent genome sequence data and the emergence of synthetic biology, we believe that research on the biochemistry and regulation of microbial lipid accumulation is necessary in order to develop competitive technologies for biodiesel production.

REFERENCES

Aguilar, P. S., J. E. Cronan, Jr., and D. de Mendoza. 1998. A *Bacillus subtilis* gene induced by cold shock encodes a membrane phospholipid desaturase. *J. Bacteriol.* 180:2194–2200.

Aguilar, P. S., A. M. Hernandez-Arriaga, L. E. Cybulski, A. C. Erazo, and D. de Mendoza. 2001. Molecular basis of thermosensing: a two-component signal transduction thermometer in *Bacillus subtilis. EMBO J.* 20:1681–1691.

Aguilar, P. S., and D. de Mendoza. 2006. Control of fatty acid desaturation: a mechanism conserved from bacteria to humans. *Mol. Microbiol.* 62:1507–1514.

Al-Feel, W., S. S. Chirala, and S. J. Wakil. 1992. Cloning of the yeast FAS3 gene and primary structure of yeast acetyl-CoA carboxylase. *Proc. Natl. Acad. Sci. USA* 89:4534–4538.

Alvarez, H. M., and A. Steinbuchel. 2002. Triacylglycerols in prokaryotic microorganisms. *Appl. Microbiol. Biotechnol.* 60:367-376.

Armbrust, E. V., J. A. Berges, C. Bowler, B. R. Green, D. Martinez, N. H. Putnam, S. Zhou, A. E. Allen, K. E. Apt, M. Bechner, M. A. Brzezinski, B. K. Chaal, A. Chiovitti, A. K. Davis, M. S. Demarest, J. C. Detter, T. Glavina, D. Goodstein, M. Z. Hadi, U. Hellsten, M. Hildebrand, B. D. Jenkins, J. Jurka, V. V. Kapitonov, N. Kroger, W. W. Lau, T. W. Lane, F. W. Larimer, J. C. Lippmeier, S. Lucas, M. Medina, A. Montsant, M. Obornik, M. S. Parker, B. Palenik, G. J. Pazour, P. M. Richardson, T. A. Rynearson, M. A. Saito, D. C. Schwartz, K. Thamatrakoln, K. Valentin, A. Vardi, F. P. Wilkerson, and D. S. Rokhsar. 2004. The genome of the diatom *Thalassiosira pseudonana*: ecology, evolution, and metabolism. *Science* 306:79–86.

Barbier, G., C. Oesterhelt, M. D. Larson, R. G. Halgren, C. Wilkerson, R. M. Garavito, C. Benning, and A. P. Weber. 2005. Comparative genomics of two closely related unicellular thermo-acidophilic red algae, *Galdieria sulphuraria* and *Cyanidioschyzon merolae*, reveals the molecular basis of the metabolic flexibility of *Galdieria sulphuraria* and significant differences in carbohydrate metabolism of both algae. *Plant Physiol.* 137:460–474.

Bouvier-Nave, P., P. Benveniste, P. Oelkers, S. L. Sturley, and H. Schaller. 2000. Expression in yeast and tobacco of plant cDNAs encoding acyl CoA:diacylglycerol acyltransferase. *Eur. J. Biochem.* 267:85–96.

Buhman, K. K., H. C. Chen, and R. V. Farese, Jr. 2001. The enzymes of neutral lipid synthesis. *J. Biol. Chem.* 276:40369–40372.

Campbell, J. W., and J. E. Cronan, Jr. 2001a. *Escherichia coli* FadR positively regulates transcription of the *fabB* fatty acid biosynthetic gene. *J. Bacteriol.* 183:5982–5990.

Campbell, J. W., and J. E. Cronan, Jr. 2001b. Bacterial fatty acid biosynthesis: targets for antibacterial drug discovery. *Annu. Rev. Microbiol.* 55:305–332.

Carman, G. M., and S. A. Henry. 1999. Phospholipid biosynthesis in the yeast *Saccharomyces cerevisiae* and interrelationship with other metabolic processes. *Prog. Lipid Res.* 38:361–399.

Cernac, A., and C. Benning. 2004. WRINKLED1 encodes an AP2/EREB domain protein involved in the control of storage compound biosynthesis in *Arabidopsis. Plant J.* 40:575–585.

Cho, H., and J. E. Cronan, Jr. 1995. Defective export of a periplasmic enzyme disrupts regulation of fatty acid synthesis. *J. Biol. Chem.* 270:4216–4219.

Cronan, J. E., Jr., and G. L. Waldrop. 2002. Multi-subunit acetyl-CoA carboxylases. *Prog. Lipid Res.* 41:407–435.

Dahlqvist, A., U. Stahl, M. Lenman, A. Banas, M. Lee, L. Sandager, H. Ronne, and S. Stymne. 2000. Phospholipid:diacylglycerol acyltransferase: an enzyme that catalyzes the acyl-CoA-independent formation of triacylglycerol in yeast and plants. *Proc. Natl. Acad. Sci. USA* 97:6487–6492.

Daniel, J., C. Deb, V. S. Dubey, T. D. Sirakova, B. Abomoelak, H. R. Morbidoni, and P. E. Kolattukudy. 2004. Induction of a novel class of diacylglycerol acyltransferases and triacylglycerol accumulation in *Mycobacterium tuberculosis* as it goes into a dormancy-like state in culture. *J. Bacteriol.* 186:5017–5030.

Davis, M. S., J. Solbiati, and J. E. Cronan, Jr. 2000. Overproduction of acetyl-CoA carboxylase activity increases the rate of fatty acid biosynthesis in *Escherichia coli. J. Biol. Chem.* 275:28593–28598.

Davis, M. S., and J. E. Cronan, Jr. 2001. Inhibition of *Escherichia coli* acetyl coenzyme A carboxylase by acyl-acyl carrier protein. *J. Bacteriol.* 183:1499–1503.

Derelle, E., C. Ferraz, S. Rombauts, P. Rouze, A. Z. Worden, S. Robbens, F. Partensky, S. Degroeve, S. Echeynie, R. Cooke, Y. Saeys, J. Wuyts, K. Jabbari, C. Bowler, O. Panaud, B. Piegu, S. G. Ball, J. P. Ral, F. Y. Bouget, G. Piganeau, B. De Baets, A. Picard, M. Delseny, J. Demaille, Y. Van de Peer, and H. Moreau. 2006. Genome analysis of the smallest free-living eukaryote *Ostreococcus tauri* unveils many unique features. *Proc. Natl. Acad. Sci. USA* 103:11647–11652.

DiRusso, C. C., and W. D. Nunn. 1985. Cloning and characterization of a gene (*fadR*) involved in regulation of fatty acid metabolism in *Escherichia coli. J. Bacteriol.* 161:583–588.

DiRusso, C. C., T. L. Heimert, and A. K. Metzger. 1992. Characterization of FadR, a global transcriptional regulator of fatty acid metabolism in *Escherichia coli*. Interaction with the *fadB* promoter is prevented by long chain fatty acyl coenzyme A. *J. Biol. Chem.* 267:8685–8691.

DiRusso, C. C., V. Tsvetnitsky, P. Hojrup, and J. Knudsen. 1998. Fatty acyl-CoA binding domain of the transcription factor FadR. Characterization by deletion, affinity labeling, and isothermal titration calorimetry. *J. Biol. Chem.* 273;33652–33659.

Dormann, P., T. A. Voelker, and J. B. Ohlrogge. 1995. Cloning and expression in *Escherichia coli* of a novel thioesterase from *Arabidopsis thaliana* specific for long-chain acyl-acyl carrier proteins. *Arch. Biochem. Biophys.* 316:612–618.

Dowhan, W. 1997. Molecular basis for membrane phospholipid diversity: why are there so many lipids? *Annu. Rev. Biochem.* 66:199–232.

Dujon, B., D. Sherman, G. Fischer, P. Durrens, S. Casaregola, I. Lafontaine, J. De Montigny, C. Marck, C. Neuveglise, E. Talla, N. Goffard, L. Frangeul, M. Aigle, V. Anthouard, A. Babour, V. Barbe, S. Barnay, S. Blanchin, J. M. Beckerich, E. Beyne, C. Bleykasten, A. Boisrame, J. Boyer, L. Cattolico, F. Confanioleri, A. De Daruvar, L. Despons, E. Fabre, C. Fairhead, H. Ferry-Dumazet, A. Groppi, F. Hantraye, C. Hennequin, N. Jauniaux, P. Joyet, R. Kachouri, A. Kerrest, R. Koszul, M. Lemaire, I. Lesur, L. Ma, H. Muller, J. M. Nicaud, M. Nikolski, S. Oztas, O. Ozier-Kalogeropoulos, S. Pellenz, S. Potier, G. F. Richard, M. L. Straub, A. Suleau, D. Swennen, F. Tekaia, M. Wesolowski-Louvel, E. Westhof, B. Wirth, M. Zeniou-Meyer, I. Zivanovic, M. Bolotin-Fukuhara, A. Thierry, C. Bouchier, B. Caudron, C. Scarpelli, C. Gaillardin, J. Weissenbach, P. Wincker, and J. L. Souciet. 2004. Genome evolution in yeasts. *Nature* 430:35–44.

Focks, N., and C. Benning. 1998. Wrinkled1: a novel, low-seed-oil mutant of *Arabidopsis* with a deficiency in the seed-specific regulation of carbohydrate metabolism. *Plant Physiol.* 118:91–101.

Galagan, J. E., S. E. Calvo, C. Cuomo, L. J. Ma, J. R. Wortman, S. Batzoglou, S. I. Lee, M. Basturkmen, C. C. Spevak, J. Clutterbuck, V. Kapitonov, J. Jurka, C. Scazzocchio, M. Farman, J. Butler, S. Purcell, S. Harris, G. H. Braus, O. Draht, S. Busch, C. D'Enfert, C. Bouchier, G. H. Goldman, D. Bell-Pedersen, S. Griffiths-Jones, J. H. Doonan, J. Yu, K. Vienken, A. Pain, M. Freitag, E. U. Selker, D. B. Archer, M. A. Penalva, B. R. Oakley, M. Momany, T. Tanaka, T. Kumagai, K. Asai, M. Machida, W. C. Nierman, D. W. Denning, M. Caddick, M. Hynes, M. Paoletti, R. Fischer, B. Miller, P. Dyer, M. S. Sachs, S. A. Osmani, and B. W. Birren. 2005. Sequencing of *Aspergillus nidulans* and comparative analysis with *A. fumigatus* and *A. oryzae*. *Nature* 438:1105–1115.

Greenberg, M. L., and J. M. Lopes. 1996. Genetic regulation of phospholipid biosynthesis in *Saccharomyces cerevisiae. Microbiol. Rev.* 60:120.

Guschina, I. A., and J. L. Harwood. 2006. Lipids and lipid metabolism in eukaryotic algae. *Prog. Lipid Res.* 45:160–186.

Han, G. S., W. I. Wu, and G. M. Carman. 2006. The *Saccharomyces cerevisiae* lipin homolog is a Mg2+-dependent phosphatidate phosphatase enzyme. *J. Biol. Chem.* 281:9210–9218.

Hasslacher, M., A. S. Ivessa, F. Paltauf, and S. D. Kohlwein. 1993. Acetyl-CoA carboxylase from yeast is an essential enzyme and is regulated by factors that control phospholipid metabolism. *J. Biol. Chem.* 268:10946–10952.

Heath, R. J., S. Jackowski, and C. O. Rock. 1994. Guanosine tetraphosphate inhibition of fatty acid and phospholipid synthesis in *Escherichia coli* is relieved by overexpression of glycerol-3-phosphate acyltransferase (*plsB*). *J. Biol. Chem.* 269:26584–26590.

Heath, R. J., and C. O. Rock. 1996a. Regulation of fatty acid elongation and initiation by acyl-acyl carrier protein in *Escherichia coli. J. Biol. Chem.* 271:1833–1836.

Heath, R. J., and C. O. Rock. 1996b. Inhibition of beta-ketoacyl-acyl carrier protein synthase III (FabH) by acyl-acyl carrier protein in *Escherichia coli. J. Biol. Chem.* 271:10996–11000.

Henry, M. F., and J. E. Cronan, Jr. 1992. A new mechanism of transcriptional regulation: release of an activator triggered by small molecule binding. *Cell* 70:671–679.

Hoja, U., S. Marthol, J. Hofmann, S. Stegner, R. Schulz, S. Meier, E. Greiner, and E. Schweizer. 2004. HFA1 encoding an organelle-specific acetyl-CoA carboxylase controls mitochondrial fatty acid synthesis in *Saccharomyces cerevisiae. J. Biol. Chem.* 279:21779–21786.

Huang, Y. S., S. L. Pereira, and A. E. Leonard. 2004. Enzymes for transgenic biosynthesis of long-chain polyunsaturated fatty acids. *Biochimie* 86:793–798.

Ishige, T., A. Tan, K. Takabe, K. Kawasaki, Y. Sakai, and N. Kato. 2002. Wax ester production from *n*-alkanes by *Acinetobacter* sp. strain M-1: ultrastructure of cellular inclusions and role of acyl coenzyme A reductase. *Appl. Environ. Microbiol.* 68:1192–1195.

Jain, R. K., M. Coffey, K. Lai, A. Kumar, and S. L. MacKenzie. 2000. Enhancement of seed oil content by expression of glycerol-3-phosphate acyltransferase genes. *Biochem. Soc. Trans.* 28:958–961.

Jako, C., A. Kumar, Y. Wei, J. Zou, D. L. Barton, E. M. Giblin, P. S. Covello, and D. C. Taylor. 2001. Seed-specific over-expression of an *Arabidopsis* cDNA encoding a diacylglycerol acyltransferase enhances seed oil content and seed weight. *Plant Physiol.* 126:861–874.

James, E. S., and J. E. Cronan. 2004. Expression of two *Escherichia coli* acetyl-CoA carboxylase subunits is autoregulated. *J. Biol. Chem.* 279:2520–2527.

Jiang, P., and J. E. Cronan, Jr. 1994. Inhibition of fatty acid synthesis in *Escherichia coli* in the absence of phospholipid synthesis and release of inhibition by thioesterase action. *J. Bacteriol.* 176:2814–2821.

Kalscheuer, R., and A. Steinbuchel. 2003. A novel bifunctional wax ester synthase/acyl-CoA:diacylglycerol acyltransferase mediates wax ester and triacylglycerol biosynthesis in *Acinetobacter calcoaceticus* ADP1. *J. Biol. Chem.* 278:8075–8082.

Kalscheuer, R., H. Luftmann, and A. Steinbuchel. 2004. Synthesis of novel lipids in *Saccharomyces erevisiae* by heterologous expression of an unspecific bacterial acyltransferase. *Appl. Environ. Microbiol.* 70:7119–7125.

Kalscheuer, R., T. Stolting, and A. Steinbuchel. 2006a. Microdiesel: *Escherichia coli* engineered for fuel production. *Microbiology* 152:2529–2536.

Kalscheuer, R., T. Stoveken, H. Luftmann, U. Malkus, R. Reichelt, and A. Steinbuchel. 2006b. Neutral lipid biosynthesis in engineered *Escherichia coli*: jojoba oil-like wax esters and fatty acid butyl esters. *Appl. Environ. Microbiol.* 72:1373–1379.

Khotimchenko, S. V., and I. M. Yakovleva. 2005. Lipid composition of the red alga *Tichocarpus crinitus* exposed to different levels of photon irradiance. *Phytochemistry* 66:73–79.

Kim, H. U., and A. H. Huang. 2004. Plastid lysophosphatidyl acyltransferase is essential for embryo development in *Arabidopsis. Plant Physiol.* 134:1206–1216.

Klaus, D., J. B. Ohlrogge, H. E. Neuhaus, and P. Dormann. 2004. Increased fatty acid production in potato by engineering of acetyl-CoA carboxylase. *Planta* 219:389–396.

Li, S. J., and J. E. Cronan, Jr. 1993. Growth rate regulation of *Escherichia coli* acetyl coenzyme A carboxylase, which catalyzes the first committed step of lipid biosynthesis. *J. Bacteriol.* 175:332–340.

Lu, Y. J., Y. M. Zhang, K. D. Grimes, J. Qi, R. E. Lee, and C. O. Rock. 2006. Acyl-phosphates initiate membrane phospholipid synthesis in Gram-positive pathogens. *Mol. Cell* 23:765–772.

Lynn, S. G., S. S. Kilham, D. A. Kreeger, and S. J. Interlandi. 2000. Effect of nutrient availability on the biochemical and elemental stoichiometry in the freshwater diatom *Stephanodiscus minutulus* (*Bacillariophyceae*). *J. Phycol.* 36:510–522.

Marini, P. E., C. A. Perez, and D. de Mendoza. 2001. Growth-rate regulation of the *Bacillus subtilis accBC* operon encoding subunits of acetyl-CoA carboxylase, the first enzyme of fatty acid synthesis. *Arch. Microbiol.* 175:234–237.

Marrakchi, H., Y. M. Zhang, and C. O. Rock. 2002. Mechanistic diversity and regulation of Type II fatty acid synthesis. *Biochem. Soc. Trans.* 30:1050–1055.

Matsuzaki, M., O. Misumi, I. T. Shin, S. Maruyama, M. Takahara, S. Y. Miyagishima, T. Mori, K. Nishida, F. Yagisawa, K. Nishida, Y. Yoshida, Y. Nishimura, S. Nakao, T. Kobayashi, Y. Momoyama, T. Higashiyama, A. Minoda, M. Sano, H. Nomoto, K. Oishi, H. Hayashi, F. Ohta, S. Nishizaka, S. Haga, S. Miura, T. Morishita, Y. Kabeya, K. Terasawa, Y. Suzuki, Y. Ishii, S. Asakawa, H. Takano, N. Ohta, H. Kuroiwa, K. Tanaka, N. Shimizu, S. Sugano, N. Sato, H. Nozaki, N. Ogasawara, Y. Kohara, and T. Kuroiwa. 2004. Genome sequence of the ultrasmall unicellular red alga *Cyanidioschyzon merolae* 10D. *Nature* 428:653–657.

Mhaske, V., K. Beldjilali, J. Ohlrogge, and M. Pollard. 2005. Isolation and characterization of an *Arabidopsis thaliana* knockout line for phospholipid:diacylglycerol transacylase gene (At5g13640). *Plant Physiol. Biochem.* 43:413–417.

Napolitano, G. E. 1994. The relationship of lipids with light and chlorophyll measurements in freshwater algae and periphyton. *J. Phycol.* 30:943–950.

Nishida, I., Y. Tasaka, H. Shiraishi, and N. Murata. 1993. The gene and the RNA for the precursor to the plastid-located glycerol-3-phosphate acyltransferase of *Arabidopsis thaliana. Plant Mol. Biol.* 21:267–277.

Oelkers, P., D. Cromley, M. Padamsee, J. T. Billheimer, and S. L. Sturley. 2002. The DGA1 gene determines a second triglyceride synthetic pathway in yeast. *J. Biol. Chem.* 277:8877–8881.

Ohlrogge, J., L. Savage, J. Jaworski, T. Voelker, and D. Post-Beittenmiller. 1995. Alteration of acyl-acyl carrier protein pools and acetyl-CoA carboxylase expression in *Escherichia coli* by a plant medium chain acyl-acyl carrier protein thioesterase. *Arch. Biochem. Biophys.* 317:185–190.

Ohlrogge, J. B., and J. G. Jaworski. 1997. Regulation of fatty acid synthesis. *Annu. Rev. Plant Physiol. Plant Mol. Biol.* 48:109–136.

Olukoshi, E. R., and N. M. Packter. 1994. Importance of stored triacylglycerols in *Streptomyces*: possible carbon source for antibiotics. *Microbiology* 140:931–943.

Overath, P., G. Pauli, and H. U. Schairer. 1969. Fatty acid degradation in *Escherichia coli*. An inducible acyl-CoA synthetase, the mapping of old-mutations, and the isolation of regulatory mutants. *Eur. J. Biochem.* 7:559–574.

Palenik, B., J. Grimwood, A. Aerts, P. Rouze, A. Salamov, N. Putnam, C. Dupont, R. Jorgensen, E. Derelle, S. Rombauts, K. Zhou, R. Otillar, S. S. Merchant, S. Podell, T. Gaasterland, C. Napoli, K. Gendler, A. Manuell, V. Tai, O. Vallon, G. Piganeau, S. Jancek, M. Heijde, K. Jabbari, C. Bowler, M. Lohr, S. Robbens, G. Werner, I. Dubchak, G. J. Pazour, Q. Ren, I. Paulsen, C. Delwiche, J. Schmutz, D. Rokhsar, Y. Van de Peer, H. Moreau, and I. V. Grigoriev. 2007. The tiny eukaryote *Ostreococcus* provides genomic insights into the paradox of plankton speciation. *Proc. Natl. Acad. Sci. USA* 104:7705–7710.

Peterfy, M., J. Phan, P. Xu, and K. Reue. 2001. Lipodystrophy in the fld mouse results from mutation of a new gene encoding a nuclear protein, lipin. *Nat. Genet.* 27:121–124.

Phan, J., and K. Reue. 2005. Lipin, a lipodystrophy and obesity gene. *Cell Metab.* 1:73–83.

Qi, B., T. Fraser, S. Mugford, G. Dobson, O. Sayanova, J. Butler, J. A. Napier, A. K. Stobart, and C. M. Lazarus. 2004. Production of very long chain polyunsaturated omega-3 and omega-6 fatty acids in plants. *Nat. Biotechnol.* 22:739–745.

Raman, N., and C. C. DiRusso. 1995. Analysis of acyl coenzyme A binding to the transcription factor FadR and identification of amino acid residues in the carboxyl terminus required for ligand binding. *J. Biol. Chem.* 270:1092–1097.

Ratledge, C., and J. P. Wynn. 2002. The biochemistry and molecular biology of lipid accumulation in oleaginous microorganisms. *Adv. Appl. Microbiol.* 51:1–51.

Ratledge, C. 2004. Fatty acid biosynthesis in microorganisms being used for single cell oil production. *Biochimie* 86:807–815.

Rock, C.O., and S. Jackowski. 1982. Regulation of phospholipid synthesis in *Escherichia coli.* Composition of the acyl-acyl carrier protein pool in vivo. *J. Biol. Chem.* 257:10759–10765.

Roessler, P. G., and J. B. Ohlrogge. 1993. Cloning and characterization of the gene that encodes acetyl-coenzyme A carboxylase in the alga *Cyclotella cryptica. J. Biol. Chem.* 268:19254–19259.

Ruuska, S. A., T. Girke, C. Benning, and J. B. Ohlrogge. 2002. Contrapuntal networks of gene expression during *Arabidopsis* seed filling. *Plant Cell* 14:1191–1206.

Salas, J. J., and J. B. Ohlrogge. 2002. Characterization of substrate specificity of plant FatA and FatB acyl-ACP thioesterases. *Arch. Biochem. Biophys.* 403:25–34.

Sasaki, Y., and Y. Nagano. 2004. Plant acetyl-CoA carboxylase: structure, biosynthesis, regulation, and gene manipulation for plant breeding. *Biosci. Biotechnol. Biochem.* 68:1175–1184.

Schujman, G. E., L. Paoletti, A. D. Grossman, and D. de Mendoza. 2003. FapR, a bacterial transcription factor involved in global regulation of membrane lipid biosynthesis. *Dev. Cell* 4:663–672.

Schujman, G. E., M. Guerin, A. Buschiazzo, F. Schaeffer, L. I. Llarrull, G. Reh, A. J. Vila, P. M. Alzari, and D. de Mendoza. 2006. Structural basis of lipid biosynthesis regulation in Gram-positive bacteria. *EMBO J.* 25:4074–4083.

Schuller, H. J., A. Hahn., F. Troster, A. Schutz, and E. Schweizer. 1992. Coordinate genetic control of yeast fatty acid synthase genes FAS1 and FAS2 by an upstream activation site common to genes involved in membrane lipid biosynthesis. *EMBO J.* 11:107–114.

Schuller, H. J., A. Schutz, S. Knab, B. Hoffmann, and E. Schweizer. 1994. Importance of general regulatory factors Rap1p, Abf1p and Reb1p for the activation of yeast fatty acid synthase genes FAS1 and FAS2. *Eur. J. Biochem.* 225:213–222.

Schweizer, E., and J. Hofmann. 2004. Microbial type I fatty acid synthases (FAS): major players in a network of cellular FAS systems. *Microbiol. Mol. Biol. Rev.* 68:501–517.

Slabas, A. R., A. White, P. O'Hara, and T. Fawcett. 2002. Investigations into the regulation of lipid biosynthesis in *Brassica napus* using antisense down-regulation. *Biochem. Soc. Trans.* 30:1056–1059.

Sorger, D., and G. Daum. 2002. Synthesis of triacylglycerols by the acyl-coenzyme A:diacyl-glycerol acyltransferase Dga1p in lipid particles of the yeast *Saccharomyces cerevisiae. J. Bacteriol.* 184: 519–524.

Stahl, U., A. S. Carlsson, M. Lenman, A. Dahlqvist, B. Huang, W. Banas, A. Banas, and S. Stymne. 2004. Cloning and functional characterization of a phospholipid:diacylglycerol acyltransferase from *Arabidopsis. Plant Physiol.* 135:1324–1335.

Stoveken, T., R. Kalscheuer, U. Malkus, R. Reichelt, and A. Steinbuchel. 2005. The wax ester synthase/acyl coenzyme A:diacylglycerol acyltransferase from *Acinetobacter* sp. strain ADP1:characterization of a novel type of acyltransferase. *J. Bacteriol.* 187:1369–1376.

Stukey, J. E., V. M. McDonough, and C. E. Martin. 1989. Isolation and characterization of OLE1, a gene affecting fatty acid desaturation from *Saccharomyces cerevisiae. J. Biol. Chem.* 264:16537–16544.

Tatsuzawa, H., E. Takizawa, M. Wada, and Y. Yamamoto. 1996. Fatty acid and lipid composition of the acidophilic green algae *Chlamydomonas* sp. *J. Phycol.* 32:598–601.

Tehlivets, O., K. Scheuringer, and S. D. Kohlwein. 2007. Fatty acid synthesis and elongation in yeast. *Biochim. Biophys. Acta* 1771: 255–270.

Uthoff, S., T. Stoveken, N. Weber, K. Vosmann, E. Klein, R. Kalscheuer, and A. Steinbuchel. 2005. Thio wax ester biosynthesis utilizing the unspecific bifunctional wax ester synthase/acyl coenzyme A:diacylglycerol acyltransferase of *Acinetobacter* sp. strain ADP1. *Appl. Environ. Microbiol.* 71:790–796.

Voelker, T. A., and H. M. Davies. 1994. Alteration of the specificity and regulation of fatty acid synthesis of *Escherichia coli* by expression of a plant medium-chain acyl-acyl carrier protein thioesterase. *J. Bacteriol.* 176:7320-7327.

Wenz, P., S. Schwank, U. Hoja, and H. J. Schuller. 2001. A downstream regulatory element located within the coding sequence mediates autoregulated expression of the yeast fatty acid synthase gene FAS2 by the FAS1 gene product. *Nucleic Acids Res.* 29:4625–4632.

Witters, L. A., and T. D. Watts. 1990. Yeast acetyl-CoA carboxylase: in vitro phosphorylation by mammalian and yeast protein kinases. *Biochem. Biophys. Res. Commun.* 169:369–376.

Wynn, J. P., A. bin Abdul Hamid, and C. Ratledge. 1999. The role of malic enzyme in the regulation of lipid accumulation in filamentous fungi. *Microbiology* 145:1911–1917.

Zhang, S., Y. Skalsky, and D. J. Garfinkel. 1999. MGA2 or SPT23 is required for transcription of the delta9 fatty acid desaturase gene, OLE1, and nuclear membrane integrity in *Saccharomyces cerevisiae. Genetics* 151:473–483.

Zhang, Y. M., H. Marrakchi, and C. O. Rock. 2002. The FabR (YijC) transcription factor regulates unsaturated fatty acid biosynthesis in *Escherichia coli. J. Biol. Chem.* 277:15558–15565.

Zhang, Y. M., H. Marrakchi, S. W. White, and C. O. Rock. 2003. The application of computational methods to explore the diversity and structure of bacterial fatty acid synthase. *J. Lipid Res.* 44:1–10.

Zheng, Z., Q. Xia, M. Dauk, W. Shen, G. Selvaraj, and J. Zou. 2003. *Arabidopsis* AtGPAT1, a member of the membrane-bound glycerol-3-phosphate acyltransferase gene family, is essential for tapetum differentiation and male fertility. *Plant Cell* 15:1872–1887.

Zou, J., V. Katavic, E. M. Giblin, D. L. Barton, S. L. MacKenzie, W. A. Keller, X. Hu, and D. C. Taylor. 1997. Modification of seed oil content and acyl composition in the *Brassicaceae* by expression of a yeast sn-2 acyltransferase gene. *Plant Cell* 9:909–923.

Zou, J., Y. Wei, C. Jako, A. Kumar, G. Selvaraj, and D. C. Taylor. 1999. The *Arabidopsis thaliana* TAG1 mutant has a mutation in a diacylglycerol acyltransferase gene. *Plant J.* 19:645–653.

Index

有 奖 征 集 反 馈 意 见

科爱传播
KE'AI COMMUNICATIONS

尊敬的读者：

科学出版社科爱森蓝文化传播有限公司（简称“科爱传播”）立足国际合作，致力于为科技专业人士提供优质的信息服务。我们很想通过自己的努力最大限度地满足您的需求，您的哪怕是一点点的建议和意见，都将成为我们改进工作的重要依据。

我们将在每年的 6 月份、12 月份各一次从半年的参与者中抽取幸运者 10 名，幸运者可以从“科爱传播”的出版物中任选价值 1000 元的图书（10 册以内）作为奖品（全部出版物信息可在我们的网站上查到）。

1. 您所购买的图书书名：《________________________________》

   您于________年____月____日在（通过）________________________购买到此书。

   您认为本书的定价：□偏高　□合适　□偏低

   您认为本书的内容有约____%对您有用。

2. 您认为我们出版物的质量：

   内容质量（学术水平、写作水平）：□很好　□较好　□一般　□较差

   译介质量（翻译水平、文字水平）：□很好　□较好　□一般　□较差

   印制质量（印制、包装）：□很好　□较好　□一般　□较差

3. 您所在的专业领域：________________________

4. 在您获取专业知识和专业信息的主要渠道中，排在前三位的是：

   1. ________　2. ________　3. ________

   A.网络　B.期刊　C.图书　D.报纸　E.电视　F.会议　G.内部交流　H.其他：________

5. 您还需要哪些类型的图书？

   □专著　□教材　□实验手册　□辞典工具书　□文集　□其他：________

   □书摘（原版书摘编）　□刊摘（国外学术期刊重要文献摘编）

6. 您还希望我们从国外引进哪些专业方向的图书（或期刊）？

7. 您建议采用何种引进形式？

☐翻译　☐影印　☐摘编（只从原书中选部分内容引进）

☐导读（原文影印加少量中文介绍）　☐注解（原文影印加大量中文介绍）

8. 您是否愿意与我们合作，参与编写、编译、翻译图书或其他科技信息？

9. 请列举您近两年看过的，您认为最有参考价值、对您帮助最大的 1~2 本书：

书名	著作者	出版社	出版日期	定价

10. 您还有什么别的意见、建议？（可另附纸）

● 请告诉我们您准确的地址和联系办法：

姓名：＿＿＿＿＿＿ 性别：＿＿＿ 生日：＿＿＿年＿月＿日

单位：＿＿＿＿＿＿＿＿＿＿＿＿职务/职称：＿＿＿＿＿

地址：＿＿＿＿＿＿＿＿＿＿＿＿＿＿＿＿＿＿＿＿

E-mail：＿＿＿＿＿＿＿＿＿＿电话：＿＿＿＿＿＿＿

传真：＿＿＿＿＿＿＿＿＿ 手机：＿＿＿＿＿＿＿＿

回邮地址（也可以通过 E-mail 反馈）：

北京东黄城根北街 16 号　科学出版社 科爱传播中心 杨 琴（收）　邮编：100717

联系电话：010-64006871；传真：010-64034056

编辑部电话：010-64034507

投稿及读者反馈：editor@kbooks.cn, keai@mail.sciencep.com

（注：本反馈单复印有效，也可以在线下载：http://www.kbooks.cn）

图　　版

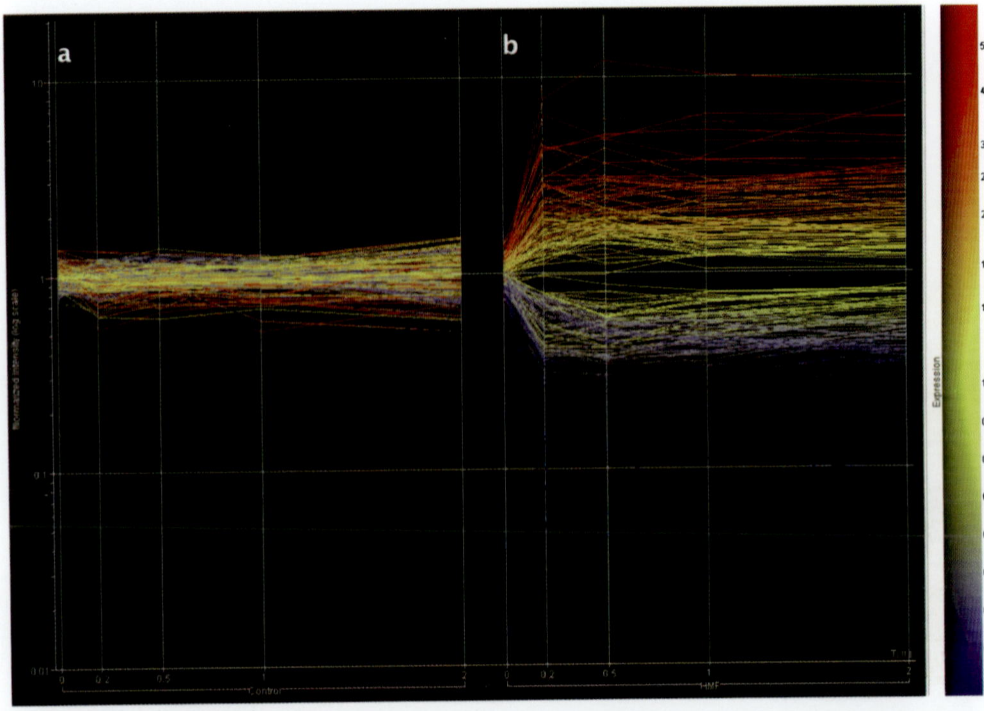

Color Plate 1 (chapter 2). The genomic adaptation to HMF stress. Expressions of selected genes of ethanologenic yeast *S. cerevisiae* are shown under a normal control condition (a) and HMF stress condition (b) from 0, 10, 30, 60, and 120 min after the treatment showing significantly induced (blue) and repressed (red) mRNA expression caused by the HMF stress on a defined medium. Yellow indicates mRNA equally expressed under different conditions. Varied colors between yellow and red or yellow and blue, as shown in a colored bar on the far right, indicate varied quantitative measurements of mRNA expression levels in a log scale. (Reprinted from Liu, 2006, with kind permission of Springer Science and Business Media.)

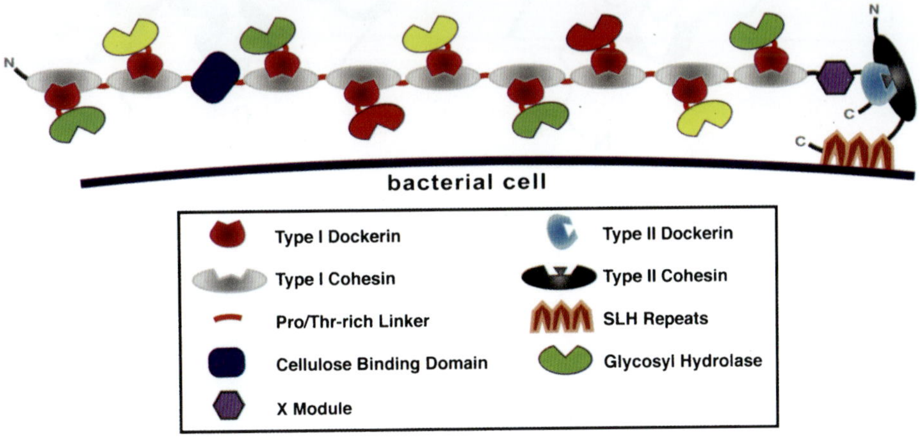

Color Plate 2 (chapter 8). Schematic diagram of the *C. thermocellum* cellulosome. The type I dockerins mediate attachment of the catalytic subunits to the scaffoldin, which is comprised of nine cohesins, a cellulose-binding domain, a hydrophilic domain of unknown function (X), and a type II dockerin. The scaffoldin likewise binds through its type II dockerin domain to a type II cohesin-containing protein on the bacterial cell surface that is thought to anchor the complex through a series of three surface layer homology (SLH) domains. Reproduced from Volkman et al., 2004, with permission of John Wiley & Sons.

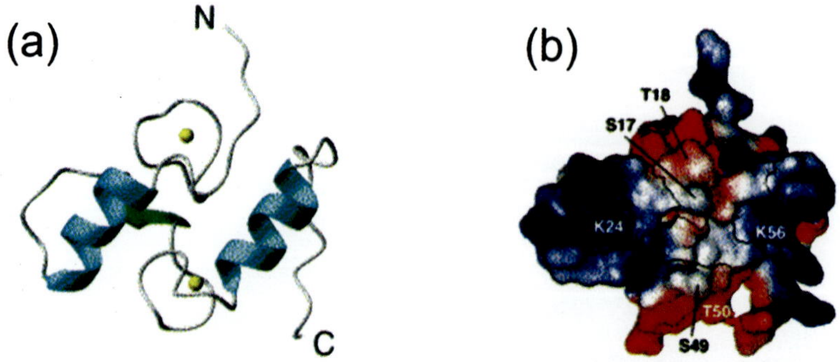

Color Plate 3 (chapter 8). Ribbon diagram structure of the cohesin-binding surface of type I dockerin from C. *thermocellum* (a) and a molecular surface representation from the same view (b). Reproduced from Lytle et al., 2001.

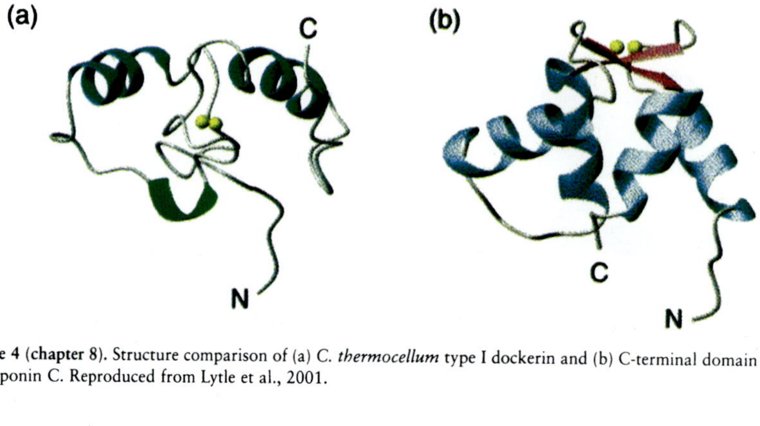

Color Plate 4 (chapter 8). Structure comparison of (a) C. *thermocellum* type I dockerin and (b) C-terminal domain of cardiac muscle troponin C. Reproduced from Lytle et al., 2001.

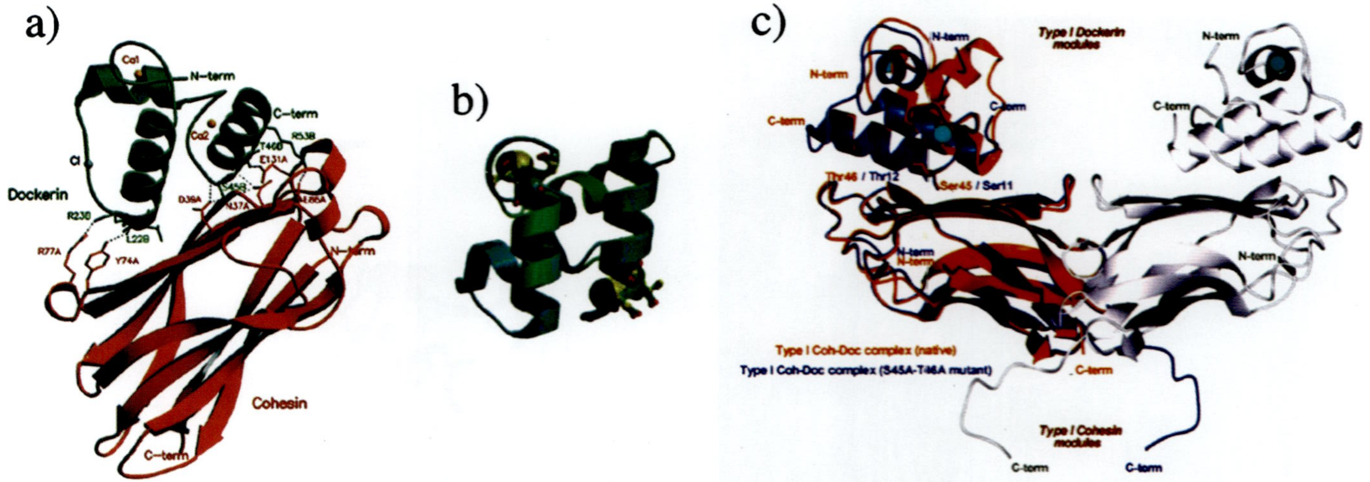

Color Plate 5 (chapter 8). Structures of type I cohesin-dockerin complex (a and c) and type I dockerin in the complex (b). (a) Complex of wild-type (WT) dockerin from *C. thermocellum* Xyn10B and cohesin 2 of *C. thermocellum* CipA. (b) Dockerin of *C. thermocellum* Xyn10B in the complex. (c) The superposition of the cohesin-dockerin complex (in orange) with its S45A-T46A mutant complex (in blue). In the mutant complex, helix-1 (containing Ser-11 and Thr-12) dominates binding, whereas in the WT complex, helix-3 (containing Ser-45 and Thr-46) plays a key role in ligand recognition. The second molecule of the mutant complex is represented by a light-gray ribbon. The Ca^{2+} ions are depicted as spheres and colored orange (the WT complex) and light blue (the mutant). Reproduced from Carvalho et al., 2003, 2007, with permission from the National Academy of Sciences.

Color Plate 6 (chapter 13). The Volta experiment in a freshwater pond. Reprinted from *Archaea: Molecular and Cellular Biology* (Ferry and Kastead, 2007) with permission of the publisher.

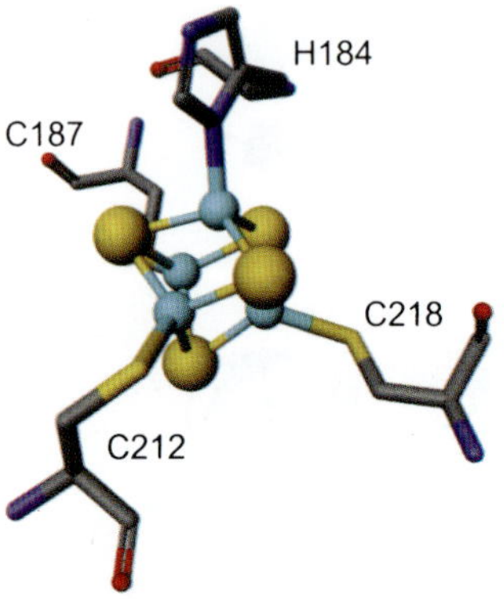

Color Plate 11 (chapter 20). Structure of the *D. fructosovorans* distal [4Fe4S] center (Rousset et al., 1998). Protein Data Bank ID 1frf. The hetero atoms are represented as spheres, and the protein ligands are represented as sticks. Iron is in cyan, sulfur is in yellow, oxygen is in red, nitrogen is in blue, and carbon is in gray.

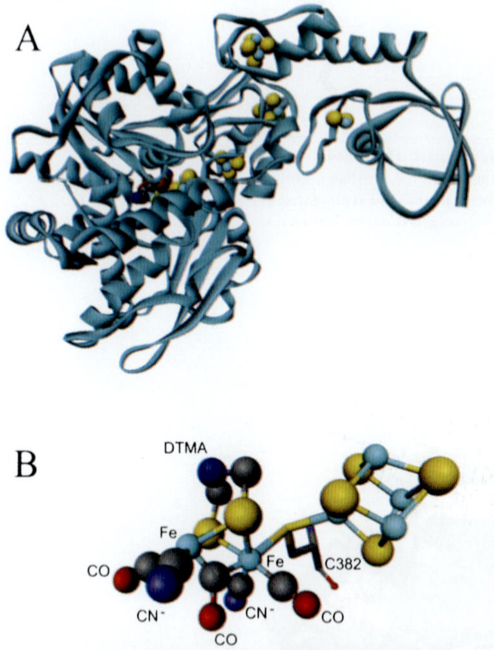

Color Plate 7 (chapter 20). [FeFe] hydrogenase from *Clostridium pasteurianum* (Peters et al., 1998). Protein Data Bank ID 1feh. (A) Three-dimensional structure of the enzyme. The dinuclear FeFe active site, the three [4Fe4S] centers, and the [2Fe2S] center are coordinated in the same subunit. (B) Structure details of the active site. The hetero atoms from the prosthetic groups are represented as spheres, and the protein ligands are represented as sticks. Iron is in cyan, sulfur is in yellow, oxygen is in red, nitrogen is in blue, and carbon is in gray. DTMA, dithiomethyl amine.

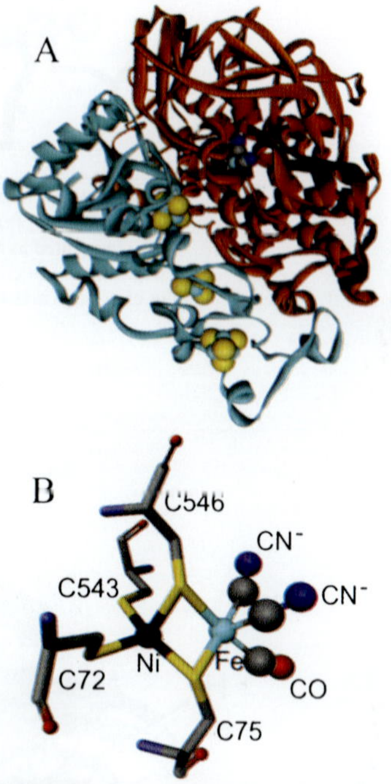

Color Plate 8 (chapter 20). [NiFe] hydrogenase from *Desulfovibrio fructosovorans* (Rousset et al., 1998). Protein Data Bank ID 1frf. (A) Three-dimensional structure of the enzyme. The dinuclear NiFe active site is coordinated in the large subunit, and the two [4Fe4S] centers and the [3Fe4S] center are coordinated in the small subunit. (B) Structure details of the active site. The hetero atoms from the prosthetic groups are represented as spheres, and the protein ligands are represented as sticks. Iron is in cyan, nickel is in dark blue, sulfur is in yellow, oxygen is in red, nitrogen is in blue, and carbon is in gray.

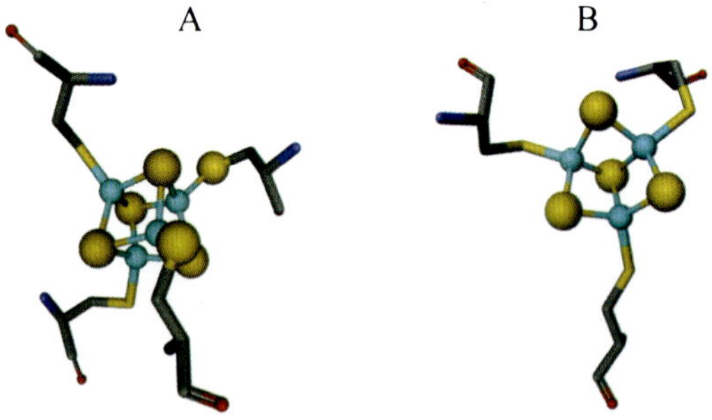

Color Plate 9 (chapter 20). Iron-sulfur centers. (A) [4Fe4S] center and (B) [3Fe4S] center. The hetero atoms from the prosthetic groups are represented as spheres, and the protein ligands are represented as sticks. Iron is in cyan, sulfur is in yellow, oxygen is in red, nitrogen is in blue, and carbon is in gray.

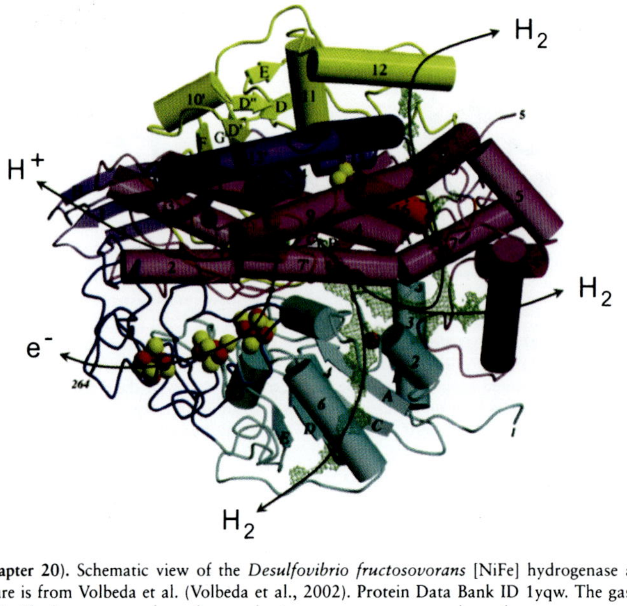

Color Plate 10 (chapter 20). Schematic view of the *Desulfovibrio fructosovorans* [NiFe] hydrogenase activity. The three-dimensional structure is from Volbeda et al. (Volbeda et al., 2002). Protein Data Bank ID 1yqw. The gas channel is represented as a wire mesh. The hetero atoms from the prosthetic groups are represented as spheres.